Microbial Interventions in Agriculture and Environment

Dhananjaya Pratap Singh
Vijai Kumar Gupta • Ratna Prabha
Editors

Microbial Interventions in Agriculture and Environment

Volume 2: Rhizosphere, Microbiome and Agro-ecology

Springer

Editors
Dhananjaya Pratap Singh
ICAR-National Bureau of Agriculturally Important Microorganisms
Maunath Bhanjan, Uttar Pradesh, India

Ratna Prabha
ICAR-National Bureau of Agriculturally Important Microorganisms
Maunath Bhanjan, Uttar Pradesh, India

Vijai Kumar Gupta
Department of Chemistry and Biotechnology
Tallinn University of Technology
Tallinn, Estonia

ISBN 978-981-13-8382-3   ISBN 978-981-13-8383-0 (eBook)
https://doi.org/10.1007/978-981-13-8383-0

This Springer imprint is published by the registered company Springer Nature Singapore Pte Ltd.
The registered company address is: 152 Beach Road, #21-01/04 Gateway East, Singapore 189721, Singapore

Preface

Microbial communities are the fundamental engines for balanced ecosystem function, natural ecological harmony, biotic and abiotic stress mitigation, nutrient recycling, acquisition and mineralization, bioconversion of complex animal and plant residues and bioremediation of soil contaminants. By such complex and crucial but continuous and extraordinary functions, they support plant growth, development and soil fertility. Plant rhizosphere, the vicinity of the root space associated with the soil, has remained among the most biologically diverse and chemically active interface for microbial activities. Microorganisms enjoy root exudates comprising sugars, organic acids, amino acids and beneficial metabolites for growth and development of their communities. They in turn secrete in the rhizosphere space valuable metabolites that function as energy sources, communicators, defence molecules and suppressor of undesirable microfauna. This helps in creating disease-suppressing environment around the root surface. The metabolites ultimately benefit agricultural crops by improving growth, development and intrinsic immunity of the plants. Microbial chemical molecules from beneficial associative communities help other microbial population to flourish in the root rhizosphere or act as antagonists to suppress non-beneficial and pathogenic organisms. This is how the mutuality among microbe-environment, microbe-microbe and microbe-plants prevails in the rhizospheric micro-environment to ultimately benefit the whole agro-ecology.

Holistically, the microbiome is crucial for the biogenesis of the earth. Microbial functions related to geochemical recycling of carbon, nitrogen, minerals and greenhouse gases are crucial for environmental processes. They fix, sequester, solubilize, mineralize, mobilize, neutralize, transform, remediate, accumulate and recycle nutrients and minerals to boost agro-ecosystem and strengthen soil fertility and plant health. Microbial communities with their functional outperformance play crucial role in disease suppression, plant growth promotion, plant immunization and induced resistance and tolerance against stresses. These are all indicative parameters for improved crop productivity. Beneficial belowground microbial composition and interaction with the rhizosphere help plants modulate their stress avoidance/tolerance mechanisms under unfavoured environmental conditions. Therefore, the benefits of beneficial microbe-rhizosphere interactions can be extended to the fields for enabling crop plants tolerate losses due to unregulated weather conditions.

The book *Microbial Interventions in Agriculture and Environment* in its volume *Rhizosphere, Microbiome and Agro-ecology* presents a detailed account of microbial mechanisms encompassing rhizosphere and agro-ecological benefits. The volume covers topics on rhizosphere engineering for improved impact of PGPRs, plant-microbiome interactions, communication network in plant-microbe interactions, inoculant applications for agricultural sustainability, functional benefits of biocontrol agents, PGPRs and PGPFs, comparative studies on the impact of chemical fertilization and inoculant use, biological treatments of biosolids, agro-waste bioconversion for rural sanitation, diversity prospects for microbial product development, microbe-mediated stress alleviation strategies in host plants, N-fertilizer for microbiome, microbial dynamics in tree ecosystem, formulation development and industrial production of inoculants. A comprehensive coverage of microbial community functions in the rhizosphere and their benefits to the agro-ecology will appraise research workers with the aspects of unexplored subject and expose problem-solving solutions based on microbial interventions for agriculture and environment. With widely covered research topics on microbial implications, we hope that this volume will attract attention of global readership of researchers, students, faculties and scientists working on the related areas. We are sure that the volume will also become a source of knowledgeable compilation for the agricultural policymakers, environmentalists, activists and advisors working at the government, industries and non-government organizations.

Maunath Bhanjan, Uttar Pradesh, India — Dhananjaya P. Singh
Tallinn, Estonia — Vijai K. Gupta
Maunath Bhanjan, Uttar Pradesh, India — Ratna Prabha

Contents

Editors and Contributors

About the Editors

Dhananjaya Pratap Singh is presently Principal Scientist in Biotechnology at ICAR-National Bureau of Agriculturally Important Microorganisms, Maunath Bhanjan, India. He did his Master's degree from G.B. Pant University of Agriculture and Technology, Pantnagar, and PhD in Biotechnology from Banaras Hindu University, Varanasi. His research interests include plant-microbe interactions, bio-prospecting of metabolites of microbial and plant origin, microbe-mediated stress management in plants, metabolomics-driven search for small molecules and bioinformatics in microbial research. He has been working on the societal implications of microbial biotechnology pertaining to microbe-mediated crop production practices and rapid composting of agro-wastes at farm and farmer's levels. The editor has successfully performed outreach of such technologies to farming community for adoption at field scale. He has been associated with the development of super-computing infrastructure for agricultural bioinformatics in microbial domain in India under the National Agricultural Bioinformatics Grid (NABG) program of ICAR. He is an Associate of the National Academy of Agricultural Sciences (NAAS), India, and has been awarded with several prestigious awards including Dr. APJ Abdul Kalam Award for Scientific Excellence. With many publications in the journals of national and international repute, the editor has also edited five books on microbial research with Springer Nature and other publishers.

Vijai Kumar Gupta is the Senior Scientist at ERA Chair of Green Chemistry, Tallinn University of Technology, Estonia. His area of research interests includes bioactive natural products, microbial biotechnology and applied mycology, bioprocess technology, biofuel and biorefinery research and glycobiotechnology of plant-microbe interaction. He is the Secretary of the European Mycological Association and Country Ambassador of the American Society of Microbiology. He is the Fellow of Linnaean Society and Mycological Society of India and Associate Fellow of the National Academy of Biological Sciences, India, and Indian Mycological Association. He has been the Editor of reputed journals of international recognition and edited 28 books with the publishers like Elsevier, Wiley-Blackwell, Frontiers, Taylor and Francis, Springer Nature, CABI and De Gruyter. To his credit, the Editor also has a vast number of research publications and review papers in internationally reputed high-impact factor journals. He also holds two IPs in the area of microbial biotechnology for sustainable product development.

Ratna Prabha is currently working as DST Women Scientist at ICAR-National Bureau of Agriculturally Important Microorganisms, India. With Doctorate in Biotechnology and Master's in Bioinformatics, she has been actively involved in different research activities. Her research interest lies in microbe-mediated stress management in plants, database development, comparative microbial analysis, phylogenomics and pan-genome analysis, metagenomics data analysis and microbe-mediated composting technology development and dissemination. She has been engaged in developing various digital databases on plants and microbes and has various edited and authored books, many book chapters and different research papers and review articles in journals of international repute.

Contributors

Iqbal Ahmad Biofilm Research Laboratory, Department of Agricultural Microbiology, Faculty of Agricultural Sciences, Aligarh Muslim University, Aligarh, India

Alejandro Alarcón Colegio de Postgraduados, Postgrado de Edafología, Texcoco, Estado de México, Mexico

Hossein Ali Alikhani Department of Soil Science, University College of Agriculture and Natural Resources, University of Tehran, Tehran, Iran

Mohd. Musheer Altaf Department of Life Science, Institute of Information Management and Technology, Aligarh, India
Department of Agricultural Microbiology, Aligarh Muslim University, Aligarh, India

Jayanthi Abraham Microbial Biotechnology Laboratory, School of Biosciences and Technology, VIT University, Vellore, Tamil Nadu, India

Galdino Andrade Department of Microbiology, Londrina State University, Londrina, Brazil

Matheus Andreata Department of Microbiology, Londrina State University, Londrina, Brazil

Desislava Angelova Department of Microbiology and Environmental Biotechnologies, Agricultural University–Plovdiv, Plovdiv, Bulgaria

Firoz Ahmad Ansari Biofilm Research Laboratory, Department of Agricultural Microbiology, Faculty of Agricultural Sciences, Aligarh Muslim University, Aligarh, India

Saumya Arora Department of Biological Sciences, Birla Institute of Technology and Science, Pilani, Pilani, Rajasthan, India

Hassan Azaizeh Institute of Applied Research (Affiliated with University of Haifa), The Galilee Society, Shefa-Amr, Israel
Department of Environmental Science, Tel Hai College, Upper Galilee, Israel

R. Bajpai School of Life and Allied Sciences, ITM University, Naya Raipur, Chhattisgarh, India

André Barazetti Department of Microbiology, Londrina State University, Londrina, Brazil

Purnima Bhandari Mehr Chand Mahajan DAV College for Women, Chandigarh, India

B. Chaudhari Ambalal School of Life Science, Kavayitri Bahinabai Chaudhari North Maharashtra University, Jalgaon, India

D. Dandi Navin School of Life Science, Kavayitri Bahinabai Chaudhari North Maharashtra University, Jalgaon, India

Mickely Liuti Dealis Department of Microbiology, Londrina State University, Londrina, Brazil

Luciana P. Di Salvo Faculty of Agronomy, Department of Applied Biology and Foods, Chair of Agricultural Microbiology, University of Buenos Aires, Buenos Aires, Argentina

M. M. Elakkya Division of Microbiology, Indian Agricultural Research Institute, New Delhi, India

Rosalba Esquivel-Cote Colegio de Postgraduados, Postgrado de Edafología, Texcoco, Estado de México, Mexico

Hassan Etesami Department of Soil Science, University College of Agriculture and Natural Resources, University of Tehran, Tehran, Iran

Inés E. García de Salamone Faculty of Agronomy, Department of Applied Biology and Foods, Chair of Agricultural Microbiology, University of Buenos Aires, Buenos Aires, Argentina

Neera Garg Department of Botany, Panjab University, Chandigarh, India

M. Gavit Pavankumar School of Life Science, Kavayitri Bahinabai Chaudhari North Maharashtra University, Jalgaon, India

Paushali Ghosh School of Biotechnology, Institute of Science, Banaras Hindu University, Varanasi, India

J. Godheja School of Life and Allied Sciences, ITM University, Naya Raipur, Chhattisgarh, India

Vijai Kumar Gupta Department of Chemistry and Biotechnology, Tallinn University of Technology, Tallinn, Estonia

Raman Gurusamy Department of Life Sciences, Yeungnam University, Gyeongsan, Gyeongbuk, Republic of Korea

Arash Hemati Department of Soil Science, University of Tabriz, Tabriz, Iran

Dulce J. Hernández-Melchor Colegio de Postgraduados, Postgrado de Edafología, Texcoco, Estado de México, Mexico

Rachana Jain Amity Food and Agriculture Foundation, Amity University, Noida, Uttar Pradesh, India

Prabhat N. Jha Department of Biological Sciences, Birla Institute of Technology and Science, Pilani, Pilani, Rajasthan, India

P. Jishma School of Biosciences, Mahatma Gandhi University, Kottayam, Kerala, India

Bhavdish N. Johri Department of Biotechnology, Barkatullah University, Bhopal, India

Samiksha Joshi Department of Microbiology, G. B. Pant University of Agriculture & Technology, Pantnagar, Uttarakhand, India

Mohd Sajjad Ahmad Khan Department of Basic Sciences, Biology Unit, Health Track, Imam Abdulrahman Bin Faisal University, Dammam, Saudi Arabia

V. Kolla School of Life and Allied Sciences, ITM University, Naya Raipur, Chhattisgarh, India

Ashok Kumar School of Biotechnology, Institute of Science, Banaras Hindu University, Varanasi, India

Indramani Kumar Department of Biotechnology, School of Life Sciences, Pondicherry University, Puducherry, India

Jay Kumar School of Biotechnology, Institute of Science, Banaras Hindu University, Varanasi, India

Gabriel Liuti Department of Microbiology, Londrina State University, Londrina, Brazil

Jean Marcos Soares Matos Department of Microbiology, Londrina State University, Londrina, Brazil

M. Mishra School of Life and Allied Sciences, ITM University, Naya Raipur, Chhattisgarh, India

D. R. Modi Department of Biotechnology, Babasaheb Bhimrao Ambedkar University, Lucknow, Uttar Pradesh, India

Fluvio Modolon Department of Microbiology, Londrina State University, Londrina, Brazil

Moumita Mondal Department of Biotechnology, School of Life Sciences, Pondicherry University, Puducherry, India

Miguel O. P. Navarro Department of Microbiology, Londrina State University, Londrina, Brazil

Erika T. G. Niekawa Department of Microbiology, Londrina State University, Londrina, Brazil

Igor Matheus Oliveira Department of Microbiology, Londrina State University, Londrina, Brazil

A. M. Pereira School of Life and Allied Sciences, ITM University, Naya Raipur, Chhattisgarh, India

Ratna Prabha ICAR-National Bureau of Agriculturally Important Microorganisms, Maunath Bhanjan, Uttar Pradesh, India

Prasanta Kumar Prusty Division of Microbiology, Indian Agricultural Research Institute, New Delhi, India

E. K. Radhakrishnan School of Biosciences, Mahatma Gandhi University, Kottayam, Kerala, India

Balasubramanian Ramakrishnan Division of Microbiology, Indian Agricultural Research Institute, New Delhi, India

Anjul Rana Division of Microbiology, Indian Agricultural Research Institute, New Delhi, India

A. Remakanthan Department of Botany, University College, Thiruvananthapuram, Kerala, India

Swati Sagar Division of Microbiology, Indian Agricultural Research Institute, New Delhi, India

Manvika Sahgal Department of Microbiology, G. B. Pant University of Agriculture & Technology, Pantnagar, Uttarakhand, India

Sakthivel Natarajan Department of Biotechnology, School of Life Sciences, Pondicherry University, Puducherry, India

Jyoti Saxena Biochemical Engineering Department, B.T. Kumaon Institute of Technology, Dwarahat, Uttarakhand, India

S. V. Sharma School of Life and Allied Sciences, ITM University, Naya Raipur, Chhattisgarh, India

S. K. Shekhar Department of Biotechnology, Babasaheb Bhimrao Ambedkar University, Lucknow, Uttar Pradesh, India

D. Shelar Rajendra Department of Microbiology, Z. B. Patil College, Dhule, India

Stefan Shilev Department of Microbiology and Environmental Biotechnologies, Agricultural University–Plovdiv, Plovdiv, Bulgaria

Dhananjaya Pratap Singh ICAR-National Bureau of Agriculturally Important Microorganisms, Maunath Bhanjan, Uttar Pradesh, India

Divya Singh School of Biotechnology, Institute of Science, Banaras Hindu University, Varanasi, India

K. Sinha School of Life and Allied Sciences, ITM University, Naya Raipur, Chhattisgarh, India

Sundarakrishnan Balakrishnan Department of Biotechnology, School of Life Sciences, Pondicherry University, Puducherry, India

Anushree Suresh Microbial Biotechnology Laboratory, School of Biosciences and Technology, VIT University, Vellore, Tamil Nadu, India

Salil K. Tewari Department of Genetics and Plant Breeding, G. B. Pant University of Agriculture & Technology, Pantnagar, Uttarakhand, India

Martha Viviana Torres Cely Department of Microbiology, Londrina State University, Londrina, Brazil

Microbial Inoculants for Sustainable Crop Management

1

Dhananjaya Pratap Singh, Ratna Prabha, and Vijai Kumar Gupta

1 Introduction

In the last few decades, agriculture has suffered at large due to industrial developments, modifications in the land use, and excessive use of chemical inputs (fertilizers and pesticides) for crop production and protection (Foley et al. 2005; Aktar et al. 2009). High-input-based agriculture has posed serious threats to the whole crop production systems that involves soil quality, crop nutrition, food safety, freshwater resources, forest cover, biodiversity, and regional microenvironment including air quality. This has caused different problems like decline in soil fertility, resistance in pests and phytopathogens, contamination of food chain, and the health of humans and animals (Horrigan et al. 2002). Since the use of chemical fertilizers and pesticides came into practice, it has been increasingly utilized for enhancing crop protection, production, safety, and food preservation. Pesticides have also been excessively employed in public health management and in domestic use. Unique intrinsic properties of these chemicals like biological specificity, target selectivity for toxicity, and lesser toxicity for nontarget organisms warrant their use as pesticides and growth regulators (Maroni et al. 2006). However, chemicalization of the whole agricultural system in terms of the excessive usage of chemical farm inputs (fertilizers; micronutrients and minerals; synthetic growth regulators; and pesticides including insecticides, herbicides, fungicides; antibiotics; etc.) is now taking the whole agroecosystem to a level at which it has started becoming nonresponsive and nonperforming in a natural way (Singh et al. 2016 Microbial inoculants). Therefore, a shift

D. P. Singh (✉) · R. Prabha
ICAR-National Bureau of Agriculturally Important Microorganisms,
Maunath Bhanjan, Uttar Pradesh, India
e-mail: dhananjaya.singh@icar.gov.in

V. K. Gupta
Department of Chemistry and Biotechnology, Tallinn University of Technology,
Tallinn, Estonia

D. P. Singh et al. (eds.), *Microbial Interventions in Agriculture and Environment*,
https://doi.org/10.1007/978-981-13-8383-0_1

of attention of the researchers, policy makers, agricultural activists, agro-industries, and farmers from high-input chemical-based farm practices to a more sustainable and environment-prone agricultural system through the usage of safer alternatives like bioorganic farm inputs is highly demanded (Weyens et al. 2009; Abhilash et al. 2013; Bhardwaj et al. 2014).

Crop production and protection system has witnessed a number of viable biological approaches in the last few decades (Philippot et al. 2013). These approaches are based on basic methodologies of natural environmental practices and balanced ecological sustenance to minimize the threats of huge chemical inputs in agricultural farms (Abhilash et al. 2013; Singh et al. 2016). The most important among all such biological options are the microbial inoculants which are now being used to promote plant health and soil fertility, improve plant resistance against biotic and abiotic agents, and reduce damage due to pests and diseases (Singh and Singh 2014; Singh et al. 2016). The prospects of enriching soils with bioorganic inputs, altering soil biology with cultural practices and manipulating plants rhizosphere by inoculating defined quantity of microbial beneficial population (bacteria, actinobacteria, cyanobacteria, mycorrhizal fungi, fungi) individually and/or as consortia, are enormous. It has shown proven impacts on plant growth and development, crop production, enhanced intrinsic plant resistance/tolerance against stresses (biotic and abiotic), plant immunization against diseases and pests, soil remediation, and improved soil fertility standard (Reddy and Wang 2011). Practically, the benefits of such approaches are seen in terms of direct reduction in the usage of chemical inputs, improved soil fertility index, decline in the disease conditions and abiotic impacts, and associated environmental risks (Nandakumar et al. 2006; Mayak et al. 2004; Nadeem et al. 2014). These approaches can be further strengthened from our understanding of biological interactions and microbial processes within communities and inhabitant plants, delivery system, edaphic conditions, stresses occurring in the rhizosphere, and viability issues. Technological reliability in the fields and adoption of technologies by the farmers will definitely pose practical requirements for the development of formulations of efficient microbial inoculants, scaling-up of viable technology, quality control management, and commercialization.

2 Microorganisms and Agriculture

Microbes are the most varied and abundant group of organisms on Earth (Fierer and Jackson 2006). Plant microbiome comprises millions of microorganisms inhabiting plants and creating complex ecological niche that directly affects plant growth by cumulative interactions and metabolic functions (Berg 2009; Lugtenberg and Kamilova 2009). Soil microbial diversity contributes enormously to valued ecological services with their diverse gene pool and metabolic functions (Nannipieri et al. 2003; Nihorimbere et al. 2010; Nihorimbere 2014) and benefit both soils and plants (Lutenberg and Kamilova 2009). Their role is fundamentally important in nutrient recycling, mineralization processes, conversion of complex agricultural residues, and remediation of soil contaminants (Sparling 2013). Close interaction mechanisms (epiphytic and endophytic both) exist in the rhizosphere with microbial communities that grow on exudates and, in turn, facilitate plants in their growth and protection attributes against biotic and abiotic stresses (Morgan et al. 2005; Berg and Smalla 2009). Thus, the task of microbial identification, characterization, utilization, and field application of these communities has a wide array of implications. Microbial communities have multifarious intrinsic traits like antagonism against phytopathogenic organisms, plant growth promotion, nutrient acquisition, carbon sequestration, and bioremediation. Such information practically help in developing efficient microbial inoculants that offer beneficial agricultural services at farm level without disturbing the soil structure, function, and ecological balance (Higa 1991; Schloter et al. 2000; Dey et al. 2014).

Over the last couple of decades, microbial biotechnology and their functional aspects have been worked out greatly due to the advancements in the molecular biology techniques, chemistry, microscopy, and analytical disciplines (Strobel and Daisy 2003; Lorenz and Eck 2005; Tringe et al. 2005; Milshteyn et al. 2014). These technologies have helped in the identification of microbial communities, structure and associations with the root system, their functional attributes, and metabolic diversity in the interaction area (phyllosphere or rhizosphere) (Rastogi and Saini 2011; Suyal et al. 2014) and facilitated identification of functional microbial communities, their potential genes, proteins, and/or metabolites linked with the functions. Advanced information is now being accumulated on plant-microbe and plant-pathogen interactions and pathogen recognition at the interface. Studies also improved our understanding on induced systemic resistance, plant's innate immunity, root rhizosphere biology, antagonistic attributes of microbial communities, impact on inoculated microbes on nontargeted organisms, and strengthened facts about the benefits of microbial inoculation in soils and/or with plants (Mei and Flinn 2010; Reddy and Saravanan 2013; Farrar et al. 2014). This article is aimed to present an in-depth account of microbes as functional communities in the soils, functional metabolic capabilities for becoming bio-inoculants, delivery and application modes, concerns about targeted and non-targeted affects, biosafety consideration, biotechnological interventions for production and commercialization, and potential benefits to the plants, soils, and the end users, i.e., the farmers. With the use of crop- and soil-friendly microbial inoculants, much can be achieved in terms of ecological sustainability, economic concerns of farming communities, and health benefits to humans and livestock.

3 Microbial Communities as Functional Components in the Soils

Soil is a dynamic but complex biological system. It is a living system by its biology that differentiates it from weathered rock (regolith). The formation of soils is a complex biological, chemical, and physical phenomenon, and microorganisms take a lead role in biological transformation of stable and labile pool of minerals (carbon, nitrogen, phosphorus, and other nutrients) to develop healthy soils that subsequently facilitate establishment of different communities of organisms including plants. Being biological in nature, soil functions result in complex multiphasic interactions with the abiotic and biotic environment (Kibblewhite et al. 2008). Of all the natural inhabitants of the soils, microbial community structure, composition, and function present the largest component that interacts widely within and between other organisms and collectively contributes to a variety of ecological services (Power 2010). Although there remain certain limitations in assessing and quantifying microbial diversity and appropriate functional activities of microbe-mediated processes in the soils (Rastogi and Saini 2011), the measures for the direct assessment of microbial composition structure and function can be used to measure ecosystem responses to anthropogenic reasons and to mark ecosystem recovery (Banning et al. 2011).

Microorganisms are the unseen, cosmopolitan, and ancient biological entities that successfully colonize all possible niches including soils. Being a prominent architect of the soil, their active presence is essential to the healthy performance of every live soil (Rajendhran and Gunasekaran 2008). Many of the ecosystem services like decomposition and biodegradation, water and air containment in soils, and carbon sequestration are directly linked to microbial activities and their functional traits that support terrestrial biology and plant production. Physicochemical and biological (organic) conditions of the soils influence microbial community composition and their function (Lombard et al. 2011). The role that microbes play to maintain dynamic equilibrium and integrity of soils is of such magnitude that continued existence of vital biological life becomes crucially and sustainably dependent on microbe-mediated processes. In soils, such microbe-mediated processes include nutrient recycling (Lloyd et al. 2009; Sparling 2013), plant growth promotion (Bais et al. 2006; Glick 2012; Compant et al. 2010; Philippot et al. 2013), disease and pest suppression (Weller et al. 2002; Mazzola 2007), fertility status upgradation (Rao 2007; Schulz et al. 2013), community structure and function (Lombard et al. 2011), remediation of contaminants (heavy metals, xenobiotics, pesticides) (Batty and Dolan 2013), and biodegradation of agricultural wastes (Bashan et al. 2012).

4 Soil as a Complex and Dynamic Biological System

Soil represents a living system characterized by different functionalities and complex interactions among all components of soil including abiotic, physical, and chemical environment. Soil, either natural or agricultural, provides habitat for various organisms which cumulatively led to a range of soil-based functionalities and specialties (Kibblewhite et al. 2008). Soil represents a complex and active biological system where it is too difficult to identify its microbial community composition. Further, determination of microbe-mediated reactions is limited because currently available assays focus on the rate measurement of the entire metabolic processes (e.g., respiration) or the activities of the enzymes like urease, protease, and phosphomonoesterase. Determined reactions do not allow enumeration of microbial species that are involved in such processes (Nannipieri et al. 2003). The essential problem caused by the relation between microbial diversity and soil function is to recognize the relations among genetic diversity and structure and also in between community structure and function. An enhanced understanding on relations between diversity of microbes and soil function needs different efficient assays for characterization of soil DNA and RNA at taxonomic and function levels and then detection of dead or active live cells in the complex soil matrix (Nannipieri et al. 2003). Soil exhibits a massive amount of interspecific relationships, including both trophic (food webs) and nontrophic (e.g. antagonism, mutualism, neutralism, commensalism, and competition) (Vasas and Jordán 2006; Knudsen et al. 2014).

5 Microbial Communities of Soil

Soil microbial community composition is a reflection of the ecosystem's reactions toward anthropogenic turbulence (van Dijk et al. 2009). This also acts as an indicator for the ecosystem functioning, loss, and recovery (Harris 2003; Lewis et al. 2010). Rather than plant communities, only little experimental information is available about predicting alterations in community composition while secondary succession (Felske et al. 2000; Kuramae et al. 2010) or even during ecosystem restoration (Gros et al. 2006; Jangid et al. 2010; Banning et al. 2011). Microorganisms respond more quickly to environmental changes and thus can serve as an early signal of the recovery trajectory (Harris 2009; Banning et al. 2011). Community structure and composition measures of soil microbes are continuously used to assess the response of ecosystem toward anthropogenic interruptions and serve as an indicator for ecosystem recovery (Harris 2003; Lewis et al. 2010; van Dijk et al. 2009). Edaphic factors which are supposed as considerable drivers behind soil microbial community structure consist of soil pH (Fierer and Jackson 2006; Rousk et al. 2010; Wakelin et al. 2008), soil carbon (C) (Bardgett et al. 1999; Carney and Matson 2005; Pennanen et al. 2004) and nitrogen (N) (Pennanen et al. 1999; Ruppel et al. 2007), soil water (Drenovsky et al. 2004; Hackl et al. 2004), texture (Carson et al. 2007), and mineralogy (Gleeson et al. 2006). These factors directly affect the structure of microbial community concurrently and result in an interactive and responsive impact (Allison et al. 2007). Therefore, microbial community structure measures might be conceptualized as a wholesome assessment of various soil and ecosystem characters (Banning et al. 2011).

Soil biota is also significantly involved in the resistance and resilience of agroecosystems toward different abiotic stresses (Brussaard et al. 2007; García-Orenes et al. 2013). Microbial communities in the soils represent the most sensitive and fast indicators for perturbations and other changes, and therefore, there is increasing interest in utilization of soil microbes and diversity as a prospective tool for soil quality evaluation (Zelles 1999; Zornoza et al. 2009; García-Orenes et al. 2013).

6 Functions Associated with the Soil Microbial Communities

Soil microbial communities owe a range of ecosystem functionalities and are crucial biotic indicators for soil health (Jackson et al. 2003; van Bruggen and Semenov 2000). Any harmful impact of agricultural management systems over soil microbial communities can damage their functionalities and influence their ecological works like nutrient cycling and crop protection (Pimentel et al. 2005). In terrestrial ecosystems, soil communities are accountable for a major proportion of decomposition and nutrient mineralization (Cregger et al. 2012). Therefore, it's worth to improve our understanding on composition, function, and dynamics of soil communities under various especially contrasting management systems.

Different assays and techniques are utilized for the study of whole soil microbial community structure, e.g., denaturing gradient gel electrophoresis (Marschner et al. 2003), phospholipid fatty acid (PLFA) analysis (Bossio et al. 1998; Zhong et al. 2010), fingerprinting-based techniques like community-level physiological profiles (Zhong et al. 2010; Zhang et al. 2008), as well as terminal restriction fragment length polymorphism (T-RFLP) (Sessitsch et al. 2001; Blackwood and Paul 2003; Blackwood et al. 2006). For microbial community analysis at phylogenetic and taxonomic level, 16S rRNA gene clone libraries (Jangid et al. 2008; Sessitsch et al. 2001), 16S rRNA-targeted oligonucleotide probes (Buckley and Schmidt 2001), and 454 pyrosequencing methods (Levine et al. 2011; Ramirez et al. 2010) have been used. Though limited information is available about ecological functions of the microbial communities in different agricultural systems, particularly functional groups of microbes, e.g., denitrifiers (Cavigelli and Robertson 2001), nitrifiers (Phillips et al. 2000), and methanotrophs (Levine et al. 2011). Particular functional genes, like *nirS*, *nosZ*, and *nifH* (Morales et al. 2010), and certain extracellular enzyme activities are analyzed, but they facilitate only with a discrete information over the ecological functions of soil microbial communities.

7 Beneficial Microorganisms and Functions

Soil microbes exert significant potential for agriculture. Numerous microbes are utilized in agricultural practices to enhance crop production. In the agriculture system, beneficial microbial inoculants majorly include plant growth-promoting bacteria and fungi. They function via various mechanisms, like as nutrient supply, plant hormone production, and the inhibition of different crop pests (Toyota and Watanabe 2013).

Plant growth promoting bacteria (PGPB) comprise a group of beneficial microbial species that inhabit plant rhizosphere, root surface, and surrounding areas. They enhance plant growth and provide resistance toward different diseases and abiotic stresses (Dimkpa and Weinanad 2009; Grover et al. 2011; Glick 2012; Souza et al. 2015). Differential mechanisms are used by PGPB for plant growth promotion. They participate in enhancing nutrient availability, nitrogen fixation, phosphate solubilization, stress alleviation, phytohormone secretion, and siderophore production. Plant and bacterial interactions are of various kinds, viz., symbiotic, endophytic, or associative, depending on the level of proximity with the roots and surrounding soil (Souza et al. 2015). Microorganisms inhabiting rhizosphere compete for space, water, carbon, and nutrients and occasionally progress their competitiveness through development of a close association with the plant (Nihorimbere et al. 2011). These microbes cooperate in the growth and ecological health of the host. An in-depth understanding about the rhizosphere microbial ecology along with the function and diversity of inhabiting microbes is needed for application of soil microbial technology in the rhizosphere (Nihorimbere et al. 2011).

8 Interactive Relations of Diversity, Community Structure, and Functions

Microbial species possess specific roles in retaining numerous ecosystem functions and services concurrently. These include primary production, nutrient cycling, climate regulation, and litter decomposition (Delgado-Baquerizo et al. 2016). Microorganisms are the central component in biogeochemical cycles of Earth, although the correlation among microbial community structure and ecosystem processes needs advanced exploration (Carney and Matson 2005; Prosser et al. 2007). Connection between complicated microbial communities and ecosystem processes are still not totally known (van der Heijden et al. 2008; Wallenstein and Hall 2012; Graham et al. 2014; Martiny et al. 2015). Current research is going on for a better understanding of carbon (C) and nitrogen (N) cycling by utilizing microbial data rather than only on environmental basis (Todd-Brown et al. 2012; Wieder et al. 2013; Reed et al. 2014; Powell et al. 2015). Although researchers are trying to improve ecosystem process models by defining parameters on microbial physiological attributes like drought tolerance (Manzoni et al. 2014), dormancy (Wang et al. 2015), growth efficiency (Hagerty et al. 2014), and turnover rates (Wieder et al. 2013). These models remain nonreproducible to include distinctions in the structure of microbial communities that can influence rate of change in ecosystem processes (Bouskill et al. 2012; Kaiser et al. 2014). There seems to be a big gap in the knowledge on an integrated understanding about the interactions that drive changes in community structure and ecosystem functions. There still lies the question that under what particular circumstances or conditions microbial communities exhibit specific functionalities. This information is needed to add extensive knowledge on critical biogeochemical cycles and their responses toward present and future adverse environmental changes. This will further allow us to identify factors that decide microbial community structure and activity in space and time (Graham et al. 2016).

With recent advancements in the area of molecular biology, knowledge and understanding of microorganisms from different habitats at genetic, taxonomic, and ecological levels is improving (Peterka et al. 2003). The extensive paradigm in ecology that function is determined by community structure currently challenged the level of complexity among microbial communities. In short, the structure and ecological functions of microbial communities is dependent on the complex interactions encompassing several environmental factors (Purahong et al. 2014).

9 Microbial Diversity and Decomposition of Organic Matter

Organic matter occupies the top 20–30 cm of most of the productive soils. It fundamentally includes different organic macromolecules comprising major combinations of carbon, hydrogen, oxygen, nitrogen, phosphorus, potassium, zinc, and sulfur. Soil organic matter (SOM) is usually calculated in terms of the quantity of organic carbon. Almost 95% biomass of the soil is occupied by bacteria and fungi. Other kinds of biomass materials comprise micro-fauna (nematodes, protozoa), meso-fauna (acari, Collembola, mites), and macro-fauna (earthworms, termites, mollusks). All types of biomass in the soils interact in a complex manner to form soil-food web systems that decide the turnover of OM (organic matter) and related bound nutrients within the environment of soils (Wardle 2002; Coleman and Wall 2007; Condron et al. 2010). Activities of bacteria and fungi are primarily responsible for the decomposition of organic carbon in the soil where almost 10–15% of soil carbon flux gets straightforwardly credited to the action of fauna (Hopkins and Gregorich 2005; Condron et al. 2010). The huge majority of soil microbes are heterotrophic in nature and depend on organic substrates from residual agricultural products and dead and decaying materials for energy and nutrients (Condron et al. 2010).

Continuous formation and breakdown of organic matter in the soil is among the foremost factors behind plant and ecosystem productivity. It facilitates different activities related to nutrient acquisition, water holding, moisture availability, and development and preservation of physical structure of soils. In the majority of soils, almost 90% of the total nitrogen and sulfur composition and more than 50% of the phosphorus content as a whole remain in the bound form. These minerals are especially associated with the microbial biomass and organic matter. Cycling and bioavailability of these key nutrients of organic matter in the soil is further regulated by the biotransformation processes which are ultimately dependent on the activity of microorganisms and fauna. The SOM is the main cause of negative charges in most of the soils and, thus, is responsible for withholding and/or availability of different nutrient cations such as potassium, calcium, magnesium, and ammonium. Organic matter is also responsible for managing water-holding capacity, aeration, and accessibility of nutrients. Thus, the enhanced relationship of SOM with secondary minerals (e.g., clay) to form aggregates that strengthen soil structure and enhance aeration and water penetration is essential for plant growth (Condron et al. 2010), and microorganisms are the key drivers in maintaining SOM in the soils.

Decomposition is a typical microbe-mediated process in which different species first become dominant and then disappear to let other species become dominant as per the steps of degradation. The actual rate and extent of decomposition are primarily affected by the climatic factors like temperature, oxygenated conditions, moisture, nitrogen content, and carbon availability along with microbivorous soil fauna (Swift et al. 1979; Janzen 2004; Johnston et al. 2009). Impact of diversity of soil microorganisms on carbon cycle-associated ecosystem processes is not very clear, but the composition of microbial diversity in the decomposed products affects soil processes. The control of microbial processes over the environmental factors is enormously challenging and is governed by the consequences of interactions among communities, edaphic behavior, and ecosystem procedures (Reed and Martiny 2007; Condron et al. 2010). Soil microorganisms with their relations to the soil fauna and involvement in different management activities play a key role in the formulation of such kind of SOM that directly facilitate plant growth and development (Condron et al. 2010). This tuning can be helpful in creating such soils that directly respond to the challenges of the plants, especially in the nutrient-deprived and abiotic-stressed conditions.

Microorganisms in the cumulative biomass of fungi, bacteria, and actinomycetes organize the breakdown of SOM and modify subsequent continued nutrient addition to the soils (Condron et al. 2010).

The breakdown of SOM is dependent on extracellular enzymes, secreted by microbes. Microorganisms exude enzymes to gain carbon (C) or restricting nutrients (Sinsabaugh et al. 2009), and to aim the most plentiful substrates (Sistla and Schimel (2012). Extracellular enzyme activities are thus frequently associated with the chemical constitution of SOM and its carbon and nitrogen content (Sinsabaugh et al. 2008; DeForest et al. 2012). In addition to SOM quality and quantity, extracellular enzymes are also found to be connected to microbial diversity or the large quantity of particular microbial groups (Strickland et al. 2009; Kaiser et al. 2010; Schnecker et al. 2014). Different environmental factors like moisture, temperature, pH, and oxygen availability, in turn, govern the microbial communities in the SOM (Eilers et al. 2012; Schnecker et al. 2014). Modifications in these features encourage precise microbial groups that exhibit improved adaptation to the new environment, although these microbial groups may slightly differ in their functional properties. This could ultimately modify microbial enzyme activities, microbial processes, and eventually decomposition of SOM (Waldrop and Firestone 2006; Talbot et al. 2013; Schnecker et al. 2014).

10 Microbial Communities and Soil Enzyme Activity

Information regarding community composition and ecological functionalities of soil microbes is not much known despite the fact that functionality and productivity of the soil are driven by its biological counterpart (Acosta-Martinez et al. 2008). Though there is immense interest in identification of global biodiversity and the role of microbes in ecosystem functionality, it is of utmost importance to gain information about the soil microbial communities (Dick-Peddie 1991; Acosta-Martinez et al. 2008). Microbial diversity is a key player behind soil quality and functioning. They modulate SOM composition, nutrient cycling, and decomposition dynamics along with the degradation of the xenobiotics. Any variations in the microbial community structure are most likely imitated in the functional properties of the soil (Insam 2001; Acosta-Martinez et al. 2008) as the microbial communities affect the capacity of soils for various enzymes like hydrolases (Kandeler et al. 1996; Acosta-Martinez et al. 2008). Significant amount of soil enzyme activity regulates SOM and nutrient recycling and influences soil management (Acosta-Martinez et al. 2008).

Levels of soil microbial enzymes are found to significantly correlate with overall organic carbon and nitrogen present in soils (Aon and Colaneri 2001). While microbial enzymes urease and phosphatase play a critical role in driving elementary soil functions like biogeochemical cycling of different nutrients (as carbon, nitrogen, and phosphorus (Garcia et al. 2002)), their levels can be utilized for evaluation of soil quality and ecosystem functioning (Dick 1997; Deng and Tabatabai 1999; Habig and Swanepoel 2015). Though microbial communities are renowned on the basis of soil type, plant species, tillage practices, cropping systems, etc. (Dick et al. 1996), stimulation of soil microbial communities with the finest agricultural systems could support the accessibility of carbon sources for microbial use. This will ultimately affect enzymatic activity and soil microbial diversity which will lead to faster nutrient recycling. In due course of time, these factors could ultimately lead to enhanced soil quality and fertility, resulting in a significant beneficial effect of sustainable agricultural productivity (Habig and Swanepoel 2015).

11 Microbes as Abiotic Stress Managers for Plants

11.1 The Plant-Microbe Association

Plant seeds harbor their own associative microflora; moreover, when a seed comes in contact with the soil, numerous indigenous microbes interact with it during as well as post germination, which involves complex cascade of events mediated by series of biologically active organic molecules released from roots of the growing seed. Being even smarter than the plants, microorganisms are likely to better utilize chemical molecules to permit the colonization of selective population in the rhizosphere region (Bais et al. 2004). Plenty of microbes actively colonize different regions of the rhizosphere, viz., ecto- and endo-rhizosphere and the rhizoplane. The colonizing population develops beneficial interactions with host and produces different biologically active molecules capable of influencing plant growth. Thus such microbes are cumulatively termed as plant growth promoters (PGPs). Microbes being omnipresent are found in both stressed and normal habitats. It is therefore quite natural to observe the above interactions in stressed habitats as well. This fact has been evident from many observations (Timmusk et al. 2014; Sorty et al. 2016; Yuan et al. 2016), relating to exploration of wild plant-associative microflora from stressed habitat and their PGP efficiency. However, the question remains whether such microbes are/will really benefit crop productivity under stress circumstances. Fortunately, the consequent studies with crop plant have established sufficient affirmative knowledge in this regard (Sorty et al. 2016; Dombrowski et al. 2017). Many observations have revealed the growth promotion and stress alleviation potential of the PGP microbial strains isolated from different habitats. Microbes elaborate different metabolites that help the plant grow smoothly under stress conditions. The metabolites predominantly include exopolysaccharides, microbial derivatives of the plant growth hormones, volatile organic compounds, different enzymes such as ACC deaminase, phytase, etc.

11.2 Microbe-Plant Interaction Under Abiotic Stressor(s)

Plants simultaneously encounter multiple kinds of abiotic stressors at a given point of time; thus, the consequences and responses generated are also equally complex (Rizhsky et al. 2004; Meena et al. 2017). Major stressors include light, salinity, drought, heat, frost, heavy metals, etc. Microbial metabolites play their counterpart under such complex situations. For instance, the ACC deaminase produced by associative microbes metabolizes the ACC, a precursor molecule involved in the pathway of ethylene biosynthesis, which plays a vital role in abiotic stress management of plants (Bleecker and Kende 2000). The lowered levels of ACC ultimately reduce the accumulation of ethylene in plant tissues, thereby preventing occurrence of the adverse effects arising due to raised levels of ethylene under the influence of various stressors of abiotic conditions (Glick et al. 1998; Mayak et al. 2004; Stearns et al. 2005). Similarly exopolysaccharides (EPS) produced by the bacteria help plants sustain under water-deficit conditions. EPS are carbohydrate polymers that serve as carbon reservoirs and water-holding agent. It can bind many folds of water by weight to its molecules. The water-holding potential of the EPS makes them extensively useful under water-deficit environments where they can ensure formation of soil aggregates and moisture retention in the rhizosphere region (Gouzou et al. 1993; Bensalim et al. 1998). Additionally, the EPS can also serve as signaling factors for microbial colonization and can stimulate development of microbial biofilms. EPS can also form a thin film on root surfaces, thereby forming sheath-like structures called "rhizofilm." The rhizofilm holds a variety of microorganisms that take part in plant growth promotion by different mechanisms. Polysaccharides can also work as osmolytes to trap the water under highly saline conditions; thus, the film of polysaccharide on the root surface can probably prevent/reduce the entry of excess salt inside the plant cells (Batool and Hasnain 2005; Sandhya et al. 2009; Arora et al. 2010).

11.3 Role of Indigenous Soil Microbiota Under Abiotic Stressor(s)

Microbial activities are possible only under the conditions of optimal colonization; exposure to the stressor(s) eventually affects important nodes in the root exudate-mediated signaling cascades during the rhizosphere colonization due to metabolic hindrance in the plant and the microorganisms in colonization (Micallef et al. 2009; Meena et al. 2017). Augmentation of microbial inoculum has been suggested under such circumstances. The physiology and metabolism of the exogenous microbes are also equally affected upon exposure to the physicochemical environment encountered in situ (O'Callaghan 2016). Advantageous utilization of the "indigenous" PGP microbes could prove beneficial owing to their well-adapted lifestyle under the stress situation. Deliberate augmentation of such microbes in rhizosphere region may help to attenuate the adverse impact of stressor. Supplement with the microbial metabolites itself could offer additional benefits, particularly due to excellent control over the production as well as the rate of augmentation in the rhizosphere region. However, simulation of the in vitro microenvironment for optimal production of metabolites by microbes represents a major challenge. Identification of the role contributed by major microbial metabolites could potentially open new avenues in the area of microbial management of abiotic stresses in plants. Keen observations have revealed that some of the metabolites produced by PGP strains contribute to a significant role in seed germination and establishment (Sorty et al. 2016). Similarly, the interactive impacts of the inoculation with multiple exogenous microbial strains also need to be studied so as to acquire the knowledge regarding interaction of the communities with the plants and influence on indigenous microflora following the invasion by the foreign inoculate. Volatile organic compounds (VOCs) from the microorganisms are of great importance under abiotic stress environment, particularly owing to their systemic action potential. Studies in this area have provided considerable insights to the potential of microbial VOCs in mitigating abiotic stressors. These volatiles rapidly diffuse throughout the plant microenvironment and can potentially act faster than other nonvolatile chemical molecules. Such studies may offer potentially desired gateways for a collective study of complex interactions between the plant and associated microbial communities.

11.4 Microbe-Mediated Metabolic Regulation Under Abiotic Stressor(s)

Both acute and chronic exposures to the stressors like salinity, drought, heat, and frost generate oxidative stress in plants. This is detectable in terms of raised levels of reactive oxygen radicals including H_2O_2. Microbes can actively enhance the ability of plants to combat increasing cellular oxidative stress under stress environment. Though the exact mode of operations is yet to be revealed, many PGPRs have been well demonstrated for their positive influence on cellular antioxidant management in plants. Microbes can enhance the antioxidant status of plants by two strategies: one being rise in the levels of free radicals/superoxide scavenging enzymes that include key enzymes having high affinity toward reactive oxygen species (ROS). Major enzymes taken into consideration include superoxide dismutase, catalase, ascorbate peroxidase, guaiacol peroxidase etc. Significant literature regarding enhancement in antioxidant enzymes followed by PGPR inoculation has been generated. These enzymes significantly reduce the levels of oxidative stress in plants by neutralizing the ROS, thus protecting the vital cellular organelles from subsequent oxidative damage. Another strategy involves nonenzymatic management of oxidative stress, where a class of reducing compounds are recruited to attack the ROS. A variety of phenolic compounds from plants have been shown to have high antioxidant potential, which have been thought to be involved in neutralization of ROS. PGPRs have been shown to efficiently enhance levels of the plant phenolic compounds, which further augment the plants' combat mechanism against the cellular oxidative stress. Though these strategies have been shown successfully mitigating oxidative stress in abiotically stressed plants, the underlying molecular mechanisms are still unclear.

11.5 Utilization of Microorganisms for Abiotic Stress Alleviation

Various soil microorganisms including the species of *Azotobacter*, *Azospirillum*, *Pseudomonas*, *Bacillus*, *Methylobacterium*, *Rhizobium*, *Pantoea*, *Bradyrhizobium*, *Sinorhizobium*, *Arthrobacter*, and *Enterobacter* are somehow associated with plant growth promotion and development under varying environmental conditions (Gouzou et al. 1993; Omar et al. 2009; Meena et al. 2012; Liu et al. 2013; Supanekar and Sorty 2013a,b; Supanekar et al. 2013; Nakbanpote et al. 2014; Sorty et al. 2016). They employ different mechanisms potentially driven by their metabolic potential. Abiotic stressors impact the nutrient-mobilizing ability of the microbes. However, a fraction of the adapted indigenous microbial population still manage to retain the ability to efficiently perform the task and may supplement the plant with adequate nutrient supply under altered soil chemical environment in the presence of stressors (e.g., salinity). Similarly, microbes adapted to lower water potential may outperform under water-deficit conditions where sustenance itself becomes a major task for the organisms lacking the ability to form resistant and/or dormant structures.

Additionally, the phyllosphere microflora harbors a handful of potential genes associated with plant growth enhancement and mitigation of different abiotic and biotic stressors. Multiple genes from culturable and nonculturable sources of the phyllosphere microflora are being explored (Chen and Pachter 2005; Kapardar et al. 2010; Oldroyd 2013). Microbial population associated with the wild plants adapted to extreme environments can potentially offer an additional package of knowledge in this regard. Efforts of utilizing genetic potential of microbes from systems-biological perspectives for mitigating abiotic stressors in plants are need of the time. Studies in this area have led to promising outcomes. Microbes are known to regulate expression of stress-responsive genes in plants with the help of different signaling molecules that they synthesize as metabolic products. For instance, inoculation with *Paenibacillus polymyxa* successfully induced the expression of ERD15 gene in arabidopsis to attenuate the adverse effects of water deficit (Timmusk and Wagner 1999). Crops including rice, beet, and tobacco have been shown to retain leaf moisture more efficiently under virus-challenged state. Deliberate exploration of molecular interplay from such regulatory cascades could potentially open new gateways to utilize genetic potential of microbes in mitigating abiotic stresses in plants.

High-throughput omic-based characterization of both the host and associated microbial communities can facilitate the rate of success. Detailed account of multi-omics strategies-mediated mitigation of abiotic stressors in plants has been reviewed recently (Meena et al. 2017).

Time also demands development of next-generation strategies involving microbial metabolic products that can attenuate the detrimental impact of abiotic stressors and permit the plant to grow well even in stressed habitat. Keen characterization of metabolome of PGPRs and their host can open new gateways in this area. Few preliminary studies concerning the same represent promising milestone (Micallef et al. 2009; Caldana et al. 2011; Pieffer et al. 2013; Jorge et al. 2015; Jia et al. 2016). This approach ensures reliable supply of the microbial product(s) to the plants in desired concentration, pertaining to their exogenous amendment, which can otherwise hardly be ensured due to metabolic hindrance in microbes caused by changing biogeochemical situations in situ. The same fact also applies to failure of the microbial inoculate under changing agro-climatic circumstances.

Microbes inhabiting in close association with the plant roots exhibit a noteworthy role in stress mitigation in crops grown in stressed soils owing to their unique properties of tolerance toward extremities, their interaction with crop plants, and potential exploitation approaches (Qin et al. 2016). Rhizobacteria, with its 2–5% population when introduced in the roots as soil inoculum, act as competitive microflora and put forth beneficial impact on plant growth. This is how they are known as plant growth-promoting rhizobacteria (PGPR) (Ahemad and Kibret 2014). Few PGPRs which are able to colonize the root system in saline conditions present significant potential as inoculants. PGPRs are advantageous bacteria that occupy plant roots in the vicinity to soil and represent one of the major beneficial microbial communities that regulate belowground growth processes. PGPRs improve plant growth due to diverse functional mechanisms like asymbiotic N_2 fixation, phosphate solubilization, plant hormone production, etc. (Meena et al. 2017). Rhizobacteria work against osmotic stress and assist plant growth.

Plants treated with rhizobacteria exhibited improved root and shoot growth, chlorophyll content, hydration, nutrient uptake, and disease resistance. Rhizobacteria improve the circulation of plant nutrients in the rhizosphere and osmolyte accumulation in plants. Furthermore, plants inoculated with rhizobacteria contain high K^+ ion concentration and also a higher K^+/Na^+ ratio that improves salinity tolerance (Qin et al. 2016). Rhizobacteria also stimulate synthesis of antioxidant enzymes in plants that degrade reactive oxygen species produced on salt shock. Presence of numerous favorable rhizobacteria like *Rhizobium*, *Bradyrhizobium*, *Azotobacter*, *Azospirillum*, *Pseudomonas*, *Bacillus*, etc. is reported from different stressed ecological niches (e.g., deserts, acid soils, saline, and alkaline areas) and are believed to be engaged in the natural reclamation process of the soil (Ahemad and Kibret 2014).

Soilborne pseudomonads have gained specific attention owing to their potential of producing metabolites, root-colonizing ability, and capability to secrete a wide range of enzymes which assist plants in adverse environments. Some PGPRs may fix nitrogen, solubilize phosphate, produce phytohormones, and acquire iron and zinc, while others exert indirect benefits to plants by protecting them against soilborne diseases (Meena et al. 2017). Beneficial microorganisms colonize the rhizosphere/endo-rhizosphere of plants and impart drought tolerance by producing exopolysaccharides (EPS), 1-aminocyclopropane-1-carboxylate deaminase (ACC deaminase), phytohormones, volatile compounds, osmolytes, and antioxidants and by regulation of stress-responsive genes and changes in root structure. Specific mechanism of plant drought tolerance by rhizobacteria routes through (i) abscisic acid (ABA), gibberellic acid, cytokinins, and indole-3-acetic acid (IAA), (ii) ACC deaminase to decrease the level of ethylene in the roots, (iii) systemic tolerance induced by bacterial compounds, and (iv) EPS accumulation (Vurukonda et al. 2016).

Commercial use of microbial inoculants for enhanced crop growth and productivity in stressful environment may become a sustainable approach in agriculture. Currently, various PGPR inoculants are commercialized that endorse growth or induce systemic resistance against pathogens or control pathogens or bring about mitigation of stress tolerance or other beneficial activities. Understanding the interactive mechanisms among consortium of microbial inoculants, soils, and plants may become advantageous for improving plant growth and increasing stress tolerance.

12 Microbial Role in Bioremediation

Increasing anthropogenic activities have resulted in the widespread pollution of the Earth (Peuke and Rennenberg 2005; Tripathi et al. 2014). Heavy metals and organic pollutants constitute the major contaminants of the soils worldwide. Therefore, there seems to exist an urgent need to focus on the development of clean-up remedies for restoration of such contaminated soil. Remediation of the contaminated soil with the current technologies is costly and relatively slow and needs to be revamped urgently as the number of contaminated site may increase worldwide (Abhilash et al. 2013). Plants and microorganisms grow and degrade pollutants from the contaminated environment. This could be exploited for developing cheaper and efficient alternative technologies for the clean-up of contaminated soils (Table 1.1).

Although the role of plant growth-promoting microorganisms has been explored in different biotic and abiotic conditions, their importance in bioremediation is still underestimated and needs extensive exploration. The microbe-mediated remediation technology also improves the soil health by enhancing the content of soil organic carbon, micro- and macronutrients, soil porosity, and permeability. However, productivity of such contaminated land is quite low and could be improved further by efficient application of microbe-assisted phytoremediation procedures. As the toxicity and bioavailability of the metals is the major problem affecting the phytoremediation process, the application of heavy metal-tolerant PGPR can improve the remediation period by metal uptake at a faster rate from the soil. Similarly, the microbes could also produce chelating agents and biosurfactants which could further enhance the bioavailability of the metals in the contaminated site (Ma et al. 2011).

Table 1.1 Details of different microorganisms with remediation capacity

Pesticide/herbicide/insecticide/environmental remediation	Organism name	References
Microbes with pesticide degradation capacity		
Allethrin	*Acidomonas* sp.	Paingankar et al. (2005)
Aniliofos	*Synechocystis* sp.	Singh et al. (2016)
Atrazine	*Pseudomonas*	Prabakaran and Peterson (2006) and Wyss et al. (2006)
Bifenthrin	*Acinetobacter calcoaceticus*, *Enterobacter aerogenes*	Tingting et al. (2012) and Lio and Xie (2009)
Carbamate	*Sphingomonas yanoikuyae*	Ouyang et al. (2008)
Carbofuran	*Bacillus* sp. and *Chryseobacterium joostei*	Foster et al. (2004)
Chlorpyrifos	*A. xylosoxidans* JCp4 and *Ochrobactrum* sp. FCp1, *Phormidium valderianum* BDU 20041	Akbar and Sultan (2016) and Palanisami et al. (2009)
Cypermethrin	*Enterobacter aerogenes*, Photosynthetic bacterium (GJ-22), *Pseudomonas putida* and *Pseudomonas mendocina*	Lio and Xie, (2009), Yin et al. (2012), and Mendoza et al. (2011)
DDT	*Sphingomonas*, *Sphingobacterium* sp., *Stenotrophomonas maltophilia*, *C. elegans*	Shunpeng and Mingxing, (2006), Fang et al. (2010), Barragán-Huerta et al. (2007), and Seo et al. (2005)
Diazinon	Rot fungi	Sagar and Singh (2011)
Difenoconazole	*Fusarium oxysporum*, *Lentinula edodes*, *Penicillium brevicompactum*, and *Lecanicillium saksenae*	Hai et al. (2012)
Endosulfan	*Stenotrophomonas maltophilia*, *Pseudomonas*, *Mortierella* sp. strains W8 and Cm1-45, *Aspergillus*	Barragán-Huerta et al. (2007), Prabakaran and Peterson (2006), Wyss et al. (2006), Kataoka et al. (2010), Bhalerao and Puranik (2007), and Javaid et al. (2016)

(continued)

Table 1.1 (continued)

Pesticide/herbicide/insecticide/environmental remediation	Organism name	References
Esbiothrin	*Acinetobacter*	Ha et al. (2009)
Ethion	*Azospirillum* and *Pseudomonas*	Zhang et al. (2007)
Fenamiphos	*Nostoc* sp. MM1, *Nostoc* sp. MM2, *Nostoc* sp. MM3, *Nostoc muscorum* and *Anabaena* sp.	Cáceres et al. (2008)
Fenpropathrin	*Sphingobium* sp. JQL4-5	Yuanfan et al. (2010)
Glyphosate	*Anabaena* sp., *Arthrospira fusiformis*, *Leptolyngbya boryana*, *Microcystis aeruginosa*, *Nostoc punctiforme*, *Spirulina platensis*, *Spirulina* sp.	Forlani et al. (2008), Lipok et al. (2009), and Lipok et al. (2007)
Hexachlorocyclohexane	*Sphingobium japonicum*	Liu et al. (2007a, b)
Imidacloprid and metribuzin	*Burkholderia cepacia* strain CH-9	Madhuban et al. (2011)
Lindane	*Bacillus* sp. and *Chryseobacterium joostei*, *Fusarium verticillioides*, *Anabaena* sp. strain PCC 7120 and *Nostoc ellipsosporum*	Foster et al. (2004), Guillén-Jiménez et al. (2012), Pinto et al. (2012), and Kuritz and Wolk (1995)
Malathion	*Bacillus thuringiensis*, *A. oryzae*, *N. muscorum*, and *S. platensis*, *Aulosira fertilissima* ARM 68 and *Nostoc muscorum* ARM 221	Zeinat et al. (2008), Ibrahim and Essa (2010), Ibrahim et al. 2014, and Subramanian et al. (1994)
Methomyl	Rot fungi	Sagar and Singh (2011)
Methyl parathion	*Bacillus* sp. and *Chryseobacterium joostei*, *Vibrio* and *Shewanella*, *Microcystis novacekii*	Foster et al. (2004), Liu et al. (2006), and Fioravante et al. (2010)
Metribuzin	*Burkholderia cepacia* strain CH-9	Madhuban et al. (2011)
Monocrotophos	*Aulosira fertilissima* ARM 68 and *Nostoc muscorum* ARM 221	Subramanian et al. (1994)
Para-nitrophenol	Rhodococcus bacteria	Zhang et al. (2009)
Pendimethalin, Terbuthylazine	*Fusarium oxysporum*, *Lentinula edodes*, *Penicillium brevicompactum*, and *Lecanicillium saksenae*	Hai et al. (2012)

(continued)

Table 1.1 (continued)

Pesticide/herbicide/insecticide/environmental remediation	Organism name	References
Permethrin	*Pseudomonas putida* and *Pseudomonas mendocina*	Mendoza et al. (2011)
Pirimicarb	*Trichoderma viridae* and *T. harzianum*	Romeh (2001)
Pyrethrin	*Sphingomonas yanoikuyae*	Ouyang et al. (2008)
Pyridine	*Paracoccus* sp. strain	Qiao and Wang (2010)
Triazophos	*Ochrobactrum*	Shunpeng and Shen (2005)
Chlorinated pesticides, herbicides, and fungicides	*Rhodobacter sphaeroides*	Harada et al. (2006)
Polychlorinated biphenyls (PCBs)	*Anabaena* PD-1	Zhang et al. (2015)
Organophosphate pesticides	*Acinetobacter johnsonii* (MA-19) strain	Xie et al. (2009)
Organo-phosphorus and organo-chlorine insecticides	*Synechococcus elongatus*, *Anacystis nidulans*, and *Microcystis aeruginosa*	Vijayakumar (2012)
Multiple (aldrin, aldicarb, alachlor, atrazine, chlordane, diuron, DDT, dieldrin, gammahexachlorocyclohexane (g-HCH), heptachlor, lindane, mirex, metalaxyl, terbuthylazine)	White-rot fungi	Das and Chandran (2011) and Nyakundi et al. (2011)
Multiple (BHC, DDT, endosulfan, HCH isomers, and 2,4-D)	*Escherichia coli*	Qiao et al. (2003), Gupta (2005), Shun-Peng et al. (2005), Chaudhary et al. (2006), Santacruz et al. (2005), and Xue-Dong et al. (2003)
Multiple (herbicide 2,4-D, endosulfan, lindane, chlorpyrifos)	*Pseudomonas* and *Alcaligenes* sp.	Mulbry and Kearney (1991), Jayashree and Vasudevan (2007a, b), Gupta et al. (2001), and Yang et al. (2005)
Microbes with environmental remediation attributes		
1-2-Dichloroethane degradation	*Xanthobacter autotrophicus* GJ10	Mena-Benitez et al. (2008)
2,4-Dichlorophenoxyacetic acid (herbicide)-contaminated land reclamation	*Pseudomonas putida* strain POPHV6	Germaine et al. (2006)
4-Chloronitrobenezene Rhizoemediation	*Comamonas* sp.	Liu et al. (2007a, b)

(continued)

Table 1.1 (continued)

Pesticide/herbicide/insecticide/environmental remediation	Organism name	References
As bioremediation	*Agrobacterium radiobacter* D14	Wang et al. (2011)
Cd removal	*Serratia nematodiphila* LRE07, *Enterobacter aerogenes* LRE17, *Enterobacter* sp. LSE04 and *Acinetobacter* sp. LSE06, *Mesorhizobium huakuii* subsp. rengei B3.	Chen et al. (2010) and Sriprang et al. (2003)
Cd and Zn bioremediation	*Paxillus involutus*, *Pseudomonas tolaasii* RP23, *Pseudomonas fluorescens* RS9	Baum et al. (2006) and Dell'Amico et al. (2005)
Cr and Pb accumulation	*Pseudomonas aeruginosa* and *P. fluorescens*	Braud et al. (2009)
Cr-contaminated land restoration	*Bacillus* species PSB10, *Pseudomonas* sp. PsA4 and *Bacillus* sp. Ba32, *Cellulosimicrobium cellulans* KUCr3	Wani and Khan (2010), Rajkumar et al. (2006), and Chatterjee et al. (2009)
Crude oil remediation	*Azospirillum lipoferum* strains, *Azospirillum brasilense* strain SR80	Muratova et al. (2005)
Cu bioremediation	*Achromobacter xylosoxidans* Ax10	Ma et al. (2009)
Fipronil and pyriproxyfen bioremediation	*Bradyrhizobium* sp. strain MRM6	Ahemad and Khan (2011)
Fly ash-contaminated soil revitalization	*Enterobacter* sp. NBRI K28 mutant NBRI K28 SD1(RS)	Kumar et al. (2008)
Multimetal Cu-, Zn-, and Cd-contaminated land phytostablization	Consortium of *Bradyrhizobium* sp., *Pseudomonas* sp. and *Ochrobactrum cytisi*)	Dary et al. (2010)
Ni and Cu solubilization	*Pseudomonas jessenii* strain PjM15, *Pseudomonas* sp. PsM6	Rajkumar and Freitas (2008)
Ni and trichloroethylene-contaminated land reclamation	*Pseudomonas putida* W619-TCE	Weyens et al. (2009)
Ni- and Cd-contaminated land restoration	*Rahnella aquatilis*, *Kluyvera ascorbata* SUD165,165/26, SUD165/26	Kumar et al. (2009) and Burd et al. (2000)
Oil-contaminated area restoration	*Paenibacillus* sp., *Acinetobacter* sp.	do Carmo et al. (2011)

(continued)

Table 1.1 (continued)

Pesticide/herbicide/insecticide/environmental remediation	Organism name	References
Organic pollutant-contaminated land reclamation	Genetically engineered *B. cepacia* L.S.2.4 with pTOM (toluene degradation plasmid of *Burkholderia cepacia* G4	Barac et al. (2004)
PAH degradation	*Pseudomonas putida*, *A. brasilense*, and *Enterobacter cloacae*, *Bacillus subtilis* BS1, *Pseudomonas* sp. KS 51, *Phanerochaete laevis* HHB-1625	Huang et al. (2004), Xiao et al. (2012), Shukla et al. (2012), and Bogan and Lamar (1996)
Pb removal	*Sinorhizobium* sp. Pb002, *Microbacterium* sp. G16	Di Gregorio et al. (2006) and Sheng et al. (2008)
Pb2+ and Cd2+ adsorption	*Azotobacter chroococcum* and *Bacillus megaterium*	Wu et al. (2009)
PCB degradation	*Phanerochaete chrysosporidium*, GM *Pseudomonas fluorescens*	Novotny et al. (1997) and de Carcer et al. (2007)
Phytostabilizing and mine tailings	*Bacillus pumilus* ES4 and RIZO1, *A. brasilense*	De Bashan et al. (2010)
Total petroleum hydrocarbon degradation	*Bacillus subtilis*, *Sphingobacterium multivorum*, *Acinetobacter radioresistens*, *Rhodococcus erythropolis*, *Pseudomonas putida*, *Pseudomonas* sp., *Gordonia* sp. S2RP-17	Tang et al. (2010), Gurska et al. (2009), Hontzeas et al. (2004), and Hong et al. (2011)
Wastewater treatment	*Pseudomonas aeruginosa* KUCd1 (RS)	Sinha and Mukherjee (2008)
Xenobiotic degradation (2,4,6-trinitrotoluene)	*Trametes versicolor*	Van Aken et al. (2000)
Zn- and Cd-contaminated land reclamation	*Enterobacter intermedius* MH8b	Płociniczaka et al., (2013)
Bioremediation of various natural and anthropogenic pollutants like p-nitrophenol, 4-chlorophenol and 4-nitroaniline, nonylphenol, polypropylene glycols, herbicides 4-(2,4-dichlorophenoxy) butyric acid, and 4-(4-chloro-2-methylphenoxy) butyric acid, Au(III) removal from contaminated wastewater, Cr (VI) resistance)	*Stenotrophomonas maltophilia*, *Stenotrophomonas rhizophila*	Ryan et al. (2009)

Soil contamination with organic pollutants has increased dramatically since industrial revolution. The injudicious use of large amount of pesticides, insecticides, chemical fertilizers, and other industrial and defense-related chemicals has resulted in severe and widespread contamination of the land with the toxic xenobiotic compounds, mostly organic in nature (Rylott et al. 2006; Rayu et al. 2012). The physicochemical methods for remediation of land contamination remain costly, inefficient, and environmentally destructive. Thus, much attention has been paid on exploiting the plant-microbe association mechanisms for the removal of organic contaminants (Wenzel 2009; Weyens et al. 2009). Various PGPRs capable of degrading or modifying the organic pollutants are reported from the rhizosphere and endosphere of the plants. These microorganisms metabolize the organic pollutants for assimilating nutrients and generating energy. Plants can also transform the toxic organic contaminants by the action of their broad-spectrum enzymes. They can also draw the pollutants toward the rhizosphere *by* transpiration pull where the rhizospheric microorganisms can degrade these contaminants. Thus, plant-microbe interaction could be applied as a sustainable low-input biotechnological tool for the remediation of organic contaminants in the soils.

13 Omics in Assessing Stress Mitigation Challenges

Omics studies can be integrated to develop a better understanding on plant-microbe interactions. Studies covering "omics" approaches and bioinformatics dedicated towards generating large-scale information will allow the understanding of molecular networks working behind stress response and tolerance. This knowledge can be further applied for prediction and validation of the process involving diverse components for generation of different pathways that regulate interactive mechanisms. Such an in silico plant will permit for predicting or modeling metabolic processes driving stress signaling with the aim of developing targeted tolerance in plants. This could lead to the creation of improved crop yield and bioenergy production. Advanced "-omics" technologies allowed to explore deeper understanding and insights on the physiological and molecular aspects of structure and function of plant-associated microbes. We know that microbes, when applied as single inoculum in the soils, usually fail to show desired increase in plant growth and stress tolerance. This may be mostly because of the competition with native inhabitants of the soils and colonization efficiency of microbial communities. Advances in the microbiome research of the rhizosphere point out the strengthening principles which causes plants to gain more benefit due to association with interacting microbial communities. It is thus worthwhile to investigate how microbes could reverse adverse impacts of abiotic stresses, strengthen stress tolerance in plants, and ultimately increase crop productivity.

14 Conclusion

It is already identified that the belowground ecological interaction network among the root, soil, and microbes plays a critical role in maintaining normal growth and defending against inappropriate conditions for both the host and its linked organisms. Billions of microorganisms occupy the plant root system and constitute an ecological community of higher complexity that affects plant growth and crop productivity through profound interactions and metabolic activities. Microbial role in support of plant growth, nutrient acquisition, and biocontrol is understood. These beneficial microbes colonize the rhizosphere or endo-rhizosphere of plants and encourage growth of the plants by direct and indirect methods. In addition, the role of microbes in management of biotic and abiotic stresses is gaining significance, and the practice could be of much practical importance in fields if applied in a judicious way.

Acknowledgment RP is thankful to DST for financial support under DST-Women Scientist Scheme-B (KIRAN Program) (Grant No. DST/WOS-B/2017/67-AAS).

References

Abhilash PC, Dubey RK, Tripathi V, Srivastava P, Verma JP, Singh HB (2013) Adaptive soil management. Curr Sci 104:2–3

Acosta-Martínez V, Acosta-Mercado D, Sotomayor-Ramírez D, Cruz-Rodríguez L (2008) Microbial communities and enzymatic activities under different management in semiarid soils. Appl Soil Ecol 38:249–260

Ahemad M, Khan MS (2011) Insecticide-tolerant and plant growth promoting Bradyrhizobium sp. (Vigna) improves the growth and yield of greengram [*Vigna radiata* (L.) Wilczek] in insecticide-stressed soils. Symbiosis 54:17–27

Ahemad M, Kibret M (2014) Mechanisms and applications of plant growth promoting rhizobacteria: Current perspective. J King Saud Univ Sci 26:1–20

Akbar S, Sultan S (2016) Soil bacteria showing a potential of chlorpyrifos degradation and plant growth enhancement. Braz J Microbiol 47(3):563–570

Aktar MW, Sengupta D, Chowdhurym A (2009) Impact of pesticides use in agriculture: their benefits and hazards. Interdiscip Toxicol 2:1–12. https://doi.org/10.2478/v10102-009-0001-7

Allison VJ, Yermakov Z, Miller RM, Jastrow JD, Matamala R (2007) Using landscape and depth gradients to decouple the impact of correlated environmental variables on soil microbial community composition. Soil Biol Biochem 39:505–516

Aon MA, Colaneri AC (2001) Temporal and spatial evolution of enzymatic activities and physicochemical properties in an agricultural soil. Appl Soil Ecol 18:255–270

Arora M, Kaushik A, Rani N, Kaushik CP (2010) Effect of cyanobacterial exopolysaccharides on salt stress alleviation and seed germination. J Environ Biol 31:701–704

Bais HP, Park SW, Weir TL, Callaway RM, Vivanco JM (2004) How plants communicate using the underground information superhighway? Trends Plant Sci 9:26–32

Bais HP, Weir TL, Perry LG, Gilroy S, Vivanco JM (2006) The role of root exudates in rhizosphere interactions with plants and other organisms. Annu Rev Plant Biol 57:233–266

Banning NC, Gleeson DB, Grigg AH, Grant CD, Andersen GL, Brodie EL, Murphy DV et al (2011) Soil Microbial community successional patterns during forest ecosystem restoration. Appl Environ Microbiol 77:6158–6164. https://doi.org/10.1128/AEM.00764-11

Barac T, Taghavi S, Borremans B, Provoost A, Oeyen L, Colpaert JV et al (2004) Engineered endophytic bacteria improve phytoremediation of water-soluble, volatile, organic pollutants. Nat Biotechnol 22:583–588

Bardgett RD et al (1999) Below-ground microbial community development in a high temperature world. Oikos 85:193–203

Barragán-Huerta BE, Costa-Pérez C, Peralta-Cruz J, Barrera-Cortés J, Esparza-García F, Rodríguez-Vázquez R (2007) Biodegradation of organochlorine pesticides by bacteria grown in microniches of the porous structure of green bean coffee. Int Biodeterior Biodegrad 59(3):239–244

Bashan Y, Salazar BG, Moreno M, Lopez BR, Linderman RG (2012) Restoration of eroded soil in the Sonoran Desert with native leguminous trees using plant growth-promoting microorganisms and limited amounts of compost and water. J Environ Manag 102:26–36

Batool R, Hasnain S (2005) Growth stimulatory effects of *Enterobacter* and *Serratia* located from biofilms on plant growth and soil aggregation. Biotechnology 4:347–353

Batty LC, Dolan C (2013) The potential use of phytoremediation for sites with mixed organic and inorganic contamination. Crit Rev Environ Sci Technol 43:217–259

Baum C, Hrynkiewicz K, Leinweber P, Meißner R (2006) Heavy metal mobilization and uptake by mycorrhizal and nonmycorrhizal willows (*Salix dasyclados*). J Plant Nutr Soil Sci 169:516–522

Bensalim S, Nowak J, Asiedu SK (1998) A plant growth promoting rhizobacterium and temperature effects on performance of 18 clones of potato. Am J Potato Res 75:145–152

Berg G (2009) Plant-microbe interactions promoting plant growth and health: perspectives for controlled use of microorganisms in agriculture. Appl Microbiol Biotechnol 84:11–18

Berg G, Smalla K (2009) Plant species and soil type cooperatively shape the structure and function of microbial communities in the rhizosphere. FEMS Microbiol Ecol 68:1–13

Bhalerao TS, Puranik PR (2007) Biodegradation of organochlorine pesticide, endosulfan, by a fungal soil isolate, Aspergillus niger. Int Biodeterior Biodegrad 59(4):315–321

Bhardwaj D, Ansari MW, Sahoo RK, Tuteja N (2014) Biofertilizers function as key player in sustainable agriculture by improving soil fertility plant tolerance and crop productivity. Microb Cell Factories 8:13–66. https://doi.org/10.1186/1475-2859-13-66

Blackwood CB, Paul EA (2003) Eubacterial community structure and population size within the soil light fraction, rhizosphere, and heavy fraction of several agricultural systems. Soil Biol Biochem 35:1245–1255

Blackwood CB, Dell CJ, Smucker AJM, Paul EA (2006) Eubacterial communities in different soil macroaggregate environments and cropping systems. Soil Biol Biochem 38:720–728

Bleecker AB, Kende H (2000) Ethylene: a gaseous signal molecule in plants. Annu Rev Cell Dev Biol 16:1–18

Bogan BW, Lamar RT (1996) Polycyclic aromatic hydrocarbon-degrading capabilities of Phanerochaete laevis HHB-1625 and its extracellular ligninolytic enzymes. Appl Environ Microbiol 62:1597–1593

Bossio DA, Scow KM, Gunapala N, Graham KJ (1998) Determinants of soil microbial communities: effects of agricultural management, season, and soil type on phospholipid fatty acid profiles. Microb Ecol 36:1–12

Bouskill NJ, Tang J, Riley WJ, Brodie EL (2012) Trait-based representation of biological nitrification: model development, testing, and predicted community composition. Front Microbiol 3:364. https://doi.org/10.3389/fmicb.2012.00364

Braud A, Jezequel K, Bazot S, Lebeau T (2009) Enhanced phytoextraction of an agricultural Cr- and Pb-contaminated soil by bioaugmentation with siderophore-producing bacteria. Chemosphere 74:280–286

Brussaard L, de Ruiter PC, Brown GG (2007) Soil biodiversity for agricultural sustainability. Agric Ecosyst Environ 121:233–244. https://doi.org/10.1016/j.agee.2006.12.013

Buckley DH, Schmidt TM (2001) The structure of microbial communities in soil and the lasting impact of cultivation. Microb Ecol 42:11–21

Burd GI, Dixon DG, Glick BR (2000) Plant growth-promoting bacteria that decrease heavy metal toxicity in plants. Can J Microbiol 46:237–245

Cáceres TP, Megharaj M, Naidu R (2008) Biodegradation of the pesticide fenamiphos by ten different species of green algae and cyanobacteria. Curr Microbiol 57(6):643–646

Caldana C, Degenkolbe T, Cuadros-Inostroza A, Klie S, Sulpice R, Leisse A, Steinhauser D, Fernie AR, Willmitzer L, Hannah MA (2011) High-density kinetic analysis of the metabolomics and transcriptomic response of Arabidopsis to eight environmental conditions. Plant J 67:869–884. https://doi.org/10.1111/j.1365-313X.2011.04640.x

Carney K, Matson P (2005) Plant communities soil microorganisms and soil carbon cycling: does altering the world belowground matter to ecosystem functioning? Ecosystems 8:928–940. https://doi.org/10.1007/s10021-005-0047-0

Carson JK, Rooney D, Gleeson DB, Clipson N (2007) Altering the mineral composition of soil causes a shift in microbial community structure. FEMS Microbiol Ecol 61:414–423

Cavigelli MA, Robertson GP (2001) Role of denitrifier diversity in rates of nitrous oxide consumption in a terrestrial ecosystem. Soil Biol Biochem 33:297–310

Chatterjee S, Sau GB, Mukherjee SK (2009) Plant growth promotion by a hexavalent chromium reducing bacterial strain, *Cellulosimicrobium cellulans* KUCr3. World J Microbiol Biotechnol 25:1829–1836

Chaudhary P, Kumar M, Khangarot BS, Kumar A (2006) Degradation and detoxification of hexachlorocyclohexane isomers by *Pseudomonas aeruginosa* ITRC-5. Int Biodeterior Biodegrad 57(2):107–113

Chen K, Pachter L (2005) Bioinformatics for whole genome shotgun sequencing of microbial communities. PLoS Comput Biol 1:106–112. https://doi.org/10.1371/journal.pcbi.0010024

Chen L, Luo S, Xiao X, Guo H, Chen J, Wan Y, Li B, Xu T, Xi Q, Rao C, Liu C, Zeng G (2010) Application of plant growth-promoting endophytes (PGPE) isolated from *Solanum nigrum* L. for phytoextraction of Cd-polluted soils. Appl Soil Ecol 46:383–389

Coleman D, Wall D (2007) Fauna: the engine for microbial activity and transport. In: Paul E (ed) Soil microbiology, ecology, and biochemistry. Academic, Amsterdam, pp 163–194

Compant S, Clément C, Sessitsch A (2010) Plant growth-promoting bacteria in the rhizo- and endosphere of plants: their role, colonization, mechanisms involved and prospects for utilization. Soil Biol Biochem 42:669–678

Condron L, Stark C, O'Callaghan M, Clinton P, Huang Z (2010) The role of microbial communities in the formation and decomposition of soil organic matter. In: Dixon G, Tilston E (eds) Soil microbiology and sustainable crop production. Springer, Dordrecht

Cregger MA, Schadt CW, McDowell NG, Pockman WT, Classen AT (2012) Response of the soil microbial community to changes in precipitation in a semiarid ecosystem. Appl Environ Microbiol 78(24):8587–8894.

Dary M, Chamber-Pérez MA, Palomares AJ, Pajuelo E (2010) "In situ" phytostabilisation of heavy metal polluted soils using *Lupinus luteus* inoculated with metal resistant plant-growth promoting rhizobacteria. J Hazard Mater 177:323–330

Das N, Chandran P (2011) Microbial degradation of petroleum hydrocarbon contaminants: an overview. Biotechnol Res Int 2011:1–13

de Bashan LE, Hernandez JP, Bashan Y, Maier MR (2010) *Bacillus pumilus* ES4: candidate plant growth-promoting bacterium to enhance establishment of plants in mine tailings. Environ Exp Bot 69:343–352

de Carcer DA, Martın M, Mackova M, Macek T, Karlson U, Rivilla R (2007) The introduction of genetically modified microorganisms designed for rhizoremediation induces changes on native bacteria in the rhizosphere but not in the surrounding soil. ISME J 1:215–223

DeForest JL, Smemo KA, Burke DJ, Elliott HL, Becker JC (2012) Soil microbial responses to elevated phosphorus and pH in acidic temperate deciduous forests. Biogeochemistry 109:189–202

Delgado BM, Maestre FT, Reich PB, Jeffries TC, Gaitan JJ, Encinar D, Berdugo M, Campbell CD, Singh BK (2016) Microbial diversity drives multifunctionality in terrestrial ecosystems. Nat Commun 28(7):10541. https://doi.org/10.1038/ncomms10541

Dell'Amico E, Amico ED, Cavalca L, Andreoni V (2005) Analysis of rhizobacterial communities inperennial Graminaceae from polluted water meadow soil, and screening of metal-resistant, potentially plant growth-promoting bacteria. FEMS Microbiol Lett 52:153–162

Deng SP, Tabatabai MA (1999) Effects of crop residue management on soil enzyme activities in soils. III. Phosphatases and arylsulphatase. Biol Fertil Soils 24:141–146

Dey R, Pal KK and Tilak KVBR (2014) Ecology diversity and application of plant growth promoting rhizobacteria. In: Singh DP, Singh HB (eds) Trends in soil microbial ecology. Studium Press LLC, USA, pp 115–140

Di Gregorio S, Barbafieri M, Lampis S, Sanangelantoni AM, Tassi E, Vallini G (2006) Combined application of Triton X-100 and *Sinorhizobium* sp Pb002 inoculum for the improvement of lead phytoextraction by *Brassica juncea* in EDTA amended soil. Chemosphere 63:293–299

Dick RP (1997) Soil enzyme activities as integrative indicators of soil health. In: Pankhurst CE, Doube BM, VVSR G (eds) Biological indicators of soil health. CAB International, Wallingford, pp 121–156

Dick RP, Breakwell DP, Turco RF (1996) Soil enzyme activities and biodiversity measurements as integrative microbiological indicators. In: Doran JW, Jones AJ (eds) Methods for assessing soil quality. Soil Science Society of America, Madison, pp 247–271

Dick-Peddie W (1991) Semiarid and arid lands: a world wide scope. In: Skujins J (ed) Semiarid lands and deserts: soil resources and reclamation. Marcel Dekker, Inc., New York, pp 3–32

Dimkpa C, Weinand T (2009) Asch F Plant-rhizobacteria interactions alleviate abiotic stress conditions. Plant Cell Environ 32:1682–1694

do Carmo FL, dos santos HF, Martins EF, van Elsas JD, Rosado AS, Peixoto RS (2011) Bacterial structure and characterization of plant growth promoting and oil degrading bacteria from the rhizospheres of mangrove plants. J Microbiol 49:535–543

Dombrowski JE, Hollenbeck VG, Martin RC (2017) Isolation and identification of bacterial endophytes from grasses along the Oregon coast. Am J Plant Sci 8:574–601. https://doi.org/10.4236/ajps.2017.83040

Drenovsky RE, Vo D, Graham KJ, Scow KM (2004) Soil water content and organic carbon availability are major determinants of soil microbial community composition. Microb Ecol 48:424–430

Eilers KG, Debenport S, Anderson S, Fierer N (2012) Digging deeper to find unique microbial communities: the strong effect of depth on the structure of bacterial and archaeal communities in soil. Soil Biol Biochem 50:58–65

Fang H, Dong B, Yan H, Tang F, Yunlong Y (2010) Characterization of a bacterial strain capable of degrading DDT congeners and its use in bioremediation of contaminated soil. J Hazard Mater 184(1–3):281–289

Farrar K, Bryant D, Cope-Selby N (2014) Understanding and engineering beneficial plant–microbe interactions: plant growth promotion in energy crops. Plant Biotechnol J 12:1193–1206

Felske A, Wolterink A, Van LR, De Vos WM, Akkermans ADL (2000) Response of a soil bacterial community to grassland succession as monitored by 16S rRNA levels of the predominant ribotypes. Appl Environ Microbiol 66:3998–4003

Fierer N, Jackson RB (2006) The diversity and biogeography of soil bacterial communities. Proc Natl Acad Sci U S A 103:626–631

Fioravante IA, Barbosa FA, Augusti R, Magalhães SM (2010) Removal of methyl parathion by cyanobacteria microcystis novacekii under culture conditions. J Environ Monit 12(6):1302–1306

Foley JA, DeFries R, Asner GP, Barford C, Bonan G, Carpenter SR, Chapin FS, Coe MT, Daily GC, Gibbs HK, Helkowski JH, Holloway T, Howard EA, Kucharik CJ, Monfreda C, Patz JA, Prentice C, Ramankutty N, Snyder P et al (2005) Global consequences of land use. Science 309:570–574. https://doi.org/10.1126/science.1111772

Forlani G, Pavan M, Gramek M, Kafarski P, Lipok J (2008) Biochemical basis for a wide spread tolerance of cyanobacteria to the phosphonate herbicide glyphosate. Plant Cell Physiol 49:443–456

Foster LJR, Kwan BH, Vancov T (2004) Microbial degradation of the organophosphate pesticide. FEMS Microbiol Lett 240(1):49–53

Garcia C, Hernandez T, Roldan A, Martin A (2002) Effect of plant cover decline on chemical and microbiological parameters under Mediterranean climate. Soil Biol Biochem 34:635–642

García-Orenes F, Morugán-Coronado A, Zornoza R, Scow K (2013) Changes in soil microbial community structure influenced by agricultural management practices in a mediterranean agro-ecosystem. PLoS ONE 8(11):e80522. https://doi.org/10.1371/journal.pone.0080522)

Germaine KJ, Liu X, Cabellos GG, Hogan JP, Ryan D, Dowling DN (2006) Bacterial endophyte-enhanced phytoremediation of the organochlorine herbicide 2,4-dichlorophenoxyacetic acid. FEMS Microbiol Ecol 57:302–310

Gleeson D, McDermott F, Clipson N (2006) Structural diversity of bacterial communities in a heavy metal mineralized granite outcrop. Environ Microbiol 8:383–393

Glick BR (2012) Plant growth-promoting bacteria: mechanisms and application. Scientifica

Glick BR, Penrose DM, Li J (1998) A model for the lowering of plant ethylene concentrations by plant growth promoting bacteria. J Theor Biol 190:63–68

Gouzou L, Burtin G, Philippy R, Bartoli F, Heulin T (1993) Effect of inoculation with *Bacillus polymyxa* on soil aggregation in the wheat Rhizosphere: preliminary examination. Geoderma 56:479–490

Graham EB, Wieder WR, Leff JW, Weintraub SR, Townsend AR, Cleveland CC et al (2014) Do we need to understand microbial communities to predict ecosystem function? A comparison of statistical models of nitrogen cycling processes. Soil Biol Biochem 68:279–282

Graham EB, Knelman JE, Schindlbacher A, Siciliano S, Breulmann M, Yannarell M, Beman JM, Abell G, Philippot L, Prosser J, Foulquier A, Yuste JC, Glanville HC, Jones DL, Angel R, Salminen J, Newton RJ, Bürgmann H, Ingram LJ, Hamer U, Siljanen HMP, Peltoniemi K, Potthast K, Bañeras L, Hartmann M, Banerjee S, Yu RQ, Nogaro G, Richter A, Koranda M, Castle SC, Goberna M, Song B, Chatterjee A, Nunes OC, Lopes AR, Cao Y, Kaisermann A, Hallin S, Strickland MS, Garcia-Pausas J, Barba J, Kang H, Isobe K, Papaspyrou S, Pastorelli R, Lagomarsino A, Lindström ES, Basiliko N, Nemergut DR (2016) Microbes as engines of ecosystem function: when does community structure enhance predictions of ecosystem processes? Front Microbiol 7:214

Gros R, Monrozier LJ, Faivre P (2006) Does disturbance and restoration of alpine grassland soils affect the genetic structure and diversity of bacterial and N2-fixing populations? Environ Microbiol 8:1889–1901

Grover M, Ali SKZ, Sandhya V, Rasul A, Venkateswarlu B (2011) Role of microorganisms in adaptation of agriculture crops to abiotic stresses. World J Microbiol Biotechnol 27:1231–1240

Guillén-Jiménez FDM, Cristiani-Urbina E, Cancino-Díaz JC, Flores-Moreno JL, Barragán-Huerta BE (2012) Lindane biodegradation by the *Fusarium verticillioides* AT-100 strain, isolated from *Agave tequilana* leaves: kinetic study and identification of metabolites. Int Biodeterior Biodegrad 74:36–47

Gupta US (2005) Biodegradation of chlorinated hydrocarbon insecticide by Pseudomonas species. Himalayan J Environ Zool 19(1):1–10

Gupta A, Kaushik CP, Kaushik A (2001) Degradation of hexachlorocyclohexane isomers by two strains of *Alcaligenes faecalis* isolated from a contaminated site. Bull Environ Contam Toxicol 66:794–800

Gurska J, Wang W, Gerhardt KE, Khalid AM, Isherwood DM, Huang XD et al (2009) Three year field test of a plant growth promoting rhizobacteria enhanced phytoremediation system at a land farm for treatment of hydrocarbon. Waste Environ Sci Technol 43:4472–4479

Ha J, Engler CR, Wild JR (2009) Biodegradation of coumaphos, chlorferon, and diethylthiophosphate using bacteria immobilized in Ca-alginate gel beads. Bioresour Technol 100(3):1138–1142

Habig J, Swanepoel C (2015) Effect of conservation agriculture and fertilization on soil microbial diversity and activity. Environment 2:358–384. https://doi.org/10.3390/environments2030358

Hackl E, Zechmeister BS, Bodrossy L, Sessitsch A (2004) Comparison of diversities and compositions of bacterial populations inhabiting natural forest soils. Appl Environ Microbiol 70:5057–5065

Hagerty SB, Groenigen KJ, Allison SD, Hungate BA, Schwartz E, Koch GW et al (2014) Accelerated microbial turnover but constant growth efficiency with warming in soil. Nat Clim Chang 4:903–906. https://doi.org/10.1038/nclimate2361

Hai FI, Modin O, Yamamoto K, Fukushi K, Nakajima F, Nghiem LD (2012) Pesticide removal by a mixed culture of bacteria and white-rot fungi. J Taiwan Inst Chem Eng 43(3):459–462
Harada N, Takagi K, Harazono A, Fujii K, Iwasaki A (2006) Isolation and characterization of microorganisms capable of hydrolysing the herbicide mefenacet. Soil Biol Biochem 38(1):173–179
Harris JA (2003) Measurements of the soil microbial community for estimating the success of restoration. Eur J Soil Sci 54:801–808
Harris J (2009) Soil microbial communities and restoration ecology: facilitators or followers? Science 325:573–574
Higa T (1991) Effective microorganisms: a new dimension for nature farming. In: Parr JF, Hornick SB, Whitman CE (eds) Proceedings of the first International Conference on Kyusei Nature Farming. US Department of Agriculture, Washington DC, pp 153–162
Hong SH, Ryu HW, Kim J, Cho KS (2011) Rhizoremediation of diesel-contaminated soil using the plant growth-promoting rhizobacterium *Gordonia* sp. S2RP-17. Biodegradation 22:593–501
Hontzeas N, Zoidakis J, Glick BR, Abu-Omar MM (2004) Expression and characterization of 1-aminocylcopropane- 1-carboxylate deaminase from the rhizobacterium *Pseudomonas putida* UW4: a key enzyme in bacterial plant growth promotion. Biochim Biophys Acta 1703:11–19
Hopkins DW, Gregorich EG (2005) Carbon as a substrate for soil organisms. In: Bardgett RD, Usher MB, Hopkins DW (eds) Biological diversity and function in soils. Cambridge University Press, Cambridge, pp 57–79
Horrigan L, Lawrence RS, Walker P (2002) How sustainable agriculture can address the environmental and human health harms of industrial agriculture. Environ Health Perspect 110:445–456
Huang X-D, El-Alawi Y, Penrose DM, Glick BR, Greenberg BM (2004) A multi-process system for removal of polycyclic aromatic hydrocarbons from contaminated soils. Environ Pollut 130:465–476
Ibrahim WM, Essa AMM (2010) Tolerance, biodegradation and utilization of malathion, an organophosphorous pesticide, by some cyanobacterial isolates. Egypt J Bot 27-29:225–240
Ibrahim WM, Karam MA, El-Shahat RM, Adway AA (2014) Biodegradation and utilization of organophosphorus pesticide malathion by cyanobacteria. Biomed Res Int
Insam H (2001) Developments in soil microbiology since the mid 1960's. Geoderma 100:389–402
Jackson LE, Calderon FJ, Steenwerth KL, Scow KM, Rolston DE (2003) Responses of soil microbial processes and community structure to tillage events and implications for soil quality. Geoderma 114:305–317
Jangid K, Williams MA, Franzluebbers AJ, Sanderlin JS, Reeves JH, Jenkins MB, Endale DM, Coleman DC, Whitman WB (2008) Relative impacts of land-use, management intensity and fertilization upon soil microbial community structure in agricultural systems. Soil Biol Biochem 40:2843–2853
Jangid K et al (2010) Development of soil microbial communities during tallgrass prairie restoration. Soil Biol Biochem 42:302–312
Janzen HH (2004) Carbon cycling in earth systems–a soil science perspective. Agric Ecosyst Environ 104:399–417
Javaid MK, Ashiq M, Tahir M (2016) Potential of biological agents in decontamination of agricultural soil. Scientifica (Cairo) 2016:1598325
Jayashree R, Vasudevan N (2007a) Organochlorine pesticide residues in ground water of Thiruvallur district, India. Environ Monit Assess 128(1–3):209–215
Jayashree R, Vasudevan N (2007b) Persistence and distribution of endosulfan under field conditions. Environ Monit Assess 131(1–3):475–487
Jia X, Sun C, Zuo Y, Li G, Li G, Ren L, Chen G (2016) Integrating transcriptomics and metabolomics to characterise the response of Astragalus membranaceus Bge var. mongolicus (Bge.) to progressive drought stress. BMC Genomics 17:188. https://doi.org/10.1186/s12864-016-2554-0
Johnston AE, Poulton PR, Coleman K, Donald LS (2009) Soil organic matter: its importance in sustainable agriculture and carbon dioxide fluxes. Adv Agron 101:1–57
Jorge TF, Rodrigues JA, Caldana C, Schmidt R, van Dongen JT, Thomas-Oates J, Antonio C (2015) Mass-spectrometry-based plant metabolomics: metabolite responses to abiotic stress. Mass Spectrom Rev 35:620–649. https://doi.org/10.1002/mas.21449

Kaiser C, Koranda M, Kitzler B, Fuchslueger L, Schnecker J et al (2010) Belowground carbon allocation by trees drives seasonal patterns of extracellular enzyme activities by altering microbial community composition in a beech forest soil. New Phytol 187:843–858
Kaiser C, Franklin O, Dieckmann U, Richter A (2014) Microbial community dynamics alleviate stoichiometric constraints during litter decay. Ecol Lett 17:680–690. https://doi.org/10.1111/ele.12269
Kandeler E, Kampichler C, Horak O (1996) Influence of heavy metals on the functional diversity of soil microbial communities. Biol Fertil Soils 23:299–306
Kapardar RK, Ranjan R, Grover A, Puri M, Sharma R (2010) Identification and characterization of genes conferring salt tolerance to Escherichia coli from pond water metagenome. Bioresour Technol 101:3917–3924. https://doi.org/10.1016/j.biortech.2010.01.017
Kataoka R, Takagi K, Sakakibara F (2010) A new endosulfan-degrading fungus, Mortierella species, isolated from a soil contaminated with organochlorine pesticides. J Pestic Sci 35(3):326–332
Kibblewhite MG, Ritz K, Swift MJ (2008) Soil health in agricultural systems. Philos Trans R Soc Lond Ser B Biol Sci 363:685–701
Knudsen GR, Louise M, Dandurand C (2014) Ecological complexity and the success of fungal biological control agents. Adv Agric: Article ID 542703. https://doi.org/10.1155/2014/542703
Kumar K, Singh N, Behl HM, Srivastava S (2008) Influence of plant growth promoting bacteria and its mutant on heavy metal toxicity in *Brassica juncea* grown in fly ash amended soil. Chemosphere 72:678–683
Kumar KV, Srivastava S, Singh N, Behl HM (2009) Role of metal resistant plant growth promoting bacteria in ameliorating fly ash to the growth of *Brassica juncea*. J Hazard Mater 170:51–57
Kuramae EE et al (2010) Microbial secondary succession in a chronosequence of chalk grasslands. ISME J 4:711–715
Kuritz T, Wolk CP (1995) Use of filamentous cyanobacteria for biodegradation of organic pollutants. Appl Environ Microbiol 61(1):234–238
Levine UY, Teal TK, Robertson GP, Schmidt TM (2011) Agriculture's impact on microbial diversity and associated fluxes of carbon dioxide and methane. ISME J 5:1683–1691
Lewis DE et al (2010) Soil functional diversity analysis of a bauxite-mined restoration chronosequence. Microb Ecol 59:710–723
Lio M, Xie X (2009) Application of enterobacteraerogenes in degrading pyrethroid pesticide residue, and preparation with enterobacteraerogenes. Faming Zhuanli Shenqing, CN102021135 A
Lipok J, Owsiak T, Młynarz P, Forlani G, Kafarski P (2007) Phosphorus NMR as a tool to study mineralization of organophosphonates – the ability of *Spirulina* spp. to degrade glyphosate. Enzym Microb Technol 41:286–291
Lipok J, Wieczorek D, Jewginski M, Kafarski P (2009) Prospects of *in vivo* ^{31}P NMR method in glyphosate degradation studies in whole cell system. Enzym Microb Technol 44:11–16
Liu J, Wang L, Zheng L, Wang X, Lee FSC (2006) Analysis of bacteria degradation products of methyl parathion by liquid chromatography/electrospray time-of-flight mass spectrometry and gas chromatography/mass spectrometry. J Chromatogr A 1137(2):180–187
Liu L, Jiang CY, Liu XY, Wu JF, Han JG, Liu SJ (2007a) Plant–microbe association for rhizoremediation of chloronitroaromatic pollutants with *Comamonas* sp. strain CNB-1. Environ Microbiol 9:465–473
Liu Z, Yang C, Qiao C-L (2007b) Biodegradation of p-nitrophenol and 4-chlorophenol by *Stenotrophomonas* sp. FEMS Microbiol Lett 277(2):150–156
Liu Y, Shi Z, Yao L, Yue H, Li H, Li C (2013) Effect of IAA produced by Klebsiellaoxytoca Rs-5 on cotton growth under salt stress. J Gen Appl Microbiol 59:59–65
Lloyd D, Soldat D, Balser T (2009) Understanding microbes helps explain nutrient cycling Turfgrass Trends. archive.lib.msu.edu/tic/golfd/article/2009feb50.pdf
Lombard N, Prestat E, Elsas JD, Simonet P (2011) Soil specific limitations for access and analysis of soil microbial com-munities by metagenomics. FEMS Microbiol Ecol 78:31–49
Lorenz P, Eck J (2005) Metagenomics and industrial applications. Nat Rev Microbiol 3:510–516
Lugtenberg B, Kamilova F (2009) Plant-growth-promoting rhizobacteria. Rev A Microbiol 63:541–556

Lutenberg B, Kamilova F (2009) Plant growth-promoting rhizobacteria. Annu Rev Microbiol:541–556
Ma Y, Rajkumar M, Freitas H (2009) Improvement of plant growth and nickel uptake by nickel resistant-plant-growth promoting bacteria. J Hazard Mater 166:1154–1161
Ma Y, Prasad MNV, Rajkumar M, Freitas H (2011) Plant growth promoting rhizobacteria and endophytes accelarate phytoremediation of metalliferous soils. Biotech Adv 29:248–258
Madhuban G, Debashis D, Das SK (2011) Biodegradation of imidacloprid and metribuzin by Burkholderia cepacia strain CH9. Pestic Res J 23(1):36–40
Manzoni S, Schaeffer S, Katul G, Porporato A, Schimel J (2014) A theoretical analysis of microbial eco-physiological and diffusion limitations to carbon cycling in drying soils. Soil Biol Biochem 73:69–83
Maroni M, Fanetti AC, Metruccio F (2006) Risk assessment and management of occupational exposure to pesticides in agriculture. Med Lav 97:430–4377
Marschner P, Kandeler E, Marschner B (2003) Structure and function of the soil microbial community in a long-term fertilizer experiment. Soil Biol Biochem 35:453–461
Martiny JB, Jones SE, Lennon JT, Martiny AC (2015) Microbiomes in light of traits: a phylogenetic perspective. Science 350:aac9323
Mayak S, Tirosh T, Glick BR (2004) Plant growth-promoting bacteria that confer resistance to water stress in tomatoes and peppers. Plant Sci 166:525–530
Mazzola M (2007) Manipulation of rhizosphere bacterial communities to induce suppressive soils. Nematol J 39:213–220
Meena KK, Kumar M, Kalyuzhnaya MG, Yandigeri MS, Singh DP, Saxena AK, Arora DK (2012) Epiphytic pink-pigmented methylotrophic bacteria enhance germination and seedling growth of wheat (*Triticum aestivum*) by producing phytohormones. Antonie Van Leeuwenhoek 101:777–786
Meena KK, Sorty AM, Bitla UM, Choudhary K, Gupta P, Pareek A, Singh DP, Prabha R, Sahu PK, Gupta VK, Singh HB, Krishnani KK, Minhas PS (2017) Abiotic stress responses and microbe-mediated mitigation in plants: the omics strategies. Front Plant Sci 8:172. https://doi.org/10.3389/fpls.2017.00172
Mei C, Flinn BS (2010) The use of beneficial microbial endophytes for plant biomass and stress tolerance improvement. Recent Pat Biotechnol 4:81–95
Mena-Benitez GL, Gandia-Herrero F, Graham S, Larson TR, McQueen-Mason SJ, French CE, Rylott EL, Bruce NC (2008) Engineering a catabolic pathway in plants for the degradation of 1,2-dichloroethane. Plant Physiol 147:1192–1198
Mendoza JC, Perea Y, Salvador JA (2011) Bacterial biodegradation of permetrina and cipermetrina pesticides in a culture assemblage. Avances en Ciencias e Ingenieria 2(3):45–55
Micallef SA, Channer S, Shiaris MP, Colón-Carmona A (2009) Plant age and genotype impact the progression of bacterial community succession in the Arabidopsis rhizosphere. Plant Signal Behav 4:777–780. https://doi.org/10.1093/jxb/erp053
Milshteyn A, Schneider JS, Brady SF (2014) Mining the metabiome identifying novel natural products from microbial communities. Chem Biol 21:1211–1223
Morales SE, Cosart T, Holben WE (2010) Bacterial gene abundances as indicators of greenhouse gas emission in soils. ISME J 4:799–808
Morgan JA, Bending GD, White PJ (2005) Biological costs and benefits to plant-microbe interactions in the rhizosphere. J Exp Bot 56:1729–1739
Mulbry W, Kearney PC (1991) Degradation of pesticides by microorganisms and the potential for genetic manipulation. Crop Prot 10:334–346
Muratova AY, Turkovskaya OV, Antonyuk LP, Makarov OE, Pozdnyakova LI, Ignatov VV (2005) Oil-oxidizing potential of associative rhizobacteria of the genus *Azospirillum*. Microbiology 74:210–215
Nadeem SM, Ahmad M, Zahir ZA, Javaid A, Ashraf M (2014) The role of mycorrhizae and plant growth promoting rhizobacteria (PGPR) in improving crop productivity under stressful environments. Biotechnol Adv 32:429–448

Nakbanpote W, Panitlurtumpai N, Sangdee A, Sakulpone N, Sirisom P, Pimthong A (2014) Salttolerant and plant growth-promoting bacteria isolated from Zn/Cd contaminated soil: identification and effect on rice under saline conditions. J Plant Interact 9:379–387

Nandakumar R, Babu S, Viswanathan R, Sheela J, Raguchander T, Samiyappan R (2006) A new bio-formulation containing plant growth promoting rhizobacterial mixture for the management of sheath blight and enhanced grain yield in rice. Biocontrol 46:493–410

Nannipieri P, Ascher J, Ceccherini MT, Landi L, Pietramellara G, Renella G (2003) Microbial diversity and soil functions. Eur J Soil Sci 54:655–670

Nihorimbere V (2014) Functional significance of microbial communities in soils. In: Singh DP, Singh HB (eds) Trends in soil microbial ecology. Studium Press LLC, USA, pp 1–21

Nihorimbere V, Ongena M, Smargiassi M, Thonart P (2010) Beneficial effect of the rhizosphere microbial community for plant growth and health. Biotechnol Agron Soc Environ 15:327–337

Nihorimbere V, Ongena M, Smargiassi M et al (2011) Beneficial effect of the rhizosphere microbial community for plant growth and health. Biotechnol Agron Soc Environ 15(2):327–337

Novotny C, Vyas BRM, Erbanova P, Kubatova A, Sasek V (1997) Removal of PCBs by various white rot fungi in liquid cultures. Folia Microbiol 42:136–140

Nyakundi WO, Magoma G, Ochora J, Nyende AB (2011) Biodegradation of diazinon and methomyl pesticides by white rot fungi from selected horticultural farms in rift valley. J Appl Technol Environ Sanit 1(2):107–124

O'Callaghan M (2016) Microbial inoculation of seed for improved crop performance: issues and opportunities. Appl Microbiol Biotechnol 100:5729–5746

Oldroyd GED (2013) Speak, friend, and enter: signalling systems that promote beneficial symbiotic associations in plants. Nat Rev Microbiol 11:252–263. https://doi.org/10.1038/nrmicro2990

Omar MNA, Osman MEH, Kasim WA, Abd El-Daim IA (2009) Improvement of salt tolerance mechanisms of barley cultivated under salt stress using Azospirillum brasiliense. Tasks Veg Sci 44:133–147. https://doi.org/10.1007/978-1-4020-9065-3_15

Ouyang ZC, Wang YH, Li XN et al (2008) Test of pesticide degradability *by Sphingomonas yanoikuyae* XJ strain. Huanan Nongye Daxue Xuebao 29(2):47–49

Paingankar M, Jain M, Deobagkar D (2005) Biodegradation of allethrin, a pyrethroid insecticide, by an *Acidomonas* sp. Biotechnol Lett 27(23–24):1909–1913

Palanisami S, Prabaharan D, Uma L (2009) Fate of few pesticide-metabolizing enzymes in the marine cyanobacterium *Phormidium valderianum* BDU 20041 in perspective with chlorpyrifos exposure. Pestic Biochem Physiol 94(2–3):68–72

Pennanen T et al (1999) Structure of the microbial communities in coniferous forest soils in relation to site fertility and stand development stage. Microb Ecol 38:168–179

Pennanen T et al (2004) Community-level responses of metabolically-active soil microorganisms to the quantity and quality of substrate inputs. Soil Biol Biochem 36:841–848

Peterka M, Tepsic K, Accetto T, Kostanjsek R, Ramsak A, Lipoglavsek L, Avgustin G (2003) Molecular microbiology of gut bacteria: genetic diversity and community structure analysis. Acta Microbiol Immunol Hung 50(4):395–406

Peuke AD, Rennenberg H (2005) Phytoremediation. EMBO Rep 6(6):497–501

Philippot L, Raaijmakers JM, Lemanceau P, van der Putten WH (2013) Going back to the roots: the microbial ecology of the rhizosphere. Nat Rev Microbiol 11:789–799

Phillips CJ, Harris D, Dollhopf SL, Gross KL, Prosser JI, Paul EA (2000) Effects of agronomic treatments on structure and function of ammonia-oxidizing communities. Appl Environ Microbiol 66:5410–5418

Pieffer JA, Spor A, Koren O, Jin Z, Green TS, Dangl JL, Buckler ES, Ley RE (2013) Diversity and heritability of the maize rhizosphere microbiome under field conditions. Proc Natl Acad Sci U S A 16:6548–6553

Pimentel D, Hepperly P, Hanson J, Douds D, Seidel R (2005) Environmental energetic and economic comparisons of organic and conventional farming systems. Bioscience 55:573–582

Pinto AP, Serrano C, Pires T et al (2012) Degradation of terbuthylazine, difenoconazole and pendimethalin pesticides by selected fungi cultures. Sci Total Environ 435-436:402–410

Powell JR, Welsh A, Hallin S (2015) Microbial functional diversity enhances predictive models linking environmental parameters to ecosystem properties. Ecology 96:1985–1993. https://doi.org/10.1890/14-1127.1

Power AG (2010) Ecosystem services and agriculture: tradeoffs and synergies. Philos Trans R Soc Lond Ser B Biol Sci 365:2959–2971. https://doi.org/10.1098/rstb.2010.0143

Prabakaran V, Peterson A (2006) Effect of Pseudomonas on biodegradation of pesticide in the fish *Cyprinus carpio*. J Ecotoxicol Environ Monit 16(5):475–479

Prosser JL, Bohannan BJ, Curtis TP, Ellis RJ, Firestone MK, Freckleton RP et al (2007) The role of ecological theory in microbial ecology. Nat Rev Microbiol 5:384–392. https://doi.org/10.1038/nrmicro1643

Purahong W, Schloter M, Pecyna MJ, Kapturska D, Däumlich V, Mital S, Buscot F, Hofrichter M, Gutknecht JL, Krüger (2014) Uncoupling of microbial community structure and function in decomposing litter across beech forest ecosystems in Central Europe. Sci Rep 4:7014. https://doi.org/10.1038/srep07014

Qiao L, Wang J (2010) Biodegradation characterization of a pyridine-degrading strain. Qinghua Daxue Xuebao 50(6):869–872

Qiao C-L, Yan Y-C, Shang HY, Zhou XT, Zhang Y (2003) Biodegradation of pesticides by immobilized recombinant *Escherichia coli*. Bull Environ Contam Toxicol 71(2):370–374

Qin Y, Druzhinina IS, Pan X, Yuan Z (2016) Microbially mediated plant salt tolerance and microbiome-based solutions for saline agriculture. Biotechnol Adv 34:1245–1259

Rajendhran J, Gunasekaran P (2008) Strategies for accessing soil metagenome for desired applications. Biotechnol Adv 26:576–590

Rajkumar M, Freitas H (2008) Influence of metal resistant-plant growth-promoting bacteria on the growth of *Ricinus communis* in soil contaminated with heavy metals. Chemosphere 71:834–842

Rajkumar M, Nagendran R, Lee KJ, Lee WH, Kim SZ (2006) Influence of plant growth promoting bacteria and Cr6+ on the growth of Indian mustard. Chemosphere 62:741–748

Ramirez KS, Lauber CL, Knight R, Bradford MA, Fierer N (2010) Consistent effects of nitrogen fertilization on soil bacterial communities in contrasting systems. Ecology 91:3463–3470

Rao DLN (2007) Microbial diversity, soil health and sustainability. J Indian Soc Soil Sci 55:1–12

Rastogi G, Saini RK (2011) Molecular techniques to assess microbial community structure function and dynamics in the environment. In: Ahmad et al (eds) Microbes and microbial technology, Agricultural and environmental applications. Springer. https://doi.org/10.1007/978-1-4419-7931-5_2

Reddy CA, Saravanan RS (2013) Polymicrobial multi-functional approach for enhancement of crop productivity. Adv Appl Microbiol 82:53–113. https://doi.org/10.1016/B978-0-12-407679-2.00003-X

Rayu S, Karpouzas DG, Singh BK (2012) Emerging technologies in bioremediation: constraints and opportunities. Biodegradation 23(6):917–926

Reddy MS, Wang Q (2011) Plant growth promoting rhizobacteria (PGPR) for sustainable agriculture. In: Proceedings of the 2nd Asian PGPR Conference. August 21–24, 2011, Beijing, China and references cited therein

Reed HE, Martiny JBH (2007) Testing the functional significance of microbial composition in natural communities. FEMS Microbiol Ecol 62:161–170

Reed DC, Algar CK, Huber JA, Dick GJ (2014) Gene-centric approach to integrating environmental genomics and biogeochemical models. Proc Natl Acad Sci U S A 111:1879–1884

Rizhsky L, Liang HJ, Shuman J, Shulaev V, Davletova S, Mittler R (2004) When defense pathways collide The response of *Arabidopsis* to a combination of drought and heat stress. Plant Physiol 134:1683–1696. https://doi.org/10.1104/pp.103.033431

Romeh AAA (2001) Biodegradation of carbosulfan, pirimicarb and diniconazole pesticides by *Trichoderma* spp. J Environ Res 3:162–172

Rousk J et al (2010) Soil bacterial and fungal communities across a pH gradient in an arable soil. ISME J 4:1340–1351

Ruppel S, Torsvik V, Daae FL, Ovreas L, Ruhlmann J (2007) Nitrogen availability decreases prokaryotic diversity in sandy soils. Biol Fertil Soils 43:449–459

Ryan RP, Monchy S, Cardinale M, Taghavi S, Crossman L, Avison MB, Berg G, van der Lelie D, Dow MJ (2009) The versatility and adaptation of bacteria from the genus Stenotrophomonas. Nat Rev Microbiol 7:514–525

Rylott EL, Jackson RG, Edwards J, Womack GL, Seth-Smith HMB, Rathbone DA, Strand SE, Bruce NC (2006) An explosive-degrading cytochrome P450 activity and its targeted application for the phytoremediation of RDX. Nat Biotechnol 24(2):216–219

Sagar V, Singh DP (2011) Biodegradation of lindane pesticide by non white- rots soil fungus Fusarium sp. World J Microbiol Biotechnol 27(8):1747–1754

Sandhya V, Ali SKZ, Grover M, Reddy G, Venkateshwarlu B (2009) Alleviation of drought stress effects in sunflower seedlings by the exopolysaccharides producing Pesudomonas Putida strain GAP-P45. Biol Fertil Soils 46:17–26

Santacruz G, Bandala ER, Torres LG (2005) Chlorinated pesticides (2,4-D and DDT) biodegradation at high concentrations using immobilized Pseudomonas fluorescens. J Environ Sci Health Part B Pest Food Contamin Agric Wastes 40(4):571–583

Schloter M, Lebuhn M, Heulin T, Heartmann A (2000) Ecology and evolution of bacterial microdiversity. FEMS Microbiol 24:647–660

Schnecker J, Wild B, Hofhansl F, Eloy Alves RJ, Barta J et al (2014) Effects of soil organic matter properties and microbial community composition on enzyme activities in cryoturbated arctic soils. PLoS ONE 9(4):e94076. https://doi.org/10.1371/journal.pone.0094076

Schulz S, Brankatschk R, Dumig A, Kogel Knabner I, Schloter M, Zeyer J (2013) The role of microorganisms at different stages of ecosystem development for soil formation. Biogeosciences 10:3983–3996

Seo J, Lee Y-G, Kim S-D, Cha C-J, Ahn J-H, Hur H-G (2005) Biodegradation of the insecticide N,N-diethyl-m-toluamide by fungi: identification and toxicity of metabolites. Arch Environ Contam Toxicol 48(3):323–328

Sessitsch A, Weilharter A, Gerzabek MH, Kirchmann H, Kandeler E (2001) Microbial population structures in soil particle size fractions of a long-term fertilizer field experiment. Appl Environ Microbiol 67:4215–4224

Sheng X-F, Xia J-J, Jiang C-Y, He L-Y, Qian M (2008) Characterization of heavy metal-resistant endophytic bacteria from rape (*Brassica napus*) roots and their potential in promoting growth and lead accumulation of rape. Environ Pollut 156:1164–1170

Shukla KP, Sharma S, Singh NK, Singh V (2012) Deciphering rhizosphere soil system for strains having plant growth promoting and bioremediation traits. Agric Res 1:251–257

Shunpeng L, Mingxing Z (2006) Sphingomonas strain for degrading chlorophenothane pesticide residue and bacteria agent containing this strain. Faming Zhuanli Shenqing Gongkai Shuomingshu

Shunpeng L, Shen LZ (2005) Pseudomonas putida and its bacterial products for degrading organophosphorus pesticide residues. Faming Zhuanli Shenqing Gongkai Shuomingshu

Shun-Peng L, Ruifu Z, Jian-Dong J et al (2005) Ochrobactrum MP-4 and its products for degrading residues of triazophos-based pesticide. Faming Zhuanli Shenqing Gongkai Shuomingshu

Singh DP, Singh HB (2014) Trends in soil microbial ecology. Studium Press LLC. USA

Singh JS, Kumar A, Rai AN, Singh DP (2016) Cyanobacteria: a precious bio-resource in agriculture, ecosystem, and environmental sustainability. Front Microbiol 2016

Sinha S, Mukherjee SK (2008) Cadmium-induced siderophore production by a high Cd-resistant bacterial strain relieved Cd toxicity in plants through root colonization. Curr Microbiol 56:55–60

Sinsabaugh RL, Lauber CL, Weintraub MN, Ahmed B, Allison SD et al (2008) Stoichiometry of soil enzyme activity at global scale. Ecol Lett 11:1252–1264

Sinsabaugh RL, Hill BH, Follstad Shah JJ (2009) Ecoenzymatic stoichiometry of microbial organic nutrient acquisition in soil and sediment. Nature 462:795–798

Sistla SA, Schimel JP (2012) Stoichiometric flexibility as a regulator of carbon and nutrient cycling in terrestrial ecosystems under change. New Phytol 196:68–78

Sorty AM, Meena KK, Choudhary K, Bitla UM, Minhas PS, Krishnani KK (2016) Effect of plant growth promoting bacteria associated with halophytic weed (*Psoralea corylifolia* L) on germination and seedling growth of wheat under saline conditions Appl. Biochem Biotechnol 180:872–882. https://doi.org/10.1007/s12010-016-2139-z

Souza R, Ambrosini A, Passaglia LM (2015) Plant growth-promoting bacteria as inoculants in agricultural soils. Genet Mol Biol 38:401–419. https://doi.org/10.1590/S1415-475738420150053

Sparling GP (2013) Soil microbial biomass activity and nutrient cycling as indicators of soil health. In: Doube CE, Gupta BM (eds) Biological indicators of soil health Pankhurst. CAB International, Wallingford, pp 97–119

Sriprang R, Hayashi M, Ono H, Takagi M, Hirata K, Murooka Y (2003) Enhanced accumulation of Cd^{2+} by a *Mesorhizobium* sp. transformed with a gene from *Arabidopsis thaliana* coding for phytochelatin synthase. Appl Environ Microbiol 69:1791–1796

Stearns JC, Saleh S, Greenberg BM, Dixon DG, Glick BR (2005) 1-aminocyclopropane-1-carboxylic acid deaminase to growth inhibition by nickel. Plant Physiol Biochem 43:701–708

Strickland MS, Lauber C, Fierer N, Bradford MA (2009) Testing the functional significance of microbial community composition. Ecology 90:441–451

Strobel G, Daisy B (2003) Bioprospecting for microbial endophytes and their natural products. Microbiol Mol Biol Rev 67:491–502

Subramanian G, Sekar S, Sampoornam S (1994) Biodegradation and utilization of organophosphorus pesticides by cyanobacteria. Int Biodeterior Biodegrad 33:129–143

Supanekar SV, Sorty AM (2013a) Catechol siderophore produced by *Klebsiella Pneumoniae* isolated from rhizosphere of *Saccharum officinarum* L. Int J Sci Res 2:423–425

Supanekar SV, Sorty AM (2013b) Siderophoregenic *Klebsiella pneumoniae* SUPII frim wheat (*Triticum aestivum* L) rhizoplane. Indian J Res 2:243–245

Supanekar S, Sorty A, Raut A (2013) Study of catechol siderophore from a newly isolated Azotobacter sp SUP-III for its antimicrobial activity. J Microbiol Biotechnol Food Sci 3:270–273

Suyal DC, Agrawal K and Goel R (2014) Identification of soil microbial communities: approaches and methodologies. In: Singh DP, Singh HB (eds) Trends in soil microbial ecology. Studium Press LLC, USA, pp 93–114

Swift MJ, Heal OW, Anderson JM (1979) Decomposition in terrestrial ecosystems. Studies in ecology. Blackwell Scientific, Oxford

Talbot JM, Bruns TD, Smith DP, Branco S, Glassman SI et al (2013) Independent roles of ectomycorrhizal and saprotrophic communities in soil organic matter decomposition. Soil Biol Biochem 57:282–291

Tang J, Wang R, Niu X, Zhou Q (2010) Enhancement of soil petroleum remediation by using a combination of ryegrass (*Lolium perenne*) and different microorganisms. Soil Tillage Res 110:87–93

Timmusk S, Gerhart E, Wagner H (1999) The plant-growth-promoting Rhizobacterium induces changes in gene expression: a possible connection between biotic and abiotic stress responses. Mol Plant-Microbe Interact 12(11):951–959

Timmusk S, Abd El-Daim IA, Copolovici L, Tanilas T, Kännaste A, Behers L, Nevo E, Seisenbaeva G, Stenstrom E, Niinemets U (2014) Drought-tolerance of wheat improved by rhizosphere bacteria from harsh environments: enhanced biomass production and reduced emissions of stress volatiles. PLoS ONE 9(5):e96086. https://doi.org/10.1371/journal.pone.0096086

Tingting L, Kunming D, Miao L et al (2012) Isolation, identification and biodegradation characteristics of a bacterial strain able to degrade bifenthrin. Nongye Huanjing KexueXuebao 31(6):1147–1152

Todd-Brown KE, Hopkins FM, Kivlin SN, Talbot JM, Allison SD (2012) A framework for representing microbial decomposition in coupled climate models. Biogeochemistry 109:19–33. https://doi.org/10.1007/s10533-011-9635-6

Toyota K, Watanabe T (2013) Recent trends in microbial inoculants in agriculture. Microbes Environ 28:403–404

Tringe SG, Mering VC, Kobayashi A, Salamov AA, Chen K, Chang HW, Podar M, Short JM, Mathur EJ, Detter JC, Bork P, Hugenholtz P, Rubin EM (2005) Comparative metagenomics of microbial communities. Science 308:554–557

Tripathi V, Dubey RK, Edrisi SA, Narain K, Singh HB, Singh N, Abhilash PC (2014) Towards the ecological profiling of a pesticide contaminated soil site for remediation and management. Ecol Eng 71:318–325

Van Aken B, Stahl JD, Naveau H, Agathos SN, Aust SD (2000) Transformation of 2,4,6-trinitrotoluene (TNT) reduction products by lignin peroxidase (H8) from the white-rot basidiomycete *Phanerochaete chrysosporium*. Bioremed J 4:135–145

Van Bruggen AHC, Semenov AM (2000) In search of biological indicators for soil health and disease suppression. Appl Soil Ecol 15:13–24

van der Heijden MGA, Bardgett RD, Straalen NM (2008) The unseen majority: soil microbes as drivers of plant diversity and productivity in terrestrial ecosystems. Ecol Lett 11:296–310

Van Dijk DJ et al (2009) Can differences in soil community composition after peat meadow restoration lead to different decomposition and mineralization rates? Soil Biol Biochem 41:1717–1725

Vasas V, Jordán F (2006) Topological keystone species in ecological interaction networks: considering link quality and non-trophic effects. Ecol Model 196:365–378

Vijayakumar S (2012) Potential applications of cyanobacteria in industrial effluents- a review. J Bioremed Biodeg 3:1–6. https://doi.org/10.4172/2155-6199.1000154

Vurukonda SS, Vardharajula S, Shrivastava M, Sk ZA (2016) Enhancement of drought stress tolerance in crops by plant growth promoting rhizobacteria. Microbiol Res 184:13–24

Wakelin SA et al (2008) Habitat selective factors influencing the structural composition and functional capacity of microbial communities in agricultural soils. Soil Biol Biochem 40:803–813

Waldrop MP, Firestone MK (2006) Response of microbial community composition and function to soil climate change. Microb Ecol 52:716–724

Wallenstein MD, Hall EK (2012) A trait-based framework for predicting when and where microbial adaptation to climate change will affect ecosystem functioning. Biogeochemistry 109:35–47. https://doi.org/10.1007/s10533-011-9641-8

Wang Q, Xiong D, Zhao P, Yu X, Tu B, Wang G (2011) Effect of applying an arsenic-resistant and plant growth–promoting rhizobacterium to enhance soil arsenic phytoremediation by *Populus deltoids* LH05-17. J Appl Microbiol 111:1065–1074

Wang G, Jagadamma S, Mayes MA, Schadt CW, Steinweg JM, Gu L et al (2015) Microbial dormancy improves development and experimental validation of ecosystem model. ISME J 9:226–237. https://doi.org/10.1038/ismej.2014.120

Wani PA, Khan MS (2010) *Bacillus* species enhance growth parameters of chickpea (*Cicer arietinum* L.) in chromium stressed soils. Food Chem Toxicol 48:3262–3267

Wardle DA (2002) Communities and ecosystems – linking the aboveground and belowground components. Princeton University Press, Princeton

Weller DM, Raaijmakers JM, Gardener BB, Thomashow LS (2002) Microbial populations responsible for specific soil suppressiveness to plant pathogens. Annu Rev Phytopathol 40:309–348

Wenzel WW (2009) Rhizosphere processes and management in plant-assisted bioremediation (phytoremediation) of soils. Plant Soil 321(1–2):385–408

Weyens N, Van der Lelie D, Taghavi S, Newman L, Vangronsveld J (2009) Exploiting plant-microbe partnerships to improve biomass production and remediation. Trends Biotechnol 27:591–598

Wieder WR, Bonan GB, Allison SD (2013) Global soil carbon projections are improved by modelling microbial processes. Nat Clim Chang 3:909–912. https://doi.org/10.1038/nclimate1951

Wu CH, Bernard SM, Andersen GL, Chen W (2009) Developing microbe–plant interactions for applications in plant-growth promotion and disease control, production of useful compounds, remediation and carbon sequestration. Microb Biotechnol 2:428–440

Wyss A, Boucher J, Montero A, Marison I (2006) Micro-encapsulated organic phase for enhanced bioremediation of hydrophobic organic pollutants. Enzym Microb Technol 40(1):25–31

Xiao X, Chen H, Si C, Wu L (2012) Influence of biosurfactant-producing strain *Bacillus subtilis* BS1 on the mycoremediation of soils contaminated with phenanthrene. Int Biodeterior Biodegradation 75:36–42

Xie S, Liu J, Li L, Qiao C (2009) Biodegradation of malathion by Acinetobacter johnsonii MA19 and optimization of cometabolism substrates. J Environ Sci 21(1):76–82

Xue-Dong W, Xiao-Ming O, Hui-Li W et al (2003) Optimized cultivation of highly-efficient bacterial strains and their biodegradation ability towards imazapyr. Nongye Huanjing Kexue Xuebao 22(1):102–105

Yang L, Zhao YH, Zhang BX, Yang CH, Zhang X (2005) Isolation and characterization of a chlorpyrifos and 3, 5, 6- trichloro-2-pyridinol degrading bacterium. FEMS Microbiol Lett 251:67–73

Yin L, Li X, Liu Y, Zhang D, Zhang S, Luo X (2012) Biodegradation of cypermethrin by rhodopseudomonas palustris GJ-22 isolated from activated sludge. Fresenius Environ Bull A 21(2):397–405

Yuan Z, Druzhinina IS, Labbe J, Redman R, Qin Y, Rodriguez R, Zhang C, Tuskan GA, Lin F (2016) Specialized microbiome of a halophyte and its role in helping non-host plants to withstand salinity. Sci Rep 6:32467. https://doi.org/10.1038/srep32467

Yuanfan H, Jin Z, Qing H, Qian W, Jiandong J, Shunpeng L (2010) Characterization of a fenpropathrin-degrading strain and construction of a genetically engineered microorganism for simultaneous degradation of methyl parathion and fenpropathrin. J Environ Manag 91(11):2295–2300. ISSN 0301-4797

Zeinat K, Nashwa AH, Ibrahim M (2008) Biodegradation and detoxification of malathion by *Bacillus thuringiensis* MOS-5. Aust J Basic Appl Sci 2:724–732

Zelles L (1999) Fatty acids patterns of phospholipids and lipopolysacharides in the characterisation of microbial communities in soil: a review. Biol Fertil Soils 29:111–129. https://doi.org/10.1007/s003740050533

Zhang X, Wu W, Zhang Y et al (2007) Screening of efficient hydrocarbon-degrading strains and study on influence factors of degradation of refinery oily sludge. Ind Eng Chem Res 46(26):8910–8917

Zhang NL, Wan SQ, Li LH, Bi J, Zhao MM, Ma KP (2008) Impacts of urea N addition on soil microbial community in a semi-arid temperate steppe in northern China. Plant Soil 311:19–28

Zhang J, Sun Z, Li Y, Peng X, Li W, Yan Y (2009) Biodegradation of p-nitrophenol by Rhodococcus sp. CN6 with high cell surface hydrophobicity. J Hazard Mater 163(2–3):723–728

Zhang H, Jiang X, Lu L, Xiao W (2015) Biodegradation of polychlorinated biphenyls (PCBs) by the novel identified cyanobacterium *Anabaena* PD-1. PLoS ONE 10(7):e0131450. https://doi.org/10.1371/journal.pone.0131450

Zhong WH, Gu T, Wang W, Zhang B, Lin XG, Huang QR, Shen WS (2010) The effects of mineral fertilizer and organic manure on soil microbial community and diversity. Plant Soil 326:511–522

Zornoza R, Guerrero C, Mataix-Solera J, Scow KM, Arcenegui V et al (2009) Changes in soil microbial community structure following the abandonment of agricultural terraces in mountainous areas of Eastern Spain. Appl Soil Ecol 42:315–323. https://doi.org/10.1016/j.apsoil.2009.05.011. PubMed: 22291451

Manufacturing and Quality Control of Inoculants from the Paradigm of Circular Agriculture

2

Inés E. García de Salamone, Rosalba Esquivel-Cote, Dulce Jazmín Hernández-Melchor, and Alejandro Alarcón

1 Introduction

Worldwide, intensive, and high-input agricultural systems have contributed to significant decreases in soil organic matter that affect the soil quality and health due to alterations in physical, chemical, and biological properties (Cassman 1999). In this regard, soil contains high genetic and functional microbial diversity that has significant influence on functionality and productivity for agricultural crops and plant communities. Soil organic matter represents an available source of energy and nutrients for heterotrophic microbial communities which control mineralization processes and nutrient cycling either in the soil matrix or in the rhizosphere (Morrisey et al. 2004; Herrera-Paredes and Lebeis 2016).

In natural systems, microbial communities are in balance and guarantee the quality and health of soils. In contrast, in agricultural systems, there are drastic alterations of such microbial balance that leads to the loss of beneficial microorganisms and the proliferation of plant pathogens. The latter has devastating effects on crop productivity and soil properties (Avis et al. 2008).

In addition, plant genetic improvement has largely ignored the role of rhizosphere microbial communities in relation to promotion of plant growth and nutrition, alleviation of abiotic or biotic stress, and maintenance and functionality of plant diversity (Morrissey et al. 2004; Rengel 2002). The first agricultural revolution in the eighteenth century introduced crop rotations to stimulate soil microbial populations, although at that time, their benefits on plant health and growth were not well known. The second revolution, named "green revolution," which began in the

I. E. García de Salamone (✉)
Faculty of Agronomy, Department of Applied Biology and Foods, Chair of Agricultural Microbiology, University of Buenos Aires, Buenos Aires, Argentina
e-mail: igarcia@agro.uba.ar

R. Esquivel-Cote · D. J. Hernández-Melchor · A. Alarcón
Colegio de Postgraduados, Postgrado de Edafología, Texcoco, Estado de México, Mexico

D. P. Singh et al. (eds.), *Microbial Interventions in Agriculture and Environment*,
https://doi.org/10.1007/978-981-13-8383-0_2

1960s, was based on traditional techniques of plant breeding and on the development of hybrids without considering the microbial processes of the rhizosphere. Currently, we are facing the era of plant genetic engineering, but production is still highly dependent on utilizing agrochemical inputs. Therefore, an increase in crop yields based on such traditional genetic improvement represents an onerous practice that has generated negative environmental and ecosystem impacts (Tilman et al. 2002). Thus, it is important to develop agronomical practices for increasing and maintaining high plant production in a more sustainable manner (Altieri and Nicholls 2000; Tilman et al. 2002).

The use of beneficial rhizosphere microorganisms to increase the yield of crops is considered as the "new green revolution" (Tilman 1999; Rengel 2005; Den Herder et al. 2010; Gewin 2010; Aeron et al. 2011). The optimization of microbial communities associated with plants offers an innovative biotechnological approach to increase crop productivity without environmental damage (Reid and Greene 2012). However, it is necessary to modify and improve the plant production systems in which either native or allochthonous beneficial microorganisms may be introduced (Malyska and Jacobi 2018; García de Salamone 2012a; Pedraza et al. 2010). This biotechnological approach is within the paradigm of circular agriculture that is a part of the circular economy model. It implies a new business model based on the economic growth of society, environmental sustainability, and the risk reduction due to the volatility and price uncertainty of raw materials and energy resources (World Economic Forum 2014).

2 Sustainability, Bioeconomy, and Circular Agriculture

Application of the circular agriculture relies on obtaining agroecosystem sustainability. Thus, sustainable agriculture is defined as a model of social and economic organization based on an equitable and participatory vision of development which recognizes the environment and natural resources as the foundation of an economic activity (Gold et al. 2017). Consequently, the agriculture is sustainable when it is ecologically friendly, economically viable, socially just, culturally appropriate, and based on a holistic scientific approach (Ikerd 1997). Sustainable and circular agriculture respects the principles of biodiversity and interdependence and uses the insights of modern science to improve the traditional knowledge accumulated over centuries by innumerable farmers around the world. These ideas are profoundly connected with the concept of bioeconomy.

Bioeconomy offers an economic model in which the production of goods and services is based on the sustainable use of biological resources – genes, bacteria, plant biomass, animals, biodiversity, and natural resources such as soil and water – and the utilization of wastes generated during their transformation, production, and consumption contributes to the global goal of decarbonizing the economy (Aguilar et al. 2018; Henry et al. 2017).

In order to feed nine billion people by the year 2050, it will be necessary to double food production. This is a big challenge especially when a third of agricultural land is threatened by desertification and soil fertility impairment due to climate change factors and excessive utilization of high agricultural inputs (Adl 2016). Therefore, the concept of bioeconomy emerges so that it represents a socio-economic model that reduces dependence on fossil resources and promotes the application of knowledge on managing natural resources and biological principles for the sustainable supply of goods and services in all economic sectors. In circular agriculture, bio-inputs constitute a fundamental part of the interface between bio-economy, environment, and agro-industry, with sustenance in prospecting and innovation (Krauss and Kuttenkeuler 2018).

For a particular territory, the successful transition toward bioeconomy will require an intensive effort to develop human resources and better mechanisms for an inclusive participation in society. It requires not only a solid technological background and a rearrangement of basic scientific skills for research and development but also producers and manufacturers capable of managing innovative processes (Aguilar et al. 2018).

Bioeconomy is a reality in several countries. In the European Union, for example, this model employs over 18 million people in the agri-food, chemical, biotechnological, and energy industries (Bell et al. 2018). In Latin America and the Caribbean, there are important developments in Argentina, Brazil, and Costa Rica (Sasson and Malpica 2018). Bioeconomy is driven by the need to ensure the availability of enough biomass feedstock for food, feed, energy, and industrial uses. Plant breeding and breeding innovations are the keystone for sustainable supply of biomass, but its demand must be properly managed in the face of several challenges including environmental issues such as biodiversity conservation and abrupt climate shifts (Małyska and Jacobi 2018).

3 Organic Farming

The paradigm of the circular agriculture also includes the organic agriculture or sustainable agriculture (Rigby and Cáceres 2001). The aims and principles of organic agriculture are declared as basic standards for production and processing by the International Federation of Organic Agriculture Movements (IFOAM 2018).

Organic agriculture is a system of holistic management of production which promotes ecosystem health without altering the biogeochemical cycles, biodiversity, morphology, fertility, and biological activity in the soils. It also optimizes the productivity of the interdependent communities of soil, plants, animals, and people through environmentally friendly methods, from the stages of production to handling and processing (El-Hage and Hattam 2003; IFOAM 2018). In organic agriculture, the ecosystem is managed avoiding the use of external agricultural inputs such as pesticides and chemical fertilizers, among others. On the contrary, the use of low inputs promotes strategies for integrated management of pests, weeds, soils, etc. and is generally oriented only to subsistence, with the aim of reducing production costs, avoiding water pollution, and reducing the presence of pesticide residues in food and the general level of risk of the farmer while increasing the profitability of exports in the short and long term (Gliessman 1998; El-Hage and Hattam 2003).

In this context, organic agriculture offers a variety of environmental, social, and economic benefits for developing countries. It has the potential to increase yields and farmers' income. Thus, it contributes to poverty reduction and sustainable rural development. From an economic point of view, the growing world markets for organic products offer interesting export opportunities for developing countries that may have some comparative advantage for organic agriculture due to relatively abundant workforces (UNCTED 2018) and the production of innocuous food (El-Hage and Hattam 2003). The techniques used in organic agriculture with low external inputs vary from the use of traditional knowledge and microbial products that substitute their synthetic equivalents. These products are regulated by different organizations worldwide. For example, in the United States, the Organic Materials Review Institute (OMRI) is an organization that reviews and evaluates the inputs destined to the production, handling, and processing of certified organic products (OMRI 2018).

4 Bio-inputs

Bio-inputs are biological products obtained from living organisms such as fungi, bacteria, plants, or their derivatives which may be directed or applied as biostimulants, biofertilizers, biocontrollers, biostabilizers, or inoculants for crop protection or nutrition and soil fertility improvers as well. Bio-inputs are also commonly referred to as biotechnological inputs and may be used for both organic agriculture and extensive agriculture.

4.1 Bio-inputs Traditionally Allowed in Organic Agriculture

The organic agriculture is mainly based on the use of locally available and renewable resources (internal inputs) to return nutrient sources to soil for regenerating its fertility and make it self-sufficient. Thus, native plants and animal classes adapted to local environments are used, as well as appropriate cultivation techniques (intercalated, multi-stratified, double excavation beds, mulching, rotations), and mixed production systems (UNCTED 2018). One of the internal inputs mostly used are natural products derived from organic wastes like crop residues, manures, sewage sludge, industrial organic waste, food processing, exploitation of the wood, and municipal garbage, which have a great significance in the maintenance of productivity in developed and developing countries.

Green manures and compost are the most used resources to soil fertilization in organic agriculture. A green manure is a crop used primarily (typically legume residues) as a soil amendment and a nutrient source for subsequent crops since residues are incorporated into soil for further microbial decomposition that allows nutrient cycling and bioavailability (Cherr et al. 2006; Guzmán-Casado and Alonso-Mielgo 2008; Guanche García 2012).

The purpose of using a green manure is to provide protection and help in the recovery of the chemical, physical, and biological conditions of the soil. This favors soil structure and facilitates the work of cultivation, increases contents of organic matter and nutrients, favors the activity of the microbial communities, and reduces the use of synthetic fertilizers. In this way, the conservation and/or recovery of the productivity of agricultural lands is promoted (FAO 2014). However, the nutritional content of organic fertilizers, as well as their content of organic matter, is variable since it depends on several factors; for example, the characteristics of the manures depend on the animal species that produces it, their age and digestive efficiency, type of feeding, and management that the manure has been subjected from its collection, maturation, and storage. In the same way, the nutritional content of a crop residue will depend on the yield potential reached by the crop, the quality of nutrition that it received, and the efficiency in its use and incorporation (INTAGRI 2018)

Composting is a process that converts organic materials (via microbial decomposition under aerobic conditions with adequate humidity) into a stable and hygienic product that favors soil structure and nutrient accumulation (Villegas-Cornelio and Laines-Canepa 2017; FAOTERM 2018). There are several methods of composting, those that use aerated piles and activated sludge and those that only use plant residues, animal manures, urban garbage, or agro-industrial byproducts (Atlas and Bartha 1997). In accordance with the paradigm of circular economy, manure composting allows the reuse of animal wastes by eliminating unpleasant odors. A good composting process of organic materials allows the generation of materials rich in available nutrients that positively impact soil quality in terms of health and fertility and the growth and productivity of crops (Labrador and Bello 2001). In addition, vermicomposting is a technique that involves the use of earthworms (mainly the California red worm *Eisenia foetida*) to get a better quality of composted organic materials. The vermicompost contains more stable compounds rich in available nutrients and is also chemically and biologically enriched by the activity of both earthworms and microbial dynamics occurring during the process (Ferrera-Cerrato and Alarcón 2001; Rivera and Cisneros-Vázquez 2008).

4.2 Bio-inputs for Global Sustainability

The FAO classified bio-inputs in accordance to their functionality as biostimulant, biofertilizer, biofungicide, bio-insecticide, bio-repellent, or inoculants. Their use in agriculture is increasing worldwide as a complement or alternative to the use of traditional agrochemicals (CAC 2017). A general description and some specific characteristics of bio-inputs are described as follows:

Biostimulants: They represent a bio-input capable of improving the efficiency of absorbing and assimilating nutrients by plants, inducing tolerance to biotic or abiotic stress, or improving some of agronomic characteristics (García 2017).

Biostimulants can be composed of humic and fulvic acids, amino acids, enzymes, or vitamins such as thiamine, mixtures of peptides, plant hormones, seaweed extracts, and/or fungi and beneficial bacteria.

Biopesticides: They are bio-based substances or mixtures of plant, animal, microbial, or mineral origin with nutritive properties utilized for preventing and controlling plant pests (insects, mites, nematodes, slugs, and snails) and diseases caused by fungi, bacteria, and viruses (IPES/FAO/RUAF Foundation 2010; FAO 2011; Nava-Pérez et al. 2012).

Biofertilizers: These are preparations containing living or latent microbial cells, as well as substances and macro- and micronutrients that promote plant growth and productivity. The utilized microorganisms can contribute to plant growth through several physiological mechanisms, for instance, nitrogen fixation; mineral solubilization; production of plant growth regulators like auxins, gibberellins, cytokinins, jasmonic acid, and ethylene; and biocontrol of pathogens (Bashan and Holguín 1998; Vessey 2003; Aguado-Santacruz 2012).

Inoculants: These are bio-inputs that contain either single or a mixture of beneficial microorganisms. The first recorded commercial inoculant was directed for legumes in 1895 (Sahoo et al. 2013). The number of registrations of commercial inoculants increases every year since OMRI was founded in 1997 (OMRI 2018). For example, in 2016, 174 microbial inoculants and 284 microbial products were registered as crop biofertilizers in North America (United States, Canada, and Mexico) (Finkel et al. 2017).

5 Inoculants

These beneficial microbial products are usually applied on seeds, in soil, and on plants. The inoculants can be classified as monovalent since they contain a single type of microorganism and as polyvalent when two or more microorganisms are included in the formulation. The use of inoculants may improve the productivity of agricultural and livestock systems and is a technology associated with the principles of sustainability when applied properly and, more importantly, if they fulfill quality controls during their production, in terms of adequate quantity of living and infective microbial cells that guarantee its effectivity.

Inoculants for agriculture have been available commercially for over 120 years (Nobbe and Hiltner 1896; Deaker et al. 2004), but have recently received increasing attention. The use of bio-inputs composed of beneficial microorganisms constitutes an environmentally sound technological strategy for extensive and intensive crops (Díaz-Franco and Mayek-Pérez 2008; Creus 2017). Inoculants are called biostimulants, biopesticides, and biofertilizers, but for being used in organic agriculture, they must avoid the use of genetically modified microorganisms in accordance to the basic norms of International Federation of Organic Agriculture Movements (IFOAM 2003) and contain any food-grade microorganism, as indicated by the Code of Federal Regulations of IFOAM.

The use of these products is increasing not only in organic agricultural production but also in traditional agriculture, at the level of large, medium, and small producers (CAC 2017). Examples of frequently used microorganisms in commercial or experimental inoculants are given in Table 2.1 and described in the following section.

Table 2.1 Examples of inoculants utilized in commercial inoculant formulations worldwide

Microorganism (inoculum)	Crop application	Country	Additional data	References
Bradyrhizobium japonicum	Soybean	Argentina	Several trademarks	Gómez et al. (1997), Benintende (2010), and SENASA (2018)
Mycorrhizal INIFAP[MR] (*Rhizophagus irregularis*), Bacterial 2709 INIFAP[MR] (*Pseudomonas* spp.), *Azospirillum brasilense* INIFAP	Tomato, barley, sorghum, rice, and maize	Mexico	Application: 1 spore mL^{-1} 1×10^6 CFU mL^{-1}	Reyes-Ramírez et al. (2014)
Pediococcus pentosaceus, *L. buchneri* 40788, *Trichoderma reesei*	Lucerne	USA	Application: 9.9 mg kg^{-1} Commercial name: Buchneri 500 inoculant	Arriola et al. (2015)
Azotobacter chroococcum, *Bacillus coagulans*, *Bacillus* sp.	*Cicer arietinum* (chickpea)	India	Commercial name: Anubhav Pravahi *Azotobacter*, Purna *Azotobacter*, Anubhav Pravahi Phosphate, Purna Phosphate Application: 10^8–10^9 CFU mL^{-1}	Ansari et al. (2015)
Azospirillum brasilense	Wheat, sorghum, barley, rice, and maize	Mexico	Application: 200×10^3 ha $año^{-1}$ Company: Biofabrica Siglo XXI S.A. de C.V.	Carrasco-Espinosa et al. (2015)
Azotobacter chroococcum, *Bacillus megaterium*, *Pseudomonas monteilii*, *Glomus intraradices*	*Andrographis paniculata* (Burm. f.) Wall. Ex Ness	India		Khan et al. (2015)

(continued)

Table 2.1 (continued)

Microorganism (inoculum)	Crop application	Country	Additional data	References
Lactobacillus buchneri NCIMB 40788, *Lactobacillus plantarum* CH 6072 y L286	Sugarcane	Brazil	Commercial name: Lalsil® sugarcane, Silobac® 5	Dos Santos et al. (2015)
			Application: 100 g/50 tons of fresh forage, and 50 g/50 tons of fresh forage, respectively	
Glomus intraradices	*Zea mays*	Canada	Application: 25 kg ha^{-1}	Owen et al. (2015)
			Commercial name: MYKE PRO SG2	
			Company: Premier Tech Biotechnologies, Canada	
Streptomyces albus	Potato, barley, lucerne, wheat	Switzerland	Application: 110 L ha^{-1}	Owen et al. (2015)
Streptomyces griseus			Commercial name: EM	
Aspergillus oryzae			Company: Bionova	
Mucor hiemalis			Hygiene GmbH, Stans, Switzerland	
Pseudomonas spp.	Wheat, rice, sorghum, sunflower, and maize	Argentina	Several trademarks	SENASA (2018)
A. brasilense Az39 *Azospirillum* spp.	Maize, wheat, sunflower, oat, rice	Argentina	Several trademarks	Cassán and Diaz-Zorita (2016) and SENASA (2018)
Bacillus amyloliquefaciens, Trichoderma virens	maize	USA	Commercial name: QuickRoots®	Parnell et al. (2016)
Trichoderma, Bradyrhizobium	soybean	USA	Commercial name: Excalibre-SA(ABM)	Parnell et al. (2016)
Trichoderma sp.	*Gossypium hirsutum*	India	Application: 300 g seeds needed per acre	Prasanna et al. (2016)

(continued)

Table 2.1 (continued)

Microorganism (inoculum)	Crop application	Country	Additional data	References
Azotobacter, *Azospirillum*, *Rhizobium*, and P-, K-, and Zn-solubilizing bacteria, arbuscular mycorrhiza, and *Acetobacter*	Rice, maize, cotton, pea, chickpea, sugarcane, sweet potato, blackberry, pomegranate, and *Catharanthus*	India	Production: 88.0 kT in 2015–2016	Sruthilaxmi and Babu (2017)
Bacteria, mycorrhizal fungi, *Trichoderma*	Green roofs	UK	Application: 6 g in 0.6 L of water m^{-2}	Rumble and Gange (2017)
			Company: Symbio Ltd. (Wormley, Surrey)	
Azospirillum brasilense Ab-V5 and Ab-V6	Legumes	Brazil	Commercial name: FORM2+P3 and FORM4+P6	Sanches-Santos et al. (2017)
Azospirillum CM1404, *Azospirillum* CM1403	Maize, wheat, sugarcane	South Africa	Commercial name: AzoBac	Tabassum et al. (2017)
Rhizobium	Legumes, soybean, lucerne, vetch and cowpea beans, wheat, maize, and rice	Pakistan	Commercial name: BioPower	Tabassum et al. (2017)
B. japonicum SCAUs36, *B. diazoefficiens*	Soybean	China	Application: 5×10^8 cells g^{-1} seeds	Thilakarathna and Raizada (2017)
SCAUs46, *Ensifer fredii* SCAUs65				
R. intraradices	Batata (*Dioscorea rotundata* Poir.)	Ivory Coast	Commercial name: MykePro	Kouadio et al. (2017)
			Application: 6,000 spores	
Azospirillum brasilense	Mezquite trees	Mexico	10^9 CFU mL^{-1} inoculum immobilized in alginate dry micropearls	Gonzalez et al. (2018)

6 Plant Growth-Promoting Rhizobacteria, Mycorrhizal Fungi, and Involved Relevant Mechanisms

The main basic crops for human nutrition have already reached their peak of productive potential through traditional genetic improvement; thus, the twenty-first century demands a new green revolution for achieving greater harvests by rationally using the available natural resources but preventing significant losses due to pest and disease incidence. To achieve this difficult objective, it is necessary to get more detailed knowledge of root and rhizosphere interactions (Lynch 2007; García de Salamone et al. 2013). Rhizosphere is one of the most dynamic habitats whose physicochemical and biological properties exhibit great spatial and temporal heterogeneity (Shen et al. 2013) since numerous microorganisms coexist (Barea et al. 2015), and many of them are beneficial for plants (Den Herder et al. 2010; Gewin 2010; Aeron et al. 2011).

Plant growth-promoting rhizobacteria (PGPR) are those bacteria that colonize the rhizosphere and produce beneficial effects on plant growth and nutrition (Bashan et al. 2004; Pedraza et al. 2010; de Souza et al. 2015; Numan et al. 2018; Finkel et al. 2017). There are several scientific evidences indicating that PGPR have a significant role in the sustainability of agroecosystems (Antoun and Prevost 2006).

Inoculation with PGPR contributes to the development and productivity of crops such as rice, wheat, and corn (Lucy et al. 2004; Siddiqui 2006; García de Salamone 2012a, b). The biological nitrogen fixation acquires ecological relevance especially for certain plant-PGPR associations but depends on the combination of plants and bacteria (García de Salamone et al. 1996, 2010; Urquiaga et al. 2004). Crop inoculation with PGPR such as *A. brasilense* must be associated with other management practices to achieve high crop yields (García de Salamone and Monzón de Asconegui 2008; Naiman et al. 2009). Some PGPR like *A. brasilense* (Cassán and García de Salamone 2008) and *P. fluorescens* (García de Salamone et al. 2001, 2012) are associated with plant species of agronomic interest and provide direct beneficial effects on both growth and nitrogen and phosphorus nutrition (Pedraza et al. 2010; García de Salamone 2012a; Naiman et al. 2009).

Under controlled conditions, certain PGPR modify the balance of cytokinins in wheat plants (García de Salamone 2000; García de Salamone et al. 2006) and *Arabidopsis thaliana* (Großkinsky et al. 2016). Furthermore, *A. brasilense* increases the concentration of auxins (Okon 1994) and gibberellins (Bottini et al. 2004), as well as the availability of phosphorus and iron (Pedraza et al. 2010). In addition, some secondary metabolites produced by fluorescent pseudomonads play a significant role in controlling phytopathogens and enhancing plant health (Mishra and Arora 2018). The action mechanisms of these PGPR are very varied (Glick 2012, Bashan and de-Bashan 2010) and are still under debate. Regardless of the mechanisms involved in plant response due to PGPR inoculation, increases in plant production are achieved, and this biotechnological application materializes a feasible strategy for sustainable agriculture (Cassán and García de Salamone 2008; García de Salamone 2012; Glick 2012).

Different crop management conditions modify the number of several microbial groups associated with rice, corn, wheat, oats, and rye (García de Salamone et al. 2010, 2012; Naiman et al. 2009; Di Salvo et al. 2013, 2018a, b; Escobar Ortega and García de Salamone 2017). In addition, under certain conditions, the PGPR inoculation modifies the activities of microorganisms linked to mineralization of organic N and biological nitrogen fixation in rice and wheat crops (D'Auria et al. 2012; García de Salamone et al. 2009; Acosta et al. 2014). Other studies have shown that the combined inoculation of *A. brasilense* and *Bradyrhizobium japonicum* increased the number of root nodules and the nitrogen content of soybean plants with respect to plants solely inoculated with *B. japonicum* (Benintende et al. 2010; Zuffo et al. 2015).

Studies about the impact of cereals on microbial communities have indicated the influence of crop rotation in different soils of the Pampean region managed under direct seeding with zero tillage. Additionally, other studies indicated that fungi and actinomycetes in soils are biological indicators for detecting seasonal variations in wheat/soybean rotations established at different locations of the Pampean region (Rorig et al. 2004; García de Salamone et al. 2006a, b). In these studies, data for the genus *Pseudomonas* showed a significant variation in the number, which indicates the capability of this bacterial group to carry out different functions in the agroecosystem by acting as environmental detoxifiers, PGPR, and controllers of other microorganisms (Großkinsky et al. 2016). The first metagenomic survey of Argentina performed to globally understand the soil microbiome of the Pampean region in agricultural systems with high input requirements (Rascovan et al. 2013) has suggested that additional efforts are needed for describing those microorganisms associated with the roots of different crops under different environments. In addition, they conclude that it is a fundamental step to understand the dynamics of rhizosphere microbial communities and certain microbe-plant associations, which can be included as agricultural practices for favoring sustainable agriculture (Schmidt et al. 2016). Similar information was reported for the rhizosphere of maize and wheat by applying culturing methods to study microbial functional diversity (Di Salvo et al. 2018a, b; Escobar Ortega and Garcia de Salamone 2017).

Mineralization of soil organic carbon and nitrogen is carried out by cellulolytic and nitrifying microorganisms that vary with the conditions imposed by rotation crops in different edaphoclimatic situations. Thus, the cultivation of wheat favors the activity of those functional groups of microorganisms, but cultivation of soybean reduces them (Rorig et al. 2004). Thus, rotation of soybean after cereal crops resulted in high rhizosphere populations of cellulolytic and nitrifier microorganisms (Escobar Ortega and García de Salamone 2017). The inoculation of *A. brasilense* resulted in alleviating stressful conditions, scavenging reactive oxygen species, and increasing the nitrate reductase activity in wheat and rice plants (Zawoznik et al. 2007a, 2009; Iannone et al. 2012, 2013; Ruiz-Sanchez et al. 2011). In addition, *Azospirillum* promotes the formation of lateral and adventitious roots and nodule development in rhizobia-legume symbiosis (Zawoznik et al. 2007b; Molina-Favero et al. 2008; Amenta et al. 2015).

Arbuscular mycorrhizal fungi (AMF) are key beneficial microorganisms for agroecosystem sustainability (Barea 2004; Johansson et al. 2004; Lone et al. 2017) since AMF form a complex symbiotic relationship with roots, called mycorrhiza (Koide and Mosse 2004). The AMF belong to the phylum *Mucoromycota*, subphylum *Glomeromycotina*, and class *Glomeromycetes*, which include several genera belonging to different families (Redecker et al. 2013; Spatafora et al. 2016). They are characterized by the production of typical structures called arbuscules, spores, and, in some cases, vesicles, throughout their life cycle (Peterson and Massicotte 2004). The AMF inoculation increases the area explored by the roots, thereby improving the efficiency of nutrient absorption from soil (Koide and Mosse 2004; Richardson et al. 2009) and contributing to soil aggregation (Rillig et al. 2002) and proper functioning of the agroecosystem (Barea et al. 2002; Jeffries et al. 2002).

Moreover, AMF are obligate biotrophs (Barea et al. 2005), and although commercial inoculants are currently formulated with these fungi, both quality and effectiveness have not yet been correctly demonstrated (Siddiqui and Kataoka 2011). For all these reasons, in the agricultural activity, the AMF inoculation is not a usual practice. However, since these fungi are present in most ecosystems (Koide and Mosse 2004, Malusá et al. 2012) and due to their ecological relevance, the quantification of natural mycorrhization constitutes an index of soil quality depending on the application of agricultural practice management (Jeffries et al. 2003; García de Salamone et al. 2006a, b). Agricultural practices such as monocropping, chemical fertilization, and excessive applications of agrochemicals could diminish or even eliminate the mycorrhizal potential in soils (Collins-Johnson et al. 2003; Oehl et al. 2003; Schalamuk et al. 2006, Willis et al. 2013).

In this regard, it was observed that neither nitrogen fertilization nor PGPR inoculation produced significant effects on AMF colonization in maize roots, but the same agricultural practices modified these variables in wheat crops (Di Salvo et al. 2014). Other authors have shown similar information for maize (Liu et al. 2000) and wheat (Covacevich et al. 2007), but inconsistencies in mycorrhizal responses to fertilization have also been reported (Collins-Johnson et al. 2003). Thus, increases in root colonization after phosphorus fertilizer addition were reported by several authors (García de Salamone et al. 2006a, b, Rubio et al. 2003), while negative responses were also reported (Blanke et al. 2005; Ellis et al. 1992; Treseder 2004). Then, it can be concluded that more research is needed to clarify this topic in order to improve the efficiency of native mycorrhiza in different agricultural ecosystems.

Legumes have an important role in maintaining the productivity of agricultural systems (Graham and Vance 2000). Grain, pulse, pasture, and fresh legumes are grown because they can form nitrogen-fixing symbioses with rhizobia which can produce root nodules (Brewin 2010). Legumes are also used in rotation systems for controlling pests, diseases, and weeds (Howieson et al. 2000b). Nowadays, there are emerging new roles for legumes in the new farming systems for these crops are very relevant to have the appropriate rhizobial inoculants. Furthermore, the agronomic practices of nitrogen fertilization and inoculation with *A. brasilense* did not modify the AMF spore diversity in the rhizosphere of maize (Zambrano et al. 2017) and wheat (Gamarnik et al. 2017) growing under field conditions in humic Hapludolls from Buenos Aires, Argentina.

7 Manufacturing of Inoculants

The use of soilborne microorganisms for increasing productivity of food crops is an attractive eco-friendly, cost-effective, and sustainable alternative to chemical fertilizers and pesticides (Manimekalai and Kannahi 2018). To do the latter, multiple beneficial mechanisms can be considered for isolating microorganisms from soils and plant tissues. Several methodological steps are involved in developing effective microbial inoculants (Fig. 2.1) for achieving consistent results in terms of grain or forage productivity under field conditions.

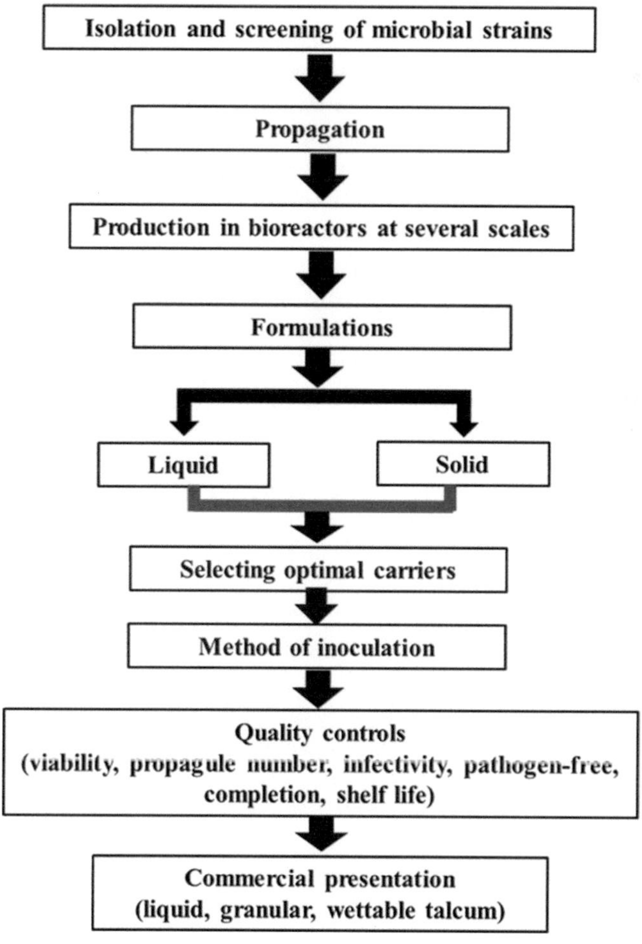

Fig. 2.1 General diagram for microbial inoculant production

The survival and maintenance of microbial activity in both rhizosphere and nonrhizosphere soils are very important for the success of any inoculation protocol. Also, it is essential the achievement of the integration of plant breeding for cultivar development with the selection of elite strains of root nodule-forming bacteria for legume crops or PGPR for nonlegume crops. This key strategy is the only way to enhance the performance of inoculants to improve both crop productivity and beneficial characteristics such as greater nitrogen-fixing ability, survival under stressful edaphic conditions, and greater competitive environment (O'Hara et al. 2002; García de Salamone et al. 2012).

The elaboration of microbial inoculants may consider a unique strain oriented toward a particular mechanism or a microbial consortium with multiple and complementary beneficial functions. The former is usually the case for root nodule-forming bacteria for legumes (Howieson et al. 2000a). For the case of PGPR, there is a trend to use polymicrobial inocula containing microorganisms for each major function involved in plant growth promotion and productivity. Consequently, inoculants can be more stable and have wider applications and wide range of crops. The understanding of biochemical and molecular mechanisms involved in plant-microbe-soil interactions would result in further advances for designing and developing inoculants with greater efficacy for several crops (Reddy and Saravanan 2013). However, for quality purposes, the inoculant must fulfill several microbial strains that can be controlled anytime during both manufacturing process and lifespan. Valverde et al. (2015) described microbiological, genetic, and agronomic tools to isolate and characterize novel *Pseudomonas* spp. from diverse environmental sources, to study the interaction with *Azospirillum* spp. in dual or multi-strain inoculants, and to evaluate the quality and effectiveness of formulated products.

In general, suspensions of PGPR are inoculated in soils and seed or root surfaces, without a proper carrier. Thus, their cell numbers decline rapidly due to predation in the soil. This makes it difficult to sustain persistence and survival of the bacteria in the rhizosphere. The major role of inoculant formulations is to provide a suitable microenvironment, combined with physical protection for a prolonged period to prevent a rapid decline in the inoculated microorganisms (Bashan and de-Bashan 2015).

8 Inoculant Formulations

One of the main limitations for using inoculants is linked to the formulations and carriers (Table 2.2), which will ultimately define the effects under field conditions (Malusa et al. 2012). The survival of any microorganism in the inoculant is fundamental to define the technology of application and product dosage, because they have to compete with native soil microbes for nutrients and niches and to survive against protozoa predation (Bashan et al. 2014; Bonkowski 2004; Xavier et al. 2004). Worldwide, the market for commercial microbial inoculants needs to develop and commercialize new inexpensive bioproducts but with high effectiveness and more stability over time.

Table 2.2 Examples of carriers for the preparation of microbial inoculants

Soil/ inorganic substrates	Plant residues	Inert materials	Liquid media
Soil, charcoal, clays, mud, and/or inorganic soils	Peat moss, vermicompost, vegetal charcoal, sawdust, wheat bran, grape bagasse, corn bagasse, cane bagasse, coconut shell powder	Polymers (alginate, chitosan, carboxymethylcellulose), vermiculite, perlite	Culture medium containing some surfactants or chemical agents (EDTA, glycerol) for improving stability, functionality, and dispersion. Liquid media (broth), mineral or organic oils, or oily water suspensions

Bashan (1998) described different techniques and carriers to prepare inoculants. Formulations obtained by these techniques are normally based on alginate or a mixture of other biopolymers and organic substances. Nevertheless, survival of microorganisms in the carrier or when applied onto the seeds always decreases significantly (Bashan et al. 2014; Cortes-Patiño and Bonilla 2015). Liquid formulations are the most commonly used by manufacturers. For example, liquid formulations of PGPR generally have higher cell numbers with higher survival rate than those formulations based on lignite or peat moss. However, further research is needed to optimize their effectiveness under field conditions. A detailed description of media to cultivate strains of *Azospirillum* was reported by Bashan and de-Bashan (2015), as well as for the phosphorus-solubilizing *Pseudomonas striata* (Mugilan et al. 2011). Valverde et al. (2015) also described several strategies to grow *Pseudomonas* spp. and to evaluate the quality and effectiveness of their formulations.

It is necessary to look for alternative functional microorganisms and for developing new carriers. Some recent studies showed that nanoparticles are promising due to their safety, low-dose application, and contribution to cell agglutination and adhesion to roots, when they are functionalized with suitable substances that interact as ligands with membrane phospholipids (Palmqvist et al. 2015). Nanoparticles have been suggested as a fundamental step to improve industrial formulations of inoculants for the new “green revolution”; then, formulations of PGPR with nanoparticles could improve bacterial growth and, consequently, the quality of microbial inoculants (Dikshit et al. 2013). Nowadays, there is a great interest to both study and compare bio-inputs with and without nanoparticles in their formulations (El-Ghamry et al. 2018)

9 Quality Control of Inoculants

The development of biofertilizers (microbial inoculants) used for agriculture contributes to the reduction or the eradication of agrochemicals and avoids environment contamination. Its application must be easy at field and industrial conditions and have low cost for producers, without detriment to traditional agricultural yields and, more importantly, without having any risk to human, animal, or plant health. Moreover, the biofertilizer must guarantee prolonged effectiveness and persistence in the plant rhizosphere. Therefore, the quality control of biofertilizers takes special relevance especially because they are based on living microorganisms. Furthermore, it is not possible for the farmer to check out the quality of the bioproduct in terms of the number of viable microbes per volume of the biofertilizer. Then, there is the need that quality control must be regulated by the corresponding government, state, or private institutions to avoid the marketing of deceptive or dubious quality products. Thus, biofertilizers should fulfill the biological requirements established by the regulatory agencies or, at least, to warrant the indications stated on the product labels.

In Argentina, in order to unify procedures, in 2005, the Inoculants Quality Control Network (REDCAI for Spanish abbreviation) was integrated with the participation of researchers from several universities, the National Agricultural Technology Institute (INTA), and the signature of some companies, through the framework of the Argentine Association of Microbiology (AAM). The REDCAI aimed to establish methodologies that would be reproducible and reliable and agreed upon registered private and public laboratories. This network works for standardization of techniques for assessing quality control of commercial inoculants based on AMF, *B. japonicum, A. brasilense*, and other PGPR (Cassán et al. 2015). Moreover, to diminish the amount of chemical fertilizers and pesticides used for enhancing crop productivity, some authors have pointed out the need for developing good inoculants and delivery systems to facilitate the environmental persistence of such bio-inputs (Perez-Montaño et al. 2014). More importantly, the impact of the inoculation on subsequent crops in relation to the buffering capacity of the plant-soil-biota is still not well documented and should be the focus of future research efforts (Trabelsi and Mhamdi 2013; Escobar Ortega and Garcia de Salamone 2017; Di Salvo et al. 2018a, b).

10 Experiences in Latin America

10.1 Argentina

In Argentina, agriculture is one of the main economic activities since not only it supplies the country, but the surplus is also destined for export. This country has an area of 34 million hectares in which the main agricultural crops are soybean (*Glycine max* L.), wheat (*Triticum* spp. L.), corn (*Zea mays* L.), sunflower (*Helianthus annuus* L.), sorghum (*Sorghum* spp. L.), and rice (*Oryza sativa* L.). However, the main crop of the agribusiness sector is soybean, an oilseed that occupies half of the land sown and that originates "the soybean chain," one of the main productive chains of the country (Aizen et al. 2009). This product is not consumed in domestic markets and is almost completely exported.

Family farming is a productive model of great importance for Argentina; this strategy contributes to national food security and sovereignty, using environmentally friendly technologies, maintaining a healthy environment and producing healthy foods for local markets (Feito 2013; Gargoloff et al. 2017). In Jujuy province, located in the north of the country, more than 12,500 peasant and indigenous families practice it by involving 18% of the productive units, which in turn represent 31% of the cultivated area and contribute 24% of the gross production value (PNUD 2018).

From a study performed in La Plata region of Buenos Aires province arises the preponderance of family agriculture, with its nuances according to the productive activity that is being analyzed. Intensive production systems based on family labor are the core of horticultural and floricultural activities, with some common and different characteristics between them. As for livestock, one of the extensive types prevails, which is only a portion of family base and, in other cases, complementary to other urban activities. Regarding the changes that have taken place in the period 2005–2015, an increase in the number of horticultural units and a conservation of the floricultural plants are seen, advancing the urbanization processes to the detriment of the bovine production (Cieza et al. 2015).

Worldwide, Argentina ranks second in certified hectares for both livestock and agricultural organic producers. The regional economies of Argentinean provinces are exporting products that are distributed to the United States (45%), to Europe (45%), and to the rest of other markets (FAO 2018). On the other hand, the seeds must comply with the so-called Good Agricultural Practices that guarantee the innocuousness and safety of workers with specific quality protocols: The National Organic Program is based on the care of soil and biodiversity and on prohibiting the use of agrochemicals and transgenic plants.

One of the technologies adopted among farmers in the organic production of soybeans is microbial inoculation. According to a survey conducted by INTA, 94% of farmers, who answered the survey, have said that they used inoculants "always or almost always" (Piccinetti et al. 2013). The "Inoculate" Project, created by INTA in 1994, in collaboration with 25 inoculant manufacturing companies, aimed to evaluate the inoculation effects in different environments for productive legumes in Argentina and disseminate annually the results.

Inoculants allow the contribution of certain selected bacteria by three essential features: competition, infectivity, and effectiveness. Thus, the inoculation of seeds or seedlings allows close contact with roots or radicles when emerging and ensures an efficient colonization; for instance, for rhizobia, close contact with root allows the formation of root nodules (Trabelsi and Mhamdi 2013).

More than 70 companies in Argentina produce commercial inoculants based on rhizobia strains, by which the symbiotic nitrogen fixation in soybean crops is encouraged. The most commonly used species in current inoculants are *B. japonicum* and *B. diazoefficiens* and *B. elkanii*. For several years, INTA recommends *B. japonicum* strain E109 which is included in most inoculants (Lodeiro 2015). Regarding the inoculation with *Azospirillum*, the best characterized strain is *A. brasilense* Az39, which has been a part of inoculants recommended for corn and wheat (García et al. 2013).

10.2 Brazil

Historically in Brazil, agriculture has been one of the main foundations of the country's economy, representing, in 2017, 5% of the gross domestic product (GDP). According to statistics and basic economic data of the Ministry of Agriculture, Livestock and Supply, in 2018, the agricultural products that obtained the highest gross value of production were, in trillions of Brazilian money, soybean (*Glycine max* L.) with 143.86, sugarcane (*Saccharum officinarum* L.) with 63.59, milho (*Panicum miliaceum* L.) with 46.96, cotton (*Gossypium herbaceum* L.) with 34.13, coffee (*Coffea* spp. L.) with 24.46, banana (*Musa x paradisiaca* L.) with 10.77, and rice (*Oryza sativa* L.) with 9.91 (MAPA 2018a).

Given the growth and danger posed using pesticides and their extensive utilization in massive crops, in the last three decades, the practice of organic agriculture in Brazil has been recognized worldwide as one of the best practices for health and environment, due to its conservationist and sustainable nature. The family farmer or "rural family entrepreneur" is one who practices activities in rural areas; thus, foresters, fish farmers, extractive people, fishermen, indigenous people, quilombolas (property of descendants of slaves), and settlers of the agrarian reform are considered family farmers (MDA 2016). In 2014, the agribusiness GDP was 1.18 trillion Brazilian money, with family agriculture representing 37% of the GDP of the agricultural sector, which added to other industries; the real GDP of family farming generated 8.5% of the GDP of Brazil (Bianchini 2016). According to data from the Brazilian Council for Organic and Sustainable Production in 2017, the country's organic sector, including food (natural and industrialized), textiles, and cosmetics, billed US$ 850 million only for domestic markets (Agencia Brasil, 2018).

According to Bortagaray (2016), Brazil is the country with the most policy instruments oriented to research, innovation, and strategic planning in Latin America. In fact, biotechnology companies are established with highly qualified professionals. An example is EMBRAPA, the Brazilian Enterprise for Agricultural Research, which was created in 1973 with the objective of diversifying agricultural production and now is responsible for developing new crops adapted to peculiar conditions of different regions of the country. One of the most outstanding scientists of EMBRAPA was the Czech-Brazilian Johanna Liesbeth Kubelka Döbereiner (1924–2000), who was a pioneer of the study of the biological nitrogen fixation (BNF) in grasses in Brazil and who discovered, for the first time, the association between the bacterium *Beijerinckia fluminensis* and sugarcane in 1958 (Döbereiner and Ruschel 1958) and the association between *Azotobacter paspali* and the grass *Paspalum notatum* in 1966 (Döbereiner 1966). The design of a semisolid culture medium allowed the isolation of two new species of *Azospirillum*: *A. lipoferum* and *A. brasilense* (Döbereiner et al. 1976). These discoveries marked the beginning of the research of biological nitrogen fixation on pastures in Brazil and worldwide. In this way, several lines of research focusing on agricultural applications were developed, thus increasing the interest in research on biological nitrogen fixation, extending it to associations between grasses and diazotrophic bacteria (Baldani and Baldani, 2005).

Research on the symbiosis between *Bradyrhizobium* and soybean (*Glycine max* L.) (Döbereiner et al. 1970) allowed the soybean crop to be revolutionized, since it was possible to reduce or eliminate the dependence of synthetic nitrogen fertilizer on crops, which currently saved between US$ 1 and 2 billion per year, which made soybeans the main national agricultural product, accounting for half of the cereal, legume, and oilseed crops and the highest export earnings of about US$ 325.3 billion, in 2016. According to the OECD and FAO Agricultural Outlook Report 2017–2016, published in July 2017, Brazil will overtake the United States as the world's largest soybean producer in the next decade (NODAL 2017). The management of biological nitrogen fixation is the main source of nitrogen for the development of soybeans. In places where soybeans are not native crops, as the case of Brazil, it is necessary to use inoculants based on bacteria of the genus *Bradyrhizobium* (Oleaginosas 2016).

10.3 Colombia

Colombian agriculture is very diverse. According to the World Bank, the percentage of participation of agriculture in GDP was 6.3% between 2011 and 2015 (The World Bank, 2018). The main crops are sugarcane (*Saccharum officinarum* L.), coffee (*Coffea arabica* L.), cotton (*Gossypium herbaceum* L.), banana (*Musa* × *paradisiaca* L.), banana (*Musa acuminata* L.), sorghum (*Sorghum* spp. L.), maize (*Zea mays* L.), rice (*Oryza sativa* L.), African oil palm (*Elaeis guineensis* L.), potato (*Solanum tuberosum* L.), cassava (*Manihot esculenta* L.), and several flowers, among others. Colombian coffee is a geographical indication protected by the European Union since 2007. Likewise, the term coffee from Colombia is a certification mark registered in the United States on July 7, 1981, and in Canada on July 6, 1990. It is also recognized as Protected Designation of Origin in other countries of the world such as Ecuador, Bolivia, and Peru. Worldwide, Colombia is the fourth coffee-producing country and the largest soft coffee producer (ICO 2018).

In Colombia, the Colombian Agricultural Institute (ICA) is the identity responsible for regulating the production and marketing of bio-inputs for agricultural use. According to Pérez-Lavalle et al. (2017), the highest production of inoculants used in the preparation of biofertilizers is concentrated in the department of Cundinamarca (central region around Bogotá), and the main microorganisms used for developing biological products correspond to AMF as *Glomus* and *Acaulospora*, as well as the nitrogen-fixing bacteria *Azotobacter chroococcum* and *B. japonicum*. Most inoculants have several biological activities and are presented in solid-type formulations. Rice is the crop where the inoculants are applied most frequently. However, in most of the departments that constitute the Caribbean region, the production of inoculants is null.

In this regard, Afanador-Barajas (2017) made a review about the situation of biofertilizers in Colombia, indicating that the research in this type of inoculants is not very large; however, it is important to highlight that there are several studies of various types of biofertilizers applied to various crops, formulated with nitrogen-fixing bacteria, phosphorus solubilizers, AMF, and other PGPR. Also, Montenegro-Gómez and Barrera-Berdugo (2014) mention that N_2-fixing bacteria such as *Azospirillum*, *Burkholderia*, *Gluconobacter*, *Azotobacter*, *Rhizobium,* and *Bradyrhizobium* are the most commonly used bacterial genera for producing nitrogen biofertilizers that help in reducing the application of synthetic fertilizers such as urea. In this way, important advances in the biofertilization of bananas with promising PGPR are also mentioned, with *Bacillus amyloliquefaciens* (BS006), *Pseudomonas fluorescens* (PS006), and *Bacillus subtilis* (EA-CB0575) (Cuéllar-Gavira 2014; Gámez et al. 2015, 2016).

10.4 Costa Rica

Agriculture is one of the traditional sectors of Costa Rica's economy. Of the GDP, 5.5% is generated by agriculture (CIA 2017). From January to September of 2018, the agricultural sector reached a growth of 1.6% with respect to 2017, due to the result of exports from the agricultural sector (2.9%) (SEPSA 2018). The most important traditional agricultural products are coffee (*Coffea arabica* L.), banana (*Musa paradisiaca* L.), sugarcane (*Saccharum officinarum* L.), cocoa (*Theobroma cacao* L.), and pineapple (*Ananas comosus*). L.) (INEC 2017; SEPSA 2018).

Within the model of sustainable agricultural development, Costa Rica has approximately 8000 hectares devoted to production without agrochemicals (MAG 2010a). In addition, this country is an example in Latin America for its advances in organic agriculture, which has allowed for approximately 11,055 hectares dedicated to this system of agricultural production to be certified by 2015. Family farming is a great variety, encompassing productive, tenure, and articulation aspects with the market (Barquero 2016).

One of the technologies carried out in the country for sustainable agricultural production is the use of beneficial microorganisms such as biofertilizers where AMF and soil bacteria such as *Rhizobium* are used. Biopesticides for the biological control of entomopathogenic fungi with the bacteria *Paecilomyces fumosoroseus*, *P. lilacinus*, and *Bacillus subtilis* and the actinomycetes *Trichoderma* spp. and *Streptomyces* are also used in this country (MAG 2010b).

However, the study and development of biofertilizers in Costa Rica is still incipient, and there are few studies reported with microorganisms that promote plant growth. After 20 years, two biological products registered in the laboratory of the center were launched on the market. Agronomic Researchers of the University of Costa Rica (UCR) have formulated Fertibiol and Degradabiol, formulated with fungi and N_2-fixing bacteria and phosphorus solubilizers which are native from different regions of Costa Rica. They comply with functions of both biofertilizers and antagonists against fungi and pathogenic bacteria that affect crops. Since 2017, both products are available to farmers thanks to an alliance between the UCR and the company Suplidora Verde© (https://suplidoraverde.com/), through the Management and Knowledge Transfer Unit (UCR 2018). With these products, it has been possible to control and manage pests such as *Fusarium*, *Rhizoctonia,* and *Mustia* in different crops such as beans (*Phaseolus vulgaris* L.) and lettuce (*Lactuca sativa* L.) as well as to fertilize soils in a sustainable way with the environment, reducing the application of agrochemicals (Chavarría-Vega 2016).

10.5 Cuba

The agricultural area of the country is just over 6.6 million hectares, of which only about 3 million are cultivated. At the beginning of the 1990s, agriculture collapsed due to the loss of commercial relations, and there was a lack of imports of fertilizers, pesticides, machinery, and fuel oil. From that moment, agriculture was reoriented, promoting organic agriculture and urban and suburban agriculture (Altieri et al. 1999). This modality included "organoponicos" (an urban organic farming system originating in Cuba), farms, and familiar patios that covered some 50,000 hectares of land (FAO 2015).

In Cuba, there is a powerful network of scientific centers dedicated to agricultural research, especially related to obtaining new varieties, growth media, soil protection, and adaptation to climate change, among others. The high capacity of human resources, whether researchers or technicians, trained in the country is one of the strengths of the agricultural sector. Urban and suburban agriculture harvests around 1.90 million tons of vegetables and employs more than 400,000 workers, and it has reached levels of 20 kg m^2 per year of usable plant material without using synthetic chemicals (FAO 2018).

One of the key elements within the alternative agricultural model that is currently being implemented in Cuba is the development and application of pest and weed management techniques and biological control, based on the reduction or elimination of synthetic pesticides (Vázquez-Moreno 2006). The most commonly used biofertilizers in Cuba are those containing *Azotobacter chroococcum*, due to the wide range of crops that can be benefited from this bacterium (Mrkovački and Milić 2001; Wani et al. 2013). Thus, selected Cuban strains can supply up to 50% of the nitrogen needs of the plants, which allows considerable savings of chemical fertilizers while reducing environmental pollution.

The production of microbial inoculants from bacteria and fungi, designed to stimulate the development and yield of vegetable species of agricultural interest while reducing the environmental impacts associated with the use and management of agrochemicals in agriculture, is a practice validated in Cuba. The isolation, characterization, and selection of bacterial strains (*Rhizobium*, *Azotobacter*, *Azospirillum*, *Bacillus*, *Gluconacetobacter*, *Bradyrhizobium*, and *Pseudomonas*) and fungi (*Trichoderma* and *Penicillium*) were performed from several ecosystems. These microorganisms have been the starting point for studying fermentative processes and obtaining formulations from different technological alternatives. The application of these bio-inputs in agricultural production systems guarantees increases in yields, between 10% and 30%, reductions of up to 50% of the affectations by diseases, and decreases of the order of 25% in the consumption of agrochemicals in certain technologies of cultivation (Tejeda-González et al. 2010).

Peña-Borrego et al. (2015) made a literature review about the scientific production on the study of biofertilizers in Cuba, and reported that sorghum (*Sorghum bicolor* L. and *S. vulgare* L.), rice (*Oryza sativa* L.), cabbage (*Brassica oleracea* L.), tomato (*Solanum lycopersicum* L.), sugarcane (*Saccharum officinarum* L.), corn (*Zea mays* L.), papaya (*Carica papaya* L.), and canavalia (*Canavalia ensiformis* L.) were the main crops in which were applied the major number of biofertilizer inoculants based on strains of *Glomus*, *Rhizobium*, *Bradyrhizobium*, *Azotobacter*, *Gluconacetobacter*, and *Pseudomonas*.

10.6 Mexico

Agriculture in Mexico is considered one of the most important economic activities, since it generates many jobs in the country; thus, it is considered as the most important productive sector from an economic, social, and environmental point of view. However, it now represents only a small percentage of Mexico's GDP (3.4% for 20017, INEGI 2018).

Mexico is one of the cradles of Mesoamerican agriculture where plants such as corn (*Zea mays* L.), beans (*Phaseolus vulgaris* L.), pepper (*Capsicum annuum* L.), tomatoes (*Solanum lycopersicum* L.), and pumpkin (*Cucurbita pepo* L.) were domesticated, in addition to avocado (*Persea americana* L.), cocoa (*Theobroma cacao* L.), and several plant species. Since the second half of the twentieth century, the Free Trade Agreement and the country's economic policies have again favored large commercial agricultural enterprises.

Organic agriculture is an economic activity with potential in the generation of employment and foreign exchange. For example, in 2017, a total of 83 organic foods were reported, which generated a market value of 6240.7 million Mexican pesos. This figure grew by 42.6% in relation to previous reports for 2016, thanks to the fact that the area destined for the sowing of organic food has grown considerably in the last 10 years. In 2017, the Ministry of Agriculture, Livestock, Rural Development, Fisheries and Food (SAGARPA, for abbreviation in Spanish) reported a total of 47,839 hectares used for growing organic crops. In Mexico, the main organic products are avocado (*Persea americana* 'Hass'), banana (*Musa* × *paradisiaca* L.), raspberry (*Rubus idaeus* L.), coffee (*Coffea arabica* L.), blackberry (*Rubus ulmifolius* L.), tomato (*Solanum lycopersicum* L.), lettuce (*Lactuca sativa* L.), strawberry (*Fragaria* sp.), and mango (*Mangifera indica* L.) (Cuevas-Valdéz 2018).

Producing food through organic agriculture is not a simple process, since among other factors, it involves obtaining technical support, which increases production costs. In Mexico, this technique is at an early stage of development, of which much is still unknown about its potential. In Mexico, the greatest impact of biofertilizers was in the 1970s and 1980s, with the biological N fixation for soybean and chickpea, where it was possible to replace nitrogen fertilization in the state of Sinaloa, which at that time was the main national producer of these legumes; thus, the use of commercial inoculants based on *Rhizobium* was a widespread practice by agricultural producers. In addition, these inoculants were recommended by research centers such as the National Research Institute in Forestry, Agriculture and Livestock (INIFAP) (Armenta-Bojórquez et al. 2010).

One of the researchers who pioneered the study and promotion of biofertilizers in Mexico was Dr. Jesús Caballero Mellado [from the Nitrogen Fixation Research Center, now the Center for Genomic Sciences (CCG) of the National Autonomous University of Mexico (UNAM)], who provided strains of *Azospirillum* and advice to produce biofertilizers directed to the crops of corn, wheat, sorghum, and barley (Rodríguez 2001). In 1999, biofertilizers were applied in around half a million hectares of corn, wheat, and other cereals, and higher yields were obtained in a range of 11–95%, with an average increase of 26% between the different crops. In 2000, the use of biofertilizers by farmers increased to about one and a half million hectares. Currently, the production of biofertilizers is carried out by small companies, by education and research institutions, and by the National Research Institute in Forestry, Agriculture and Livestock (INIFAP) and the International Maize and Wheat Improvement Center (CIMMYT), supported by the federal government through the PROGRO program of SAGARPA (SAGARPA 2018). In fact, SAGARPA announced a new project that focused on promoting the use of biofertilizers developed by these institutions (Curiel 2018). However, the distribution and application on a large scale have had serious difficulty due to problems of promotion and distribution.

On the other hand, for the last 13 years, the company Biofabrica Siglo XXI has been developing together with the UNAM biofertilizers based on *A. brasilense* and *Rhizobium etli* which help in absorbing soil nutrients for plants. This endeavor has been supported by the National Council on Science and Technology (CONACYT) (CONACYT 2016).

11 Regulations and Legislations for Microbial Inoculants

Worldwide, procedures for the regulation, development, and registration of commercial inoculants must be settled. Every country has its regulatory requirements which also vary depending on the characteristics of the new inoculant or bio-input. These regulations apply for the whole production cycle, and some of them include restrictions for the utilization of natural microorganisms.

Some other ministries have issued complementary regulations to ensure that biological products are safe for living organisms and environment and for being properly disposed. Examples of Latin American countries that have a regulation of the use of biofertilizers in agricultural practices are Argentina, Brazil, Colombia, and Mexico.

In Argentina, the registration and regulation of biological fertilizers were originally issued by Resolution No. 1131 of 12/29/1988 of the Ministry of Agriculture, Livestock and Fisheries (SAGyP), which established the characteristics that biological fertilizers must accomplish. The specific norm on the procedure of qualification and control of quality of commercial inoculants is regulated by the Law 20.466; Decree 4.830/1973; decree 1624/1980; Resolution SAG 6673; and Resolutions

SAGyP 310/1994 and 422/2004. Furthermore, the laboratory of the National Service of Agri-Food Health and Quality (SENASA, for its abbreviation in Spanish) is established as the official body for quality control and, at the same time, the minimum requirements for such control by private organisms, which are listed as an annex to Resolution No. 310/1994. Recently created within the scope of the National Advisory Commission on Agricultural Biotechnology (CONABIA for its abbreviation in Spanish) by Resolution 29/2016, the Advisory Committee on Bio-inputs for Agricultural Use (CABUA, for its abbreviation in Spanish) functions to advise on the technical requirements of quality, efficacy, and biosecurity that must be met by bio-inputs for their release to agroecosystems, as well as to propose new standards and issue opinions regarding the regulation and promotion of these products. The use of bio-inputs as biological control agents or phytotherapeutic metabolites is regulated by the Resolution No. 350/1999 of SENASA.

Brazil is a country that has a significant number of regulations on fertilizers, inoculants, and breeders available in the Ministry of Agriculture, Livestock and Food Supply (MAPA 2018b), which are listed below:

- Decree No. 4.954 of 01/14/2004 with modifications of Decree No. 8.384/2014 which approves the regulation of Law No. 6894 of December 16, 1980 that provides the inspection and control of production and trade in fertilizers, improvers, inoculants or biofertilizers, remineralizers, and substrates for plants for agriculture
- Normative Instruction No. 53 of 10/23/2013 with the modifications of the IN No. 6 of 03/10/2016, which establishes the provisions and criteria for the definitions, classification, registration and renewal of the registration of establishment, product registration, marketing authorization and use of secondary materials, registration and renewal of registration of storage service providers, packaging, labeling and advertising of products, alterations or cancellations of registration of establishment, product and registration, and the procedures to be adopted in the inspection and inspection, laboratory analysis, companies that generate secondary materials and suppliers of minerals, production, import, export, and trade of fertilizers, improvers, inoculants, biofertilizers, remineralizers, and substrate for plants and secondary materials; and accreditation of private research institutions
- The Normative Instruction No. 13 of 03/24/2011 approves the standards on specifications, guarantees, registration, packaging and labeling of inoculants destined for agriculture, as well as the relationships of authorized and recommended microorganisms for the production of inoculants in Brazil.
- Normative Instruction No. 30 of 12/11/2010 establishes the official methods for the analysis of inoculants, their counting, identification, and purity analysis.
- Normative Instruction No. 25 of 07/23/2009 approves the norms on specifications and the guarantees, tolerances, registration, packaging and labeling of simple, mixed, compound, organomineral, and biofertilizing organic fertilizers destined to the farming.

- Normative Instruction No. 27 of 05/06/2006, modified by the IN SDA 07 of 12/04/2016, establishes the limits of maximum allowed concentrations for phytotoxic agents, pathogens to humans, animals and plants, toxic heavy metals, pests and weeds to produce, import or market fertilizers, inoculants, and biofertilizers.
- Normative Instruction No. 14 of 10/16/2003 establishes the rules for registration in the Integrated Foreign Trade System (SISCOMEX) for imports of fertilizers, improvers, inoculants, and biofertilizers and their respective raw materials.
- Normative Instruction No. 08 of 02/07/2003 mentions in the Ministry of Agriculture, Livestock and Food Supply, MAPA, the fertilizers, improvers, and inoculants imported directly by the final consumer, for their own use.

In Colombia, the Ministry of Agriculture and Rural Development establishes the regulation for the primary production, processing, packaging, labeling, storage, certification, importation, and commercialization of ecological agricultural products through Regulation No. 0074 of 2002, in which is included in *Annex I. Fertilizers, fertilizers and soil conditioners* the use of products such as microbiological broths (rhizosphere microbial broth), soil biological inoculants, mycorrhiza, *Rhizobium*, *Azotobacter*, *Azospirillum*, *Nitrosomonas*, and *Nitrobacter*, as biological soil inoculants (SAC 2002).

In Mexico, SAGARPA, through the National Service of Health, Food Safety and Agro-Food Quality (SENASICA, for abbreviation in Spanish), is responsible for establishing the requirements and specifications for carrying out effectiveness studies of the inputs of plant nutrition, through the Official Mexican Standard NOM-077-FITO-2000 (https://www.gob.mx/senasica/documentos/nom-077-fito-2000), where the tests of biological effectiveness must be fulfilled in order to commercialize the biofertilizers, besides presenting the required documentation by the Federal Commission for the Protection against Sanitary Risks (COFEPRIS).

12 Challenges of Organic Agriculture

Organic agriculture is a development strategy that tries to change some of the limitations found in conventional production. More than a production technology, the organic agriculture is a strategy based not only on better soil management and promotion of the use of local inputs but also on a greater added value and fairer marketing chain. The organic agriculture is not the universal panacea for all producers in all circumstances. It also has limitations and own challenges, for example, to know the proportion of nitrogen released from the organic material and how much quantity to be applied in accordance to the nutritional requirements of the plants. Nitrogen is the most required element for plants and is more likely depleted in both extensive and organic agriculture. In most inorganic fertilizers, nitrogen is immediately available, but in most organic fertilizers, its availability is slower. In the same way, the amounts of phosphate and potassium, generally, are not in proportion one to the other. Also, the nutritional composition of the organic material depends on the origin of ingredients used for its preparation, which can change depending on the season, the type of organic waste, the mixtures, and the time they undergo composting (Hernández -Rodríguez et al. 2013).

Climate change, emerging pests and weeds, increased resistance to pesticides, malnutrition, famine, and the need for greater production of essential foods are the great challenge for both extensive and organic agriculture. Thus, it is necessary that governments around the globe schedule more programs that involve evenness and inclusion, in favor of sustainable development in the economic, social, and environmental spheres.

13 Future Perspectives of the Use of Inoculants

After more than a century of research associated with technological advances and the need for sustainable crop yield increases, the study of PGPR-plant interactions is being focused on plant microbiomes. Thus, soil biodiversity is being characterized and up to certain functionality is being deciphered. The microbial inputs may contribute to plant growth promotion, disease control, and resistance to pursue the goal for achieving sustainable agriculture worldwide (Finkel et al. 2017).

It is expected that by the year 2030, the effect of the increase in the average annual temperature and the decrease in rainfall will generate significant negative impacts on world agriculture. This is a challenge for Latin America countries, because many farming areas that support exports and food security in rural areas will be affected in the future, and the capacity of this population to adapt to these changes will depend on their access to basic services, information, technology, and maintenance of ecosystems.

To study the complex problem that will arise, it is proposed to carry out interdisciplinary research focused on the practice of crop inoculation using selected strains of beneficial microorganisms, in general, and PGPR, in particular, to obtain formulations of good quality and effectiveness. These biotechnological developments require studies that generate a detailed knowledge of physiological and structural changes that occur on the microbial communities that establish associations with roots. Nevertheless, it is necessary to clarify the role of the plant when interacting with PGPR and soils. This approach could reduce the risk of environmental contamination due to the excessive use of fertilizers in agricultural systems but achieving more efficient use of the available nutrients in the rhizosphere soil.

As our understanding of the complex environment of the rhizosphere, PGPR mechanisms of action, and the practical aspects of inoculant formulations, we can expect to see new PGPR products becoming available. The success of these bioinputs will depend on our ability to manage the rhizosphere to enhance survival and competitiveness of such beneficial microorganisms (Mhlongo et al. 2018). This rhizosphere management will require reconsiderations of soil and crop cultural practices as well as inoculant formulation and delivery. Genetic enhancement of PGPR strains to favor colonization and effectiveness may involve the addition of one or more traits associated with plant growth promotion. Genetic manipulation of crops for root-associated traits to enhance establishment and proliferation of beneficial microorganisms is the way to follow (García de Salamone 2012a; Kroll et al. 2017; Małyska and Jacobi 2018). The use of multi-strain inoculant of PGPR with known functions is of interest as these formulations may favor consistencies in the field (Bashan et al. 2014; Parnell et al. 2016). They offer the potential to address multiple modes of action, multiple pathogens, and temporal or spatial variability. The application of molecular tools is enhancing our ability to understand and manage the rhizosphere and will lead to new products with improved effectiveness (Arora and Mishra 2016; Finkel et al. 2017).

Bacterial inoculation resulted in the stimulation of the native bacteria, actinomycetes, and a group of N_2-fixing free-living bacteria in rhizosphere while suppressing fungal pathogenic populations (Trivedi et al. 2005). Also, the positive effects of bacterial inoculation on the growth of maize are attributed to the stimulation of native microflora (Kumar et al. 2007; Perez-Montaño et al. 2014). In this regard, the rhizosphere microbial communities for several crops were altered after seed inoculation with two bacterial strains like *P. fluorescens* and *A. brasilense* (Escobar Ortega and García de Salamone 2017; Di Salvo et al. 2018a, b); other beneficial microorganisms are related to the phosphorus mobilization such as phosphorus-solubilizing bacteria and AMF (Sharma et al. 2013). In addition, the inoculation with *Bacillus subtilis* enhanced the efficacy of the symbiosis between *Rhizobium* and *Lens esculenta* (Rinu and Pandey 2009). Also, the *B. subtilis* strain NRRL B-30408 has resulted in a significant decrease in AMF colonization in roots (Trivedi et al. 2012).

However, proper guidelines for the production and commercialization of inoculants should be framed and standardized in order to promote the use of such bio-inputs for maintaining the sustainability of agricultural ecosystems and taking care of required safety measures associated with the use of living microbial cultures. Biotechnological and molecular approaches can be applied to develop complementary knowledge about the mode of action of bacterial inoculants and thus to reach successful plant-microbe interactions. Finally, it is essential to combine sustainable management practices with a circular economy approach to meet the environmental, economic, and social needs.

References

Acosta-Andocilla C, Di Salvo LP, Escobar-Ortega JS et al (2014) Nitrógeno potencialmente mineralizable y producción de trigo fertilizado e inoculado con rizobacterias. In: Paper presented at the XXIV Congreso Argentino de la Ciencia del Suelo, Universidad Nacional de Río Cuarto, Argentina, 27 June–1 July 2016

Adl S (2016) Rhizosphere, food security and climate change: a critical role for plant-soil research. Rhizosphere 1:1–3

Aeron A, Kumar S, Pandey P et al (2011) Emerging role of plant growth promoting rhizobacteria in agrobiology. In: Maheshwari DK (ed) Bacteria in agrobiology: crop ecosystems. Springer, Berlin/Heidelberg, pp 1–36

Afanador-Barajas LN (2017) Biofertilizantes: conceptos, beneficios y su aplicación en Colombia. Universidad Central. http://editorial.ucentral.edu.co/ojs_uc/index.php/Ingeciencia/article/view/2353/2177. Accessed 25 Aug 2018

Agencia Brasil (2018) La producción orgánica crece en Brasil. http: / /agenciabrasil.ebc.com.br/en/economia/noticia/2018-08/la-produccion-organica-crece-en-brasil. Accessed 20 Sept 2018
Aguado-Santacruz GA (2012) Uso de microorganismos como biofertilizantes. In: Aguado-Santacruz GA (ed) Introducción al uso y manejo de los biofertilizantes en la agricultura. INIFAP/SAGARPA, México, pp 35–78
Aguilar A, Wohlgemuth R, Twardowsk T (2018) Preface to the special issue bioeconomy. New Biotechnol 40:1–4
Aizen MA, Garibaldi LA, Dondo M (2009) Expansión de la soja y diversidad de la agricultura argentina. Ecol Aust 19:45–54
Altieri MA, Nicholls CI (2000) Applying agroecological concepts to development of ecologically based pest management strategies. In: National Research Council (ed) Professional societies and ecologically based pest management: proceedings of a workshop. National Academy Press, Washington, DC, p 60
Altieri MA, Companioni N, Cañizares K et al (1999) The greening of the "barrios": urban agriculture for food security in Cuba. Agric Hum Values 16(2):131–140
Amenta M, Molina-Favero C, Creus CM et al (2015) Nitric oxide in *Azospirillum* and related bacteria: production and effects. In: Cassán F, Okon Y, Creus C (eds) Handbook for *Azospirillum*. Springer, Cham, pp 155–180
Ansari MF, Tipre DR, Dave SR (2015) Efficiency evaluation of commercial liquid biofertilizers for growth of *Cicer aeritinum* (chickpea) in pot and field study. Biocatal Agric Biotechnol 4(1):17–24
Antoun H, Prevost D (2006) Ecology of plant growth promoting rhizobacteria. In: Siddiqui ZA (ed) PGPR: biocontrol and biofertilization. Springer, Dordrecht, pp 1–38
Armenta-Bojórquez AD, García-Gutiérrez C, Camacho-Báez JR et al (2010) Biofertilizantes en el desarrollo agrícola de México. Ra Ximahai 6(1):51–56
Arora NK, Mishra J (2016) Prospecting the roles of metabolites and additives in future bioformulations for sustainable agriculture. Appl Soil Ecol 107:405–407
Arriola KG, Queiroz OC, Romero JJ et al (2015) Effect of microbial inoculants on the quality and aerobic stability of bermudagrass round-bale haylage. J Dairy Sci 98:478–485
Atlas RM, Bartha R (1997) Microbial ecology. Fundamentals and applications, 4th edn. Benjamin/Cummings, San Francisco, p 694
Avis TJ, Gravel V, Antoun H et al (2008) Multifaceted beneficial effects of rhizosphere microorganisms on plant health and productivity. Soil Biol Biochem 40(7):1733–1740
Baldani JI, Baldani VLD (2005) History on the biological nitrogen fixation research in graminaceous plants: special emphasis on the Brazilian experience. Ann Acad Bras Ciênc 77(3):549–579
Barea JM (2004) Impacto de las micorrizas en la calidad del suelo y la productividad vegetal en sistemas agrícolas y espacios naturales. In: Monzón de Asconegui MA, García de Salamone IE, Miyazaki SS (eds) Biología del Suelo. Transformaciones de la materia orgánica, usos y biodiversidad de los organismos edáficos. FAUBA, Universidad de Buenos Aires, pp 7–11
Barea JM, Azcón R, Azcón-Aguilar C (2002) Mycorrhizosphere interactions to improve plant fitness and soil quality. Antonie Van Leeuwenhoek 81:343–351
Barea J-M, Pozo MJ, Azcón R, Azcón-Aguilar C (2005) Microbial co-operation in the rhizosphere. J Exp Bot 56(417):1761–1778
Barea JM, Pozo MJ, Azcón R et al (2015) Microbial co-operation in the rhizosphere. J Exp Bot 56(417):1761–1778
Barquero M (2016) Terrenos para cultivo orgánico crecieron en dos últimos años. La nación. https://www.nacion.com/economia/agro/terrenos-organicos-en-costa-rica-repuntaron-en-dos-ultimos-anos/6652KW7OSJBOZFCGANEUYMNQVQ/story/. Accessed 20 Sept 2018
Bashan Y (1998) Inoculants of plant growth-promoting bacteria for use in agriculture. Biotechnol Adv 16(4):729–770
Bashan Y, de Bashan LE (2010) How plant growth-promoting bacterium *Azospirillum* promotes plant growth – a critical assessment. Adv Agron 108:77–136
Bashan Y, de-Bashan LE (2015) Inoculant preparation and formulations for *Azospirillum* spp. In: Cassán FD, Okon Y, Creus CM (eds) Handbook for *Azospirillum*. Technical issues and protocols. Springer, Geneva, pp 469–485

Bashan Y, Holguín G (1998) Proposal for the division of plant growth-promoting rhizobacteria into two classifications: biocontrol PGPB (plant growth-promoting bacteria) and PGPB. Soil Biol Biochem 30:1225–1228

Bashan Y, Holguin G, de-Bashan LE (2004) *Azospirillum*-plant relationships: agricultural, physiological, molecular and environmental advances (1997–2003). Can J Microbiol 50:521–577

Bashan Y, de-Bashan LE, Prabbu SR et al (2014) Advances in plant growth-promoting bacterial inoculant technology: formulations and practical perspectives (1998–2013). Plant Soil 378(1–2):1–33

Bell J, Paula L, Dodd T et al (2018) EU ambition to build the world's leading bioeconomy – uncertain times demand innovative and sustainable solutions. New Biotechnol 40:25–30

Benintende S (2010) Calidad de inoculantes comerciales para el cultivo de soja en la Argentina: concentración de rizobios viables y presencia de contaminantes. Rev Argent Microbiol 42(2):129–132

Benintende S, Uhrich W, Herrera M et al (2010) Comparación entre coinoculación com *Bradyrhizobium japonicum* y *Azospirillum brasilense* e inoculación simple con *Bradyrhizobium japonicum* en la nodulación, crecimiento y acumulación de N en el cultivo de soja. Agriscientia 23(2):71–77

Bianchini V (2016) Vinte anos do PRONAF, 1995–2015. Avanços e desafios. Ministerio de Desarrollo Agrario. 113 pp. Retrieved from http://www.mda.gov.br/sitemda/sites/sitemda/files/ceazinepdf/PRONAF_20_ANOS_VALTER_BIANCHINI.pdf. Accessed 12 Sept 2018

Blanke V, Renker C, Wagner M et al (2005) Nitrogen supply affects arbuscular mycorrhizal colonization of *Artemisia vulgaris* in a phosphate-polluted field site. New Phytol 166:981–992

Bonkowski M (2004) Protozoa and plant growth: the microbial loop in soil revisited. New Phytol 162(3):617–631

Bortagaray I (2016) Políticas de ciencia, tecnología e innovación sustentable e inclusiva en América Latina. UNESCO. http://www.unesco.org/new/fileadmin/MULTIMEDIA/FIELD/Montevideo/pdf/PolicyPapersCILAC-InnovacionEmpresarial.pdf. Accessed 2 Aug 2018

Bottini R, Cassán F, Piccoli P (2004) Gibberellin production by bacteria and its involvement in plant growth promotion and yield increase. Appl Microbiol Biotechnol 65:497–503

Brewin NJ (2010) Root nodules (Legume–*Rhizobium* Symbiosis). In: eLS. Wiley, Chichester. https://doi.org/10.1002/9780470015902.a0003720.pub2

CAC (2017) Biopesticides, biofertilizers, biostimulants. Joint FAO/WHO Food Standards Programme Codex Alimentarius Commission 40th Session CICG, Geneva, Switzerland 17–22 July 2017. Accessed 1 Oct 2018

Carrasco-Espinosa K, García-Cabrera RI, Bedoya-López A et al (2015) Positive effect of reduced aeration rate on growth and stereospecificity of dl-malic acid consumption by *Azospirillum brasilense*: improving the shelf life of a liquid inoculant formulation. J Biotechnol 195:74–81

Cassán F, Diaz-Zorita M (2016) *Azospirillum* sp. in current agriculture: from the laboratory to the field. Soil Biol Biochem 103:117–130

Cassán FD, García de Salamone IE (2008) *Azospirillum* sp.: cell physiology, plant interactions and agronomic research in Argentina. In: Paper presented at the international workshop on *Azospirillum*: cell physiology, plant response and agronomic research in Argentina. Córdoba, Argentina, 12–13 October 2007

Cassán FD, Okon Y, Creus CM (eds) (2015) Handbook for *Azospirillum*. Technical issues and protocols. Springer, Geneva. https://doi.org/10.1007/978-3-319-06542-7_24

Cassman KG (1999) Ecological intensification of cereal production systems: yield potential, soil quality, and precision agriculture. Proc Natl Acad Sci U S A 96:5952–5959

Chavarría-Vega M (2016) El uso de microorganismos benéficos: Biofertilizantes y Biocontroladores. CIA. INTA. ACCS. http://www.mag.go.cr/biblioteca_virtual_ciencia/biofertilizantes_biocontroladores.pdf. Accessed 1 Oct 2018

Cherr CM, Scholberg JMS, McSorley R (2006) Green manure approaches to crop production. Agron J 98(2):302–319

CIA (2017) The world factbook. https://www.cia.gov/library/publications/the-world-factbook/fields/2012.html#84. Accessed 20 Nov 2018

Cieza RI, Ferraris G, Seibane C et al (2015) Aportes a la caracterización de la agricultura familiar en el partido de La Plata. Rev Fac Agron 114(3):129–142
Collins-Johnson N, Rowland DL, Corkidi L et al (2003) Nitrogen enrichment alters mycorrhizal allocation at five mesic to semiarid grasslands. Ecology 84:1895–1908
CONACYT (2016). http://www.conacytprensa.mx/index.php/tecnologia/biotecnologia/9822-biofertilizantes-alternativa-ecologica-y-confiable. Accessed 25 Oct 2018
Cortes-Patiño S, Bonilla RR (2015) Polymers selection for a liquid inoculant of *Azospirillum brasilense* based on the *Arrhenius thermodynamic* model. Afr J Biotechnol 14(33):2547–2553
Covacevich F, Echeverría HE, Aguirrezabal LAN (2007) Soil available phosphorus status determines indigenous mycorrhizal colonization at field and glasshouse-grown spring wheat from Argentina. Appl Soil Ecol 35:1–9
Creus CM (2017) Inoculantes microbianos: piezas de un rompecabezas que aún requiere ser ensamblado. Rev Argent Microbiol 49(3):207–209
Cuéllar-Gavira TZ (2014) Evaluación de la promoción de crecimiento de *Bacillus subtilis* EA-CB0575 en cultivos de banano, crisantemo y café. Universidad EAFIT, Colombia, p 135
Cuevas-Valdéz J (2018) Agricultura orgánica, oportunidades de crecimiento en México (II). https://www.eleconomista.com.mx/opinion/Agricultura-organica-oportunidades-de-crecimiento-en-Mexico-II-20180726-0093.html. Accessed 20 Oct 2018
Curiel R (2018) Anuncia SAGARPA biofertilizantes para agricultores de PROGRO Productivo que recibirán asesoría de CIMMYT. (https://masagro.mx/es/inicio/39-boletines/boletines-2018/170-anuncia-sagarpa-biofertilizantes-para-agricultores-de-progro-productivo-que-recibiran-asesoria-de-cimmyt). Accessed 20 Oct 2018
D'Auria F, Di Salvo LP, García de Salamone IE (2012) Microbiología de la fitorremediación de suelos contaminados con hidrocarburos. Paper presented at XIX Congreso Latinoamericano de la Ciencia del Suelo. XXIII Congreso Argentino de la Ciencia del Suelo, Mar del Plata, Argentina, 16–20 April 2012
de Souza R, Ambrosini A, Passaglia LMP (2015) Plant growth-promoting bacteria as inoculants in agricultural soils. Genet Mol Biol 38(4):401–419
Deaker R, Roughley RJ, Kennedy IR (2004) Legume seed inoculation technology – a review. Soil Biol Biochem 36:1275–1288
Den Herder G, Van Isterdael G, Beeckman T et al (2010) The roots of a new green revolution. Trends Plant Sci 15:600–607
Di Salvo LP, Cellucci GC, Carlino ME et al (2013) Inocular el cultivo de maíz con *Azospirillum brasilense* y fertilizar con urea incrementa el rendimiento y altera temporalmente los microorganismos rizosféricos asociados a los ciclos del carbono y el nitrógeno. Simposio IPNI, Rosario, Argentina, May 22–23.
Di Salvo LP, López Rondó GR, Radio Brandoni P et al (2014) Influence of *Azospirillum brasilense* inoculation and fertilization practices on the colonization of wheat by natural arbuscular mycorrhiza. In: Paper presented at the II Taller Latinoamericano sobre Rizobacterias Promotoras del Desarrollo Vegetal. Córdoba, Argentina, 21–26 September 2014
Di Salvo LP, Cellucci GC, Carlino ME et al (2018a) Plant growth promoting rhizobacteria inoculation and nitrogen fertilization increase maize grain yield and modified rhizosphere microbial communities. Appl Soil Ecol 126:113–120
Di Salvo LP, Ferrando L, Fernández-Scavino A et al (2018b) Microorganisms reveal what plants do not: wheat growth and rhizosphere microbial communities after application of *Azospirillum brasilense* and nitrogen fertilizer under field conditions. Plant Soil 424:405–417
Díaz-Franco A, Mayek-Pérez N (2008) La biofertilización como tecnología sostenible. Consejo Nacional de Ciencia y Tecnología y Consejo Tamaulipeco de Ciencia y Tecnología de México. Plaza y Valdés, S.A. de C.V. México, p 260
Dikshit A, Shukla SK, Mishra RK (2013) Exploring manomaterials with PGPR in current agriculture scenario. LAP Lambert Academic Publishing, Saarbrucken, p 78
Döbereiner J (1966) *Azotobacter paspali* sp.n., uma bacteria fixadora de nitrogenio na rizosfera de *Paspalum*. Pesq Agropec Bras 1:357–365
Döbereiner J, Ruschel AP (1958) Uma nova especie de *Beijerinckia*. Rev Biol 1:261–272

Döbereiner J, Francos AA, Cuzmán I (1970) Estirpes ide *Rhizobium japonicum* de excepcional eficiência. Pesq Agropec Bras 5:155–161
Döbereiner J, Marriel IE, Nery M (1976) Ecological distribution of *Spirillum lipoferum* Beijerinck. Can J Microbiol 22:1464–1473
Dos Santos WCC, do Nascimento WG, Magalhães ALR et al (2015) Nutritive value, total losses of dry matter and aerobic stability of the silage from three varieties of sugarcane treated with commercial microbial additives. Anim Feed Sci Technol 204:1–8
El-Ghamry AM, Mosa AA, Alshaal TA et al (2018) Nanofertilizers vs. biofertilizers: new insights. Environ Biodivers Soil Security 2:51–72
El-Hage S, Hattam C (ed) (2003) La agricultura orgánica, ambiente y seguridad alimentaria. Colección FAO: Ambiente y Recursos Naturales No. 4. http://www.fao.org/docrep/005/y4137s/y4137s03.htm#bm03.1. Accessed 18 Oct 2018
Ellis JR, Roder W, Mason SC (1992) Grain sorghum soybean rotation and fertilization influence on vesicular arbuscular mycorrhizal fungi. Soil Sci Soc Am J 56:789–794
Escobar-Ortega JS, García de Salamone IE (2017) Dynamics of rhizosphere microbial communities of cover crops dried with glyphosate. In: Singh DP et al (eds) Plant-microbe interactions in agro-ecological perspectives. Springer Nature, pp 17–34. https://link.springer.com/content/pdf/bfm%3A978-981-10-5813-4%2F1.pdf
FAO (2011) FAO enseña a combatir plagas y enfermedades con sustancias naturales. http://www.fao.org/americas/noticias/ver/es/c/230667/. Accessed 25 Aug 2018
FAO (2014) Organic Agriculture. http://www.fao.org/organicag/oa-home/en/. Accessed 25 Aug 2018
FAO (2015) La Habana. Agricultura urbana y periurbana en América Latina y el Caribe. http://www.fao.org/ag/agp/greenercities/Es/CMVALC/la_habana.html
FAO (2018) FAOSTAT: Agricultura bajo agricultura orgánica certificada. Organización de las Naciones Unidas para la Alimentación y la Agricultura. http://www.fao.org/faostat/es/#search/Agricultura%20bajo%20agricultura%20org%C3%A1nica%20certificada. Accessed 30 Aug 2018
FAOTERM–Organic Agriculture (2018) Manual del compostaje del agricultor. Experiencias en América Latina. http://termportal.fao.org/faooa/main/start.do. Accessed 25 Sept 2018
Feito MC (2013) Agricultura familiar con enfoque agroecológico en zonas periurbanas. Análisis de una experiencia de intervención para el desarrollo rural en Lujan (Buenos Aires, Argentina) Rev Electron Geogr Austral. http://revistanadir.yolasite.com/resources/agricultura%20fmiliar%20FEITO.pdf. Accessed December 2018
Ferrera-Cerrato R, Alarcón A (2001) La microbiología del suelo en la agricultura sostenible. Revista Ciencia Ergo Sum. Universidad Autónoma del Estado de México 8(22):175–183
Finkel OM, Castrillo G, Herrera-Paredes S et al (2017) Understanding and exploiting plant beneficial microbes. Curr Opin Plant Biol 38:155–163
Gamarnik M, Zambrano-Soledispa A, Di Salvo LP et al (2017) Diversidad de esporas de micorrizas arbusculares modificada por ciertas prácticas agronómicas del cultivo de trigo. In: Paper presented at the XI Reunión Nacional de Biología de Suelos REBIOS. Corrientes, Argentina, October 25–27
Gámez RM, Rodríguez F, Bernal JF et al (2015) Genome sequence of the banana plant growth-promoting rhizobacterium *Bacillus amyloliquefaciens* BS006. Genome Announc 3(6):e01391–e01315
Gámez RM, Rodríguez F, Ramírez S et al (2016) Genome sequence of the banana plant growth-promoting rhizobacterium *Pseudomonas fluorescens* PS006. Genome Announc 4(3):e00329–e00316
García SD (2017) Bioestimulantes Agrícolas, Definición, Principales Categorías y Regulación a Nivel Mundial. Serie Nutrición Vegetal Núm. 94. INTAGRI. México. https://www.intagri.com/articulos/nutricion-vegetal/bioestimulantes-agricolas-definicion-y-principales-categorias. Accessed 25 Sept 2018
García de Salamone IE (2000) Direct beneficial effects of cytokinin-producing rhizobacteria on plant growth. University of Saskatchewan, Canada. www.usask.ca

García de Salamone IE (2012a) Use of soil microorganisms to improve plant growth and ecosystem sustainability. In: Caliskan M (ed) The molecular basis of plant genetic diversity. INTECH, Rijeka, pp 233–258. https://www.intechopen.com/books/the-molecular-basis-of-plant-genetic-diversity/use-of-soil-microorganisms-to-improve-plant-growth-and-ecosystem-sustainability.
García de Salamone IE (2012b) Microorganismos promotores del crecimiento vegetal. Informaciones Agronómicas de Hispanoamérica (IPNI) 5:12–16
García de Salamone IE, Monzón de Asconegui M (2008) Ecofisiología de la respuesta a la inoculación con *Azospirillum* en cultivos de cereales. In: Cassán FD, García de Salamone IE (eds) *Azospirillum* sp.: cell physiology, plant interactions and agronomic research in Argentina. Asociación Argentina de Microbiología, Buenos Aires, pp 209–226
García de Salamone IE, Dobereiner J, Urquiaga S et al (1996) Biological nitrogen fixation in *Azospirillum* strain-maize genotype associations as evaluated by the 15N isotope dilution technique. Biol Fertil Soils 23:249–256
García de Salamone IE, Hynes RK, Nelson LM (2001) Cytokinin production by plant growth promoting rhizobacteria and selected mutants. Can J Microbiol 47:404–411
García de Salamone IE, Hynes RK, Nelson LM (2006a) Role of cytokinins in plant growth promotion by rhizosphere bacteria. In: Siddiqui ZA (ed) PGPR: biocontrol and biofertilization. Springer, Dordrecht, pp 173–195
García de Salamone IE, Michelena R, Rodríguez A et al (2006b) Ocurrencia de micorrizas vesículo arbusculares en plantas de maíz, soja y trigo en sistemas de siembra directa. Revista de la Facultad de Agronomía, Universidad de Buenos Aires 26(1):67–72
García de Salamone IE, Escobar-Ortega JS, Gatica M et al (2009) Effect of *Azospirillum* inoculation on N-cycling microorganisms associated with rice and wheat crops. In: Paper presented at the 16th Nitrogen Workshop, Turin, Italy, June 28th–July 1st 2009.
García de Salamone IE, Di Salvo LP, Escobar-Ortega JS et al (2010) Field response of rice paddy crop to inoculation with *Azospirillum*: physiology of rhizosphere bacterial communities and the genetic diversity of endophytic bacteria in different parts of the plants. Plant Soil 336:351–362
García de Salamone IE, Funes JM, Di Salvo LP et al (2012) Inoculation of paddy rice with *Azospirillum brasilense* and *Pseudomonas fluorescens*: Impact of plant genotypes on the rhizosphere microbial communities and field crop production. Appl Soil Ecol 61:198–204
García de Salamone IE, Vázquez S, Penna C, Cassán FD (2013) Rizosfera, biodiversidad y agricultura sustentable. Paper publicated at the Taller Internacional sobre Rizosfera, Biodiversidad y Agricultura Sustentable 2010. Buenos Aires, Argentina. 21–22 October 2010
García JE, Puente ML, Maronichea GA et al (2013) Estudio de *Azospirillum* como tecnología aplicable en los cultivos de trigo y maíz. Microbiología Agrícola. Un aporte de la Investigación en Argentina, pp 351–366. https://www.researchgate.net/publication/307513653_Estudio_del_Azospirillum_como_tecnologia_aplicable_en_los_cultivos_de_trigo_y_maiz. Accessed December 2018
Gargoloff NA, Blandi ML, and Sarandón SJ (2017) The importance of the family horticultural history on the knowledge and ecological management of agrobiodiversity. A case study in the green belt of La Plata, Argentina. VI Congreso Latino-Americano de Agroecologia, Brasilia, Brazil. September 12–15. file:///C:/Users/inese/Documents/SPRINGER%202018/family%20Agriculture%20Argentina%202018.pdf. Accessed December 2018.
Gewin V (2010) Food: An underground revolution. Nature 466:552–553
Glick BR (2012) Plant growth-promoting bacteria: mechanisms and applications. Scientifica. https://doi.org/10.6064/2012/963401
Gliessman SR (1998) Agroecology: ecological processes in sustainable agriculture. Ann Arbor Press, Ann Arbor, p 357
Gold S, Kunz N, Reine G (2017) Sustainable global agrifood supply chains: exploring the barriers. J Ind Ecol 21(2):249–260
Gómez M, Silva N, Hartmann A et al (1997) Evaluation of commercial soybean inoculants from Argentina. World J Microbiol Biotechnol 13:167–173

Gonzalez EJ, Hernandez JP, de-Bashan LE (2018) Dry micro-polymeric inoculant of *Azospirillum brasilense* is useful for producing mesquite transplants for reforestation of degraded arid zones. Appl Soil Ecol 129:84–93
Graham PH, Vance CP (2000) Nitrogen fixation in perspective: an overview of research and extension needs. Field Crop Res 65:93–106
Großkinsky DK, Tafner R, Moreno MV et al (2016) Cytokinin production by *Pseudomonas fluorescens* G20-18 determines biocontrol activity against *Pseudomonas syringae* in *Arabidopsis*. Sci Rep 6:23310. https://doi.org/10.1038/srep23310
Guanche-García A (2012) Los abonos verdes. Información técnica. http://www.agrocabildo.org/publica/Publicaciones/agec_454_abonos_verdes.pdf
Guzmán-Casado GI, Mielgo AM (2008) Buenas prácticas en producción ecológica. Uso de abonos verdes. Ministerio de Medio Ambiente y Medio Rural y Marino. España, p 22 http://www.agroecologia.net/recursos/publicaciones/bppe/Uso_de_Abonos_Verdes_tcm7-187426.pdf
Henry G, Hodson E, Aramendis R et al (2017) Bioeconomy: An engine for integral development of Colombia. file:///C:/Users/inese/Documents/SPRINGER%202018/Articulos%20revisados/Bioeconomy_An_engine_for_the_integral_development_of_Colombia.pdf
Hernández-Rodríguez OA, Hernández-Tecorral A, Rivera-Figueroa C et al (2013) Calidad Nutrimental de cuatro abonos orgánicos producidos a partir de residuos vegetales y pecuarios. Terra Latinoamericana 31:35–46
Herrera Paredes S, Lebeis SL (2016) Giving back to the community: microbial mechanisms of plant–soil interactions. Funct Ecol 30:1043–1052
Howieson JG, Malden J, Yates RJ et al (2000a) Techniques for the selection and development of elite inoculant strains of *Rhizobium leguminosarum* in southern Australia. Symbiosis 28:33–48
Howieson JG, O'Hara GW, Carr SJ (2000b) Changing roles for legumes in Mediterranean agriculture: developments from an Australian perspective. Field Crop Res 65:107–122
Iannone MF, Rosales EP, Groppa MD et al (2012) Reactive oxygen species formation and cell death in catalase-deficient tobacco leaf discs exposed to paraquat. Biol Trace Elem Res 146:246–255
Iannone MF, Rosales EP, Groppa MD et al (2013) H_2O_2 Involvement in Polyamine-induced cell death in tobacco leaf discs. J Plant Growth Regul 32:745–757
ICO (2018). www.ico.org. Accessed 28 Sept 2018
IFOAM (2003) Normas para la producción y procesado orgánico. Die Deutsche Bibliothek, p 154
IFOAM (2018) Definition of organic agriculture. https://www.ifoam.bio/en/organic-landmarks/definition-organic-agriculture. Accessed 1 Oct 2018
Ikerd J (1997) Toward an economics of sustainability. Dept of Agricultural Economics, University of Missouri. http://www.ssu.missouri.edu/faculty/JIkerd/papers/econ-sus.htm. Accessed November 2018.
INEC (2017) Encuesta Nacional Agropecuaria. http://inec.cr/encuesta-nacional-agropecuaria. Accessed 12 Nov 2018
INEGI (2018). http://www.beta.inegi.org.mx/contenidos/saladeprensa/boletines/2018/ pib_pconst/pib_pconst2018_08.pdf. Accessed 1 Oct 2018
INTAGRI (2018) Los abonos orgánicos beneficios tipos y contenidos nutrimentales. https://www.intagri.com/articulos/agricultura-organica/los-abonos-organicos-beneficios-tipos-y-contenidos-nutrimentales. Accessed 28 Sept 2018
IPES/FAO/RUAF Foundation (2010) Biopreparados para el manejo sostenible de plagas y enfermedades en la agricultura urbana y periurbana. http://www.fao.org/3/a-as435s.pdf. Accessed 30 Aug 2018
Jeffries P, Craven-Griffiths A, Barea JM et al (2002) Application of arbuscular mycorrhizal fungi in the revegetation of desertified Mediterranean ecosystems. In: Gianinazzi S, Schuepp H, Barea JM et al (eds) Mycorrhizal technology in agriculture. Birkhauser verlag, Basel, pp 151–174
Jeffries P, Gianinazzi S, Perotto S, Turnau K, Barea JM (2003) The contribution of arbuscular mycorrhizal fungi in sustainable maintenance of plant health and soil fertility. Biol Fertil Soils 37:1–16
Johansson JF, Paul LR, Finlay RD (2004) Microbial interactions in the mycorrhizosphere and their significance for sustainable agriculture. FEMS Microbiol Ecol 48:1–13

Khan K, Pankaj U, Verma SK et al (2015) Bio-inoculants and vermicompost influence on yield, quality of *Andrographis paniculata*, and soil properties. Ind Crop Prod 70:404–409
Koide RT, Mosse B (2004) A history of research on arbuscular mycorrhiza. Mycorrhiza 14(3):145–163
Kouadio ANM-S, Nandjui J, Krou SM et al (2017) A native arbuscular mycorrhizal fungus inoculant outcompetes an exotic commercial species under two contrasting yam field conditions. Rhizosphere 4:112–118
Krauss JB, Kuttenkeuler D (2018) Intellectual property rights derived from academic research and their role in the modern bioeconomy-a guide for scientist. New Biotechnol 40:133–139
Kroll S, Agler MT, Kemen E (2017) Genomic dissection of host-microbe and microbe-microbe interactions for advance plant breeding. Curr Opin Plant Biol 36:71–78
Kumar B, Trivedi P, Pandey A (2007) *Pseudomonas corrugate*: a suitable bacterial inoculant for maize grown under rainfed conditions of Himalayan region. Soil Biol Biochem 39:3093–3100
Labrador MJ, Bello A (2001) La materia orgánica en los agroecosistemas. 2ª. Ed. Mundi-Prensa. España, p 293
Liu A, Hamel C, Hamilton RI et al (2000) Mycorrhizae formation and nutrient uptake of new corn (*Zea mays* L.) hybrids with extreme canopy and leaf architecture as influenced by soil N and P levels. Plant Soil 221:157–166
Lodeiro AR (2015) Interrogantes en la tecnología de la inoculación de semillas de soja con *Bradyrhizobium* spp. Rev Argent Microbiol 47(3):261–273
Lone R, Shuab R, Khan S et al (2017) Arbuscular mycorrhizal fungi for sustainable agriculture. In: Kumar V, Kumar M, Sharma S et al (eds) Probiotics and plant health. Springer, Singapore, pp 553–577
Lucy M, Reed E, Glick BR (2004) Applications of free-living plant growth-promoting rhizobacteria. Review. Antonie van Leewenhoek 86:1–25
Lynch JP (2007) Roots of the second green revolution. Aust J Bot 55(5):493–512
MAG (2010a). Unidad de prensa oficial del Ministerio de Agricultura y Ganadería de Costa Rica. http://prensamag.blogspot.com/2010/09/costa-rica-cuenta-con-mas-de-8-mil.html. Accessed 25 Sept 2018
MAG (2010b) Guía técnica para la difusión de tecnología de producción agropecuaria sostenible. http://www.mag.go.cr/biblioteca_virtual/bibliotecavirtual/a00192.pdf. Accessed 25 Sept 2018
Malusá E, Sas-Paszt L, Ciesielska J (2012) Technologies for beneficial microorganisms inocula used as biofertilizers. Sci World J 491206:1–12
Małyska A, Jacobi J (2018) Plant breeding as the cornerstone of a sustainable bioeconomy. New Biotechnol 40:129–132
Manimekalai G, Kannahi M (2018) Evaluation of low cost liquid formulation of PGPR inoculants with protective substances. Int J Recent Sci Res 9(6):27330–27335. https://doi.org/10.24327/ijrsr.2018.0906.2237
MAPA (2018a) Estadísticas e dados básicos de economía agrícola. http://www.agricultura.gov.br/assuntos/politica-agricola/todas-publicacoes-de-politica-agricola/estatisticas-e-dados-basicos-de-economia-agricola/pasta-de-setembro-2018.pdf. Accessed 12 Nov 2018
MAPA (2018b) Legislações. http://www.agricultura.gov.br/assuntos/insumos-agropecuarios/insumos-agricolas/fertilizantes/legislacoes. Accessed 10 Oct 2018
MDA (2016) Ministerio de Desarrollo Agrario Brasil. O que é a agricultura familiar. http://www.mda.gov.br/sitemda/noticias/o-que-%C3%A9-agricultura-familiar. Accessed 28 Sept 2018
Mhlongo MI, Piater LA, Madala NE et al (2018) The chemistry of plant-microbe interactions in the rhizosphere and the potential for metabolomics to reveal signaling related to defense priming and induced systemic resistance. Front Plant Sci 9(112):1–19. https://doi.org/10.3389/fpls.2018.00112
Mishra J, Arora NK (2018) Secondary metabolites of fluorescent pseudomonads in biocontrol of phytopathogens for sustainable agriculture. Appl Soil Ecol 125:35–45
Molina-Favero C, Creus CM, Simontacchi M et al (2008) Aerobic nitric oxide production by *Azospirillum brasilense* Sp245 and its influence on root architecture in tomato. Mol Plant-Microbe Interact 21(7):1001–1009

Montenegro-Gómez SP, Barrera-Berdugo SE (2014) Biofertilización nitrogenada como aporte a la sustentabilidad de la agricultura colombiana. Revista de Investigación Agraria y Ambiental 5(2):135–144

Morrissey JP, Dow JM, Mark GL et al (2004) Are microbes at the root of a solution to world food production? Rational exploitation of interactions between microbes and plants can help to transform agriculture. EMBO Rep 5:922–926

Mrkovački N, Milić V (2001) Use of *Azotobacter chroococcum* as a potentially useful in agricultural application. Ann Microbiol 51:145–158

Mugilan I, Gayathri P, Elumalai EK et al (2011) Studies on improve survivability and shelf life of carrier using liquid inoculation of *Pseudomonas striata*. Int J Pharm Biol Arch 2(4):1271–1275

Naiman AD, Latronico AE, García de Salamone IE (2009) Inoculation of wheat with *Azospirillum brasilense* and *Pseudomonas fluorescens*: impact on the production and rhizospheric microflora. Eur J Soil Biol 45:44–51. https://doi.org/10.1016/j.ejsobi.2008.11.001.ISSN:11645563

Nava-Pérez E, García-Gutiérrez C, Camacho-Báez JR et al (2012) Bioplaguicidas: Una opción para el biocontrol biológico de plagas. Ra Ximhai 8(3):17–29

Nobbe F, Hiltner L (1896) U.S. Patent 570 813. Inoculation of the soil for cultivating leguminous plants.

NODAL (2017) La soja cambia el mapa de Brasil y su estatus agrícola en el mundo. https://www.nodal.am/2017/07/la-soja-cambia-mapa-brasil-estatus-agricola-mundo/. Accessed 28 Sept 2018

Numan M, Bashira S, Khana Y et al (2018) Plant growth promoting bacteria as an alternative strategy for salt tolerance in plants: a review. Microbiol Res 209:21–32

O'Hara G, Yates R, Howieson J (2002) Selection of strains of root nodule bacteria to improve inoculant performance and increase legume productivity in stressful environments. In: Herridge D (ed) Inoculants and nitrogen fixation of legumes in Vietnam. ACIAR proceedings 109e

Oehl F, Sieverding E, Ineichen K et al (2003) Impact of land use intensity on the species diversity of arbuscular mycorrhizal fungi in agroecosystems of central Europe. Appl Environ Microbiol 69:2816–2824

Okon Y (1994) *Azospirillum*/plant associations. CRC Press, Florida, p 192

Oleaginosas (2016) La producción de soya en Brasil. http://www.oleaginosas.org/art_374. Accessed 12 Nov 2018

OMRI (2018) Organic Materials Review Institute. https://www.omri.org. Accessed 30 Aug 2018

Owen D, Williams AP, Griffith GW et al (2015) Use of commercial bio-inoculants to increase agricultural production through improved phosphorous acquisition. Appl Soil Ecol 86:41–54

Palmqvist NGM, Bejai S, Meijer J et al (2015) Nano titania aided clustering and adhesion of beneficial bacteria to plant roots to enhance crop growth and stress management. Sci Rep 5:10146. https://doi.org/10.1038/srep10146

Parnell JJ, Berka R, Young HA et al (2016) From the lab to the farm: an industrial perspective of plant beneficial microorganisms. Front Plant Sci 7:1110. https://doi.org/10.3389/fpls.2016.01110

Pedraza RO, Teixeira KRS, Scavino AF et al (2010) Microorganismos que mejoran el crecimiento de las plantas y la calidad de los suelos. Revista Corpoica – Ciencia y Tecnología Agropecuaria 11(2):155–164

Peña-Borrego MD, de Zayas-Pérez MR, Rodríguez-Fernández RM (2015) Scientific production about biofertilizers in Cuba in the 2008-2012 period: a bibliometric analysis of Cuban journal. Cultivos Tropicales 36(1):43–52

Perez-Lavalle L, Bolivar Anillo HJ, Diaz Perez A (2017) Biofertilizantes en Colombia. In: Estrada-Lopez HH, Saumett-Espana HG, Iglesias-Navas MA et al (eds) Productos de confitería nutracéutica. Una opción empresarial para cultivadores de frutas y hortalizas. Universidad Simon Bolivar, Colombia, pp 179–222. https://www.researchgate.net/publication/326989814_Biofertilizantes_en_Colombia

Pérez-Montano F, Alías-Villegas C, Bellogín RA et al (2014) Plant growth promotion in cereal and leguminous agricultural important plants: from microorganism capacities to crop production. Microbiol Res 169:325–336

Peterson RL, Massicotte HB (2004) Exploring structural definitions of mycorrhizas, with emphasis on nutrient-exchange interfaces. Can J Bot 82:1074–1088

Piccinetti C, Arias N, Ventimiglia L et al (2013) Efectos positivos de la inoculación de soja sobre la nodulación, la FBN y en los parámetros de producción del cultivo. In: Albanesi AS (ed) Microbiología agrícola. Un aporte de la investigación argentina. Magna Publicaciones, Tucumán, pp 283–297

PNUD (2018) Un plan de agricultura familiar para Jujuy. Programa de las Naciones Unidas para el desarrollo. http://www.ar.undp.org/content/argentina/es/home/ ourwork/environmentandenergy/successstories/PlanAgriculturaFamiliar.html. Accessed November 2018

Prasanna R, Ramakrishnan B, Ranjan K et al (2016) Microbial inoculants with multifaceted traits suppress *Rhizoctonia* populations and promote plant growth in cotton. J Phytopathol 164(11–12):1030–1042. https://doi.org/10.1111/jph.12524

Rascovan N, Carbonetto B, Revale S et al (2013) The PAMPA datasets: a metagenomic survey of microbial communities in Argentinean pampean soils. Microbiome 1:21. https://doi.org/10.1186/2049-2618-1-21

Reddy CA, Saravanan RS (2013) Polymicrobial multi-functional approach for enhancement of crop productivity. Adv Appl Microbiol 82:53–113

Redecker D, Schüßler A, Stockinger H et al (2013) An evidence-based consensus for the classification of arbuscular mycorrhizal fungi (Glomeromycota). Mycorrhiza 23:515–531

Reid A, Greene S (2012) How microbes can help feed the world. Report on an American Academy of Microbiology Colloquium Washington, DC. http://ofrf.org/sites/ofrf.org/files/FeedTheWorld_0.pdf. Accessed December 2018

Rengel Z (2002) Breeding for better symbiosis. Plant Soil 245:147–162

Rengel Z (2005) Breeding crops for adaptation to environments with low nutrient availability. In: Ashraf M, Harris PJC (eds) Abiotic stresses: plant resistance through breeding and molecular approaches. The Haworth Press, New York, pp 239–276

Reyes-Ramírez A, López-Arcos M, Ruiz-Sánchez E et al (2014) Efectividad de inoculantes microbianos en el crecimiento y productividad de chile habanero (*Capsicum chinense* Jacq.). Agrociencia 48(3):285–294

Richardson AE, Barea JM, McNeill AM et al (2009) Acquisition of phosphorus and nitrogen in the rhizosphere and plant growth promotion by microorganisms. Plant Soil 321:305–339

Rigby D, Caceres D (2001) Organic farming and the sustainable agricultural systems. Agric Syst 68:21–40

Rillig MC, Wright SF, Eviner VT (2002) The role of arbuscular mycorrhizal fungi and glomalin in soil aggregation: comparing effects of five plant species. Plant Soil 238:325–333

Rinu K, Pandey A (2009) *Bacillus subtilis* NRRL B-30408 inoculation enhances the symbiotic efficiency of *Lens esculenta* Moench at a Himalayan location. J Plant Nutr Soil Sci 172:134–139

Rivera JS, Cisneros-Vázquez JM (2008) Micronutrimentos solubles en vermicomposta. Revista Chapingo Series Zonas Áridas 7(1):29–35

Rodríguez O (2001) Biofertilizante hecho en México. http://www.comoves.unam.mx/assets/revista/29/ojodemosca_29.pdf. Accessed 25 Sept 2018

Rorig M, Alderuccio S, Malcolm V et al (2004) Estimaciones del número de microorganismos, producción de nitratos y actividad de fosfatasa alcalina en un suelo Argiudol vértico de la localidad de Ramírez, Entre Ríos, bajo siembra directa y durante la rotación trigo-soja. In: Monzón de Asconegui M, García de Salamone IE, Miyazaki S (eds) Biología del Suelo. FAUBA, Buenos Aires, pp 137–142

Rubio R, Borie F, Schalchli C et al (2003) Occurrence and effect of arbuscular mycorrhizal propagules in wheat as affected by the source and amount of phosphorus fertilizer and fungal inoculation. Appl Soil Ecol 23:245–255

Ruiz-Sanchez M, Armada E, Muñoz Y et al (2011) *Azospirillum* and arbuscular mycorrhizal colonization enhance rice growth and physiological traits under well-watered and drought conditions. J Plant Physiol 168:1031–1037

Rumble H, Gange AC (2017) Microbial inoculants as a soil remediation tool for extensive green roofs. Ecol Eng 102:188–198

SAC (2002) No. 074 de 2002 Reglamentación para productos agropecuarios. https://www.sac.org.co/es/ambito-juridico/resoluciones/136-no-0074-de-2002-reglamentacion-para-productos-agropecuarios.html. Accessed 15 Nov 2018
SAGARPA (2018) PROAGRO. https://www.sagarpa.gob.mx/proagro. Accessed 1 Oct 2018
Sahoo RK, Bhardwaj D, Tuteja N (2013) Biofertilizers: a sustainable eco-friendly agricultural approach to crop improvement. In: Tuteja N, Singh Gill S (eds) Plant acclimation to environmental stress. Springer, New York, pp 403–432
Sanches-Santos M, Hungria M, Nogueira MA (2017) Production of polyhydroxybutyrate (PHB) and biofilm by *Azospirillum brasilense* aiming at the development of liquid inoculants with high performance. Afr J Biotechnol 16(37):1855–1862
Sasson A, Malpica C (2018) Bioeconomy in Latin America. New Biotechnol 40:40–45
Schalamuk S, Velázquez S, Chidichimo H et al (2006) Fungal spore diversity of arbuscular mycorrhizal fungi associated with spring wheat: effects of tillage. Mycologia 98:16–22
Schmidt JE, Bowles TM, Gaudin ACM (2016) Using ancient traits to convert soil health into crop yield: impact of selection on maize root and rhizosphere function. Front Plant Sci 7:373
SENASA (2018) Listados oficiales de registros. Servicio Nacional de Sanidad y Calidad Alimentaria. Argentina. http://www.senasa.gob.ar/cadena-animal/bovinos-y-bubalinos/industria/establecimiento-industrializador/registros-y-1. Accessed December 2018.
SEPSA (2018) Informe comercio exterior del sector agropecuario. http://www.sepsa.go.cr/docs/2018-023_Informe_Comercio_Exterior_enero-setiembre_2017-2018.pdf. Accessed 15 Nov 2018
Sharma SB, Sayyed RZ, Trivedi MH et al (2013) Phosphate solubilizing microbes: sustainable approach for managing phosphorus deficiency in agricultural soils. Springerplus 2:587. https://doi.org/10.1186/2193-1801-2-587
Shen J, Li C, Mi G et al (2013) Maximizing root/rhizosphere efficiency to improve crop productivity and nutrient use efficiency in intensive agriculture of China. J Exp Bot 64:1181–1192
Siddiqui ZA (2006) PGPR: biocontrol and biofertilization. Springer, Dordrecht, p 318
Siddiqui ZA, Kataoka R (2011) Mycorrhizal inoculants: progress in inoculant production technology. In: Ahmad I, Ahmad F, Pichtel J (eds) Microbes and microbial technology. Springer, New York, pp 489–506
Spatafora JW, Chang Y, Benny GL et al (2016) A phylum-level phylogenetic classification of zygomycete fungi based on genome-scale data. Mycologia 108(5):1028–1046. https://doi.org/10.3852/16-042
Sruthilaxmi CB, Babu S (2017) Microbial bio-inoculants in Indian agriculture: ecological perspectives for a more optimized use. Agric Ecosyst Environ 242:23–25
Tabassum B, Khan A, Tariq M et al (2017) Bottlenecks in commercialisation and future prospects of PGPR. Appl Soil Ecol 121:102–117
Tejeda-González G, Martínez-Viera R, Arozarena-Daza N et al (2010) Biofertilizantes microbianos para una agricultura sostenible: sistema cubano de gestión tecnológica para las actividades de investigación, desarrollo, innovación y producción. http://agris.fao.org/agris-search/search.do?recordID=CU2011000127. Accessed 20 Oct 2018
The World Bank (2018). http://wdi.worldbank.org/table/4.2. Accessed 15 Nov 2018
Thilakarathna MS, Raizada MN (2017) A meta-analysis of the effectiveness of diverse rhizobia inoculants on soybean traits under field conditions. Soil Biol Biochem 105:177–196
Tilman D (1999) Global environmental impacts of agricultural expansion: the need for sustainable and efficient practices. PNAS 96(11):5995–6000
Tilman D, Cassman KG, Matson PA et al (2002) Agricultural sustainability and intensive production practices. Nature 418:671–677
Trabelsi D, Mhamdi R (2013) Microbial inoculants and their impact on soil microbial communities: a review. BioMed Research International. Hindawi Publishing Corporation. https://doi.org/10.1155/2013/863240
Treseder KK (2004) A meta-analysis of mycorrhizal responses to nitrogen, phosphorus, and atmospheric CO_2 in field studies. New Phytol 164:347–355

Trivedi P, Pandey A, Palni LMS (2005) Carrier based formulations of plant growth promoting bacteria suitable for use in the cooler regions. World J Microbiol Biotechnol 21(6-7):941–945
Trivedi P, Pandey A, Palni LMS (2012) Bacterial inoculants for field applications under mountain ecosystem: present initiatives and future prospects. In: Maheshwari D (ed) Bacteria in agrobiology: plant probiotics. Springer, Berlin/Heidelberg, pp 15–44
UCR (2018) Biofertilizantes desarrollados en la UCR ya están al alcance de los agricultores. https://www.ucr.ac.cr/noticias/2018/02/16/biofertilizantes-desarrollados-en-la-ucr-ya-estan-al-alcance-de-los-agricultores.html. Accessed 25 Sept 2018
UNCTED (2018) Organic agriculture. http://unctad.org/en/Pages/DITC/Trade-and-Environment/Organic-Agriculture.aspx. Accessed 25 Sept 2018
Urquiaga S, Jantalia CP, Alves BJR et al (2004) Importancia de la FBN en el secuestro de carbono en el suelo y en la sustentabilidad agrícola. In: Monzón de Asconegui MA, García de Salamone IE, Miyazaki SS (eds) Biología del Suelo. FAUBA, Buenos Aires, pp 1–6
Valverde C, Gonzalez Anta G, Ferraris G (2015) *Pseudomonas* and *Azospirillum*. In: Cassán F, Okon Y, Creus C (eds) Handbook for *Azospirillum*. Technical issues and protocols. Springer, Geneva, pp 389–409
Vázquez-Moreno L (2006) La lucha contra las plagas agrícolas en Cuba. De las aplicaciones de plaguicidas químicos por calendario al manejo agroecológico de plagas. Fitosanidad 10(3):221–242
Vessey JK (2003) Plant growth promoting rhizobacteria as biofertilizers. Plant Soil 255:571–586
Villegas-Cornelio VM, Laines-Canepa JR (2017) Vermicompostaje: I avances estrategias en el tratamiento de residuos orgánicos. Rev Mex Ciencias Agrícolas 8(2):393–406
Wani SA, Chand S, Ali T (2013) Potential use of *Azotobacter chroococcum* in crop production: an overview. Curr Agric Res J 1(1):35–38
Willis A, Rodrigues BF, Harris PJC (2013) The ecology of arbuscular mycorrhizal fungi. Crit Rev Plant Sci 32:1–20
World Economic Forum (2014) Towards the circular economy: accelerating the scale-up across global supply chains. World Economic Forum, Geneva
Xavier IJ, Holloway G, Leggett M (2004) Development of rhizobial inoculant formulations. Crop Manag 3(1). https://doi.org/10.1094/CM-2004-0301-06-RV
Zambrano-Soledispa A, Gamarnik M, Di Salvo LP et al (2017) Diversidad de hongos micorrícicos arbusculares nativos del cultivo de maíz bajo distintas prácticas agronómicas. V Congreso CONEBIOS. Lujan, Argentina, November 5–8
Zawoznik M, Groppa MD, Benavides MP (2007a) Nitric oxide and salt stress tolerance in wheat-*Azospirillum* association. XLIII Reunión Anual de la Sociedad Argentina de Investigación en Bioquímica y Biología Molecular. November 2–15
Zawoznik MS, Groppa MD, Tomaro ML et al (2007b) Endogenous salicylic acid potentiates cadmium-induced oxidative stress in *Arabidopsis thaliana*. Plant Sci 173:190–197
Zawoznik MS, Rosales EP, Benavides MP et al (2009) Colonización radical con *Azospirillum* como factor mitigador del estrés salino. Influencia de la cepa microbiana. VII Simposio Nacional de Biotecnología (REDBIO- Argentina) y II Congreso Internacional-REDBIO-Argentina. Lugar: Rosario, Santa Fe, Argentina. April 20–24
Zuffo AM, Rezende PM, Bruzi AT et al (2015) Co-inoculation of *Bradyrhizobium japonicum* and *Azospirillum brasilense* in the soybean crop. Rev Ciênc Agrár 38(1):87–93

Microbial Biological Control of Diseases and Pests by PGPR and PGPF

3

Miguel O. P. Navarro, André Barazetti, Erika T. G. Niekawa, Mickely Liuti Dealis, Jean Marcos Soares Matos, Gabriel Liuti, Fluvio Modolon, Igor Matheus Oliveira, Matheus Andreata, Martha Viviana Torres Cely, and Galdino Andrade

1 Introduction

The rhizosphere is the area around the roots influenced for exudates released by plant root and is highly colonized by several microorganisms. The exudates improve the nutrient uptake and make the environment highly nutritious, maintaining high diversity of microbial community. Due to the conditions provided by plants, the rhizosphere is a niche with high competition among them. During the evolution, different kinds of interactions were established among microorganism-plant and microorganism-microoganism; the relationship among plants and microorganisms is beneficial or deleterious for plants. Similarly, the microorganism-microorganism interactions may be symbiotic or not (Zhang et al. 2014; Allard-Massicotte et al. 2016).

The bacteria and fungi are capable to promoting plant growth and are called PGPR (plant growth-promoting rhizobacteria) and PGPF (plant growth-promoting fungi), respectively. These interactions and the benefits resulted from them have been extensively studied in the last decades, mainly because these microorganisms may decrease the use of pesticides. The PGPR and PGPF not only promote plant growth but also protect plant against phytopathogenic microorganisms (Etesami and Maheshwari 2018).

Biocontrol agents can act by two different ways, by direct mechanisms of biological control and by induction of systemic resistance (ISR) in plants. Direct mechanisms of biocontrol against pathogens may vary for: (i) production lytic enzymes which degrade cell wall, (ii) antimicrobial production (Fig. 3.1), (iii) production of iron siderophores chelated by microorganisms, and (iv) strategies for competition of nutrients in the rhizosphere and specific sites for colonization (Borrero et al. 2009).

M. O. P. Navarro · A. Barazetti · E. T. G. Niekawa · M. L. Dealis · J. M. S. Matos · G. Liuti · F. Modolon · I. M. Oliveira · M. Andreata · M. V. T. Cely · G. Andrade (✉)
Department of Microbiology, Londrina State University, Londrina, Brazil
e-mail: andradeg@uel.br

D. P. Singh et al. (eds.), *Microbial Interventions in Agriculture and Environment*,
https://doi.org/10.1007/978-981-13-8383-0_3

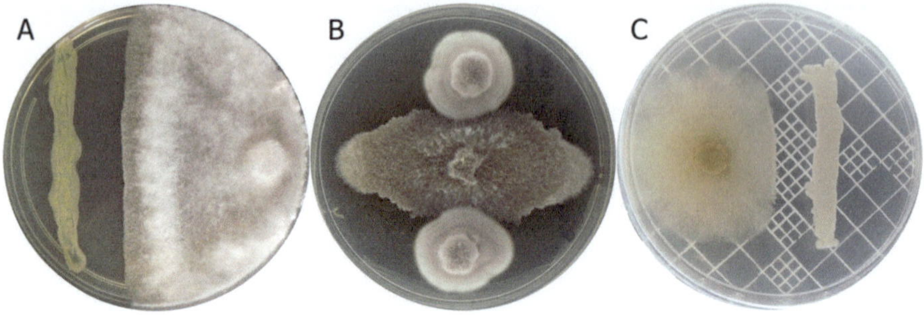

Fig. 3.1 In vitro biocontrol by antimicrobial production. (**a**) *Pseudomonas aeruginosa* strain LV against *Sclerotinia sclerotiorum*, (**b**) *Paecilomyces* sp. against *Rhizoctonia solani*, and (**c**) *Burkholderia cepacia* strain RV7S3 against *R. solani*

Meanwhile, in the ISR, the elicitors activate plant innate defense, reducing disease symptoms after subsequent infections by pathogen.

The biocontrol management is carried out by introducing PGPR and/or PGPF under natural conditions, improving crop health. According **to** Chalfoun (2010), good biological agent atributes are the following: (i) not pathogenic to the host, (ii) genetically stable, (iii) efficient against a wide series of pathogens, (iv) able to develop in diverse environment conditions, (v) effective in low concentrations, (vi) grow satisfactorily in cheap media, (vii) easily transported and stored, (viii) not produce secondary metabolites that can be harmful to humans, (ix) resistant to pesticides, and (x) compatible with other chemical and physical treatments of the product. The mechanism of protection such as antibiosis and ISR used as biocontrol agents, PGPR, PGPF, and their interactions will be discussed in the present chapter.

2 Plant Growth-Promoting Rhizobacteria (PGPR)

The main bacterial genera of PGPR found in agricultural soils are *Bacillus*, *Paenibacillus*, *Streptomyces*, *Frankia*, *Rhizobium*, *Bradyrhizobium*, *Azospirillum*, *Acetobacter*, *Burkholderia*, *Enterobacter*, *Pantoea*, *Serratia*, and *Pseudomonas* (Tariq et al. 2017). The PGPR are present in the soil in planktonic form or they colonize the root of plant. They are move by flagella in the rhizosphere and are attracted by exudates excreted by the roots through positive chemotaxis. The exudates are composed of organic acids such as citrate, malate, succinate, pyruvate, fumarate, oxalate, and acetate (Zhang et al. 2014; Ma et al. 2018). Other important components are sugars such as glucose, xylose, fructose, maltose, sucrose, galactose, and ribose, vitamins, amino acids, and flavonoids. Root exudates control the composition of microbial community in the rhizosphere and have an important role in the communication signal to establish and for the survival of PGPR (Dutta and Podile 2010; Saleem et al. 2018).

These microorganisms provide many benefits for plant host because they may induce a systemic resistance and are producers of antimicrobial and insecticide compounds (Table 3.1) (Mhlongo et al. 2018; Tariq et al. 2017). The most common antimicrobials produced by PGPR are 2,4-diacetylphloroglucinol (DAPG), pyrrolnitrin, pyoluteorin, phenazines, and lipopeptides (LPs) (Zhou et al. 2014). Also, PGPRs can produce plant hormones, mainly auxins and gibberellins, promoting branching and lateral root growth, increasing nutrient absorption surface, and expanding plant habitat (Ahemad and Kibret 2014).

2.1 Gram-Positive Species

The *Bacillus* genus is well known for its PGP ability; it exhibits strong antagonistic activity against phytopathogenic microorganisms by producing several antibiotics (Gerst et al. 2018; Mnif et al. 2015) and improves plant health by producing phytohormones (Shao et al. 2015) and compounds associated with ISR, controlling fungi, bacteria, viruses, root-knot nematodes, and other pathogens (Gond et al. 2015; Borriss 2011, 2015; Kloepper et al. 2004).

This genus of gram-positive bacterium is very versatile, producing a wide range of compounds correlated with PGP characteristics. In recent years, researchers have identified many molecules that are produced in the secondary metabolism of *Bacillus* species. This genus can produce several antimicrobial compounds that possess antibacterial and antifungal properties, and they generally show peptide characteristics. Usually, these peptides compounds produced by secondary metabolism of *Bacillus* are classified in two distinct families: non-ribosomally synthesized products, as inturins (Besson et al. 1976), fengycins (Vanittanakom et al. 1986), surfactins (Nakano et al. 1988), maltacines (Hagelin et al. 2004), mycobacillin (Majumdar and Bose 1958), and ribosomally synthesized products, as lantibiocs types (Nagao et al. 2006). However, non-peptide compounds are related to polyketides and phospholipids with antimicrobial effects (Wang et al. 2015).

Bacillus amyloliquefaciens FZB42 strain has five gene clusters directly related to nonribosomal synthesis of the cyclic lipopeptides: surfactin, bacillomycin, fengycin, and unknown peptide. It also produces iron siderophore, bacillibactin, which suppresses pathogens (Chen et al. 2009). Others studies analyzed the effect of volatile organic compounds (VOC) produced by *B. amyloliquefaciens* UCMB5113 strain to supress fungi growth *such as Botrytis cinerea*, *Alternaria brassicicola*, *A. brassicae*, and *Sclerotinia sclerotiorum* , and also promoted plant growth (Asari et al. 2016).

Auxin, phenylacetic acid, produced by *Bacillus fortis* IAGS162 strain has a key role to play in inducing systemic resistance in tomato plant, and it has contributed to the improvement of plants' resistance to wilt diseases caused by *Fusarium* species (Akram et al. 2016). *Bacillus* is well known as a producer of IAA, and *B. altitudinis* WR10, a strain highly resistant to iron when inoculated in wheat, decreased iron stress by the regulation of IAA-inducing ferritin-encoded genes in roots (Sun et al. 2017).

Table 3.1 PGPR and PGPF used as biological control against phytopathogens

Microorganism	Pathogens	Plants	Mechanism	References
PGPR				
Gram-positive species				
Bacillus subtilis RB14	*Rhizoctonia solani*	Tomato	Antimicrobial Inturin A	Zohora et al. (2016).
B. amyloliquefaciens FZB4	*Xanthomonas oryzae*	Rice	Antibacterial: difficidin and vacilysin	Wu et al. (2015a, b)
B. amyloliquefaciens LJ02	Powdery mildew disease (many fungi)	Cucurbits	ISR-SA mediated	Li et al. (2015)
Bacillus sp. *Strain* B25	*Fusarium verticillioides*	Maize	Antibiosis by chitinases, glycoside hydrolases, siderophores and antibiotics	Douriet-Gámez et al. (2018)
Paenibacillus polymyxa NSY50	*F. oxysporum*	Cucumber	ISR	Du et al. (2017)
Paenibacillus ehimensis KWN38	*F.oxysporum*	Tomato	Antibiosis by unknown enzyme extracts	Naing et al. (2015)
Actinobacteria				
Streptomyces spp.	*Colletotrichum gloeosporioides, Curvularia* spp., *Aspergillus niger, Helminthosporium* spp., *Fusarium* spp. *Alternaria* spp., *Phy tophthora capsici, Colletotrichum* sp., *Scleotinia* sp., and *R. solani*.	Chili, ginger, tomato, tobacco, pepper, maize, banana, cherry, soybean, chickpea, cucumber, Raspberry	Antibiosis, increasing activity of defense-related enzymes and synthesis of defense related chemicals in plants	Vurukonda et al. (2018)
Streptomyces spp.	*Rhizoctonia solani, Alternaria alternata Macrophomina phaseolina* (Tassi) Goid.	Wheat, tomato, rice, neem, chickpea.	Siderophores, ammonia production, phosphate solubilization, nitrogen fixation, hydrolytic enzyme production, IAA, induced accumulation of plant phenolics, enzyme production, hydrocyanic acid	Singh et al. (2018)

(continued)

Table 3.1 (continued)

Microorganism	Pathogens	Plants	Mechanism	References
Streptomyces spp., *Saccharopolyspora* spp., *Actinopolyspora* spp., *Nocardia* spp. *Thermobifida* sp.	*A. niger, A. brassicicola, Chaetomium globosum, F. oxysporum, Phytophthora dresclea, R. solani, Botrytis cinerea*	Cow pea, wheat, sorghum, rice, chickpea , clover plants,	IAA and siderophore production, antifungal activity, biocontrol direct, antibiosis	Saif et al. (2014)
Gram-negative species				
Mesorrhizobium ciceri A13 and CR24	*Fusarium* sp.	*Cicer arietinum* (Chickpea)	Antibiosis by enzymatic activity (phenylalanine ammonia lyase, peroxidase, and polyphenol oxidase)	Das et al. (2017)
Azotobacter spp.	*Helminthosporium* sp., *Macrophomina* sp., *Fusarium* sp.	*Zea mays* (Maize)	Antagonism by siderophores and NCN production	Nagaraja et al. (2016)
Burkholderia cenocepacia 869T2	*F. oxysporum* f. sp. cubense	Banana	Antagonism by pyrrolnitrin and pyrroloquinoline quinone	Ho et al. (2015)
Burkholderia spp.	*A. niger*	Agave sisalana	Unkown antibiosis mechanisms	Magalhães et al. (2017)
Burkholderia cepacia JBK9	*P. capsici, F. oxysporum,* and *R. solani*	Pepper	Pyrrolnitrin	Jung et al. (2018)
B. seminalis TC3.4.2R3	*F. oxysporum*	*Saccharum officinarum*	Siderophore pyochelin and rhamnolipid Rha-Rha-C15-C14	Araújo et al. (2017)
B. contaminans KNU17BI1	*R. solani, F. graminearum, F. moniliforme, F. oxysporum, Pythium graminicola, Alternaria alternata, A. solani, Stemphylium botryosum, Colletotrichum dematium, and Stemphylium lycopersici*	*Zea mays* (Maize)	Unknown antibiosis mechanisms	Tagele et al. (2018)

(continued)

Table 3.1 (continued)

Microorganism	Pathogens	Plants	Mechanism	References
Pantoea agglomerans strain ENA1	*Macrophomina phaseolina*	*Glycine max* (L.) Merrill	Antibiosis by pyrrolnitrin	Vasebi et al. (2015)
P. agglomerans strain P10c	*Erwinia amylovor*	*Pyrus communis*; *Malus domestica*	Direct biocontrol	Ait Bahadou et al. (2016)
Serratia phymuthica strain A30	*Pectobacterium* sp., *Ralstonia* sp., *Dickeya* sp.	*Solanum tuberosum* L.	Direct biocontrol	Carpenter and Maloney (2015)
S. marscescens	*R.. solani*	*Solanum tuberosum* L.	Direct biocontrol	Khaldi et al. (2015)
Pseudomonas fluorescens Strain CZ	Tobacco mosaic virus (TMV)	*Nicotiana glutinosa* L.; *Nicotiana tabacum*	Direct biocontrol	Shen et al. (2014)
P. fluorescens	*Xanthomonas campestris* pv. campestris	*Brassica oleracea* var. capitata	Induced systemic resistance (ISR)	Umesha and Roohie (2017)
P. fluorescens Strain PTA-CT2	*B. cinerea*	Grapevine	Induced systemic resistance (ISR)	Gruau et al. (2015)
P. aeruginosa strain LV	*Xanthomonas citri* subsp.*Citri*	*C. sinensis* cv. *Valencia*	Antibiosis by organocopper compound	de Oliveira et al. (2016)
P. aeruginosa strain LV	*Pectobacterium carotovorum* subsp. *carotovorum*	*Solanum lycopersicum* L.	Antibiosis by the purified fraction F4a	Munhoz et al. (2017)
P. aeruginosa strain LV	*Xanthomonas axonopodis* pv. *Malvacearum*; *X. axonopodis* pv. *Phaseoli*; *X. axonopodis* pv. *citri*	*Gossypium hirsutum; Phaseolus vulgaris; C. sinensis cv. Valencia*	Antibiosis by the purified fraction VLC4f	Spago et al. (2014)
P. aeruginosa strain LV	*Xanthomonas arboricola* pv. *pruni*	*Prunus persica* L. Batsch	Antibiosis by the purified fraction F3	Silva Vasconcellos et al. (2014)
P. putida strain PCI2	*F. oxysporum*	*Solanum lycopersicum* L.	Induced systemic resistance (ISR)	Pastor et al. (2016)
PGPF				
Oomycetes				
Pythium oligandrum	*Phaeomoniella chlamydospora*	Grape	Induced resistance by colonization	Yacoub et al. 2016

(continued)

Table 3.1 (continued)

Microorganism	Pathogens	Plants	Mechanism	References
P. oligandrum	*Botrytis cinerea*	Tomato	Induced resistance by Elicitin-like proteins Oli-D1 and Oli-D2	Ouyang et al. 2015
P. oligandrum	*Phytophthora* spp.	Pepper	Mycoparasitism	Yin and Yuan (2017)
P. oligandrum	*Aphanomyces cochlioides*	Sugar beet	Nutrient and/or space competition	Takenaka and Ishikawa (2013)
Ascomycetes				
Trichoderma spp.	*Colletotrichum lindemuthianum*	Bean	Antibiosis, SAR.	Pedro et al. (2012)
Penicillium citrinum LWL4, *Aspergillus terreus* LWL5	*Sclerotium rolfsii*	Sunflower	SAR	Waqas et al. (2015)
Funneliformis mosseae – Fusarium equiseti (GF18-3 and GF19-1)	*Colletotrichum orbiculare*	Cucumber	Not determined	Saldajeno and Hyakumachi (2011)
AMF				
Gigaspora margarita and *Acaulospora scrobiculata*	*R. solani* and *Colletotrichum falcatum*	5.1.1.1. In vitro	Glomalin production	Sharma et al. (2017)
Acaulospora spinosa, Glomus mosseae ,Glomus fasciculatum	*Fusarium* sp.	Chickpea	Mycorrhiza-induced resistance	Singh et al. (2013)
Glomus mosseae	*Meloidogyne incognita, Pratylenchus penetrans*	Tomato	Mycorrhiza-induced resistance	Vos et al. (2012a, b)
Acaulospora lacunosa	*Cercospora arachidicola*	Groundnut	Improve nutrient absorption	Hemavani and Thippeswamy (2014)
Gigaspora margarita,	*Cylas puncticollis*	Sweet potato	Improve nutrient absorption	Issa et al. (2017)
Rhizophagus irregularis	*Ralstonia solanacearum*	Tomato	Mycorrhiza-induced resistance	Chave et al. (2017)

(continued)

Table 3.1 (continued)

Microorganism	Pathogens	Plants	Mechanism	References
Rhizophagus irregularis	*Phytophthora infestans*	Potato	Mycorrhiza-induced resistance	Alaux et al. (2018)
Glomus fasciculatum	*Sclerotium rolfsii*	Groundnut	Mycorrhiza-induced resistance	Doley et al. (2017)
Rhizophagus irregulari	*Clavibacter michiganensis subsp. michiganensis*	Tomato	Mycorrhiza-induced resistance	Hong and Katalin (2018)
Gigaspora margarita and *Bacillus* sp.*; Bacillus thuringiensis* and *Paenibacillus rhizospherae*	*F. oxysporum* f. sp. lactucae MAFF 744088, *Rosellinia necatrix, R. solani* MAFF 237426, and *Pythium ultimum* NBRC 100123	5.1.1.2. In vitro	Not determined	Cruz and Ishii (2018)

Moreover, *B. subtilis* is a well-known species largely applied in biological control traits against several pathogens in agriculture. For example, *B. subtilis* strain QST 713 utilized as a large-spectrum biological fungicide – approved for use in the European Union (Reg. (EC) No 839/2008) – shows great activity in the control of yellow rust disease of wheat caused by *Puccinia striiformis* (Reiss and Jørgensen 2017). *B subtilis* strains are known for the control of other fungal infections in plants, as *Botrytis cinerea* and *Pseudomonas syringae* on grown tomato plants (Hinarejos et al. 2016) and *Phytophthora* crown and root rot of pistachio (Moradi et al. 2018). Furthermore, other trials with *B. subtilis* show the high capacity of this bacterium in biocontrol of diseases caused by *Rhizoctonia solani* (Ma et al. 2015; Asaka and Shoda 1996), *Fusarium* species (Chaurasia et al. 2005), and *Alternaria alternata* (Chaurasia et al. 2005), among other phytopathogenic microorganisms.

Paenibacillus is another gram-positive bacterium that shows PGPR properties, but low study was carried, some species was related as N-fixing (Goswami et al. 2016) and showed strong antagonism against *Fusarium* spp. (Lounaci et al. 2017). Others authors, identified the capacity to solubilizetri-calcium-phosphate, and producers of indole-3-acetic acid, ammonia and siderophore production by the PGPR *P. mucilaginosus* N3 strain (Goswami et al. 2015a, b).

2.1.1 Actinobacteria

Plant-associated actinobacteria are common in soils and represent a high proportion of the rhizosphere microbial community, and they are very efficient in colonizing root systems (Sousa et al. 2008; Bulgarelli et al. 2013). Some species are symbionts and endophytes such as *Frankia, Streptomyces, Micromonospora Nocardia, and Microbispora.* These species are nonpathogenic and live inside the plant while improving plant growth (Tokala et al. 2002; Taechowisan et al. 2003; Roy et al.

2007). Others species are soil saprophytes with important role in nutrient cycling, especially in the organic matter turnover due to their capacity to degrade complex molecules and recalcitrant substances such as cellulose, lignocellulose, xylan, and lignin (Sousa et al. 2008; Zhou et al. 2009).

Actinomycetes are recognized by their ability to produce auxin (plant growth regulators) that promotes root growth and root hair proliferation, improving absorption of water and nutrients from soil solution. Additionally, actinomycetes are able to solubilize phosphates, whereas almost 95–99% of soil phosphate is adsorbed and cannot be used by plants. Lack of phosphate is one of the main limitations of plant growth. The ability of actinomycetes to solubilize phosphate into soluble form is mediated by the production of organic acids.

Actinomycetes have great potential as biocontrol agents in agricultural systems, they produce ionophores (increase the linked nutrients including cations) and enzymes having antimicrobial activity. The most common enzymes are chitinases that can be used as a biocontrol mechanisms, especially against fungi. Other enzymes are catalase, amylase, and lipase, which are important for plant growth. On the other hand, cellulase and xylanase act in the decomposition of organic matter and increasing soil quality and plant growth. Additionally, they are involved in the induction of resistance against some plant-pathogen systems and produce siderophores, which can solubilize and chelate iron from the soil and thus inhibit pathogen growth.

The most common species is the genus *Streptomyces,* which forms pseudomycelia with a complex multicellular life cycle and propagates by sporulation (the spores release from the aerial mycelium). *Streptomyces* are important as producers of several biotechnological products, including an extensive variety of important antimicrobials, as well as a wide range of enzymes with industrial application.

In addition, *Streptomyces are* widely used for the biological control of pests and as PGPR due to its ability to improve plant health and produce extracellular proteases, IAA, siderophores, and antibiotics**.** Some species of *Streptomyces* are described as PGPR in chickpea, eucalyptus (Salla et al. 2014), pine (Dalmas et al. 2011), beans (Nassar et al. 2003), pea (Tokala et al. 2002), rice (Gopalakrishnan et al. 2015), tomato (Dias et al. 2017), and wheat (Sadeghi et al. 2012). They induce a systemic response in the modulation of enzymes related to plant defense and/or production of bioactive secondary metabolites. In contrast, *Frankia* is the most commonly known symbiotic N-fixing bacteria of nonleguminous plants, called actinorhizal symbiosis promoting plant growth through and can suppress root diseases (Gopinathan 1995).

On the other hand, *Micromonospora* sp. is an important bioactive genera of actinobacteria, and is considered as PGPR and biocontrol agent by two ways, inducing plant defense (plant immunity) and producing antifungal compounds. Additionally, *Micromonospora* spore formation occurs directly on the substrate mycelium, and their activity is sustained in time, features applicable for its use as bioinoculants, since it allows maintaining the microorganism without a remarkable loss of viability for prolonged periods of time (Martínez-Hidalgo et al. 2015; Barka et al. 2016).

Other genera of the order Actinomycetales are *Actinoplanes, Amorphosporangium, Microbispora,* and *Streptosporangium,* which showed great potential as PGPR and/or biocontrol agents against plant pathogens (El-Tarabily and Sivasithamparam 2006), but a more detailed study needs to be carried out.

2.2 Gram-Negative Species

The family of Rhizobiaceae comprises the main genera of root-forming endophytic PGPR, which induce directly or indirectly the growth of several plants, mainly leguminous plants (Bhattacharyya and Jha 2012), but growth is also seen in nonleguminous plants (Antoun et al. 1998). These microorganisms are mainly PGPR biofertilizer because they are strongly characterized by high BNF capacity and synthesize *nod* factors (Geurts and Bisseling 2002; Hadri and Bisseling 1998), increasing the bioavailability of nitrogen (N) for plants.

Many species are considered as PGPR, such as *Rhizobium giardinii, R. gallicum* (Amarger et al. 1997), *R. ciceri* (Nour et al. 1994), *R. galegae* (Lindström 1989), *R. etli* (Segovia et al. 1993), *R. fredii* (Scholla and Elkan 1984), *R. ecuadorense* (Ribeiro et al. 2015), *R. sophorae*, *R. sophoriradicis* (Jiao et al. 2015), *Allorhizobium undicola* (de Lajudie et al. 1998a) *Bradyrhizobium japonicum* (Guerinot and Chelm 1984), *B. vignae* (Grönemeyer et al. 2016), *B. tropiciagri*, *B. embrapense* (Delamuta et al. 2015), *Azorhizobium caulinodans* (Dreyfus et al. 1988), *Mesorhisobium chacoense* (Velázquez et al. 2001), *M. pluriforium* (de Lajudie et al. 1998b), *Sinorhizobium arboris* (Nick et al. 1999), *S. fredii* (Chen et al. 1988), and *S. medicae* (Rome et al. 1996).

In addition, members of this taxon also show antagonism against phytopathogens and are considered also as biopesticides. *M. ciceri* acts as an antagonist for *Fusarium* sp. by forming biofilm in in vitro experiments, suggesting that it is a potential PGPR and a biocontrol agent (Das et al. 2017). Also, the strain producing HCN, ammonia, IAA, and hydrolytic enzymes can also solubilize inorganic phosphates. Other authors described *R. radiobacter* as a producer of polypeptides having antibacterial and antifungal activities from secondary metabolism (Rozi et al. 2018).

Some authors suggest that *Gluconacetobacter* (Acetobacteriaceae family) is an important PGPR which shows phosphate and zinc solubilization (Intorne et al. 2009), nitrogen fixation (Fuentes-Ramíres et al. 2001), and antagonism against plant pathogens such as *Xanthomonas albilineans* (Blanco et al. 2005) and *F. oxysporum* (Logeshwarn et al. 2011). The compounds related with antibiosis are 2,4-diacetylphloroglucinol, pyrrolnitrin, and pyoluteorin (Logeshwarn et al. 2011). Some authors observed that when *Gluconacetobacter* and *Burkholderia* were associated the P solubilization increased when compared in plants inoculated with each strain alone (Stephen et al. 2015). Other mechanism described is compounds can immobilize Fe and N, that is the key of some PGPR to suppress soil diseases especially fungi (Lareen et al. 2016, Salomon et al. 2017).

In field conditions, inoculation of *A. brasiliense* and *A. lipoferum* increases the yield of maize up to 27% and of wheat up to 31% (Hungria et al. 2010). Many authors performed co-inoculation assays with *Azospirillum* and *Bradyrhizobium* in soybean under both greenhouse conditions and field conditions, and it was observed that early root growth occurred because of the presence of *Azospirillum*, and early nodule formation increased total nodule biomass (Chibeba et al. 2015). The authors concluded that the co-inoculation of these two microorganisms was the key role to supply deficiency of N in soil with low fertility specially N during soybean growth.

In addition, *A. lipoferum* is antagonist against many bacterial and fungal isolates due to the production of siderophores such as salicylic acid, 2,3-dihydroxybenzoic acid (DHBA), and 3,5-DHBA (Shah et al. 1992). Other free-living N fixing who show PGPR is *Azotobacter* genus, and is antagonist against phytopathogens such as many *Fusarium* species (Bjelić et al. 2015; Nagaraja et al. 2016) *Apergillus flavus*, and *Cercospora* sp. (Ponmurugan et al. 2012).

2.2.1 *Burkholderia*

The *Burkholderia* genus comprises more than 100 species (Bochkareva et al. 2018; Depoorter et al. 2016) present in the most diverse environments and habitats and can be found especially in the soil and even isolated from infection disease of animal and human. Many of them are known for their ability to promote growth in the most diverse plant species. For the taxonomic and epidemiological purposes, based on the partial sequencing of the 16S rRNA and the current ribosomal multilocus sequence typing (rMLST) scheme, the species are distributed into four groups: *B. cepacia* complex group (Bcc), *B. pseudomallei* group, *B. glathei* group, and *B. xenovorans* group (Depoorter et al. 2016).

The Bcc group comprises the main strains of PGPR of the genus, and they also include some animal and human pathogenic species, and also few plant pathogens (Rojas-Rojas et al. 2018). The pathogenicity observed in some species is a problem and limits the use of these bacteria as biocontrol agents and PGPR in the field. In the USA, for example, the use of any *Burkholderia* strain was banned for agro-technological purposes (Estrada-de los Santos et al. 2016). Therefore, in recent years, the scientific community has been the harsh critic of the classification methodologies currently accepted for the genus. Estrada-de los Santos et al. (2016) proposed a phylogenetic reclassification of the *Burkholderia* group, dividing it into two groups (A and B); in group A, only plant-beneficial-environmental strains are included, and in group B, strains that can be harmful to animal and plant health (including human) are included. The phylogeny proposed by the authors is based on multilocus sequence analysis (MLSA) scheme. Zuleta et al. (2014) made a phylogenetic analysis of 545 housekeeping genes from 15 different *Burkholderia* species and also suggested the separation of the genus into 2 groups, one of them with only plant-beneficial-environmental strains. Other works have already proposed a review of the classification (Gyaneshwar et al. 2011; Sawana et al. 2014), suggesting that plant-beneficial-environmental strains should be included in a different genus.

However, the metabolites produced by *Burkholderia* confer PGPR capacity and suppression of plant pathogens. Many compounds with antimicrobial activity with low molecular weight are produced such as pyrrolnitrin (El-Banna and Winkelmann 1998), phenazines (Pierson and Pierson 2010), and siderophores (Darling et al.

1998) and also some compounds of peptide as xylocandins (Bisacchi et al. 1987), occidiofungins (Lu et al. 2009), and burkholdins (Thomson & Dennis 2012). Also, Burkholderia also produces indole acetic acid (IAA) (Pandey et al. 2005), rhamnolipids (Irorere et al. 2018), cepafungins (Shoji et al. 1990), and hydrogen cyanide (Gilchrist et al. 2013).

B. pyrrocinia and *B. cepacia* are well known for producing the antimicrobial pyrrolnitrin, which has antibacterial and strong antifungal activities against several microorganisms known for phytopathogenicity (El-Banna and Winkelmann 1998; Jung et al. 2018). Jung et al. (2018) demonstrated the PGP potential of *B. cepacia* JBK9 strain, while the bacterium produces pyrrolnitrin by secondary metabolism. The authors compared the cell colonization and cell motility of JBK9 in red pepper roots with the ability of *B. cepacia* ATCC 25416 and *B. pyrrocinia* KCTC 2973, in addition to the antifungal activity of each strain and correlation with the ability to control *Phytophthora capsici*.

Araújo et al. (2017) demonstrated in vitro mechanisms of antagonism between *B. seminalis* TC3.4.2R3 and *F. oxysporum*, both isolated from sugarcane. Coculture assays were performed with the microorganisms using mass spectrometry imaging capabilities; thus spatial and temporal distribution of metabolites could be analyzed simultaneously. The compound exhibiting antifungal activity was identified to be pyochelin, a siderophore. In addition, the presence of bikaverin and fusarin C, mycotoxins produced by the fungus in response to the presence of the bacterium, was identified in the trials.

The antifungal properties and their importance in the metabolism of *Burkholderia* as a biocontrol agent and PGPR is largely known. Bevivino et al. (1998) described how pyochelin, in addition to other siderophores produced by 14 distinct isolates of a *B. cepacia* population, is essential for the control of phytopathogens and growth promotion of maize. The authors carried out tests with all the isolates separately and evaluated the shoot fresh weight and root weight of maize plants after 21 days and 55 days of treatment. In all treatments, except one, increased fresh weight compared with noninoculated plants was observed. In other experiment performed, plants were infected with *F. proliferatum* ITEM-381 and *F. moniliforme* ITEM-504 strains. The potential of 11 isolates of *Burkholderia* to suppress fungal infection was evaluated. Only one strain decreased root fresh weight and in the others treatments, were increased shoot and/or root weight, and suppressing phytopathogen growth, promoting maize growth.

B. contaminans KNU17BI1 in in vitro and in vivo experiments showed antifungal activity against *R. solani*, *F. graminearum*, *F. moniliforme*, *F. oxysporum*, *Pythium graminicola*, *Alternaria alternata*, *A. solani*, *Stemphylium botryosum*, *Colletotrichum dematium*, and *Stemphylium lycopersici* (Tagele et al. 2018). The strain KNU17BI1 also produces plant growth compounds as IAA and ammonia, phosphate and zinc solubilization, hydrolytic and 1-aminocyclopropane-1-carboxylate (ACC) deaminase activity and fixintg-N. In greenhouse conditions, infection suppression of *R. solani* in two maize cultivars was observed. In addition, the PGPR capacity was tested and the strain increased in shoot and root dry weight of 52.3% when compared with control (Tagele et al. 2018).

2.2.2 *Pseudomonas* sp.

The genus Pseudomonas consists of a group of gram-negative bacteria which are ubiquitous and are of great clinical and environmental importance. Some species, such as *Pseudomonas aeruginosa*, are opportunistic pathogens often associated with respiratory infections in immunosuppressed individuals. In the soil, where they are found in greater abundance, species such as *P. fluorescens* and *P. putida* participate in bioremediation processes, degrading oily compounds such as petroleum, diesel, and kerosene, among other agricultural machinery residues (Wasi et al. 2013).

This genus also performs important interaction with the vegetables, being the PGPR group that is widely studied. The benefits promoted by Pseudomonas in plants can be direct, through the promotion of growth, or indirect, by the production of antimicrobial compounds. Six classes of antimicrobials produced by *Pseudomonas* are phenazines, phloroglucinols, pyoluteorin, pyrrolnitrin, cyclic lipopeptides, and hydrogen cyanide (Haas and Défago 2005). Several studies have demonstrated that the application of cellular suspensions or purified compounds of *Pseudomonas* offers antimicrobial activity against nematodes, fungi, oomycetes, and bacteria (Fernando et al. 2006).

The phenazine produced by *Pseudomonas* has as its precursor salicylic acid, produced by the shikimate route. The first formed phenazine is the phenazine-1-carboxylic acid (PCA), from the modification of PCA in reactions mediated by modifying enzymes; other types of phenazines are formed. The most studied phenazines with antimicrobial activity are phenazine-1-carboxylic acid (PCA), phenazine-1-carboxamide (PCN), 1-hydroxyphenazine (OHP), and 5-Nmethyl-1-hydroxyphenazinium betaine (PYO); the latter produced exclusively by *P. aeruginosa* (Blankenfeldt et al. 2004). PCA has excellent in vitro activity against several phytopathogens of great economic importance, such as *Xanthomonas oryzae*, *Rhizoctonia solani*, and *Fusarium oxyporum* (Upadhyay and Srivastava 2011; Xu et al. 2015; Zhou et al. 2016). In a study conducted by Simionato and coworkers, the phenazine-1-carboxylic acid (PCA) extracted and purified from *Pseudomonas aeruginosa* presented reduction of mycelial growth, in vitro, of *Botritys cinerea* at a minimum concentration of 25 μg mL^{-1} (Simionato et al. 2017). In another study, tomato plant roots colonized by PCA-producing *Pseudomonas chrlororaphis* were shown to inhibit *Fusarium oxysporum* infection (Chin-Woeng et al. 2000).

The antifungal activity of PCN was evaluated by Xiang and his colleagues. In this work, it was observed that small concentrations of PCN, 18 μg/mL, cause morphological changes in hyphae of *Rhizoctonia solani*. Through the physiological and biochemical results obtained in the real-time PCR test, the researchers observed that PCN affected the cell wall by inhibiting chitin synthesis and decreased mitochondrial activity, affecting the I complex of the electron transport chain (Xiang et al. 2017). *Pseudomonas* produce and release siderophores, molecules with high affinity for metal ions, mainly Fe^{3+} and Fe^{2+} (Chu et al. 2010; Rakin et al. 2012). These micronutrients are indispensable to living beings because they are essential to the performance of various metabolic processes, such as cofactors in enzymatic reactions, nucleic acid biosynthesis, cellular respiration, and photosynthesis. The sequestration of these ions by siderophores restricts the supply of these micronutrients to competing organisms, suppressing the development of phytopathogenic agents due to nutritional deprivation (Pahari and Mishra 2017). Salicylic acid (SA) is a molecule produced by *Pseudomonas* and is an important inducer of systemic resistance in plants. SA leads to the activation of SA-dependent defense mechanisms. Plants with resistance mechanism activated are able to react faster to the attack of several phytopathogens and predation of herbivores (Bernsdorff et al. 2015). However, the induction mechanisms of systemic resistance and the involvement of siderophores have not yet been well elucidated.

In 2002, Kris Audenaert and colleagues evaluated the influence of Pch, SA, and the phenycin compound pyocianin (PYO) produced by *Pseudomonas aeruginosa* on the induction of systemic resistance against *Botrytis cinerea* in tomato plants. The PHZ1 (PYO−, Pch+, AS+) and 7NSK2-562 (Pch−, AS−, PYO+) strains when inoculated alone had no induction of systemic resistance. However, the characteristic was restored when co-inoculated, evidencing that in this case, synergy occurred between PYO, SA, and Pch in the ISR. Contrary to what was imagined, SA did not induce resistance to this plant, but a combination of Pch and PYO. Therefore, in this work, the author evidenced that SA is important for the induction of systemic resistance against *B. cinerea* in tomato plants because it is the precursor of Pch (Audenaert et al. 2002).

Maria Péchy-Tarr reported, in 2008, that *Pseudomonas* were capable of producing an insecticidal toxin. In this work, the researchers identified the genomic locus Fit (*P. fluorescens* insecticidal toxin) in *P. protegens* (formerly called *P. fluorescens*). In this study, researchers injected a solution containing 3×10^4 cells of *Pseudomonas fluorescens* carrying the Fit locus into the hemocoel of larvae, and all of the insects died within 24 h. When using defective bacteria for this locus, about 60–95% survived infection (Péchy-Tarr et al., 2008). In 2015, Chen and colleagues, through genomic analysis, determined that *Pseudomonas chlororaphis*, another rhizobacterium, is also able to biosynthesize the insecticide toxin Fit (Chen et al. 2015). In addition, lipopeptides from the secondary *Pseudomonas* metabolite may exhibit activity against some insects. Kim et al. (2011) demonstrated that rhamnolipids, a biosurfactant produced by *Pseudomonas aeruginosa* during the degradation of diesel oil, exhibit in vivo insecticidal activity against *Myzus persicae*, an aphid commonly known as the green peach aphid.

2.2.3 Other Enterobacteriales

The *Serratia* genus produces a wide variety of bioactive secondary metabolites such as the antifungal and bactericidal pyrrolnitrin; the antifungal and antioomycete oocydin A; the antibiotic polyamino of broad spectrum and nematicide zeamine; and acetyl-CoA carboxylase (ACC) inhibitor andrimid, a potent antibiotic with high selectivity for prokaryotic ACC (Singh and Jha 2016). In a study carried out in vitro, *S. plymuthica* strain AS3 and *S. proteamaculans* strain S4 promoted morphological changes and growth inhibition of *Rhizoctonia solani* by the production of chitinases, proteases, and antifungal toxins (Gkarmiri et al. 2015).

The *S. plymuthica* strains 3Rp8 isolated from rhizosphere of *Brassica napus* L. and *S. plymuthica* strains 3Re4-18 isolated from endorhiza of *Solanum tuberosum* L. showed activity, in vitro, against soil-borne fungi *Verticillium dahliae*, *R. solani*, and *S. sclerotiorum* (Adam et al. 2016). Also, *S. plymuthica* A153 strain, isolated from the wheat rhizosphere, was able to kill the nematode *Caenorhabditis elegans* in a few hours. To find out if the nematicidal compound produced by this strain was zeamine, the researchers promoted a disruption in the cluster gene of this compound, which culminated in a decline in nematicidal activity. This result indicated that zeamine biosynthesis plays an important role in the nematicidal activity of *Serratia* sp. (Hellberg et al. 2015).

Likewise, *Enterobacter* sp. strain SA187 isolated from root nodules of *Indigofera argentea*, a plant found in the desert of the Kingdom of Saudi Arabia. The genomic sequencing identified the phzF and ubiC genes, enzymes involved in the biosynthesis of phenazine and 4-hydroxybenzoate, compounds with recognized activity against phytopathogenic bacteria, as well as six other genes known as chitinase coding with antifungal and insecticidal activity (Paulsen et al. 2005).

Similarly, some species of the genus *Pantoea* produce the antibiotic tripeptide Pantocin A, often related to the control of *Erwinia amylovora*, a bacterium that causes fire blight of pear and apple flowers. The most known commercially available strains are *Pantoea agglomerans* strain Eh252, EH318, and P10c and *Pantoea vagans* strain C9-1 (Klein et al. 2017).

3 Plant Growth-Promoting Fungi (PGPF)

Plant growth can also be promoted by the groups of fungi such as ascomycetes (*Trichoderma*, *Fusarium*, *Penicillium*), oomycetes (*Pythium*, *Phythophthora*), and arbuscular mycorrhizal fungi (AMF – *Glomus*, *Funneliformis*, and *Rhizophagus*) which have the capacity to either colonize the roots of many plants and induce ISR or protect the plant directly against pathogens (biological control). Interestingly, biocontroller-PGPF may be a nonvirulent strain of phytopathogenic fungi (Bent 2006). Thus, PGPF are nonpathogenic fungi that present great biotechnological potential, which could provide important benefits to agriculture to protect large crops against diseases through direct biocontrol or ISR (Table 3.1).

3.1 Ascomycetes

The large number of interactions among plant and microorganisms occur in the rhizosphere, ranging from harmful relationship as parasitism to complex beneficial symbioses. The relationships established by growth promoting bacteria in plants (PGPR), between plants and growth promoting fungi (PGPF), in addition to mycorrhizal associations are the examples. The most important PGPF are *Trichoderma*, *Fusarium*, and *Penicillium*. The relationships among plants and microorganisms involve the molecular recognition between both symbionts through a signaling network, mediated by plant hormones, as salicylic acid (AS), jasmonic acid (AJ), and ethylene (ET). While AJ and ET are described as signaling molecules involved in induced systemic resistance (ISR), AS is described as a signaling molecule of systemic acquired resistance (SAR) (Hermosa et al. 2012).

ISR is effective to protect plants against many types of pathogens, the induction carried out by microorganisms does not cause diseases or lesions in the host, unlike what happens in SAR, where is inducing when plant is in contact with a pathogen, resulting in a hypersensitivity response, and often results in necrotic lesions. SAR increases plant resistance against attacks of pathogens. ISR and SAR, plant recognize the early attach that leads to the triggering of cell response by elicitors. The elicitors are proteins, glycoprotein, peptides, chitin, glucan, polysaccharides, lipids, and secondary metabolites produced by fungi and bacteria, which induce the synthesis of compounds such as phytoalexins, defensins, phenolic compounds, flavonoids, and proteins that directly attack the agents pathogens (Navarro et al. 2017).

3.1.1 *Trichoderma*

Trichoderma generally live in soil, are nonpathogenic, and are found in many ecosystems, and some of the strains have the ability to reduce plant severity by inhibiting plant pathogens in soil and roots for antagonism and/or mycoparasitism. The presence in soil is stimulated by the availability of nutrients released by root exudate, favoring the establishment in the rhizosphere. Different strains of *Trichoderma* demonstrated direct effect on plants, increasing growth, nutrient absorption, germination rate, and stimulating plant defenses against biotic and abiotic factors by ISR (Hermosa et al. 2012).

T. harzianum can solubilize nutrients, and in experiments with cucumber inoculated with *T. asperellum* increased P and Fe availability to the plants and increased dry mass, length, and leaf area were observed. In maize, root colonized with *T. virens*, was observed that photosynthetic rates and CO2 absorption enhanced in leaves. The authors suggested that *Trichoderma* inoculation has its effects on auxins' and other phytohormones' production; they stimulate plant growth and root development (Hermosa et al. 2012).

Species of *Trichoderma* are able to inhibit phytopathogens by different mechanisms, such as direct competition by growth factors, by parasitism of mycelia or spores, and by secondary metabolite production. The use of *Trichoderma* isolates also improved seed germination, the development of lateral roots, dry matter, and plant height, probably by phytohormones production, increase in nutrient acquisition, and resistance against biotic and abiotic stress (Machado et al. 2012).

Trichoderma may also act as mycoparasites; the mylcelia detect hyphae of the host through chemical signals, once in contact, they form appressory and penetrates, after which the *Trichoderma* digests the hyphae of the host. Trichoderma also compete for nutrients, water, light, space, growth factors, oxygen, and other factors that suppress phytopathogenic microorganisms in the rhizosphere. *Trichoderma* species are still capable to produce several secondary metabolites with antimicrobial activity, such as antibiotics and enzymes that inhibit and destroy infectious fungal propagules. *T. harzianum* produce a protease that hydrolyzes pathogen enzymes that destroy the cell wall of plants, reducing the ability of the pathogen to infect the plant host (Machado et al. 2012).

The properties discussed above make the *Trichoderma* an effective biocontrol agent to be used as a biofungicides. There are several commercial products based in Trichoderma spores to use in biological control of fungi.

3.1.2 *Fusarium*

The rhizosphere is a highly competitive environment where one of the many strategies for survival is the production of antimicrobial. The high biodiversity provides several kinds of antimicrobial molecules. *Fusarium* is a soil-borne fungi living in the rhizosphere of several plants, establishing both parasitic and growth-promoting relationships; this depends on several factors such as the availability of nutrients and environmental conditions. Phytopathogenic species affect a wide range of hosts, causing diseases such as root rot, vascular wilt, yellowing, and foliar necrosis. On the other hand, saprophyte species have the ability to survive degrading lignin, survive complex carbohydrates, use the exudates as nutrient source, or colonize roots protecting plants against other diseases (Islam et al. 2014).

Organisms from different kingdoms use volatile metabolites for inter- and intraspecific communication and in some cases as factor of sexual modulation and reproduction, controlling physiology and growth, suppressing other organisms, and driving symbioses between microorganism and plant (Bitas et al. 2015).

In some cases, *F. oxysporum* living in association with a consortium of ectosymbiotic bacteria produces β-caryophyllene, a volatile sesquiterpene that apparently increases lettuce plant growth, and negatively influences mycelial growth of other *F. oxysporum* strain suppressing the virulence genes expressions. However, when the symbiotic bacteria is removed, the strain stops producing β-caryophyllene and becomes pathogenic; thus it stops promoting plant growth (Minerdi et al. 2011). The reverse transcription PCR analysis showed significant differences, an increase seven times the level of gene expression of expansin in leaves and four times in the roots, when compared with F. oxysporum not associated with symbiotic bacteria.

The volatile compounds have important functions like increasing lateral root density, which facilitates root penetration by phytopathogenic fungi. The increase in the number of lateral roots enhanced the infection level due to the presence of more entry points through the root. The auxin transport and signaling promoted by volatile compounds of *F. oxysporum* are related of growth promotion. Chemical inhibition of auxin efflux blocks growth promotion mediated by volatile compounds in *Arabidopsis thaliana* and tobacco. Also, plant with AUX1 or TIR1 genes silient which is related with auxin transport and reception, respectively, inhibit growth by volatile compounds (Bitas et al. 2015).

Penicillium sp. is found in several environments and has an important role in the dynamics of carbon cycle especially in organ mater turnover. Most of the species are saprophytic but some species can cause rot in, in addition release a large variety of mycotoxins. On the other hand, several strains act as beneficial organisms and are used in the production of cheese and sausages, and others are penicillin producers.

PGPFs perform essential functions in soil, which are well known, such as the production of a large number of secondary metabolites (IAA, siderophores, ammonia, organic acids, antibiotics, extracellular enzymes) which increase plant growth, crop productivity, and the soil fertility. *Penicillium* species have shown a greater potential in solubilizing inorganic phosphates more than bacteria, since they secrete more organic acids resulting in a greater P solubilization (Altaf et al. 2018).

When the effect of *Penicillium* sp. EU0013 strain on the wilt caused by pathogenic *Fusarium* in tomato and cabbage was evaluated, the presence of inhibition zones indicated disease control. In tomato, the effects of wilt were observed in greenhouse conditions. Seedlings were immersed in conidial suspensions of different concentrations, and the disease symptoms were reduced significantly in the highest conidial concentration of *Penicillium*, suggesting the potential of this species as a biocontrol agent (Alam et al. 2011)

The downy mildew disease caused by *Sclerospora graminicola* (Sacc.) Schroeter is the main cause of millet damage (*Pennisetum glaucum* (L.) R. Brown) in India and other countries, reducing productivity up to 80% (Murali and Amruthesh 2015). The effect of *P. oxalicum* fungi clearly increased seed germination and seedling vigor, induced systemic resistance against diseases, and increased nutrient absorption. The response varied according to the interval time of exposure between the inducer and the pathogen, showing no protective response between 1 and 2 days, and conferring protection from the 3rd day, inhibiting pathogen growth and suppressing mildew disease.

In another study, with *P. menonorum*, Babu et al. (2015) observed that the strain was able to produce IAA, an important siderophore, which substantially suppresses iron deficiency, as well as indirectly depriving the fungal pathogens of iron and of chelating Mg^{2+}, Ca^{2+} and Al^{3+} in the soil solution. It also increased the biosynthesis of starch and proteins and chlorophyll content in the yield culture of cucumber.

Trichodherma sp., *Fusarium* sp., and *Penicillium* sp. are a few examples of fungi which promote plant growth, enhance productivity and nutrient absorption, and also controls diseases. Similar effects was observed with other PGPF, such as *Aspergillus* sp. (Angel et al. 2011; Waqas et al. 2015; Wang et al. 2018), *Paecilomyces* sp. (Cavello et al. 2015; Siddiqui and Akhtar 2008; Siddiqui and Futai 2009), and *Heteroconium* sp. (Narisawa et al. 1998; Usuki and Narisawa 2007).

3.2 Oomycetes

Oomycetes is a group of heterotrophic microorganisms, containing 90 genera with 600 species, among them are pathogens of plants, animals, and algae. The group is ecologically important in the organic-matter turnover and the rhizosphere equilibrium (Margulis and Schwartz, 2000; Moore et al. 2011). The filamentous morphology, similar to Fungi Kingdom individuals, these microorganisms were previously classified in this group. The evolution of molecular biology and modern biochemical analyses found that the oomycetes are phylogenetically related with diatoms and brown algae, and now are classified as *Stramenopila* group (Gunderson et al. 1987; Jiang and Tyler 2012; Lamour and Sophie 2009).

Due to the great epidemic on potato yield in Europe between 1845 and 1849 by *Phythophthora infestans*, an oomycete, killed more than 1 million people of hunger (Erwin et al. 1996); the genus *Phythophthora* is the most studied among the oomycetes. The last decades increasing the interest in oomycete species who can promoting plant growth, highlighting whether the genera *Pythium* and *Phythophthora*.

The *Phythophthora* species can produce extracellular protein, named elicitins, capable of inducing hypersensitive (HR) cell death in tobacco (Ricci 1997; Bonnet et al. 1994); more than 30 *Phytophthora* species were described as elicitin excretors. Initially were observed in *P. cryptogea* and *P. capisici* in tabacco to stimulate plant natural defenses followed the presence of leave necrosis. After the appearance of HR the systemic acquired resistance (SAR) was discovered (Ricci et al. 1989; Bonnet and Rousse 1985). Later, it was verified that the improvement of the immune system was not directly linked to foliar necrosis, once the HR induction is not observed in all plants (Ricci et al. 1989; Kamoun et al. 1993, 1997; Keller et al. 1996).

Recent studies have demonstrated that ELR receptors present on the cell surface are able to mediate the recognition of elicitin, since they are bound to co-receptors of the immune system (Du et al. 2015; Domazakis et al. 2014), but knowledge about the regulation mechanisms in oomycetes is poor. Another interesting feature about the synthesis of this molecule is the diversity of forms presented; some species express a single class of elicitin, while others are capable of producing more complex patterns (Ricci 1997).

Other crops showed a response to elicitins, including tomato, potato and pepper (Solanaceae), pigeon (Fabaceae), grapevine (Vitaceae), citrus (Rutaceae), oak (Fagaceae), as well as some radicots and radicchio (Brassicaceae) (Vleeshouwers et al. 2006; Dalio et al. 2011; Akino et al. 2014; Oßwald et al. 2000).

With the increase in the number of studies of biocontrol of oomycetes, the soil species *P. oligandrum* is highlighted as promising (Rey et al. 2008), although other nonpathogenic species of *Pythium* genus also presented biocontrol activities. *P. oligandrum* has a worldwide distribution, which facilitates its study and application. This *P. oligandrum* colonizes the rhizosphere of several crops and shows positive activity on plant, promoting growth and protection against abiotic stresses and biocontrol of pathogens. The success of this microorganism in the control of pathogens is due to the synergism by several mechanisms of action in soil, including direct competition for nutrients, production of antimicrobials, and mycoparasitism (Benhamou et al. 2012).

There are various "guns" to compete for nutrients in the rhizosphere; mycoparasitism is a fundamental part of their success in the antagonistic process (Benhamou et al. 1999). The species *P. oligandrum* exhibit an atypical characteristic that make it a strong competitor in the rhizosphere, parasitizing species closely related, including other species of *Pythium* (Bahramisharif et al. 2013). Apparently, it is the result of the evolution of specific mechanisms, not yet understood, of *P. oligandrum* to recognize and degrade the wall of the host. So far, it was also observed in the mycoparasitism of ascomycetes (Benhamou et al. 1999), and resistance structure as sclerotia (Rey et al. 2005), basidiospores (Ikeda et al. 2012), and pathogenic oomycetes (Le Floch et al. 2005).

The nonpathogenic *P. oligandrum* still demonstrates the ability to increase plant resistance by stimulating the immune system, penetrating and sensitizing the plant by the production of pathogen-associated molecular pattern (PAMPs), and among these standards include wall fractions and oligandrine. The *P. oligandrum*-mediated resistance is associated with the changes in host metabolism, which trigger a series of physical and biochemical responses involved in the protection against penetration and development of pathogens in host tissues, conferring wall reinforcement or increased antimicrobial activity (Benhamou et al. 1996).

The oomycetes and pathogenic bacteria have been described as resistance-inducing fungi, they were found in tomato plants interrupting bacterial infection process (Masunaka et al. 2009). The effect on nutrient content by the fungus was also observed in the control of gray-eared caused by *Botrytis cinerea* (Le Floch et al. 2009; Mohamed et al. 2007).

As we already know, plant growth is highly dependent on the production of phytoregulators, such as auxin (Zeiger and Taiz 2017). The growth promotion by microorganisms involved the production of secondary metabolites and phytohormones (Helman et al. 2011; Hermosa et al. 2012). The ability of *P. oligandrum* to produce compounds similar to auxin has been described (Le Floch et al. 2003), when was add auxin precursors in the culture medium, producing tryptamine (TNH2), an auxin-like molecule, was observed. This pathway is common for nonpathogenic fungi; what distinguishes this group from others is that this species does not present the pathway to convert TNH2 to indole-3-acetic acid (IAA). In addition, it was also observed that TNH2 was rapidly adsorbed by roots which promote an improve secondary root growth increasing plant biomass.

3.3 Arbuscular Mycorrhizal Fungi (AMF)

Microorganisms are an important component of natural systems. In terrestrial ecosystems, soil microorganisms play several functions in nutrient dynamics and regulation of relationships between microbial communities, which is fundamental in sustainable agriculture (Bender et al. 2016). One of the most important points to approach in this context are the interactions among soil microorganisms and plant roots in the rhizospheric environment. The microorganisms can influence plant growth positively by biochemical nutrient transformation and absorption, or negatively, if some microorganisms in the rhizospheric community are phytopathogens (Giri et al. 2005).

The biotic interactions in the rhizosphere are also important for plant health and plant production (Bender et al. 2016). An example is the arbuscular mycorrhizal (AM) fungi which are a group of obligate biotrophic microorganisms that establish symbiotic relationship with plant root and make the interaction in 90% of most plant species (Giovannetti et al. 2010). The AM association is very important in the present and in the future of sustainable agriculture due to their relevant influence in plant growth and health. The AM association promotes plant growth directly by nutrient uptake (Neumann and George 2010) and indirectly by other mechanisms such as tolerance against pathogens (Pozo et al. 2010; Jung et al. 2012) and abiotic stresses (Porcel et al. 2006; Ruiz-Lozano and Aroca 2010).

Currently, one of the most interesting approaches of AM association is the plant protection against microbial disease and pest attack (Dar and Reshi 2017). Improved nutrient absorption is being cited as one of the mechanisms whereby the AM symbiosis assists plants to act against pathogens. However, the data in this point are inconclusive. Some experiments show that the effect of *Glomus intraradices* biocontrol of *Alternaria solani* in tomato can be regulated by phosphorous (P) availability, showing effective disease control at lower levels of P and reducing the effect of the biocontrol by AM at high P levels, but this is a result of modulation of P levels in AM colonization (Fritz et al. 2006). On the other hand, the inoculation of *G. mosseae* into tomato and eggplant contrasts the pathogenic effect of *Verticillium dahliae* in the meantime that increases the P and N uptake (Karagiannidis et al. 2002). In soybean, it was observed that *Entrophospora infrequens* colonization inhibits the infection of *Pseudomonas syringae* pv. glycinea; this effect is probably observed by N increase in mycorrhizated plants (Malik et al. 2016).

The interaction between AM fungi and soil-borne pathogens in the rhizosphere can modify the root exudates, promoting the biocontrol. In tomato roots, the *G. musseae* colonization and interaction with *F. oxysporum* f.sp. *lycopersici* increase the release of chlorogenic acid, reducing the spore pathogen germination (Hage-Ahmed et al. 2013). In tomato, transformed roots, in vitro conolized by *G. intraradices* modify the exudates composition showed repulsion of *Phytophthora nicotianae* zoospores (Lioussanne et al. 2007), similarly root exudates from banana plants previously colonized by *G. mosseae* reduce the chemotaxis of plant-parasitic nematode *Radopholus similis* (Vos et al. 2012a, b). The infection of other parasitic nematodes is controlled by AM fungi (Tchabi et al. 2016; Marro et al. 2014; Anjos et al. 2010), but the mechanism of this biocontrol effect is not clearly described (Schouteden et al. 2015). In addition, competitions for infection sites, described for root pathogens how fungus that colonized the root cortical cells can to leave at the biocontrol mediated by AM fungi. *G. mosseae* colonization reduces the infection loci of *Phytophthora parasitica* in tomato plants (Vigo et al. 2000) and *G. etunicatum* colonization reduces the disease severity caused by *Verticillium dahliae* in cotton plants (Kobra et al. 2009).

Other mechanism, most commonly described in AM biocontrol, is the induction of plant defense. The plant defense response depend on the levels of several phytohormones such as salicylic acid (SA), jasmonic acid (JA), ethylene (ET) and abscisic acid (ABA); they can be activated by AM root colonization (Dar and Reshi 2017; Pozo et al. 2010). Early stages of AM colonization are regulated by auxins and cytokinins; these are involved in recognition and root architecture modifications for mycorrhizal infections; in later stages, the fungal arbuscular formation into the roots is modulated by ABA and JA levels, and the rate of AM colonization is given as a response to the levels of ET and SA (Ludwig-Müller 2010). The induction of ET and JA production by AM colonization can trigger an induced systemic resistance (ISR) which leads to the control of pathogens in distant parts of roots, as in leafs (Pozo et al. 2010). In in vitro culture conditions, the mycorrhization of banana plantlets with *Rhizophagus irregularis* reduced the symptoms of black sigatoka (*Mycosphaerella fijiensis*) in leafs (Anda et al. 2015). The same ISR effect by JA can protect the plants of herbivorous insects attack (Minton et al. 2016; Song et al. 2013)

As described by Burketova et al. (2015), the plant immunity is enabled by the recognition of microbe-associated molecular patterns (MAMPs) or pathogen-associated molecular patterns (PAMPs) in plant cell tissues. MAMPs can be the common molecules from bacteria or fungi, such as flagellin, chitin, glucans, or lipids. These typical microbial molecules are recognized by pattern-recognition receptors in plasmatic membrane and produce a cellular response that can include production of reactive oxygen species (ROS) and nitric oxide (NO) and activation of mitogen-activated and calcium-dependent protein kinases (MAPKs and CDPKs) that are considered to be transcriptional regulators for some genes related to plant defense. After genetic transcriptional activation of plant defense response, some molecules such as pathogenesis-related proteins (PRP) (β-1, 3-glucanases and chitinases), antimicrobial compounds (phytoalexins), enzymes of the phenylpropanoid pathway (i.e., phenylalanine ammonia lyase (PAL), related with phenolic and flavonoid production) or deposition of high levels of lignin to the cell wall all counteract the microbial/pathogen colonization.

Modifications in concentrations of chitinase, chitosanase, and β-1, 3-glucanase in tomato roots by *G. mosseae* and *G. intraradices* association led to a reduction of infection by *Phytophtora parasitica* (Pozo et al. 2002). High levels of PAL activity have been related with the control *of Fusarium solani* symptoms by *G. intraradices*, *G. hoi*, *Gigaspora margarita*, and *Scutellospora gigantean* in common bean (*Phaseolus vulgaris* L) under greenhouse conditions (Eke et al. 2016). In the same way, *G. deserticola* induces the defense of pepper against *Verticillium dahliae* (Garmendia et al. 2006).

4 Interaction of PGPR and PGPF

4.1 Arbuscular Mycorrhizal Fungi: Gram Positive

As discussed previously, AMF are common in the rhizosphere and interact with PGPR, thereby enhancing plant growth, improving nutrient uptake, phytohormones production, or controlling pathogens. Pathogen biocontrol by PGPR can happen by production of antimicrobial compounds (antagonist relationship) or by inducing plant resistance (Beneduzi et al. 2012). The gram-positive bacteria of genera *Bacillus* and *Paenibacillus* are ubiquitous in the rhizosphere environment, and they enhance plant growth and induce resistance to several plants species (Won-Il et al. 2011). The synergistic effect of both AMF (*Gigaspora margarita*) and gram-positive PGPR (*P. rhizospherae*) is efficient in the control of white root rot caused by

Rosellinia necatrix in seeding fruits; the results suggest a mechanisms of induction resistance or new root formation (Cruz et al. 2014). The root rot produced by *Macrophomina phaseolina* infection and *Meloidogyne incognita* root invasion was controlled by *G. intraradices* and *Bacillus pumilus* inoculation, in chickpea (Akhtar and Siddiqui 2007). However, the effectiveness of these combinations depends on the species and the origin of microorganisms. In experiments with papaya (*Carica papaya* L.) to control root-knot nematode (*M. incognita*), the effect of AM biocontrol was observed isolates (*G. mosseae* or *G. manihotis*) but not in interaction with *Bacillus* consortia (Jaizme-Vega et al. 2005). On the other hand, Bakhtiar et al. (2012) report that the mycorrhizal endosymbiotic bacteria (*B. subtilis* B10) have synergic effect with AM fungi consortia in control of *Ganoderma boninense* in oil palm seedlings by improving phosphorous uptake.

4.2 Arbuscular Mycorrhizal Fungi: Gram Negative

Naturally, AMF and PGPR coexist in the rhizosphere, and studies show that the inoculation of these genera into plants of agricultural interest shows excellent results regarding phytopathogen suppression (Miransari 2011). In a greenhouse study in tomato plants, it was observed that the inoculation of AMF *Rhizophagus irregularis* and PGPR gram-negative *P. jassenii* and *P. synxantha* promoted the reduction of infection caused by the root-knot nematode *Meloidogyne incognita* (Sharma et al. 2017). The combination of *G. intraradices* with *P. fluorescens*, *P. putida*, and *Enterobacter cloacae* reduced the infection caused by *F. oxysporum*. Co-inoculation of one to three strains combined with AMF was evaluated. The results showed that the triple inoculation presented better results against the pathogen when compared to the inoculation of only one strain (Akköprü 2005). The results of another study evaluating the combination of biocontrol agents against three phytopathogens *Pyricularia grisea*, *Bipolaris oryzae*, and *Gerlachia oryzae* also presented better results for inoculants with more than one strain combined. In this work, rice seeds were treated with bacterial suspensions containing different associations, from one to four strains, being *P. synxantha* strain DFs185; *P. fluorescens* strain DFs223; two strains of *Bacillus* sp DFs416 and DFs418; and an unidentified strain DFs306 (de Souza et al. 2017).

4.3 Arbuscular Mycorrhizal Fungi: Actino

The microorganisms help the plant to acquire nutrients from the soil by symbiotic interaction. The fungal-bacterial interactions (FBI) are an integral component of soil and plant health. Mycorrhiza helper bacteria (MHB) assist mycorrhizal fungi in the establishment of a mycorrhizal association with the plant (Duponnois and Garbaye 1991; Frey-Klett et al. 2007). Mechanisms involved in this interaction include the fungal spore germination, increased root colonization, production of factors that stimulate the growth of mycelia, and reduction of stress by detoxification of substances that are antagonistic (Haq et al. 2014).

Actinomycetes have been associated with ectomycorrhizal fungi and have been discussed as modulators of plant symbiosis (Choudhary et al. 2017). Immunofluorescence microscopy showed that MHB *Streptomyces* promoting hyphal growth and morphological changes in ectomycorrhizal fungi act on basic cell growth processes, cap of fungal hyphae, and symbiosis formation. The use of MHB actinobacteria and AMF as bio-inoculum for growth promotion and biocontrol of disease caused by plant pathogen emerged like an alternate solution to the use of chemical fertilizers and pesticides. For instance, when actinobacteria isolates, separately and along with *G. fasiculatum*, are screened against *Fusarium* sp. in tomato, it is found that plants treated with actinobacterial isolates in association with *G. fasiculatum* showed higher efficiency (no disease incidence), whereas the plants treated with the only pathogen as expected showed 100% disease symptom (Krishnaraj 2017).

Poovarasan et al. (2013) elucidated the role that actinobacteria-associated mycorrhizae play in promoting plant growth and their effectiveness as antibacterial agents in controlling *Xanthomonas axonopodis* pv. *punicae*, causing blight disease in pomegranate. Thus, actinomycetes enhance the beneficial role of mycorrhizal fungi in plant growth promotion by producing growth regulators, phosphate solubilization, siderophore production, and as an antimicrobial agent (Mohandas et al. 2013). Large-scale utilization of FBI, MHB actinomycetes, and AMF on farm as "fertilizers" and "pesticides" is much less demanding than chemical products. Mykorrhiza soluble (Glückspilze, Innsbruck, Austria) is a microbial-consortia-based products containing *Pisolithus tinctorius*, *Rhizopogon* spp., *Scleroderma* spp., *Suillus* spp., *Laccaria* spp., *Glomus* spp., *Streptomyces griseus*, and *S. lydicus*. It is further proposed that these properties in microbial consortium development and commercialization have to be explored.

4.4 Arbuscular Mycorrhizal Fungi

Arbuscular mycorrhizal fungi (AMF) establish interactions with most plants, and in this complex symbiosis, a series of molecular signal exchanges takes place. In this communication, the plant can develop early and improved defense responses to pathogen attack, after establishment of the AMF-plant symbiosis, in a phenomenon known as the priming effect (Jung et al. 2012; Selosse et al. 2014). Among these responses to colonization are the production of auxins, cytokinins, abscisic acid (ABA), and jasmonic acid (JA) (Ludwig-Müller 2010; Pozo et al. 2015). Induction of defense in plants is correlated with the synthesis of PR proteins, such as 1,3-b-glucanases and chitinases (Pozo et al. 2010).

As seen earlier in this chapter, the ability of both AMFs and PGPFs to promote plant growth, as biocontrol agents, and their potential to induce resistance are widely documented, with interesting results in the most diverse cultures, therefore, to seek a better understanding of the relationships between these microorganisms in soil is important, since these benefits can be stimulated in co-inoculation, as well as competition for space and nutrients, and one of those involved in symbiosis can be repressed.

Martinez-Medina et al. (2011), for example, evaluated the effect of co-inoculation of two AMFs (*Glomus intraradices* and *G. mosseae*) with PGPF *Trichoderma harzianum* and observed that the inoculation with AMF alone did not result in fresh weight differences in the plants when compared to the control, while inoculation with *T. harzianum* alone increased the fresh weight by up to 20%. They also observed that co-inoculation resulted in fresh weight gain when compared to plants inoculated with AMF alone. When evaluating the root/shoot ratio, they verified that co-inoculation presented values up to 25% higher when compared to the control, and they were higher in the treatments in which the inoculation of the microorganisms was done separately. In this same study, it was verified that *T. harzianum* fungus increased mycorrhizal colonization; in contrast, the number of colonies of *T. harzianum* was not affected by AMF inoculation. Co-inoculated plants still had the lowest rates of disease, being more effective than the treatments in which inoculation was done separately. As for hormone production, no synergistic effects were observed for IAA, ACC, or ABA production, which presented lower values than those presented in plants inoculated with *T. harzianum* alone.

In another study, Chandanie et al. (2006) sought to evaluate the interaction between *Penicillium simplicissimum* and *Phoma* sp. with AMF *G. mosseae* and to verify if it was neutral, negative, or positive for ISR against anthracnose caused by *C. orbiculare* in cucumber. They observed that the anthracnose was successfully suppressed by fungal isolates (*P. simplicissimum* and *Phoma* sp.) when separately inoculated; however, when co-inoculated with *Phoma* sp. and *G. mosseae*, there was a reduction in the resistance induction capacity, and when the co-inoculation with *P. simplicissimum* and *G. mosseae* was performed, there was no difference. When evaluating root colonization of cucumber plants, they found that co-inoculation inhibited colonization by *Phoma* sp., since it was higher when PGPF was inoculated individually and co-inoculation did not affect colonization for *P. simplicissimum*. Thus, it was possible to observe the relationship between root colonization by PGPF and suppression of anthracnose.

Elsharkawy et al. (2012), still with cucumber plants, evaluated the effect of co-inoculation of *G. mosseae* and *Fusarium equiseti* on the induction of resistance against cucumber mosaic virus, and observed that both co-inoculation, and the inoculation only with *F. equiseti* demonstrated a potential for reducing disease severity; however, the inoculation with AMF alone did not demonstrate this potential.

Many studies report the potential for the biocontrol of AMF and PGPF when inoculated individually; however, few studies aim to address the relationship between symbionts, and even in these studies, what is observed is that the responses can be varied depending on the species of mycorrhizal fungi, PGPF, and on the plants involved, indicating that this is a field that still has to be studied.

5 Field Experiences

To prove the effectiveness of biocontrol agents, field studies are a necessary step to develop the biotechnology and formulation of commercial products. Currently, 149 microorganisms are commercialized as biopesticides (Anwer 2017). Despite the potential of AMF in suppressing plant pathogens under controlled conditions is widely known (Baum et al. 2015) and many of the mechanisms involved have already been clarified (Jacott et al. 2017), recent studies on the effect of AMF inoculation in the control of plant diseases in field conditions are scarce when compared to the number of studies under controlled conditions (Hohmann and Messmer 2017). In most studies inoculation with AMF is held in greenhouse or in nursery stage before transplantation when held in perennial plants.

In a field study carried out by Neeraj and Singh (2011), two AMF species, *Glomus sinuosum* and *Gigaspora albida*, were separately inoculated into bean plants (Phaseolus Vulgaris L.) in order to control *Rhizoctonia solani*. Both AMF species were able to reduce the negative effects of *R. solani,* resulting in greater development of both root and shoot even under pathogen presence. The severity index and disease incidence also decreased with AMF inoculation, and the grain weight was also higher compared to uninoculated plants. In another study, inoculation of groundnut with *Glomus fasciculatum* reduced the severity of stem rot caused by *Sclerotinia rolfsii* (Doley and Jite 2013) and increased plant development even under pathogen attack. Zachée et al. (2010) using a substrate containing *Glomus* Sp. and *Gigaspora* Sp. spores to inoculate groundnut plants were able to reduce the severity of the rosette virus disease by 38%, and the severity of leaf spot caused by *Cercospora arachidicola* and *C. personatum* was reduced by 50%; leaf spot is the major diseases that affect groundnut. The results show that, although it does not guarantee the total protection of the plant, the AMF inoculation can significantly reduce the disease damage.

Inoculation of AMF in seedlings at nursery stage for later transplant to field is the most used technique for field testing. Using this methodology, Abo-Elyousr et al. (2014) inoculated tomato plants with *Glomus mosseae* (current *Funneliformis mosseae*) and successfully reduced the damage of bacteria wilt disease. Inoculation with *G. mosseae* reduced the disease incidence by 35% and provided an increase of 46% in productivity under natural conditions of disease infestation. Pepper plants infested with *Phytophthora capsici*, causal agent of the Phytophthora blight, have also presented a reduction in disease severity when inoculated with *G. mosseae,* although the increase in productivity had been only significant with the inoculation of *G. fasciculatum* and *Gigaspora margarita,* which presented less disease control (Ozgonen and Erkilic 2007). On the other hand, preinoculation of potato plants with *Rhizophagus irregularis* was able to reduce the damage caused by *Phytophthora infestans* only in periods of low pathogen pressure, and it did not have a positive effect on final production or tubers quality (Alaux et al. 2018). Nursery inoculation of melon plants with *R. srregularis* before transplant to the field demonstrated a satisfactory control of the disease complex *Monosporascus* root rot and vine decline (MRRVD), caused by soil fungus Monosporascus *cannonball* (Aleandri et al. 2015), while inoculation with *Rizophagus constrictum, R. claroideum, R. intraradices*, and *G. mosseae* considerably decreased the incidence of wilt caused by *F. oxysporum* (Martínez-Medina et al. 2011). The preinoculation of *Carica papaya* also demonstrated to reduce the incidence of stem diseases caused by pathogenic fungi (Sukhada et al. 2011; Olawuyi et al. 2014).

Although the biocontrol effect of AMF on plant parasitic nematode (PPN) under controlled conditions has been widely reported in the literature (Schouteden et al. 2015), few works are performed in open field or under conditions similar to commercial cultivation. Protective effects against PPN by AMF were obtained in cultures such as cowpea (Odeyemi et al. 2010), banana (Olaniyi 2014), tomato (Gómez et al. 2008), and peach (Calvet et al. 2001). Affokpon et al. (2011) inoculated tomato plants with isolated *Acaulospora scrobiculata*, *Kikolspora kentinensis*, and *Glomus etunicatum* and assessed its protective effect on the attack of *Meloidogyne* spp, causal agent of root-knot. Inoculation with the AMF did not reduce the symptoms of root-knot in a significant way but was able to ensure an increase in the final productivity of tomato fruits. The most interesting, though, was that, after the tomato harvest, carrot plants were transplanted over the plots, but this time without any inoculation with AMF. Plots which received inoculated tomato plants presented the highest productivity of carrots. These results demonstrate the long-term benefits from the inoculation of plants with AMF, proving the potential of these microorganisms for large-scale use in commercial crops.

In addition to the AMF, another group of PGPF widely studied as biocontrol agent are those belonging to the genus *Trichoderma*. The use of this fungi as a biocontrol agent was reported for the first time in 1932 (Kumar et al. 2017), and today more than 60% of biopesticides registered have *Trichoderma* in its formulation (Gangwar and Singh 2018). The great success of the use of *Trichoderma* for biological control of plant diseases is in the ability of this group of fungi to survive under unfavorable conditions, its high reproduction capacity, efficacy in the use of nutrients, the ability to promote plant development, and its aggressiveness against phytopathogenic fungi (Kumar et al. 2017).

The success of the use of *Trichoderma* in different cultures has driven industrial production of inoculated substrates, whether solid or liquid. Techniques for applying living spores include foliar application, seed treatments, post-pruning treatment, and incorporation in the soil or in irrigation system, among other (Fraceto et al. 2018). In soybean, for example, seed inoculation with *T. viride* before sowing reduced the incidence of *Colletotrichum truncatum,* a damping-off causing pathogen (Begum et al. 2010), while the spraying of *T. harzianum* in soil covering after planting reduced the incidence and sclerotia number of *Sclerotinia sclerotiorum* (Zeng et al. 2012). Görgen et al. (2009), also evaluating the control of *S. sclerotiorum*, concluded that application of *T. harzianum* on the soil before and after soybean seeding decreases disease incidence as *T. harzianum* doses increase. The reduction in more than 90% of disease incidence and 30% of disease severity of *Macrophomina phaseolina*, causal agent of charcoal rot disease in soybean, also were obtained with seed inoculation with *T. harzianum* (Barari and Foroutan 2016). In chickpeas (*Cicer Arietinum* L.), seed inoculation with commercial products based on *T. harzianum* reduced the incidence of *F. oxysporum* (Marzani et al. 2017). In a 3-year-long experiment, Xue et al. (2017) tested six isolated *Trichoderma* spp. against *Fusarium* spp. in wheat, achieving a decrease of at least 50% disease severity and a variation of 6–11% of productivity increment.

In addition to controlling soil diseases, foliar application of *Trichoderma* also shows positive results, controlling above-ground diseases. In wheat, foliar application of *T. harzianum* reduced the incidence and severity of spot blotch at 15% and 26%, respectively. This reduction, however, was not greater than that provided by chemical fungicides, which reduced more than 50% of the disease incidence compared with the control (Yadav et al. 2015). Similar results were obtained by Sharkawy et al. (2015), where the application of chemical fungicides presented a better control of *Puccinia triticina*, compared to *T. harzianum* application, which decreased at least 25% of disease severity. Despite presenting a lesser disease control, these studies show that the use of *Trichoderma* as biocontrol agents can serve as an alternative in disease-integrated control, aiming mainly at the reduction of pathogens' resistance to chemical control.

Hydroponic cultivations are susceptible to pathogen infestation, mainly belonging to the genera *Pythium* and *Phytophthora* (Watanabe et al. 2008). In this way, the use of *Trichoderma* seems to be a great control strategy, mainly in organic cultivation. The application of *Trichoderma* in the hydroponic nutritional solution demonstrated to be successful in controlling infections caused by *Pythium aphanidermatum* in lettuce plants (Patekoski and Pires-Zottarelli 2010). Alternatives to chemical compounds can also be obtained through pasty spore-based formulations which can be used as a biocontrol in the healing of perennial crops after pruning or grafting. Sanjay et al. (2008) evaluated mixed culture of *T. harzianum* spores and hyphae, with commercial talc, carboxymetil cellulose (0.5%), and distilled water, forming a paste which was then applied in pruning cuts in tea plants, and in canker injuries caused by *Phomopsis theae*. The application of the *T. harzianum* paste resulted in a larger number of buds per branch, besides increasing the weight of the buds and the production of leaves, parameters that were larger than the ones obtained with the application of chemical protectors. The length and the width of cankers produced by *Phomopsis* showed a considerable decrease under the application of the paste. A paste formulated with *Gliocladium virens* produced similar results. *Trichoderma*-based products can also be used to prevent possible infections of perennial crops during the grafting phase, as demonstrated by Kumar et al. (2017).

Fungi of the genus *Penicillium* also demonstrate positive results in disease control in field conditions. De Cal et al. (2008) successfully reduced powdery mildew severity in strawberry through the foliar application of a solution containing *Penicillium oxalicum* conidia. The results were obtained from seven different strawberry genotypes and repeated for 3 consecutive years, proving the beneficial effect of applying *P. oxalicum*. The authors state the needs of field experiments to attest the efficacy of the biocontrol agents, since under controlled conditions, there may be favoring to the growth of the pathogen, which would not happen in the open fields. *P. oxalicum* also controlled *Fusarium* wilt in watermelon (De Cal et al. 2009) and tomato (De Cal and Melgarejo 2001) and was capable of inducing resistance in pearl millet against *Sclerospora grassbird* (Murali and Amruthesh 2015). *Fusarium* root rot in okra was also controlled with an isolated *Penicillum* sp. in seed treatment (Ullah et al. 2015). Control of *Cercospora* leaf spot on sugar beet was obtained through the foliar application of *Penicillium regulars*, both as preventative as curative (El-fawy et al. 2018). In the same experiment, different adhesive compounds were tested at 1% concentration together with a suspension *P. Regulars* spores, being the best control results obtained with agar, starch flour, and white glue. Currently, several studies have focused on the use of *Penicillium* on fruit coating, seeking to reduce postharvest losses by decomposing fungi, thereby ensuring a longer shelf time (Marín et al. 2017).

Other fungi species are used on a smaller scale, like *Purpureocillium lilacinum (syn: Paecilomyces lilacinus),* which presents great success in the control of root nematodes (Abd-Elgawad 2016). Nonpathogenic isolates of *F. oxysporum* are effective in reducing nematode population on infested areas (Waweru et al. 2014). Gliocladium catenulatum also demonstrates to control phytopathogenic fungi in field conditions.

Among the PGPR, great prominence is given to the genera *Bacillus* and *Pseudomonas*, which are basis of several commercial products, as summarized by Shaikh and Sayyed (2015) and Velivelli et al. (2014). In the genus *Bacillus*, the most commonly used species are *B. subtilis*, *B. amyloliquefaciens*, and *B. thuringiensis* (Yao et al. 2006; Velivelli et al. 2014; Wu et al. 2015a, b). *B. subtilis* is used to control a wide variety of phytopathogens in open fields (Hinarejos et al. 2016) like *Botrytis cinera*, *Pseudomonas syingae*, *Bremia lactucae* (Hinarejos et al. 2016), *R. solani* (Ma et al. 2015), *Puccinia striiformis* (Reiss and Jørgensen 2017), and S *sclerotiorum* (Hu et al. 2014). Similarly, control of a wide range of pathogens is reported using *B. amyloliquefaciens* (Chowdhury et al. 2015; Beibei et al. 2016; Gotor-Vila et al. 2017; Kulimushi et al. 2017a, b).

The genus *Pseudomonas* is known for its metabolic versatility and adaptability; therefore, several nonpathogenic species possess the ability of promoting plant growth and protection against biotic stress (Goswami et al. 2015). *P. fluorescens*, when added to soil, was able to increase the weight of potato tubers, as it reduced the symptoms of scab coverage caused by *Streptomyces* sp. (Arseneault et al. 2015). Guyer et al. (2015), also working with potato, decreased the incidence of *P. infestans* through foliar application of *Pseudomonas* isolates. The cell viability test, however, demonstrated that after 8 days of application of the biocontrol agent, the number of UFC on surface of potato leaves fell drastically, indicating the need for continuous applications for maintenance of pathogen control. Isolated *Pseudomonas* sp. were also able to induce the resistance in tomatoes, decreasing the severity of leaf curl virus (ToLCV) by more than 80%, besides promoting an increase in the fruit number per plant and fruit weight (Mishra et al. 2014). Erdogan and Benlioglu (2010) succeeded in controlling *Verticillium* wilt in cotton by inoculating seeds with *Pseudomonas* sp., achieving up to 10% increase on final productivity (Erdogan and Benlioglu 2010). In sugarcane, *P. fluorescens* drenched in the soil around the roots resulted in a decrease of at least 50% in the severity of *Glomerella tucumanensis*, causal agent of red rot disease (Hassan et al. 2011), while in rice, it controlled the root-knot nematode and enhances yield (Seenivasan et al. 2012).

Studies using bacteria as biocontrol agents in field conditions are not limited only to the use of *Pseudomonas* and *Bacillus*. The use of *Paenibacillus dendritiformis* in irrigation water demonstrated to be able to reduce the damage of *Pectobacterium carotovorum* in potato, resulting in higher productivity in relation to control (Lapidot et al. 2015). The addition of organic substrate inoculated with *Paenibacillus polymyxa* in transplant of watermelon seedlings reduced the attack of *F. oxysporum* and resulted in the increment of plant development (Ling et al. 2010).

6 Conclusions

New agriculture practices, aimed **at** reducing the use of agrochemicals by ensuring productivity, are essential for the near future. The use of biological agents, such as bacteria and fungi, or products of their metabolism, can become an excellent strategy to overcome serious environmental problems such as soil and water contamination and the selection of resistant microorganisms.

In this chapter, several microorganisms with potential to be used as biocontrol agents were presented by different action model, such as the production of volatile compounds, siderophores, and induction of resistance (ISR and SAR), including some already being marketed (see Table 3.1). However, the relationships between microorganisms and host plants are not yet fully understood, and the choice of a fungus, bacterium, or even a combination of both will in many cases elicit different responses depending on a series of factors such as host plant choice, competition for space and efficiency in root colonization against pathogens, and compatibility of selected agents, among others. Therefore, new studies aimed at elucidating these communications between microorganisms and plants are extremely important for the development of new products and technologies.

References

Abd-Elgawad MMM (2016) Biological control agents of plant-parasitic nematodes. Egypt J Biol Pest Control 26:423–429

Abo-Elyousr KAM, Seleim MEA, Abd-El-Moneem KMH, Saead FA (2014) Integrated effect of *Glomus mosseae* and selected plant oils on the control of bacterial wilt disease of tomato. Crop Prot 66:67–71. https://doi.org/10.1016/j.cropro.2014.07.022

Adam E, Mïuller H, Erlacher A, Berg G (2016) Complete genome sequences of the *Serratia plymuthica* strains 3Rp8 and 3Re4-18, two rhizosphere bacteria with antagonistic activity towards fungal phytopathogens and plant growth promoting abilities. Stand Genomic Sci 11:1–8. https://doi.org/10.1186/s40793-016-0185-3

Affokpon A, Coyne DL, Lawouin L et al (2011) Effectiveness of native West African arbuscular mycorrhizal fungi in protecting vegetable crops against root-knot nematodes. Biol Fertil Soils 47:207–217. https://doi.org/10.1007/s00374-010-0525-1

Ahemad M, Kibret M (2014) Mechanisms and applications of plant growth promoting rhizobacteria: current perspective. J King Saud University - Science 26(1):1–20

Ait Bahadou S, Ouijja A, Tahiri A (2016) Evaluation of biological control agent Pantoea agglomerans P10c against fire blight in Morocco. African J Agric Res 11:1661–1667. https://doi.org/10.5897/AJAR2016.10986

Akhtar MS, Siddiqui ZA (2007) *Glomus intraradices*, *Pseudomonas alcaligenes*, and *Bacillus pumilus*: effective agents for the control of root-rot disease complex of chickpea (*Cicer arietinum* L.). J Gen Plant Pathol 74:53–60

Akino S, Takemoto D, Hosaka K (2014) *Phytophthora infestans*: a review of past and current studies on potato late blight. J Gen Plant Pathol 80:24–37

Akköprü A, Demir S (2005) Biological control of Fusarium wilt in tomato caused by Fusarium oxysporum f. sp. lycopersici by AMF Glomus intraradices and some Rhizobacteria. J Phytopathol 153(9):544–550

Akram W, Anjum T, Ali B (2016) Phenylacetic acid is ISR determinant produced by *Bacillus fortis* IAGS162, which involves extensive re-modulation in metabolomics of tomato to protect against Fusarium wilt. Front Plant Sci 7:498. https://doi.org/10.3389/fpls.2016.00498

Alam SS, Sakamoto K, Inubushi K (2011) Biocontrol efficiency of *Fusarium* wilt diseases by a root-colonizing fungus *Penicillium* sp. Soil Sci Plant Nutr 57:204–212

Alaux PL, César V, Naveau F et al (2018) Impact of *Rhizophagus irregularis* MUCL 41833 on disease symptoms caused by *Phytophthora infestans* in potato grown under field conditions. Crop Prot 107:26–33. https://doi.org/10.1016/j.cropro.2018.01.003

Aleandri MP, Martignoni D, Reda R, Chilosi G (2015) Effects of preconditioning through mycorrhizal inoculation on the control of melon root rot and vine decline caused by *Monosporascus cannonballus*. J Phytopathol 163:898–907. https://doi.org/10.1111/jph.12389

Allard-Massicotte R, Tessier L, Lécuyer F et al (2016) Bacillus subtilis early colonization of *arabidopsis thaliana* roots involves multiple chemotaxis receptors. MBio 7(6):e01664–e01616. https://doi.org/10.1128/mBio.01664-16

Altaf MM, Imran M, Abulreesh HH et al (2018) Diversity and applications of *Penicillium* spp. in plant-growth promotion. In: Gupta V, Rodrigues-Couto S (eds) New and future developments in microbial biotechnology and bioengineering: *Penicillum* system properties and applications. Elsevier, Amsterdam, pp 261–276

Amarger N, Macheret V, Laguerre G (1997) *Rhizobium gallicum* sp. nov. and *Rhizobium giardinii* sp. nov., from Phaseolus vulgaris nodules. Int J Syst Evol Microbiol 47(4):996–1006. https://doi.org/10.1099/00207713-47-4-996

Anda CCO, Boulois HD, Declerck S (2015) The arbuscular mycorrhiza fungus *Rhizophagus irregularis* MUCL 41833 decreases disease severity of black sigatoka on banana c.v. Grande naine, under in vitro culture conditions. Fruits 70(1):37–46

Angel S-M, Silva-Flores MA, Cervantes-Badillo MG et al (2011) The plant growth-promoting fungus *Aspergillus ustus* promotes growth and induces resistance against different lifestyle pathogens in *Arabidopsis thaliana*. J Microbiol Biotechnol 21:686–696

Anjos ECT, Cavalcante UMT, Gonçalves DMC et al (2010) Interactions between an arbuscular mycorrhizal fungus (*Scutellospora heterogama*) and the root-knot nematode (*Meloidogyne incognita*) on sweet passion fruit (*Passiflora alata*). Braz Arch Biol Technol 53(4):801–809

Antoun H, Beauchamp CJ, Goussard N et al (1998) Potential of *Rhizobium* and *Bradyrhizobium* species as plant growth promoting rhizobacteria on non-legumes: effect on radishes (Raphanus sativus L.). In: Hardarson G, Broughton WJ (eds) Molecular microbial ecology of the soil, Developments in plant and soil sciences, 83. Springer, Dordrecht. https://doi.org/10.1007/978-94-017-2321-3_5

Anwer A (2017) Status of biopesticides and biocontrol agents in agriculture: an overview. In: Anwer A (ed) Biopesticides and bioagents: novel tools for pest management, 1st edn. Apple Academic Press, New York, p 418

Araújo FDS, Araújo WL, Eberlin MN (2017) Potential of *Burkholderia seminalis* TC3. 4.2 R3 as biocontrol agent against Fusarium oxysporum evaluated by mass spectrometry imaging. J Am Soc Mass Spectrom 28(5):901–907. https://doi.org/10.1007/s13361-017-1610-6

Arseneault T, Goyer C, Filion M (2015) *Pseudomonas fluorescens* LBUM223 increases potato yield and reduces common scab symptoms in the field. Phytopathology 105:1311–1317

Asaka O, Shoda M (1996) Biocontrol of Rhizoctonia solani damping-off of tomato with *Bacillus subtilis* RB14. Appl Environ Microbiol 62(11):4081–4085

Asari S, Matzén S, Petersen MA et al (2016) Multiple effects of *Bacillus amyloliquefaciens* volatile compounds: plant growth promotion and growth inhibition of phytopathogens. FEMS Microbiol Ecol 92(6). https://doi.org/10.1093/femsec/fiw070

Audenaert K, Pattery T, Cornelis P, Höfte M (2002) Induction of systemic resistance to *Botrytis cinerea* in tomato by *Pseudomonas aeruginosa* 7NSK2: role of salicylic acid, pyochelin, and pyocyanin. Mol Plant-Microbe Interact 15:1147–1156. https://doi.org/10.1094/MPMI.2002.15.11.1147

Babu AG, Kim SW, Yadav DR et al (2015) *Penicillium menonorum*: a novel fungus to promote growth and nutrient management in cucumber plants. Mycobiology 43:49–56

Bahramisharif A, Lamprecht SC, Calitz F et al (2013) Suppression of *Pythium* and *Phytophthora* damping-off of rooibos by compost and a combination of compost and nonpathogenic *Pythium* taxa. Plant Dis 97:1605–1610

Bakhtiar Y, Yahya S, Sumaryono W et al (2012) Adaptation of oil palm seedlings inoculated with arbuscular mycorrhizal fungi and mycorrhizal endosymbiotic bacteria Bacillus subtilis B10 towards biotic stress of pathogen *Ganoderma boninense*. Microbiology 6(4):157–164

Barka EA, Vatsa P, Sanchez L, Gaveau-Vaillant N et al (2016) Taxonomy, physiology, and natural products of Actinobacteria. Microbiol Mol Biol Rev 80:1–43. https://doi.org/10.1128/MMBR.00019-15

Barari H, Foroutan A (2016) Biocontrol of soybean charcoal root rot disease by using *Trichoderma* spp. Cercet Agron Mold 49:41–51. https://doi.org/10.1515/cerce-2016-0014

Baum C, El-Tohamy W, Gruda N (2015) Increasing the productivity and product quality of vegetable crops using arbuscular mycorrhizal fungi: a review. Sci Hortic (Amsterdam) 187:131–141. https://doi.org/10.1016/j.scienta.2015.03.002

Begum MM, Sariah M, Puteh AB et al (2010) Field performance of bio-primed seeds to suppress *Colletotrichum truncatum* causing damping-off and seedling stand of soybean. Biol Control 53:18–23. https://doi.org/10.1016/j.biocontrol.2009.12.001

Beibei W, Zongzhuan S, Fengge Z et al (2016) *Bacillus amyloliquefaciens* strain W19 can promote growth and yield and suppress fusarium wilt in banana under greenhouse and field conditions. Pedosphere 26:733–744. https://doi.org/10.1016/S1002-0160(15)60083-2

Bender FS, Wagg C, Van der Heijden MGA (2016) An underground revolution: biodiversity and soil ecological engineering for agricultural sustainability. Trends Ecol Evol 31(6):440–451

Beneduzi A, Ambrosini A, Passaglia LMP (2012) Plant growth-promoting rhizobacteria (PGPR): their potential as antagonists and biocontrol agents. Genet Mol Biol 35(4):1044–1051

Benhamou N, Rey P, Chérif M et al (1996) Treatment with the mycoparasite *Pythium oligandrum* triggers induction of defense-related reactions in tomato roots when challenged with *Fusarium oxysporum* f.sp. radicis-lycopersici. Phytopathology 87(1):108–122

Benhamou N, Rey P, Picard K, Tirilly Y (1999) Ultrastructural and cytochemical aspects of the interaction between the mycoparasite. *Pythium oligandrum* and soilborne plant pathogens. Phytopathology 89(6):506–517

Benhamou N, le Floch G, Vallance J et al (2012) *Pythium oligandrum*: an example of opportunistic success. Microbiology (UK) 158:2679–2694

Bent E (2006) Induced systemic resistance mediated by plant growth-promoting rhizobacteria (PGPR) and Fungi (PGPF). In: Tuzun S, Bent E (eds) Multigenic and induced systemic resistance in plants. Springer, New York, pp 225–258

Bernsdorff F, Doering A-C, Gruner K et al (2015) Pipecolic acid orchestrates plant systemic acquired resistance and defense priming via salicylic acid-dependent and independent pathways. Plant Cell 28:TPC2015–00496–RA. https://doi.org/10.1105/tpc.15.00496

Besson F, Peypoux F, Michel G, Delcambre L (1976) Characterization of iturin A in antibiotics from various strains of *Bacillus subtilis*. J antibiot 29(10):1043–1049. https://doi.org/10.7164/antibiotics.29.1043

Bevivino A, Sarrocco S, Dalmastri C et al (1998) Characterization of a free-living maize-rhizosphere population of *Burkholderia cepacia*: effect of seed treatment on disease suppression and growth promotion of maize. FEMS Microbiol Ecol 27(3):225–237. https://doi.org/10.1111/j.1574-6941.1998.tb00539.x

Bhattacharyya PN, Jha DK (2012) Plant growth-promoting rhizobacteria (PGPR): emergence in agriculture. World J Microbiol Biotechnol *28*(4):1327–1350. https://doi.org/10.1007/s11274-011-0979-9

Bisacchi GS, Hockstein DR, Koster WH et al (1987) Xylocandin: a new complex of antifungal peptides. J Antibiot 40(11):1520–1529. https://doi.org/10.7164/antibiotics.40.1520

Bitas V, McCartney N, Li N et al (2015) *Fusarium oxysporum* volatiles enhance plant growth via affecting auxin transport and signaling. Front Microbiol 6:1248. https://doi.org/10.3389/fmicb.2015.01248

Bjelić DĐ, Marinković JB, Tintor BB et al (2015) Screening of *Azotobacter* isolates for PGP properties and antifungal activity. Zbornik Matice srpske za prirodne nauke 129:65–72. https://doi.org/10.2298/ZMSPN1529065B

Blanco Y, Blanch M, Piñón D et al (2005) Antagonism of *Gluconacetobacter diazotrophicus* (a sugarcane endosymbiont) against *Xanthomonas albilineans* (pathogen) studied in alginate-immobilized sugarcane stalk tissues. J Biosci Bioeng 99(4):366–371. https://doi.org/10.1263/jbb.99.366

Blankenfeldt W, Kuzin AP, Skarina T et al (2004) Structure and function of the phenazine biosynthetic protein PhzF from *Pseudomonas fluorescens*. Proc Natl Acad Sci U S A 101:16431–16436. https://doi.org/10.1073/pnas.0407371101

Bochkareva OO, Moroz EV, Davydov II, Gelfand MS (2018) Genome evolution in *Burkholderia* spp. bioRxiv. https://doi.org/10.1101/319723

Bonnet P, Rousse G (1985) Réactions différentielles du tabac à 9 espèces de Phytophthora. Agronomie *5*(9):801–808

Bonnet P, Lacourt I, Venard P et al (1994) Diversity in pathogenicity to tobacco and in elicitin production among isolates of *Phytophthora parasitica*. J Phytopathol 141:25–37

Borrero C, Trillas MI, Avilés M (2009) Carnation *Fusarium* wilt suppression in four composts. Eur J Plant Pathol 123:425–433

Borriss R (2011) Use of plant-associated *Bacillus* strains as biofertilizers and biocontrol agents in agriculture. In: Maheshwari D (ed) Bacteria in agrobiology: plant growth responses. Springer, Berlin/Heidelberg. https://doi.org/10.1007/978-3-642-20332-9_3

Borriss R (2015) *Bacillus*, a plant-beneficial bacterium. In: Lugtenberg B (ed) Principles of plant-microbe interactions. Springer, Cham. https://doi.org/10.1007/978-3-319-08575-3_40

Bulgarelli D, Schlaeppi K, Spaepen S et al (2013) Structure and functions of the bacterial microbiota of plants. Annu Rev Plant Biol 64:807–838. https://doi.org/10.1146/annurev-arplant-050312-120106

Burketova L, Trda L, Ott PG, Valentova O (2015) Bio-based resistance inducers for sustainable plant protection against pathogens. Biotechnol Adv 33:994–1004

Calvet C, Pinochet J, Hernández-Dorrego A et al (2001) Field microplot performance of the peach-almond hybrid GF-677 after inoculation with arbuscular mycorrhizal fungi in a replant soil infested with root-knot nematodes. Mycorrhiza 10:295–300. https://doi.org/10.1007/PL00009998

Carpenter RD, Maloney K (2015) (12) Patent Application Publication (10) Pub. No.: US 2006 / 0222585 A1 Figure 1. 002:354. https://doi.org/10.1037/t24245-000

Cavello IA, Crespo JM, Garcia SS et al (2015) Plant growth promotion activity of keratinolytic fungi growing on a recalcitrant waste known as "Hair Waste". Biotechnol Res Int 2015:952921. https://doi.org/10.1155/2015/952921

Chalfoun SM (2010) Biological control and bioactive microbial metabolites: a coffee quality perspective. Ciência e Agrotecnologia 34(5):1071–1085

Chandanie WA, Kubota M, Hyakumachi M (2006) Interactions between plant growth promoting fungi and arbuscular mycorrhizal fungus *Glomus mosseae* and induction of systemic resistance to anthracnose disease in cucumber. Plant Soil 286:209–217

Chaurasia B, Pandey A, Palni LMS et al (2005) Diffusible and volatile compounds produced by an antagonistic *Bacillus subtilis* strain cause structural deformations in pathogenic fungi in vitro. Microbiol Res 160(1):75–81. https://doi.org/10.1016/j.micres.2004.09.013

Chave M, Crozilhac P, Deberdt P et al (2017) Rhizophagus irregularis MUCL 41833 transitorily reduces tomato bacterial wilt incidence caused by *Ralstonia solanacearum* under in vitro conditions. Mycorrhiza 27:719–723

Chen WX, Yan GH, Li JL (1988) Numerical taxonomic study of fast-growing soybean rhizobia and a proposal that Rhizobium fredii be assigned to Sinorhizobium gen. nov. J Syst Evol Micr 38(4):392–397. https://doi.org/10.1099/00207713-38-4-392

Chen XH, Koumoutsi A, Scholz R, Borriss R (2009) More than anticipated–production of antibiotics and other secondary metabolites by *Bacillus amyloliquefaciens* FZB42. J Mol Microbiol Biotechnol 16(1-2):14–24. https://doi.org/10.1159/000142891

Chen Y, Shen X, Peng H et al (2015) Comparative genomic analysis and phenazine production of *Pseudomonas chlororaphis*, a plant growth-promoting rhizobacterium. Genomics Data 4:33–42. https://doi.org/10.1016/j.gdata.2015.01.006

Chibeba AM, de Fátima Guimarães M, Brito OR et al (2015) Co-inoculation of soybean with *Bradyrhizobium* and *Azospirillum* promotes early nodulation. Am J Plant Sci 6(10). https://doi.org/10.4236/ajps.2015.610164

Chin-Woeng TFC, Bloemberg GV, Mulders IHM et al (2000) Root colonization by phenazine-1-carboxamide-producing bacterium *Pseudomonas chlororaphis* PCL1391 is essential for biocontrol of tomato foot and root rot. Mol Plant-Microbe Interact 13:1340–1345. https://doi.org/10.1094/MPMI.2000.13.12.1340

Choudhary DK, Varma A, Tuteja N (2017) Mycorrhizal helper bacteria: sustainable approach. In: Mycorrhiza – function, diversity, state of the art. Springer, Cham, pp 61–74

Chowdhury SP, Hartmann A, Gao XW, Borriss R (2015) Biocontrol mechanism by root-associated *Bacillus amyloliquefaciens* FZB42 – a review. Front Microbiol 6:1–11

Chu BC, Garcia-Herrero A, Johanson TH et al (2010) Siderophore uptake in bacteria and the battle for iron with the host; a bird's eye view. BioMetals 23:601–611. https://doi.org/10.1007/s10534-010-9361-x

Cruz AF, Ishii T (2018) Arbuscular mycorrhizal fungal spores host bacteria that affect nutrient biodynamics and biocontrol of soil-borne plant pathogens. Biol Open. https://doi.org/10.1242/bio.2011014

Cruz AF, Soares WRO, Blum LEB (2014) Impact of the arbuscular mycorrhizal fungi and bacteria on biocontrol of white root rot in fruit seedlings. J Plant Physiol Pathol. https://doi.org/10.4172/2329-955X.1000114

Dalio RJ, Fleischmann F, Osswald W (2011) Localiazation of *Phytophthora plurivora* effector protein citricolin in *Fagus sylvatica* roots by light and fluorescence laser scanning microscopy. Phytopathology 101:S40

Dalmas FR, Pereira TCB, Bogo MR, Astarita LV (2011) Autochthonous Streptomyces regulate the metabolism of seedlings of *Araucaria angustifolia* (Coniferales) during root colonisation. Aust J Bot 59:118. https://doi.org/10.1071/BT10175

Dar MH, Reshi ZA (2017) Vesicular arbuscular mycorrhizal (VAM) fungi- as a major biocontrol agent in modern sustainable agriculture system. Microbiology 43(2):138–143

Darling P, Chan M, Cox AD, Sokol PA (1998) Siderophore production by cystic fibrosis isolates of Burkholderia cepacia. Infect Immun 66(2):874–877

Das K, Rajawat MVS, Saxena AK, Prasanna R (2017) Development of *Mesorhizobium ciceri*-based biofilms and analyses of their antifungal and plant growth promoting activity in chickpea challenged by *Fusarium* wilt. Indian J Microbiol 57(1):48–59. https://doi.org/10.1007/s12088-016-0610-8

De Cal A, Melgarejo P (2001) Repeated applications of *Penicillium oxalicum* prolongs biocontrol of fusarium wilt of tomato plants: biological control of fusarium wilt in tomato plants. Eur J Plant Pathol 107:805–811. https://doi.org/10.1023/A:1012432404886

De Cal A, Redondo C, Sztejnberg A, Melgarejo P (2008) Biocontrol of powdery mildew by *Penicillium oxalicum* in open-field nurseries of strawberries. Biol Control 47:103–107. https://doi.org/10.1016/j.biocontrol.2008.07.010

De Cal A, Sztejnberg A, Sabuquillo P, Melgarejo P (2009) Management fusarium wilt on melon and watermelon by *Penicillium oxalicum*. Biol Control 51:480–486. https://doi.org/10.1016/j.biocontrol.2009.08.011

De Lajudie P, Laurent-Fulele E, Willems A et al (1998a) Allorhizobium undicola gen. nov., sp. nov., nitrogen-fixing bacteria that efficiently nodulate Neptunia natans in Senegal. J Syst Evol Microbiol 48(4):1277–1290. https://doi.org/10.1099/00207713-48-4-1277

De Lajudie P, Willems A, Nick G et al (1998b) Characterization of tropical tree rhizobia and description of Mesorhizobium plurifarium sp. nov. J Syst Evol Micr 48(2):369–382. https://doi.org/10.1099/00207713-48-2-369

De Oliveira AG, Spago FR, Simionato AS et al (2016) Bioactive organocopper compound from *Pseudomonas aeruginosa* inhibits the growth of Xanthomonas citri subsp. citri. Front Microbiol 7:1–12. https://doi.org/10.3389/fmicb.2016.00113

de Souza Júnior IT, Schafer JT, Corrêa BO, Funck GD, Moura AB (2017) Expansion of the biocontrol spectrum of foliar diseases in rice with combinations of rhizobacteria. Rev Ciênc Agron 48(3):513–522. https://doi.org/10.5935/1806-6690.20170060

Delamuta JRM, Ribeiro RA, Ormeño-Orrillo E et al (2015) *Bradyrhizobium tropiciagri* sp. nov. and *Bradyrhizobium embrapense* sp. nov., nitrogen-fixing symbionts of tropical forage legumes. Int J Syst Evol Microbiol 65(12):4424–4433. https://doi.org/10.1099/ijsem.0.000592

Depoorter E, Bull MJ, Peeters C et al (2016) *Burkholderia*: an update on taxonomy and biotechnological potential as antibiotic producers. Appl Microbiol Biotechnol 100(12):5215–5229. https://doi.org/10.1007/s00253-016-7520-x

Dias MP, Bastos MS, Xavier VB et al (2017) Plant growth and resistance promoted by *Streptomyces* spp. in tomato. Plant Physiol Biochem 118:479–493. https://doi.org/10.1016/j.plaphy.2017.07.017

Doley K, Jite PK (2013) Management of stem-rot of groundnut (*Arachis hypogaea* L.) cultivar in field. Notulae Scientia Biologicae 5:316–324

Doley K, Dudhane M, Borde M (2017) Biocontrol of *Sclerotium rolfsii* in groundnut by using microbial inoculants. Notulae Scientia Biologicae 9(1):124–130

Domazakis E, Du J, Liebrand TWH et al (2014) Potato SOBIR1 and SOBIR1-like interact with the elicitin-response receptor (ELR) of potato and are involved in the response to INF1 elicitin of *Phytophthora infestans*. In: Book of abstracts XVI international congress on molecular plant-microbe interactions, pp 117–117

Douriet-Gámez NR, Maldonado-Mendoza IE, Ibarra-Laclette E et al (2018) Genomic analysis of *Bacillus* sp. Strain B25, a biocontrol agent of maize pathogen *Fusarium verticillioides*. Curr Microbiol 75(3):247–255. https://doi.org/10.1007/s00284-017-1372-1

Dreyfus B, Garcia JL, Gillis M (1988) Characterization of *Azorhizobium caulinodans* gen. nov., sp. nov., a stem-nodulating nitrogen-fixing bacterium isolated from *Sesbania rostrata*. Int J Syst Evol Microbiol 38(1):89–98. https://doi.org/10.1099/00207713-38-1-89

Du J, Verzaux E, Chaparro-garcia A et al (2015) Elicitin recognition confers enhanced resistance to *Phytophthora infestans* in potato. Nat Plants 1:1–5

Du N, Shi L, Yuan Y et al (2017) Isolation of a potential biocontrol agent *Paenibacillus polymyxa* NSY50 from vinegar waste compost and its induction of host defense responses against *Fusarium* wilt of cucumber. Microbiol Res 202:1–10. https://doi.org/10.1016/j.micres.2017.04.013

Duponnois R, Garbaye J (1991) Mycorrhization helper bacteria associated with the Douglas fir-*Laccaria laccata* symbiosis: effects in aseptic and in glasshouse conditions. Ann des Sci For 48:239–251. https://doi.org/10.1051/forest:19910301

Dutta S, Podile AR (2010) Plant Growth Promoting Rhizobacteria (PGPR): the bugs to debug the root zone. Crit Rev Microbiol 36:232–244. https://doi.org/10.3109/10408411003766806

Eke P, Chatue GC, Wakam LN et al (2016) Mycorrhiza consortia suppress the fusarium root rot (*Fusarium solani* f. sp. *Phaseoli*) in common bean (*Phaseolus vulgaris* L.). Biol Control 103:240–250

El-Banna N, Winkelmann G (1998) Pyrrolnitrin from Burkholderia cepacia: antibiotic activity against fungi and novel activities against streptomycetes. J Appl Microbiol 85(1):69–78

El-fawy MM, El-sharkawy RMI, Abo-elyousr KAM (2018) Evaluation of certain *Penicillium frequentans* isolates against Cercospora leaf spot disease of sugar beet. Egypt J Biol Pest Control 28:49. https://doi.org/10.1186/s41938-018-0053-0

Elsharkawy MM, Shimizu M, Takahashi H et al (2012) The plant growth-promoting fungus *Fusarium equiseti* and the arbuscular mycorrhizal fungus *Glomus mosseae* induce systemic resistance against *Cucumber mosaic virus* in cucumber plants. Plant Soil 361:397–409

El-Tarabily KA, Sivasithamparam K (2006) Non-streptomycete actinomycetes as biocontrol agents of soil-borne fungal plant pathogens and as plant growth promoters. Soil Biol Biochem 38:1505–1520. https://doi.org/10.1016/j.soilbio.2005.12.017

Erdogan O, Benlioglu K (2010) Biological control of verticillium wilt on cotton by the use of fluorescent *Pseudomonas* spp. under field conditions. Biol Control 53:39–45. https://doi.org/10.1016/j.biocontrol.2009.11.011

Erwin DC et al (1996) Phytophthora diseases worldwide. American Phytopathological Society (APS Press), St Paul

Estrada-De Los Santos P, Rojas-Rojas FU, Tapia-García EY et al (2016) To split or not to split: an opinion on dividing the genus Burkholderia. Ann Microbiol 66(3):1303–1314. https://doi.org/10.1007/s13213-015-1183-1

Etesami H, Maheshwari DK (2018) Use of plant growth promoting rhizobacteria (PGPRs) with multiple plant growth promoting traits in stress agriculture: action mechanisms and future prospects. Ecotoxicol Environ Saf 156:225–246

Fernando WGD, Nakkeeran S, Zhang Y (2006) Biosynthesis of antibiotics by PGPR and its relation in biocontrol of plant diseases. PGPR Biocontrol Biofertil:67–109. https://doi.org/10.1007/1-4020-4152-7_3

Fraceto LF, Maruyama CR, Guilger M et al (2018) *Trichoderma harzianum* -based novel formulations: potential applications for management of Next-Gen agricultural challenges. J Chem Technol Biotechnol 28:303–325. https://doi.org/10.1002/jctb.5613

Frey-Klett P, Garbaye J, Tarkka M (2007) The mycorrhiza helper bacteria revisited. New Phytol 176:22–36. https://doi.org/10.1111/j.1469-8137.2007.02191.x

Fritz M, Jakobsen I, Lyngkjaer MF, Thordal-Christensen H, Pons-Kühnemann J (2006) Arbuscular mycorrhiza reduces susceptibility of tomato to *Alternaria solani*. Mycorrhiza 16:413–419

Fuentes-Ramírez LE, Bustillos-Cristales R, Tapia-Hernández A et al (2001) Novel nitrogen-fixing acetic acid bacteria, *Gluconacetobacter johannae* sp. nov. and *Gluconacetobacter azotocaptans* sp. nov., associated with coffee plants. Int J Syst Evol Microbiol 51(4):1305–1314. https://doi.org/10.1099/00207713-51-4-1305

Gangwar OP, Singh APK (2018) *Trichoderma* as an efficacious bioagent for combating biotic and abiotic stresses of wheat – a review. Agric Rev 39:48–54. https://doi.org/10.18805/ag.R-1770

Garmendia I, Aguirreolea J, Goicoechea N (2006) Defence-related enzymes in pepper roots during interactions with arbuscular mycorrhizal fungi and/or *Verticillium dahliae*. Biol Control 51:293–310

Gerst MM, Yesil M, Yousef AE (2018) Draft genome sequence of *Bacillus velezensis* OSY-S3, a producer of potent antimicrobial agents active against bacteria and fungi. Genome Announc 6(3):e01465–e01417. https://doi.org/10.1128/genomeA.01465-17

Geurts R, Bisseling T (2002) Rhizobium Nod factor perception and signalling. Plant Cell 14(1):239–249. https://doi.org/10.1105/tpc.002451

Gilchrist FJ, Sims H, Alcock A et al (2013) Is hydrogen cyanide a marker of *Burkholderia cepacia* complex? J Clin Microbiol JCM-02157

Giovannetti M, Avio L, Sbrana C (2010) Fungal spore germination and pre-symbiotic mycelial growth – physiological and genetic aspects. In: Koltai H, Kapulnik Y (eds) Arbuscular mycorrhizas: physiology and function. Springer, New York, pp 3–31

Giri B, Giang PH, Kumari R et al (2005) Microorganisms in soils: roles in genesis and functions. In: Buscot F, Varma A (eds) Soil biology. Springer, New York, pp 19–43

Gkarmiri K, Finlay RD, Alström S et al (2015) Transcriptomic changes in the plant pathogenic fungus *Rhizoctonia solani* AG-3 in response to the antagonistic bacteria *Serratia proteamaculans* and Serratia plymuthica. BMC Genomics 16:1–17. https://doi.org/10.1186/s12864-015-1758-z

Gómez L, Rodríguez M, de la Noval B et al (2008) Interacción entre el ecomic® y una población cubana de *Meloidogyne incognita* en tomate. Rev Protección Veg 23:90–98

Gond SK, Bergen MS, Torres et al (2015) Endophytic *Bacillus* spp. produce antifungal lipopeptides and induce host defence gene expression in maize. Microbiol Res 172:79–87. https://doi.org/10.1016/j.micres.2014.11.004

Gopalakrishnan S, Srinivas V, Alekhya G, Prakash B (2015) Effect of plant growth-promoting *Streptomyces* sp. on growth promotion and grain yield in chickpea (*Cicer arietinum* L). 3 Biotech 5:799–806. https://doi.org/10.1007/s13205-015-0283-8

Gopinathan S (1995) Biological control of *Rhizoctonia* sp. root rot of *Casuarina equisetifolia* seedlings by *Frankia* spp. strains. Biol Fertil Soils 20:221–225. https://doi.org/10.1007/BF00336081

Görgen CA, da Silveira Neto AN, Carneiro LC et al (2009) Controle do mofo-branco com palhada e *Trichoderma harzianum* 1306 em soja. Pesqui Agropec Bras 44:1583–1590. https://doi.org/10.1590/S0100-204X2009001200004

Goswami D, Patel K, Parmar S et al (2015a) Elucidating multifaceted urease producing marine *Pseudomonas aeruginosa* BG as a cogent PGPR and bio-control agent. Plant Growth Regul 75:253–263. https://doi.org/10.1007/s10725-014-9949-1

Goswami D, Parmar S, Vaghela H et al (2015b) Describing *Paenibacillus mucilaginosus* strain N3 as an efficient plant growth promoting rhizobacteria (PGPR). Cogent Food Agric 1(1). https://doi.org/10.1080/23311932.2014.1000714

Goswami D, Dhandhukia P, Thakker JN (2016) Expanding the horizons for the use of Paenibacillus species as PGPR for sustainable agriculture. In: Islam M, Rahman M, Pandey P, Jha C, Aeron A (eds) Bacilli and agrobiotechnology. Springer, Cham, pp 281–307. https://doi.org/10.1007/978-3-319-44409-3_12

Gotor-Vila A, Teixidó N, Casals C et al (2017) Biological control of brown rot in stone fruit using *Bacillus amyloliquefaciens* CPA-8 under field conditions. Crop Prot 102:72–80. https://doi.org/10.1016/j.cropro.2017.08.010

Grönemeyer JL, Hurek T, Bünger W, Reinhold-Hurek B (2016) *Bradyrhizobium vignae* sp. nov., a nitrogen-fixing symbiont isolated from effective nodules of *Vigna* and *Arachis*. Int J Syst Evol Microbiol 66(1):62–69. https://doi.org/10.1099/ijsem.0.000674

Gruau C, Trotel-Aziz P, Villaume S et al (2015) *Pseudomonas fluorescens* PTA-CT2 triggers local and systemic immune response against Botrytis cinerea in grapevine. Mpmi 28:1117–1129. https://doi.org/10.1094/MPMI-04-15-0092-R

Guerinot ML, Chelm BK (1984) Isolation and expression of the *Bradyrhizobium japonicum* adenylate cyclase gene (cya) in Escherichia coli. J Bacteriol 159(3):1068–1071

Gunderson JH, Elwood H, Ingold H, Kindle A, Sogin ML (1987) Phylogenetic relationships between chlororphytes, chrysophytes, and oomycetes. Proc Natl Acad Sci U S A 84:5823–5827

Guyer A, De Vrieze M, Bönisch D et al (2015) The anti-phytophthora effect of selected potato-associated *Pseudomonas* strains: from the laboratory to the field. Front Microbiol 6:1–13. https://doi.org/10.3389/fmicb.2015.01309

Gyaneshwar P, Hirsch AM, Moulin L et al (2011) Legume-nodulating betaproteobacteria: diversity, host range, and future prospects. Mol Plant Microb 24(11):1276–1288. https://doi.org/10.1094/MPMI-06-11-0172

Haas D, Défago G (2005) Biological control of soil-borne pathogens by fluorescent pseudomonads. Nat Rev Microbiol 3:307–319. https://doi.org/10.1038/nrmicro1129

Hadri AE, Bisseling T (1998) Responses of the plant to nod factors. In: Spaink HP, Kondorosi A, Hooykaas PJJ (eds) The rhizobiaceae. Springer, Dordrecht. https://doi.org/10.1007/978-94-011-5060-6_21

Hage-Ahmed K, Moyses A, Voglgruber A, Hadacek F, Steinkellne S (2013) Alterations in root exudation of intercropped tomato mediated by the arbuscular mycorrhizal fungus *Glomus mosseae* and the soil borne pathogen *Fusarium oxysporum* f.sp. *lycopersici*. J Phytopathol. https://doi.org/10.1111/jph.12130

Hagelin G, Oulie I, Raknes Å, Undheim K, Clausen OG (2004) Preparative high-performance liquid chromatographic separation and analysis of the Maltacine complex–a family of cyclic peptide antibiotics from *Bacillus subtilis*. J Chromatogr B 811(2):243–251. https://doi.org/10.1016/j.jchromb.2004.09.008

Haq IU, Zhang M, Yang P, van Elsas JD (2014) The interactions of bacteria with fungi in soil: emerging concepts. Adv Appl Microbiol 89:185–215. https://doi.org/10.1016/B978-0-12-800259-9.00005-6

Hassan MN, Afghan S, Hafeez FY (2011) Biological control of red rot in sugarcane by native pyoluteorin-producing *Pseudomonas putida* strain NH-50 under field conditions and its potential modes of action. Pest Manag Sci 67:1147–1154. https://doi.org/10.1002/ps.2165

Hellberg JEEU, Matilla MA, Salmond GPC (2015) The broad-spectrum antibiotic, zeamine, kills the nematode worm *Caenorhabditis elegans*. Front Microbiol 6:1–13. https://doi.org/10.3389/fmicb.2015.00137

Helman Y, Burdman S, Okon Y (2011) Plant growth promotion by rhizosphere bacteria through direct effects. In: Beneficial microorganisms in multicellular life forms. Springer, Berlin, pp 89–103

Hemavani C, Thippeswamy B (2014) Effect of arbuscular mycorrhizal fungus, *Acaulospora lacunosa* on growth of groundnut disease caused by *Cercospora arachidicola*. Int J Res App Nat S Sci 2(4):57–60

Hermosa R, Viterbo A, Chet I, Monte E (2012) Plant-beneficial effects of *Trichoderma* and of its genes. Microbiology 158:17–25

Hinarejos E, Castellano M, Rodrigo I et al (2016) *Bacillus subtilis* IAB/BS03 as a potential biological control agent. Eur J Plant Pathol 146:597–608. https://doi.org/10.1007/s10658-016-0945-3

Ho YN, Chiang HM, Chao CP et al (2015) In planta biocontrol of soilborne Fusarium wilt of banana through a plant endophytic bacterium, *Burkholderia cenocepacia* 869T2. Plant Soil 387(1–2):295–306. https://doi.org/10.1007/s11104-014-2297-0

Hohmann P, Messmer MM (2017) Breeding for mycorrhizal symbiosis: focus on disease resistance. Euphytica 213:113. https://doi.org/10.1007/s10681-017-1900-x

Hong DN, Posta K (2018) Mycorrhiza-induced alleviation of plant disease caused by *Clavibacter michiganensis* subsp. michiganensis and role of ethylene in mycorrhiza-induced resistance in tomato. Acta Biol Hung 69:170–181. https://doi.org/10.1556/018.69.2018.2.6

Hu X, Roberts DP, Xie L et al (2014) Formulations of *Bacillus subtilis* BY-2 suppress *Sclerotinia sclerotiorum* on oilseed rape in the field. Biol Control 70:54–64. https://doi.org/10.1016/j.biocontrol.2013.12.005

Hungria M, Campo RJ, Souza EM, Pedrosa FO (2010) Inoculation with selected strains of *Azospirillum brasilense* and *A. lipoferum* improves yields of maize and wheat in Brazil. Plant Soil 331(1-2):413–425. https://doi.org/10.1007/s11104-009-0262-0

Ikeda S, Shimizu A, Shimizu M, Takahashi H, Takenaka S (2012) Biocontrol of black scurf on potato by seed tuber treatment with Pythium oligandrum. Biol Control 60(3):297–304

Intorne AC, de Oliveira MVV, Lima ML et al (2009) Identification and characterization of *Gluconacetobacter diazotrophicus* mutants defective in the solubilization of phosphorus and zinc. Arch Microbiol 191(5):477–483. https://doi.org/10.1007/s00203-009-0472-0

Irorere VU, Smyth TJ, Cobice D et al (2018) Fatty acid synthesis pathway provides lipid precursors for rhamnolipid biosynthesis in *Burkholderia thailandensis* E264. Appl Microbiol Biotechnol:1–12. https://doi.org/10.1007/s00253-018-9059-5

Islam S, Akanda AM, Prova A et al (2014) Growth promotion effect of *Fusarium* spp. PPF1 from bermudagrass (*Cynodon dactylon*) rhizosphere on Indian spinach (*Basella alba*) seedlings are linked to root colonization. Arch Phytopathol Plant Protect 47:2319–2331

Issa M, Ngakou A, Haouvangh CL, Nukenine EN (2017) Potentials of arbuscular mycorrhiza fungi (AMF) and Neem (*Azadirachta indica* A. Juss.) leaves extract as biological control agents against the sweet potato weevil (*Cylas puncticollis* Boh.) in two agro-ecological zones of Cameroon. J Exp Agric Int 17(1):1–13

Jacott C, Murray J, Ridout C (2017) Trade-offs in arbuscular mycorrhizal symbiosis: disease resistance, growth responses and perspectives for crop breeding. Agronomy 7:75. https://doi.org/10.3390/agronomy7040075

Jaizme-Vega MC, Rodríguez-Romero AS, Núñez LAB (2005) Effect of the combined inoculation of arbuscular mycorrhizal fungi and plant growth-promoting rhizobacteria on papaya (*Carica papaya* L.) infected with the root-knot nematode *Meloidogyne incognita*. Fruits 61:151–162

Jiang RHY, Tyler BM (2012) Mechanisms and evolution of virulence in oomycetes. Annu Rev Phytopathol 50:295–318

Jiao YS, Yan H, Ji ZJ et al (2015) *Rhizobium sophorae* sp. nov. and *Rhizobium sophoriradicis* sp. nov., nitrogen-fixing rhizobial symbionts of the medicinal legume *Sophora flavescens*. Int J Syst Evol Microbiol 65(2):497–503. https://doi.org/10.1099/ijs.0.068916-0

Jung SC, Martinez-Medina A, Lopez-Raez JA, Pozo MJ (2012) Mycorrhiza-induced resistance and priming of plant defenses. J Chem Ecol 38:651–664

Jung BK, Hong SJ, Park GS et al (2018) Isolation of *Burkholderia cepacia* JBK9 with plant growth-promoting activity while producing pyrrolnitrin antagonistic to plant fungal diseases. Appl Biol Chem 61(2):173–180. https://doi.org/10.1007/s13765-018-0345-9

Kamoun S, Young M, Glascock CB, Tyler BM (1993) Extracellular protein elicitors from Phytophthora: host-specificity and induction of resistance to bacterial and fungal phytopathogens. Mol Plant-Microbe Interact 6:15–25

Kamoun S, vanWest P, deJong AJ, deGroot KE, Vleeshouwers VGAA, Govers F (1997) A gene encoding a protein elicitor of *Phytophthora infestans* is down-regulated during infection of potato. Mol Plant-Microbe Interact 10:13–20

Karagiannidis N, Bletsos F, Stavropoulos N (2002) Effect of Verticillium wilt (Verticillium dahliae Kleb.) and mycorrhiza (Glomus mosseae) on root colonization, growth and nutrient uptake in tomato and eggplant seedlings. Sci Hortic 94:145–159

Keller H, Blein J, Bonnet P, Ricci P (1996) Physiological and molecular characteristics of elicitin-induced systemic acquired resistance in tobacco. Plant Physiol 110(2):365–376

Khaldi RE, Daami-remadi M, Hamada W et al (2015) Plant Pathology & Microbiology the Potential of Serratia marcescens : an indigenous strain isolated from date palm compost as biocontrol agent of Rhizoctonia solani on potato. Plant Pathol Microbiol 3:3–5. https://doi.org/10.4172/2157-7471.S3-006

Kim SK, Kim YC, Lee S et al (2011) Insecticidal activity of rhamnolipid isolated from *Pseudomonas* sp. EP-3 against green peach aphid (*Myzus persicae*). J Agric Food Chem 59:934–938. https://doi.org/10.1021/jf104027x

Klein JM, Loper JE, Stockwell VO (2017) Nutritional environment influences transcription of a Pantocin a biosynthesis gene in *Pantoea vagans* strain C9-1. J Plant Pathol 99:99–103

Kloepper JW, Ryu CM, Zhang S (2004) Induced systemic resistance and promotion of plant growth by *Bacillus* spp. Phytopathology 94(11):1259–1266. https://doi.org/10.1094/PHYTO.2004.94.11.1259

Kobra N, Jalil K, Youbert G (2009) Effects of three glomus species as biocontrol agents against *Verticillium*-induced wilt in cotton. J Plant Protect 49(2):185–189

Krishnaraj PU (2017) Effect of Actinobacteria and *Glomus fasiculatum* against *Fusarium oxysporum* f. sp. lycopersici in tomato plant. Int J Curr Microbiol App Sci 6:488–498. https://doi.org/10.20546/ijcmas.2017.609.059

Kulimushi PZ, Basime GC, Nachigera GM et al (2017a) Efficacy of *Bacillus amyloliquefaciens* as biocontrol agent to fight fungal diseases of maize under tropical climates: from lab to field assays in south Kivu. Environ Sci Pollut Res:1–14. https://doi.org/10.1007/s11356-017-9314-9

Kulimushi PZ, Arias AA, Franzil L et al (2017b) Stimulation of fengycin-type antifungal lipopeptides in *Bacillus amyloliquefaciens* in the presence of the maize fungal pathogen *Rhizomucor variabilis*. Front Microbiol 8. https://doi.org/10.3389/fmicb.2017.00850

Kumar G, Maharshi A, Patel J et al (2017) *Trichoderma*: a potential fungal antagonist to control plant diseases. SATSA (MB)

Lamour K, Sophie K (2009) Oomycete genetics and genomics: diversity, interactions and research tools. Wiley-Blackwell, Hoboken

Lapidot D, Dror R, Vered E et al (2015) Disease protection and growth promotion of potatoes (*Solanum tuberosum* L.) by *Paenibacillus dendritiformis*. Plant Pathol 64:545–551. https://doi.org/10.1111/ppa.12285

Lareen A, Burton F, Schäfer P (2016) Plant root-microbe communication in shaping root microbiomes. Plant Mol Biol 90(6):575–587. https://doi.org/10.1007/s11103-015-0417-8

Le Floch G, Rey P, Déniel F et al (2003) Enhancement of development and induction of resistance in tomato plants root colonization by a biocontrol agent, *Pythium oligandrum*. Plant Physiol Biochem 43:1–11

Le Floch G, Vallance J, Benhamou N, Rey P (2009) Combining the oomycete *Pythium oligandrum* with two other antagonistic fungi: root relationships and tomato grey mold biocontrol. Biol Control 50:288–298

Le Floch G, Benhamou N, Mamaca E et al (2005) Characterisation of the early events in atypical tomato root colonisation by a biocontrol agent, Pythium oligandrum. Plant Physiol Biochem 43(1):1–11

Li Y, Gu Y, Li J et al (2015) Biocontrol agent *Bacillus amyloliquefaciens* LJ02 induces systemic resistance against cucurbits powdery mildew. Front Microbiol 6:883. https://doi.org/10.3389/fmicb.2015.00883

Lindström K (1989) *Rhizobium galegae*, a new species of legume root nodule bacteria. Int J Syst Evol Microbiol 39(3):365–367. https://doi.org/10.1099/00207713-39-3-365

Ling N, Xue C, Huang Q et al (2010) Development of a mode of application of bioorganic fertilizer for improving the biocontrol efficacy to fusarium wilt. BioControl 55:673–683. https://doi.org/10.1007/s10526-010-9290-1

Lioussanne L, Jolicoeur M, St-Arnaud M (2007) Mycorrhizal colonization with *Glomus intraradices* and development stage of transformed tomato roots significantly modify the chemotactic response of zoospores of the pathogen *Phytophthora nicotianae*. Soil Biol Biochem 40:2217–2224

Logeshwarn P, Thangaraju M, Rajasundari K (2011) Antagonistic potential of *Gluconacetobacter diazotrophicus* against *Fusarium oxysporum* in sweet potato (Ipomea batatus). Arch Phytopathol Plant Protect 44(3):216–223. https://doi.org/10.1080/03235400902952707

Lounaci L, Guemouri-Athmani S, Boureghda et al (2017) Suppression of crown and root rot of wheat by the rhizobacterium *Paenibacillus polymyxa*. Phytopathol Mediterr 55(3):355–365. https://doi.org/10.14601/Phytopathol_Mediterr-18055

Lu SE, Novak J, Austin et al (2009) Occidiofungin, a unique antifungal glycopeptide produced by a strain of *Burkholderia contaminans*. Biochemist 48(35):8312–8321. https://doi.org/10.1021/bi900814c

Ludwig-Müller J (2010) Hormonal responses in host plants triggered by arbuscular mycorrhizal fungi. In: Koltai H, Kapulnik Y (eds) Arbuscular mycorrhizas: physiology and function. Springer, New York, pp 169–198

Ma X, Wang X, Cheng J et al (2015) Microencapsulation of *Bacillus subtilis* B99-2 and its biocontrol efficiency against *Rhizoctonia solani* in tomato. Biol Control 90:34–41. https://doi.org/10.1016/j.biocontrol.2015.05.013

Ma L, Chao S, Ti Z et al (2018) Effect of nicotine from tobacco root exudates on chemotaxis , growth , biocontrol efficiency, and colonization by *Pseudomonas aeruginosa* NXHG29:1237–1257. https://doi.org/10.1007/s10482-018-1035-7

Machado DFM, Parzianello FR, Silva ACF et al (2012) *Trichoderma* no Brasil: o fungo e o bioagente. Revista de Ciências Agrárias 26:274–288

Magalhães VC, Barbosa LO, Andrade JP et al (2017) *Burkholderia* isolates from a sand dune leaf litter display biocontrol activity against the bole rot disease of *Agave sisalana*. Biol Control 112:41–48. https://doi.org/10.1016/j.biocontrol.2017.06.005

Majumdar SK, Bose SK (1958) Mycobacillin, a new antifungal antibiotic produced by B. subtilis. Nature 181(4602):134. https://doi.org/10.1038/181134a0

Malik RJ, Dixon MH, Bever JD (2016) Mycorrhizal composition can predict foliar pathogen colonization in soybean. Biol Control 103:46–53

Margulis L, Schwartz KV (2000) Five kingdoms: an illustrated guide to the phyla of life on earth. W. H. Freeman and Co., New York

Marín A, Atarés L, Chiralt A (2017) Improving function of biocontrol agents incorporated in antifungal fruit coatings: a review. Biocontrol Sci Tech 27:1220–1241. https://doi.org/10.1080/09583157.2017.1390068

Marro N, Lax P, Cabello M et al (2014) Use of the arbuscular mycorrhizal fungus *Glomus intraradices* as biological control agent of the nematode *Nacobbus aberrans* parasitizing tomato. Braz Arch Biol Technol 57(5):668–674

Martínez-Hidalgo P, García JM, Pozo MJ (2015) Induced systemic resistance against Botrytis cinerea by *Micromonospora* strains isolated from root nodules. Front Microbiol 6:1–11. https://doi.org/10.3389/fmicb.2015.00922

Martinez-Medina A, Roldán A, Albacete A et al (2011) The interaction with arbuscular mycorrhizal fungi or *Trichoderma harzianum* alters the shoot hormonal profile in melon plants. Phytochemistry 72:223–229

Martínez-Medina A, Roldán A, Pascual JA (2011) Interaction between arbuscular mycorrhizal fungi and *Trichoderma harzianum* under conventional and low input fertilization field condition in melon crops: Growth response and fusarium wilt biocontrol. Appl Soil Ecol 47:98–105. https://doi.org/10.1016/j.apsoil.2010.11.010

Marzani QA, Fatah KM, Mustafa MH (2017) Field efficacy of the commercial formulation of the antagonistic *Trichoderma harzianum* on chickpea wilt caused by *Fusarium oxysporum*. Sci J Univ Zakho 5:64–69

Masunaka A, Nakaho K, Sakai M, Takahashi H, Takenaka S (2009) Visualization of *Ralstonia solanacearum* cells during biocontrol of bacterial wilt disease in tomato with Pythium oligandrum. J Gen Plant Pathol 75(4):281–287

Mhlongo MI, Piater LA, Madala NE et al (2018) The chemistry of plant–microbe interactions in the rhizosphere and the potential for metabolomics to reveal signaling related to defense priming and induced systemic resistance. Front Plant Sci 9:1–17. https://doi.org/10.3389/fpls.2018.00112

Minerdi D, Bossi S, Maffei ME et al (2011) *Fusarium oxysporum* and its bacterial consortium promote lettuce growth and expansin A5gene expression through microbial volatile organic compound (MVOC) emission. FEMS Microbiol Ecol 76:342–351

Minton MM, Barber NA, Gordon L (2016) Effects of arbuscular mycorrhizal fungi on herbivory defense in two *Solanum* (Solanaceae) species. Plant Ecol Evol 149(2):157–164

Mishra S, Jagadeesh KS, Krishnaraj PU, Prem S (2014) Biocontrol of tomato leaf curl virus (ToLCV) in tomato with chitosan supplemented formulations of *Pseudomonas* sp. under field conditions. Aust J Crop Sci 8:347–355

Miransari, M. (2011) Interactions between arbuscular mycorrhizal fungi and soil bacteria. Appl Microbiol Biotechnol 89: 917. https://doi.org/10.1007/s00253-010-3004-6

Mnif I, Hammami I, Triki MA et al (2015) Antifungal efficiency of a lipopeptide biosurfactant derived from *Bacillus subtilis* SPB1 versus the phytopathogenic fungus, Fusarium solani. Environ Sci Pollut Res 22(22):18137–18147. https://doi.org/10.1007/s11356-015-5005-6

Mohamed N, Lherminier J, Farmer M et al (2007) Defense responses in grapevine leaves against *Botrytis cinerea* induced by application of a *Pythium oligandrum* strain or Its elicitin, oligandrin, to Roots. Phytopathology 97(5):611–620

Mohandas S, Poovarasan S, Panneerselvam P et al (2013) Guava (*Psidium guajava* L.) rhizosphere *Glomus mosseae* spores harbor actinomycetes with growth promoting and antifungal attributes. Sci Hortic (Amsterdam) 150:371–376. https://doi.org/10.1016/j.scienta.2012.11.019

Moore D, Robson GD, Trinci AP (2011) 21st century guidebook to fungi with CD. Cambridge University Press, Cambridge

Moradi M, Nejad FJ, Bonjar GHS et al (2018) Efficacy of *Bacillus subtilis* native strains for biocontrol of *Phytophthora* crown and root rot of pistachio in Iran. Trop. Plant Pathol 43:306. https://doi.org/10.1007/s40858-018-0226-0

Munhoz LD, Fonteque JP, Santos IMO et al (2017) Control of bacterial stem rot on tomato by extracellular bioactive compounds produced by *Pseudomonas aeruginosa* LV strain. Cogent Food Agric 3:1–16. https://doi.org/10.1080/23311932.2017.1282592

Murali M, Amruthesh KN (2015) Plant growth-promoting fungus *Penicillium oxalicum* enhances plant growth and induces resistance in pearl millet against downy mildew disease. J Phytopathol 163:743–754. https://doi.org/10.1111/jph.12371

Nagao JI, Asaduzzaman SM, Aso Y et al (2006) Lantibiotics: insight and foresight for new paradigm. J Biosci Bioeng 102(3):139–149. https://doi.org/10.1263/jbb.102.139

Nagaraja H, Chennappa G, Rakesh S et al (2016) Antifungal activity of *Azotobacter nigricans* against trichothecene-producing *Fusarium* species associated with cereals. Food Sci Biotechnol 25(4):1197–1204. https://doi.org/10.1007/s10068-016-0190-8

Naing KW, Nguyen XH, Anees M et al (2015) Biocontrol of Fusarium wilt disease in tomato by *Paenibacillus ehimensis* KWN38. World J Microbiol Biotechnol 31(1):165–174. https://doi.org/10.1007/s11274-014-1771-4

Nakano MM, Marahiel MA, Zuber P (1988) Identification of a genetic locus required for biosynthesis of the lipopeptide antibiotic surfactin in *Bacillus subtilis*. J Bacteriol 170(12):5662–5668. https://doi.org/10.1128/jb.170.12.5662-5668.1988

Narisawa K, Tokumasu S, Hashiba T (1998) Suppression of clubroot formation in Chinese cabbage by the root endophytic fungus, *Heteroconium chaetospira*. Plant Pathol 47:206–210

Nassar AH, El-Tarabily KA, Sivasithamparam K (2003) Growth promotion of bean (*Phaseolus vulgaris* L.) by a polyamine-producing isolate of *Streptomyces griseoluteus*. Plant Growth Regul 40:97–106. https://doi.org/10.1023/A:1024233303526

Navarro MOP, Simionato AS, Barazetti AR et al (2017) Disease-induced resistance and plant immunization using microbes. In: Singh DP, Singh HB, Prabha R (eds) Plant-microbe interactions in agro-ecological perspectives, vol 1. Springer, Singapore, pp 447–465

Neeraj, Singh K (2011) Organic amendments to soil inoculated arbuscular mycorrhizal fungi and *Pseudomonas fluorescens* treatments reduce the development of root-rot disease and enhance the yield of *Phaseolus vulgaris* L. Eur J Soil Biol 47:288–295. https://doi.org/10.1016/j.ejsobi.2011.07.002

Neumann E, George E (2010) Nutrient uptake: the arbuscular mycorrhiza fungal symbiosis as a plant nutrient acquisition strategy. In: Koltai H, Kapulnik Y (eds) Arbuscular mycorrhizas: physiology and function. Springer, New York, pp 137–167

Nick G, de Lajudie P, Eardly BD et al (1999) *Sinorhizobium arboris* sp. nov. and *Sinorhizobium kostiense* sp. nov., isolated from leguminous trees in Sudan and Kenya. J Syst Evol Micr 49(4):1359–1368. https://doi.org/10.1099/00207713-49-4-1359

Nour SM, Fernandez MP, Normand P, Cleyet-Marel JC (1994) *Rhizobium ciceri* sp. nov., consisting of strains that nodulate chickpeas (*Cicer arietinum* L.). Int J Syst Evol Microbiol 44(3):511–522. https://doi.org/10.1099/00207713-44-3-511

Odeyemi I, Afolami S, Sosanya O (2010) Effect of *Glomus Mosseae* (arbuscular mycorrhizal fungus) on host – parasite relationship of *Meloidogyne incognita* (southern root-knot nematode) on four improved cowpea varieties. J Plant Prot 50:321–325. https://doi.org/10.2478/v10045-010-0055-8

Olaniyi MO (2014) Effects of mycorrhizal inoculant and organic mulches on nematode damage to cooking banana. J Biol Agric Health 4:81–86

Olawuyi OJ, Odebode AC, Oyewole IO et al (2014) Effect of arbuscular mycorrhizal fungi on *Pythium aphanidermatum* causing foot rot disease on pawpaw (*Carica papaya* L.) seedlings. Arch Phytopathol Plant Protect 47:185–193. https://doi.org/10.1080/03235408.2013.806079

Oßwald W, Fleischmann F, Rigling D et al (2000) Cytological effects of cellulases in the parasitism of *Phytophthora parasitica* by *Pythium oligandrum*. Appl Environ Microbiol 66:4305–4314

Ouyang Z, Li X, Huang L, Hong Y, Zhang Y, Zhang H, Song F (2015) Elicitin-like proteins Oli-D1 and Oli-D2 from *Pythium oligandrum* trigger hypersensitive response in *Nicotiana benthamiana* and induce resistance against *Botrytis cinerea* in tomato. Mol Plant Pathol 16(3):238–250

Ozgonen H, Erkilic A (2007) Growth enhancement and phytophthora blight (*Phytophthora capsici Leonian*) control by arbuscular mycorrhizal fungal inoculation in pepper. Crop Prot 26:1682–1688. https://doi.org/10.1016/j.cropro.2007.02.010

Pahari A, Mishra BB (2017) Antibiosis of siderophore producing bacterial isolates against phytopathogens and their effect on growth of okra. Int J Curr Microbiol App Sci 6:1925–1929

Pandey P, Kang SC, Maheshwari DK (2005) Isolation of endophytic plant growth promoting *Burkholderia* sp. MSSP from root nodules of *Mimosa pudica*. Curr Sci:177–180

Patekoski S, Pires-Zottarelli CLA (2010) Patogenicidade de *Pythium aphanidermatum* a alface cultivada em hidroponia e seu biocontrole com *Trichoderma*. Pesq Agropec Bras 45:805–810

Pastor N, Masciarelli O, Fischer S et al (2016) Potential of Pseudomonas putida PCI2 for the protection of tomato plants against fungal pathogens. Curr Microbiol 73:346–353. https://doi.org/10.1007/s00284-016-1068-y

Paulsen IT, Press CM, Ravel J et al (2005) Complete genome sequence of the plant commensal *Pseudomonas fluorescens* Pf-5. Nat Biotechnol 23:873–878. https://doi.org/10.1038/nbt1110

Péchy-Tarr M, Bruck DJ, Maurhofer M et al (2008) Molecular analysis of a novel gene cluster encoding an insect toxin in plant-associated strains of *Pseudomonas fluorescens*. Environ Microbiol 10:2368–2386. https://doi.org/10.1111/j.1462-2920.2008.01662.x

Pedro EAS, Harakava R, Lucon CMM et al (2012) Promoção do crescimento do feijoeiro e controle da antracnose por *Trichoderma* spp. Pesq Agropec Bras 47:1589–1595

Pierson LS, Pierson EA (2010) Metabolism and function of phenazines in bacteria: impacts on the behavior of bacteria in the environment and biotechnological processes. Appl Microbiol Biotechnol 86(6):1659–1670. https://doi.org/10.1007/s00253-010-2509-3

Ponmurugan K, Sankaranarayanan A, Al-Dharbi NA (2012) Biological activities of plant growth promoting *Azotobacter* sp. isolated from vegetable crops rhizosphere soils. J Pure Appl Microbiol 6(4):1–10

Poovarasan S, Mohandas S, Paneerselvam P et al (2013) Mycorrhizae colonizing actinomycetes promote plant growth and control bacterial blight disease of pomegranate (*Punica granatum* L. cv Bhagwa). Crop Prot 53:175–181. https://doi.org/10.1016/J.CROPRO.2013.07.009

Porcel R, Aroca R, Azcón R, Ruiz-Lozano JM (2006) PIP aquaporin gene expression in arbuscular mycorrhizal *Glycine max* and *Lactuca sativa* plants in relation to drought stress tolerance. Plant Mol Biol 60:389–404

Pozo MJ, Cordier C, Dumas-Gaudot E et al (2002) Localized versus systemic effect of arbuscular mycorrhizal fungi on defense to *Phytophthora infections* in tomato plants. J Exp Bot 53(368):525–534

Pozo MJ, Jung SC, López-Ráez JA, Azcón-Aguilar C (2010) Impact of arbuscular mycorrhizal symbiosis on plant response to biotic stress: the role of plant defense mechanisms. In: Koltai H, Kapulnik Y (eds) Arbuscular mycorrhizas: physiology and function. Springer, New York, pp 193–208

Pozo MJ, Lopez-R JÁ, Azcon-Aguilar C, Garcia-Garrido JM (2015) Phytohormones as integrators of environmental signals in the regulation of mycorrhizal symbioses. New Phytol 205:1431–1436

Rakin A, Schneider L, Podladchikova O (2012) Hunger for iron: the alternative siderophore iron scavenging systems in highly virulent Yersinia. Front Cell Infect Microbiol 2:1–7. https://doi.org/10.3389/fcimb.2012.00151

Reiss A, Jørgensen LN (2017) Biological control of yellow rust of wheat (*Puccinia striiformis*) with Serenade®ASO (*Bacillus subtilis* strain QST713). Crop Prot 93:1 8. https://doi.org/10.1016/j.cropro.2016.11.009

Rey P, Le Floch G, Benhamou N, Salerno MI, Thuillier E, Tirilly Y (2005) Interactions between the mycoparasite Pythium oligandrum and two types of sclerotia of plant-pathogenic fungi. Mycol Res 109(7):779–788

Rey P, Agro BS, Le Floch G (2008) *Pythium oligandrum* biocontrol: its relationships with fungi and plants. Plant Microbe-Interact:47–53

Ribeiro RA, Martins TB, Ormeño-Orrillo E et al (2015) *Rhizobium ecuadorense* sp. nov., an indigenous N2-fixing symbiont of the Ecuadorian common bean (*Phaseolus vulgaris* L.) genetic pool. Int J Syst Evol Microbiol 65(9):3162–3169. https://doi.org/10.1099/ijsem.0.000392

Ricci P (1997) Induction of the hypersensitive response and systemic acquired resistance by fungal proteins: the case of elicitins. Plant-Microbe Interact 3:53–75

Ricci P, Bonnet P, Huet J-C et al (1989) Structure and activity of proteins from pathogenic fungi Phytophthora eliciting necrosis and acquired resistance in tobacco. Eur J Biochem 183:555–563

Rojas-Rojas FU, López-Sánchez D, Meza-Radilla G et al (2018) El controvertido complejo Burkholderia cepacia, un grupo de especies promotoras del crecimiento vegetal y patógenas de plantas, animales y humanos. Rev Argent Microbiol. https://doi.org/10.1016/j.ram.2018.01.002

Rome S, Fernandez MP, Brunel B et al (1996) *Sinorhizobium medicae* sp. nov., isolated from annual *Medicago* spp. J Syst Evol Micr 46(4):972–980. https://doi.org/10.1099/00207713-46-4-972

Roy S, Khasa DP, Greer CW (2007) Combining alders, frankiae, and mycorrhizae for the revegetation and remediation of contaminated ecosystems. Can J Bot 85:237–251. https://doi.org/10.1139/B07-017

Rozi P, Rasulov BA, Waili A et al (2018) Isolation, characterization and antimicrobial activity of extracellular polypeptides from diazotrophic strain *Rhizobium radiobacter*. Chem Nat Compd 54(2):330–333. https://doi.org/10.1007/s10600-018-2334-y

Ruiz-Lozano JM, Aroca R (2010) Host response to osmotic stresses: stomatal behaviour and water use efficiency of arbuscular mycorrhizal plants. In: Koltai H, Kapulnik Y (eds) Arbuscular mycorrhizas: physiology and function. Springer, New York, pp 239–254

Sadeghi A, Karimi E, Dahaji PA et al (2012) Plant growth promoting activity of an auxin and siderophore producing isolate of *Streptomyces* under saline soil conditions. World J Microbiol Biotechnol 28:1503–1509. https://doi.org/10.1007/s11274-011-0952-7

Saif S, Khan MS, Zaidi A, Ahmad E (2014) Role of phosphate-solubilizing actinomycetes in plant growth promotion: current perspective. In: Phosphate solubilizing microorganisms. Springer, Cham, pp 137–156

Saldajeno MGB, Hyakumachi M (2011) The plant growth-promoting fungus *Fusarium equiseti* and the arbuscular mycorrhizal fungus *Glomus mosseae* stimulate plant growth and reduce severity of anthracnose and damping-off diseases in cucumber (*Cucumis sativus*) seedlings. Ann Appl Biol 159:28–40. https://doi.org/10.1111/j.1744-7348.2011.00471.x

Saleem M, Law AD, Sahib MR et al (2018) Impact of root system architecture on rhizosphere and root microbiome. Rhizosphere 6:47–51. https://doi.org/10.1016/j.rhisph.2018.02.003

Salla TD, Ramos da Silva T, Astarita LV, Santarém ER (2014) Streptomyces rhizobacteria modulate the secondary metabolism of Eucalyptus plants. Plant Physiol Biochem 85:14–20. https://doi.org/10.1016/j.plaphy.2014.10.008

Salomon MV, Funes Pinter I, Piccoli P, Bottini R (2017) Use of plant growth-promoting rhizobacteria as biocontrol agents: induced systemic resistance against biotic stress in plants. In: Kalia V (ed) Microbial applications, 2. Springer, Cham. https://doi.org/10.1007/978-3-319-52669-0_7

Sanjay R, Ponmurugan P, Baby UI (2008) Evaluation of fungicides and biocontrol agents against grey blight disease of tea in the field. Crop Prot 27:689–694. https://doi.org/10.1016/j.cropro.2007.09.014

Sawana A, Adeolu M, Gupta RS (2014) Molecular signatures and phylogenomic analysis of the genus *Burkholderia*: proposal for division of this genus into the emended genus *Burkholderia* containing pathogenic organisms and a new genus Paraburkholderia gen. nov. harboring environmental species. Front Genet:5. https://doi.org/10.3389/fgene.2014.00429

Scholla MH, Elkan GH (1984) *Rhizobium fredii* sp. nov., a fast-growing species that effectively nodulates soybeans. Int J Syst Evol Microbiol 34(4):484–486. https://doi.org/10.1099/00207713-34-4-484

Schouteden N, De Waele D, Panis B, Vos CM (2015) Arbuscular mycorrhizal fungi for the biocontrol of plant-parasitic nematodes: a review of the mechanisms involved. Front Microbiol 6:1–12. https://doi.org/10.3389/fmicb.2015.01280

Seenivasan N, David PMM, Vivekanandan P, Samiyappan R (2012) Biological control of rice root-knot nematode, *Meloidogyne graminicola* through mixture of *Pseudomonas fluorescens* strains. Biocontrol Sci Tech 22:611–632. https://doi.org/10.1080/09583157.2012.675052

Segovia L, Young JPW, Martínez-Romero E (1993) Reclassification of American *Rhizobium leguminosarum* biovar phaseoli type I strains as *Rhizobium etli* sp. nov. Int J Syst Evol Microbiol 43(2):374–377. https://doi.org/10.1099/00207713-43-2-374

Selosse MA, Bessis A, Pozo MJ (2014) Microbial priming of plant and animal immunity: symbionts as developmental signals. Trends Microbiol 22:607–613

Shah S, Karkhanis V, Desai A (1992) Isolation and characterization of siderophore, with antimicrobial activity, from *Azospirillum lipoferum* M. Curr Microbiol 25(6):347–351. https://doi.org/10.1007/BF01577233

Shaikh SS, Sayyed RZ (2015) Role of plant growth-promoting rhizobacteria and their formulation in biocontrol of plant diseases. In: Arora NK (ed) Plant microbes symbiosis: applied facets. Springer, New Delhi

Shao J, Xu Z, Zhang N et al (2015) Contribution of indole-3-acetic acid in the plant growth promotion by the rhizospheric strain *Bacillus amyloliquefaciens* SQR9. Biol Fertil Soils 51(3):321–330. https://doi.org/10.1007/s00374-014-0978-8

Sharkawy HHAE, Tohamey S, Khalil AA (2015) Combined effects of *Streptomyces viridosporus* and *Trichoderma harzianum* on controlling wheat leaf rust caused by *Puccinia triticina*. Plant Pathol J 14:182–188. https://doi.org/10.3923/ppj.2015.182.188

Sharma S, Prasad R et al (2017) Glycoprotein associated with *Funneliformis coronatum*, *Gigaspora margarita* and *Acaulospora scrobiculata* suppress the plant pathogens *in vitro*. Asian J Plant Pathol 11(4):199–202

Shen L, Wang F, Yang J et al (2014) Control of tobacco mosaic virus by *Pseudomonas fluorescens* CZ powder in greenhouses and the field. Crop Prot 56:87–90. https://doi.org/10.1016/j.cropro.2013.11.020

Shoji JI, Hinoo H, Kato et al (1990) Isolation of cepafungins I, II and III from *Pseudomonas* species. J Antibiot 43(7):783–787. https://doi.org/10.7164/antibiotics.43.783

Siddiqui Z, Akhtar S (2008) Synergistic effects of antagonistic fungi and a plant growth promoting rhizobacterium, an arbuscular mycorrhizal fungus, or composted cow manure on populations of *Meloidogyne incognita* and growth of tomato. Biocontrol Sci Tech 18:279–290

Siddiqui Z, Futai K (2009) Biocontrol of *Meloidogyne incognita* on tomato using antagonistic fungi, plant-growth-promoting rhizobacteria and cattle manure. Pest Manag Sci 65:943–948

da Silva Vasconcellos FC, de Oliveira AG, Lopes-Santos L et al (2014) Evaluation of antibiotic activity produced by *Pseudomonas aeruginosa* LV strain against Xanthomonas arboricola pv. Pruni. Agric Sci 05:71–76. https://doi.org/10.4236/as.2014.51008

Simionato AS, Navarro MOP, de Jesus MLA et al (2017) The effect of phenazine-1-carboxylic acid on mycelial growth of *Botrytis cinerea* produced by *Pseudomonas aeruginosa* LV strain. Front Microbiol 8. https://doi.org/10.3389/fmicb.2017.01102

Singh RP, Jha PN (2016) The multifarious PGPR *Serratia marcescens* CDP-13 augments induced systemic resistance and enhanced salinity tolerance of wheat (*Triticum aestivum* L.). PLoS One 11:1–24. https://doi.org/10.1371/journal.pone.0155026

Singh PK, Singh M et al (2013) Arbuscular mycorrhizal fungi: biocontrol against Fusarium wilt of chickpea. Int J Sci Res 3(1):38–42

Singh DP, Patil HJ, Prabha R et al (2018) Actinomycetes as potential plant growth-promoting microbial communities. Crop Improv Microb Biotechnol:27–38. https://doi.org/10.1016/B978-0-444-63987-5.00002-5

Song YY, Ye M, Li CY et al (2013) priming of anti-herbivore defense in tomato by arbuscular mycorrhizal fungus and involvement of the Jasmonate pathway. J Chem Ecol 39:1036–1044

Sousa CS, ACF S, Garrido MS (2008) Characterization of streptomycetes with potential to promote plant growth and biocontrol. Sci Agric 65:50–55. https://doi.org/10.1590/S0103-90162008000100007

Spago FR, Ishii Mauro CS, Oliveira AG et al (2014) *Pseudomonas aeruginosa* produces secondary metabolites that have biological activity against plant pathogenic Xanthomonas species. Crop Prot 62:46–54. https://doi.org/10.1016/j.cropro.2014.04.011

Stephen J, Shabanamol S, Rishad KS, Jisha MS (2015) Growth enhancement of rice (*Oryza sativa*) by phosphate solubilizing *Gluconacetobacter* sp. (MTCC 8368) and *Burkholderia* sp. (MTCC 8369) under greenhouse conditions. 3. Biotech 5(5):831–837. https://doi.org/10.1007/s13205-015-0286-5

Sukhada M, Manjula R, Rawal RD (2011) Evaluation of arbuscular mycorrhiza and other biocontrol agents against *Phytophthora parasitica* var. *nicotianae* infecting papaya (*Carica papaya* cv. Surya) and enumeration of pathogen population using immunotechniques. Biol Control 58:22–29. https://doi.org/10.1016/j.biocontrol.2011.03.013

Sun Z, Liu K, Zhang J et al (2017) IAA producing *Bacillus altitudinis* alleviates iron stress in Triticum aestivum L. seedling by both bioleaching of iron and up-regulation of genes encoding ferritins. Plant Soil 419(1-2):1–11. https://doi.org/10.1007/s11104-017-3218-9

Taechowisan T, Peberdy JF, Lumyong S (2003) Isolation of endophytic actinomycetes from selected plants and their antifungal activity. World J Microbiol Biotechnol 19:381–385. https://doi.org/10.1023/A:1023901107182

Tagele SB, Kim SW, Lee, et al. (2018) Effectiveness of multi-trait *Burkholderia* contaminans KNU17BI1 in growth promotion and management of banded leaf and sheath blight in maize seedling. Microbiol Res. https://doi.org/10.1016/j.micres.2018.05.004

Takenaka S, Ishikawa S (2013) Biocontrol of sugar beet seedling and taproot diseases caused by *Aphanomyces cochlioides* by *Pythium oligandrum* treatments before transplanting. Jpn Agric Res Q 47(1):75–83

Tariq M, Noman M, Ahmed T et al (2017) Antagonistic features displayed by Plant Growth Promoting Rhizobacteria (PGPR): a review. J Plant Sci Phytopathol 1:38–43. https://doi.org/10.29328/journal.jpsp.1001004

Tchabi A, Hountondji FCC, Ogunsola B et al (2016) The influence of arbuscular mycorrhizal fungi inoculation on micro-propagated hybrid yam (*Dioscorea* spp.) growth and root knot nematode (*Meloidogyne* spp.) suppression. Int J Curr Microbiol App Sci 5(10):267–281

Thomson EL, Dennis JJ (2012) A *Burkholderia cepacia* complex non-ribosomal peptide-synthesized toxin is hemolytic and required for full virulence. Virulence 3(3):286–298. https://doi.org/10.4161/viru.19355

Tokala RK, Strap JL, Jung CM et al (2002) Novel plant-microbe rhizosphere interaction involving *Streptomyces lydicus* WYEC108 and the pea plant (*Pisum sativum*). Appl Environ Microbiol 68:2161–2171. https://doi.org/10.1128/AEM.68.5.2161-2171.2002

Ullah Z, Syed SA, Abbas A et al (2015) Evaluation of *Penicillium* sp. Eu0013 for management of root rot disease of okra. Int J Biosci 7:11–15

Umesha S, Roohie RK (2017) Role of Pseudomonas fluorescens and INA against black rot of cabbage. J Phytopathol 165:265–275. https://doi.org/10.1111/jph.12558

Upadhyay A, Srivastava S (2011) Phenazine-1-carboxylic acid is a more important contributor to biocontrol *Fusarium oxysporum* than pyrrolnitrin in *Pseudomonas fluorescens* strain Psd. Microbiol Res 166:323–335. https://doi.org/10.1016/j.micres.2010.06.001

Usuki F, Narisawa K (2007) A mutualistic symbiosis between a dark septate endophytic fungus, *Heteroconium chaetospira*, and a nonmycorrhizal plant, Chinese cabbage. Mycologia 99:175–184

Vanittanakom N, Loeffler W, Koch U, Jung G (1986) Fengycin-a novel antifungal lipopeptide antibiotic produced by *Bacillus subtilis* F-29-3. J antibiot 39(7):888–901. https://doi.org/10.7164/antibiotics.39.888

Vasebi Y, Alizadeh A, Safaie N (2015) Pantoea agglomerans ENA1 as a biocontrol agent of Macrophomia phaseolina and growth enhancer of soybean. J Crop Prot 4:43–57

Velázquez E, Igual JM, Willems A et al (2001) Mesorhizobium chacoense sp. nov., a novel species that nodulates *Prosopis alba* in the Chaco Arido region (Argentina). Int J Syst Evol Microbiol 51(3):1011–1021. https://doi.org/10.1099/00207713-51-3-1011

Velivelli SLS, De Vos P, Kromann P et al (2014) Biological control agents: From field to market, problems, and challenges. Trends Biotechnol 32:493–496

Vigo C, Norman JR, Hooker JE (2000) Biocontrol of the pathogen *Phytophthora parasitica* by arbuscular mycorrhizal fungi is a consequence of effects on infection loci. Plant Pathol 49:509–514

Vleeshouwers VGAA, Driesprong JD, Kamphuis LG et al (2006) Agroinfection based high throughput screening reveals specific recognition of INF elicitins in Solanum. Mol Plant Pathol 7:499–510

Vos CM, Tesfahuna A, Panisa N et al (2012a) Arbuscular mycorrhizal fungi induce systemic resistance in tomato against the sedentary nematode *Meloidogyne incognita* and the migratory nematode *Pratylenchus penetrans*. Appl Soil Ecol 61:1–6

Vos C, Broucke DD, Lombi FM et al (2012b) Mycorrhiza-induced resistance in banana acts on nematode host location and penetration. Soil Biol Biochem 47:60–66

Vurukonda SSKP, Giovanardi D, Stefani E (2018) Plant growth promoting and biocontrol activity of streptomyces spp. As endophytes. Int J Mol Sci 19. https://doi.org/10.3390/ijms19040952

Wang T, Liang Y, Wu M et al (2015) Natural products from *Bacillus subtilis* with antimicrobial properties. Chin J Chem Eng 23(4):744–754. https://doi.org/10.1016/j.cjche.2014.05.020

Wang X, Wang C, Sui J et al (2018) Isolation and characterization of phosphofungi, and screening of their plant growth-promoting activities. AMB Express 8:63. https://doi.org/10.1186/s13568-018-0593-4

Waqas M, Khan AL, Hamayun M et al (2015) Endophytic fungi promote plant growth and mitigate the adverse effects of stem rot: an example of *Penicillium citrinum* and *Aspergillus terreus*. J Plant Interact 10:280–287

Wasi S, Tabrez S, Ahmad M (2013) Use of Pseudomonas spp. for the bioremediation of environmental pollutants: a review. Environ Monit Assess 185(10):8147–8155

Watanabe H, Kageyama K, Taguchi Y et al (2008) Bait method to detect *Pythium* species that grow at high temperatures in hydroponic solutions. J Gen Plant Pathol 74:417–424. https://doi.org/10.1007/s10327-008-0116-2

Waweru B, Turoop L, Kahangi E et al (2014) Non-pathogenic *Fusarium oxysporum* endophytes provide field control of nematodes, improving yield of banana (*Musa* sp.). Biol Control 74:82–88. https://doi.org/10.1016/j.biocontrol.2014.04.002

Won-Il K, Cho WK, Su-Nam K et al (2011) Genetic diversity of cultivable plant growth-promoting rhizobacteria in Korea. J Microbiol Biotechnol 21(8):777–790

Wu L, Wu HJ, Qiao J et al (2015a) Novel routes for improving biocontrol activity of *Bacillus* based bioinoculants. Front Microbiol 6:1–13

Wu L, Wu H, Chen L et al (2015b) Difficidin and bacilysin from *Bacillus amyloliquefaciens* FZB42 have antibacterial activity against *Xanthomonas oryzae* rice pathogens. Sci Rep 5:12975. https://doi.org/10.1038/srep12975

Xiang Y, Zhang Y, Wang C et al (2017) Effects and inhibition mechanism of phenazine-1-carboxamide on the mycelial morphology and ultrastructure of *Rhizoctonia solani*. Pestic Biochem Physiol 147:32–39. https://doi.org/10.1016/j.pestbp.2017.10.006

Xu S, Pan X, Luo J et al (2015) Effects of phenazine-1-carboxylic acid on the biology of the plant-pathogenic bacterium *Xanthomonas oryzae* pv. oryzae. Pestic Biochem Physiol 117:39–46. https://doi.org/10.1016/j.pestbp.2014.10.006

Xue AG, Guo W, Chen Y et al (2017) Effect of seed treatment with novel strains of *Trichoderma* spp. on establishment and yield of spring wheat. Crop Prot 96:97–102. https://doi.org/10.1016/j.cropro.2017.02.003

Yacoub A, Gerbore J, Magnin N, Chambon P, Dufour MC, Corio-Costet MF, Rey P (2016) Ability of *Pythium oligandrum* strains to protect *Vitis vinifera* L., by inducing plant resistance against *Phaeomoniella chlamydospora*, a pathogen involved in Esca, a grapevine trunk disease. Biol Control 92:7–16

Yadav B, Singh R, Kumar A (2015) Management of spot blotch of wheat using fungicides, bioagents and botanicals. Afr J Agric Res 10:2494–2500. https://doi.org/10.5897/AJAR2013.8196

Yao A, Bochow H, Karimov S et al (2006) Effect of FZB 24® *Bacillus subtilis* as a biofertilizer on cotton yields in field tests. Arch Phytopathol Plant Protect 39:323–328. https://doi.org/10.1080/03235400600655347

Yin J, Yuan L (2017) Phytophthora Disease Control and Growth Promotion of Pepper by Pythium oligandrum. *Acta Hortic Sinica 12*:009

Zachée A, Bekolo N, Bime et al (2010) Effect of mycorrhizal inoculum and urea fertilizer on diseases development and yield of groundnut crops (*Arachis hypogaea* L.). Afr J Biotechnol 7:2823–2827. https://doi.org/10.5897/AJB07.443

Zeiger E, Taiz L (2017) Fisiologia e desenvolvimento vegetal, 6th edn. Artmed, Porto Alegre

Zeng W, Kirk W, Hao J (2012) Field management of sclerotinia stem rot of soybean using biological control agents. Biol Control 60:141–147. https://doi.org/10.1016/j.biocontrol.2011.09.012

Zhang N, Wang D, Liu Y et al (2014) Effects of different plant root exudates and their organic acid components on chemotaxis, biofilm formation and colonization by beneficial rhizosphere-associated bacterial strains. Plant Soil 374:689–700. https://doi.org/10.1007/s11104-013-1915-6

Zhou Q, Li K, Jun X, Bo L (2009) Role and functions of beneficial microorganisms in sustainable aquaculture. Bioresour Technol 100:3780–3786. https://doi.org/10.1016/j.biortech.2008.12.037

Zhou TT, Li CY, Chen D et al (2014) PhlF- mutant of *Pseudomonas fluorescens* J2 improved 2,4-DAPG biosynthesis and biocontrol efficacy against tomato bacterial wilt. Biol Control 78:1–8. https://doi.org/10.1016/j.biocontrol.2014.07.006

Zhou L, Jiang HX, Sun S et al (2016) Biotechnological potential of a rhizosphere *Pseudomonas aeruginosa* strain producing phenazine-1-carboxylic acid and phenazine-1-carboxamide. World J Microbiol Biotechnol 32:1–12. https://doi.org/10.1007/s11274-015-1987-y

Zohora US, Ano T, Rahman MS (2016) Biocontrol of *Rhizoctonia solani* K1 by iturin A producer *Bacillus subtilis* RB14 seed treatment in tomato plants. Adv Microbiol 6(06):424. https://doi.org/10.4236/aim.2016.66042

Zuleta LFG, Cunha CD, De Carvalho FM et al (2014) The complete genome of *Burkholderia phenoliruptrix* strain BR3459a, a symbiont of *Mimosa flocculosa*: highlighting the coexistence of symbiotic and pathogenic genes. BMC Genomics 15(1). https://doi.org/10.1186/1471-2164-15-535

PGPR Inoculation and Chemical Fertilization of Cereal Crops, How Do the Plants and Their Rhizosphere Microbial Communities' Response?

4

Luciana P. Di Salvo and Inés E. García de Salamone

1 Introduction

Wheat (*Triticum aestivum* L.) and maize (*Zea mays* L.) are two cereal crops commonly used in any crop rotation extensive scheme. With rice, they provided more than 60% of the calories of the human diet (FAOSTAT 2012). Wheat and maize, along with soybean and sunflower, constitute the main crops of Argentina's Pampas, the principal productive region of this country in which highest grain yields were produced.

Food production increases based on "green revolution" paradigm of the twentieth century can generate negative environmental impact to the ecosystem (Tilman et al. 2002). Thus, new agricultural practices are important to increase production levels in a more sustainable way (Altieri and Nicholls 2000). In this regard, nowadays, many authors propose that a better understanding of plant-soil interactions and the management of beneficial soil microorganisms constitutes a "new green revolution" (Den Herder et al. 2010; Gewin 2010). Thus, cereal crops are capable to associate with many beneficial bacteria, commonly named as plant growth-promoting rhizobacteria (PGPR). In these associations, PGPR can produce beneficial direct and indirect effects on plant growth (Pliego et al. 2011; Verma et al. 2010) by the synthesis of indolic compounds (Pedraza et al. 2004), cytokinins (García de Salamone et al. 2001), and siderophores (Pedraza et al. 2007), by the inhibition of ethylene synthesis (Glick 1995), and by the improvement of plant nutrition by phosphorus solubilization and biological nitrogen fixation (Kennedy et al. 2004; García de Salamone et al. 1990, 1996; Sarig et al. 1990; Freitas et al. 1997), among other mechanisms. In the context of agricultural production, PGPR inoculation constitutes an economical and ecological alternative to increase plant growth and grain

L. P. Di Salvo · I. E. García de Salamone (✉)
Faculty of Agronomy, Department of Applied Biology and Foods, Chair of Agricultural Microbiology, University of Buenos Aires, Buenos Aires, Argentina
e-mail: igarcia@agro.uba.ar

D. P. Singh et al. (eds.), *Microbial Interventions in Agriculture and Environment*,
https://doi.org/10.1007/978-981-13-8383-0_4

yield that is spreading worldwide (Bashan et al. 2004; Cassán and García de Salamone 2008; Cassán and Díaz Zorita 2016). PGPR can be used as bioinsumes which mostly are named inoculants or biofertilizers, and they are becoming a crucial input of organic farming and a major player for bioeconomy and circular agricultural production on a global scale (Vejan et al. 2016). The rhizosphere is of paramount importance for ecosystem services, such as nutrient and water cycling and crop production, but global climate change will affect rhizosphere ecology and hence ecosystem function, through a variety of direct and indirect ways (Adl 2016). The knowledge of above-belowground ecology is important for the understanding of how plant interactions with beneficial microorganisms, decomposers, and enemies affect crop production, biodiversity, and response to global changes (Ramirez et al. 2018). Due to the complexity of interactions in the microbial ecology of rhizosphere, which is causing inconsistent responses (Cassán and Díaz Zorita 2016), it is necessary to improve the knowledge about those interactions in order to improve grain production. In this chapter, results of different field experiments performed in different locations of the province of Buenos Aires, Argentina, are presented. Table 4.1 summarizes the edapho-climatic description of sites, crop managements, experimental designs, treatments, and sampling dates of each field experiment. Wheat and maize agronomic responses (Sect. 2) and those of rhizosphere microbial communities (Sect. 3) to PGPR inoculation and chemical fertilization were evaluated. Results were compared with information previously described by the literature in order to advance knowledge related to the microbial ecology of crop rhizosphere. This knowledge can be used to improve PGPR inoculation response in pursuit of more sustainable agricultural production.

2 Agronomic Response to PGPR Inoculation and Chemical Fertilization

It has been established that agronomic response due to inoculation of PGPR bacteria such as *A. brasilense* has shown high variability at field conditions (Cassán and Díaz Zorita 2016). This is due to complex interactions between plant genotype, inoculated strains, and environmental conditions which determine crop response to *A. brasilense* inoculation (García de Salamone 2012a). In this regard, some authors demonstrated significant interaction between inoculated strains and plant genotype in maize (García de Salamone and Döbereiner 1996), rice (García de Salamone et al. 2012), and wheat (García de Salamone et al. 2009) experiments.

Previously, it has been demonstrated that crop inoculation with 40M and 42M *A. brasilense* strains increased grain yield of rice from 8% (García de Salamone et al. 2010) to 20% (Gatica et al. 2009), grain yield of wheat up to 30% (García de Salamone and Monzón de Asconegui 2008), and grain yield of maize more than 70% (Salvaré 1995). Recently, five field experiments showed positive responses in grain yield and aerial biomass production of wheat and maize due to 40M + 42M inoculation, two of which showed significant responses (Table 4.2). Increases in aerial biomass production involve higher volumes of plant residues after crop

Table 4.1 Features and experimental design of the five field experiments

	Experiment 1	Experiment 2	Experiment 3	Experiment 4	Experiment 5
Cereal crop	Wheat	Maize	Maize	Maize	Wheat
	Baguette 19 (Nidera™)	AX886 MG (Nidera™)	DK190 MG RR2 (Dekalb™)	DK190 MG RR2 (Dekalb™)	Klein Castor (Klein™)
Location	30 de Agosto, Buenos Aires	Pehuajó, Buenos Aires	30 de Agosto, Buenos Aires	30 de Agosto, Buenos Aires	Villa Moll, Buenos Aires
Soil texture	Sandy loam	Silty loam	Sandy loam	Sandy loam	Sandy loam
pH[1] (1:2.5 soil:water)	6.6	5.8	5.3	5.3	6.1
Electrical conductivity[a] (dS m^{-1})	0.19	0.92	0.78	0.78	0.45
Organic matter[a] (%)	2.9	3.2	2.9	2.9	3.9
Nitrogen (N)[a]	52.3 ppm (N-NO_3)	0.19% (organic N)	0.16% (organic N)	0.16% (organic N)	59.7 ppm (N-NO_3)
Available phosphorus (P)[a] (ppm)	8.35	8.54	11.8	11.8	5.37
Preceding crop	Soybean	Soybean	Soybean	Soybean	Soybean
Sowing date	May 31, 2011	September 30, 2010	October 1, 2010	October 1, 2010	June 18, 2009
Fertilization	None	20 kg P ha^{-1} as monoammonium phosphate	24 kg P ha^{-1} as triple superphosphate (TSP)	None	20 kg P ha^{-1} as TSP
Experiment design	Split-plot design with factorial arrangement	Completely randomized block design with a factorial arrangement	Completely randomized block design with a factorial arrangement	Completely randomized block design with a factorial arrangement	Completely randomized block with factorial arrangement

(continued)

Table 4.1 (continued)

	Experiment 1	Experiment 2	Experiment 3	Experiment 4	Experiment 5
Treatments	-2 levels of inoculation[b]: Control (no inoculation) inoculated with both strains (40M + 42M)	-5 levels of inoculation[c]: Control (no inoculation) inoculated with commercial inoculant[d] inoculated with 40M strain (40M) inoculated with 42M strain (42M) inoculated with 40M + 42M	-2 levels of inoculation[d]: Control (no inoculation) inoculated with 40M + 42M	-2 levels of inoculation[d]: Control (no inoculation) inoculated with 40M + 42M	-4 levels of inoculation[e]: Control (no inoculation) inoculated with 40M inoculated with 42M inoculated with 40M + 42M
	-3 levels of N fertilization: 0, 75, and 150 kg urea ha^{-1}	-3 levels of N fertilization[f]: 0, 90 and 180 kg urea ha^{-1}	-3 levels of N fertilization: 0, 100 and 200 kg urea ha^{-1}	-3 levels of N fertilization: 0, 60 and 120 kg TSP ha^{-1}	-2 levels of N fertilization[g]: 0 and 46 kg N ha^{-1}
	−3 levels of P fertilization: 0, 60 and 120 kg TSP ha^{-1}				
Sampling	Tillering (118 DDS)	V5 (62 DDS)	V5 (61 DDS)	V5 (61 DDS)	Jointing (88 DDS)
	Grain-filling (182 DDS)	R3 (132 DDS)[h]	R3 (129 DDS)[h]	R3 (129 DDS)[h]	Grain-filling (133 DDS)

[a]Upper soil layer (0–20 cm)
[b]The dose per each kg of seed was 12 ml of inoculant containing 4.3×10^9 CFU ml^{-1}
[c]The dose per each kg of seed was 10 ml of the 40M, 42M, and 40M + 42M inoculants containing 10^{10} CFU ml^{-1}
[d]The dose per each kg of seed was 2.0×10^{11} CFU ml^{-1}
[e]The dose per each kg of seed was 15 ml of the 40M, 42M, and 40M + 42M inoculants containing 6.6×10^8, 2.0×10^8, and 3.7×10^8 CFU ml^{-1}, respectively. Both strains were previously isolated from maize rhizosphere (García de Salamone and Döbereiner 1996), identified (García de Salamone et al. 2010) and vastly characterized (Di Salvo et al. 2014)
[f]N fertilization was performed at V4 stage (Ritchie and Hanway, 1982)
[g]N fertilizer was Solmix™ (PASA Fertilizantes Petrobras; 28% N and 2.6% S)
[h]V5 and R3 are two phenological stages as described by Ritchie and Hanway (1982)
[i]Commercial formulation of *A. brasilense* and *Pseudomonas fluorescens* (Rhizoflo Premium Maíz®, Laboratorios CKC®, Argentina). The dose per each kg of seed was 5 ml of commercial inoculant

Table 4.2 Agronomic response due to inoculation with 40M + 42M in five field experiments (see Table 4.1)

	Experiment 1	Experiment 2	Experiment 3	Experiment 4	Experiment 5
Grain yield	−1%	11%[a]	9%	18%	0%
Aerial biomass at a vegetative stage	−8% (20%[a])[b]	30%	37%	0%	3%
Aerial biomass at a reproductive stage	−14% (55%[a])[b]	24%	5%	−9%	23%

[a]Significant differences with Tukey's test vs. control plants without inoculation ($p \leq 0.05$)
[b]Percentages in parentheses indicate inoculation response of fertilized plants with 120 kg of triple superphosphate ha^{-1}

harvest, which allow the accumulation of soil organic matter. Besides, and taking into account the low cost of the inoculation practice, increases in grain production improve economic benefit according to the paradigm of conservation agriculture.

The genus *Azospirillum* is one of the most studied PGPR. It could produce phytohormones (Bottini et al. 1989; Dobbelaere et al. 1999), fix atmospheric nitrogen (García de Salamone et al. 1996), and increase the amount and length of both radical hairs and adventitious roots (Okon and Vanderleyden 1997), improving water and nutrient absorption (Bashan et al. 2007). Regarding the growth plant promotion mechanisms of this bacterium, the additive hypothesis suggests that various mechanisms are playing simultaneously or successively during the plant cycle. These mechanisms will determine the observable effects on plant growth and yield (Bashan et al. 2004). Due to the complex interaction in the rhizosphere, such as unfavorable environmental conditions to crop growth, one or more mechanisms could be inactive. Thus, the benefits of plant-bacterium association could not be so evident (Bashan and Levanony 1990; Kaushik et al. 2002). This could explain no significant inoculation responses observed in some field experiments (Table 4.2).

It has been demonstrated that co-inoculation of two different *A. brasilense* strains produced better inoculation response of several cereal crops than the inoculation of each strain individually (García de Salamone et al. 1996, 2010; García de Salamone 2012b). According to this, maize crop in the experiment 2 showed higher co-inoculation response (Table 4.2) than individual 40M and 42M inoculation responses, with an average of 5% grain yield increase (Di Salvo et al. 2018a). On the contrary, this was not observed in wheat crop (experiment 5) because neither individual inoculation nor co-inoculation of both strains modified grain yield production.

Fertilization must meet crop nutrient demands but without affecting soil nutrient availability and reducing environmental negative impacts (Tilman et al. 2002), such as groundwater contamination risk. In Argentina, chemical fertilizer consumption was doubled in the last 20 years (Fertilizar 2018a). During the 2015–2016 campaign, 90% and 77% of the total cultivated area of wheat and maize crops were fertilized, with average doses of 131 and 186 kg ha^{-1}, respectively (Fertilizar 2018b). Wheat and maize are the two crops that have the highest percentage of nutrient reposition in all the campaigns, taking into account that the percentage of reposition corresponds to the relation between nutrient application and crop removal after harvest. During the 2011–2012 campaign, the percentage of nutrient reposition was 95% and 80% in wheat and maize crops, respectively (González Sanjuan et al. 2013). This demonstrates that fertilization does not always meet crop nutrient demands, in large part due to market fluctuations regarding chemical fertilizers and grain costs. Besides, under certain environmental conditions, recovery efficiency could be diminished (Steinbach 2005). Due to this, it is necessary to increase fertilizer use efficiency (Tilman et al. 2002).

In this context, biological nitrogen fixation constitutes an essential and potential nitrogen source in sustainable agroecosystems (Urquiaga et al. 2004). It has been demonstrated that certain *Azospirillum* spp. strains can fix atmospheric nitrogen in association with maize at levels of 100 kg ha^{-1} of nitrogen (García de Salamone et al. 1996). Besides, increases in root growth due to inoculation improve soil exploration and maximize nutrient uptake (Dardanelli et al. 2004; Dobbelaere et al. 2001). This combination of mechanisms constitutes an opportunity to improve fertilizer use efficiency of chemical fertilizers in several crops, under different environmental conditions (Caballero-Mellado 2004; Dobbelaere et al. 2001). In this regard, inoculation with a commercial inoculant formulated with *A. brasilense* increases wheat grain yield in interaction with nitrogen fertilization (Naiman et al. 2009). In rice crop, inoculation with 40M + 42M *A. brasilense* strains increases grain yield in interaction with micronutrient fertilization (García de Salamone et al. 2010). In the five field experiments of wheat and maize, reported here, only wheat crop in experiment 1 showed interaction between *A. brasilense* inoculation and phosphorus fertilization (Table 4.2) while, interestingly, in the other field experiments, inoculation did not interact with chemical fertilization. Furthermore, two of the five field experiments showed a significant positive response to chemical fertilization (Table 4.3). Environmental conditions during crop development have explained differences in fertilization response. Maize experiments 3 and 4 were under high hydric stress. Despite the fact that precipitation levels during the critical period of grain yield setting were normal, the first 3 months of this maize campaign corresponded to a severe drought that affects plant growth and productivity. Thus, average grain yield of these maize field experiments was 5800 kg ha^{-1}, while the same hybrid in a field

Table 4.3 Agronomic response due to fertilization in five field experiments (see Table 4.1)

	Experiment 1	Experiment 2	Experiment 3	Experiment 4	Experiment 5
Dose and type of fertilizer (see Table 4.1)	120P	180 N	200 N	120P	46 N
Grain yield	40%[a]	40%[a]	0%	10%	7%
Aerial biomass at a vegetative stage	97%[a]	−12%	47%	0%	0%
Aerial biomass at a reproductive stage	29%[a]	27%	13%	0%	21%

[a]Significant differences with Tukey's test vs. control plants without fertilization ($p \leq 0.05$)

with similar soil characteristics showed 9400 kg ha^{-1} of grain yield in the previous campaign which had better levels of precipitations than the 2010–2011 campaign. Regarding wheat in experiment 5, nitrate availability at sowing was high (Table 4.1).

Independently of the agronomic response to chemical fertilization and *A. brasilense* inoculation, it is important to note that wheat and maize field experiments with significant inoculation response showed that inoculation with 40M + 42M strains increased the fertilizer use efficiency of phosphorus and nitrogen fertilizers by 11% and 38%, respectively. In this sense, it has been established that *A. brasilense* inoculation improves the use efficiency of chemical fertilizers (Caballero-Mellado 2004; Hayat et al. 2012; Kennedy et al. 2004). Despite the fact that no significant response due to fertilization was observed in the other three field experiments, it is important to point out that chemical fertilization, especially phosphorus fertilization, is applied not only to increase grain yield but also to achieve nutrient reposition in productive soils. Thus, soil analyses after harvest showed that phosphorus fertilization not only increased wheat grain yield and aerial biomass production (Table 4.3) but also increased soil available phosphorus to 11 ppm and 18 ppm with the 60 and 120 kg ha^{-1} of triple superphosphate fertilization doses, respectively. Regarding maize crop (experiment 4), phosphorus fertilization did not increase grain yield or aerial biomass production (Table 4.3) but increased soil available phosphorus to 17 ppm and 27 ppm with the 60 and 120 kg ha^{-1} of triple superphosphate fertilization doses, respectively.

It has been shown that *A. brasilense* inoculation response increases under hydric or nutritional stress conditions (Creus et al. 2008; Okon and Labandera-Gonzalez 1994; Rodríguez-Cáceres et al. 1996). This could explain the lack of inoculation response in field experiment 5 (Table 4.2) due to that wheat crop grew under good hydric and nutritional conditions (Table 4.1). However, this could not explain the lack of inoculation response in maize experiments (Table 4.2). According to what has been discussed previously, these maize crops grew under several hydric stresses, which determined decreases in plant growth, biomass production, and grain yield. Differences in water availability between maize field experiment in Pehuajó (experiment 2) and maize field experiments in 30 de Agosto (experiments 3 and 4), both performed in the same campaign (Table 4.1), could explain their different inoculation responses (Table 4.2). However, it is important to note that differences in crop management – regarding decisions on agricultural practices, crop rotation, etc. – between both productive establishments can also affect the level of agronomic response of field experiments. Taking this into account, it is expected that hydric stress condition has defined the interaction between maize plants and soil microorganisms (Abril et al. 2006). This could have inactivated one or more plant growth-promoting mechanisms of *A. brasilense* and masked inoculation-positive effects (Creus et al. 2008) on the agronomic response of these maize field experiments. Besides, nitrogen fertilization response is highly dependent on water availability during the whole crop cycle, while phosphorus fertilization response is determined by nitrogen availability (Alvarez 2005a, b). This could explain, to a large extent, both crop responses and the lack of them to chemical fertilization in field experiments (Table 4.3). Other authors have discussed that variability in the inoculation response is due to the reflection of environmental heterogeneity, including different soil types and native microflora, which determined the level of inoculation response (Aeron et al. 2011; Babalola 2010). Regarding the heterogeneity of native microflora, it is important to evaluate the effects of chemical fertilization or PGPR inoculation on rhizosphere microbial communities in order to characterize the environmental risks of these agricultural practices for a more sustainable agricultural production. Results are presented in the following section.

3 Rhizosphere Microbial Communities' Response to PGPR Inoculation and Chemical Fertilization

The rhizosphere is the small volume of soil around plant roots under its direct influence (Morgan et al. 2005). It is one of the most dynamic environments with the highest diversity on the Earth (Ahkami et al. 2017; Hisinger et al. 2009; Pliego et al. 2011). It has been established that changes in microbial diversity by any microbial inoculation are temporary due to the fact that soil is a "biological buffer" (Bashan 1999). However, many biological processes that occur in the rhizosphere (Hisinger et al. 2009) and microbial communities responsible for carrying them out (Kent and Triplett 2002) are still unknown. Thus, it is necessary to improve the knowledge about rhizosphere microbial ecology (García de Salamone 2012a; Minz and Ofek 2011) in order to achieve, mainly, two aims. On the one hand, microorganisms play a predominating role in many ecosystem functions (Winding et al. 2005) and both abundance and biological diversity support a sustainable agricultural production (Altieri 1999; Zhu et al. 2012). Thus, it is necessary to study if crop inoculation is effectively either harmless to the native soil microbial diversity or that modifications generate by inoculation are temporary. On the other hand, an inoculant is applied into a certain environment with an established native microbiota (Bashan 1999), which constitute an ecological competition for the inoculant (Cummings 2009). It has been suggested that microbial colonization reorganize and change plant spaces and resources, so they define their niches inside the plant and determine in consequence microbial diversity (Kroll et al. 2017). Thus, any contribution to the microbial ecology of the rhizosphere, which frequently is the place where the first contact is established, will provide new information to improve crop response to PGPR inoculation. Results regarding the physiology of rhizosphere microbial communities of wheat and maize crops under two commonly adopted agronomic practices, chemical fertilization and PGPR inoculation, showed in this section constitute a significant contribution to the knowledge of microbial ecology.

As previously discussed (Sect. 2), agronomic response to inoculation also depends on native microbial communities, among them, microorganisms with biological nitrogen fixation capacity. As *A. brasilense* belongs to this microbial community, the aim was to evaluate the number of microaerophilic diazothrophic bacteria, a particular group of bacteria with biological nitrogen-fixing capacity, by the use of NFb semisolid medium which was formulated to *A. brasilense* isolation and characterization (Döbereiner 1998) due to its typical growth (Di Salvo and

García de Salamone 2018). However, this medium allows other microorganisms than *A. brasilense* to grow fixing atmospheric nitrogen (Di Salvo et al. 2014). Taking this into account, it was expected that plants inoculated with any *A. brasilense* strain showed a higher count of microaerophilic nitrogen-fixing bacteria than control plant without inoculation. This would be in accordance with results shown by other authors for rice (García de Salamone et al. 2010; Pedraza et al. 2009) and maize (Abril et al. 2006; Cappelletti et al. 2004; Casaretto and Labandera 2008). In the five field experiments, described in this chapter, inoculation with 40M + 42M *A. brasilense* strains did not modify the number of microaerophilic nitrogen-fixing bacteria in comparison with control plants without inoculation in none of the two phenological stages (Table 4.4). Only in the experiment 2, maize plants inoculated with 40M + 42M showed a 17% more number of this bacterial community than plants inoculated with 40M strain at reproductive stage. These results suggest a competitive advantage of co-inoculation (Di Salvo et al. 2018a), in addition to the advantages of this mixture inoculant to increase grain yield and biomass production (Sect. 2). No changes in the number of this diazothrophic community due to any PGPR inoculation were observed in the other four field experiments (Table 4.4; Di Salvo et al. 2018b). Regarding this, some authors have been demonstrated that many field experiments did not show differences in the number of this microbial community in relationship with culture media (Abril et al. 2006). In NFb semisolid medium, the growth of bacteria other than *A. brasilense* and the growth of native strains of this bacterium could explain these results. Besides, related to that the benefits of plant-bacterium association could not be so evident under unfavorable environmental conditions to crop growth (Sect. 2), it has been demonstrated that crops growing under low precipitation levels showed lower number of microaerophilic nitrogen fixers in their rhizospheres than crops growing under good water availability (Abril et al. 2006; Reis et al. 2004). By the contrary, under simulated drought conditions, wheat grain yield decreased but the number of diazotrophic bacteria has not been modified (Creus et al. 2004). Same results were observed in stressed maize field experiments (experiments 3 and 4) (Table 4.4).

Regarding the effect of chemical fertilization response on diazotrophic bacteria, in the five field experiments, the maximum dose of nitrogen and phosphorus fertilizers did not modify the number of microaerophilic nitrogen-fixing bacteria in comparison with control plants without fertilization in none of the two phenological stages (Table 4.5). Only in the experiment 2 (Table 4.1), maize plants fertilized with 90 kg ha^{-1} of urea showed a 9% higher number of this bacterial community than plants fertilized with 180 kg ha^{-1} of urea at vegetative stage. It is known that nitrogen availability inhibits biological nitrogen fixation (Cocking 2003), but not necessarily the growth of heterotrophic microaerophilic nitrogen-fixing microorganisms or crop inoculation response (Bashan and Levanony 1990). However, a lower competitive capacity of diazotrophic bacteria than other heterotrophic microorganisms could explain the decreases in the number of this microbial community with the dose of 180 kg ha^{-1} of urea. In this sense, some authors showed that a better nitrogenous nutrition of the plant increases carbon compound secretion through the root exudates, generating the proliferation of other heterotrophic microorganisms which

Table 4.4 Rhizosphere microorganisms' response due to inoculation with 40M + 42M in five field experiments (see Table 4.1)

Phenological stage of crop		Experiment 1	Experiment 2	Experiment 3	Experiment 4	Experiment 5
Vegetative	MNF bacteria	−3%	1%	1%	7%	0%
	Cellulolytic microorganisms	−2%	−1%	−1%	5%	nd
	Nitrifying microorganisms	1%	−8%	−3%	−2%	nd
	Shannon's index	1%	0%	0%	1%	2%
Reproductive	MNF bacteria	2%	11%	−7%	5%	−11%
	Cellulolytic microorganisms	3%	−1%	1%	−3%	nd
	Nitrifying microorganisms	2%	7%	3%	−6%	nd
	Shannon's index	1%	0%	1%	1%	0%

MNF microaerophilic N_2 fixing (also diazotrophic bacteria), *nd* not determined
No significant differences vs. control plants without inoculation were observed with Tukey's test ($p \leq 0.05$)

Table 4.5 Rhizosphere microorganisms' response due to fertilization in five field experiments (see Table 4.1)

		Experiment 1	Experiment 2	Experiment 3	Experiment 4	Experiment 5
Phenological stage of crop	Dose and type of fertilizer (see Table 4.1)	120P	180N	200N	120P	46N
Vegetative	MNF bacteria	1%	5%	3%	1%	4%
	Cellulolytic microorganisms	4%	6%	3%	2%	nd
	Nitrifying microorganisms	7%	12%[a]	9%	0%	nd
	Shannon's index	3%	1%	0%	0%	0%
Reproductive	MNF bacteria	3%	3%	3%	9%	9%
	Cellulolytic microorganisms	4%	3%	4%	2%	nd
	Nitrifying microorganisms	5%	6%	15%	3%	nd
	Shannon's index	3%	0%	0%	0%	0%

MNF microaerophilic N_2 fixing (also diazotrophic bacteria), *nd* not determined
[a]Significant differences with Tukey's test vs. control plants without fertilization ($p \leq 0.05$)

can compete with the microaerophilic nitrogen-fixing microorganisms (Reis et al. 2000a; Dobbelaere et al. 2002).

Despite the fact that most of field experiments did not show differences in the number of diazotrophic bacteria due to PGPR inoculation or chemical fertilization, counting of this microbial community was high in all field experiments reported here, with an average of 6.29 Log MPN g^{-1} of dry root. Interestingly, field experiments with significant agronomic response due to PGPR inoculation (Table 4.2) did not show significant differences in the number of diazotrophic bacteria between phenological stages (Table 4.6). Although it was not evaluated in this work, this lack of response does not imply that the PGPR inoculation does not change the structure of the community of microorganisms with biological nitrogen fixation capacity, according to previously demonstrated (Soares et al. 2006). By the contrary, differences between phenological stages were observed in three of the five field experiments. At the grain-filling stage of these crops, the number of diazotrophic bacteria was lower than the number observed at vegetative stages (Table 4.6). These results are in accordance with other authors (Cappelletti et al. 2004; García de Salamone et al. 2010; Reis et al. 2000b). The higher crop-demand of nitrates promotes biological nitrogen fixation, which increases during flowering (Kapulnik et al. 1981).

Cultivable soil microorganisms are the most active even though they represent a small fraction of the total soil microbial diversity (Ellis et al. 2003). Thus, human impact on soil microbial communities was vastly evaluated with cultivable-dependent techniques (Chessa et al. 2016). For instance, it has been proposed shredded pruning litters as an organic amendment after the evaluation of changes in nitrogen and organic carbon soil content and microbial properties (Pramanik et al. 2017). Thus, some culture media can be used to evaluate key microbial communities responsible for soil nutrient cycles, such as carbon and nitrogen which are essential macronutrients for sustainable crop production. Cellulolytic and nitrifying microbial communities can be evaluated by the inoculation of soil samples into specific culture media (Di Salvo et al. 2018a).

It has been reported that the wheat and rice inoculation with *A. brasilense* increases both the number of nitrifying microorganisms and the amount of potentially mineralizable nitrogen (García de Salamone et al. 2009). However, in the five field experiments, crop inoculation with 40M + 42M strains did not modify the

Table 4.6 Rhizosphere microorganisms' variation during crop development in five field experiments (see Table 4.1)

	Experiment 1	Experiment 2	Experiment 3	Experiment 4	Experiment 5
MNF bacteria	−2%	4%	−28%[a]	−26%[a]	−11%[a]
Cellulolytic microorganisms	3%	−9%[a]	−27%[a]	−27%[a]	nd
Nitrifying microorganisms	1%	−1%	−23%[a]	−24%[a]	nd
Shannon's index	9%[a]	4%[a]	1%	−1%	−2%[a]

MNF microaerophilic N_2 fixing (also diazotrophic bacteria), *nd* not determined
[a]Significant differences between vegetative and reproductive stages with Tukey's test ($p \leq 0.05$)

rhizospheric communities of cellulolytic and nitrifiers microorganisms (Table 4.4). Regarding chemical fertilization, it has been suggested that in long-term fertilization experiments, nitrogen fertilization increases the abundance of nitrifying microorganisms and modifies microbial communities' composition (Geisseler and Scow 2014). Nitrogen fertilization modified the number of nitrifying (Table 4.5) and cellulolytic microorganisms (Di Salvo et al. 2018a) only in the experiment 2 at vegetative stage. An extensive discussion of these results is presented in Di Salvo et al. (2018a). Briefly, nitrogen fertilization at V4 stage would be promoting organic matter degradation as "positive priming effect," which could increase counts and activity of cellulolytic microorganisms. Thus, it could increase ammonium immobilization in detriment of nitrifying microorganisms due to a lower competitive capacity of this functional group compared to heterotrophic microorganisms.

Differences between phenological stages were observed in three of the five field experiments. At grain-filling stage of maize crops, the number of nitrifying microorganisms was lower than the number observed at vegetative stages (Table 4.6) due probably to greater remaining crop residues at this phenological stage than reproductive stage (Di Salvo et al. 2018a). Regarding wheat crop, residues of the predecessor crop are available as a substrate for cellulolytic microorganisms. However, climatic conditions during this winter crop are unfavorable for the activity of any mesophilic microorganisms, including cellulolytics. Thus, it would be expected a higher number of this functional group in more advanced stages of the crop, under more favorable temperature conditions. However, the rhizospheric effect (Bais et al. 2006; Doornbos et al. 2012), which stimulates other heterotrophic microorganisms' growth on root exudates, would explain the lack of significant differences in the number of cellulolytic microorganisms in wheat crops (Table 4.6). Thus, it could increase ammonium immobilization in detriment of nitrifying microorganisms due to a lower competitive capacity of this functional group compared to heterotrophic microorganisms.

Nitrate demand of a crop is greater during vegetative stage, until flowering, than during grain-filling stage (Kapulnik et al. 1981), where the greatest plant nitrogen remobilization occurs. However, it is important to note that maize field experiments 3 and 4 (Table 4.1) were under severe hydric stress (Sect. 2), which could have reduced nitrate assimilation by these crops and their agronomic responses to chemical fertilization (Table 4.3). Besides, increases in the residues degradation at V5 stages of maize crops imply greater nitrogen immobilization in microbial biomass. However, some authors have demonstrated that under certain conditions, nitrifying microorganisms are more competitive for the ammonium added to the soil by fertilization than heterotrophic microorganisms (Shi and Norton 2000). These could explain the higher number of nitrifying microorganisms at vegetative stage than reproductive stages observed in the field maize experiments 3 and 4 (Table 4.6). This plant regulation of the ammonium use by nitrifier microorganisms has not been before studied in relationship with the combination of agricultural practices like PGPR inoculation and chemical fertilization. This is a relevant knowledge due to its impact on agroecosystem sustainability (Vejan et al. 2016).

Community-level physiological profiling (CLPP) of rhizosphere microbial communities constitutes a valuable and widely used tool to evaluate the potential catabolism of cultivable microorganisms present in a microbial community (Di Salvo and García de Salamone 2012). Several authors have demonstrated the usefulness of Biolog® commercial microplates for the CLPP analyses of microbial communities from different environments or in a certain environment under different treatments (Campbell et al. 1997; Gómez et al. 2004; Preston-Mafham et al. 2002; Zak et al. 1994). Some authors have showed changes in microbial communities due to bacterial inoculation (Conn and Franco 2004; García de Salamone et al. 2010, 2012; Minz and Ofek 2011; Naiman et al. 2009). Because of its advantages, Di Salvo and García de Salamone (2012) standardized this technique using microplates prepared in the laboratory in order to obtain similar results than using commercial microplates. Briefly, this technique consists in the inoculation of soil dilutions in microplates with 23 sole carbon sources. After incubation, microbial growth due to the carbon source consumption could be determined by absorbance measurements. Physiological profiles can be analyzed by multivariate analyses. Also, absorbance values can be used to calculate ecological indexes, such as Shannon's diversity (Di Salvo et al. 2018a). Shannon's diversity index is an ecological index which combines richness and evenness in the distribution of metabolic activity. Besides, absorbance values can be used to perform mean comparisons among different treatments (Di Salvo and García de Salamone 2012). Using this methodology, it was possible to analyze the effect on microbial communities of different of soil compaction levels (García de Salamone et al. 2004), different bioremediation treatments (Di Salvo and García de Salamone 2012), and different agriculture managements of rice and wheat crops (García de Salamone et al. 2010, 2012; Naiman et al. 2009).

We analyzed CLPP of rhizosphere microbial communities in samples obtained from the five field crop experiments at two phenological stages performing both principal component and discriminant analyses. Previously, it has been demonstrated that using different methods for the analysis of the CLPP data allows to maximize information of microbial functional diversity and arrive at complementary conclusions (Di Salvo et al. 2012). Principal component analysis of CLPP of the five field experiments (Fig. 4.1) showed differences in the physiology of rhizosphere microbial communities due mainly to experiment features (Table 4.1) and phenological stage of each crop. Axes 1 and 2 explained the 58% of the total variation. Rhizosphere microbial communities are clustered on the Axis 1 mainly by lactic acid, maltose, putrescine, mannitol, and arginine. This axis clustered maize and wheat samples, the latter depending on phenological stage of the crop (Fig. 4.1). Rhizosphere microbial communities are clustered on the Axis 2 mainly by tween 20 and oxalic acid. On this axis, two clusters can be seen (Fig. 4.1). One of these is on the right top of the figure and corresponds to some samples of experiment 5 at grain-filling stage. Considering that these samples are all from one of the blocks of this experiment, it could expect that differences in the microbial physiology are due to any different soil characteristic in this block compared to the other three blocks. These differences could interact with plant development at this phenological stage and thereby could modify the physiology of their rhizosphere microbial

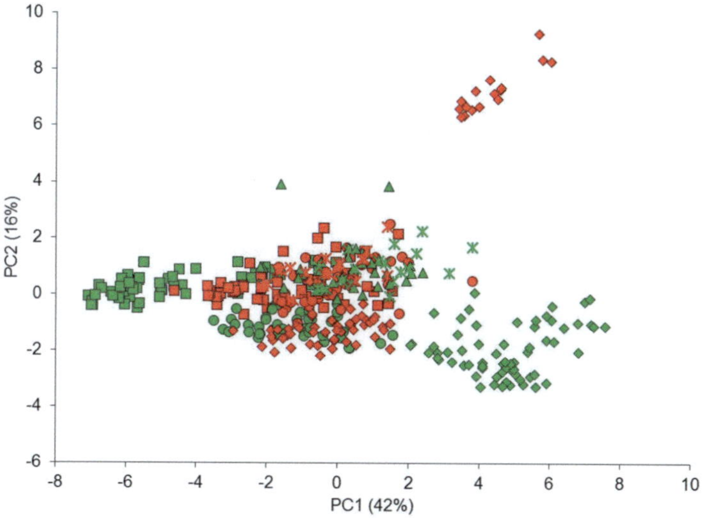

Fig. 4.1 Principal component analysis of the physiological profiles of rhizosphere microbial communities from wheat (*squares for experiment 1; diamonds for experiment 5*) and maize (*circles for experiment 2; triangles for experiment 3; crosses for experiment 4*) crops at vegetative stage (*green symbols*) and reproductive stage (*red symbols*). Analysis was performed with absorbance values at 72 h of incubation. Total explained variance by each axis is in parenthesis

communities. However, no differences in physicochemical parameters were observed in this block in comparison with the other, which could explain differences in the principal component analysis. Because of this, Table 4.1 shows average values of soil physicochemical parameters of the whole experimental site.

Discriminant analysis of CLPP of the data from the five field experiments (Fig. 4.2a) showed similar differences than the principal component analysis due mainly to experiment features (Table 4.1) and phenological stage of each crop. Axes 1 and 2 explained the 68% of the total variation. Rhizosphere microbial communities are clustered on the Axis 1 mainly by oxalic acid and lactic acid. This axis clustered wheat field experiments and wheat phenological stages (Fig. 4.2a). Rhizosphere microbial communities are clustered on the Axis 2 mainly by cellobiose, mannitol, and proline. This axis clustered maize and wheat field experiments (Fig. 4.2a).

Means comparisons of absorbance values showed that wheat and maize rhizosphere microbial communities used differentially 20 of the 23 carbon sources depending on the interaction between field experiment and phenological stage (Fig. 4.3). Besides, three of the 23 carbon sources showed differences among field experiments (Fig. 4.4a). It is important to note that differences among field experiments showed by means comparisons and multivariate analyses cannot be explained only by differences in plant genotypes. Conversely, differences in environmental conditions that affect crop growth and its associated rhizosphere microbial communities, such as soil type, rainfall, and temperature, should be considered.

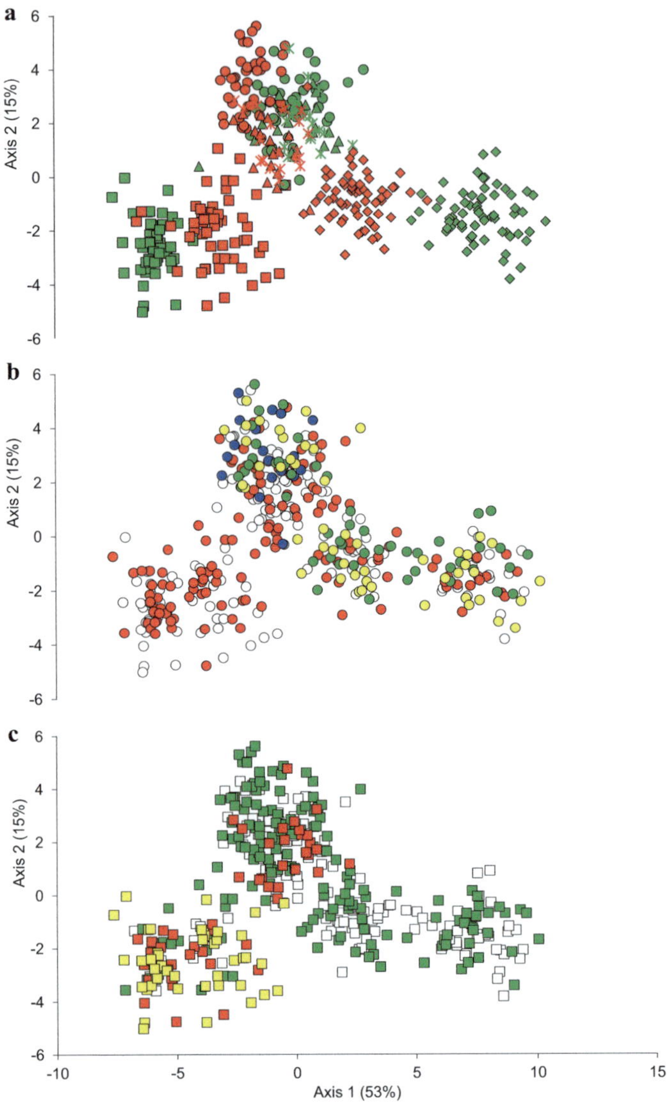

Fig. 4.2 Discriminant analysis of the physiological profiles of rhizosphere microbial communities from (**a**) wheat (*squares for experiment 1; diamonds for experiment 5*) and maize (*circles for experiment 2; triangles for experiment 3; crosses for experiment 4*) crops at vegetative stage (*green symbols*) and reproductive stage (*red symbols*), (**b**) wheat and maize crops under different inoculation treatments (*white for control plants without inoculation; green for 40M inoculation; yellow for 42M inoculation; red for 40M + 42M inoculation; blue for commercial inoculant*), (**c**) wheat and maize crops under different fertilization treatments (*white for plants without any fertilization; green for plants with nitrogen fertilization; red for plants with phosphorus fertilization; yellow for plants with nitrogen and phosphorus fertilization*). More details of inoculation and fertilization treatments are available in Table 4.1. Analysis was performed with absorbance values at 72 h of incubation. Total explained variance by each axis is in parenthesis

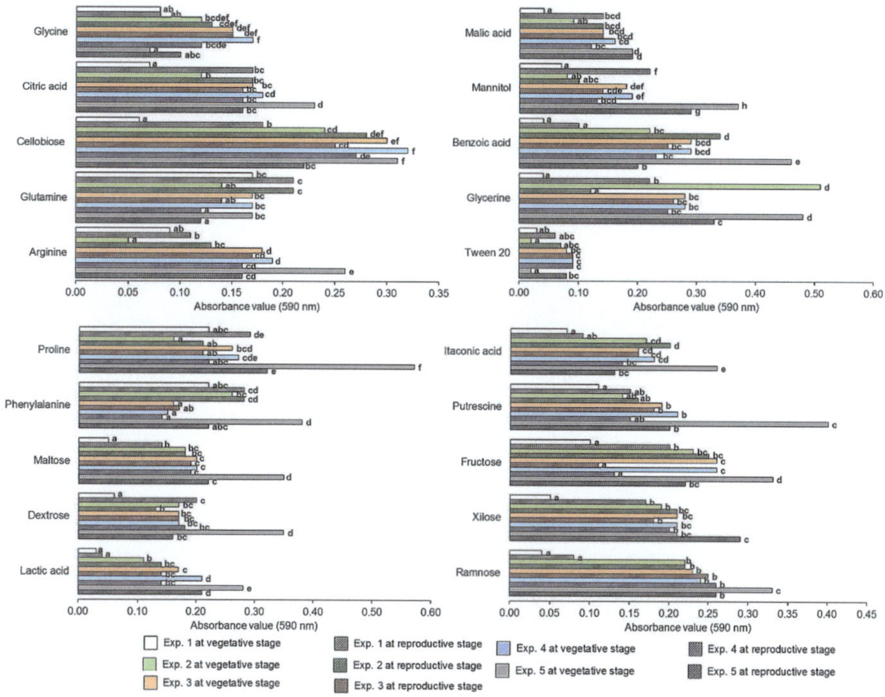

Fig. 4.3 Differential use of carbon sources from rhizosphere microbial communities of wheat and maize crops at different phenological stages. Means comparisons were performed with absorbance values at 72 h of incubation. Different letters indicate significant difference for each carbon source, with Tukey's test ($p \leq 0.05$)

Regarding this, some authors showed differences in the microbial communities of the rhizosphere among different soil types (Bossio et al. 2005; Minz and Ofek 2011). Instead of this, other authors showed that plant genotype defined more clearly the associated microbial communities than the effect of soil type on these communities (Miethling et al. 2000; Roesti et al. 2006; Wieland et al. 2001), due to differences in root architecture and the composition of root exudates (Ahkami et al. 2017; Houlden et al. 2008; Rengel 2002).

In accordance with the results of the multivariate analyses (Figs. 4.1 and 4.2a), the same three carbon sources showed differences between phenological stages of the crops. Microbial communities associated with the rhizosphere of wheat and maize crops at vegetative stage used less histidine and oxalic acid and more salicylic acid than microbial communities associated with the rhizosphere of wheat and maize crops at reproductive stage (Fig. 4.4b). Regarding this, other authors showed changes in the physiology of microbial communities during crop cycles due mainly to differences in quantity and quality of root exudates (Benizri and Amiaud 2005; Houlden et al. 2008; Ortiz-Castro et al. 2009; Wieland et al. 2001).

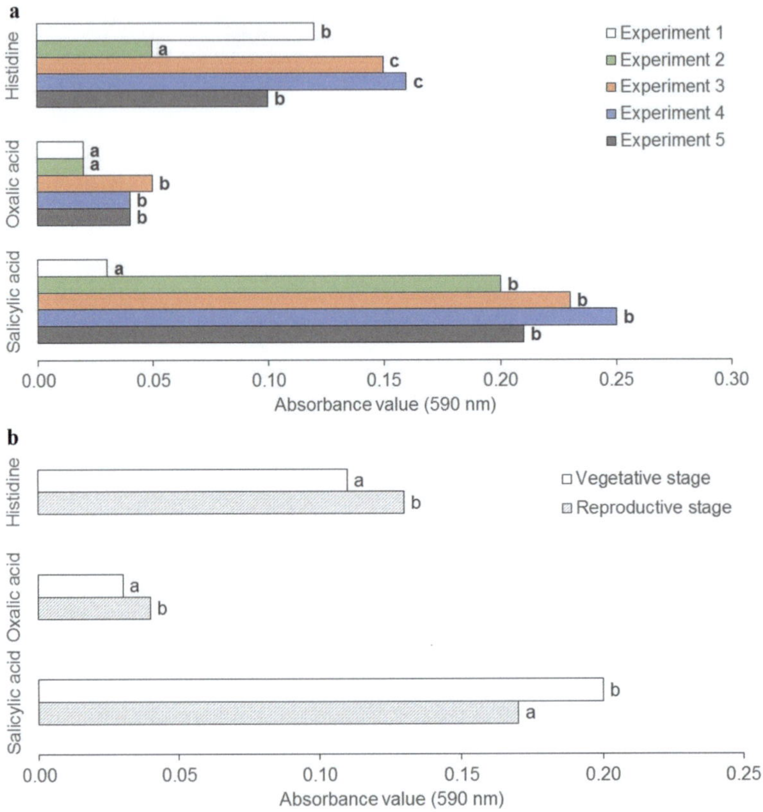

Fig. 4.4 Differential use of carbon sources from rhizosphere microbial communities of (**a**) wheat and maize crops at (**b**) different phenological stages. Means comparisons were performed with absorbance values at 72 h of incubation. Different letters indicate significant difference for each carbon source, with Tukey's test ($p \leq 0.05$)

Regarding agricultural practices, changes in the physiology of rhizosphere microbial communities due to PGPR inoculation (Fig. 4.2b) or chemical fertilization (Fig. 4.2c) were less significant than changes in the CLPP due to plant-environment-phenology interaction (Fig. 4.2a). This knowledge could help to increase the efficacy of benign microbes to promote the development of beneficial traits in plants. Regarding this, other authors demonstrated the effect of plant genotype and its phenology on the functional diversity of the rhizosphere microbial communities (Baudoin et al. 2003; Grayston et al. 1998; Houlden et al. 2008; Kent and Triplett 2002; Kristin and Miranda 2013; Söderberg et al. 2002). It has been pointed out that due to altered gene expression, plants subsequently release an array of primary and secondary metabolites that define relationships to establish vital root interactions with rhizosphere microorganisms (Mhlongo et al. 2018; Rosier et al. 2016).

Under controlled conditions, it has been reported that *A. brasilense* inoculation did not modify the structural diversity of maize rhizosphere (Herschkovitz et al. 2005). Besides, other authors did not find differences in functional diversity due to land-use and agricultural practices (Bossio et al. 2005), while other authors did show differences in this index among soil types (Øvreås and Torsvik 1998) and land-use changes (Xue et al. 2008). Although the results observed in the multivariate analyses (Figs. 4.1 and 4.2) and means comparisons of the absorbance values (Figs. 4.3 and 4.4), no differences in Shannon's diversity index were observed in any of the five field experiments due to PGPR inoculation (Table 4.4) or chemical fertilization (Table 4.5). However, in accordance with the other results (Figs. 4.2a and 4.4b.), wheat and maize phenology modified Shannon's diversity index. Functional diversity at reproductive stages was higher than functional diversity at vegetative stages in two of the five field experiments (Table 4.6). These results are in accordance with changes in structural and functional diversity in maize rhizosphere microbial communities at different phenological stages reported by Baudoin et al. (2002). According to what was previously discussed, as the crop develops, the quantity and quality of root exudates are modified (Aulakh et al. 2001; Kamilova et al. 2006), which could change function or composition of microbial communities associated with the rhizosphere (Coskun et al. 2017; Kristin and Miranda 2013; Houlden et al. 2008; Marschner et al. 2002; Meier et al. 2017). Thus, soil pH is the edaphic variable that best explains diversity and structural profiles of microbial communities on a continental scale (Fierer and Jackson 2006), while both floristic composition and carbon availability explain changes in microbial diversity at local scale (Bais et al. 2006; Benizri and Amiaud 2005).

Here, it is shown that functional diversity at reproductive stages was lower than functional diversity at vegetative stages in one of the five field experiments (Table 4.6). Differences in Shannon's index between wheat field experiments do not surprise due to the fact that samplings at vegetative stage of both field experiments were performed at different phenological stages (Table 4.1). Regarding field experiments 3 and 4, no differences in Shannon's index between phenological stages were observed. Both field experiments showed an average Shannon's index value of 3.02 and 3.03, respectively. Despite the fact that no means comparison can be used in order to compare this index among field experiments due to the characteristics of this ecological index, it is interesting to note that these field experiments showed Shannon's index values higher than Shannon's index values of the other three field experiments which were 2.94 for the experiment 2 and 2.74 and 2.92 for the experiments 1 and 5, respectively. These functional diversity values could be indicating an environmental stress situation (Degens et al. 2001) that is in accordance with what was previously discussed in relationship with hydric stress of these maize crops in the experiments 3 and 4. Also, higher Shannon's index values could be associated with higher amounts of root exudates (Eisenhauer et al. 2017), and this occurs when plants are coping with different types of stresses (Vejan et al. 2016).

4 Conclusions

Crop response to inoculation with 40M and 42M strains of *A. brasilense* was determined by an interaction between both plant and bacteria genotypes' environmental conditions. Interaction between plant genotype and environmental conditions also determined crop responses to phosphorus and nitrogen fertilization. In some cases, this inoculation response even exceeded the level of response to commercial PGPR inoculation. In other cases, strain-mixed inoculant improved plant growth and grain yield better than single-strain inoculants. Besides, *A. brasilense* inoculation and chemical fertilization did not affect most of the evaluated bacterial communities. Furthermore, changes observed in the rhizosphere microbial diversity of wheat and maize due to these agricultural practices were less significant than changes due to plant phenology. In summary, 40M and 42M *A. brasilense* strains have showed its potential to be used as biofertilizers in maize and wheat production. The PGPR inoculation and chemical fertilization, at least at the levels evaluated here, constitute good agricultural practices for food production in a sustainable way. The knowledge related to microbial ecology of crop rhizosphere can be used to improve PGPR inoculation response in pursuit of more sustainable agricultural production in alignment with the bioeconomy paradigm.

Acknowledgments This work was partially supported by FONCYT 2008 PICT1864 from the MINCyT, UBACyT project 20020090100255, Universidad de Buenos Aires in Argentina. We are grateful to Perdoménico's family and the personal of both "El Correntino," 30 de Agosto, and "El Coronel," Pehuajó, Buenos Aires, Argentina. We are also grateful to Lic. Florencia D'Auria, Ma. Laura Beldoménico, Ma. Eugenia Carlino, Marcos Falabella, Ing. Agr. Gabriel C. Cellucci, and Ing. Agr. Claudio Acosta Andocilla for helping during the field experiments and determinations. We are grateful to Laboratorios CKC, Buenos Aires, Argentina, for supplying the commercial inoculant used in this work. We would like to dedicate this work to the memory of Dr. Yoav Bashan, who will always be in our hearts.

References

Abril A, Biasutti C, Maich R, Dubbini L, Noe L (2006) Inoculacion con *Azospirillum* spp. en la región semiarida-central de la Argentina: factores que afectan la colonización rizosférica. Cienc Suelo 24:11–19

Adl S (2016) Rhizosphere, food security and climate change: a critical role for plant-soil research. Rhizosphere 1:1–3

Aeron A, Kumar S, Pandey P, Maheshwari DK (2011) Emerging role of plant growth promoting rhizobacteria in agrobiology. In: Maheshwari DK (ed) Bacteria in agrobiology: crop ecosystems. Springer, Berlin Heidelberg, pp 1–36

Ahkami AH, Allen White R, Handakumburaa PP, Jansson C (2017) Rhizosphere engineering: enhancing sustainable plant ecosystem productivity. Rhizosphere 3:233–243

Altieri MA (1999) The ecological role of biodiversity in agroecosystems. Agric Ecosyst Environ 74:19–31

Altieri MA, Nicholls CI (2000) Applying agroecological concepts to development of ecologically based pest management strategies. In: National Research Council (ed) Professional societies and ecologically based pest management: proceedings of a workshop. National Academy Press, Washington DC

Álvarez R (2005a) Fertilización de trigo. In: Álvarez R (coord) Fertilización de cultivos de granos y pasturas. Diagnóstico y recomendación en la Región Pampeana, Editorial Facultad de Agronomía, Buenos Aires, Argentina, pp 55–80

Álvarez R (2005b) Fertilización de maíz. In: Álvarez R (coord) Fertilización de cultivos de granos y pasturas. Diagnóstico y recomendación en la Región Pampeana, Editorial Facultad de Agronomía, Buenos Aires, Argentina, pp 91–110

Aulakh MS, Wassmann R, Bueno C, Kreuzwieser J, Rennenberg H (2001) Characterization of root exudates at different growth stages of the rice (*Oryza sativa* L.) cultivars. Plant Biol 3:139–148

Babalola OO (2010) Beneficial bacterial of agricultural importance. Biotechnol Lett 32:1559–1570

Bais HP, Weir TL, Perry LG, Gilroy S, Vivanco JM (2006) The role of root exudates in rhizosphere interactions with plants and other organisms. Annu Rev Plant Biol 57:233–266

Bashan Y (1999) Interactions of *Azospirillum* spp. in soils: a review. Biol Fertil Soils 29:246–256

Bashan Y, Levanony H (1990) Current status of *Azospirillum* inoculation technology: *Azospirillum* as a challenge for agriculture. Can J Microbiol 36:591–600

Bashan Y, Holguin G, de-Bashan LE (2004) *Azospirillum*-plant relationships: physiological, molecular, agricultural and environmental advances (1997–2003). Can J Microbiol 50:521–577

Baudoin E, Benizri E, Guckert A (2002) Impact of growth stage on the bacterial community structure along maize roots, as determined by metabolic and genetic fingerprinting. Appl Soil Ecol 19:135–145

Baudoin E, Benizri E, Guckert A (2003) Impact of artificial root exudates on the bacterial community structure in bulk soil and maize rhizosphere. Soil Biol Biochem 35:1183–1192

Benizri E, Amiaud B (2005) Relationship between plants and soil microbial communities in fertilized grasslands. Soil Biol Biochem 37:2055–2064

Bossio DA, Girvan MS, Verchot L, Bullimore J, Borelli T, Albrecht A, Scow KM, Ball AS, Pretty JN, Osborn AM (2005) Soil microbial community response to land use change in an agricultural landscape of Western Kenya. Microb Ecol 49:50–62

Bottini R, Fulchieri M, Pearce D, Pharis RP (1989) Identification of gibberellins A1, A3, and Iso-A3 in cultures of *Azospirillum lipoferum*. Plant Physiol 90:45–47

Caballero-Mellado J (2004) Uso de *Azospirillum* como alternativa tecnologica viable para cultivos de cereales. In: Monzón de Asconegui MA, García de Salamone IE, Miyazaki SS (eds) Biología del suelo. Transformaciones de la materia orgánica, usos y biodiversidad de los organismos edáficos. Editoral Facultad de Agronomía, Buenos Aires, pp 45–49

Campbell CD, Grayston SJ, Hirst DJ (1997) Use of rhizosphere carbon sources in sole carbon source tests to discriminate soil microbial communities. J Microbiol Methods 30:33–41

Cappelletti JP, García de Salamone IE, Ventimiglia L (2004) Colonización rizosférica con bacterias microaerofílicas, longitud de raíces y rendimiento en grano de plantas *Zea mays* inoculadas con dos inoculantes comerciales de *Azospirillum* spp. In: Monzón de Asconegui MA, García de Salamone IE, Miyazaki SS (eds) Biología del suelo. Transformaciones de la materia orgánica, usos y biodiversidad de los organismos edáficos. Editoral Facultad de Agronomía, Buenos Aires, pp 201–205

Casaretto E, Labandera C (2008) Respuesta de Maíz a la inoculación con *Azospirillum*. In: Cassán FC, García de Salamone IE (eds) *Azospirillum* sp.: cell physiology, plant interactions and agronomic research in Argentina. Asociación Argentina de Microbiología, Buenos Aires, pp 261–268

Cassán FD, Díaz-Zorita M (2016) *Azospirillum* sp. in current agriculture: from the laboratory to the field. Soil Biol Biochem 103:117–130

Cassán FD, García de Salamone IE (eds) (2008) *Azospirillum* sp.: cell physiology, plant interactions and agronomic research in Argentina. Asociación Argentina de Microbiología, Buenos Aires

Chessa L, Pusino A, Garau G, Mangia NP, Pinna MV (2016) Soil microbial response to tetracycline in two different soils amended with cow manure. Environ Sci Pollut Res 23:5807–5817

Cocking EC (2003) Endophytic colonization of plant roots by nitrogen-fixing bacteria. Plant Soil 252:169–175

Conn VM, Franco CMM (2004) Analysis of the endophytic actinobacterial population in the roots of wheat (*Triticum aestivum* L.) by terminal restriction fragment length polymorphism and sequencing of 16S rRNA clones. Appl Environ Microbiol 70:1787–1794

Coskun D, Britto D, Shi W, Kronzucker HJ (2017) How plant root exudates shape the nitrogen cycle. Trends Plant Sci 22:661–673

Creus CM, Sueldo RJ, Barassi CA (2004) Water relations and yield in *Azospirillum*-inoculated wheat exposed to drought in the field. Can J Bot 82:273–281

Creus CM, Pereyra MA, Molina-Favero C, Ramella NA, Casanovas EM, Pereyra CM, Arruebarrena-Di Palma A, Lamattina L, Sueldo RJ, Barassi CA (2008) La pared celular como target en la promoción del crecimiento de las plantas por *Azospirillum*. In: Cassán FC, García de Salamone IE (eds) *Azospirillum* sp.: cell physiology, plant interactions and agronomic research in Argentina. Asociacion Argentina de Microbiología, Buenos Aires, pp 97–112

Cummings SP (2009) The application of plant growth promoting rhizobacteria (PGPR) in low input and organic cultivation of graminaceous crops; potential and problems. Environ Biotechnol 5:43–50

Dardanelli J, Collino D, Otegui ME, Sadras OV (2004) Bases funcionales para el manejo del agua en los sistemas de producción de los cultivos de grano. In: Satorre E, Benech-Arnold R, Slafer GA, de la Fuente EB, Miralles DJ, Otegui ME, Savin R (eds) Producción de granos. Bases funcionales para su manejo, Editorial Facultad de Agronomía, Buenos Aires, pp 377–440

de Bashan LE, Holguin G, Glick BR, Bashan Y (2007) Bacterias promotoras de crecimiento en plantas para propósitos agrícolas y ambientales. In: Ferrera-Cerrato R, Alarcón A (eds) Microbiología Agrícola: Hongos, bacterias, micro y macrofauna, control biológico y planta-microorganismo. Editorial Trillas, México DF, pp 161–215

de Freitas JR, Banerjee MR, Germida JJ (1997) Phosphate-solubilizing rhizobacteria enhance the growth and yield but not phosphorus uptake of canola (*Brassica napus* L.). Biol Fertil Soils 24:358–364

Degens BP, Schipper LA, Sparling GP, Duncan LC (2001) Is the microbial community in a soil with reduced catabolic diversity less resistant to stress or disturbance? Soil Biol Biochem 33:1143–1153

Den Herder G, Van Isterdael G, Beeckman T, De Smet I (2010) The roots of a new green revolution. Trends Plant Sci 15:600–607

Di Salvo LP, García de Salamone IE (2012) Evaluation of soil-microbial communities by their CLPP. Standardization of a laboratory technique to replace commercial available microplates. Ecol Austral 22:129–136

Di Salvo LP, García de Salamone IE (2018) Veil-like pellicle development by *Azospirillum brasilense* in semisolid NFb medium. Rev Argent Microbiol 51:184–185. https://doi.org/10.1016/j.ram.2018.04.002

Di Salvo LP, Carlino ME, Cellucci GC, García de Salamone IE (2012) Métodos de análisis complementarios para maximizar la información obtenida sobre la diversidad funcional de comunidades microbianas rizosféricas. In: Abstracts of the 25° Reunión Argentina de Ecología, Universidad Nacional de Luján, Buenos Aires, Argentina, 24–28 September 2012

Di Salvo LP, Silva E, Teixeira KR, Esquivel-Cote R, Pereyra MA, García de Salamone IE (2014) Physiological and biochemical characterization of *Azospirillum brasilense* strains commonly used as plant growth-promoting rhizobacteria. J Basic Microbiol 54:1310–1321

Di Salvo LP, Cellucci GC, Carlino ME, García de Salamone IE (2018a) Plant growth-promoting rhizobacteria inoculation and nitrogen fertilization increase maize grain yield and modified rhizosphere microbial communities. Appl Soil Ecol 126:113–120

Di Salvo LP, Ferrando L, Fernández Scavino A, García de Salamone IE (2018b) Microorganisms reveal what plants do not: wheat growth and rhizosphere microbial communities after application of *Azospirillum brasilense* and nitrogen fertilizer under field conditions. Plant Soil 424:405–417

Dobbelaere S, Croonenborghs A, Thys A, VandeBroek A, Vanderleyden J (1999) Phytostimulatory effect of *Azospirillum brasilense* wild type and mutant strains altered in IAA production on wheat. Plant Soil 212:155–164

Dobbelaere S, Croonenborghs A, Thys A, Ptacek D, Vanderleyden J, Dutto P, Labandera-Gonzalez C, Caballero-Mellado J, Aguirre J, Kapulnik Y, Brener S, Burdman S, Kadouri D, Sarig S, Okon Y (2001) Responses of agronomically important crops to inoculation with *Azospirillum*. Aust J Plant Physiol 28:871–887

Dobbelaere S, Croonenborghs A, Thys A, Ptacek D, Okon Y, Vanderleyden J (2002) Effect of inoculation with wild type *Azospirillum brasilense* and *A. irakense* strains on development and nitrogen uptake of spring wheat and grain maize. Biol Fertil Soils 36:284–297

Döbereiner J (1998) Isolation and identification of aerobic nitrogen-fixing bacteria from soil and plants. In: Alef K, Nannipieri P (eds) Methods in applied soil microbiology and biochemistry. Academic, London

Doornbos RF, van Loon LC, Bakker PAHM (2012) Impact of root exudates and plant defense signaling on bacterial communities in the rhizosphere. A review. Agron Sustain Dev 32:227–243

Eisenhauer N, Lanoue A, Strecker T, Scheu S, Steinauer K, Thakur MP, Mommer L (2017) Root biomass and exudates link plant diversity with soil bacterial and fungal biomass. Sci Rep 7:44641

Ellis RJ, Morgan P, Weightman AJ, Fry JC (2003) Cultivation-dependent and -independent approaches for determining bacterial diversity in heavy-metal-contaminated soil. Appl Environ Microbiol 69:3223–3230

FAOSTAT (2012) Food and agricultural commodities production. http://faostat.fao.org/site/339/default.aspx. Accessed 2 Mar 2015

Fertilizar Asociación Civil (2018a). https://www.fertilizar.org.ar/subida/Estadistica/Evolucion%20de%20Consumo%201990%202013/EvolucionConsumo1990-2017.pdf

Fertilizar Asociación Civil (2018b). https://www.fertilizar.org.ar/subida/Estadistica/Detalle%20de%20Consumo%20de%20Fertilizantes%20Extensivos/Consumo%20Fertilizantes%20Campa%C3%B1a%202015-2016.pdf. Accessed 6 Sept 2018

Fierer N, Jackson RB (2006) The diversity and biogeography of soil bacterial communities. Proc Natl Acad Sci U S A 103:626–631

García de Salamone IE (2012a) Use of soil microorganisms to improve plant growth and ecosystem sustainability. In: Caliskan M (ed) The molecular basis of plant genetic diversity. INTECH, Rijeka, pp 233–258. Open access: http://www.intechopen.com/articles/show/title/use-of-soil-microorganisms-to-improve-plant-growth-and-ecosystem-sustainability

García de Salamone IE (2012b) Microorganismos promotores del crecimiento vegetal. Informaciones Agronómicas de Hispanoamérica (IPNI) 5:12–16

García de Salamone IE, Döbereiner J (1996) Maize genotype effects on the response to *Azospirillum* inoculation. Biol Fertil Soils 21:193–196

García de Salamone IE, Monzón de Asconegui MA (2008) Ecofisiología de la respuesta a la inoculación con *Azospirillum* en cultivo de cereales. In: Cassán FC, García de Salamone IE (eds) *Azospirillum* sp.: cell physiology, plant interactions and agronomic research in Argentina. Asociación Argentina de Microbiología, Buenos Aires, pp 209–226

García de Salamone IE, Laballos N, Monzón de Asconegui MA (1990) Effect of inoculation with *Azospirillum* of wheat cv. Buck Pucará growing in field conditions. In: AIANBA (eds) Actas del II Congreso Nacional de Trigo, Pergamino, Buenos Aires, pp 101–110

García de Salamone IE, Döbereiner J, Urquiaga S, Boddey RM (1996) Biological nitrogen fixation in *Azospirillum* strain-maize genotype associations as evaluated by the 15N isotope dilution technique. Biol Fertil Soils 23:249–256

García de Salamone IE, Hynes RK, Nelson LM (2001) Cytokinin production by plant growth promoting rhizobacteria and selected mutants. Can J Microbiol 47:404–411

García de Salamone IE, Rorig M, Bordato F, Michelena R (2004) Actividad microbiana luego de la aplicación de compactacion sobre un suelo Haplustol éntico bajo siembra directa. In: Abstracts of the XVI y XII Congresos latinoamericano y colombiano de la ciencia del suelo, Cartagena de Indias, Colombia, 26 September–1 October 2004

García de Salamone IE, Escobar Ortega J, Gatica M, Di Salvo L, Vilches MV, Zubillaga M, Urquiaga S (2009) Effect of *Azospirillum* inoculation on N-cycling microorganisms associated with rice and wheat crops. In: Abstracts of the 16th Nitrogen Workshop, Torino, Italy, 28 July–1 July 2009

García de Salamone IE, Di Salvo LP, Escobar Ortega JS, Boa Sorte PMF, Urquiaga S, Teixeira KRS (2010) Field response of rice paddy crop to *Azospirillum* inoculation: physiology of rhizosphere bacterial communities and the genetic diversity of endophytic bacteria in different parts of the plants. Plant Soil 336:351–362

García de Salamone IE, Funes JM, Di Salvo LP, Escobar Ortega JS, D'Auria F, Ferrando L, Fernandez Scavino A (2012) Inoculation of paddy rice with *Azospirillum brasilense* and *Pseudomonas fluorescens*: impact of plant genotypes on the rhizosphere microbial communities and field crop production. Appl Soil Ecol 61:196–204

Gatica SM, Anzovini MB, Di Salvo LP, Escobar Ortega JS, García de Salamone IE (2009) Inoculación de arroz con *Azospirillum brasilense*: incremento en rendimiento con impacto reversible sobre las comunidades microbianas nativas. In: Abstracts of the VII Reunión Nacional Científico Técnica de Biología de Suelo y Fijación Biológica del Nitrógeno, San Miguel de Tucumán, Tucumán, Argentina, 1–4 July 2009

Geisseler D, Scow KM (2014) Long-term effects of mineral fertilizers on soil microorganisms – a review. Soil Biol Biochem 75:54–63

Gewin V (2010) An underground revolution. Nature 466:552–553

Glick BR (1995) The enhancement of plant growth by free-living bacteria. Can J Microbiol 41:109–117

Gómez E, Garland J, Conti M (2004) Reproducibility in the response of soil bacterial community-level physiological profiles from a land use intensification gradient. Appl Soil Ecol 26:21–30

González Sanjuan MF, Grasso AA, Bassi J (2013) Fertilizantes en Argentina – Análisis del Consumo. In: García FO, Correndo AA (eds) Simposio Fertilidad 2013: nutrición de cultivos para la intensificación productiva, https://www.fertilizar.org.ar/?page_id=612. Accessed 27 Sept 2018

Grayston SJ, Wang S, Campbell CD, Edwards AC (1998) Selective influence of plant species on microbial diversity in the rhizosphere. Soil Biol Biochem 30:369–378

Hayat R, Ahmed I, Sheirdil RA (2012) An overview of plant growth promoting rhizobacteria (PGPR) for sustainable agriculture. In: Ashraf M, Ozturk M, Ahmad MSA, Aksoy A (eds) Crop production for agricultural improvement. Springer, Dordrecht, pp 557–579

Herschkovitz Y, Lerner A, Davidov Y, Rothballer M, Hartmann A, Okon Y, Jurkevitch E (2005) Inoculation with the plant-growth-promoting rhizobacterium *Azospirillum brasilense* causes little disturbance in the rhizosphere and rhizoplane of maize (*Zea mays*). Microb Ecol 50:277–288

Hisinger P, Glyn Bengough A, Vetterlein D, Young IM (2009) Rhizosphere: biophysics, biogeochemistry and ecological relevance. Plant Soil 321:117–152

Houlden A, Timms-Wilson TM, Day MJ, Bailey MJ (2008) Influence of plant development stage on microbial community structure and activity in the rhizosphere of three field crops. FEMS Microbiol Ecol 65:193–201

Kamilova F, Kravchenko LV, Shaposhnikov AI, Azarova T, Makarova N, Lugtenberg B (2006) Organic acids, sugars, and L-tryptophane in exudates of vegetables growing on stonewool and their effects on activities of rhizosphere bacteria. Mol Plant-Microbe Interact 19:250–256

Kapulnik Y, Okon Y, Kigel J, Nur I, Henis Y (1981) Effects of temperature, nitrogen fertilization, and plant age on nitrogen fixation by *Setaria italica* inoculated with *Azospirillum brasilense* (strain cd). Plant Physiol 68:340–343

Kaushik R, Saxena AK, Tilak KVBR (2002) Can *Azospirillum* strains capable of growing at a sub-optimal temperature perform better in field-grown-wheat rhizosphere. Biol Fertil Soils 35:92–95

Kennedy IR, Choudhury ATMA, Kecskes ML (2004) Non-symbiotic bacterial diazotrophs in crop farming systems: can their potential for plant growth promotion be better exploited? Soil Biol Biochem 36:1229–1244

Kent AD, Triplett EW (2002) Microbial communities and their interactions in soil and rhizosphere ecosystems. Annu Rev Microbiol 56:211–236

Kristin A, Miranda H (2013) The root microbiota -a fingerprint in the soil? Plant Soil 370:671–686
Krol S, Agler MT, Kemen E (2017) Genomic dissection of host-microbe and microbe-microbe interactions for advanced plant breeding. Curr Opin Plant Biol 36:71–78
Marschner P, Neumann G, Kania A, Weiskopf L, Lieberei R (2002) Spatial and temporal dynamics of the microbial community structure in the rhizosphere of cluster roots of white lupin (*Lupinus albus* L.). Plant Soil 246:167–174
Meier IC, Finzi AC, Phillips RP (2017) Root exudates increase N availability by stimulating microbial turnover of fast-cycling N pools. Soil Biol Biochem 106:119–128
Mhlongo MI, Piater LA, Madala NE, Labuschagne N, Dubery IA (2018) The chemistry of plant-microbe interactions in the rhizosphere and the potential for metabolomics to reveal signaling related to defense priming and induced systemic resistance. Front Plant Sci 9:112
Miethling R, Wieland G, Backhaus H, Tebbe CC (2000) Variation of microbial rhizosphere communities in response to crop species, soil origin, and inoculation with *Sinorhizobium meliloti* L33. Microb Ecol 41:43–56
Minz D, Ofek M (2011) Rhizosphere microorganisms. In: Rosenberg E, Gophna U (eds) Beneficial microorganisms in multicellular life forms. Springer, Berlin, pp 105–121
Morgan JAW, Bending GD, White PJ (2005) Biological costs and benefits to plant-microbe interactions in the rhizosphere. J Exp Bot 56:1729–1739
Naiman AD, Latrónico A, García de Salamone IE (2009) Inoculation of wheat with *Azospirillum brasilense* and *Pseudomonas fluorescens*: impact on the production and culturable rhizosphere microflora. Eur J Soil Biol 45:44–51
Okon Y, Labandera-Gonzalez CA (1994) Agronomic applications of *Azospirillum*: an evaluation of 20 years worldwide field inoculation. Soil Biol Biochem 26:1591–1601
Okon Y, Vanderleyden J (1997) Root-associated *Azospirillum* species can stimulate plants. ASM News 63:366–370
Ortiz-Castro R, Contreras-Cornejo HA, Macías-Rodríguez L, López-Bucio J (2009) The role of microbial signals in plant growth and development. Plant Signal Behav 4:701–712
Øvreås L, Torsvik V (1998) Microbial diversity and community structure in two different agricultural soil communities. Microb Ecol 36:303–315
Pedraza RO, Ramirez-Mata A, Xiqui ML, Baca BE (2004) Aromatic amino acid aminotransferase activity and indole-3-acetic acid production by associative nitrogen-fixing bacteria. FEMS Microbiol Lett 233:15–21
Pedraza RO, Motok J, Tortora ML, Salazar SM, Diaz Ricci JC (2007) Natural occurrence of *Azospirillum brasilense* in strawberry plants. Plant Soil 295:169–178
Pedraza RO, Bellone CH, Carrizo de Bellone S, Fernandes Boa Sorte PM, Teixeira KRS (2009) *Azospirillum* inoculation and nitrogen fertilization effect on grain yield and on the diversity of endophytic bacteria in the phyllosphere of rice rainfed crop. Eur J Soil Biol 45:36–43
Pliego C, Kamilova F, Lugtenberg B (2011) Plant growth-promoting bacteria: fundamentals and exploitation. In: Maheshwari DK (ed) Bacteria in agrobiology: plant growth responses. Springer, Verlag, Berlin, pp 295–343
Pramanik P, Safique S, Zahan A, Phukan M, Ghosh S (2017) Cellulolytic microorganisms control the availability of nitrogen in microcosm of shredded pruning litter treated highly acidic tea-growing soils of Assam in Northeast India. Appl Soil Ecol 120:20–34
Preston-Mafham J, Boddy L, Randerson PF (2002) Analysis of microbial community functional diversity using sole-carbon-source utilisation profiles – a critique. FEMS Microbiol Ecol 42:1–14
Ramirez KS, Geisen S, Morriën E, Snoek BL, van der Putten WH (2018) Network analyses can advance above-belowground ecology. Trends Plant Sci 23:759–768
Reis FBD, Da Silva LG, Reis VM, Dobereiner J (2000a) Ocorrência de bactérias diazotróficas em diferentes genótipos de cana-de-açúcar. Pesq Agrop Brasileira 35:985–994
Reis FBD, Reis VM, Urquiaga S, Dobereiner J (2000b) Influence of nitrogen fertilisation on the population of diazotrophic bacteria *Herbaspirillum* spp. and *Acetobacter diazotrophicus* in sugarcane (*Saccharum* spp.). Plant Soil 219:153–159

Reis FBD, Silva MF, Teixeira KRS, Urquiaga S, Reis VM (2004) Identificaçao de isolados de *Azospirillum amazonense* associados a *Brachiaria* spp., em diferentes épocas e condiçoes de cultivo e produçao de fitormônio pela bactéria. Rev Bras Cienc Solo 28:103–113

Rengel Z (2002) Genetic control of root exudation. Plant Soil 245:59–70

Ritchie SW, Hanway JJ (1982) How a corn plant develops. Iowa State University. Special Report No. 48.

Rodríguez-Cáceres EA, Ciocco CD, Pacheco-Basurco JC (1996) Influencia de la inoculación con *Azospirillum brasilense* en trigo cultivado en suelos de la provincia de La Pampa, Argentina. Cienc Suelo 14:110–112

Roesti D, Gaur R, Johri BN, Imfeld G, Sharma S, Kawaljeet K, Aragno M (2006) Plant growth stage, fertilizer management and bio-inoculation of arbuscular mycorrhizal fungi and plant growth promoting rhizobacteria affect the rhizobacterial community structure in rain-fed wheat fields. Soil Biol Biochem 38:1111–1120

Rosier A, Bishnoi U, Lakshmanan V (2016) A perspective on inter-kingdom signaling in plant-beneficial microbe interactions. Plant Mol Biol 90:537–548

Salvaré FD (1995) Evaluación del rendimiento y la nutrición nitrogenada del cultivo de maíz (*Zea mays* L.) inoculado con bacterias del género *Azospirillum* spp. Trabajo de intensificación para optar por el título de ingeniero agrónomo, Facultad de Agronomía, Universidad de Buenos Aires, Argentina

Sarig S, Okon Y, Blum A (1990) Promotion of leaf area development and yield in *Sorghum bicolor* inoculated with *Azospirillum brasilense*. Symbiosis 9:235–245

Shi W, Norton JM (2000) Microbial control of nitrate concentrations in an agricultural soil treated with dairy waste compost or ammonium fertilizer. Soil Biol Biochem 32:1453–1457

Soares RA, Roesch LFW, Zanatta G, Oliveira Camargo FA, Passaglia LMP (2006) Occurrence and distribution of nitrogen fixing bacterial community associated with oat (*Avena sativa*) assessed by molecular and microbiological techniques. Appl Soil Ecol 33:221–234

Söderberg KH, Olsson PA, Baath E (2002) Structure and activity of the bacterial community in the rhizosphere of different plant species and the effect of arbuscular mycorrhizal colonisation. FEMS Microbiol Ecol 40:223–231

Steinbach HS (2005) Funciones de producción. In: Álvarez R (coord) Fertilización de cultivos de granos y pasturas. Diagnóstico y recomendación en la Región Pampeana, Editorial Facultad de Agronomía, Buenos Aires, Argentina, pp 13–25

Tilman D, Cassman KG, Matson PA, Naylor R, Polasky S (2002) Agricultural sustainability and intensive production practices. Nature 418:671–677

Urquiaga S, Jantalia CP, Alves BJR, Boddey RM (2004) Importancia de la FBN en el secuestro de carbono en el suelo y en la sustentabilidad agrícola. In: Monzón de Asconegui MA, García de Salamone IE, Miyazaki SS (eds) Biología del suelo. Transformaciones de la materia orgánica, usos y biodiversidad de los organismos edáficos. Editoral Facultad de Agronomía, Buenos Aires, pp 1–6

Vejan P, Abdullah R, Khadiran T, Ismail S, Nasrulhaq Boyce A (2016) Role of plant growth promoting rhizobacteria in agricultural sustainability-a review. Molecules 21:573–590

Verma JP, Yadav J, Tiwari KN, Lavakush SV (2010) Impact of plant growth promoting rhizobacteria on crop production. Int J Agric Res 5:954–983

Wieland G, Neumann R, Backhaus H (2001) Variation of microbial communities in soil, rhizosphere, and rhizoplane in response to crop species, soil type, and crop development. Appl Environ Microbiol 67:5849–5854

Winding A, Hund-Rinke K, Rutgers M (2005) The use of microorganisms in ecological soil classification and assessment concepts. Ecotoxicol Environ Saf 62:230–248

Xue D, Yao HY, Ge DY, Huang CY (2008) Soil microbial community structure in diverse land use systems: a comparative study using Biolog, DGGE, and PLFA analyses. Pedosphere 18:653–663

Zak HC, Willig MR, Moorhead DL, Wildman HG (1994) Functional diversity of microbial communities: a quantitative approach. Soil Biol Biochem 26:1101–1108

Zhu W, Wang S, Caldwell CD (2012) Pathways of assessing agroecosystem health and agroecosystem management. Acta Ecol Sin 32:9–17

Biological Treatment: A Response to the Accumulation of Biosolids

5

Stefan Shilev, Hassan Azaizeh, and Desislava Angelova

1 Introduction

The public concern about environmental issues has been constantly increasing in the last few decades. The implementation of concepts such as sustainable development and zero-waste trends is widely accepted and has led to recent economic and social responsibilities. One of the most significant challenges in waste management is wastewater treatment. During the wastewater treatment process, liquids and solids are separated. After several stages of purification, the liquids, which cover the legislation criteria, are discharged into suitable aquatic environments or are collected and used for irrigation of certain crops. The solid part represents a mix of organic and inorganic components. During the wastewater treatment process, the solids are separated to screenings, grit, and sludge and removed for further treatment and disposal. The organic part of solids (sludge) is an inevitable and dangerous by-product generated in significant volumes. It is a heterogeneous mass where the individual components can vary considerably depending on the technological level of the wastewater treatment system (available stages of purification and ways of sludge stabilization). The sewage sludge is potentially dangerous to the environment (all environmental components: soil, air, water) and has certain risks for human health. It may contain harmful levels of toxic metals such as zinc, cadmium, mercury, copper, etc. The sludge has heavy metal content of nearly 0.5–2% on a dry weight basis that may become as high as 6% in some cases (Lester et al. 1983).

S. Shilev (✉) · D. Angelova
Department of Microbiology and Environmental Biotechnologies, Agricultural University–Plovdiv, Plovdiv, Bulgaria
e-mail: stefan.shilev@au-plovdiv.bg

H. Azaizeh
Institute of Applied Research (Affiliated with University of Haifa), The Galilee Society, Shefa-Amr, Israel

Department of Environmental Science, Tel Hai College, Upper Galilee, Israel

D. P. Singh et al. (eds.), *Microbial Interventions in Agriculture and Environment*,
https://doi.org/10.1007/978-981-13-8383-0_5

Besides heavy metals in sewage sludge are found high levels of environmentally persistent chemicals (such as polychlorinated biphenyls and dioxins) and also high levels of fecal pathogens, eggs of parasites, etc.

Agricultural utilization of raw sewage sludge, which is a common practice nowadays, brings certain environmental risks. Once applied to the farmland, the heavy metals are accumulated at the surface of the soils and can remain there for longer duration. Some metals like zinc and cadmium are quite toxic and, therefore, may disturb various biological mechanisms that normally occur in the healthy soils. The significant presence of toxic metals in soil may lead to contamination of the planted crops (Shilev et al. 2014). The level of contamination varies widely depending on the different uptake for each crop. Dowd et al. (2000) reported that when applied to land, raw sewage sludge can introduce pathogens to the spreading sites. This may also create contamination of the groundwater, wells, and surface water, or can even contaminate the food chain. A recent study at the University of Arizona indicated a high risk of pathogen contamination in a significant area of 10-km radius due to the spread of sludge fields.

However, biosolids contain a significant amount of macronutrients (such as nitrogen, phosphorus, potassium, sulfur) and also trace elements of some micronutrients that are very important to plants growth and development. The biosolids appear to be a significant source of phosphorus. According to the report on critical raw materials for the EU (2014), phosphate is listed as one of the 20 critical resources for the European Union. The use of sewage sludge directly to the agricultural lands results in improved physical, chemical, and biological properties of soil (Beck et al. 1996).

The contemporary concept of wastewater management needs to meet the requirements of sustainability and efficiency, recycling, and utilization without supplying harmful substances to the environment. The selection of utilization method varies from country to country and depends on economic, geographical, cultural, legal, and political factors. Globally, besides the long-standing practices, such as landfill disposal, incineration, land reclamation, and agricultural utilization, there are new ones such as energy and fuel production, gasification, cement manufacture, etc. Recycling of farmlands, land restoration, and reclamation strategies have become the main disposal routes, and disposal to landfills has been decreased significantly.

The amount of sewage sludge in the last 15 years is increasing mainly as a result of investments in new wastewater treatment plants, especially in new EU members. Lately, the amount of generated sludge per year in the EU countries is over ten million tons, of which 8.7 million tons are generated in the 15 old EU members and the rest, around 1.2 million tons, came from the 12 new members of EU (Milieu Ltd., WRc and RPA, 2010).

The anaerobic and aerobic stabilization are considered to be the most widely used methods for sludge stabilization in EU. The aerobic degradation is mainly applied in regions with smaller wastewater treatment facilities. Composting, chemical stabilization, or lime stabilization are used in several EU countries. On the other hand, over the past years, important changes have been made in terms of sludge utilization and recycling conducted to approximately 39% of the sludge generated in the EU to be used in agriculture. Besides agricultural purposes, sludge is also used in forestry, for reclamation of disturbed areas (closed mines or landfills), etc. The reuse strategies of the sludge including direct agricultural application after composting had been the predominant choice for its utilization in the EU countries (53% of produced sludge), followed by incineration (21% of produced sludge). However, new EU members prefer to use sludge through landfilling. Although the total amount of sludge utilized in agriculture has been increasing since 1995, some of the EU members (Swiss and The Netherlands) ceased this practice due to the growing public concerns about its safety. In addition, some regions already examined this opportunity (Flanders, Bavaria, parts of Austria) because of the increasing public concern and food safety. The main alternative in old EU members is incineration and residual ash disposal, while in the new members it is still landfilling and agricultural usage.

Wastewater treatment in Bulgaria is an area that has been growing rapidly over the past few years. Finding safe ways of utilizing the increasing sludge production is a relatively new challenge for the new members of the EU. With the implementation of the European Urban Waste Water Treatment Directive (91/271/EEC) concerning the introduction of secondary treatment for all urban agglomerations with more than 2000 inhabitants by the end of 2016, 87% of the population in Bulgaria is covered by the sewer system. The number of urban wastewater treatment plants (WWTP) increased rapidly from 79 in 2010 to 89 in 2014 and up to 174 in 2016. The sludge production for 2014 was nearly 55,000 tons dry weight, and according to the National Strategic Plan for sludge management of urban WWTPs, the amount of sewage sludge generated is expected to reach 160,000 tons in 2020. About 41% of this quantity was temporarily stored, 30% was used in agriculture, and 15% was landfilled by 2014. In addition, according to the recent data of Ministry of environment and waters (MOEW) from 2016, 65,800 tons of sludge dry weight was generated in Bulgaria, while half of it was produced in the capital of Sofia (33,000 tons). In this rapidly growing sector, the fate of sewage sludge is modified with respect to 2014, increasing the agricultural utilization till 40% in the state of the stored quantities, while the recultivation of disturbed areas with biosolids increases till 17.5%. The recycled quantities of sewage sludge in 2016 were 62% of the generated with the objective to reach 65% in 2020 and 35% with energy recovery.

2 Strategies for Biosolids Management

The management of biosolids differs all over the world. It varies a lot even in EU member states according to recent investigations (Zhen et al. 2017). Here, we present the current status of the most effective and used technologies for sewage sludge treatment.

2.1 Anaerobic Digestion of Biosolids for Biogas Production

Biological processes of biosolids like activated sludge (AS) have become common to treat municipal and industrial wastewater. Sludge produced during wastewater treatments may contain different substances such as organics, trace organic compounds such as pharmaceuticals and hormones, polycyclic aromatic compounds, nutrients, and pathogens among which most may affect human health and the environment (Roy et al. 2011). In addition to this, it may also contain various heavy metals, foul odor, and other contaminants; therefore, proper treatment and disposal is a problem (Aldin et al. 2011; Fytili and Zabaniotou 2008). However, despite that activated sludge is efficient in wastewater treatment, the application includes the production of bulk amount of waste-activated sludge (WAS) that needs to be suitably managed, properly treated, and safely disposed of. Inappropriate sludge management may create environmental issues concerning with their odor, soil, and groundwater contamination, sanitation, and greenhouse gas emissions. Therefore, sludge processing, treatment, and disposal are the most essential parts of wastewater treatment systems. Primary and secondary sludge are two major types of sludge produced at conventional biological wastewater treatment plants using AS (Diagram 1). Total solids (TS) in the primary sludge (PS) constitute biodegradable carbohydrates and fats in comparison to AS fraction. Primary sludge is mainly comprised of 20–30% proteins, 5–8% fats and grease, and 8–15% cellulose (Miron et al. 2000; Tchobanoglous et al. 2014). About 50–60% of the operating cost of any wastewater treatment plant is spent for sludge management (Mininni et al. 2015; Tomei et al. 2016). The main problem of sludge stabilization by biological (aerobic and anaerobic) processes is the long retention period (15–30 days) and low digestion efficiency. This may be caused partly by the inability of microorganisms to degrade organic components efficiently. The sludge consists of more than 95% water; therefore, dewatering is made difficult mainly due to the colloidal particles and the gel-like flocculated systems (Dursun and Dentel 2009). Effective dewatering requires microbial action through their presence as the aggregates like films, flocs, and sludges which are embedded in a matrix of extracellular polymeric substances (EPS) that comprises proteins and polysaccharides predominant organic components (Dignac et al. 1998). The sludge has a high affinity for water (Wingender et al. 1999). It has a large percentage of the WAS mass (Chen et al. 2015) and, therefore, dewaterability of WAS may get improved through EPS degradation.

The biodegradation of organic compounds using anaerobic digestion (AD) is the most widely used technology for the production of biogas mainly methane and hydrogen. The effectiveness of the process depends on the four stages of the biodegradation process: hydrolysis, acidification, acetogenesis, and methanogenesis. The AD process is a preferable stabilization procedure in comparison to the aerobic digestion because of low cost, low energy, and moderate performance for stabilization. However, in many cases pretreatment is required that makes the process expensive in many cases. The main sludge treatment techniques used as a pretreatment to anaerobic digestion include biological (largely thermal phased AD), thermal hydrolysis, mechanical, chemical with oxidation (mainly ozonation), and alkali treatments (Carrère et al. 2010). The first three treatment techniques are the most widespread, where they have a great impact on the resulting sludge properties, on the potential energy production and on their application at industrial scale plants. The thermal biological process provides a moderate performance advantage compared to the mesophilic digestion process, with moderate energetic input. It is very important to emphasize that pretreatment methods increase the cost of the anaerobic process and should be evaluated before applying the industrial AD process (Carrère et al. 2010).

Anaerobic digestion (Fig. 5.1) is a well-established method of sludge treatment and stabilization for recovering energy through biogas production. The specific production of biogas from the sludge ranges from 0.75 to 1.12 m^3/kg VS biodegraded, or 0.5 to 0.75 m^3/kg VS into the reactor, or 0.03 to 0.04 m^3/person/day. Biogas comprises methane (65–70%), carbon dioxide (30–35%) and traces of nitrogen, hydrogen, hydrogen sulfide, water vapor, and other gases. The calorific value of the sludge may range from 21 to 25 MJ/m^3 (Appels et al. 2008). The major challenge associated with the AD of wastewater sludge is the slow rate of biodegradation, which requires very long retention (20–30 days) and, thus, need large reactors. This increases the cost of the treatment as high capital investment is required for achieving low solid biodegradation (Mustafa et al. 2014). The performance of 11 different full-scale AD sewage sludge digesters was studied. The results showed that VS degradation efficiency varied from 40 to 65% and was relatively low (Shang et al. 2005). Hydrolysis is the major rate-limiting factor for this process. Waste-activated sludge has low biodegradability due to the presence of different microbial life forms, organic and inorganic components, and extracellular polymeric constituents. Compared to the WAS of the AS, the primary sludge is digested effectively with relative ease. Mixing of WAS with primary sludge decreases the digestion rate

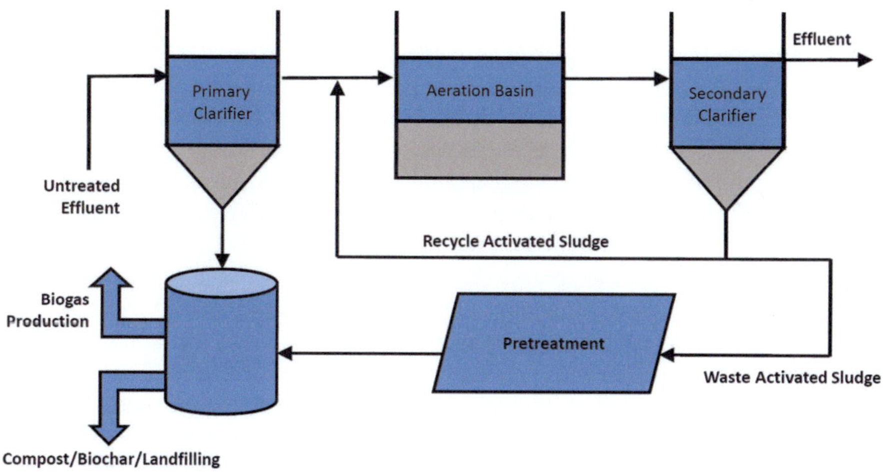

Fig. 5.1 Scheme of wastewater treatment plan with directions for sewage sludge treatment and reuse

under anaerobic conditions and deteriorates the solid–liquid separation quality of digested primary sludge (Tchobanoglous et al. 2014). An anaerobic Digestion Model No. 1 (ADM1) was developed as the state-of-the-art structured mathematical model describing AD bioprocesses (Batstone et al. 2002). This is being used for the implementation of anaerobic method simulation for the treatment of various wastewaters/slurries. The model has shown promising results with considerable accuracy at high-scale plants (Jeong et al. 2005; Lohani et al. 2016; Beline et al. 2017; Nordlander et al. 2017; Aboulfotoh 2018).

Due to the increasing demand for renewable energy, the energy efficiency of AD can be utilized. Primary sludge and WAS organic fractions from municipal wastewater, industrial wastewater treated biosolids, and other food and beverage wastes are becoming the prime sources for degradation (Iacovidou et al. 2012). AD is a cost-effective and reliable technology for waste management but is not effective for food wastes (FW). Food wastes comprise of high organics and volatile fatty acid (VFA) accumulation and, thus, suitable biochemical inputs are required for optimal production of biogas. To overcome these challenges, FW co-digestion with complementary organic waste such as sewage sludge (SS) mixed are now being used for maintaining suitable C:N ratio (Mehariya et al. 2018). The Food and Agriculture Organization (FAO) of the United Nations projected that ~2.2 billion tons of FW produced worldwide by 2025 will need suitable waste management practices (Ariunbaatar et al. 2014, 2015). Anaerobic co-digestion (AcoD) becomes the process of better efficiency for better product yield, nutrients availability, bulk density, lower feed volume, substrate variability, toxicity dilution, synergism, divers, and robust microbiome. However, there are even more pronounced challenges in individual AD operation along with the mixing of FW and SS or SS alone (Chakraborty et al. 2017; Li et al. 2017). More often AD from food waste may not meet the theoretical CH_4 yield of ~ 550 m^3 CH_4 per ton of volatile solid due to limited nutrients (Lisboa and Lansing 2013; Chakraborty et al. 2017; Li et al. 2017).

The co-digestion is also affected by various operating parameters (temperature, pH, system configuration, and feeding modes) (Schievano et al. 2012). Total solid (TS) and moisture contents become the crucial parameters that greatly influence the whole AD process (Iacovidou et al. 2012; Yi et al. 2014; Krishnan et al. 2017; Li et al. 2017). The beneficial effects of FW as co-substrate for sludge anaerobic digestion involve improved methane yield and enhanced methane production (Iacovidou et al. 2012; Koch et al. 2016). The average C/N ratio of different FW samples may not be suitable for the optimum required for the AD process. Thus, the addition of another waste improves the process (Kim et al. 2015). The acceleration of biogas production from AcoD in comparison to mono-digestion was found to be synergistic and attributed to promoted microbial enzyme production (Insam and Markt 2016; Koch et al. 2016). In an interesting work, five semi-continuous flow AD treating a mixture of FW and municipal biosolids at solid retention time (SRT) of 20 days, at different mixing ratios, were operated in order to investigate the AcoD performance (Kim et al. 2017). The synergistic effects of co-digestion of food wastes and wastewater biosolids showed a 37% increase in primary sludge and 50%

thickened waste-activated sludge PS. The TWAS degradation rate was attributed primarily to the biodegradability rather than to the COD/N ratio (Kim et al. 2017).

Recently it is shown that photo-Fenton pretreatment followed by AD of WAS became very effective and yielded 75.7% total VS reduction, 81.5% COD removal, and 0.29–0.31 m^3/kg VS•d biogas production rate in comparison to 40.7% total VS reduction, 54.7% COD removal, and 0.12–0.17 m^3/kg VS•d biogas production rate (Heng et al. 2017). Thus, photo-Fenton can be considered a useful pretreatment step in sludge management; however, more research is still required along with the cost-benefit assessment before being applied.

Poor digestibility of the algal cell wall biomass is the major obstacle towards methane generation by the AD process (Passos et al. 2014). Several reports discussed improving the biodegradability of both micro and macroalgal biomass using AD processes through different approaches including physicochemical as well as biological pretreatment methods (Mendez et al. 2014; Passos et al. 2014). Moreover, attempts were made worldwide for the development of low-cost biological algal biomass pretreatment, where algae grow in close association with a native microorganism of wastewater with the inter-species transfer of CO_2 and O_2 (Ramanan et al. 2016). Moreover, most of the wastewater grown microorganisms, especially bacteria, are hydrolytic (Krah et al. 2016). Hence, it is possible that the natural storage of wastewater-grown algae could lead to a higher degree of algal cell breakdown due to the presence of hydrolytic bacteria. Methane production was assessed from the algal biomass grown in a continuous photobioreactor using sewage. The algal biomass reached up to 1.69 ± 0.35 g L^{-1} in 12 days. Algae naturally colonized low-nutrient effluent water in a wetland treatment system, and the results showed that the algae grow in wastewater (Dalrymple et al. 2013). This study showed that the potential biogas production was estimated to be above 415,000 kg/yr., equivalent to providing the power > 500 homes for one year (Dalrymple et al. 2013).

2.2 Pyrolysis of Biosolids for Biochar Production for Soil Amendment

The Sewage Sludge Directive 86/278/EEC (EC 2001) has a purpose to promote the use of biosolids and prevent undesirable side impacts on soil, groundwater, air, vegetation, animals, and humans. It restricts the use of untreated biosolids, e.g., sludge directly on agricultural soils for reducing undesired effects. Treated sludge is defined as having undergone “biological, chemical or heat treatment, long-term storage or any other appropriate process in order to significantly reduce its biodegradability and the health hazards resulting from its use” (EC 2001). Once treated, sludge can be recycled or disposed of through the use including the main routes: reuse (soil application, biogas production such as AD), incineration including pyrolysis or landfilling. Reuse (including land application and composting) seemed to be the predominant choice for the management of the biosolids in the EU-15 (biosolids 53% of the produced) followed by incineration (biosolids 21%) (Kelessidis and Stasinakisa 2012). This contrasted with practices in the 12 countries which joined the EU after 2004, where the main use was and still mainly landfilling which is considered as the least appropriate biosolid treatment.

Pyrolysis of the biosolids adds several benefits to the product in comparison to the traditional landfilling, incineration, or land application approach. This includes gas emissions of high potentials to obtain energy and a solid product called biochar, which is being used for amending the soils and removing heavy metals from contaminated water and industrial wastewater (Abdelhadi et al. 2017). Pyrolysis can be a potentially promising method for biosolids management compared to other alternatives used, where pyrolysis has a lower carbon footprint and production of biochar for further uses (Miller-Robbie et al. 2015). Pyrolysis is advantageous for delivering lesser gas that could be managed easily than in the process of incineration. This also diminishes the amount of acidic gases and dioxins and helps in producing a higher yield of py-gas with high hydrogen content (Dominguez et al. 2008). It is well known that temperature increase during the pyrolysis enhances the yield of gaseous fraction and decreases biochar production (Inguanzo et al. 2002; Abdelhadi et al. 2017). At low temperature (300–400 °C), the produced biochar becomes acidic while an alkaline product can be obtained at higher temperatures (700 °C), which is good for soil amendment. However, this needs high energy and makes the overall process less economic (Hossain et al. 2011). Agronomic and physicochemical properties of biosolids biochar produced at 400 and 600 °C showed that the volatile matter content of biochar decreased at higher pyrolysis temperatures (Méndez et al. 2013). The BET surface area, pH, porosity, and total concentration of several metals in biosolids biochar increased with temperature, whereas electrical conductivity (EC) and cation exchange capacity of biochar were lower than the original biosolids and were further reduced at higher temperatures (Chen et al. 2014). Another study demonstrated that nutrient content of biochars obtained from biosolids was high for phosphorus, an important constituent for plant nutrition (Antunes et al. 2017). Biochar usage for growing non-edible plants like forest trees could mitigate growing media concerns. Biosolids can also be blended with pre-pyrolysis materials having a lower content of heavy metals for dilution purposes such as lignocellulose from olive oil production (Abdelhadi et al. 2017).

Sewage sludge is an important organic matter source having a good content of essential plant nutrients like P and N (Eid et al. 2017; Wollmann et al. 2017). SS production is increasing continuously because of the availability of treated wastewater worldwide (Eid et al. 2017). Soil amendment using SS is also increasing because of its importance to provide essential plant minerals that enhance soil properties, crop growth, and yield (Antoniadis et al. 2010; Bai et al. 2017). However, SS often contains toxic metals which varied with the types and origin of SS (Samara et al. 2017; Xiong et al. 2018). The heavy metals are plant toxicants that impact plant growth and development by different ways (Rizwan et al. 2016, 2017). Therefore, many countries have specified standards and limits of heavy metals in the SS for field application (Udayanga et al. 2018). In addition, SS is also a secondary source of P and the level of availability increases with the improvements in wastewater management technology (Li et al. 2015). The biochar can be used as an amendment in the soils to increase soil fertility and stability, water holding capacity, and carbon level enhancement. However, this depends on the type of feedstock used for biochar production and pyrolysis conditions (Abbas et al. 2017; Ali et al. 2017). Biochar produced from SS decreases the volume and weight of SS and reduces its toxic compound content (Song et al. 2014), and thus improves soil properties (Yue et al. 2017). Due to the increasing demand of P fertilizers, recyclable P from biosolids is now the main attention for research. Different SS and biochars produced from them were evaluated on soil properties and P uptake in wheat (*Triticum aestivum* L.) with and without P fertilizer. Results indicated that the plant biomass and grain yield were significantly increased due to the application of SS and their biochars (Rehmana et al. 2018). High P accumulation was reported in populated area sludge applied-soil than the disposal sludge and their biochars. Some of the biochars produced from the SS could be the efficient alternate sources of P to enhance plant productivity and implement organic farming systems (Rehmana et al. 2018). Such additional research is still needed to test the produced biochar from SS as a soil amendment for different plant crops and for long-term applications in different soils and regions.

2.3 Biotreatments of Biosolids

The biodegradation of organic matter can be distinguished in two different processes by the nature of decomposition (aerobic and anaerobic). Under the conditions of limited or absent oxygen supply, the anaerobic microorganisms predominate, while metabolizing the nutrients, break down the organic compounds through a process of reduction and developing of intermediate compounds, including gases (such as carbon dioxide and methane), organic acids, etc. In the conditions of oxygen absence, these intermediate compounds are not metabolized further, which may deliver a strong odor and phytotoxicity of the product. The energy released in the anaerobic process is less than released in the aerobic decomposition process (composting). As a low-temperature process, the anaerobic decomposition takes longer and can leave weed seeds and pathogens intact (Shilev et al. 2007). On the other hand, the composting is a technology for the biodegradation of organic matter under aerobic conditions by producing stable, safe, and rich in humic substances material called "compost". The reaction is exothermic, and the temperature of composting material rises fast from the very beginning of the process. The aerobic composting is a dynamic system where various biological forms (such as bacteria, actinomycetes, fungi, protozoa, etc.) and also macroorganisms are actively involved. The relative prevalence of one or other microorganisms depends on constantly changing energy sources, temperature, and conditions. Different organisms are responsible for different decomposition stage, which occur in the composting piles—thermophilic bacteria which are mainly responsible for the breakdown of proteins, fungi, and actinomycetes play an important role in the decomposition of lignin, cellulose, etc. (Shilev et al. 2007). During the initial phase of the process, the mesophilic forms predominate in the composting pile, but when the temperature rises up to 45–60 °C the microorganism species change to thermophilic. The latter stage is associated with a decrease in temperature and the formation of humic compounds.

In general, the composting process could be divided into two stages—the active degradation stage and the maturing stage. The first one is characterized by a high rate of oxygen consumption, carbon dioxide and energy release due to intense decomposition, breaking the chemical bonds of the various organic components. This phase, purely thermophilic, could last from several weeks to more than a month, depending on the characteristics of the substrate and the composting technique that is being applied. The thermophilic phase leads to sanitizing the final product by destroying harmful microorganisms, weed seeds, etc.

With the disappearance of easily biodegradable components, which being metabolized during the first phase, decomposition already affects more complex compounds and require slower processes. Due to the shortages of food, a large part of the microbial population collapse, the temperature rapidly drops, and this leads to a significant change in the type of microbial populations—from thermophilic to mesophilic. During this phase, temperatures are in a range of 40–45 °C and begin to balance progressively with the ambient temperature. This maturing stage can continue for several months. During its mesophilic period, actinomycetes, which actively degrade cellulose and lignin, appear. The activity of this physiological group of microorganisms is of major importance for humus formation. Their presence in the composted mass can be noticed by the specific smell of forest soil they give to the end product. There is also an intense disintegration of the material by very small animals (rainworms, ticks, and centipedes), which also contribute to the fragmentation and mixing of inorganic and organic compounds. Compost stability (maturity) is shown to affect the successful application of the compost in agricultural purposes (Iannotti et al. 1993; Inbar et al. 1990; Mathur et al. 1993). Insufficient stability or maturity of compost may damage plant growth and overall productivity of crops due to competing interest of organisms for oxygen and carbon that may create phytotoxicity (Brodie et al. 1994; He et al. 1995; Keeling et al. 1994).

The most important factor for controlled biodegradation is the small particle size that yields homogeneous compost mixture and improves the temperature regime of the composting system. However, too small particles may prevent entering air freely within the composting mass. The C/N ratio considers the available carbon as well as the available nitrogen. The optimal C:N ratio at the beginning of the process should be between 25–30:1, but actually ratios 20:1 and 35:1 are also acceptable. Luck or excess of carbon may lead to significant loss of nitrogen, decreasing the activity of microorganisms and respectfully elongation of the composting time. The C:N ratio in the finished product has to be between 10 and 15:1. The sewage sludge is nitrogen-rich material; therefore, achieving the adequate C/N ratio requires the sludge to be co-composted with carbon-rich material (such as straw and wood chips). However, because of its compacted structure, significant water content, and low C/N ratio, municipal sludge needs to be supplied with other biodegradable materials (Banegas et al. 2007). Co-composting of municipal sludge and other materials, including municipal solid waste, sawdust, etc., are applied in many cases (Lu et al. 2009; Yousefi et al. 2013; Angelova et al. 2016).

The microorganisms inhabiting a compost pile need enough water for their survival. The optimal water content needed to support microbial metabolic activities ranges from 40 to 65%. If the pile is too dry, the processes occur more slowly. If the moisture content is over 65%, the composting piles may develop anaerobic conditions. The desirable moisture could be achieved by appropriate selection of raw materials because the content is different and some source materials, such as sewage sludge, contain more moisture than others. Besides, microorganisms require a certain range of temperature for their metabolic activities. Both water content and temperature requirements have their fundaments in biological processes and organic compounds that need to be accomplished when biological systems work. The optimal temperature that promotes rapid decomposition, destroying pathogens and the weed seeds is between 50 and 60°C for at least a week (Shilev et al. 2007, Angelova et al. 2016). Aeration helps to produce an odorless end-product, while high temperatures are required for the inactivation of pathogens and weeds, and removal of waste gases, excess heat, and moisture (Rynk 1992; Tambari and Stentiford 1990).

Aerobic organisms involved in the composting process need of about 1.6 kg of oxygen to process 1 kg of organic matter with a minimum oxygen concentration of 5% that is essential for aerobic decomposition. Aeration has been shown to decrease the active decomposition time (Sartaj et al. 1997; Feinstein et al. 1980); therefore it is necessary to ensure that oxygen is supplied, so metabolic activity is maintained. Aeration provides oxygen in a composting system through various means such as physical mechanical or not mechanical overturning (dynamic models), natural convection or forced aeration (positive or negative modes).

2.3.1 Composting Techniques: Advantages and Disadvantages

Aerated Static Pile (ASP)

The usage of aerated static pile composting technique is adequate almost for any type of organic wastes, including sewage sludge from urban WWTP. Under this technique, the organic wastes are mixed in a large pile. The height of the aerated static piles can be between 150 and 250 m depending on the implemented aeration technique. Once the pile is properly formed and the air supply is sufficient, the turning does not occur in the composting mass. If the supply of oxygen is not adequate for the process, the growth of aerobic microorganisms will be limited which will result in slower decomposition. Moreover, the good aeration decreases the risk of overheating of the composting piles. For proper pile aeration and to ensure porosity especially in sludge treatment, due to the specific properties of the material, are needed layers of loosely piled bulking agents (e.g., wood chips, grass clippings, straw, etc.), so the air can circulate through up and down movement in the piles. The aerated static piles can be supplied with the oxygen needed for the aerobic process by natural convection (passive aerated static piles) or forced aeration (active aerated static piles). The main difference between the two is that the forced aeration piles are placed over a net of pipes and blowers to deliver air and ensure its circulation. The forced aeration technique appears to be more suitable for sludge treatment, by a significant reduction of composting time and bad odors occurrence due to anaerobic conditions. The aerated static piles composting technique (both passive and active) requires precise monitoring of temperature to ensure that it is the same near to the surface of the pile as well as in the core zone. The monitoring can be accomplished by sensors placed inside of each pile, due to the lack of physical turning. The moisture content must be maintained precisely under this composting technique as well. In warm, dry conditions, the aerated static piles are recommended to be covered under a shelter to prevent fast evaporation of water. Excessive drying of the composting mater may decrease microbiological activities. Therefore, during winter the piles can be larger in order to retain the heat for the process. The most common way to alleviate some odors, which may appear, is to use a layer of finished compost on the top surface of piles.

Windrow Composting

Aerated (turned) windrow composting is suitable for large volumes of diverse organic waste, including sewage sludge, which can be composted through this method. Under this technique, the organic wastes are mixed together and formed into rows of long piles called "windrows". The aeration is achieved by mechanical turning of the windrows periodically. The proper temperature regime of the windrow composting piles could be easily achieved by controlling the physical quality of the materials, especially piles and the particle size, moisture content, and ensuring the adequate frequency of turning. The pile has to be large enough to generate sufficient heat, therefore, to ensure the thermophilic stage of the process, but not too large—the oxygen flow to the windrow's core must be ensured. The ideal pile height is between 1.20 and 1.80 m, and the pile width is between 4 and 5 m. Windrows should be located over an impervious surface, so the surface rainwater and the infiltrate, which occur during the composting process, should be properly collected and treated in order to prevent local soil, ground, and surface water contamination.

In-Vessel Composting

This kind of composting can treat large amounts of waste, including biosolids, using less space compared to the windrow method. It may utilize any type of organic waste (animal manure, biosolids, food residue, municipal wastes). The method involves mixing the organic components into a drum. The process allows perfect control of environmental factors like temperature, moisture, and air. The material should be turned mechanically even mixing in order to ensure aeration. The size of the vessel often varies. This method can produce compost in just a few weeks.

2.3.2 Vermicomposting of Biosolids and Soil Amendment

Biochar is a carbon-rich solid material obtained through thermal bioconversion of plant lignocellulose and animal biomass without the oxygen, and the process is known as pyrolysis (Lehmann and Joseph 2009; Abdelhadi et al. 2017). Biochar has characters of binding with the contaminants like heavy metals and PAH. This can retain nutrients and water, enhance microbial activities (Khan et al. 2013; Zhang et al. 2013; Abdelhadi et al. 2017), and can supplement composting material (Malinska et al. 2014; Czekała et al. 2016). Biochar is a potential soil amendment agent in vermicomposting of various organic wastes including SS (Malinska et al. 2016). Bio-products like vermicompost could be applied to the soils to supply organic matter (Smith et al. 2014). Vermicomposting is considered as an ecofriendly approach to recycle organic waste materials for conditioning the soils and/or amending their properties (Sharma and Garg 2017). The mutual reaction of earthworms and aerobic microorganisms available in the biosolids accelerates the biodegradation of organic materials (Lim et al. 2016; Rorat et al. 2015).

It is well reported that worms can utilize different varieties of organic matter and convert them into the quantity equal to their body's weight per day. They thereby provide nitrate, phosphorus, potassium, calcium, and magnesium-rich bioorganic products (El-Haddad et al. 2014). However, the quality of the product may vary depending on environmental factors like the type of wastes, aeration condition, humidity, pH, temperature, and earthworm species. Vermicomposting of municipal SS is reported (Wang et al. 2013; He et al. 2016). However, the management of SS through this process has various challenges, especially because of the content of contaminants in the SS (heavy metals and calcitrant compounds like PAH, pharmaceuticals, etc.) (Garg and Kaushik 2005; He et al. 2016). The contamination of heavy metals in the SS can inhibit earthworms activity and result in reduced growth, reproduction, and mortality. So, vermicomposting of SS may require additional supplementary materials to be added to support the growth and development of worms for ensuring optimal characters of earthworms. The supplementary materials may include bulking agents like different crop straw, wood chips or sawdust (Wang et al. 2013), coal ash from tea factory (Goswami et al. 2014), rock phosphate (Wang et al. 2013), and biochar (Malinska et al. 2016). These materials may be added before or after composting, and the mixture may then be subjected to vermicomposting. The addition of supplementary materials straw or sawdust improves C/N ratio and accelerate SS stabilization (Nayak et al. 2013; He et al. 2016).

Anthropogenic activities using compounds such as antibiotics and hormones found in wastewater are difficult to be eliminated during the biological wastewater treatment process (Mao et al. 2015), where many of these compounds are transferred from wastewater to the excess sludge, a major byproduct of the municipal wastewater treatment plants, and the abundant and diverse microorganisms in the sludge could facilitate the formation and spread of resistant bacteria. Consequently, excess sludge is becoming the hotspot and reservoir for the existence of antibiotic resistance genes (ARGs), where the diverse ARGs present in SS are difficult to be eliminated using the conventional sludge treatment processes. Little is known about their fate during the vermicomposting of sludge and their effect on the bioprocess. Wastewater treatment plants receive sewage from different sources, making them a hotspots and rising the human concern for antibiotic-resistant bacteria (ARB) and ARGs (Guo et al. 2017; Rizzo et al. 2013).

As the by-product of WWTP, the high abundances of diverse ARB and ARGs in the SS are recorded in dewatered SS (Li et al. 2013; Guo et al. 2017; Karkman et al. 2017). The removal efficiency of ARGs in sewage treatment is a less efficient process (Yang et al. 2014). Moreover, the SS with high organic matter and diversified microbial population enables the ARGs to flourish and disseminate among bacterial communities through horizontal gene transfer (HTG) (Li et al. 2013; Guo et al. 2017). Vermicomposting significantly decreased the abundances of tetracycline and fluoroquinolone resistance genes and *intI* with complete removal for *parC* gene (Huang et al. 2018). Variations in ARGs were found to be associated with different conditions like HTG, bacterial community composition, and earthworms. In addition to this, earthworms are strongly affected by the inhabiting host bacteria encoding *ARGs* and *IntI* abating the pathogenic bacteria during compost product formation (Huang et al. 2018).

Sewage sludge derived biochar (SSDB) was used as supplementary material for municipal SS and wood chips mixtures (WC) for use in coupled composting and vermicomposting processes (Malinska et al. 2017). The SSDB addition before composting results in higher reproduction rate of the microbial population. Amendment of the SS mixtures with biochar before composting had a significant effect on the activity of *Eisenia fetida* vermicomposting at six weeks (Malinska et al. 2017). Municipal sludge (MS) and vermicomposted sludge (VS) were evaluated for their toxic potential by *Allium cepa* in order to understand the effect of vermicomposting on the reduction of toxicity, if any (Srivastava et al. 2005), where the morphological studies of *A. cepa* roots showed coiled and wavy roots on exposure to MS but no root abnormality were observed in VS. Based on different studies, it was evident that vermicomposting could be an important tool to reduce the toxicity of MS as evidenced by the results of genotoxicity and phytotoxicity (Srivastava et al. 2005). Thus, vermitechnology could be an excellent technique for the recycling of MS; however, the quality of SS used in the vermicomposting process should be tested for its heavy metal content and other toxic compounds before being applied. A safe vermicomposting process, mixing of SS with FW or algae biomass, could be a good solution to dilute the heavy metal content originated from SS where further studies still needed.

2.4 Landfilling of Biosolids as a Waste Management

Conventional treatments like landfilling, incineration, or land application of biosolids (Brisolara and Qi 2013) are operationally convenient and economical. However, such treatments face obstacles due to legislative and public perceptions. The construction and operation of landfills is now restricted in many countries especially in the EU due to legislation (Werle and Wilk 2010; Fytili and Zabaniotou 2008), due to the leachates containing heavy metals and other toxic compounds (Fytili and Zabaniotou 2008; Singh and Agrawal 2008), despite providing technologically simple means of biosolid disposal for many decades. Landfilling encounters issues of public acceptability and greenhouse gas emissions as well as groundwater contamination (Wang et al. 2008) Land application, similar to landfilling except typical agricultural usage (Brisolara and Qi 2013), provides an alternative for biosolids disposal. However, it is problematic due to contaminants like carcinogenic organic compounds such as PAH and heavy metals contaminating water, soil, and crops (Clarke and Cummins 2015; Cincinelli et al. 2012; Singh and Agrawal 2008; Elissen et al. 2010). Land application is, thus, not considered to be a sustainable option for sludge disposal especially that contaminated with high levels of heavy metals and pharmaceuticals. Approaches need to be made to minimize sludge disposal for beneficial reuse such as biochar and compost.

The European Union continued to encourage governments for waste reduction (EC, 2012; EEA, 2013). This is because the landfilling and waste incineration are not good approaches for municipal solid waste (MSW) management issues. The emissions from MSW landfill are biogas to produce energy and leachate which are considered pollutants and can persist for a long period of time. Many studies on the risks due to leachate contamination demonstrate correlations between the exposition of the human population to leachates and occurrence of pathogens and diseases (Butt et al. 2014). Due to both contamination risk and the lack of space for disposal, policies on waste treatment in large cities often focus on thermal treatment facilities for the solid (plastic and paper) and AD or composting from the wet organic fraction. Techniques based on health risk assessment are important and can help decision making during and after an accident due to waste release into the environment (Butt et al. 2016; Davoli et al. 2010; Mishra et al. 2016). The risk for human health due to the air, surface water, groundwater, and soil contamination demands for the waste-to-energy management solutions which consider thermal treatment or gasification of the dry fraction coupled with AD of the wet portion (Paladino and Massab 2017). Studies showed unacceptable HI values found due to groundwater contamination, while HIs were due to river pollution under the threshold (Paladino and Massab 2017).

Landfills are considered as sinks for resources and can potentially be used for resource recovery. Incineration of wastes before landfilling is important to reduce volume and sometimes also for energy recovery; in addition, it had been shown that solid waste, and its incineration residues, in particular, contains as much P as does the SS (Kalmykova et al. 2012; Ott and Rechberger 2012; Kalmykova and Fedje 2013). Therefore, incineration of MSW and SS for P recovery needs more research for cost effectiveness where the remaining solid will be landfilled. The landfilling of MSW as well as the SS should be considered as the last option after the recovery of energy, minerals, biochar, and other beneficial products, in order to minimize leaching of these resources.

3 Case Study

During the last few years, we conducted several experiments in order to investigate the co-composting of different proportions sewage sludge and other biodegradable wastes from agriculture and landscape activities (straw, wood chips, grass clippings) and livestock production (cow manure) to obtain a quality and safe product, which could be used successfully for agricultural purposes.

This studies were also referred to: optimization of the speed of composting process; estimation the relevant amount and type of biodegradable waste for the type of treated sludge; the correct adaptation of selected organisms to the substrate in order to obtain a higher quality of the final product; establishment the presence or absence of harmful substances (especially heavy metals and pathogens) in the final product (compost and vermicompost); study of the accomplishment of the requirements of the Bulgarian legislation for compost production; monitoring the effect of the compost and vermicompost upon test plants development—absence or presence of phytotoxicity; and establishment of the accumulation of heavy metals in the tissues of the test plants.

3.1 Experimental Design

For the present study, we used sewage sludge provided by WWTP—Plovdiv supplemented with different biodegradable wastes. In the first experiment, the wastes used were cow manure and straw, delivered by local farmers with the purpose to utilize the rests from agriculture for the production of quality compost but at the same time to reduce the greenhouse gases that could be produced if self-degradation of waste is performed. On the other hand, in the second experiment, we used biodegradable wastes from gardens of the city of Plovdiv, such as wood chips, grass clippings, and leaves. The composting process was performed outside on a concrete surface. The experiments were performed as per windrow composting technology.

Five shaped piles were made adding different percentage of sewage sludge and cow manure: treatment 1 (75:25), treatment 2 (50:50), treatment 3 (25:75), treatment 4 (100:0), and treatment 5 (0:100). A relevant quantity of straw was added to each case to ensure the appropriate C/N ratio for the composting process—30:1. The dimensions of the piles were width 2 × 2 m and height of about 1.5 m, in order to ensure the process with the minimum amount of material for its normal course. The piles were turned on with a wheel loader twice a week.

During the composting process, the temperature dynamics in both experiments were monitored daily with a borer at a depth between 50 and 100 cm several times into the composting piles. In order to provide and ensure more rapid and complete aerobic decomposition, the whole profile of the piles was supplemented with oxygen by mechanical turning twice a week. The optimal moisture for the process was established to be between 55 and 60%. The adjustment of the adequate moisture content, temperature, and aeration was accomplished by turning and watering of the composting piles. At the end of composting, the volume was reduced by 50%. The biodegradation of organic matter was followed by the formation of leaching from the piles, which was recollected and measured for elemental content. The duration of composting process was 19 weeks, after which composting mixtures were transferred to so-called "beds", with dimensions 3.50 m length, 0.60 m width, and depth of about 0.40 m. To each of those "beds" were added 1 m^2 (30–40,000) worms of mixed species Lumbricus rubellus and Eisenia fetida. The duration of vermicomposting was 12 weeks.

Besides the monitoring of temperature variations into the composting piles, aiming to establish the end of the composting and the maturity of final compost, lab analyses of EC, pH, soil respiration were also performed. The same tests were repeated after the completion of vermicomposting. Finally, with the purpose to define the applicability of the end-product (vermicompost) for agriculture purposes, biotests on the field were carried out. As test cultures were used pepper plants, Capsicum annuum L. The biotests were done using three 20 l pots per treatment, which were filled with agricultural soil from layer 0–30 cm with corresponding addition of vermicompost 25% or 50% (v:v), according to the Regulation of compost (2017).

3.2 Results and Discussion

The standard temperature variation was reported in the phase of active degradation during the composting process. In all five treatments of the present study, the expected rapid increase in temperature (within a range of 36–58 °C) in the first week was observed (Fig. 5.2). Despite pile mixing, the temperature remained extremely high (in the range of 58–70 °C) for the next 4 weeks. Sixth and seventh weeks were characterized with high temperatures still (over 50 °C). During the next weeks, the temperature falls to 25–29 °C in week 19. Relatively higher temperatures are noted in the treatment with cow manure only. Deviation of about 3–4 °C during the composting process was determined by the change of ambient temperature, pile mixing, precipitation, etc.

The exposure to high-temperature levels was sufficient to destroy weed seeds and sanitize the composting piles (Table 5.1). One of the major objectives of the study was to obtain a valuable product, respecting EU and BG legislation that permits to utilize it in agriculture or gardening (Regulation of compost 2017). In this sense, the maximum permissible concentrations (MPC) of some important parameters in the initial SS were over the limits notified in the corresponding BG regulation of the quality of compost produced by SS from WWTP: *Salmonella* sp., heavy metals (Cu, Cd, Cr).

The main macronutrients such as carbon, nitrogen, and phosphorus were converted into different forms, but a certain amount of them was lost. There was a

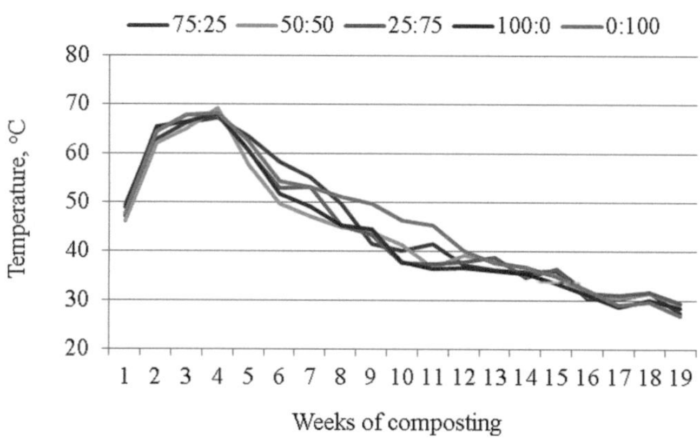

Fig. 5.2 Changes of compost pile temperatures in studied treatments during the process. The results represent the mean of three replicates, while the standard error is in the range of 5%

Table 5.1 Analyzed sewage sludge (source material), composts, and vermicomposts comparing to the MPC according to Bulgarian regulations

	MPC compost	MPC *soil conditioner*	Sewage sludge	Compost					Vermicompost				
				75:25	50:50	25:75	100:0	0:100	75:25	50:50	25:75	100:0	0:100
Salmonella spp.	Presence in 25 g	Presence in 25 g	Absence	Absence	Absence	Absence	Absence	Absence	Absence	Absence	Absence	Absence	Absence
E coli titer	<100 cfu/g fw	<100 cfu/g fw	0.001	**1.0**	**1.0**	**1.0**	1.0	1.0	**1.0**	**1.0**	**1.0**	1.0	Absence
Cl. Perfringens	–	–	0.001	0.001	0.001	0.001	0.001	0.001	0.001	0.001	0.001	0.001	Absence
Organic content, %	>15	>15	66.81	44.40	32.20	34.47	32.03	24.57	34.15	24.60	26.10	27	24
Total N, %	–	–	48	1.87	1.42	1.6	1.86	1.10	1.80	1.13	1.4	1.4	1.13
Cu	250	400	447	237.88	159.58	109.75	154.13	53	235.3	147	152.3	196.4	42.7
Cd	2	3	3.69	2.50	2.30	**1.97**	2.94	0.05	**1.1**	**0.3**	**0.2**	**0.3**	0
Pb	180	250	312	**158.04**	**107.88**	**83.88**	107.89	1.95	137.1	**92.5**	**127.2**	173.1	52.7
Hg	1	2	<0.05	<0.05	<0.05	<0.05	<0.05	<0.05	<0.05	<0.05	<0.05	<0.05	<0.05
Cr	100	200	136	65	45.625	34.5	46.05	17.85	84	53.7	55	60.4	25.6
Ni	80	100	72.7	45.13	32.38	24.78	30.95	12.93	51.9	35.3	33.3	36.8	15.8
Zn	800	1200	1115	533.25	354.50	275.38	354.56	135.16	777.2	496.8	502.3	543.2	199.9
Respiration (CO_2)			–	0.091	0.111	0.147	0.107	0.08	0.078	0.081	0.086	0.079	0.08

significant reduction of the total N concentration into the compost treatments compared to the N content into the sludge. In both compost and vermicompost, the higher values were found in treatment 1 (75:25). Generally, the total nitrogen content was reduced in vermicompost compared to the corresponding treatment in the compost stage. Organic matter expressed as total organic carbon decreased since the beginning of the degradation. After the composting, the highest content of carbon was found in treatment 1 (75:25), respectfully the lowest value was noticed in treatment 5 (0:100), while in the other treatments there was not any statistically significant difference. The decreasing is attributed to the changes that occur during the composting process. It is clear that during the vermicomposting process decomposition of organic matter continues, resulting in a decrease in organic content (Table 5.1).

In addition, during the composting process, the part of the carbon is converted to CO_2 (assuming carbon losses) and the C/N ratio decreases, with the values of final product typically close to 10–15:1. Authors from Pennsylvania State University reported different reduction rates (from 12.3 to 20.9) for C/N ratio during composting (Cekmecelioglu et al. 2005). In our case for both (compost and in the vermicompost), the C/N ratios remained higher than the desired with values between 17 and 23 (Fig. 5.3).

In composting stage, the concentrations of the different heavy metals were reduced, which is attributed to an increase in the initial volume with the addition of biodegradable waste, but also to the leaching of element during the biodegradation. After vermicomposting the amount of heavy metals remained practically constant with some fluctuations due to the heterogeneity of the material for analyses.

The decomposition of organic matter may conduct to a higher content of salts in the substrate during the active phase of the composting. In the matured compost, the content of salts is decreasing. Higher values of EC may have an inhibitory effect

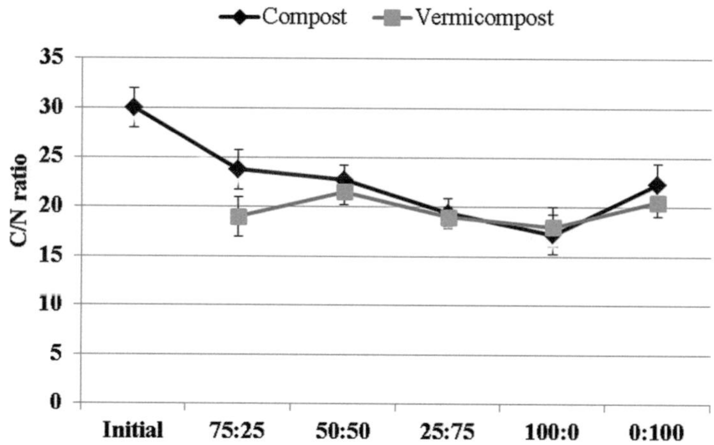

Fig. 5.3 C/N ratio of compost and vermicompost by treatments. The results represent the mean of three replicates and the standard error

about the germination and development of various plant seeds. The electrical conductivity of compost solution can be considered as a measure of its maturity. According to BG Regulation (Regulation of compost, 2017), compost produced with SS from WWTP and designated to be used as fertilizer in gardening or as substrate for plants should possess EC equal or lower than 3 mS/cm^2. In our study, the highest value of the EC was found in sewage sludge before the composting (more than 8 mS/cm^2, Fig. 5.4). After the composting, EC of composting piles was higher in treatment with 75% sludge, while lower was found in treatment with 25% SS and a moderate quantity of cow manure and straw. In the rest of composting piles, the EC data were slightly higher. In this moment of biological treatment of SS, the legislative requirements concerning EC are not fulfilled, so this kind of compost is not sufficiently treated to be used legally in gardening. The subsequent treatment (vermicomposting) ensures that the rate of EC in all treatment is lower than the limitation of 3 mS/cm^2, thus significantly decreases compared with the compost from the same treatment. In the first treatment, the reduction of EC was with 77%, followed by Treatments 4 and 5 with 61 and 64.7%, respectively, while in the second and third treatment the reduction was with 52 and 37%, respectively. It seems that a higher initial concentration leads to higher reduction after vermicomposting.

The changes in the pH into the test plies can evaluate the progress of composting, and also the level of decomposition of organic matter. The pH of the SS at the beginning of the process was found to be 6.98, i.e., neutral. After 19 weeks of composting in the resulted composts, reduction in this value was reported, that was most significant in treatment (80:20) and most negligible in (40:60). In Treatment 5 (fully manure), even a small increase in pH —7.26—was noted. After the vermicomposting, the rate of pH does not present important fluctuations. In that sense, Li et al. (2001) reported that during composting of SS, the pH was increasing in the thermophilic phase and decreased after that. The same authors described that the EC decreased throughout the composting time.

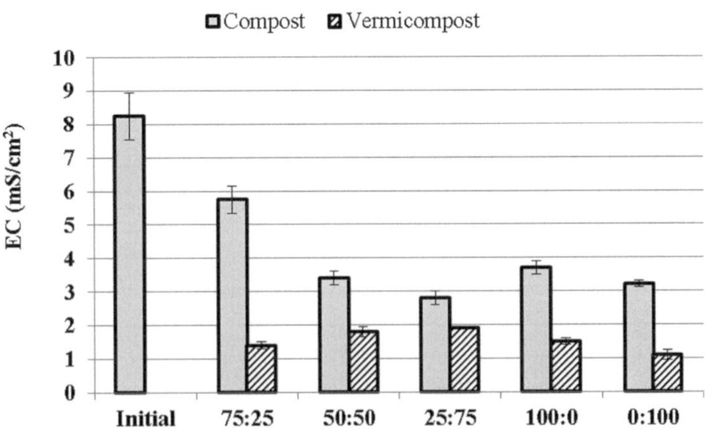

Fig. 5.4 Electrical conductivity of compost and vermicompost by treatment. The results represent the mean of three replicates and the standard error

The concentration of heavy metals into the compost (in all five treatments) was reduced during the process. The most significant decrease was noticed in the third treatment (25% sludge:75% cow manure and 47% of straw from the total quantity, adjusting the C/N ratio). Obviously, besides the influence of biodegradation processes and high temperature of composting, a significant impact on the obtained results had different sludge concentration in composting piles. Heavy metals content in the vermicompost from each treatment was significantly decreased compared to the compost from the same one. Various studies emphasized the need for the implementation of vermicomposting for obtaining high-quality and safe product. According to several authors, earthworms possess the ability to accumulate heavy metals into their tissues during the vermicomposting of sewage sludge (Saxena 1998). The potential of vermicomposting, as a useful approach for sewage sludge management, stabilization, and pathogens removal is discussed from Hartenstein et al. (1979), Neuhauser et al. (1998), and Bogdanov (1998). All these studies correspond with Eastman et al. (2001) that carried out a large-scale laboratory experiment reporting a significant reduction in populations of pathogens, such as *Salmonella enteritidis*, *Escherichia coli*, and helminth eggs during the vermicomposting of sewage sludge from wastewater treatment.

One of the most important indicators is the presence or absence of *Salmonella* ssp. In accordance with the analysis of sludge, composts, and vermicomposts, the presence of this pathogen is not found. In addition, *Escherichia coli* titer is another indicator showing the minimum volume, in which is presented the corresponding bacterium. It was established that there is the presence of *Escherichia coli* in 0.001 kg into the used SS, while after composting and vermicomposting the minimum volume in which the pathogen is present is noticed to increase up to 1 kg in all five treatments. In the presence of *Clostridium perfringens*, of a pathogen in 0.001 kg into the sludge was revealed, as well as into the final composts and vermicomposts (in all five treatments). There were not any changes in this parameter in compost compared to the vermicompost.

Biotests were performed in order to finalize the evaluation of the end-product for correspondence with the legislation. At the end of the study, we measured root and stem length and leaf fresh weight of pepper (*Capsicum annuum L.*). The results showed a similarity between treatments from 1 to 3 in both studied concentration (Fig. 5.5).

Slightly reduced growth was found in the case of fourth and fifth treatments regarding some of the parameters. Nevertheless, the pepper plants in all treatments and all vermicompost concentrations showed much higher growth in folds compared to the control treatment (without vermicompost). The results clearly confirm the necessity of addition of compost in soils in order to overcome the lack of nutrients and to improve soil structure. Regarding the requirements of the legislation, in the treatments with compost no reduction of growth was found. Fig. 5.6 shows the aspect of plants from different treatments. On the other hand, Pb and Cd were not found in the tissues of the plants, while Cu concentration was below 1 mg/kg DW that was similar to the concentration in the plants grown in control treatment (results not shown).

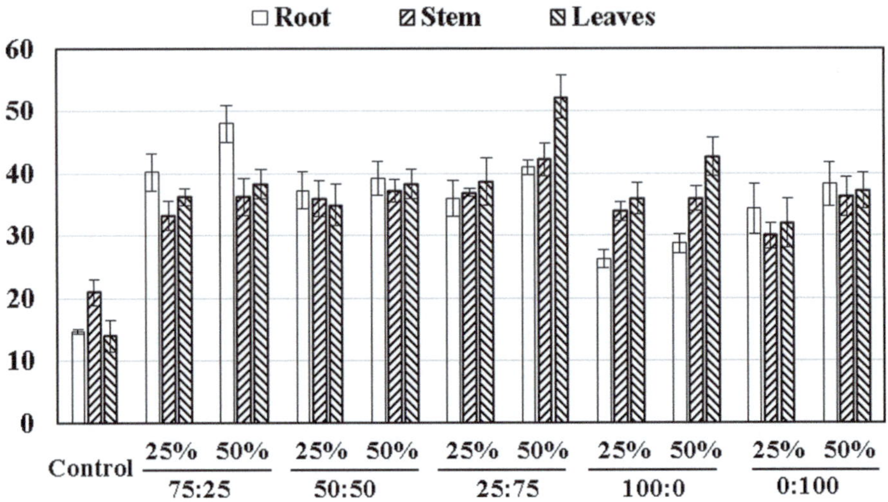

Fig. 5.5 Growth parameters of pepper plants [root and stem length (cm) and leaves' fresh weight(g)] grown in soil supplemented with vermicompost in the concentration of 25 and 50%. The results represent the mean of three replicates and the standard error

Fig. 5.6 Pictures of representative plants of each treatment comparing consecutively control treatment (without compost) to the other ones (concentration of compost 25 and 50%)

4 Conclusion

Sewage sludge is one of the most important sources of nutrient elements. It could be a successful partial substitute of phosphorus in agricultural soils. However, in the world different strategic decisions exist concerning the fate of sewage sludge. The major public concern is attributed to the contaminants in the sludge that could be very important depending on the source. Excessive amounts of heavy metals or pathogens are successfully reduced by different kinds of treatments, including biological one, decreasing significantly the volume of contaminated waste, thus decreases the environmental load of contaminants. These considerations are taken into account in the limitations of legislation for the utilization of treated sludge in agriculture and are very important from the point of view of food safety. Although sewage sludge treatment could not give a full guaranty for safety product, here is the role of legislation and controlling bodies, even self-control of the landowner to not permit a minimum doubt for food chain translocation of contaminants. The technology development in last years may support in-situ analysis of sewage sludge, by-products, and soils with the purpose to help to farmers, water associations, and controlling bodies to be more exact and to fulfill the requirements of the legislation.

Acknowledgments This paper was partially supported by the Bulgarian Ministry of Education and Science under the National Research Programme "Healthy Foods for a Strong Bio-Economy and Quality of Life" approved by DCM # 577 / 17.08.2018.

Authors' Contribution Study concepts, design, and editing: S. Shilev; Abstract and Introduction: S. Shilev; Sects. 5.2.1, 5.2.2, 5.2.3.2, and 5.2.4: H. Azaizeh; Sect. 5.2.3: D. Angelova and S. Shilev; Sect. 5.3: D. Angelova and S. Shilev; Conclusion: S. Shilev; Manuscript final version approval: All authors read and approved the manuscript.

References

Abbas T, Rizwan M, Ali S, Zia-ur-Rehman M, Farooq Qayyum M, Abbas F, Hannan F, Rinklebe J, Sik Ok Y (2017) Effect of biochar on cadmium bioavailability and uptake in wheat (*Triticum aestivum* L.) grown in a soil with aged contamination. Ecotoxicol Environ Saf 140:37–47

Abdelhadi OS, Dosoretz GC, Rytwo G, Gerchman Y, Azaizeh H (2017) Production of biochar from olive mill solid waste for heavy metal removal. Bioresour Technol 244:759–767

Aboulfotoh AM (2018) Anaerobic digestion model no. 1 simulation of high solids anaerobic digestion with feasibility study for El Gabal El Asfar water resource recovery facility. Water Environ Res 90(3):197–205

Aldin S, Tu F, Nakhla G, Ray MB (2011) Simulating the degradation of odor precursors in primary and waste-activated sludge during anaerobic digestion. Appl Biochem Biotech 164:1292–1304

Ali S, Rizwan M, Qayyum MF, Ok YS, Ibrahim M, Riaz M et al (2017) Biochar soil amendment on alleviation of drought and salt stress in plants: a critical review. Environ Sci Pollut Res 24:12700–12712

Angelova D, Shilev S, Naydenov M (2016) Composting of sewage sludge at large scale for subsequent utilization in agriculture. In: Filcheva, Stefanova, Ilieva (Eds), 4th Nat. conf. of BHSS with 8–10 September, 2016, Sofia, ISBN 978-619-90189-2-7, pp 285–295

Antoniadis V, Tsadilas CD, Samaras V (2010) Trace element availability in a sewage sludge-amended cotton grown Mediterranean soil. Chemosphere 80:1308–1313

Antunes E, Schuman J, Brodie G, Jacob MV, Schneider PA (2017) Biochar produced from biosolids using asingle-mode microwave: characterization of its potential for phosphorus removal. J Envir Manag 196:119–126

Appels L, Baeyens J, Degreve J, Dewil R (2008) Principles and potential of the anaerobic digestion of waste-activated sludge. Prog Energy Combust 34:755–781

Ariunbaatar J, Panico A, Esposito G, Pirozzi F, Lens PNL (2014) Pretreatment methods to enhance anaerobic digestion of organic solid waste. Appl Energy 123:143–156

Ariunbaatar J, Panico A, Yeh DH, Pirozzi F, Lens PNL, Esposito G (2015) Enhanced mesophilic anaerobic digestion of food waste by thermal pretreatment: substrate versus digestate heating. Waste Manag 46:176–181
Bai Y, Zang C, Gu M, Gu C, Shao H, Guan Y, Wang X, Zhou X, Shan Y, Feng K (2017) Sewage sludge as an initial fertility driver for rapid improvement of mudflat salt-soils. Sci Total Environ 578:47–55
Banegas V, Moreno JL, Moreno JI, García C, León G, Hernández T (2007) Composting anaerobic and aerobic sewage sludges using two proportions of sawdust. Waste Manag 27:1317–1327
Batstone DJ, Keller J, Angelidaki I, Kalyuzhnyi SV, Pavlostathis SG, Rozzi A, Sanders WTM, Siegrist H, Vavilin VA (2002) Anaerobic digestion model no.1, scientific and technical report no.13. IWA Publishing, London
Beck AJ, Johnson DL, Jones KC (1996) The form and bioavailability of non-ionic organic chemicals in sewage sludge-amended agricultural soils. Sci Total Environ 185(1–3):125–149
Beline F, Rodriguez-Mendez R, Girault R, Le Bihan Y, Lessard P (2017) Comparison of existing models to simulate anaerobic digestion of lipid-rich waste. Bioresour Technol 226(99):107
Bogdanov P (1998) Casting Call 1:2. http://vermico.com/news
Brisolara KF, Qi Y (2013) Biosolids and sludge management. Water Environ Res 85:1283–1297
Brodie HL, Francis RG, Lewis EC (1994) What makes good compost? Biocycle 35:60–68
Butt TE, Gouda HM, Baloch MI, Paul P, Javadi AA, Alam A (2014) Literature review of baseline study for risk analysis the landfill leachate case. Environ Int 63:149–162
Butt TE, Javadi AA, Nunns MA, Beal CD (2016) Development of a conceptual framework of holistic risk assessment landfill as a particular type of contaminated land. Sci Total Environ 569570:815–829
Carrère H, Dumas C, Battimelli A, Batstone DJ, Delgenès JP, Steyer JP, Ferrer I (2010) Pretreatment methods to improve sludge anaerobic degradability: a review. J Hazard Mat 183:1–15
Cekmecelioglu D, Demirci A, Graves RE, Davitt NH (2005) Optimization of windrow food waste composting to inactivate pathogenic microorganisms. Trans ASAE 48(5):2023–2032
Chakraborty D, Karthikeyan OP, Selvam A, Wong JWC (2017) Co-digestion of food waste and chemically enhanced primary treated sludge in a continuous stirred tank reactor. Biomass Bioenergy 111:232. https://doi.org/10.1016/j.biombioe.2017.06.002
Chen T, Zhang Y, Wang H, Lu W, Zhou Z, Zhang Y, Ren L (2014) Influence of pyrolysis temperature and heavy metal adsorptive performance of biochar derived from municipal sewage sludge. Bioresour Technol 164:47–54
Chen Z, Zhang W, Wang D, Ma T, Bai R (2015) Enhancement of activated sludge dewatering performance by combined composite enzymatic lysis and chemical re-flocculation with inorganic coagulants: kinetics of enzymatic reaction and re-flocculation morphology. Water Res 83:367–376
Cincinelli A, Martellini T, Misuri L, Lanciotti E, Sweetman A, Laschi S, Palchetti I (2012) PBDEs in Italian sewage sludge and environmental risk of using sewage sludge for land application. Environ Pollut 161:229–234
Clarke RM, Cummins E (2015) Evaluation of classic and emerging contaminants resulting from the application of biosolids to agricultural lands: a review. Hum Ecol Risk Assess 21:492–513
Czekała W, Malinska K, Cáceres R, Janczak D, Dach J, Lewicki A (2016) Co-composting of poultry manure mixtures amended with biochar - the effect of biochar on temperature and C-CO_2 emission. Bioresour Technol 200:921–927
Dalrymple OK, Halfhide T, Udom I, Gilles B, Wolan J, Zhang Q, Ergas S (2013) Wastewater use in algae production for generation of renewable resources: a review and preliminary results. Aquatic Biosys 9:1–11
Davoli E, Fattore E, Paiano V, Colombo A, Palmiotto M, Rossi AN, Il Grande M, Fanelli R (2010) Waste management health risk assessment: a case study of a solid waste landfill in South Italy. Waste Manag 30(89):1608–1613
Dignac MF, Urbain V, Rybacki D, Bruchet A, Snidaro D, Scribe P (1998) Chemical description of extracellular polymers: implication on activated sludge floc structure. Water Sci Technol 38:45–53

Dominguez A, Fernandez Y, Fidalgo B, Pis JJ, Menendez JA (2008) Bio-syngas production with low concentrations of CO_2 and CH_4 from microwave-induced pyrolysis of wet and dried sewage sludge. Chemosphere 70:97–403

Dowd SF, Gerba CP, Pepper IL, Pillai SD (2000) Ecological risk assessment: bioaerosol transport modeling and risk assessment in relation to biosolid placement. J Environ Qual 29:343–348

Dursun D, Dentel SK (2009) Toward the conceptual and quantitative understanding of biosolids conditioning: the gel approach. Water Sci Technol 59:1679–1685

Eastman BR, Kane PN, Edwards CA, Trytek L, Gunadi B, Stermer AL, Mobley JR (2001) The effectiveness of Vermiculture in human pathogen reduction for USEPA biosolids stabilization. Compost Sci Util 9(1):38–49

Eid EM, Alrumman SA, El-Bebany AF, Hesham AEL, Taher MA, Fawy KF (2017) The effects of different sewage sludge amendment rates on the heavy metal bioaccumulation, growth and biomass of cucumbers (*Cucumis sativus* L.). Environ Sci Pollut Res 24:16371–16382. https://doi.org/10.1007/s11356-017-9289-6

El-Haddad M, Zayed MS, El-Sayed G, Hassanein M, El-Satar AA (2014) Evaluation of compost, vermicompost and their teas produced from rice straw as affected by addition of different supplements. Annals Agric Sci 59:243–251

Elissen HJH, Mulder WJ, Hendrickx TLG, Elbersen HW, Beelen B, Temmink H, Buisman CJN (2010) Aquatic worms grown on biosolids: biomass composition and potential applications. Bioresour Technol 101:804–811

Environmental, economic and social impacts of the use of sewage sludge on land – final report, RPA, Milieu Ltd and WRc for the European Commission. Report on critical raw materials for the EU, published 2014

European Communities (2001) Disposal and recycling routes for sewage sludge part 3-scientific and technical report; Office for Official Publications of the European Communities: Luxembourg, p 70, ISBN 92-894-1800-1

European Communities (2008) Directive 2008/98/EC of the European Parliament and of the council of 19 November 2008 on waste and repealing certain directi.ves. Off J Eur Union L 312, 330, <http://eur-lex.europa.eu/legal-content/EN/TXT/PDF/?uri=CELEX:32008L0098&from =EN

Feinstein MS et al. (1980) Sludge composting and utilization rational approach to process control. Final Report, Project No. C-340-678-01-1;

Fytili D, Zabaniotou A (2008) Utilization of sewage sludge in EU application of old and new methods e a review. Renew. Sust Energ Rev 12:116–140

Garg VK, Kaushik P (2005) Vermistabilization of textile mill sludge spiked with poultry droppings by an epigeic earthworm *Eisenia foetida*. Bioresour Technol 96:1063–1071

Goswami L, Sarkar S, Mukherjee S, Das S, Barman S, Raul P, Bhattacharyya P, Mandal NC, Bhattacharya S, Bhattacharya SS (2014) Vermicomposting of tea factory coal ash: metal accumulation and metallothionein response in *Eisenia fetida* (Savigny) and Lampito mauritii (Kinberg). Bioresour Technol 166:96–102. https://doi.org/10.1016/j.biortech.2014.05.032

Guo J, Li J, Chen H, Bond PL, Yuan Z (2017) Metagenomic analysis reveals wastewater treatment plants as hotspots of antibiotic resistance genes and mobile genetic elements. Water Res 123:468–478

Hartenstein R, Neuhauser EF, Kaplan DL (1979) Reproductive potential of the earthworm Eisenia foetida. Oecologia 43(3):329–340

He XT, Logan TJ, Traina SJ (1995) Physical and chemical characteristics of selected U.S. municipal solid-waste composts. J Environ Qual 24:543–552

He X, Zhang Y, Shen M, Zeng G, Zhou M, Li M (2016) Effect of vermicomposting on concentration and speciation of heavy metals in sewage sludge with additive materials. Bioresour Technol 218:867–873

Heng GC, Hasnain Isa M, Lim JW, Hol YC, Zinatizadeh A-K (2017) Enhancement of anaerobic digestibility of waste activated sludge using photo-Fenton pretreatment. Environ Sci Pollut Res DOI 24:27113. https://doi.org/10.1007/s11356-017-0287-5

Hossain MK, Strezov V, Chan KY, Ziolkowski A, Nelson PF (2011) Influence of pyrolysis temperature on production and nutrient properties of wastewater sludge biochar. J Environ Manag 92:223–228

Huang K, Xia H, Wu Y, Chen J, Cui G, Li F, Chen Y, Wu N (2018) Effects of earthworms on the fate of tetracycline and fluoroquinolone resistance genes of sewage sludge during vermicomposting. Bioresour Technol 259:32–39. https://doi.org/10.1016/j.biortech.2018.03.021

Iacovidou E, Ohandja D-G, Voulvoulis N (2012) Food waste co-digestion with sewage sludge - realising its potential in the UK. J Environ Manag 112:267–274

Iannotti DA, Pang T, Toth BL et al (1993) A quantitative respirometric method for monitoring compost stability. Compost Sci Utiliz 1:5245

Inbar Y, Chen Y, Hadar Y (1990) Humic substances formed during the composting of organic matter. Soil Sci Soc Am J 54(5):1316–1323

Inguanzo M, Dominguez A, Menendez JA, Blanco CG, Pis JJ (2002) On the pyrolysis of sewage sludge: the influence of pyrolysis conditions on solid, liquid and gas fractions. J Anal Appl Pyrolysis 63:209–222

Insam H, Markt R (2016) Comment on "synergistic co-digestion of solid-organicwaste and municipal-sewage-sludge: 1 plus 1 equals more than 2 in terms of biogas production and solids reduction" [Water Res 87:416–423]. Water Res 95:392–393

Jeong HS, Suh CW, Lim JL, Lee SH, Shin HS (2005) Analysis and application of ADM1 for anaerobic methane production. Bioprocess Biosyst Eng 27(2):81–89

Kalmykova Y, Fedje KK (2013) Phosphorus recovery from municipal solid waste incineration fly ash. Waste Manag 33:1403–1410

Kalmykova Y, Harder R, Borgestedt H, Svanäng I (2012) Pathways and management of phosphorus in urban areas. J Ind Ecol 6:928–939

Karkman A, Do TT, Walsh F, Virta MPJ (2017) Antibiotic-resistance genes in waste water. Trends Microbiol 26(3):220–228

Keeling AA, Paton IK, Mullett JAJ (1994) Germination and growth of plants in media containing unstable refuse derived compost. Soil Biol Biochem 26:767–772

Kelessidis A, Stasinakisa AS (2012) Comparative study of the methods used for treatment and final disposal of sewage sludge in European countries. Waste Manag 32(6):1186–1195

Khan S, Wang N, Reid BJ, Freddo A, Cai C (2013) Reduced bioaccumulation of PAHs by Lactuca satuva L. grown in contaminated soil amended with sewage sludge and sewage sludge derived biochar. Environ Pollut 175:64–68

Kim M, Chowdhury MMI, Nakhla G, Keleman M (2015) Characterization of typical household food wastes from disposers: fractionation of constituents and implications for resource recovery at wastewater treatment. Bioresour Technol 183:61–69

Kim M, Chowdhury MMI, Nakhla G, Keleman M (2017) Synergism of co-digestion of food wastes with municipal wastewater treatment biosolids. Waste Manag 61:473–483

Koch K, Plabst M, Schmidt A, Helmreich B, Drewes JE (2016) Co-digestion of food waste in a municipal wastewater treatment plant: comparison of batch tests and full-scale experiences. Waste Manag 47(Part A):28–33

Krah D, Ghattas A-K, Wick A, Bröder K, Ternes TA (2016) Micropollutant degradation via extracted native enzymes from activated sludge. Water Res 95:348–360. https://doi.org/10.1016/j.watres.2016.03.037

Krishnan S, Singh L, Sakinah M, Thakur S, Wahid ZA, Ghrayeb OA (2017) Role of organic loading rate in bioenergy generation from palm oil mill effluent in a two-stage up-flow anaerobic sludge blanket continuous-stirred tank reactor. J Clean Prod 142:3044–3049

Lehmann J, Joseph S (2009) Biochar for environmental management: an introduction. In: Lehmann J, Joseph S (eds) Biochar for environmental management. Earthscan, London

Lester JN, Sterrit RM, Kirk PWW (1983) Significance and behaviour of heavy metals in wastewater treatment process. II. Sludge treatment and disposal. Sci Total Environ 30:45–83

Li Z, Ryan JA, Chen J-L, Al-Abed SR (2001) Adsorption of cadmium on biosolids-amended soils. J Environ Qual 30(3):903

Li W, Shi Y, Gao L, Liu J, Cai Y (2013) Occurrence, distribution and potential affecting factors of antibiotics in sewage sludge of wastewater treatment plants in China. Sci Total Environ 445-446:306–313

Li R, Zhang Z, Li Y, Teng W, Wang W, Yang T (2015) Transformation of apatite phosphorus and non-apatite inorganic phosphorus during incineration of sewage sludge. Chemosphere 141:57–61. https://doi.org/10.1016/j.chemosphere.2015.05.094

Li Q, Li H, Wang G, Wang X (2017) Effects of loading rate and temperature on anaerobic co-digestion of food waste and waste activated sludge in a high frequency feeding system, looking in particular at stability and efficiency. Bioresour Technol 237:231–239

Lim SL, Lee LH, Wu TY (2016) Sustainability of using composting and vermicomposting technologies for organic solid waste biotransformation: recent overview, greenhouse gases emissions and economic analysis. J Cleaner Product 111(Part A):262–278

Lisboa MS, Lansing S (2013) Characterizing food waste substrates for co-digestion through biochemical methane potential (BMP) experiments. Waste Manag 33:2664–2669

Lohani SP, Wang S, Lackner S, Horn H, Khanal SN, Bakke R (2016) ADM1 modeling of UASB treating domestic wastewater in Nepal. Renew Energy 95:263–268

Lu Y, Wu X, Guo J (2009) Characteristics of municipal solid waste and sewage sludge co-composting. Waste Manag 29(3):1152–1157

Malinska K, Zabochnicka-Swiatek M, Dach J (2014) Effects of biochar amendment on ammonia emission during composting of sewage sludge. Ecological Eng 71:474–478

Malinska K, Zabochnicka-Swiatek M, Cáceres R, Márfa O (2016) The effect of precomposted sewage sludge mixtures amended with biochar on the growth and reproduction of *Eisenia fetida* during laboratory vermicomposting. Ecological Eng 90:35–41

Malinska K, Golanska M, Cáceres R, Rorat A, Weisser P, Ewelina S (2017) Biochar amendment for integrated composting and vermicomposting of sewage sludge – The effect of biochar on the activity of *Eisenia fetida* and the obtained vermicompost. Bioresour Technol 225:206–214. https://doi.org/10.1016/j.biortech.2016.11.049

Mao D, Yu S, Rysz M, Luo Y, Yang F, Li F, Hou J, Mu Q, Alvarez PJJ (2015) Prevalence and proliferation of antibiotic resistance genes in two municipal wastewater treatment plants. Water Res 85:458–466

Mathur SP, Owen G, Dinel H, Schnitzer M (1993) Determination of compost biomaturity I. Literature review. Biol Agric Hortic 10:1065–1085

Mehariya S, Patel AK, Obulisamy PK, Punniyakotti E, Wong JWC (2018) Co-digestion of food waste and sewage sludge for methane production: Current status and Perspective. Biores Technol 265:519–531. https://doi.org/10.1016/j.biortech.2018.04.030

Méndez A, Terradillos M, Gascó G (2013) Physicochemical and agronomic properties of biochar from sewage sludge pyrolysed at different temperatures. J Anal Appl Pyrolysis 102:124–130

Mendez L, Mahdy A, Demuez M, Ballesteros M, González-Fernández C (2014) Effect of high pressure thermal pretreatment on *Chlorella vulgaris* biomass: organic matter solubilisation and biochemical methane potential. Fuel 117:674–679. https://doi.org/10.1016/j.fuel.2013.09.032

Miller-Robbie L, Ulrich BA, Ramey DF, Spencer KS, Herzog SP, Cath TY, Stokes JR, Higgins CP (2015) Life cycle energy and greenhouse gas assessment of the co-production of biosolids and biochar for land application. J Clean Prod 91:118–127

Mininni G, Laera G, Bertanza G, Canato M, Sbrilli A (2015) Mass and energy balances of sludge processing in reference and upgraded wastewater treatment plants. Environ Sci Pollut Res 22:7203–7215

Miron Y, Zeeman G, van Lier JB, Lettinga G (2000) The role of sludge retention time in the hydrolysis and acidification of lipids, carbohydrates and proteins during digestion of primary sludge in CSTR systems. Water Res 34:1705–1713

Mishra H, Karmakar S, Kumar R, Singh J (2016) A framework for assessing uncertainty associated with human health risks from MSW landfill leachate contamination. Risk Analysis, Version online: 24 SEP 2016 | https://doi.org/10.1111/risa.12713

Mustafa N, Elbeshbishy E, Nakhla G, Zhu J (2014) Anaerobic digestion of municipal wastewater sludges using anaerobic fluidized bed bioreactor. Bioresour Technol 172:461–466

National strategic plan for management of sludge from WWTP in Republic of Bulgaria for the period 2014–2020, MOEW, Sofia, 2014
Nayak AK, Varma S, Kalamdhad AS (2013) Effects of various C/N ratios during vermicomposting of sewage sludge using *Eisenia fetida*. J Environ Sci Technol 6(2):63–78
Neuhauser EF, Loehr RC, Malecki MR (1998) The potential of earthworms for managing sewage sludge. In: Edwards CA, Neuhauser EF (eds) Earthworms in waste and environmental management. SPB Academic Publishing, Netherlands, pp 9–20
Nordlander E, Thorin E, Yan J (2017) Investigating the possibility of applying an ADM1 based model to a full-scale co-digestion plant. Biochem Eng J 120:73–83
Ott C, Rechberger H (2012) The European phosphorus balance. Resour Conserv Recycl 60:159–172
Paladino O, Massab M (2017) Health risk assessment as an approach to manage an old landfill and to propose integrated solid waste treatment: a case study in Italy. Waste Manag 68:344–354. https://doi.org/10.1016/j.wasman.2017.07.021
Passos F, Uggetti E, Carrère H, Ferrer I (2014) Pretreatment of microalgae to improve biogas production: a review. Bioresour Technol 172:403–412. https://doi.org/10.1016/j.biortech.2014.08.114
Ramanan R, Kim BH, Cho DH, Oh HM, Kim HS (2016) Algae-bacteria interactions: evolution, ecology and emerging applications. Biotechnol Adv 34:14–29. https://doi.org/10.1016/j.biotechadv.2015.12.003
Regulation for separate recollection of biowaste and treatment of biodegradable wastes. Decree of Council of Ministers № 20/25.01.2017
Regulation from 1.08.2008 for the way of utilization of sludge from wastewater purification through their use in agriculture. SG, 112 from 23.12.2004
Rehmana RA, Rizwanb M, Qayyuma MF, Alib S, Zia-ur-Rehmanc M, Zafar-ul-Hyea M, Hafeezd F, Iqbal MF (2018) Efficiency of various sewage sludges and their biochars in improving selected soil properties and growth of wheat (*Triticum aestivum*). J Environ Manag 223:607–613
Report of MOEW to the 2nd national conference "Utilization of sewage sludge from WWTP", Bulgarian water association, Starosel village, 2018
REPORT ON CRITICAL RAW MATERIALS FOR THE EU, Report of the Ad hoc Working Group on defining critical raw materials, May 2014
Rizwan M, Ali S, Adrees M, Rizvi H, Zia-ur-Rehman M, Hannan F, Qayyum MF, Hafeez F, Ok YS (2016) Cadmium stress in rice: toxic effects, tolerance mechanisms, and management: a critical review. Environ Sci Pollut Res 23(18):17859–17879
Rizwan M, Ali S, Qayyum MF, Ok YS, Zia-ur-Rehman M, Abbas Z, Hannan F (2017) Use of maize (Zea mays L.) for phytomanagement of Cd-contaminated soils: a critical review. Environ Geochem Health 39:259–277
Rizzo L, Manaia C, Merlin C, Schwartz T, Dagot C, Ploy MC, Michael I, Fatta-Kassinos D (2013) Urban wastewater treatment plants as hotspots for antibiotic resistant bacteria and genes spread into the environment: a review. Sci Total Environ 447:345–360
Rorat A, Suleiman H, Grobelak A, Grosser A, Kacprzak M, Płytycz B, Vandenbulcke F (2015) Interactions between sewage sludge-amended soil and earthworms-comparison between *Eisenia fetida* and *Eisenia andrei* composting species. Environ Sci Pollut Res 23(4):3026–3035
Roy MM, Dutta A, Corscadden K, Havard P, Dickie L (2011) Review of biosolids management options and co-incineration of a biosolid-derived fuel. Waste Manag 31:2228–2235
Rynk R (1992) On-farm composting handbook. Natural Resource, Agriculture, and Engineering Service, Ithaca
Samara E, Matsi T, Balidakis A (2017) Soil application of sewage sludge stabilized with steelmaking slag and its effect on soil properties and wheat growth. Waste Manag 68:378–387
Sartaj M, Fernandes L, Patni N (1997) Performance of forced, passive, and natural aeration methods for composting manure slurries. Trans ASAE 40:457–463. https://doi.org/10.13031/2013.21273
Saxena M, Chauhan A (1998) Flyash vermicomposting from non-organic wastes. Pollut Res 17:5–11
Schievano A, Tenca A, Scaglia B, Merlino G, Rizzi A, Daffonchio D, Oberti R, Adani F (2012) Two-stage vs single-stage thermophilic anaerobic digestion: comparison of energy production and biodegradation efficiencies. Environ Sci Technol 46:8502–8510

Shang Y, Johnson R, Sieger R, Forbes R (2005) Evaluation of the IWA anaerobic digestion model (ADM1) for simulating full-scale anaerobic sewage sludge digestion. Water Sci Technol 52:487–492

Sharma K, Garg V (2017) Management of food and vegetable processing waste spiked with buffalo waste using earthworms (*Eisenia fetida*). Environ Sci Pollut Res Int 24(8):7829–7836. https://doi.org/10.1007/s11356-017-8438-2

Shilev S, Naydenov M, Vancheva V, Aladjadjiyan A (2007) Composting of food and agricultural wastes. In: Oreopoulou V, Russ W (Eds.), utilization of by-products and treatment of waste in the food industry", series: integrating safety and environmental knowledge into food studies towards European Sustainable Development, Vol. 3, ISBN-10 0-387-33511-0, ISBN-13 978-0-387-33511-7, Springer, pp. 283–302

Shilev S, Naydenov M, Gachev V, Rangova I, Babrikov T (2014) Compost incorporation in contaminated soil affects heavy metal mobility and accumulation in spinach. In: Industrial, medical and environmental applications of microorganisms: current status and trends, proc. V international conference of environmental, industrial and applied microbiology, Madrid, Spain, 2-4 October 2013. Wageningen Academic Publishers

Singh RP, Agrawal M (2008) Potential benefits and risks of land application of sewage sludge. Waste Manag 28:347–358

Smith J, Abegaz A, Matthews RB, Subedi M, Orskov ER, Tumwesige V, Smith P (2014) What is the potential for biogas digesters to improve soil carbon sequestration in Sub-Saharan Africa? Comparison with other uses of organic residues. Biomass Bioenergy 70:73–86

Song C, Li M, Jia X, Wei Z, Zhao Y, Xi B, Zhu C, Liu D (2014) Comparison of bacterial community structure and dynamics during the thermophilic composting of different types of solid wastes: anaerobic digestion residue, pig manure and chicken manure. Microb Biotechnol 7:424–433

Srivastava R, Kumar D, Gupta SK (2005) Bioremediation of municipal sludge by vermitechnology and toxicity assessment by *Allium cepa*. Bioresour Technol 96:1867–1871

Tambari L, Stentiford E (1990) Control of aeration in static pile composting. Waste Manag Res 8(4):299–306

Tchobanoglous G, Stensel HD, Tsuchihashi R, Burton F (2014) Wastewater engineering: treatment and resource recovery, 5th edn. McGraw-Hill, Singapore

Tomei MC, Bertanza G, Canato M, Heimersson S, Laera G, Svanström M (2016) Techno-economic and environmental assessment of upgrading alternatives for sludge stabilization in municipal wastewater treatment plants. J Clean Prod 112:3106–3115

Udayanga WC, Veksha A, Giannis A, Lisak G, Chang VWC, Lim TT (2018) Fate and distribution of heavy metals during thermal processing of sewage sludge. Fuel 226:721–744

Wang H, Brown SL, Magesan GN, Slade AH, Quintern M, Clinton PW, Payn TW (2008) Technological options for the management of biosolids. Environ Sci Pollut R 15:308–317

Wang L, Zhang Y, Lian J, Chao J, Gao Y, Yang F, Zhang L (2013) Impact of fly ash and phosphatic rock on metal stabilization and bioavailability during sewage sludge vermicomposting. Bioresour Technol 136:281–287

Werle S, Wilk RK (2010) A review of methods for the thermal utilization of sewage sludge: the polish perspective. Renew Energ 35:1914–1919

Wingender J, Jäger KE, Flemming HC (1999) Interaction between extracellular polysaccharides and enzymes. In: Wingender J, Neu TR (eds) Microbial extracellular polymeric substances. Springer, Heidelberg, pp 1–19

Wollmann I, Gauro A, Müller T, Möller K (2017) Phosphorus bioavailability of sewage sludge, based recycled fertilizers. J Plant Nutr Soil Sci, vol 181, p 158. https://doi.org/10.1002/jpln.201700111

Xiong Q, Zhou M, Liu M, Jiang S, Hou H (2018) The transformation behaviors of heavy metals and dewaterability of sewage sludge during the dual conditioning with Fe^{2+}- sodium persulfate oxidation and rice husk. Chemosphere 208:93–100

Yang Y, Li B, Zou S, Fang HH, Zhang T (2014) Fate of antibiotic resistance genes in sewage treatment plant revealed by metagenomic approach. Water Res 62:97–106

Yi J, Dong B, Jin J, Dai X (2014) Effect of increasing total solids contents on anaerobic digestion of food waste under mesophilic conditions: performance and microbial characteristics analysis. PLoS One 9:7

Yousefi J, Younesi H, Ghasempoury SM (2013) Co-composting of municipal solid waste with sawdust: improving compost quality, CLEAN – soil air water 41(2), February 2013

Yue Y, Cui L, Lin Q, Li G, Zhao X (2017) Efficiency of sewage sludge biochar in improving urban soil properties and promoting grass growth. Chemosphere 173:551–556

Zhang W, Mao S, Chen H, Huang L, Qiu R (2013) Pb(II) and Cr(VI) sorption by biochars pyrolyzed from the municipal wastewater under different heating conditions. Bioresour Technol 147:545–552

Zhen G, Lu X, Kato H, Zhao Y, Li Y (2017) Overview of the pretreatment strategies for enhancing sewage sludge disintegration and subsequent anaerobic digestion: current advances, full scale application and future perspectives. Renew Sust Energ Rev 69:559–577

Microbial Bioconversion of Agricultural Wastes for Rural Sanitation and Soil Carbon Enrichment

6

Hassan Etesami, Arash Hemati, and Hossein Ali Alikhani

1 Introduction

Million tons of organic municipal and agricultural solid wastes in the world are produced annually through human activities. The lack of proper management of these materials causes severe environmental damage, such as greenhouse gas emissions, municipal waste accumulation, and pollution of water and soil (Ahmad et al. 2007). Correct management of wastes is one of the determining factors in urban cleaning and maintaining public health (Saha et al. 2010; Sharholy et al. 2008). The rapid expansion of cities and the massive migration from rural to urban centers have led to a significant increase in per capita production of solid municipal wastes. Increased solid waste production, limited landfill space, and more stringent environmental regulations for landfill and waste incineration facilities have increased waste disposal costs, especially in developing countries. Therefore, municipalities and local governments are under heavy pressure to find sustainable and affordable strategies for solid waste management. On the other hand, management of waste and plant residues and returning them to agricultural and horticultural land is one of the most important ways to manage soil fertility for sustainable agriculture (Saha et al. 2010; Sharholy et al. 2008; Tosun et al. 2008). One of the basic principles for managing organic wastes is the production of compost. Composting is the biological decomposition and stabilization of organic matter under special conditions (humidity and aeration) that is stabilized by heat generation (thermophilic conditions), and the final product is also free from pathogens and plant seeds (Kluczek-Turpeinen et al. 2003).

H. Etesami (✉) · H. A. Alikhani
Department of Soil Science, University College of Agriculture and Natural Resources, University of Tehran, Tehran, Iran
e-mail: hassanetesami@ut.ac.ir

A. Hemati
Department of Soil Science, University of Tabriz, Tabriz, Iran

D. P. Singh et al. (eds.), *Microbial Interventions in Agriculture and Environment*,
https://doi.org/10.1007/978-981-13-8383-0_6

2 Compost Production

In the preparation of compost, the Chinese were among the nations who 4000 years ago made a suitable fertilizer from plant and human wastes and used it for soil fertilization. Since then, the issue of the use of wastes in agriculture has been a major activity in different countries. The first compost experiments were carried out during 1926–1941 by Waksman (1936). Among the many scholars in the world who have done extensive research in the field of composting or, principally, the use of municipal waste materials, the Gotaas name is very familiar to most specialists in the field (Gotaas 1956). From 1950 to 1952, he conducted several studies on compost that are noteworthy. In recent years, the mechanization of composting in Europe and the United States has produced many results, which are being used now. Other scientists in Italy and in the United States have made innovations in this field that recorded them in their own name. DANO's system in Denmark and then the a mechanical silo- type digester known as the Bio-stabilizer developed in this country. However, the continuation of research and operations on composting were carried out to the extent that in 1975 the number of compost-producing factories in Europe was reported up to more than 200 factories (Epstein 1994).

The compost is derived from the microbial degradation of organic residues (including agricultural and urban wastes) under proper aeration and temperature conditions. In other words, composting is a biological process which converts heterogeneous organic wastes into humus-like substances by a group of microorganisms such as fungi, bacteria, and actinomyces at mesophilic and thermophilic temperatures, suitable moisture, and intermittent aeration. During the composting process, due to the production of antibiotics, relative pasteurization is also carried out (Insam et al. 2013; Kutzner 1999).

The compost has a high percentage of humus. Humus is a soil amendment that improves the biological conditions and performance of soil organisms. The important point is that humus contains a lot of nitrogenous substances that are gradually released in the soil and are then taken up by plants. Compost is a valuable resource for the supply of organic matter required for most agricultural soils (Hargreaves et al. 2008). The compost produced from a variety of wastes and organic residues can be used for a variety of soils and various products. But the amount of compost consumption is also very important in this regard. The amount of organic fertilizer application can vary according to the type of crop, soil type, available compost type, and climatic conditions. For example, the type of desired crop is one of the most important factors in determining the amount of consumed compost. In this case, the amount of nutrients needed for the crop cultivation should be considered and the amount of compost used should be such that it does not damage seed germination.

Adding organic fertilizers to heavy-textured soils like clay soils increases soil porosity and permeability, and if the soil texture is sandy and light (the light-textured sandy soils), organic fertilizers increase water holding capacity in the soil and reduce the amount of using water. Generally, the amount of using compost in the soil varies according to soil texture, gradient of land, and groundwater depth. In areas where rainfall is an average of 1250 mm per year, compost consumption is recommended to be about 12.5 tons per hectare, and in low rainfall areas with an average rainfall of 500 mm and drylands, 5 tons per hectare and 2.5 tons per hectare, respectively, are recommended (Kutzner 1999).

2.1 Compost Production Process

From a microbiological point of view, composting is an extremely complex process in which many microorganisms participate (Ryckeboer et al. 2003). The main groups of microorganisms involved in the composting process are bacteria and fungi (Hultman et al. 2010). Fungi play a more important role in this process, which is due to the ability of fungi to survive dry and acidic environments and to decompose organic matter with very low nitrogen, in which bacteria have a low efficiency (Ryckeboer et al. 2003). Some fungal strains have the highest potential for decomposition of lignin and organic compounds. A number of thermophilic fungi that have the ability to decompose these hardly degradable materials are presented in some reports (Kutzner 1999; Ryckeboer et al. 2003). Among the various stages of composting, the thermophilic stage reflects the maximum activity of the microbial population, which leads to degradation of organic matter increasingly through successive activities (Kutzner 1999). Despite the importance of heat-tolerant and mesophilic fungi in compost production, little research has been done in this field (Langarica-Fuentes et al. 2014). In thermophilic conditions, degradation-resistant materials are also decomposed. Lignin is one of these hardly degradable materials (Tuomela et al. 2000).

The process of composting is the decomposition of organic wastes in the presence of oxygen, which produces products such as water, ammonia, carbon dioxide, and heat. On the other hand, anaerobic processes lead to the decomposition of waste materials in an oxygen-free environment, the final product of which is methane, carbon dioxide, ammonia, and other gases (Suárez-Estrella et al. 2008). The final compost contains three main substances: humus, microbial mass, and ash. Humus contains ligninous compounds and degradation-resistant organic compounds, as well as the humic acids caused by biological degradation. Microbial mass (biomass) also includes dead and living microorganisms inside the compost pile. Ash contains minerals in the compost pile, which are a stable component of the compost pile (Marhuenda-Egea et al. 2007).

The materials used to produce compost should have the following four characteristics: (i) energy, (ii) food, (iii) water, and (iv) proper structure. The first three parameters are important for the activity and growth of microorganisms, and the last parameter is needed for proper aeration, which is also essential for the microorganisms in the compost pile (Hargreaves et al. 2008).

2.2 Composting Steps

Preparation of compost, which usually lasts from a few weeks to several months, consists of three main stages, namely, pre-active, active, and mature. The pre-active stage of compost occurs under mesophilic conditions. At this stage, microorganisms that act at the mesophilic temperature are propagated by the use of simple sugars and some proteins, resulting in a gradual increase in temperature in the range of 50–55 °C (Nelson and Babie 2005).

In the active stage of the compost, known as the thermophilic stage, part of organic compounds such as sugars, proteins, etc. is consumed by the activity of the mesophilic microorganisms, and the produced heat increases the thermophilic population (mainly fungi, actinomycetes, and a group of bacteria). The time to reach this stage is 3–15 days, and the temperature of the substrate is 46–60 °C. Fast increase in temperature, significant pH changes, complex organic matter decomposition, increase in CO_2 production, and a decrease in the C/N ratio are the characteristics of this step (Hargreaves et al. 2008).

Compost maturity stage is the longest stage in the preparation of compost, because, at this stage, compost maturity and changes leading to humidification occur. The onset of this phase is accompanied by a drop in temperature and in bioactivity. Under these conditions, some microorganisms that can decompose cellulosic and lignin compounds continue their activity. Decrease in toxic compounds, stabilization of pH and C/N ratio, increase of cation exchange capacity (CEC), and synthesis of humic substances also occur at this stage. Fulvic acid (FA) initially and humic acid (HA) are formed at the middle of this stage, which continue until the end of the composting time (Hargreaves et al. 2008).

Different temperature phases for the production of compost, in which microorganisms are active, include the mesophilic and thermophilic temperature range and the cooling of the compost pile (Galai et al. 2009; Tuomela et al. 2000).

In the mesophilic phase, which begins from the very first day, mesophilic microorganisms begin to grow and function inside the compost pile. Microorganisms that operate in the temperature range of 20–45 °C are part of this group, namely, mesophilic microorganisms. Most of the active microorganisms inside the compost pile are from this group (Xiao et al. 2009). Microorganisms known as thermophilic microorganisms are active in the temperature range of 45–65 °C (Awasthi et al. 2014). In the cooling phase, which occurs at the very last stages of the composting process, the temperature is around the ambient temperature. In this phase, microorganisms are also active (Xiao et al. 2009). When the temperature of the compost pile is above 60 °C, the diversity and activity of the bacteria are heavily influenced and reduced. The main reason for the low level of decomposition at high temperatures is not only due to the low population of microbes but also because of the low activity of the microbial population (Jurado et al. 2014).

2.3 Factors Affecting Composting Process

2.3.1 Temperature

Temperature is an important factor at different stages of composting. It indicates the intensity of the microbial activity of the compost, which is due to the use of simple compounds and energy liberation (self-heating phenomenon). This action leads to the release of energy and heat production. The generated heat makes environment conditions suitable for the activity of other microorganisms. In cold regions where energy exchange is rapid, the self-heating phenomenon overcomes composting problems (Haug 2018; Langarica-Fuentes et al. 2014).

Optimum temperature in the published reports has a wide range due to differences in the physical and chemical properties of compost, composting process type, and equipment and available facilities. According to some researchers, temperatures over 55 °C have inhibitory effects on some of the cellulose-degrading microorganisms, which lead to a decrease in the severity of degradation (Haug 2018). Pietro and Paola (2004) showed that the produced heat during composting is directly related to the physicochemical composition of the raw material of the compost. In this research, heat in several formulations such as wheat straw compost and chicken manure was studied. The results showed that 37.4% of the generated heat was the result of the bio-oxidation of the existing materials, and the rest was the energy generated from incomplete organic material oxidation. Therefore, maintaining the produced heat in the active phase has a significant effect on the development of the stage of maturation of the compost. In general, the pile temperature reaches 60–65 °C. The temperatures above the thermophilic temperatures effectively reduce the decomposition pathway within the compost pile. New research has shown that the proper temperature for compost production is determined by the activity of microorganisms. The temperature of the compost pile is regulated by adjusting the aeration rate and moisture content inside the compost pile. In some circumstances, suitable coatings are used to maintain the temperature inside the compost pile (Haug 2018; Pietro and Paola 2004).

2.3.2 Moisture

The amount of moisture in the compost plays an important role in the activity of microorganisms. Excessive moisture has an adverse effect on activity of aerobic microorganisms, which causes the production of immature compost with a low quality. Low moisture reduces the biological activity of microorganisms and reduces organic matter decomposition. The most suitable moisture for most composts is 55–65% by weight (Miller 1989).

2.3.3 Aeration

Due to bioactivity, large volumes of CO_2 are released during composting process. Due to not being exited CO_2 from the compost pile or the anaerobic conditions, there is an incomplete oxidation of organic matter that results in the production of intermediate compounds or types of organic acids. Therefore, the amount of aeration affects the quality of compost (Haug 2018) .

2.3.4 pH

pH plays a crucial role in all biological processes. The inappropriate pH will slow down or stop the chemical reactions. pH control in the range of 7–8 accelerates the decomposition of organic matter. pH changes depend on the activity of microorganisms in the composting process and the presence of sugar and protein compounds in the aqueous phase. In the logarithmic phase, the microbial activity in the neutral pH is the best. The logarithmic phase results from the microbial activity from mesophilic stage to thermophilic stage. Therefore, the decreased pH in the acid range causes delayed composting steps, especially from the mesophilic to thermophilic phase (Kutzner 1999; Sundberg et al. 2004).

2.3.5 Electrical Conductivity (EC)

Electrical conductivity is one of the most important factors in biological interactions. The EC of 5–25 dS/m for microorganisms is a good range. The process of changes in the various composting stages shows that in the active phase, an increase in EC occurs due to the degradation of organic matter, but in the late stages of compost, these changes are minor (Kutzner 1999).

2.3.6 Carbon to Nitrogen (C/N) Ratio

The ratio of C/N is often referred to as an index of compost stability and an estimate of compost quality. According to Hirai et al. (1983) and Bernai et al. (1998), the C/N ratio of 20 is the most appropriate ratio for determining the maturation of compost. The C/N ratio at the active stage of compost is decreased significantly due to organic matter decomposition and CO_2 exit, but in the maturity phase, the process of change is slowed down. Thus, the C/N ratio shows the intensity of activity of microorganisms decomposing available organic matter and compost stability over time. One of the effective factors at reducing the C/N ratio is the optimum temperature conditions for the activity of microorganisms. C/N ratio is one of the important parameters in the composting process, and this parameter affects other parameters such as microorganism's activity, nitrogen losses, and so on. For appropriate nutrient uptake by microorganisms, there must be a good ratio between carbon and nitrogen, which is between 20 and 30. If the amount of nitrogen is low, it directly affects the activity of microorganisms, and the organic matter decomposition will be slower. In this case, due to the excessive use of carbon by microorganisms, the amount of CO_2 production is increased. On the contrary, if nitrogen in the compost pile is high, the activity of microorganisms is increased, but on the other hand, the problem makes the smell more acute because oxygen is consumed heavily by microorganisms, and if the aeration is not done well, anaerobic conditions are created. On the other hand, the amount of excess nitrogen in the form of ammonia gas is removed from the compost pile. Also, in these conditions, less C/N ratio shortens duration of the thermophilic phase (Bernai et al. 1998; Haug 2018; Hirai et al. 1983).

If the C/N ratio is higher than 30 (though the ratio of 50–30 is within the permissible limits), the following problems are created:

(i) The time required to prepare the compost is increased because much time is needed for microorganisms to oxidize the excess carbon and remove carbon dioxide.
(ii) Adding the compost with a high C/N ratio to the soil causes the plant to be deprived of nitrogen, because microorganisms need nitrogen to oxidize excess carbon. In order to meet this need, microorganisms compete with plants to consume soil-soluble nitrogen. The result of this competition is to stop plant growth until a large amount of carbon is oxidized.
(iii) If a C/N ratio is higher than 30, due to the high carbon content and nitrogen deficiency, some microorganisms die to provide the nitrogen needed by other microorganisms. Generally, this decrease in nitrogen reduces the number of microorganisms and, as a result, slows down the fermentation process (Bernai et al. 1998; Haug 2018; Hirai et al. 1983).
(iv) If the C/N ratio is less than 15, the following problems occur:
    1. At first, microbial growth and decomposition are accelerated, but after a while, oxygen is reduced and the process is continued as an anaerobic status. Under these conditions, nitrogen, which is the most important component in the final product of the compost, is released as ammonia out of the environment and is caused an unpleasant odor.
    2. When nitrogen levels are high, high concentrations of ammonia are produced, and this high concentration of N for the microbial population is toxic and will be an inhibitor.

The easiest way to adjust the C/N ratio is to add the materials with the appropriate carbon and nitrogen values to the input materials. Table 6.1 shows the ratio of C/N of some compostable materials which on addition to the compost results in an optimal C/N value (Epstein 1994; Haug 2018).

Table 6.1 C/N ratio in different materials

Materials	C/N
Sewage sludge	6
Wood	700
Sawdust	500
Leaves of trees	40–80
Fruit wastes	35
Decomposed manure	20
Sugarcane remnants	50
Wheat straw	80
Dried alfalfa	12
Humus	10
Alfalfa straw	50

The main method of increasing C/N is to add cellulosic material with a high C/N ratio, such as straw and sawdust, and the most important ways to reduce C/N ratio are, respectively, by:

(i) If possible, removing materials with very high C/N from the wastes. For example, if the wastes contain sawdust or wood chips, C/N can be reduced to an optimal value by sieving the wastes and removing the large wood chips from the wastes that are also being reused.
(ii) Adding organic fertilizers such as poultry manure, as well as mineral fertilizers such as urea and ammonium nitrate, which increase nitrogen levels.
(iii) Use of compost products with C/N ratio of 15 or less to compensate N during the process. It should be noted that in municipal wastes, nitrogen is usually sufficient and sometimes more than needed.
(iv) In a situation where the C/N ratio is stabilized by increasing nitrogen, this can be estimated by mixing the sludge with the waste. In these cases, sludge must not contain harmful pollutants. The amount of the added sludge is determined by the ratio of the initial C/N present in the mixture.

In practice, accurate calculation and proper adjustment of the C/N ratio are difficult. The main reasons are:

(i) A part of the organic carbon such as cellulose and lignin has high resistance to biological degradation, and their decomposition requires a long time.
(ii) Some substances, such as some proteins, are not effectively exposed to microorganisms. and
(iii) Measuring carbon content is very difficult and usually is not accurate.

During the production of compost, if the waste material is heterogeneous, it is necessary to determine the ratio of C/N material according to the carbon and nitrogen content of each of the raw materials (Bernai et al. 1998; Haug 2018; Hirai et al. 1983).

Usually, carbon and nitrogen values of each substance are already known (Table 6.1), and if these values are not known, these values can be determined in the laboratory.

2.3.7 Compost Pile Size

In all composting systems, the appropriate pile is in equilibrium between the heat produced by microorganisms and thermal drops through convection, conductivity, and radiation. In passive systems where aeration is dependent on diffusion and convection, the compost pile is usually massive (having many surfaces) to help moving air into the mass (Epstein 1994; Haug 2018). On the other hand, this increase in surface leads to heat losses through radiation and conduction. Heat losses in these systems are largely dependent on compost pile levels, while microbial-mediated heat production is considered a function of compost pile volume. Larger compost piles usually have a lower level than their volume, in which case the compost pile temperature is very high. In active systems, heat loss is increased through convection by increasing the aeration of compost piles. This factor reduces the average temperature of the compost pile. In large-size compost piles, the sites close to the air entrances are very dry and have a low temperature, but other sites have higher temperatures (Kutzner 1999; Rasapoor et al. 2009).

2.4 Organic Matter in the Compost Pile

Waste materials are initially complex organic compounds. Due to the activity of microorganisms, part of the organic material in the compost piles is converted into inorganic forms, and some is also wasted. It has been shown that lignin compounds are the most suitable indicators for the identification of organic matter residues at the end of the composting process (Langarica-Fuentes et al. 2014; Tuomela et al. 2000).

Carbon is also one of the elements needed for the activity of microorganisms. At the beginning of the composting process, the organic carbon content of the compost pile is high but its significant percentage is reduced during the first month of compost production due to thermophilic temperatures in the period. Thermophilic bacteria easily decompose organic compounds including organic carbon and produce carbon dioxide, water, and heat. In later stages, that is, at the maturity stage, the percentage of decreased organic carbon is lower than the initial stages, due to the presence of more resistant carbon compounds such as cellulose and lignin. The decomposable carbon in the compost pile is found in two states, some of which are organic carbon that can easily be decomposed (this part of the organic carbon is higher in the food materials) and the others are organic carbon resistant to degradation like lignin and cellulose. Carbon losses occur between 46% and 62% due to oxidation. Available carbon plays an important role in stabilizing nitrogen in the compost pile. At the time of decomposition of organic compounds, microorganisms lose about 60–70% of carbon as carbon dioxide, and only about 30–40% carbon in their cellular components are stored. Therefore, in a combination of waste products with a C/N ratio of 30, only 12 units of carbon become available for each unit of nitrogen (Bernai et al. 1998; Haug 2018; Hirai et al. 1983; Zaved et al. 2008).

3 Organic Matter Compounds in Organic Wastes

The chemical composition of the cell wall of the plants varies in layers, but cellulose molecules are the main constructors of the cell wall. In addition to cellulose, there are other compounds such as pectin, hemicellulose, and lignin in the cell wall. Some cell walls have fatty substances such as suberin and waxes. Pectin and hemicellulose are polysaccharide compounds. Hemicellulose and other polysaccharides can be converted into simple sugars through the use of chemicals or bacteria and fungi.

Lignin is a complex organic material that is formed in the wall of the vascular cells. In these cells, lignin is seen not only in the lateral wall but also in other layers of the cell wall. In fact, the ligninization phenomenon begins in the intercellular layers and progresses internally, so that the middle blade and the first cell wall have greater lignin than the lateral wall. Lignin increases the strength and hardness of the cell wall. In general, cellulose, hemicellulose, and lignin are the main components of the cell wall of plants (Alexander 1977; Pietro and Paola 2004; Zaved et al. 2008), which are discussed below.

3.1 Cellulose

Cellulose is an important carbon component of the structure of plants and the most abundant organic matter in nature. Because large amounts of this material are added to the soil as different parts of the plants (e.g., the roots, the leaves, and the stems) every year, its decomposition and degradation play a special role in the biological cycle of carbon in life network. As a result, special attention has been paid to the cellulose-degrading microorganisms. Cellulose is present in the structure of grain crops, many fungi and even in cysts of some protozoans. This polysaccharide is not visible as simple chains in the cell wall of the organisms, but is seen as small rod-shaped units, called the micelle, under a microscope. These micelles are organized as larger units called microfibrils and have specific connections to other polysaccharides (i.e., hemicelluloses) and lignin. Except for living organisms, it is said that cellulose is also found in the organic part of the soil. The soil organic matter or humus consists of two parts of humic substances and non-humic substances. Cellulose is located in the non-humic section of the soil humus (Gupta and Sowden 1964; Schwarz 2001).

The high cellulose content of plants during their lifetime is never fixed, but varies with the type and age of the plant. The amount of this polymer in woody parts of plants (i.e., straw, wood shavings, and leaves) is much higher than other parts of the plant. The juicy plant tissues are generally poor in terms of cellulose content, but this increases with the growth and development of the plant. For example, in herbaceous plants and vegetables, the cellulose content makes up only 15% of the dry weight of these plants, but this content in woody plants is higher than 50%. The cellulose content is between 15% and 45% in most common plants. Interestingly, starch and cellulose are both polymers with the same constituent units, glucose. But the differences in the glycosylated bonds of the two types of polymers (α-1,6-glycosidic bonds in starch and β-1,4-glycosidic bonds in cellulose) give specific properties to any type of molecule by which microbial attack on starch is easy, while the cellulose resistance to enzymatic microbial degradation is much higher. In addition, due to the specific construction difference maintained above, the two carbohydrates attract quite different microbial populations (Alexander 1977; Schwarz 2001).

3.2 Hemicellulose

Hemicellulose is one of the structural polysaccharides that were first isolated from the cell wall of the plants. Hemicellulose is a complex of polymeric carbohydrates with xylene as its major compound (Shallom and Shoham 2003). Hemicellulose macromolecules often include polymers of pentose (xylose, arabinose, and hexoses) and a number of glucuronic acids. Hemicelluloses are usually classified according to the main sugar in their building, and the xylenes and manases are the main groups in their continuous chain. Hemicelluloses have a higher solubility than cellulose. Alkaline extractants completely extract hemicelluloses (Maijala et al. 2012; Sjostrom 2013).

3.3 Lignin

Lignin consists of monomeric units of phenylpropane and is a highly irregular and insoluble polymer that includes phenylpropanoid subunits. Lignin is made from an aromatic ring and a tri-carbon chain. Unlike cellulose and hemicellulose, there are no chains with the repeated subunits in lignin. Therefore, the enzymatic hydrolysis of this polymer is extremely difficult (Malherbe and Cloete 2002). Lignin is a hydrophobic aromatic polymer and is not amorphous in plant cell walls. Its chemical structure is basically changed under conditions of high temperature and acidic conditions. At temperatures above 200 °C, it can be seen as dense form and in smaller sizes apart from cellulose. Lignin acts as a physical and non-penetrating agent to prevent penetration of solutions and enzymes and also acts as an obstacle to biological attacks, including tackling cellulose-degrading microorganisms (Xie et al. 2000). Lignin is the most abundant aromatic polymer and the second organic polymer containing cellulose on the Earth. Lignin is produced after cellulose at 60 million tons per year, accounting for about 30% of the organic carbon on the Earth. It seems that a limited number of microorganisms can completely break it down. Structurally, biosynthesis and biological degradation of the lignin is complex (Xie et al. 2000). The cell wall of plants consists of cellulose, hemicellulose, and lignin, all of which are in the form of microbial degradation-resistant crystalline fibers, and the process of their decomposition is very complex and requires a wide variety of enzymes. Lignin is degraded by chemical reactions and oxidative enzymes such as phenol oxidases, laccase, and peroxidases (Kamimura et al. 2017).

4 Organic Waste Decomposers

The microorganisms are classified according to the use of oxygen as follows.

(i) *Obligate aerobes*: These types of microorganisms require oxygen for their survival and proper activity.

(ii) *Obligate anaerobes*: Types of microorganisms that cannot function in the presence of oxygen.
(iii) *Facultative anaerobes*: Types of microorganisms that can operate in both aerobic and anaerobic environments, but their activity in oxygen-enriched environments is higher than in non-oxygenated environments.

The various types of active microorganisms in a compost pile include bacteria (i.e., actinomycetes), fungi, protozoa, and rotifers (Alexander 1961; Langarica-Fuentes et al. 2014).

4.1 Bacteria

The number of bacteria in the compost pile is very high, so that about 80–90% of active microorganisms in the compost pile constitute bacteria. The bacteria are unicellular and have a higher resistance to high temperatures than other microorganisms. The most important bacteria are of genus *Bacillus*, which is active in the mesophilic and thermophilic temperatures (Schwarz 2001; Tuomela et al. 2000). Because of their small size, the bacteria have a high surface-to-volume ratio, which quickly absorbs soluble materials. Some bacteria, such as *Bacillus*, are capable of creating thick-walled endospores that are highly resistant to heat, irradiation, and chemical disinfection. Indicator bacteria of the thermophilic stage are species of *Bacillus* (*B. subtilis*, *B. licheniformis*, and *B. circulans*) (Tuomela et al. 2000). Generally, only a small number of each bacterial species are actively cellulolytic. The ability in the trait (to be cellulolytic) in relation to oxygen, temperature, and tolerance to EC is different among bacteria (Schwarz 2001). Aerobic cellulolytic bacteria and fungi have high cellular growth potentials that produce the cellular protein from the cellulose residues (Singh et al. 1991; Xi et al. 2015).

Similar to fungi, actinomycetes are also multicellular, but functionally are similar to bacteria. These types of microorganisms in compost pile play an important role in complex organic matter decomposition. Most of these microorganisms are active in the cooling phase of the compost pile, in which only organic matter resistant to decomposition is found (Ryckeboer et al. 2003).

4.2 Fungi

In the compost pile, fungi are remarkable for their high activity in lignin decomposition and function well under acidic conditions. They can decompose compounds or residual organic matter, which are very dry, acidic, or low in nitrogen. Fungi do not have high resistance to high temperatures and are the most sensitive types of microorganisms to temperature increase (Awasthi et al. 2014; Maheshwari et al. 2000). To be thermophilic in fungi is not as high as bacteria or algae. Only a handful among more than 50,000 fungi species can grow at high temperatures, and the known thermophilic fungi belong to the genera of phycomycetes, ascomycetes, and imperfect fungi (Mouchacca 1997). These fungi cause composting and humification of the materials; but all mechanisms of the organic matter decomposition are not well known. It is thought that these fungi are a good source of enzymes that can destroy the cell walls of the plants. Thermophilic fungi can be found in the genus *Chaetomium*, which is active at an appropriate temperature of 45–55 °C and at a maximum temperature of 58–61 °C. Of the mesophilic fungi, it can be referred to two species of *Trichoderma* including *Trichoderma reesei* and *T. viride* that are active at an optimum temperature of 30 °C (Langarica-Fuentes et al. 2014; Maheshwari et al. 2000).

Trichoderma is a group of imperfect fungi in the soil and is located on rotten wood and herbaceous litters. Among imperfect fungi, the fungus *Trichoderma* is the dominant genus. This superiority may be interpreted by the various metabolic ability and effective competitive nature of this species. *T. harzianum* strains have been considered in recent years due to their potential use as a biological control agent. For example, *Trichoderma* species are capable of producing biological control factors against soilborne plant pathogens (Cho and Lee 1999).

4.3 Protozoa

Protozoa are eukaryotic unicellular organisms that are active in the water droplets inside the compost pile. This type of organisms plays a very limited role in decomposition of organic matter (Gupta and Sowden 1964; Ilyin et al. 2005). These microorganisms feed on living bacteria and play an important role in liberating the nutrients from bacterial biomass and in accelerating the nutrient cycle.

4.4 Rotifers

Rotifers are multi-cellular organisms that are found in water droplets. In addition to organic matter, they feed on bacteria and fungi (Gupta and Sowden 1964; Ilyin et al. 2005).

5 Value of Microbial Inoculation for Organic Waste Decomposition

There are only a few farm-level reports in which microbial inoculations have been successful (Verstraete and Top 1999). One of the key factors of failure may be the inability to survive and compete with the natural microflora in a highly competitive and microbial environment in a sample of a compost system. For this reason, the value of microbial inoculation as a way to improve the quality of compost remains unclear. The importance of lignin-degrading fungi has been determined in sterile conditions, but its effect in non-sterile conditions is still ambiguous (Romero et al. 1999).

Microbial inoculation has been reported to have a positive and significant effect on composting processes. For example, Shin et al. (1999) reported that inoculation of garbage with specific microorganisms increased microbial activity and reduced pH. In general, the decomposition efficiency was 1.7–2.1 times higher. Following the inoculation of compost with *Trichoderma viride*, the ratio of humic acids with high molecular weight decreased. Awasthi et al. (2014) showed that inoculation of fungi *Trichoderma viride*, *Aspergillus niger*, and *Aspergillus flavus* improved the quality of the produced compost and accelerated composting process. The inoculation with lignocellulose-degrading fungus can potentially increase the degradation of organic matter. *Trichoderma viride*, *Aspergillus niger*, and white rot fungi can reduce the volume of waste through cellulose decomposition and ligninization (Huang et al. 2006; Vargas-Garcia et al. 2010; Yu et al. 2007). These microorganisms significantly reduce the time of composting. The results of the research showed that the inoculation in several stages increased the thermophilic time during the composting process. By multiphase inoculation of bacteria and fungi, the thermophilic temperature was longer for 5 days than single-stage inoculation at the beginning of the process. Also, multistage inoculation induced the increased microbial diversity indices (Xi et al. 2015). In the study, the primary materials were converted to compost with continuous heating. The results showed that following increasing the heating time, the quality of compost production increased (Varma and Kalamdhad 2015). Microbial activity plays a major role in the changes made during humification processes. Therefore, the ability of microorganisms in lignocellulose decomposition is a key factor in the recycling of waste, especially in the process of composting. The presence of an adequate microbial population promoting the proper degradation of organic matter is affected by environmental conditions. Inoculation with useful microorganisms activates organic biodegradation and improves the final composting properties (Gaind and Pandey 2005). Some compounds, such as polyphenols, sugars, and amine compounds, are caused by microbial activity, which seems to contribute to the formation of humus (Ji et al. 2006).

Microbial inocula are added to the compost pile to produce high-value-added compost or reduce the amount of composting time. But due to the inability to survive some of the microbial inocula in a competitive environment, the effects of such treatments may sometimes be unknown. Therefore, the selection of beneficial microorganisms as useful inocula for successful treatments is necessary (del Carmen Vargas-García et al. 2006). Nakasaki et al. (1994) reported that a thermophilic bacterium (*B. licheniformis*) can effectively break down the protein and prevent the loss of initial pH levels during composting so it can stimulate the activity of other thermophilic bacteria.

Ohtaki et al. (1998) showed that microbial inocula increase the microbial population and cause the production of desired enzymes, leading to fundamental changes and transformations, and reducing the emission of odorous gasses. Shin et al. (1999) studied the increase in composting efficiency by increasing the solid and liquid inoculum and showed that inoculum efficiency is usually affected by the combination of native microorganisms. Many studies have shown that inoculated microbial populations should be used continuously to cause changes in the various composting stages. Few attempts have been made to investigate the inoculation processes at multiple stages, which would lead to further improvements in this field by conducting needed research (Xi et al. 2005). del Carmen Vargas-García et al. (2006) reported that biodegradation of organic matter is a complex process that is affected by several factors, among which microbial activity is one of the most important factors. The addition of microorganisms with the ability to break down lignocellulose leads to production of the compost with higher quality or faster process progress. On the other hand, the efficiency of inoculum in the composting process depends on the conditions in which the process occurs, including the raw material and the inoculum. Xi et al. (2005) developed a method for improving the composting efficiency by inoculation with inoculum A (*Bacillus azotofixans*, *B. megaterium*, and *B. mucilaginosus*), inoculum B (a combination of cellulolytic strains *Trichoderma koningii* and *Streptomyces cellulose*), and inoculum C (a combination of inoculum A and B). The results showed that the inoculum containing cellulolytic and white rot fungi species (inoculum B and C) could significantly accelerate the composting process, especially inoculum C, because it contained large amounts of inoculum A and B. As a result, inoculum C, which contained *Bacillus* species, cellulolytic species, and white rot fungi, was very effective at accelerating decomposition, reducing the release of odorous gases, and stabilizing compost products.

Chen and Stamler (2006) showed that inoculation of the compost pile with cellulolytic and heat-resistant microorganisms could lead to rapid degradation, improvement of the quality of the compost, and decrease in the maturity time of the compost. del Carmen Vargas-García et al. (2006), in their study, showed that inoculation in the composting process can be considered as a useful tool for increasing the degree of humification in the final product. Therefore, in order to improve the quality of the compost and achieve higher levels of composting stability and maturity, especially when wastes included lignocellulose materials, high-performance microorganisms should be used at an early stage to make the humification process faster and more complete. In the study of microbiology of food and sewage sludge decomposition by Wang et al. (2004), the biological conversion process of municipal waste and food waste was carried out in a cylindrical reactor at 60 °C. In this study, *B. amilourans* was used as a starter in the reactor. The treatment continued for 10–13 days. In this study, the organic matter decomposition rate was determined based on the intensity of carbon dioxide production. Biodegradation of organic matter decreased from 3.8 to 1.3 mg CO_2 per day per mg organic matter after 10 days. That is, the bulk of the mixture of food and urban waste was easily decomposed for 10 days. To understand the biological conversion mechanisms of organic waste to organic fertilizer (compost), it is very important to study the microbial communities in compost (Fang et al. 2000). In a study, to isolate cellulose-degrading bacteria from kitchen waste at appropriate temperature and pH, the 21 bacterial strains were isolated on culture medium. Of these, four strains were found to be more effective than the rest at degrading cellulose, and the most suitable temperatures and pH for them were found to be 35 °C and 7, respectively. Generally, the role of cellulose-degrading bacteria is to transform the complex polymers into simpler products such as sugars, proteins, antibiotics, and so on (Singh et al. 1991). Biodegradation involves the breakdown of organic waste via the extracellular enzymes produced by the microorganisms that these decomposed organic matters act as the primary food and energy for them (Maier et al. 2000).

The waste management strategy in the future involves the achievement of human health, respect for the environment and ecology, and adaptation to biological systems. The advantages of this method can be to reduce the amount of organic waste, to control the biological damage of the waste and to adapt the system to other systems. In a study where the decomposition process for organic and fermentative waste was analyzed, the number of *Lactobacillus* and *Clostridium* was increased up to 100 times, and protozoan, *Enterobacteria*, and *Staphylococcus* pathogens were eliminated during the process (Ilyin et al. 2005). In the study of the efficiency of microorganisms in the management of organic debris by Singh et al. (2004), by adding several microbial species, decomposition time decreased and lateral products were also produced. In the study, the microorganisms with ability to produce enzymes such as lipase, protease, amylase, and cellulase were used. The results showed that three types of microorganisms prevented the rapid degradation of the debris, and some of them led to intensification of decomposition and composting, so that after 28 days the maximum decomposition was in a way that led to 60% reduction in weight and 67% decrease in mass volume, while in the control treatment, these values were 40% and 38%, respectively. Maximum temperature in treatment with bacteria number 12 was 58 and in control treatment was 45 °C. According to Raynal et al. (1998), organic waste components contain about 75% sugar and hemicellulose, 9% cellulose, and 5% lignin, carbohydrates, amino acids, peptides, proteins, volatile acids, fatty acids, and esters. In isolation and determination of metabolic parameters of bacteria, 80 strains were isolated from 35 culture media that produced enzymes of interest. The 35 bacterial strains were used in combination. In the evaluation of the organic waste degradation efficiency, the combination of bacteria 3, 5, and 12 had about 50% of the waste degradation potential; the remaining bacteria had less than 50% of decomposition ability. The weight loss of 25 kg was 48, 54, and 60%, but in control treatment, it was 40%. The maximum temperature increase was observed in all treatments from day 14 to 21. Microorganisms in the compost pile can absorb soluble elements and water of the masses and create a uniform environment in the compost bed. For appropriate composting conditions, suitable humidity is 40–60% by weight (Gotaas 1956). According to Miller (1959), the best temperature for maximum microbial decomposition is 55 °C. At 70 °C, degradation is minimal. High temperature in the compost pile leads to increased evaporation of excess water, control and escape of vermin, and stimulation of composting. On the other hand, high temperatures provide better ventilation and reduce the unpleasant smell of compost. In a study conducted by Zaved et al. (2008), three strains of bacteria that were effective at decomposing waste were isolated. Characterization of these isolates, by recognizing apparent colonies and biochemical tests, showed that they belonged to genera of *Xanthomonas*, *Bacillus*, and *Pseudomonas*. Changes of color, odor, and weight loss and reduction in volume, temperature, and pH in each compost pile were studied. To observe the effect of various additives on the rate of organic matter decomposition, concentrations of 5, 10, and 15% sucrose and molasses solution were used in bacterial suspension, in which the best result was observed at 15% concentration. The decomposed waste as organic fertilizer was used to observe its effects on potato dry matter production. The result of this study showed that inoculation with beneficial bacteria could be effective for decomposing solid organic matter. Harper et al. (2002) reported that the weight of the compost pile, the initial weight of 26 kg, reached 3.1 kg after composting for 2 weeks. In this study, the average weight loss per day was about 6% of the initial wastes. In this assay, molasses treatments along with bacteria significantly decreased the volume and weight of wastes. The reason for this greater increase may be due to the variety of sugary compounds in molasses relative to sucrose, which can aggravate bacterial growth and proliferation. Bacterial growth is an important factor in the decomposition of organic wastes. In the composting pile, the temperature increases after 4–6 days, and after 15–24 days the temperature reaches the maximum and decreases after 28–30 days. The pH changed from acid to alkaline from the beginning of composting to the next in all types of organic wastes and treatments. The proper pH for the growth of the bacteria was 6–7.5. In the first 2–7 days, pH decreased due to the production of organic acids and then increased to 7.5–8.5. The increase in temperature led to the decomposition of organic compounds and increased pH.

Research on microbial communities in the process of municipal solid waste composting showed that 44% of the microbial populations were bacteria, 32% actinomycetes, and 23% fungi. The dominant bacterial species were *Bacillus*, *Streptomyces*, *Actinomyces*, *Pseudomonas*, and *Azospirillum*. Bacteria were a dominant group that was able to decompose a large amount of organic matter by producing a wide range of enzymes. On the other hand, bacteria are responsible for the primary decomposition and heat production in the compost. In the process of composting, pH decreases from 1.8 to 7.7 (Wang et al. 2001). Moisture content reduces due to heat generation (Tiquia et al. 2002). Gradually, the compost becomes more complex and the population of microorganisms will decrease over time, except for actinomycetes. The increase in the population of the fungi and bacteria in the mesophilic stage is affected by temperature and pH. The food and vegetable residues often have an initial pH of about 4.4–5, which causes fungal proliferation (Ryckeboer et al. 2003). The high level of the surface to volume of bacteria leads to the rapid transfer of soluble material into their cell. The high temperature leads to the degradation of lignocellulose organic matter and the elimination of pathogens. Microorganisms with high hydrolysis ability, spore-forming microorganisms (such as *Bacillus*), and missile-forming microorganisms (i.e., *Streptomyces*, *Actinomyces*, and fungi) are of cellulolytic microorganisms.

6 Suitable Microorganisms for Organic Waste Decomposition

6.1 Cellulose-Degrading Microorganisms

About three decades ago, research on cellulose microbial decomposition was conducted. Isolation of cellulose-degrading bacteria was done by Stewart and Leatherwood (1976) on various carbonaceous media. Cellulose-degrading enzymes are produced by a large number of microorganisms. Cellulose-degrading microorganisms are found among fungi, actinomycetes, true bacteria, gliding bacteria, and myxobacteria. The enzyme hydrolysis of the natural cellulose is catalyzed by a group of enzymes that act synergistically. The cellulose-degrading enzyme system is often different in different microorganisms. In addition, many microorganisms that do not have the ability to grow on natural cellulose produce cellulose-degrading enzymes. But their enzymatic systems lack some of the enzymes necessary for the hydrolysis of natural cellulose. Often, they can use cellulose that has been partially disintegrated to grow. Such microorganisms, despite this defect, may be of particular interest in the industrial production of cellulose-degrading enzymes. Some microorganisms secrete large amounts of cellulose-degrading enzymes into their culture medium. While the other group, despite the growth on the medium including cellulose, secretes no enzymes or a small amount of enzymes into their culture medium. The question of whether cellulose-degrading enzymes are truly extracellular or intracellular is not yet clear. In many cases, the high activity of cellulose degradation is found only in the logarithmic phase of growth in a liquid medium. Hence, it can be said that the enzymes responsible for this action are released by the autolysis of microorganisms outside the cells (Enari 1983).

6.2 Hemicellulose-Degrading Microorganisms

Among the hemicellulase-producing fungi, the *Trichoderma* and *Aspergillus* groups can be mentioned. The bacterial secretion of hemicellulases in *Bacillus polymyxa* and *B. subtilis* has been reported. Hemicellulase enzymes in *B. subtilis* are endo-α-1,5-arabinose and endo-galactanase activity, which are active at temperatures of 60 °C and 48 °C and pH 6.8 and 6, respectively. But the activity of β-glucosidase in *B. polymyxa* has not been reported (Howard et al. 2003).

6.3 Lignin-Degrading Microorganisms

Few microorganisms are capable of decomposing lignin polymers. Among these microorganisms, white rot fungi (i.e., *Basidiomycetes*) are the most efficient in the decomposition, leading to significant mineralization of lignin (Hatakka 2001). However, the microbial population in the compost pile are usually free of Basidiomycetes fungi (Tuomela et al. 2000). A number of compost microorganisms that are capable of decomposing the polymer structure of lignin in pure culture have been reported in some studies (Tuomela et al. 2000). Among them, there are filamentous fungi belonging to ascomycetes or deuteromycetes. *Aspergillus*, *Penicillium*, *Fusarium*, *Chaetomium*, *Trichoderma*, and *Paecilomyces* are also lignin-degrading microorganisms (Tuomela et al. 2000). In a study of lignin degradation by *Paecilomyces inflatus*, laccase was the only detectable lignin-degrading enzyme isolated from the compost (Kluczek-Turpeinen et al. 2003). Ascomycetes were the dominant class in this compost (90% of all fungal mycelia). In thermophilic conditions, the number of fungi from the orders Sordariales and Eurotiales was high. Molecular studies revealed the presence of thermophilic fungi (i.e., *Scytalidium thermophilum* and *Myriococcum thermophilum*) and heat-resistant fungi (i.e., *Pseudallescheria boydii*, *Verrucosus corynascus*, and *Coprinopsis* sp.) in the compost. Few fungi of *Aspergillus fumigatus* and *Mycocladus corymbifer* were also observed in this study (Langarica-Fuentes et al. 2014). Lignin-degrading bacteria have also been reported, including *Rhodococcus* (Ahmad et al. 2011), *Pseudomonas*, and *Amycolatopsis* (Brown et al. 2012; Brown et al. 2011). These observations indicate that bacteria are also part of the lignin decomposers in nature (Lladó et al. 2017). Generally, bacteria are thought to play an important role in the decomposition of low molecular weight aromatic compounds (Masai et al. 2007).

7 Lignocellulase Enzymes Involved in the Decomposition of Organic Waste

Lignocelluloses consist of three major structural polymers: cellulose, hemicellulose, and lignin. The carbohydrates are composed of units of β-(1, 4′) pyranosyl and monomeric sugars. Glucose in cellulose, xylose in xylenes, and mannose and glucose in glucomans are some these monomers. The hydrolysis of these carbohydrates is accomplished through the decomposition of β-1,4 glycosidic bonds in the polysaccharide structure by cellulases, xylanases, mannases, and laccases. The enzymes such as α-galactosidase, α-glucuronidase, and α-arabinosidase are required for the decomposition of heteropolymeric mannases and xylenes. In addition to these enzymes, esterases are required to remove acetyl groups from xylan in woody and annual plants (Polizeli et al. 2005). Hydrolysis of plant cell wall polysaccharides has been the subject of study for decades (Aro et al. 2005). However, the efficient enzymatic hydrolysis to convert lignocelluloses into simpler compounds is still a topic for further research. Depending on the carbon source, variable proportions of heat-resistant hydrolytic enzymes are required. All thermophilic strains have the ability to secrete the xylan- and mannose-degrading enzymes (Maijala et al. 2012). Aeration and microbial inoculation are important factors affecting composting. Mixing different types of enzymes such as cellulase, protease, amylase, and lipase is an effective strategy to improve composting (Echeverria et al. 2012). Microorganisms often produce a wide range of hydrolytic enzymes (in particular, cellulase, xylanases, proteases, lipases, and phosphatases) during composting, which can provide information about the process of decomposition of organic matter and stability of the produced compost (Ben-David et al. 2011; Portillo et al. 2011). The enzymes of cellulase and β-glucosidase, proteases and urease, phosphatases, and arylsulfatase are related to carbon mineralization, the nitrogen cycle, the phosphorus cycle, and the sulfur cycle in this process, respectively. Results have shown that in the first stage of composting (oxidative biologic phase), the enzymatic activity is caused by the availability of degradable organic compounds. These activities were observed with less intense between the second and third months of composting (mesophilic phase). Chemical parameters and humification have been very similar and close in during 119–189 days. There is a significant relationship between enzymatic activities and also between enzyme activity and other parameters such as C/N ratio, cation exchange capacity, extractable carbon, and evolution of humic material, which show that both types of indices can be used as a reliable tool for determining the degree of stability and maturation of compost (Jurado et al. 2014). In improving hydrolysis technology, the thermal stability of enzymes is considered as one of the main goals for the conversion and decomposition of biomass (Maijala et al. 2012). Evaluations of genomic and proteomic data from several fungi have shown that these microorganisms are a very important source for polysaccharide-degrading hydrolytic enzymes (Cantarel et al. 2008).

8 Quality Criteria of Produced Compost for Use in Agriculture

The quality of compost for agricultural use is usually determined by its degree of repining and stability. Parameters such as temperature and thermophilic phase duration, C/N ratio, cation exchange capacity, extractable carbon, and humic material evolution are used to evaluate the maturity and stability of the compost (Ben-David et al. 2011; Portillo et al. 2011). Temperature is one of the important parameters that are measured in composting to assess the quality of compost. Thermophilic temperature is very essential for the elimination of microorganisms pathogenic to animal and plant. According to the US Environmental Protection Agency (EPA, 2003), a 15-day high temperature of 55 °C or at least 5 consecutive days at a temperature above 55 °C is sufficient to eliminate these microorganisms (Rasapoor et al. 2009). The ratio of C/N is strongly influenced by the level of microbial activity. By increasing the microbial activity, the amount of carbon and non-nitrogenous substances, such as carbohydrates, is reduced. But in the final months, due to the reduced decomposition of the biodegradable material, the microbial population is decreased, and the carbon material decomposition is also reduced (Rasapoor et al. 2009). It is also reported that available easy materials for microorganisms are reduced in the final stages. Therefore, at this stage, amine groups and sugary compounds and materials such as fulvic acids are used, which reduce the quality of the produced compost (Aparna et al. 2008; Veeken et al. 2000). Another quality criterion for producing compost quality is pH, which should be neutralized in the mature compost. In the early months, pH is increased, which is due to the process of ammonification and ammonium production, but with approaching the final stages, the pH is decreased, which is due to the nitrification and H^+ production processes (Rasapoor et al. 2009). An important indicator, which is usually the criterion for the quality of compost, is the measurement of the ratio of NO_3^--N to NH_4^+-N. Usually, the amount of this ratio in the final stages is greater than its value at the initial stages of composting. The importance of NO_3^--N is due to its easy absorption by plants. The production of NO_3^--N is more likely at the stage of compost production fulfillment. Nitrifying bacteria have slow growth and are inactive at temperatures above 40 °C and are therefore activated when the organic matter decomposition reactions are complete (Rasapoor et al. 2009). EC is another indicator of compost quality, which is usually increased during the production of compost. An increase in EC can be due to the decomposition of organic matter and the enhancement of absorbable nutrients (Campbell Jr et al. 1997). Other characteristics important in compost quality are indicators of humification and amounts of humic substances. Increased humification indexes are associated with increasing the stability of organic matter (humus) and thus increasing the composting effect (Amir et al. 2008; Bustamante et al. 2008). Fulvic acid has amino acids, polysaccharides, sugars, and more compounds than humic acid, which can be used in the final stages of organic matter decomposition for microorganisms. Therefore, in the final stages of composting, the amount of fulvic acid is decreased, but humic acid is increased (Amir et al. 2008; Bustamante et al. 2008). The polymerization of the ratio of HA/FA represents the formation of more complex molecules (i.e., HA) from simpler molecules (FA). This ratio is known as a compost maturity index. Other indicators (humification index and humification ratio) indicate the amount of compost maturity. Humic acid analysis in different stages of composting showed that with increasing time, functional groups in humic acid increased (Amir et al. 2008; Sánchez-Monedero et al. 2002). Also, the results of the analysis of FTIR spectra showed that, during composting, little changes were made in the structure and functional groups of humic acid and in amide groups. However, the decrease in these groups was observed with increasing time. In addition, aromatic compounds increased strongly, and carboxylic, aliphatic, and phenolic groups also enhanced (Miikki et al. 1997). Ethereal and peptide groups are increased in humic acid, and amide and aliphatic groups, which are rapidly degradable compounds, are decreased (Amir et al. 2010). Also, during the production of compost, the elemental composition of humic acid is changed and the ratio of H/C and O/C and the ratio of C/N are increased and are decreased, respectively (Amir et al. 2010; Sánchez-Monedero et al. 2002).

The cation exchange capacity is also one of the qualitative criteria for compost. According to the standard of the sale of compost products of the Department of Ecology, Washington, CEC of compost should be above 100 meq 100 g^{-1}. Also, according to the same institution, the produced compost should be free of seeds of live weeds and have a germination index of more than 80% (seed of *Lepidium sativum*) (Zazouli et al. 2009). One of the disadvantages of premature compost is the presence of its toxic compounds. These compounds, which are produced due to the incomplete oxidation of organic matter and the formation of various organic acids, alcohols, and aldehydes and other compounds, have a deterrent effect on root growth and seed germination. Therefore, reducing or eliminating these compounds indicates that the composting process is complete.

9 Conclusions

The production of compost from urban and rural organic wastes, in addition to the necessity of increasing the organic matters of soils and reducing soil and water pollution, can also be important in order to generate income in rural communities. For the production of compost, many factors are important, such as temperature, humidity, pH, EC, organic matter, active microorganisms, compost pile size, etc. Active microorganisms in the compost pile with the production of lignocellulase enzymes decompose organic matter during three stages (pre-active, active, and mature) and convert them into stable forms. The pre-active stage of compost occurs in mesophilic conditions. Maximum changes in organic matter and microbial mass of the compost occur during the active or thermophilic stage, and at the maturity stage, changes in humification of compost occur. The lignin compounds are highly degraded, and their amount in organic matter determines the time of composting. Finally, in order to evaluate the quality of the produced compost, the maturity factors such as the temperature and duration of the thermophilic phase, the C/N ratio, the cation exchange capacity, and the evolution of humic material should be standard.

Acknowledgments We wish to thank University of Tehran for providing the necessary facilities for doing this study.

References

Ahmad M, Roberts JN, Hardiman EM, Singh R, Eltis LD, Bugg TDH (2011) Identification of DypB from Rhodococcus jostii RHA1 as a lignin peroxidase. Biochemistry 50:5096–5107

Ahmad R, Jilani G, Arshad M, Zahir ZA, Khalid A (2007) Bio-conversion of organic wastes for their recycling in agriculture: an overview of perspectives and prospects. Ann Microbiol 57:471–479

Alexander M (1961) Introduction to soil microbiology. Wiley, New York

Alexander M (1977) Introduction to soil microbiology-2

Amir S, Benlboukht F, Cancian N, Winterton P, Hafidi M (2008) Physico-chemical analysis of tannery solid waste and structural characterization of its isolated humic acids after composting. J Hazard Mater 160:448–455

Amir S, Jouraiphy A, Meddich A, El Gharous M, Winterton P, Hafidi M (2010) Structural study of humic acids during composting of activated sludge-green waste: elemental analysis, FTIR and 13C NMR. J Hazard Mater 177:524–529

Aparna C, Saritha P, Himabindu V, Anjaneyulu Y (2008) Techniques for the evaluation of maturity for composts of industrially contaminated lake sediments. Waste Manag 28:1773–1784

Aro N, Pakula T, Penttilä M (2005) Transcriptional regulation of plant cell wall degradation by filamentous fungi. FEMS Microbiol Rev 29:719–739

Awasthi MK, Pandey AK, Khan J, Bundela PS, Wong JWC, Selvam A (2014) Evaluation of thermophilic fungal consortium for organic municipal solid waste composting. Bioresour Technol 168:214–221

Ben-David EA, Zaady E, Sher Y, Nejidat A (2011) Assessment of the spatial distribution of soil microbial communities in patchy arid and semi-arid landscapes of the Negev Desert using combined PLFA and DGGE analyses. FEMS Microbiol Ecol 76:492–503

Bernai MP, Paredes C, Sanchez-Monedero MA, Cegarra J (1998) Maturity and stability parameters of composts prepared with a wide range of organic wastes. Bioresour Technol 63:91–99

Brown ME, Barros T, Chang MCY (2012) Identification and characterization of a multifunctional dye peroxidase from a lignin-reactive bacterium. ACS Chem Biol 7:2074–2081

Brown ME, Walker MC, Nakashige TG, Iavarone AT, Chang MCY (2011) Discovery and characterization of heme enzymes from unsequenced bacteria: application to microbial lignin degradation. J Am Chem Soc 133:18006–18009

Bustamante MA, Paredes C, Marhuenda-Egea FC, Pérez-Espinosa A, Bernal MP, Moral R (2008) Co-composting of distillery wastes with animal manures: carbon and nitrogen transformations in the evaluation of compost stability. Chemosphere 72:551–557

Campbell AG Jr, Folk RL, Tripepi RR (1997) Wood ash as an amendment in municipal sludge and yard waste composting processes. Compost Sci Util 5:62–73

Cantarel BL, Coutinho PM, Rancurel C, Bernard T, Lombard V, Henrissat B (2008) The Carbohydrate-Active EnZymes database (CAZy): an expert resource for glycogenomics. Nucleic Acids Res 37:D233–D238

Chen Z, Stamler JS (2006) Bioactivation of nitroglycerin by the mitochondrial aldehyde dehydrogenase. Trends Cardiovasc Med 16:259–265

Cho C-F, Lee W-C (1999) Formulation of a biocontrol agent by entrapping biomass of Trichoderma viride in gluten matrix. J Biosci Bioeng 87:822–824

del Carmen Vargas-García M, Suárez-Estrella FF, López MJ, Moreno J (2006) Influence of microbial inoculation and co-composting material on the evolution of humic-like substances during composting of horticultural wastes. Process Biochem 41:1438–1443

Echeverria MC et al (2012) Microbially-enhanced composting of wet olive husks. Bioresour Technol 104:509–517

Enari TM (1983) Microbial cellulase; microbial enzymes and biotechnology. Applied Science Publishers, London, pp 183–223

Epstein E (1994) Composting and bioaerosols. BioCycle

Fang A, Pierson DL, Mishra SK, Demain AL (2000) Growth of Streptomyces hygroscopicus in rotating-wall bioreactor under simulated microgravity inhibits rapamycin production. Appl Microbiol Biotechnol 54:33–36

Gaind S, Pandey AK (2005) Biodegradation study of crop residues as affected by exogenous inorganic nitrogen and fungal inoculants. J Basic Microbiol 45:301–311

Galai S, Limam F, Marzouki MN (2009) A new Stenotrophomonas maltophilia strain producing laccase. Use in decolorization of synthetics dyes. Appl Biochem Biotechnol 158:416–431

Gotaas HB (1956) Composting. Sanitary disposal and reclamation of organic wastes composting sanitary disposal and reclamation of organic wastes

Gupta UC, Sowden FJ (1964) Isolation and characterization of cellulose from soil organic matter. Soil Sci 97:328–333

Hargreaves JC, Adl MS, Warman PR (2008) A review of the use of composted municipal solid waste in agriculture. Agric Ecosyst Environ 123:1–14

Harper JW, Burton JL, Solomon MJ (2002) The anaphase-promoting complex: it's not just for mitosis any more. Genes Dev 16:2179–2206

Hatakka A (2001) Biopolymers. Biology, chemistry, biotechnology, applications. Lignin, humic substances and coal 1:129–180

Haug R (2018) The practical handbook of compost engineering. Routledge, London

Hirai MF, Chamyasak V, Kubota H (1983) Standard measurement for compost maturity. BioCycle 24:54–56

Howard RL, Abotsi E, Van Rensburg ELJ, Howard S (2003) Lignocellulose biotechnology: issues of bioconversion and enzyme production. Afr J Biotechnol 2:602–619

Huang GF, Wu QT, Wong JWC, Nagar BB (2006) Transformation of organic matter during co-composting of pig manure with sawdust. Bioresour Technol 97:1834–1842

Hultman J, Kurola J, Rainisalo A, Kontro M, Romantschuk M (2010) Utility of molecular tools in monitoring large scale composting. In: Microbes at Work. Springer, pp 135–151

Ilyin VK, Korniushenkova IN, Starkova LV, Lauriniavichius KS (2005) Study of methanogenesis during bioutilization of plant residuals. Acta Astronautica 56:465–470

Insam H, Riddech N, Klammer S (2013) Microbiology of composting. Springer Science & Business Media, Berlin

Ji P, Campbell HL, Kloepper JW, Jones JB, Suslow TV, Wilson M (2006) Integrated biological control of bacterial speck and spot of tomato under field conditions using foliar biological control agents and plant growth-promoting rhizobacteria. Biol Control 36:358–367

Jurado MM, Suárez-Estrella F, Vargas-García MC, López MJ, López-González JA, Moreno J (2014) Evolution of enzymatic activities and carbon fractions throughout composting of plant waste. J Environ Manag 133:355–364

Kamimura N, Takahashi K, Mori K, Araki T, Fujita M, Higuchi Y, Masai E (2017) Bacterial catabolism of lignin-derived aromatics: new findings in a recent decade: update on bacterial lignin catabolism. Environ Microbiol Rep 9:679–705

Kluczek-Turpeinen B, Tuomela M, Hatakka A, Hofrichter M (2003) Lignin degradation in a compost environment by the deuteromycete Paecilomyces inflatus. Appl Microbiol Biotechnol 61:374–379

Kutzner HJ (1999) The microbiology of composting. Wiley-VCH, New York

Langarica-Fuentes A, Handley PS, Houlden A, Fox G, Robson GD (2014) An investigation of the biodiversity of thermophilic and thermotolerant fungal species in composts using culture-based and molecular techniques. Fungal Ecol 11:132–144

Lladó S, López-Mondéjar R, Baldrian P (2017) Forest soil bacteria: diversity, involvement in ecosystem processes, and response to global change. Microbiol Mol Biol Rev 81:e00063–e00016

Maheshwari R, Bharadwaj G, Bhat MK (2000) Thermophilic fungi: their physiology and enzymes. Microbiol Mol Biol Rev 64:461–488

Maier RM, Pepper IL, Gerba CP (2000) Terrestrial environments. Environ Microbiol:61–80

Maijala P, Kango N, Szijarto N, Viikari L (2012) Characterization of hemicellulases from thermophilic fungi. Antonie Van Leeuwenhoek 101:905–917

Malherbe S, Cloete TE (2002) Lignocellulose biodegradation: fundamentals and applications. Rev Environ Sci Biotechnol 1:105–114

Marhuenda-Egea FC, Martínez-Sabater E, Jordá J, Moral R, Bustamante MA, Paredes C, Pérez-Murcia MD (2007) Dissolved organic matter fractions formed during composting of winery and distillery residues: evaluation of the process by fluorescence excitation–emission matrix. Chemosphere 68:301–309

Masai E, Katayama Y, Fukuda M (2007) Genetic and biochemical investigations on bacterial catabolic pathways for lignin-derived aromatic compounds. Biosci Biotechnol Biochem 71:1–15

Miikki V, Senesi N, Hänninen K (1997) Characterization of humic material formed by composting of domestic and industrial biowastes: Part 2 spectroscopic evaluation of humic acid structures. Chemosphere 34:1639–1651

Miller FC (1989) Matric water potential as an ecological determinant in compost, a substrate dense system. Microb Ecol 18:59–71

Miller GL (1959) Use of dinitrosalicylic acid reagent for determination of reducing sugar. Anal Chem 31:426–428

Mouchacca J (1997) Thermophilic fungi: biodiversity and taxonomic status. Cryptogam Mycol 18:19–70

Nakasaki K, Fujiwara S, Kubota H (1994) A newly isolated thermophilic bacterium, Bacillus licheniformis HA1 to accelerate the organic matter decomposition in high rate composting. Compost Sci Util 2:88–96

Nelson R, Babie B (2005) An experimental study of the stability of a four-vortex system, p 4852

Ohtaki A, Akakura N, Nakasaki K (1998) Effects of temperature and inoculum on the degradability of poly-ε-caprolactone during composting. Polym Degrad Stab 62:279–284

Pietro M, Paola C (2004) Thermal analysis for the evaluation of the organic matter evolution during municipal solid waste aerobic composting process. Thermochim Acta 413:209–214

Polizeli M, Rizzatti ACS, Monti R, Terenzi HF, Jorge JA, Amorim DS (2005) Xylanases from fungi: properties and industrial applications. Appl Microbiol Biotechnol 67:577–591

Portillo MC, Villahermosa D, Corzo A, Gonzalez JM (2011) Microbial community fingerprinting by differential display-denaturing gradient gel electrophoresis. Appl Environ Microbiol 77:351–354

Rasapoor M, Nasrabadi T, Kamali M, Hoveidi H (2009) The effects of aeration rate on generated compost quality, using aerated static pile method. Waste Manag 29:570–573

Raynal J, Delgenes JP, Moletta R (1998) Two-phase anaerobic digestion of solid wastes by a multiple liquefaction reactors process. Bioresour Technol 65:97–103

Romero MD, Aguado J, González L, Ladero M (1999) Cellulase production by Neurospora crassa on wheat straw. Enzym Microb Technol 25:244–250
Ryckeboer J, Mergaert J, Coosemans J, Deprins K, Swings J (2003) Microbiological aspects of biowaste during composting in a monitored compost bin. J Appl Microbiol 94:127–137
Saha JK, Panwar N, Singh MV (2010) An assessment of municipal solid waste compost quality produced in different cities of India in the perspective of developing quality control indices. Waste Manag 30:192–201
Sánchez-Monedero MA, Cegarra J, García D, Roig A (2002) Chemical and structural evolution of humic acids during organic waste composting. Biodegradation 13:361–371
Schwarz W (2001) The cellulosome and cellulose degradation by anaerobic bacteria. Appl Microbiol Biotechnol 56:634–649
Shallom D, Shoham Y (2003) Microbial hemicellulases. Curr Opin Microbiol 6:219–228
Sharholy M, Ahmad K, Mahmood G, Trivedi RC (2008) Municipal solid waste management in Indian cities – a review. Waste Manag 28:459–467
Shin HS, Hwang EJ, Park BS, Sakai T (1999) The effects of seed inoculation on the rate of garbage composting. Environ Technol 20:293–300
Singh A, Abidi AB, Agrawal AK, Darmwal NS (1991) Single cell protein production by Aspergillus niger and its evaluation. Zentralblatt für mikrobiologie 146:181–184
Singh S, Sinha S, Saxena R, Pandey K, Bhatt K (2004) Translocation of metals and its effects in the tomato plants grown on various amendments of tannery waste: evidence for involvement of antioxidants. Chemosphere 57:91–99
Sjostrom E (2013) Wood chemistry: fundamentals and applications. Elsevier, Saint Louis
Stewart BJ, Leatherwood JM (1976) Derepressed synthesis of cellulase by Cellulomonas. J Bacteriol 128:609–615
Suárez-Estrella F, Vargas-Garcia MC, López MJ, Moreno J (2008) Changes in carbon fractions during composting of plant wastes and the influence of a humic extract on soil microorganism growth dynamic biochemistry process. Biotechnol Mol Biol 2:90–95
Sundberg C, Smårs S, Jönsson H (2004) Low pH as an inhibiting factor in the transition from mesophilic to thermophilic phase in composting. Bioresour Technol 95:145–150
Tiquia SM, Wan HC, Tam NFY (2002) Microbial population dynamics and enzyme activities during composting. Compost Sci Util 10:150–161
Tosun I, Gönüllü MT, Arslankaya E, Günay A (2008) Co-composting kinetics of rose processing waste with OFMSW. Bioresour Technol 99:6143–6149
Tuomela M, Vikman M, Hatakka A, Itävaara M (2000) Biodegradation of lignin in a compost environment: a review. Bioresour Technol 72:169–183
Vargas-Garcia MC, Suárez-Estrella F, Lopez MJ, Moreno J (2010) Microbial population dynamics and enzyme activities in composting processes with different starting materials. Waste Manag 30:771–778
Varma VS, Kalamdhad AS (2015) Evolution of chemical and biological characterization during thermophilic composting of vegetable waste using rotary drum composter. Int J Environ Sci Technol 12:2015–2024
Veeken A, Nierop K, de Wilde V, Hamelers B (2000) Characterisation of NaOH-extracted humic acids during composting of a biowaste. Bioresour Technol 72:33–41
Verstraete W, Top EM (1999) Soil clean-up: lessons to remember. Int Biodeterior Biodegradation 43:147–153
Waksman SA (1936) Humus origin, chemical composition, and importance in nature. The Williams and Wilkins Company, Baltimore
Wang GK et al (2001) Regulation of the cycling of timeless (tim) RNA. J Neurobiol 47:161–175
Wang P, Changa CM, Watson ME, Dick WA, Chen Y, Hoitink HAJ (2004) Maturity indices for composted dairy and pig manures. Soil Biol Biochem 36:767–776
Xi B et al (2015) Effect of multi-stage inoculation on the bacterial and fungal community structure during organic municipal solid wastes composting. Bioresour Technol 196:399–405
Xi B, Zhang G, Liu H (2005) Process kinetics of inoculation composting of municipal solid waste. J Hazard Mater 124:165–172

Xiao Y, Zeng G-M, Yang Z-H, Shi W-J, Huang C, Fan C-Z, Xu Z-Y (2009) Continuous thermophilic composting (CTC) for rapid biodegradation and maturation of organic municipal solid waste. Bioresour Technol 100:4807–4813

Xie Y, Yasuda S, Wu H, Liu H (2000) Analysis of the structure of lignin-carbohydrate complexes by the specific 13 C tracer method. J Wood Sci 46:130–136

Yu H et al (2007) Microbial community succession and lignocellulose degradation during agricultural waste composting. Biodegradation 18:793–802

Zaved HK, Rahman MM, Rahman MM, Rahman A, Arafat SMY, Rahman MS (2008) Isolation and characterization of effective bacteria for solid waste degradation for organic manure. Curr Appl Sci Technol 8:44–44

Zazouli MA, Ardebilian MB, Ghahramani E, Alahabad GM (2009) Principles of compost production technology. Khaniran, Tehran, p 25

Plant Growth-Promoting Rhizobacteria (PGPRs): Functions and Benefits

7

Divya Singh, Paushali Ghosh, Jay Kumar, and Ashok Kumar

1 Introduction

The term "plant growth-promoting rhizobacteria (PGPRs)" was first used to define a group of soil bacteria that colonizes around or in/on the root surface of plants and exerts various beneficial effects on its growth and development (Kloepper and Schroth 1978). Rhizosphere generally represents a narrow zone of soil around the root system and is highly rich in nutrients than the rest of the soil due to the release of a vast array of plant exudates including sugars and amino acids. These exudates serve as reservoir of nutrients and energy which supports the growth and metabolism of various microorganisms (Gray and Smith 2005). This in turn probably leads to higher population of bacteria in the rhizosphere in comparison to other regions of the soil (Lynch and Whipps 1990). Bacteria inhabiting the rhizosphere are called rhizobacteria, and these may be either symbiotic or nonsymbiotic depending on their mode of interaction with the plants (Kundan et al. 2015). PGPRs usually boost plant growth and development via indirect and direct ways as illustrated in Fig. 7.1. Plant growth is promoted indirectly by these rhizobacteria due to the inhibition or prevention of the detrimental effects of phytopathogens through synthesis of antagonistic substances and induction of resistance against pathogen (Fig. 7.2). In the case of direct promotion, plant growth-promoting effects are manifested by the synthesis of growth-promoting compounds such as phytohormones (IAA, cytokinin, ethylene, etc.), vitamins, enzymes, etc. as well as nutrient acquisition from natural resources such as fixed nitrogen, iron, and phosphate (Glick 2012).

There are two main classes of PGPRs, i.e., extracellular plant growth-promoting rhizobacteria (ePGPR) and intracellular plant growth-promoting rhizobacteria (iPGPR) (Viveros et al. 2010). ePGPR typically colonize the rhizosphere or the spaces on surface of the root cortex. Bacterial genera belonging to ePGPR include *Serratia*, *Azospirillum*, *Azotobacter*, *Bacillus*, *Chromobacterium*, *Caulobacter*,

D. Singh · P. Ghosh · J. Kumar · A. Kumar (✉)
School of Biotechnology, Institute of Science, Banaras Hindu University, Varanasi, India

D. P. Singh et al. (eds.), *Microbial Interventions in Agriculture and Environment*,
https://doi.org/10.1007/978-981-13-8383-0_7

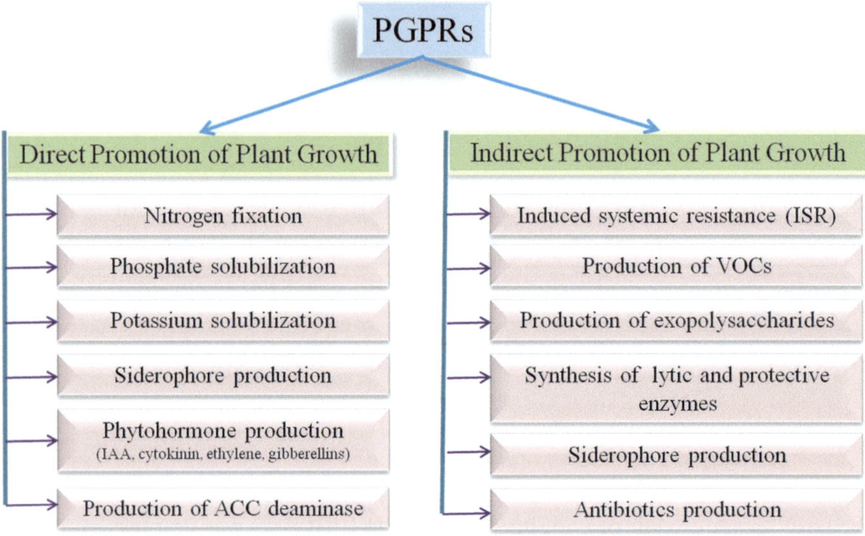

Fig. 7.1 Direct and indirect promotion of plant growth by PGPRs

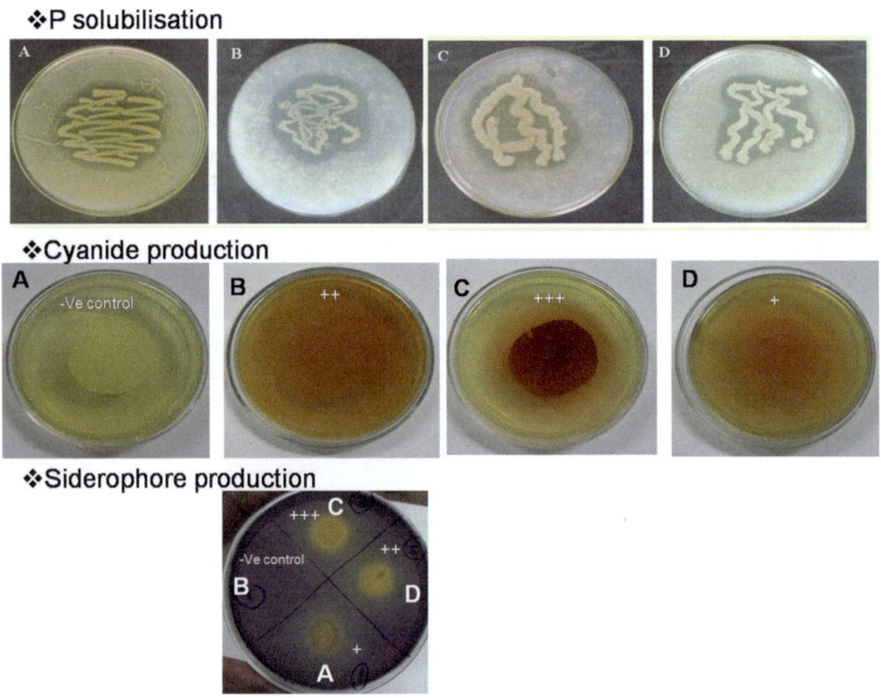

Fig. 7.2 Photographs of plates showing phosphate solubilization and cyanide and siderophore production by PGPR

Agrobacterium, *Erwinia*, *Pseudomonas*, *Flavobacterium*, *Micrococcus*, *Arthrobacter*, and *Burkholderia*. On the other hand, iPGPR mostly reside in the specific nodulated parts of root cells, e.g., *Rhizobium*, *Bradyrhizobium*, *Allorhizobium*, *Mesorhizobium*, etc. (Bhattacharyya and Jha 2012; Viveros et al. 2010). PGPRs perform different activities in the rhizosphere, thereby rendering the soil dynamic and fertile for enhanced crop production (Gupta et al. 2015). They colonize plant roots and promote growth and crop yield via various processes such as nitrogen fixation, mineral phosphate solubilization, production of siderophores and indole-3-acetic acid (IAA), and synthesis of 1-amino-cyclopropane-1-carboxylate (ACC) deaminase enzyme and hydrogen cyanate (Ahemad and Khan 2012; Glick 2012; Jahanian et al. 2012; Liu et al. 2016). Additionally, PGPRs are known to synthesize different types of antibiotics and lytic enzymes which degrade a variety of environmental pollutants and kill/reduce pathogenic bacteria (Xie et al. 2016). Furthermore, certain PGPRs possess certain novel attributes which help in heavy metal detoxification (bioremediation), salinity tolerance, and biological control of insects and other pathogens (Egamberdieva and Lugtenberg 2014).

Presently, the choice for biological alternatives for achieving improved crop production in a sustainable way is receiving much attention among agriculturalists and environmentalists. In fact, attempts are being made to screen and exploit a large number of PGPRs which can be used in place of synthetic fertilizers and other harmful agrochemicals so as to maintain the fertility and good health of soil. To this effect, several symbiotic (*Bradyrhizobium*, *Rhizobium*, *Mesorhizobium*) and non-symbiotic (*Bacillus*, *Pseudomonas*, *Klebsiella*, *Azospirillum*, *Azotobacter*, *Azomonas*) rhizobacteria are widely used as biofertilizers for promoting plant growth and yield in different agroclimatic conditions (Ahemad and Khan 2012; Egamberdieva and Lugtenberg 2014). However, the processes by which these microbes promote growth of plants are poorly understood; nevertheless, PGPR-mediated growth-promoting effects have been well documented (Khan et al. 2009). As such, interactions of PGPRs with crop plants such as wheat, oat, maize, peas, canola, barley, soy, potatoes, lentils, tomatoes, radicchio, cucumber, etc. are now exploited commercially (Gray and Smith 2005). This chapter briefly highlights the functions and benefits of plant growth-promoting rhizobacteria in different agroecosystems with an aim to expand their applicability and benefits for attaining sustainable crop production.

2 Functions of PGPRs

PGPRs assist growing plant due to its specific functions which include nutrient uptake through nitrogen fixation, solubilization of nutrients including organic compounds, and production of growth hormones and siderophores (Bhardwaj et al. 2014). Furthermore, PGPRs also elevate plant growth by conferring resistance against different biotic (phytopathogens) and abiotic (drought, salinity, heat, temperature) stresses through induction of systemic resistance and synthesis of volatile organic compounds (VOCs), exopolysaccharides (EPSs), protective enzymes, hydrolytic enzymes (cellulases, chitinases, and proteases), various antibiotics, etc. (Nivya 2015). Some of the well-known functions of PGPRs which are beneficial for growth and development of plants are described below.

2.1 Nitrogen Fixation

Nitrogen is an essential nutrient for various metabolic processes of plants governing growth and development. N_2 constitutes around 78% of the atmospheric gases but still not used by the plants since plants lack N_2-fixing machinery. However, the biological nitrogen fixation (BNF) process carried out mostly by certain bacteria and cyanobacteria converts atmospheric N_2 into forms usable by plants with the help of the enzyme nitrogenase (Kim and Rees 1994). Both symbiotic and nonsymbiotic PGPRs have been reported to fix N_2. Symbiotic PGPR strains, namely, *Rhizobium*, *Sinorhizobium*, *Bradyrhizobium*, *Mesorhizobium*, *Beijerinckia*, *Azoarcus*, *Pantoea*, and *Klebsiella*, are the most widely reported N_2 fixers in the soil. Symbiotic N_2 fixation is a result of mutualistic relationship between bacteria and leguminous/nonleguminous plants wherein the microbes first colonize the root and form nodules in which N_2 is converted to ammonia and/or its product that are supplied to the plants as nitrogen source (Ahemad and Kibret 2014). Microbes, in return, obtain carbon sources from the plants specifically in the form of dicarboxylates. Continued supply of dicarboxylates and its metabolism enable bacteria to drive the highly energetic process of nitrogen fixation (Udvardi and Poole 2013). Nonsymbiotic nitrogen fixation is mostly performed by the free-living diazotrophic PGPR including the genera *Azotobacter*, *Azospirillum*, *Pseudomonas*, *Acetobacter*, *Enterobacter*, *Burkholderia*, and *Gluconacetobacter* and cyanobacteria (Bhattacharyya and Jha 2012).

Both symbiotic and free-living rhizobacteria harbor *nif* genes which encode the nitrogenase enzyme responsible for nitrogen fixation. Nitrogenase enzyme comprises of two components: dinitrogenase containing both iron (Fe) and molybdenum (Mo) as its cofactor and dinitrogenase reductase containing iron (Fe) as its cofactor. Both these components function in a coordinated manner to reduce N_2 to NH_3 (Santi et al. 2013). Several workers have reported certain other types of nitrogenase complexes on the basis of different cofactors associated with the enzyme complex (Ahemad and Kibret 2014). Inoculation of nitrogen-fixing PGPRs to crop plants serves as an integrated strategy for growth stimulation, disease suppression, and maintenance of nitrogen level in the soil of agricultural fields (Damam et al. 2016).

2.2 Phosphate Solubilization

Phosphorus (P) is the second vital nutrient other than nitrogen required for optimum growth of the plants. It plays an important role in various metabolic activities of plants, namely, energy transfer and generation, respiration, photosynthesis, signal transduction, and macromolecular synthesis (Anand et al. 2016). P occurs in the soil both in the inorganic form like apatite and organic forms, viz., inositol phosphate, phosphomonoesters, and phosphotriesters (Glick 2012). Despite the vast abundance of phosphorus in the soil, its utilizable form available for the plants is usually very low. This is mainly due to the fact that more than 95% of P exists in insoluble and/or immobilized forms in the soil which are not absorbed by the plants. In general, plants absorb P if it is available in the form of monobasic ($H_2PO_4^-$) and dibasic ($H_2PO_4^{-2}$) ions (Bhattacharyya and Jha 2012). To overcome this problem, PGPRs have developed novel mechanism(s) to convert insoluble soil P into soluble forms which become readily available to the plants. The processes include (a) solubilization of inorganic P by the synthesis of organic acids (gluconic and citric acid) that lowers the pH of soil leading to the release of bound forms of P and (b) mineralization of phosphorus through the production of phosphatases which break down phosphoric esters (Glick 2012; Zaidi et al. 2009). Important genera among PGPRs involved in P solubilization include *Azotobacter*, *Arthrobacter*, *Burkholderia*, *Bacillus*, *Beijerinckia*, *Enterobacter*, *Erwinia*, *Flavobacterium*, *Microbacterium*, *Mesorhizobium*, *Pseudomonas*, *Rhizobium*, *Rhodococcus*, and *Serratia* (Bhattacharyya and Jha 2012). P-solubilizing rhizobacteria are generally considered as potent biofertilizers because they promote plant growth and productivity by supplying phosphorus in usable forms in environment-friendly and inexpensive mode as compared to the chemical fertilizers (Zaidi et al. 2009).

2.3 Potassium Solubilization

Potassium, an essential macronutrient, also plays an important role in growth and development of plants. However, the amount of soluble potassium in the soil is always low; approximately 90% is present as insoluble rocks and minerals (Parmar and Sindhu 2013). Its deficiency results in poor crop productivity because of its adverse effects on root development and overall growth of the plants. Henceforth, search for other possible sources of potassium which can sustain plant growth by maintaining the required concentration of potassium in the soil without causing any harmful effects to the environment is essential (Kumar and Dubey 2012). To this effect, PGPRs have emerged as one of the most reliable and efficient candidates because of its ability to solubilize rocks and minerals containing potassium by producing organic acids. Certain PGPRs including *Bacillus mucilaginosus*, *Bacillus edaphicus*, *Acidithiobacillus ferrooxidans*, *Paenibacillus* sp., *Burkholderia* sp., and *Pseudomonas* sp. are known to solubilize potassium in a manner similar to P solubilization (Liu et al. 2012). It is presumed that the application of potassium-solubilizing PGPRs as biofertilizer to crop plants shall not only reduce dependency on hazardous and costly chemical fertilizers but may support sustainable crop production.

2.4 Siderophore Production

Certain PGPRs produce a low molecular weight iron-chelating compound, siderophore especially under iron-limiting conditions which facilitate the uptake of iron.

Almost all the living organisms require iron to drive different biological processes. However, iron (Fe) occurs predominantly as Fe^{3+} which in turn forms insoluble hydroxides and oxyhydroxides, thereby rendering its unavailability for uptake and assimilation by plants and microorganisms (Rajkumar et al. 2010). Siderophores having high specificity for iron form complexes with Fe^{3+} and eventually bind to the bacterial cell membrane. Inside the cell, Fe^{3+} is enzymatically reduced to Fe^{2+} which has low affinity for siderophore and is thus released from the siderophore into the cell where it assists in microbial growth and metabolism (Boukhalfa and Crumbliss 2002). Based on the presence of specific functional groups, siderophores have been classified into three major groups, i.e., catecholates, hydroxamates, and carboxylates. Rhizobacteria generally differ in their ability to cross-utilize siderophore; some of them are capable of utilizing only homologous siderophore (synthesized by the same genus), while others can efficiently use heterologous siderophores (synthesized by different genera) (Khan et al. 2009).

Role of siderophores produced by PGPRs has been implicated in both direct and indirect promotion of plant growth. It has been reported that siderophores directly benefit plants since plants absorb iron from the bacterial siderophores through different modes such as direct uptake of siderophore-Fe complexes, chelation and release of iron, and ligand exchange reaction (Schmidt 1999). Crowley and Kraemer (2007) reported that bacterial siderophores transfer iron to oat plants which possess siderophore-mediated iron transport system. Likewise, siderophore-Fe complexes produced by *Pseudomonas fluorescens* C7 have been reported to be used by *Arabidopsis thaliana* which resulted in an increased growth of the plants and higher iron content in different tissues (Vansuyt et al. 2007). Studies conducted by several researchers have demonstrated that plants are proficient in capturing iron bound to siderophore produced by PGPRs including *Aeromonas*, *Azotobacter*, *Burkholderia*, *Bacillus*, *Rhizobium*, *Pseudomonas*, *Serratia*, and *Streptomyces* sp. (Sujatha and Ammani 2013). Siderophore also provides benefits to the plants indirectly. For example, PGPRs enhance plant growth due to the iron sequestering efficiency of siderophore; this process deprives pathogens from iron nutrition and thus serves as a biocontrol agent resulting in increased crop yield (O'Sullivan and O'Gara 1992). Furthermore, siderophores have been reported to form complexes with certain heavy metals such as Cd, Al, Ga, Cu, Pb, In, and Zn. Complex formation of siderophores with heavy metals results in increased concentration of soluble metals and thus helps the plants in alleviating heavy metal stress (Neubauer et al. 2000; Rajkumar et al. 2010). Kloepper et al. (1980) have demonstrated the ability of siderophore-producing *Pseudomonas putida* B10 to suppress growth of *Fusarium oxysporum* in iron-deficient soil. However, such effect was abolished if the soil was amended with iron most probably due to the inhibition of siderophore production by the bacterium.

2.5 Phytohormone Production

PGPRs play an important role in promoting growth and development of plants by the production of various growth-stimulating hormones such as indole-3-acetic acid (IAA), gibberellins (GAs), cytokinins, and ethylene that have been well documented. These phytohormones are produced by bacteria including *Azotobacter chroococcum*, *Enterobacter asburiae*, *Klebsiella oxytoca*, *Mesorhizobium ciceri*, *Pseudomonas aeruginosa*, *Pseudomonas putida*, *Paenibacillus polymyxa*, *Rhizobium leguminosarum*, *Stenotrophomonas maltophilia*, etc. (Ahemad and Kibret 2014). Most importantly, above hormones induce root proliferation resulting in overproduction of lateral roots and root hairs which lead to improved water and nutrient uptake (Arora et al. 2013). Brief details of selected hormones are described below:

1. *Indole Acetic Acid (IAA):* Several researchers have documented that around 80% of the rhizospheric microorganisms produce and secrete auxins as secondary metabolites. Among these, IAA, a type of auxin, plays an important role in plant developmental processes because absorption of IAA secreted by the rhizobacteria modifies the intrinsic pool of plant IAA (Glick 2012; Spaepen et al. 2007). IAA produced by PGPRs exerts broad effects on various processes of plants including cell division and differentiation, photosynthesis, and pigment production, invigorates the rate of xylem development, enhances seed and tuber germination, induces adventitious and lateral root formation, regulates vegetative growth, facilitates responses to light and gravity, and confers resistance to various types of stresses (Spaepen and Vanderleyden 2011). Moreover, IAA assists in establishing rhizobacteria-plant interactions. It also functions as a bacterial signaling molecule and exerts several direct effects on bacterial physiology (Spaepen et al. 2007). As per the report of Glick (2012), rhizobacterial IAA slackens cell wall of plants, thereby enabling a large amount of root exudates that provides surplus nutrients to sustain the growth of rhizobacteria. Most commonly occurring amino acid in root exudates is tryptophan, and it has been recognized as the main progenitor molecule in the biosynthetic pathway of IAA in bacteria (Zaidi et al. 2009). The role of tryptophan also became evident from the work of Spaepen and Vanderleyden (2011) who observed increase in IAA production by rhizobacteria on supplementing culture media with tryptophan. Biosynthesis of IAA in PGPRs involving tryptophan as precursor shows high degree of similarity with plants. However, certain PGPRs such as *Erwinia herbicola*, *Agrobacterium*, *Azospirillum*, *Pseudomonas*, *Bradyrhizobium*, *Rhizobium*, *Enterobacter*, and *Klebsiella* typically synthesize IAA via the formation of indole-3- pyruvic acid and indole-3-acetic aldehyde as intermediates (Patten and Glick 1996; Spaepen and Vanderleyden 2011).
2. *Cytokinins and Gibberellins:* Cytokinins and gibberellins also play an indispensable role as phytohormones that govern virtually all the processes of growth and development in plants. Certain strains of *Rhizobium* sp., *Azotobacter* sp., *Azospirillum* sp., *Pseudomonas fluorescens*, *Bacillus subtilis*, *Bradyrhizobium* sp., *Pantoea agglomerans*, *Rhodospirillum rubrum*, and *Paenibacillus polymyxa* have been reported to synthesize cytokinins and gibberellins (Glick 2012). Cytokinins are basically involved in cell division, root induction, and enlargement of root surface area of plants through increased lateral and adventitious

root formation (Salamone et al. 2005). However, when cytokinins are applied in high concentration, it enhances shoot development instead of root development (Skoog and Miller 1957). *Azospirillum* sp. has been described to metabolize and absorb gibberellins in association with higher plants both under in vitro and in vivo conditions (Bottini et al. 2004). Gene clusters, namely, CYP112, CYP114, and CYP117 from *Bradyrhizobium japonicum* involved in gibberellin synthesis, have been identified and sequenced. Interestingly, the structure of these gene clusters showed similarity with the plant *ent*-kaurene synthase (KS) genes which is responsible for gibberellin synthesis (Tully et al. 1998). The role of gibberellins in promoting growth and yield of plants has been demonstrated by several workers. For example, Dobert et al. (1992) inoculated *Phaseolus lunatus* plants with *Bradyrhizobium* sp. and found marked internode elongation. Furthermore, inoculation of wheat and maize roots with gibberellin-producing *Azospirillum* sp. resulted in increased ^{15}N uptake and promotion of root growth, respectively (Kucey 1988). Also, reversal of the dwarf phenotype (both genetic and inhibitor-induced) has been observed by inoculating rice and maize seedlings with *Azospirillum* sp. (Lucangeli and Bottini 1996).

3. *Ethylene:* Ethylene, an important plant hormone, shows a broad range of biological roles in plant growth, development, and survival. Its function comprises promotion of fruit ripening, leaf abscission, flower wilting, and activation of certain other phytohormones (Glick et al. 2007a). In addition to its role as growth regulator, its role has been also implicated in stress tolerance as its production increases endogenously in almost all the plants under different stress conditions including drought, salinity, waterlogging, and pathogen infection. However, its higher concentration leads to defoliation and changes in other cellular processes resulting in reduced crop productivity (Bhattacharyya and Jha 2012; Saleem et al. 2007). Synthesis of ethylene occurs by using 1-aminocyclopropane-1-carboxylate (ACC) as a precursor. Several PGPRs including diverse genera *Achromobacter*, *Alcaligenes*, *Acinetobacter*, *Agrobacterium*, *Azospirillum*, *Burkholderia*, *Bacillus*, *Enterobacter*, *Pseudomonas*, *Rhizobium*, *Ralstonia*, *Serratia*, etc. produce the enzyme 1-aminocyclopropane-1-carboxylate (ACC) deaminase which destroys ACC by converting it into 2-oxobutanoate and NH_3, thereby inhibiting ethylene levels and elevating growth and development of plants by conferring drought resistance and salt tolerance in plants (Nadeem et al. 2009; Zahir et al. 2008). ACC deaminase-producing PGPRs have been reported to confer resistance to plants against different stresses including biotic (arising from phytopathogenic microorganisms like viruses, bacteria, and fungi) and abiotic (temperature, high light intensity, radiation, insect predation, polyaromatic hydrocarbons, wounding, and flooding) (Glick 2012). Gene encoding ACC deaminase has been identified, cloned, and sequenced, and attempts are underway to transfer this trait to other useful microbes and plants (Singh et al. 2015).

2.6 Stress Management

Stresses (biotic or abiotic) negatively affect the plant by increasing reactive oxygen species (ROS) production such as H_2O_2, O^{2-}, and OH^- radicals. ROS generation results in oxidative stress causing oxidation of membrane lipids, proteins, nucleic acids, and photosynthetic pigments and damages to other biologically important macromolecules in the plant cells (Ramegowda and Senthil-Kumarb 2015). Certain PGPRs help the plants to grow under environmental stress conditions through the following strategies:

2.6.1 Induced Systemic Resistance (ISR)

ISR is defined as physiological condition of improved defense ability generated against certain environmental stimuli. PGPRs activate defense system in many plants by inducing systemic resistance against several biotic stresses involving phytopathogens (Prathap and Ranjitha 2015). In most cases, PGPRs manifest systemic resistance response in plants through phytohormones jasmonic acid and ethylene (Zamioudis and Pieterse 2012). However, a few rhizobacteria induce systemic response via the salicylic acid pathway (van de Mortel et al. 2012); others such as *Bacillus cereus* AR156 use signaling pathways for the activation of ISR (Niu et al. 2011). Upon exposure to pathogens, signals generated induce a defense mechanism through the vascular system resulting in the activation of a vast array of enzymes which potentiate host plant defense responses against phytopathogens. ISR is not specific to a particular pathogen; rather, it aids the plant in controlling a number of diseases (Kamal et al. 2014). Additionally, different rhizobacterial components such as siderophores, lipopolysaccharides, 2,4-diacetylphloroglucinol, cyclic lipopeptides, homoserine lactones, and volatile compounds like 2,3-butanediol and acetoin have been reported to evoke induced systemic response (Berendsen et al. 2015).

2.6.2 Production of Volatile Organic Compounds (VOCs)

VOCs are low molecular weight compounds (MW <300 Da) most commonly produced by several rhizospheric bacteria. These compounds have high vapor pressure which makes them diffusible through pores in the soil and have been demonstrated to promote plant growth by inhibiting bacterial and fungal phytopathogens and inducing ISR (Insam and Seewald 2010). VOCs are produced by selected bacterial species of genera, viz., *Arthrobacter*, *Bacillus*, *Pseudomonas*, *Serratia*, and *Stenotrophomonas* (Santoro et al. 2016). Important VOCs include benzene, benzene (1-methylnonadecyl), 2-(benzyloxy)ethanamine, decane, dodecane, tetradecane, dotriacontane, methyl, cyclohexane, 1-(N-phenylcarbamyl)-2-morpholinocyclohexene, dotriacontane, 1-chlorooctadecane, 2,6,10-trimethyl, 11-decyldocosane, 2,3-butanediol, and acetoin (Kanchiswamy et al. 2015). Among these, 2,3-butanediol and acetoin are regarded as the most potent VOCs in preventing fungal growth and enhancing growth of the plants (Santoro et al. 2016). Both the above compounds are secreted by *Bacillus* sp. and proved to be the main agent for promoting growth and inducing ISR in *Arabidopsis* (Lucy et al. 2004). Recently,

Chernin et al. (2011) reported that a VOC, namely, dimethylsulfide, hinders the process of quorum sensing in phylogenetically distinct pathogenic bacteria by inhibiting the transcription of N-acyl-homoserine lactone synthase gene. However, quorum-sensing compounds released by PGPRs have been reported to provoke plant defense responses by activating different defense-related genes such as *MPK6*, *MPK3*, *WRKY29*, *WRKY22*, and *Pdf1.2* (Hartmann and Schikora 2012).

2.6.3 Production of Exopolysaccharides (EPSs)

Exopolysaccharides, the high molecular weight biopolymer and composed of monosaccharide residues, are synthesized by a wide range of PGPRs. These compounds play a vital role in enhancing the soil fertility and crop yield. *Azotobacter vinelandii*, *Agrobacterium* sp., *Bacillus drentensis*, *Rhizobium leguminosarum*, *Enterobacter cloacae*, and *Xanthomonas* sp. are some of the examples of EPS-producing PGPRs (Mahmood et al. 2016). EPSs act as an effective signal molecule that elicits defense response in plants during the course of infection (Parada et al. 2006). Additionally, it has the ability to bind cation such as Na^+ and thus helps in mitigating salt stress by lowering the concentration of Na^+ present in the soil (Arora et al. 2013). EPSs secreted by rhizobacteria have the capability to sequester, remove, and recover heavy metals in the rhizospheric soil by inducing the formation of biofilms which not only provide protective sheath to bacteria but also convert toxic metal ions into nontoxic forms after adsorption (Gupta and Diwan 2017). Moreover, these biofilms act as a microenvironment which retains water and dehydrates slowly as compared to the external environment. In fact, biofilms provide an excellent protective strategy to the rhizobacteria and host plants against desiccation (Hepper 1975).

2.6.4 Production of Lytic and Protective Enzymes

A number of PGPRs are known to produce cell wall-degrading enzymes, viz., cellulase, hemicellulase, chitinase, protease, and glucanase, which are able to hydrolyze cellulose, hemicellulose, chitin, proteins, and glucans, respectively. Production of above enzymes by PGPRs indirectly stimulates growth of the plants by suppression of fungal phytopathogens including *Rhizoctonia solani*, *Sclerotium rolfsii*, *Pythium ultimum*, *Fusarium oxysporum*, *Botrytis cinerea*, and *Phytophthora* sp. (Nadeem et al. 2013; Upadyay et al. 2012). For example, chitinase-producing *Pseudomonas fluorescens* acts as a competent biological control agent against black root rot disease of tobacco plant caused by the infection of the fungus, *Thielaviopsis basicola* (Voisard et al. 1989). Besides the production of lytic enzymes, several PGPRs also synthesize and secrete antioxidative (protective) enzymes such as catalase, superoxide dismutase, glutathione reductase, and ascorbate peroxidase which scavenges ROS produced during abiotic stresses (Kaushal and Wani 2015).

2.6.5 Production of Antibiotics

Antibiotics, the low molecular weight, heterogeneous organic compounds, are well known to inhibit growth of microorganisms at low concentrations by suppressing/blocking different metabolic activities (Duffy 2003). The production of antibiotics by PGPRs has emerged as the most studied and useful mechanism of biocontrol against various phytopathogens (Glick et al. 2007b). Of all the PGPRs, *Bacillus* and *Pseudomonas* species are considered to be the most efficient in inhibiting growth and proliferation of phytopathogens through antibiotic production (Ulloa-Ogaz et al. 2015). As per the report of Haas and Défago (2005), antibiotics can be classified into six classes on the basis of their ability to control root diseases: phloroglucinols, phenazines, pyoluteorin, cyclic lipopeptides, pyrrolnitrin (diffusible antibiotics), and hydrogen cyanide (volatile antibiotic). For example, 2,4-diacetylphloroglucinol synthesized by *Pseudomonas* sp. was found to control disease caused by the fungus *Gaeumannomyces graminis* var. *tritici* in wheat plant (de Souza et al. 2003). Similarly, *P. fluorescens* BL915 produced an antibiotic, pyrrolnitrin, which protected cotton plant from the infection of *Rhizoctonia solani* under damping-off conditions (Hill et al. 1994). Another type of antibiotic, phenazines, produced by pseudomonads is also known to have antagonistic activity against phytopathogens, such as *Fusarium oxysporum* and *Gaeumannomyces graminis* (Chin-A-Woeng et al. 2003). Likewise, lipopeptide and polyketide production by *Bacillus amyloliquefaciens* is well known for their biocontrol potential against soilborne pathogens (Ongena and Jacques 2008). Besides the synthesis of diffusible antibiotics, PGPRs also produce volatile antibiotic known as hydrogen cyanide (HCN) which is responsible for preventing black root rot of tobacco caused by *Thielaviopsis basicola* (Sacherer et al. 1994).

3 Benefits of PGPRs

In order to attain sustainable crop production without compromising with the biodiversity, agroecosystem, and environmental quality, there has been a growing dependency on microbial inoculants in the form of biofertilizers, plant growth promoters, biopesticides, soil health managers, etc. A large number of bacterial strains belonging to different classes and genera showing multifaceted plant growth-promoting attributes are actively considered for their role in improving plant growth and yield in modern agriculture. These PGPRs have gained considerable attention among researchers, agriculturists, farmers, and policy makers. Besides the role of PGPRs in improving plant health and crop yield, their roles in nutrient uptake and conferring resistance to environmental stress have also been established (Malhotra and Srivastava 2009). PGPRs promote growth and yield of crops by stimulating nitrogen transformation; increased availability of phosphorous; iron acquisition; mineral solubilization; synthesis of phytohormones such as IAA, cytokinins, and gibberellin; synthesis of various volatile compounds; and offering protection to plants against pathogens through the production of various antibiotics (Ryu et al. 2003; Vessey 2003). They also produce antioxidative enzymes for scavenging reactive oxygen species under stress and decrease the overaccumulation of stress hormone ethylene in plants by the bulk synthesis of ACC-degrading enzyme ACC deaminase (Glick et al. 1998; Penrose et al. 2001). Important benefits of PGPRs to the plants are briefly described below.

3.1 Biofertilization

The necessity of feeding an escalating world population without damaging soil health and level of salinity and deforestation has become a challenging problem for the scientists working in the area of soil productivity research. To overcome above problems, there is a need to develop suitable biotechnological techniques not only to enhance crop productivity but also to manage soil health through establishing and improving interaction of plants with soil microorganisms (Lugtenberg et al. 2002). The rhizospheric soil exhibits varied types of PGPR community, which have a vital role in sustainable agriculture. These PGPRs promote plant growth and development by several biological processes and seem much superior compared to synthetic pesticides, insecticides, and fertilizers. For example, inoculation with PGPRs enhances root growth, P (phosphorous) and K (potassium) solubilization, and uptake of N (nitrogen) and promotes symbioses of the host plant, thereby acting as principal contributors in agroecosystems. Microbial biofertilizers are excellent source of organic farming as they provide up to 65% of the nitrogen source to agricultural crops globally (Babalola 2010). PGPR formulations, containing useful microbial strains, are cheaper, nontoxic, and eco-friendly and contain readily available carrier material. Formulation is the most important step for the bacterial biofertilizer as it shall decide the success or failure of the inoculants in the field (Domenech et al. 2006). A successful bacterial biofertilizer strain should have an efficient colonization potential under different agroclimatic conditions and must retain and exhibit its growth-promoting features once applied in the field. Equally important parameter relates to its ability to retain viability under different storage conditions.

3.2 PGPRs as "Helper" Bacteria

The beneficial effects of PGPRs as a biofertilizer involve synergism with a third-party microorganism. In such cases, PGPRs helping other host-symbiotic interactions are considered as "helper" bacteria. Although different modes for PGPR-mediated legume-rhizobia symbioses occur, phytohormone-induced (IAA) stimulation of root growth (Vessey and Buss 2002) is the most common. Burdman et al. (1996) reported that *Azospirillum brasilense* stimulate node formation in common bean leading to an enhanced synthesis of flavonoids by the host. Subsequently, flavonoids induce *nod* genes in rhizobia which in turn lead to the legume-rhizobia symbiosis (Schultze and Kondorosi 1998). PGPRs as a biofertilizer often increase plant growth indirectly by augmenting association between the host plant and the beneficial rhizospheric fungi such as arbuscular mycorrhizae (AM). Ratti et al. (2001) tested the effects of AM fungus *Glomus aggregatum* and PGPRs *Bacillus polymyxa* and *Azospirillum brasilense* on *Cymbopogon martini* grown with inorganic phosphate and found enhanced P content and biomass of the plant. However, the interactions between PGPRs and AM are not always positive. Certain studies reported that PGPR inoculants may show inhibitory effects on the symbiotic relationship of host plants and AM. This is also evident from the experiments of Walley and Germida (1997), who have observed variable results of co-inoculation of PGPR strains of *P. cepacia* and *P. putida* and AM fungi in spring wheat.

3.3 Growth Enhancement

PGPRs are a boon in agriculture as they have tremendous potential in meeting ever-increasing global demand for food without affecting soil health. They are known to increase overall growth of plant including root and shoot biomass, seed emergence, and yield of crops (Minorsky 2008). PGPRs have been reported to affect several pathways of hormone synthesis resulting in marked changes in elongation rate of lateral roots and overall structure of root system (Kapulnik et al. 1985). Inoculation of PGPRs has been reported to enhance growth traits such as leaf area, chlorophyll content, and dry weight of root system and the aerial parts of the plants leading to enhanced flowering and development (Dobbelaere et al. 2001). Several PGPR strains such as *Achromobacter xylosoxidans*, *Bacillus subtilis*, *B. licheniformis*, and *Pseudomonas putida* play an indispensable role in cell elongation due to increased activity of ACC deaminase (Sgroy et al. 2009). It has been reported that PGPRs evoke changes in root cortex due to bulk division of cells in the root tips of wheat and maize seedlings (Baset Mia et al. 2010). Inoculation of a mixture of PGPRs and rhizobia to the seeds of certain crops and ornamental plants before planting resulted in increased growth and disease resistance (Zehnder et al. 2001). Ahanthem and Jha (2007) have reported changes in rice plants challenged with AM fungi *Glomus* sp. and PGPR *Azotobacter chroococcum* in soils differing in nitrogen concentrations and recorded maximized shoot biomass and shoot phosphorous and nitrogen availability.

3.4 Microbial Antagonists

Rhizobacteria often exhibit antagonistic activity such as (1) synthesis of hydrolytic enzymes like chitinase, cellulase, β-1,3-glucanase, protease, and lipase which can break fungal cell walls and suppress the growth of deleterious microbes, (2) favoring active colonization of niches at the root surface creating nutrient competition (Kamilova et al. 2005), (3) alteration in the level of ethylene by ACC deaminase activity in host plants exposed to abiotic stress (Glick and Bashan 1997; van Loon 2007), and (4) production of siderophores by PGPRs which inhibit the proliferation of pathogenic microbes allowing better growth of plants (Castignetti and Smarelli 1986). That the uptake of iron (Fe^{3+}) by the iron-siderophore complex prevents proliferation of pathogenic microbes in the rhizospheric region has been reported by a number of researchers. For example, *Pseudomonas putida* mutant strain overproducing siderophore was found more efficient in inhibiting a pathogenic strain of *Fusarium oxysporum*, to tomato as compared to the wild type. In addition to siderophores, production of low molecular weight metabolite such as HCN by certain PGPRs also promotes growth of plants by acting as biocide. Production of antibiotics by *Bacillus* sp. and other species of PGPR also inhibits growth of the residential deleterious fungi and bacteria around rhizosphere by exhibiting antagonistic activity (Principe et al. 2007; Choudhary and Johri 2009).

3.5 Biocontrol Agents

PGPRs have the ability to decrease the prevalence and/or magnitude of diseases caused by pathogenic microbes. Potential resistance-inducing compounds, viz., lipopolysaccharides, salicylic acid, biosurfactants, N-acyl-homoserine lactones (AHL), N-alkylated benzylamines, antibiotics, and exopolysaccharides (EPSs), have been reported from several species of PGPRs (De Vleesschauwer and Höfte 2009). ISR evokes an exemplary defensive ability against various diseases. ISR is equivalent to systemic acquired resistance (SAR) which is pathogen induced and renders resistance to uninfected plant parts against plant pathogens (van Wees et al. 1999). ISR is stimulated most commonly by the species of *Pseudomonas* and *Bacillus* sp. residing in the rhizosphere (Kloepper et al. 2004). Notably, combined action of ISR and SAR is more effective against pathogens as compared to acting alone (van Wees et al. 1999).

3.6 Stress Relievers

Benefits of PGPR including enhanced growth are more or less similar under unstressed natural environment, but during stress conditions, certain strains become inefficient due to their incapability to sustain in the unfavorable environmental conditions. However, certain PGPR strains not only cope up with these stress conditions but also show better plant growth. Since plant growth and development are affected by a variety of biotic and abiotic stresses, PGPRs alleviate the stress-mediated harmful effects on physiological and biochemical processes of plants by various mechanisms (Evelin et al. 2009; Saharan and Nehra 2011). The deleterious outcomes of abiotic stresses include enhanced synthesis of ethylene in the root, ionic imbalance, and hyperosmotic shock in plants (Mayak et al. 2004a, b; Niu et al. 1995; Zhu et al. 1997). Some abiotic stresses like salinity, drought, and waterlogging elevate the levels of ethylene due to enhanced production of ACC, which is a precursor of ethylene biosynthesis (Zapata et al. 2007). This results in the retardation of root extension and photosynthesis, thereby causing changes in chlorophyll content as well as damage to photosynthesis framework (Iturbe-Ormaetxe et al. 1998) and inhibition of nitrogenase activity (Rai and Tiwari 1999). Although many agrochemicals like amino ethoxy vinyl glycine (AVG), silver ion (Ag^+), and cobalt ion (Co^{2+}) have been applied as ethylene inhibitors, however, they are expensive and toxic for human as well as soil health (Dodd et al. 2004). Roles of PGPRs in alleviating various stresses are briefly mentioned below:

1. *Salinity:* High salinity results in nutritional imbalance due to elevated levels of sodium (Na^+), thereby causing ion toxicity (Ashraf 1994). Certain strains of PGPR produce exopolysaccharides to protect the plants from the deleterious effects of high Na^+ level. The exopolysaccharides bind to Na^+ and render it unavailable for plant uptake. This enables the plants to adjust high K^+/Na^+ ratio for osmotic tolerance, thereby allowing plant's survival even under high saline conditions (Ashraf et al. 2004; Khodair et al. 2008).
2. *Drought stress:* The role of exopolysaccharides on growth of plants under water-limiting conditions has been studied by Sandhya et al. (2009). As such, water deficit condition causes negative impacts on both plants and microbial flora. Under such circumstances, exopolysaccharides produced by PGPR not only safeguard bacteria but also plants from desiccation and allow growth even under drought condition. Plant growth is also inhibited under water deficit conditions due to enhanced level of ethylene. It has been also reported that the level of chlorophyll content decreases during severe drought condition (Reddy and Rao 1968). Mayak et al. (2004b) demonstrated that inoculation of *Achromobacter piechaudii* possessing ACC deaminase activity showed increase in dry and fresh weight of pepper and tomato seedlings subjected to short-term water stress. Similarly, inoculation of *P. fluorescens* strain TDK1 having ACC deaminase activity increased drought resistance in root rot pathogen (*Macrophomina phaseolina*)-infected plants (Grichko and Glick 2001).
3. *Flood:* Other than the stresses of salinity and drought, there are reports that PGPR strains are capable to negate the impacts of flooding stress by reducing the level of ethylene accumulation (Grichko and Glick 2001). Tomato plants showed significant tolerance to flooding stress after the inoculation with PGPRs possessing ACC deaminase activity suggesting the key role of ACC deaminase in conferring tolerance to flooding by reducing the level of ethylene (Grichko and Glick 2001).
4. *Temperature:* Frequent fluctuations in ambient temperature (either increase or decrease mode) possess a serious threat to agriculture worldwide. The phenomenon of global warming may be a big threat in agriculture sector if the trend in temperature increase continues. There is a general consensus that temperature changes can cause hormonal imbalance resulting in marked effects on growth and development of plants. Increased ethylene production under high and chilling temperature in plants has been reported (Strzelczyk et al. 1994). One of the approaches to overcome the above problem would be to apply rhizobacteria possessing higher level of ACC deaminase activity in the field so as to block the accumulation of ethylene in plants. This approach seems promising considering the report wherein *Burkholderia phytofirmans* PsJN, a rhizobacterium endowed with high ACC deaminase activity when inoculated to potato clones grown under high temperature, showed normal growth as was evident from stem length and shoot and root biomass (Bensalim et al. 1998).
5. *Heavy metals:* Although several metals are essential for growth and development of plants, they may impart toxicity when present in higher concentrations (Ernst 1998). Several researchers have reported that PGPRs displaying ACC

deaminase activity can lower the level of ethylene production in plants subjected to metal stress (Safronova et al. 2006; Arshad et al. 2008). Seeds of canola inoculated with PGPR *Kluyvera ascorbata* SUD165 possessing ACC deaminase activity conferred resistance to plants to higher concentrations of nickel chloride (Burd et al. 1998, 2000).

3.7 Bioremediation

Phytoremediation demands the utility of plants to extract or balance hazardous substances present in the soil; however, adding PGPRs is an additional benefit as it accelerates contaminant uptake and plant nutrition and health. Whiting et al. (2001) reported that bacterial inoculation in *Zea mays* and *Thlaspi caerulescens* crops enhanced the uptake of heavy metals. PGPRs applications also help in decreasing the bioavailability of toxicity in heavy metal-contaminated soils by detoxifying chemicals, nutrient cycling, and maintaining soil structure (Denton 2007). Pairing of transgenic plants with PGPRs is now emerging as a novel approach to exploit the process of phytoremediation. Canola plants exposed to arsenate showed significant increase in root and shoot biomass by the combined action of the PGPR, *Enterobacter cloacae* CAL2, and canola plants due to the expression of stress tolerance gene ACC deaminase (Nie et al. 2002). PGPRs together with AM fungi have been recently employed in the management of nutrient-deficient agricultural soils to enhance the solubility of heavy metals and accomplish rhizoremediation (Bhattacharyya and Jha 2012).

3.8 Integrated Nutrient Management (INM)

Soil fertility and management of plant nutrients are an integral venture for increasing food production to meet the demand of the escalating population. Sustainable agriculture incorporates the idea to generate large output and income without depleting natural resources. Persistent use of chemical fertilizers including pesticides has degraded the fertility of soil and affected the microbial diversity. An integrated approach using PGPRs as microbial inoculants has been applied to reduce the use of fertilizers for enhancing growth and yield of crop plants. The goal of INM is to maintain soil as storehouses of plant nutrients and integrate all man-made and natural deposit of nutrients for achieving the desired level of crop yield in a sustainable manner. Henceforth, uses of PGPRs as alternative source of fertilizers are encouraged as a crucial constituent of an integrated plant nutrient system (IPNS). The synergistic effects of PGPRs and INM have been tested on several crops like groundnut, sunflower, soybean, mustard, etc. (Dubey and Maheshwari 2011).

4 Conclusion and Future Perspectives

This chapter deals with the multifaceted nature of PGPRs in the rhizospheric region of the soil with emphasis on their functions and benefits for sustainable agriculture. PGPRs represent a substitute to chemical fertilizers which otherwise increase the crop yield but are expensive, deplete nonrenewable energy, make crops more susceptible to diseases, reduce soil fertility, and cause an irreplaceable damage to the agroecosystem. It is now a proven fact that certain PGPRs exhibit more than one attribute of plant growth promotion, e.g., the production of enzymes, antibiotics, bioactive compounds, growth promoters, siderophores, etc. Since PGPRs have great potential through their effective biocontrol mechanism and potent eco-friendly strategy, judicious application of these microbial populations should be encouraged so that they can be utilized in a sustainable way. One essential aspect rests in the evaluation of the aforesaid characters under field conditions. Unfortunately, earlier studies have been mostly performed under laboratory conditions, and significant variability has been reported in field performance probably due to a number of environmental factors which influence PGPRs. There is a need to develop efficient biofertilizer strains which retain and perform all the beneficial characters in the field.

Undoubtedly, PGPRs may be a boon for sustainable agriculture, but certain limitations which limit the use of PGPRs for improved crop production need to be addressed. The main limitation is the quality control of the PGPR inoculums in terms of their proper functioning in the farmer's field. Future prospects lie in the search of bacterial strains/traits which are beneficial and suitable for different plants in terms of their performance in different agroclimatic conditions. Additionally, it will be also useful to make a detailed study on how different bacterial strains can work together as a consortium for the synergistic effects on plant growth. A proper understanding of the factors that increase the environmental persistence of the PGPR strains would be equally useful. For the quality assurance of PGPRs, it is necessary that inoculant strains be labelled with different reporter genes (e.g., *lux* or green fluorescent protein (*gfp*) genes or antibiotic markers), so that they can be readily detected in the field after their release. During the past few years, successful attempts have been made to produce transgenic plants encoding stress tolerance genes of desired traits of PGPRs, and trials have been made under laboratory conditions, but extensive field-based studies are also necessary to monitor the proper functioning in fluctuating environmental conditions. To this effect, proper knowledge and detailed analysis of the genes responsible for stress tolerance of PGPRs would be desirable to reveal the molecular biology of the tolerance mechanism. One of the most important constraints pertaining to the large-scale use of PGPRs as biofertilizer is the lack of awareness among the users. To exploit the benefits of PGPRs as biofertilizer in the farmer's field, it would be necessary to create awareness and provide adequate technical skills among farmers/stockholders for the production of PGPRs as biofertilizer and its application.

Acknowledgments DS is grateful to DST-INSPIRE, Department of Science and Technology, New Delhi, for the award of Junior Research Fellowship (DST/INSPIRE Fellowship/2014/296, IF140707). PG and JK are thankful to University Grants Commission (UGC), New Delhi, for providing financial assistance as JRF. Research in the area of PGPRs is partly supported by a research grant sanctioned to AK by the Indian Council of Agricultural Research, Government of India, New Delhi (NBAIM/AMAAS/2014-17/PF/4).

References

Ahanthem S, Jha DK (2007) Response of rice crop inoculated with arbuscular mycorrhizal fungi and plant growth promoting rhizobacteria to different soil nitrogen concentrations. Mycorrhiza News 18:15–20

Ahemad M, Khan MS (2012) Evaluation of plant-growth promoting activities of rhizobacterium *Pseudomonas putida* under herbicide stress. Ann Microbiol 62:1531–1540

Ahemad M, Kibret M (2014) Mechanisms and applications of plant growth promoting rhizobacteria: current perspective. J King Saud Univ Sci 26:1–20

Anand K, Kumari B, Mallick MA (2016) Phosphate solubilizing microbes: an effective and alternative approach as bio-fertilizers. Int J Pharm Sci 8:37–40

Arora NK, Tewari S, Singh R (2013) Multifaceted plant-associated microbes and their mechanisms diminish the concept of direct and indirect PGPRs. In: Arora NK (ed) Plant microbe symbiosis: fundamentals and advances. Springer, New Delhi, pp 411–449

Arshad M, Sharoona B, Mahmood T (2008) Inoculation with *Pseudomonas* sp. containing ACC deaminase partially eliminate the effects of drought stress on growth, yield and ripening of pea (*Pisum sativum* L.). Pedosphere 18:611–620

Ashraf M (1994) Breeding for salinity tolerance in plants. Crit Rev Plant Sci 13:17–42

Ashraf M, Berge SH, Mahmood OT (2004) Inoculating wheat seedling with exopolysaccharide producing bacteria restricts sodium uptake and stimulates plant growth under salt stress. Biol Fertil Soils 40:157–162

Babalola OO (2010) Beneficial bacteria of agricultural importance. Biotechnol Lett 32:1559–1570

Baset Mia MA, Shamsuddin ZH, Wahab Z et al (2010) Effect of plant growth promoting rhizobacterial (PGPR) inoculation on growth and nitrogen incorporation of tissue cultured *Musa* plantlets under nitrogen-free hydroponics condition. Aust J Crop Sci 4:85–90

Bensalim S, Nowak J, Asiedu SK (1998) A plant growth promoting rhizobacterium and temperature effects on performance of 18 clones of potato. Am J Potato Res 75:145–152

Berendsen RL, Verk MCV, Stringlis IA et al (2015) Unearthing the genomes of plant-beneficial *Pseudomonas* model strains WCS358, WCS374 and WCS417. BMC Genomics 16:539

Bhardwaj D, Ansari MW, Sahoo RK et al (2014) Biofertilizers function as key player in sustainable agriculture by improving soil fertility, plant tolerance and crop productivity. Microbial Cell Fact 13:1–10

Bhattacharyya PN, Jha DK (2012) Plant growth-promoting rhizobacteria (PGPR): emergence in agriculture. World J Microbiol Biotechnol 28:1327–1350

Bottini R, Cassán F, Piccoli P (2004) Gibberellin production by bacteria and its involvement in plant growth promotion and yield increase. Appl Microbiol Biotechnol 65:497–503

Boukhalfa H, Crumbliss AL (2002) Chemical aspects of siderophore mediated iron transport. BioMetals 15:325–339

Burd GI, Dixon DG, Glick BR (1998) A plant growth-promoting bacterium that decreases nickel toxicity in seedlings. Appl Environ Microbiol 64:3663–3668

Burd GI, Dixon DG, Glick BR (2000) Plant growth-promoting bacteria that decrease heavy metal toxicity in plants. Can J Microbiol 46:237–245

Burdman S, Volpin H, Kigel J et al (1996) Promotion of nod gene inducers and nodulation in common bean (*Phaseolus vulgaris*) roots inoculated with *Azospirillum brasilense* Cd. Appl Environ Microbiol 62:3030–3033

Castignetti D, Smarelli J (1986) Siderophores, the iron nutrition of plants, and nitrate reductase. FEBS Lett 209:147–151

Chernin L, Toklikishvili N, Ovadis M et al (2011) Quorum-sensing quenching by rhizobacterial volatiles. Environ Microbiol Rep 3:698–704

Chin-A-Woeng TF, Bloemberg GV, Lugtenberg BJ (2003) Phenazines and their role in biocontrol by *Pseudomonas* bacteria. New Phytol 157:503–523

Choudhary DK, Johri BN (2009) Interactions of *Bacillus* sp. and plants-with special reference to induced systemic resistance (ISR). Microbiol Res 164:493–513

Crowley DE, Kraemer SM (2007) Function of siderophores in the plant rhizosphere. In: Pinton R et al (eds) The rhizosphere, biochemistry and organic substances at the soil-plant interface. CRC Press, Boca Raton, pp 73–109

Damam M, Kaloori K, Gaddam B et al (2016) Plant growth promoting substances (phytohormones) produced by rhizobacterial strains isolated from the rhizosphere of medicinal plants. Int J Pharm Sci Rev 37:130–136

de Souza JT, Weller DM, Raaijmakers JM (2003) Frequency, diversity and activity of 2, 4-diacetylphloroglucinol producing fluorescent *Pseudomonas* sp. in Dutch take-all decline soils. Phytopathology 93:54–63

De Vleesschauwer D, Hofte M (2009) Rhizobacteria-induced systemic resistance. Adv Bot Res 51:223–281

Denton B (2007) Advances in phytoremediation of heavy metals using plant growth promoting bacteria and fungi. Basic Biotechnol 3:1–5

Dobbelaere S, Croonenborghs A, Thys A et al (2001) Responses of agronomically important crops to inoculation with *Azospirillum*. Aust J Plant Physiol 28:871–879

Dobert RC, Rood SB, Blevins DG (1992) Gibberellins and the legume-Rhizobium symbiosis. I. Endogenous gibberellins of lima bean (*Phaseolus lunatus* L.) stems and nodules. Plant Physiol 98:221–224

Dodd IC, Belimov AA, Sobeih WY et al (2004) Will modifying plant ethylene status improve plant productivity in water-limited environments? In: Proceedings of the 4th International Crop Science Congress, Brisbane, Australia, 26 Sept–1 Oct 2004

Domenech J, Reddy MS, Kloepper JW et al (2006) Combined application of the biological product LS213 with *Bacillus*, *Pseudomonas* or *Chryseobacterium* for growth promotion and biological control of soil-borne diseases in pepper and tomato. Bio Control 51:245–258

Dubey DC, Maheshwari DK (2011) Role of PGPR in integrated nutrient management of oil seed crops. In: Maheshwari DK (ed) Bacteria in agrobiology: plant nutrient management. Springer-Verlag, Heidelberg, pp 1–17

Duffy B (2003) Pathogen self-defense: mechanisms to counteract microbial antagonism. Annu Rev Phytopathol 41:501–538

Egamberdieva D, Lugtenberg B (2014) Use of plant growth-promoting rhizobacteria to alleviate salinity stress in plants. In: Miransari M (ed) Use of microbes for the alleviation of soil stresses. Springer, New York, pp 73–96

Ernst WHO (1998) Effects of heavy metals in plants at the cellular and organismic level. In: Schuurmann G, Markert B (eds) Ecotoxicology. Wiley, Heidelberg, pp 587–620

Evelin H, Kapoor R, Giri B (2009) Arbuscular mycorrhizal fungi in alleviation of salt stress: a review. Ann Bot 104:1263–1280

Glick BR (2012) Plant growth-promoting bacteria: mechanisms and applications. Hindawi Publishing Corporation Scientifica. https://doi.org/10.6064/2012/963401, Article ID 963401

Glick BR, Bashan Y (1997) Genetic manipulation of plant growth-promoting bacteria to enhance biocontrol of fungal phytopathogens. Biotechnol Adv 15:353–378

Glick BR, Penrose DM, Li J (1998) A model for the lowering of plant ethylene concentrations by plant growth-promoting bacteria. J Theor Biol 190:3–68

Glick BR, Cheng Z, Czarny J et al (2007a) Promotion of plant growth by ACC deaminase-producing soil bacteria. Eur J Plant Pathol 119:329–339

Glick BR, Todorovic B, Czarny J et al (2007b) Promotion of plant growth by bacterial ACC deaminase. Crit Rev Plant Sci 26:227–242

Gray EJ, Smith DL (2005) Intracellular and extracellular PGPR: commonalities and distinctions in the plant-bacterium signaling processes. Soil Biol Biochem 37:395–412

Grichko VP, Glick BR (2001) Amelioration of flooding stress by ACC deaminase containing plant growth-promoting bacteria. Plant Physiol Biochem 39:11–17

Gupta P, Diwan B (2017) Bacterial exopolysaccharide mediated heavy metal removal: a review on biosynthesis, mechanism and remediation strategies. Biotechnol Rep 13:58–71

Gupta G, Parihar SS, Ahirwar NK et al (2015) Plant growth promoting rhizobacteria (PGPR): current and future prospects for development of sustainable agriculture. J Microbiol Biochem 7:96–102

Haas D, Défago G (2005) Biological control of soil-borne pathogens by fluorescent pseudomonads. Nat Rev Microbiol 3:307–319

Hartmann A, Schikora A (2012) Quorum sensing of bacteria and trans-kingdom interactions of N-acyl homoserine lactones with eukaryotes. J Chem Ecol 38:704–713

Hill DS, Stein JI, Torkewitz NR et al (1994) Cloning of genes involved in the synthesis of pyrrolnitrin from *Pseudomonas fluorescens* and role of pyrrolnitrin synthesis in biological control of plant disease. Appl Environ Microbiol 60:78–85

Hepper CM (1975) Extracellular polysaccharides of soil bacteria. In: Walker N (ed) Soil microbiology, a critical review. Wiley, New York, pp 93–111

Insam H, Seewald MSA (2010) Volatile organic compounds (VOCs) in soils. Biol Fertil Soils 46:199–213

Iturbe-Ormaetxe I, Escuredo PR, Arrese-Igor C et al (1998) Oxidative damage in pea plants exposed to water deficit or paraquat. Plant Physiol 116:173–181

Jahanian A, Chaichi MR, Rezaei K et al (2012) The effect of plant growth promoting rhizobacteria (PGPR) on germination and primary growth of artichoke (*Cynara scolymus*). Int J Agric Crop Sci 4:923–929

Kamal R, Gusain YS, Kumar V (2014) Interaction and symbiosis of fungi, actinomycetes and plant growth promoting rhizobacteria with plants: strategies for the improvement of plants health and defense system. Int J Curr Microbial Appl Sci 3(7):564–585

Kamilova F, Goel V, Azarova T et al (2005) Enrichment for enhanced competitive plant root tip colonizers selects for a new class of biocontrol bacteria. Environ Microbiol 7:1809–1817

Kanchiswamy CN, Malnoy M, Maffei ME (2015) Chemical diversity of microbial volatiles and their potential for plant growth and productivity. Front Plant Sci 6:151

Kapulnik Y, Okon Y, Henis Y (1985) Changes in root morphology of wheat caused by *Azospirillum* inoculation. Can J Microbiol 31:881–887

Kaushal M, Wani SP (2015) Plant-growth-promoting rhizobacteria: drought stress alleviators to ameliorate crop production in drylands. Ann Microbiol 66:35–42

Khan MS, Zaidi A, Wani PA et al (2009) Role of plant growth promoting rhizobacteria in the remediation of metal contaminated soils. Environ Chem Lett 7:1–19

Khodair TA, Galal GF, El-Tayeb TS (2008) Effect of inoculating wheat seedlings with exopolysaccharide-producing bacteria in saline soil. J Appl Sci Res 4:2065–2070

Kim J, Rees DC (1994) Nitrogenase and biological nitrogen fixation. Biochemistry 33:389–397

Kloepper JW, Schroth MN (1978) Plant growth-promoting rhizobacteria on radishes. In: Proceedings of the IVth international conference on plant pathogenic bacteria, France, vol 2, pp 879–882

Kloepper JW, Leong J, Teintze M et al (1980) *Pseudomonas* siderophores: a mechanism explaining disease- suppressive soils. Curr Microbiol 4:317–320

Kloepper JW, Ryu CM, Zhang SA (2004) Induced systemic resistance and promotion of plant growth by *Bacillus* sp. Phytopathology 94:1259–1266

Kucey RMN (1988) Plant growth-altering effects of *Azospirillum brasilense* and *Bacillus* C-11-25 on two wheat cultivars. J Appl Bacteriol 64:187–196

Kumar P, Dubey RC (2012) Plant growth promoting rhizobacteria for biocontrol of phytopathogens and yield enhancement of *Phaseolus vulgaris*. J Curr Perspect Appl Microbiol 1:6–38

Kundan R, Pant G, Jado N et al (2015) Plant growth promoting rhizobacteria: mechanism and current prospective. J Fertilizers Pesticides 6:2

Liu D, Lian B, Dong H (2012) Isolation of *Paenibacillus* sp. and assessment of its potential for enhancing mineral weathering. J Geomicrobiol 29:413–421
Liu W, Wang Q, Hou J et al (2016) Whole genome analysis of halotolerant and alkalotolerant plant growth-promoting rhizobacterium *Klebsiella* sp. D5A. Sci Rep 6:26710
Lucangeli C, Bottini R (1996) Reversion of dwarfism in dwarf-1 maize (*Zea mays* L.) and dwarf-x rice (*Oryza sativa* L.) mutants by endophytic *Azospirillum* sp. Biocell 20:223–228
Lucy M, Reed E, Glick BR (2004) Applications of free living plant growth-promoting rhizobacteria. Anton Leeuw 86:1–25
Lugtenberg BJ, Chin-A-Woeng TF, Bloemberg GV (2002) Microbe-plant interactions: principles and mechanisms. Anton Leeuw 81:373–383
Lynch JM, Whipps JM (1990) Substrate flow in the rhizosphere. Plant Soil 129:1–10
Mahmood S, Daur I, Al-Solaimani SG et al (2016) Plant growth promoting rhizobacteria and silicon synergistically enhance salinity tolerance of mung bean. Front Plant Sci 7:1–14
Malhotra M, Srivastava S (2009) Stress-responsive indole-3-acetic acid biosynthesis by *Azospirillum brasilense* SM and its ability to modulate plant growth. Eur J Soil Biol 45:73–80
Mayak S, Tirosh T, Glick BR (2004a) Plant growth-promoting bacteria that confer resistance in tomato plants to salt stress. Plant Physiol Biochem 42:565–572
Mayak S, Tirosh T, Glick BR (2004b) Plant growth-promoting bacteria that confer resistance to water stress in tomato and pepper. Plant Sci 166:525–530
Minorsky PV (2008) On the inside. Plant Physiol 146:323–324
Nadeem SM, Zahir ZA, Naveed M et al (2009) Rhizobacteria containing ACC-deaminase confer salt tolerance in maize grown on salt-affected fields. Can J Microbiol 55:1302–1309
Nadeem SM, Naveed M, Zahir ZA et al (2013) Plant-microbe interactions for sustainable agriculture: fundamentals and recent advances. In: Arora NK (ed) Plant microbe symbiosis: fundamentals and advances. Springer, New Delhi, pp 51–103
Neubauer U, Furrer G, Kayser A et al (2000) Siderophores, NTA, and citrate: potential soil amendments to enhance heavy metal mobility in phytoremediation. Int J Phytoremediation 2:353–368
Nie L, Shah S, Rashid A et al (2002) Phytoremediation of arsenate contaminated soil by transgenic canola and the plant growth-promoting bacterium *Enterobacter cloacae* CAL2. Plant Physiol Biochem 40:355–361
Niu X, Bressan RA, Hasegawa PM et al (1995) Ion homeostasis in NaCl stress environments. Plant Physiol 109:735–742
Niu DD, Liu HX, Jiang CH et al (2011) The plant growth-promoting rhizobacterium *Bacillus cereus* AR156 induces systemic resistance in *Arabidopsis thaliana* by simultaneously activating salicylate and jasmonate/ethylene-dependent signaling pathways. Mol Plant Microbe Interact 24:533–542
Nivya RM (2015) A study on plant growth promoting activity of the endophytic bacteria isolated from the root nodules of *Mimosa pudica* plant. Int J Innov Res Sci Er Technol 4:6959–6968
O'Sullivan DJ, O'Gara F (1992) Traits of fluorescent *Pseudomonas* sp. involved in suppression of plant root pathogens. Microbiol Rev 56:662–676
Ongena M, Jacques P (2008) *Bacillus* lipopeptides: versatile weapons for plant disease biocontrol. Trends Microbiol 16:115–125
Parada M, Vinardell J, Ollero F et al (2006) *Sinorhizobium fredii* HH103 mutants affected in capsular polysaccharide (KPS) are impaired for nodulation with soybean and *Cajanus cajan*. Mol Plant Microbe Interact 19:43–52
Parmar P, Sindhu SS (2013) Potassium solubilisation by rhizosphere bacteria: influence of nutritional and environmental conditions. J Microbial Res 3:25–31
Patten CL, Glick BR (1996) Bacterial biosynthesis of indole-3-acetic acid. Can J Microbiol 42:207–220
Penrose DM, Moffatt BA, Glick BR (2001) Determination of 1-aminocyclopropane-l-carboxylic acid (ACC) to assess the effects of ACC deaminase-containing bacteria on roots of canola seedlings. Can J Microbiol 47:77–80
Prathap M, Ranjitha KBD (2015) A critical review on plant growth promoting rhizobacteria. J Plant Pathol Microbiol 6:1–4

Principe A, Alvarez F, Castro MG et al (2007) Biocontrol and PGPR features in native strains isolated from saline soils of Argentina. Curr Microbiol 55:314–322
Rai AK, Tiwari SP (1999) Response to NaCl of nitrate assimilation and nitrogenase activity in the cyanobacterium *Anabaena* sp. PCC 7120 and its mutants. J Appl Microbiol 87:877–883
Rajkumar M, Ae N, Prasad MNV et al (2010) Potential of siderophore-producing bacteria for improving heavy metal phytoextraction. Trends Biotechnol 28:142–149
Ramegowda V, Senthil-Kumarb M (2015) The interactive effects of simultaneous biotic and abiotic stresses on plants: mechanistic understanding from drought and pathogen combination. J Plant Physiol 176:47–54
Ratti N, Kumar S, Verma HN et al (2001) Improvement in bioavailability of tricalcium phosphate to *Cymbopogon martinii* var. motia by rhizobacteria, AMF and *Azospirillum* inoculation. Microbiol Res 156:145–149
Reddy AJ, Rao IM (1968) Influence of induced water stress on chlorophyll components of proximal and distal leaflets of groundnut plants. Curr Sci 5:118–121
Ryu CM, Farag MA, Hu CH et al (2003) Bacterial volatiles promote growth in *Arabidopsis*. Proc Natl Acad Sci 100:4927–4932
Sacherer P, Défago G, Haas D (1994) Extracellular protease and phospholipase C are controlled by the global regulatory gene *gacA* in the biocontrol strain *Pseudomonas fluorescens* CHA0. FEMS Microbiol Lett 116:155–160
Safronova VI, Stepanok VV, Engqvist GL et al (2006) Root-associated bacteria containing 1-aminocyclopropane-1-carboxylate deaminase improve growth and nutrient uptake by pea genotypes cultivated in cadmium supplemented soil. Biol Fertil Soil 42:267–272
Saharan BS, Nehra V (2011) Plant growth promoting rhizobacteria: a critical review. Life Sci Med Res 21:1–30
Salamone IEG, Hynes RK, Nelson LM (2005) Role of cytokinins in plant growth promotion by rhizosphere bacteria. In: Siddiqui ZA (ed) PGPR: biocontrol and biofertilization. Springer, Dordrecht, pp 173–195
Saleem M, Arshad M, Hussain S et al (2007) Perspective of plant growth promoting rhizobacteria (PGPR) containing ACC deaminase in stress agriculture. J Indian. Microbiol Biotechnol 34:635–648
Sandhya V, Ali SKZ, Grover M et al (2009) Alleviation of drought stress effects in sunflower seedlings by the exopolysaccharides producing *Pseudomonas putida* strain GAP-P45. Biol Fertil Soils 46:17–26
Santi C, Bogusz D, Franche C (2013) Biological nitrogen fixation in non-legume plants. Ann Bot 111:743–767
Santoro MV, Bogino PC, Nocelli N et al (2016) Analysis of plant growth promoting effects of fluorescent *Pseudomonas* strains isolated from *Mentha piperita* rhizosphere and effects of their volatile organic compounds on essential oil composition. Front Microbiol 7:1–17
Schmidt W (1999) Mechanisms and regulation of reduction-based iron uptake in plants. New Phytol 141:1–26
Schultze M, Kondorosi A (1998) Regulation of symbiotic root nodule development. Annu Rev Gen 32:33–57
Sgroy V, Cassan F, Masciarelli O et al (2009) Isolation and characterization of endophytic plant growth-promoting (PGPB) or stress homeostasis-regulating (PSHB) bacteria associated to the halophyte *Prosopis strombulifera*. Appl Microbiol Biotechnol 85:371–381
Singh RP, Shelke GM, Kumar A, Jha PN (2015) Biochemistry and genetics of ACC deaminase: a weapon to "stress ethylene" produced in plants. Front. Microbiol 6:937
Skoog F, Miller CO (1957) Chemical regulation of growth and organ formation in plant tissue cultures in vitro. Symp Soc Exp Biol 11:118–131
Spaepen S, Vanderleyden J (2011) Auxin and plant-microbe interactions. Cold Spring Harb Perspect Biol 3:a001438
Spaepen S, Vanderleyden J, Remans R (2007) Indole-3-acetic acid in microbial and microorganism-plant signaling. FEMS Microbiol Rev 31:425–448

Strzelczyk E, Kampert M, Pachlewski R (1994) The influence of pH and temperature on ethylene production by mycorrhizal fungi of pine. Mycorrhiza 4:193–196

Sujatha N, Ammani K (2013) Siderophore production by the isolates of fluorescent Pseudomonads. Int J Cur Res Rev 5:1–7

Tully RE, van Berkum P, Lovins KW, Keister DL (1998) Identification and sequencing of a cytochrome P450 gene cluster from *Bradyrhizobium japonicum*. Biochim Biophys Acta 1398:243–255

Udvardi M, Poole PS (2013) Transport and metabolism in legume-rhizobia symbioses. Annu Rev Plant Biol 64:781–805

Ulloa-Ogaz AL, Munoz-Castellanos LN, Nevarez-Moorillon GV (2015) Biocontrol of phytopathogens: Antibiotic production as mechanism of control. In: Mendez-Vilas A (ed) The battle against microbial pathogens: basic science, technological advances and educational programs. Formatex, Badajoz, pp 305–309

Upadyay SK, Maurya SK, Singh DP (2012) Salinity tolerance in free living plant growth promoting rhizobacteria. Ind J Sci Res 3:73–78

van de Mortel JE, de Vos RCH, Dekkers E et al (2012) Metabolic and transcriptomic changes induced in *Arabidopsis* by the rhizobacterium *Pseudomonas fluorescens* SS101. Plant Physiol 160:2173–2188

van Loon LC (2007) Plant responses to plant growth-promoting rhizobacteria. Eur J Plant Pathol 119:243–254

Van Wees SCM, Luijendijk M, Smoorenburg I et al (1999) Rhizobacteria-mediated induced systemic resistance (ISR) in *Arabidopsis* is not associated with a direct effect on expression of known defense-related genes but stimulates the expression of the jasmonate-inducible gene *Atvsp* upon challenge. Plant Mol Biol 41:537–549

Vansuyt G, Robin A, Briat JF et al (2007) Iron acquisition from Fe-pyoverdine by *Arabidopsis thaliana*. Mol Plant Microbe Interact 20:441–447

Vessey JK (2003) Plant growth promoting rhizobacteria as biofertilizers. Plant Soil 255:571–586

Vessey JK, Buss TJ (2002) *Bacillus cereus* UW85 inoculation effects on growth, nodulation, and N accumulation in grain legumes: controlled-environment studies. Can J Plant Sci 82:282–290

Viveros OM, Jorquera MA, Crowley DE et al (2010) Mechanisms and practical considerations involved in plant growth promotion by rhizobacteria. J Soil Sci Plant Nutr 10:293–319

Voisard C, Keel C, Haas D et al (1989) Cyanide production by *Pseudomonas fluorescens* helps suppress black root rot of tobacco under gnotobiotic conditions. EMBO J 8:351–358

Walley FL, Germida JJ (1997) Response of spring wheat (*Triticum aestivum*) to interactions between *Pseudomonas* sp. and *Glomus clarum* NT4. Biol Fertil Soils 24:365–371

Whiting SN, de Souza MP, Terry N (2001) Rhizosphere bacteria mobilize Zn for hyperaccumulation by *Thlaspi caerulescens*. Environ Sci Technol 35:3144–3150

Xie J, Shi H, Du Z et al (2016) Comparative genomic and functional analysis reveals conservation of plant growth promoting traits in *Paenibacillus polymyxa* and its closely related species. Sci Rep 6:21329

Zahir ZA, Munir A, Asghar HN et al (2008) Effectiveness of rhizobacteria containing ACC-deaminase for growth promotion of pea (*Pisum sativum*) under drought conditions. J Microbiol Biotechnol 18:958–963

Zaidi A, Khan MS, Ahemad M et al (2009) Plant growth promotion by phosphate solubilizing bacteria. Acta Microbiol Immunol Hung 56:263–284

Zamioudis C, Pieterse CMJ (2012) Modulation of host immunity by beneficial microbes. Mol Plant Microbe Interact 25:139–150

Zapata PJ, Botella MA, Pretel MT et al (2007) Responses of ethylene biosynthesis to saline stress in seedlings of eight plant species. Plant growth Regul 53:97–106

Zehnder GW, Murphy JF, Sikora EJ et al (2001) Application of rhizobacteria for induced resistance. Eur J Plant Pathol 107:39–50

Zhu JK, Hasegawa PM, Bressan RA (1997) Molecular aspects of osmotic stress in plants. Crit Rev Plant Sci 16:253–277

Functional Diversity of Plant Growth-Promoting Rhizobacteria: Recent Progress and Future Prospects

8

Mohd. Musheer Altaf, Mohd Sajjad Ahmad Khan, and Iqbal Ahmad

1 Introduction

Agriculture serves as the biggest source of income since the beginning of human civilization. Nearly 7.7 billion people are currently living on this globe, of which 60% population directly earn their livelihood by agriculture. In agreement to the Food and Agriculture Organization of the United Nations (FAO), earth terrestrial environment serves as the source of majority of food for world population. Every year 84.21 million people are added to this earth; to feed this growing population, there has been a constant increment in the requirement of food with a concurrent shortage in the availability (Paustian et al. 2016; Gouda et al. 2018; https://ourworldindata.org/world-population-growth). In India about 60% land is under agricultural use. The fertility of agricultural soil is directly influenced by moisture, volume of organic carbon, phosphorous, nitrogen, potassium, and further biotic and abiotic parameters. On the other hand, agriculture in this century confronted with severe challenges like loss of soil fertility, indiscriminate use of synthetic fertilizers, pesticides, aridity, high salt concentration, irregular climatic situations, and growing attack on crops by pathogens and pest (Gomiero et al. 2011; Gupta et al. 2015). This is the reality that the extreme external effort-based agriculture, which was a component of our Green Revolution policy, has exhausted soil fertility significantly in our

M. M. Altaf (✉)
Department of Life Science, Institute of Information Management and Technology, Aligarh, India

Department of Agricultural Microbiology, Aligarh Muslim University, Aligarh, India

M. S. A. Khan
Department of Basic Sciences, Biology Unit, Health Track, Imam Abdulrahman Bin Faisal University, Dammam, Saudi Arabia

I. Ahmad
Biofilm Research Laboratory, Department of Agricultural Microbiology, Faculty of Agricultural Sciences, Aligarh Muslim University, Aligarh, India

D. P. Singh et al. (eds.), *Microbial Interventions in Agriculture and Environment*,
https://doi.org/10.1007/978-981-13-8383-0_8

main agricultural systems. This is reflected in FAO report; 38.47% of the total world area is under agriculture, of which only about 4% is used for producing food. The condition will further become bad as 20–25% of agricultural area around the globe is being ruined every year (Abhilash et al. 2016).

Agricultural sustainability, environment protection, and increased crop yield can be obtained by using ecologically sustainable techniques like the use of bioinoculant and biopesticides (Bhardwaj et al. 2014; Kamkar 2016). The positive impact of bacterial inoculants, chiefly plant growth-promoting rhizobacteria (PGPR), necessitates the demand for increased research and their utility in modern agriculture.

Plant growth-promoting rhizobacteria (PGPR) are the diverse class of microbes which inhabit the plant root as ectophytes or endophytes, enhancing the host plant growth through definite and indefinite methods. PGPR can stimulate plant development by increasing the plant nutrition by several mechanisms like the production of plant growth augmenting hormones, nitrogen fixation, siderophore production, and phosphate solubilization (Glick 2012; Gabriela et al. 2015). Maintaining soil productivity, environmental safety, as well as increased crop production is the major task which is to be addressed in modern agriculture. The use of PGPR for enhancing crop productions and soil and plant health had been in practice since long. Scientist around the globe has considerably enhanced our understanding of the methods utilized by diverse PGPR in the last few decades (Glick 2014; Perez et al. 2016). Further, in-depth additional understanding of the fundamental mechanism of their plant-bacteria interaction and their beneficial association will probably accelerate the receiving efficacy of these microorganisms as appropriate and valuable component to agriculture practices. Therefore, new dimension and fundamental mechanism must be explored before increasing practical use of these microorganisms in crop production approaches. Microbes that can directly boost crop development preferentially by enhancing absorption of nitrogen and phosphorus along with other important minerals or changing the level of phytohormones and obliquely by diminishing the harmful impact of various plant pathogens in the form of biocontrol are designated as plant growth-promoting rhizobacteria (PGPR) (Kloepper et al. 1989).

The application of different PGPR in the form of single or mixed bioinoculants for minimizing the dependence on chemical fertilizers without decreasing the productivity is currently an important subject of research in the field of crop productivity. The authors in this chapter attempted to review the recent progress associated with functional diversity of PGPR which possibly assist in the transformation to bioinoculant exhibiting multiple PGP characters that could be employed in different agricultural environments.

2 Functional Diversity of Plant Growth-Promoting Rhizobacteria

PGPR are free-living rhizobacteria that colonizes plant roots and improves the plant growth. The PGPR improves the plant proliferation through the secretion of plant hormones and nitrogen fixation along with phosphate solubilization or by improving the absorption of water and nutrients. These activities lead to the enhanced root length and suppression of phytopathogens (Fig. 8.1).

These capabilities of PGPR take part in a significant task in agriculture productivity and dependence on chemical fertilizers. Various mechanisms used by different PGPR have been well documented as reviewed in Table 8.1.

2.1 Biofertilization

Biofertilization of crops by means of plant growth encouraging microbes is regarded as the most efficient eco-friendly strategy for crop production with minimum use of inorganic fertilizers. Kloepper and Schroth (1981) reported that PGPR-based plant growth improvement takes place via making changes in the entire microbial population of rhizosphere by the secretion of a mixture of compounds. Rhizobacteria that improve plant development via increasing the nutrient absorption are designated as biofertilizers. These microbes comprise a role in enhancing the nourishment condition of host plants by fixing nitrogen, improving the accessibility of nutrients, increase in root surface area, and enhancement in beneficial symbiosis of the host plant (Pérez-Montaño et al. 2014; Berger et al. 2015). Generally, growth is the result of the combined effects of these traits. Plants are capable of absorbing nitrogen (N),

Fig. 8.1 Mechanism of plant growth promotion by PGPR. (Adapted from Altaf 2016)

Table 8.1 Plant growth-promoting substances secreted by rhizobacterial isolates

Plant growth-promoting rhizospheric bacteria	Plant growth-promoting characters	References
Pseudomonas fluorescens	I, P, Ac	Jeon et al. (2003) and Cheng et al. (2007)
Pseudomonas putida and *Bacillus amyloliquefaciens*	I, S, P, Ac, biofilm formation, motility	Kumar et al. (2016)
Pseudomonas aeruginosa and *Bacillus subtilis*	I, S, H, P, Ac, antifungal activity	Adesemoye et al. (2008)
Azotobacter, *fluorescent Pseudomonas*, *Mesorhizobium*, and *Bacillus* sp.	I, S, A, H, P, antifungal activity	Ahmad et al. (2008) and Altaf and Ahmad (2017)
Paenibacillus durus	I, P, S, A, antifungal activity	Ahmad et al. (2016)
Klebsiella pneumoniae	I, P, salinity stress, nitrogenase activity	Sachdev et al. (2009), Bhattacharyya and Jha (2012), Sharma et al. (2013), Kuan et al. (2016), and Sapre et al. (2018)
Klebsiella sp.		
Azotobacter sp.	I, P, S	Farajzadeh et al. (2012) and Kasa et al. (2015)
Bacillus sp.	I, S, P, H, Ac, lytic enzyme	Kumar et al. (2012) and Shakeel et al. (2015)
Agrobacterium, *Burkholderia*, *Enterobacter*, *Pseudomonas*, *Pseudacidovorax*	I, S, P, biological nitrogen fixation	Souza et al. (2013)
Enterobacter sp.	I, S, P, Ac	Kumar et al. (2008) and Sharma et al. (2013)
Pantoea NII-186, BRM17, *Exiguobacterium* NII-0906	I, S, H, P, Ac, EPS, antifungal activity	Dastager et al. (2009, 2010) and Trifi et al. (2017)

Adapted from Altaf (2016)
I IAA, *A* Ammonia, *P* Phosphate solubilization, *S* Siderophores, *H* Hydrogen cyanide, *AC* ACC deaminase, *EPS* Exopolysaccharides

a key plant source of nourishment, from the earth as nitrite, nitrate, or ammonia (Pérez-Montaño et al. 2014).

Most of the agricultural soils are deficient in these forms of nitrogen, and the nitrogen applied in the form of chemical fertilizers is normally wiped out by rainfall or by mineral leaching. Atmospheric nitrogen-fixing microbes like *Rhizobium* and *Bradyrhizobium* are capable of creating symbiosis producing nodules on the root surface of legumes, for example, chickpea, soybean, pea, peanut, and alfalfa, where they convert nitrogen to ammonia, that could be utilized through plants as a resource of nitrogen (Murray et al. 2011; Mus et al. 2016). However, this method is basically restricted to leguminous plants. Several nonsymbiotic microbes have been identified as free-living nitrogen fixers such as *Azotobacter*, *Azospirillum*, *Azoarcus*, *Bacillus polymyxa*, *Burkholderia*, *Gluconacetobacter*, and *Herbaspirillum*. These

PGPR are capable of colonizing many significant agricultural crops like wheat, sorghum, maize, rice, soybean, chickpea, and sugarcane (Mäder et al. 2011; Pérez-Montaño et al. 2014). The application attributed to these PGPR normally improves plant's dry mass, flowering, and grain yield. But the enhancement in production as a result of application of PGPR could be due to the improvement in root growth, which increases the water and mineral absorption (Okon et al. 1998; Sureshbabu et al. 2016).

Another important plant nutrient is phosphorus (P). Although it is present in soil in large amount, mostly it is not available to plants, which hindered their growth significantly. Many PGPR can solubilize the unavailable form of phosphorus through acidification, chelation, or enzymatic activity (Hameeda et al. 2008; Richardson et al. 2009). Microorganisms belonging to genera *Azospirillum*, *Bacillus*, *Burkholderia*, *Erwinia*, *Pseudomonas*, *Rhizobium*, *Rhodococcus*, *Serratia*, *Bradyrhizobium*, *Salmonella*, *Sinomonas*, and *Thiobacillus* are accounted as phosphate solubilizers (Sharma et al. 2013; Alori et al. 2017a). Moreover, application of PGPR could improve plant absorption of numerous nutrient elements, for instance, calcium, potassium, iron, copper, manganese, and zinc. This absorption normally takes place throughout acidification of rhizospheric soil by the discharge of natural acids or by the activation of proton pump ATPase (Mantelin and Touraine 2004).

2.2 Nitrogen Fixation

The nitrogen fixation (important nutrient) plays a significant role in plant development and health. The earth atmosphere contains about 78% nitrogen, but it is not available directly to plants. This nitrogen is to be transformed into plant functional form through biological nitrogen fixation (BNF), in which soil bacteria using nitrogenase system convert atmospheric nitrogen to ammonia. BNF takes place at low temperature, by nitrogen-fixing bacteria (Raymond et al. 2004). Moreover, BNF serves as economically favorable and eco-friendly alternative to chemical fertilizers (Ahemad and Kibret 2014). Nitrogen-fixing microbes normally categorize into two groups. The first group consists of root-associated symbiotic bacteria, which fix nitrogen through nodule formation like *Rhizobium*. The second group which does not possess any specificity contains free existing nitrogen-fixing bacteria such as *Azospirillum*, *Azotobacter*, *Burkholderia*, *Herbaspirillum*, *Bacillus*, and *Paenibacillus* (Bhattacharyya and Jha 2012).

Free existing nitrogen-fixing bacteria do not work as endophytes, but their association is so close that the atmospheric nitrogen fixed by these microbes is easily consumed by the host plant directly or after the death of the cell organisms. This kind of association is known as non-specific and loose symbiosis. It is estimated that approximately 20–30 kg nitrogen hectare^{-1} year^{-1} was fixed by means of biological nitrogen fixation process (Stacey et al. 1992). Bacterial species associated with the genera *Azotobacter* and *Azospirillum* are reported to be extensively utilized under field conditions. The earliest information of their use came in 1902 and they are in use till date (Bhattacharyya and Jha 2012). Besides the isolates of *Azotobacter* and

Azospirillum, which received attention for nitrogen fixation, other bacteria, for instance, *Bacillus* and *Paenibacillus*, are also documented to possess the *nif* genes, responsible for nitrogen fixation. Species of *Bacillus* such as *Bacillus azotofixans*, *Bacillus macerans*, and *Bacillus polymyxa* were reported to be as nitrogen fixers, as evident by the presence of nitrogenase properties (Seldin et al. 1984); on the other hand, these strains were grouped in *Paenibacillus* genus after reclassification. *Paenibacillus odorifer*, *P. graminis*, *P. peoriae*, *P. brasilensis*, *P. azotofixans*, *P. macerans*, and *P. polymyxa* have been categorized as nitrogen fixers. Many other workers also reported the nitrogen fixation ability of *Bacillus* and *Paenibacillus* (Ding et al. 2005; Beneduzi et al. 2008; Fernandes et al. 2014). Therefore, nitrogen fixation ability of microorganism is accepted as a key trait of PGPR, which influences the plant growth directly.

2.3 Phosphate Solubilization

Phosphorus (P) is considered as the second vital macronutrient implicated in boosting plant maturation, after nitrogen. It has an important function in entirely all the important metabolic methods within plants, for instance, energy transfer, signal transmission, and respiration along with photosynthesis (Plaxton and Tran 2011; Anand et al. 2016). However, a significant amount of phosphorus is available in soils equally as organic and inorganic mode. This is the main nutrient that hindered the plant growth because of its very limited bioavailability. To mitigate the problem of phosphate deficiency, phosphatic fertilizers are used in agriculture soils, but majority of the applied phosphate is precipitated and only a marginal amount is accessible to plant (Sharma et al. 2013). Moreover, the constant and careless use of synthetic phosphatic fertilizers not only causes depletion of soil fertility, but it also upsets the diversity of microorganisms (Gyaneshwar et al. 2002; Alori et al. 2017a). Plants can take up on their own only mono- and dibasic phosphate, but the organic and insoluble forms of phosphate require to be solubilized by plant growth-promoting rhizospheric bacteria (Richardson et al. 2009; Ramaekers et al. 2010; Sharma et al. 2013).

Concerning India, 98% of agriculture land is poor in available forms of soil phosphorus. Intensive agriculture practices used to increase crop production during Green and White Revolution has also produced large-scale phosphorus scarcity (Sharma et al. 2013). To mitigate the problem of phosphorus deficiency, the use of microbial inoculants possessing phosphate-solubilizing capability could be considered as an eco-friendly and cost-effective alternative to reduce dependence on chemical phosphatic fertilizers (Gaur 1990; Hassen et al. 2016). A number of investigations have been performed to test the effectiveness of phosphate-solubilizing bacteria (PSB) by soil inoculation methods that resulted in increased crop yield. However, the phosphate-solubilizing activity of these PSB is largely hampered by various environmental factors of soil like salinity, pH, moisture, and temperature (Sharma et al. 2013).

Isolates associated with genera *Pseudomonas, Bacillus, Mesorhizobium, Rhizobium Agrobacterium, Flavobacterium, Rhodococcus, Burkholderia,* and *Erwinia* are found to be the main phosphate solubilizers which increase plant growth and yield (Yu et al. 2011; Oteino et al. 2015). The key methods pertaining to phosphate solubilization employed by these bacteria include (a) secretion of mineral breaking and combining substances, for instance, organic acids, siderophores, protons, hydroxyl ions, as well as carbon dioxide, (b) secretion of enzymes, and (c) discharge of phosphate amid substrate decaying (McGill and Cole 1981). Microbes inhabit a vital position in the entire key methods of soil phosphate cycle (i.e., dissolution-precipitation, sorption-desorption, and mineralization-immobilization). The most important means of P solubilization by microorganisms is by the release of organic acids through lessening the pH, by means of boosting chelation of the cations connected with P, by challenging with P for adsorption spot, and by producing soluble substances by means of metal ions related through insoluble P and lastly liberating it (Sharma et al. 2013).

The additional means of phosphate solubilization is through the procedure of mineralization in which organic phosphorus was solubilized by the release of various enzymes like non-specific acid phosphatases (Nannipieri et al. 2011), phytases (Richardson 1994) and phosphonatases and C-P lyases (Rodriguez et al. 2006). Therefore, it is clear that P solubilization takes place through various methods, and microbes also showed considerable dissimilarity in this characteristic.

2.4 Potassium Solubilization

Potassium is also a vital macronutrient for plant development. About 90% of potassium is present in the shape of insoluble rock and silicate reserves; the level of soluble potassium is typically extremely small in soil (Parmar and Sindhu 2013). Shortage of potassium turns out to be a main restriction in crop yield. The deficiency of potassium causes poor root development, decreased seed yield, and diminished growth. Therefore to mitigate the problem of low potassium, PGPR can be used as environmentally sustainable methods. The potential of PGPR to dissolve potassium rock by releasing organic acids has being extensively studied. Potassium-dissolving PGPR, such as *Acidithiobacillus* sp., *Bacillus edaphicus, Ferrooxidans* sp., *Bacillus mucilaginosus, Pseudomonas* sp., *Burkholderia* sp., and *Paenibacillus* sp., have been documented to make potassium bioavailable to plants originating at soil (Liu et al. 2012).

2.5 Siderophore Production

Siderophore secretion is a key attribute exhibited by the PGPR to improve plant growth. Siderophore augments plant development together by the direct and indirect methods. The direct method includes the delivery of iron to plants and indirectly by combining by the iron powerfully making them unavailable to plant pathogens (Glick 2012). Roughly every microorganism with the exemption of few lactobacilli forcibly requires iron (Neilands 1995). Iron is an important source of nourishment for plants. It is involved as cofactor in numerous enzymes which are essential for many important physiological procedures like respiration, photosynthesis, and nitrogen fixation. Fe^{3+} is the main type of iron in soil, but it reacts to create insoluble oxides and hydroxides, not available to plants and microorganisms. In order to stay alive, microorganisms produce low molecular weight substances ($\approx$1 kDa) termed as siderophores that composed of functional groups that attach to iron, as a result making them accessible to microorganisms (Glick 2012). The main functional groups are hydroximates and catechols. The intensity of siderophores existing in soil is approximately 10–30 M. Normally, siderophore-producing microbes fit in the genus *Pseudomonas*, in which *Pseudomonas fluorescens* and *Pseudomonas aeruginosa* are extensively evaluated for their siderophore (pyochelin and pyoverdine type) (Haas and Défago 2005). These substances were produced by bacteria to enhance their combativeness, since these substances have antibiotic potency and augment the iron absorption by plants (Glick 1995). In addition, fastening to iron, siderophores in addition are recognized to manufacture constant complexes with diverse heavy metals and other radionuclides that help in alleviation of heavy metal stress for plants in polluted soil (Neubauer et al. 2000; Rajkumar et al. 2010).

2.6 Plant Growth Regulator Produced by Rhizobacteria

Rhizobacteria are acknowledged to secrete various kinds of plant hormones, for instance, auxins, gibberellins, cytokinins, ethylene, as well as abscisic acid (Patten and Glick 1996; Arshad and Frankenberger 1998). Plant retorts to several phytohormones within the rhizosphere which are supplied from outside or formed by rhizospheric bacteria. These phytohormones decide several vital procedures like plant cell elongation and division and growth of root structure (Patten and Glick 1996; Glick 2014).

2.7 Indole Acetic Acid

Microbial products in the form of plant growth regulators are commonly recognized for their involvement in improving plant health (Arshad and Frankenberger 1993). Indole acetic acid (IAA) is the single most important physiologically dynamic auxin produced by PGPR. IAA is a result of L-tryptophan metabolism produced via a number of microbes particularly PGPR (Lynch 1985). This metabolite is the secondary product of several tryptophan-reliant and tryptophan-free mechanisms in microbes and plants. The microorganisms are accounted to possess in excess of one mechanism (Patten and Glick 1996). IAA remunerate the plants by improving the root length having increased root hairs and growth of lateral root structure that enhances the nutrition absorption (Datta and Basu 2000). Furthermore, IAA promote cell elongation by escalating osmotic content of the root cell, growing the water permeability into the cell, declining cell wall pressure, and increasing the growth of cell wall production. Likewise it participates in an important function such as gametogenesis and embryogenesis, delays abscission of leaves, and promotes seedling development, flowering, and fruiting (Zhao 2010). It has been found that besides these physiological roles, IAA also employed as signaling compounds and control gene interpretation within microorganisms (Spaepen and Vanderleyden 2011).

IAA production in rhizospheric bacteria takes place by means of L-tryptophan-dependent and L-tryptophan-independent mechanisms. Three L-tryptophan-dependent pathways have been accounted. The bulk of rhizobacteria employed L-tryptophan, as originator, secreted by plants as root discharge for IAA production. Nevertheless, the exact pathway and enzymes implicated in L-tryptophan-independent IAA production are still not recognized (Spaepen et al. 2007; Jha and Saraf 2015). Furthermore, microbial species of genera *Rhizobium*, *Bradyrhizobium*, and *Azospirillum* employ indole-3-pyruvic acid (IPyA) route for producing IAA (Patten and Glick 1996), while several bacteria such as *Agrobacterium tumefaciens*, *Pseudomonas syringae*, *Pantoea agglomerans*, *Rhizobium*, *Bradyrhizobium*, and *Erwinia herbicola* make IAA primarily via indole-3-acetamide (IAM) pathway (Dobbelaere et al. 2003). In contrast *Bacillus subtilis*, *B. licheniformis*, and *B. megaterium* make IAA via tryptamine pathway (Goswami et al. 2016).

2.8 Other Plant Growth Regulators

Rhizospheric bacteria are capable of secreting various kinds of phytohormones, for instance, auxins, cytokinins, gibberellins, ethylene, and abscisic acid, that help in several developmental procedures (Glick 2014; Tabassum et al. 2017). Crop plants inoculated with cytokinin-secreting PGPR demonstrated an enhanced growth as compared to control (Amara et al. 2015). Cytokinin is a part pertaining to plant growth controller which is associated with shoot development, suppression of root enlargement, and enhanced cell splitting and root growth (Porcel et al. 2014; Jha and Saraf 2015). Cytokines of different kinds are secreted via PGPR wherein zeatin and kinetin are dominant forms. PGPR manufactures zeatin via couple of special methods, namely, certain and uncertain. The certain method includes the production pertaining to dimethylallyldiphosphate (DMAPP) and N6-isopentenyladenosine monophosphate (i6 AMP), whereas uncertain method includes cis-zeatin having tRNA to discharge cytokinins. Several species of *Pseudomonas*, *Proteus*, *Azospirillum*, *Bacillus*, *Escherichia*, *Klebsiella*, and *Xanthomonas* existed to possess the power to secrete cytokinin (Castillo et al. 2015; Maheshwari et al. 2015).

One more important group of plant hormones secreted by PGPR is gibberellins that are associated with several growth and developmental procedures like seed germination, flowering, and stem lengthening, along with fruit fixing in vascular plants (Saleem et al. 2015a, b). Research has revealed certain plants having gibberellin-secreting PGPR in their root system have improved development percentage (Poupin et al. 2013; Vacheron et al. 2014). Merely heptad belonging to growth-promoting rhizobacteria has been documented as gibberellin-secreting PGPR and comprises *Bacillus amyloliquefaciens* (Shahzad et al. 2016) and *Sphingomonas* sp. (Khan et al. 2014).

3 Other Mechanisms of Plant Growth

The application of microorganisms as biocontrol instrument is a cheap and environment-friendly approach. The key indirect method of plant growth improvement employed by PGPR is through functioning as biocontrol agent (Glick 2012). Usually the PGPR act as biocontrol agent via contesting for source of nourishment, niche, induced systemic resistance, as well as discharge of antifungal and antibacterial substances (Lugtenberg and Kamilova 2009). Numerous plant root-connected bacteria existed to secrete antifungal intermediate products similar to hydrogen cyanide, phenazines, pyrrolnitrin, 2,4-diacetylphloroglucinol, pyoluteorin, and viscosinamide in addition to tensin (Bhattacharyya and Jha 2012).

3.1 1-Aminocyclopropane-1-Carboxylate Deaminase

Usually, ethylene acts as an important substance intended for the development pertaining to crop plant. This compound is produced by most of the plants; in addition it is also formed by numerous living and nonliving processes in soil, and it is associated with several biological procedures. Ethylene is capable of regulating the development of plants in several dissimilar ways such as encouraging root initiation, hindering root lengthening, stimulating fruit ripening, encouraging flower sagging, exciting seed germination, promoting leaf abscission, activating the production of new plant hormones, hampering *Rhizobia* spp. nodule production, hindering mycorrhizae-plant communication, and reacting to both living and nonliving stresses (Glick 2012). On the other hand, in stress situations, for example, dryness, saltiness, heavy metals, and waterlogging, organic substances in addition to infection from plant pathogens enhance the ethylene concentration that harmfully influences the plant growth through inducing defoliation and additional key biological procedures (Glick 2014). The rhizospheric bacteria having the enzyme 1-aminocyclopropane-1 -carboxylate (ACC) deaminase could alleviate these tribulations connected to plant improvement, through lessening the ethylene level (Glick 2012). Several rhizobacterial species (*Acinetobacter*, *Achromobacter*, *Agrobacterium*, *Alcaligenes*, *Azospirillum*, *Bacillus*, *Burkholderia*, *Enterobacter*, *Pseudomonas*, *Ralstonia*, *Serratia*, and *Rhizobium*) have been documented to possess ACC deaminase activity (Kang et al. 2010). This enzyme split the plant ethylene precursor, 1-aminocyclopropane-1-carboxylate, into ammonia and α-ketobutyrate (Honma and Shimomura 1978), therefore reducing the plant ethylene level, which in high concentration can cause immature plant growth and even death (Glick et al. 2007; Glick 2014).

3.2 Vitamins

Generally, plants in good conditions create adequate amount of vitamins to maintain their growth and development, whereas under substandard or in stress situations, plants can experience vitamin shortage, and as a result microbes which secrete vitamins are competent of encouraging plant development in stress conditions. Few rhizobacteria associated with the genus *Bacillus* sp. generate vitamin B complex, like pantothenic acid, thiamine, and riboflavin, along with biotin, that can be taken up by plant (Alexandre et al. 2000). Still, added research is required to establish the involvement of vitamins in plant growth promotion (Dobbelaere et al. 2003).

3.3 Diacetylphloroglucinol (2,4-DAPG)

Diacetylphloroglucinol is the most effective antimicrobial that is associated with phytopathogens (Fernando et al. 2006) that could be secreted via several strains of *Pseudomonas*, which are regularly detected in the rhizosphere of crops (Picard et al. 2000; Couillerot et al. 2009). The 2,4-DAPG is of broad range; it is found to be antifungal (Loper and Gross 2007), antibacterial, and antihelminthic (Rezzonico et al. 2007). In agricultural soils, it represses the development pertaining to wheat pathogenic fungus *Gaeumannomyces graminis* var. *tritici*. Raaijmakers and Weller (1998) demonstrated the secretion of 0.62 ng 2, 4-DAPG per 10^5–10^7 CFU g/1 root by *Pseudomonas fluorescens*, strain Q2-87.

3.4 Volatile Organic Compounds (VOC)

Small molecular weight substances, for example, hydrogen cyanide, a biocide, with antifungal property, are secreted in a mixture of volatile organic compounds like 2,3-butanediol and acetoin which promotes the development of *Arabidopsis thaliana*. In an assessment pertaining to the importance of HCN and DAPG to *Pseudomonas fluorescens* CHA0, HCN, not suppressed by fusaric acid, take part in a very important function in disease suppression compared to DAPG (Duffy et al. 2004). The production of HCN was a common attribute of *Pseudomonas* sp. (88.89%) (Ahmad et al. 2008). Riedlinger et al. (2006) documented other substances like pyoluteorin and auxofuran.

3.5 Induced Systemic Resistance (ISR)

Connection of numerous rhizobacteria to the plant roots generates protection among plants against plant pathogens. This incident is described as induced systemic resistance (ISR) (Lugtenberg and Kamilova 2009). ISR includes jasmonate along with ethylene communication within the crop plant, and these hormones support the owner plant's protection reply in opposition to plant pathogens (Glick 2012). A variety of specific bacterial substances stimulate ISR, like lipopolysaccharides (LPS), flagella, siderophores, cyclic lipopeptides, 2,4-diacetylphloroglucinol, homoserine lactones, and volatiles such as acetoin and 2,3-butanediol (Lugtenberg and Kamilova 2009). A few PGPR strains have been marketed due to their biocontrol capacity (mainly *Pseudomonas* and *Bacillus*). On the other hand, unreasonable application of these strains leads to the growth of refusal in the disease-causing microbes (Raaijmakers et al. 2002; Haas and Keel 2003; Compant et al. 2005).

Rhizobacterial stimulated ISR was examined on carnation (*Dianthus caryophyllus*) amid low vulnerability to wilt created by *Fusarium* sp. and on cucumber (*Cucumis sativus*) amid low vulnerability to foliar disease created by *C. orbiculare* (Viswanathan and Samiyappan 1999) for the first time. The consequence of ISR is reliant on the mixture of owner plant and bacterial strain. Free-living rhizospheric bacteria are thoroughly evaluated for its ISR-promoting capability; however lately endophytic PGPR also assessed its ISR initiation. Likely, ISR was stimulated by *Pseudomonas fluorescens* EP1 against red rot created by *C. falcatum* on sugarcane, *Burkholderia phytofirmans* PsJN toward *Botrytis cinerea* on grapevine and *Verticillium dahliae* on tomato, *P. denitrificans* 1-15 and *P. putida* 5-48 against *Ceratocystis fagacearum* on oak, *P. fluorescens* 63-28 against *F. oxysporum* f. sp. *radicis-lycopersici* on tomato and *Pythium ultimum* and *F. oxysporum* f. sp. *pisi* on pea roots, and *Bacillus pumilus* SE34 in opposition to *F. oxysporum* f. sp. *pisi* on pea roots and *F. oxysporum* f. sp. *vasinfectum* associated with cotton roots (Stephane et al. 2005).

3.6 Competition

Competition related to appropriate habitat is considered a key method for eliminating disease-causing microbes associated with the rhizosphere through PGPR. Secretions from plant roots are component of the accessible source of nourishment for rhizosphere microorganisms. These nourishment loaded regions draw microbes, together with disease-causing bacteria through chemotaxis against organic acids, sugars, and amino acids (Miller and Oldroyd 2012). Plant root-associated bacteria travel in the direction of the root through flagellar movement and attach itself in a biphasic procedure (De Weert et al. 2002). At the time of settlement, possible necessary positions for pathogens are engaged, eliminating possible harmful microbes as of the plant root. While the nourishment content together with the accessibility of the rhizosphere is highly reliant on root secretions, microbes having a flexible metabolism and high-affinity transporters are very much supported, and the majority of rhizobacteria can employ a large diversity of compounds such as carbon and nitrogen supply in a well-organized mode. In this fashion they can hunt nutrients and decrease the propagation of disease-causing bacteria in the root region, thus defending the crop plants from soil produced disease-causing microbes (Bach et al. 2016). A number of experiments have attempted to recognize the microbial ecology in these soils to see deeper inside the microbial machinery concerned in inhibition of disease (Mazzola 2004; Shameer and Prasad 2018).

4 Application of PGPR in Bioremediation of Heavy Metals

A broad variety of heavy metal-resistant microbes and plant root-associated bacteria/fungi comprises the ability to augment plant development under metal stress situations. These microorganisms use several methods to survive under metal stress conditions, for instance, efflux, impermeability to metals, volatilization, EPS sequestration, and metal complex and enzymatic detoxification. Additionally, root attached microorganisms encourage plant development and maturity by diminishing ethylene level and secretion of growth monitors, for example, IAA and ACC deaminase, and repress disease (Glick 2010). Besides these, nitrogen fixation, nutrient recruitment, and siderophore and phosphate solubilization improve both plant development and elimination of heavy metals (Verma et al. 2013; Kumar and Verma 2018). The majority of the heavy metals is poisonous to plants; however, microbes are capable in reducing the outcome of the metal poisoning through combining on its negatively charged functional groups above the entire cell membrane which provides ample communication by way of positive ions, particularly, metal ions, and connects them; this phenomenon is called as metal biosorption (Syed and Chinthala 2015).

Microorganisms belonging to different genera with substantial metal resistance have revealed immobilization of poisonous metals once inserted into the polluted earth, thus decreasing its poisonous effect against plants (Kong and Glick 2017). Microbes belonging to *Bacillus* and *Pseudomonas* are excellent biosorbents of a diversity of heavy metals (Syed 2016). An investigation done by Prapagdee et al. (2013) in cadmium-polluted earth revealed cadmium-tolerant plant growth-promoting bacteria (*Micrococcus* sp. MU1 and *Klebsiella* sp. BAM1) increase cadmium mobilization and encourage lengthening of root and development of plant. Another experiment by Fatnassi et al. (2015) ascertained that over 1 mM level of copper (Cu) hampers development of *Vicia faba*; however, application of rhizobia in addition to plant root-associated bacteria decreased their harmful outcomes. Hashem et al. (2016) reported that arbuscular mycorrhiza fungi mitigate harmful outcome pertaining to cadmium toxicity through decreasing malonaldehyde and hydrogen peroxide.

5 Potential of PGPR in Bioremediation of Organic Pollutants

On the contrary for inorganic substances, microbes could degenerate and even calcify natural substances associated with plants (Saleh et al. 2004). Therefore, finding pertaining to efficient methods for degeneration and mineralization of organic substances might take part in key function in the coming future. The application of root-associated beneficial bacteria enhances the elimination related to organic contaminants like polycyclic aromatic hydrocarbon and creosote, possibly by increasing development of plant, accumulating biomass of root plus existence within highly contaminated soils (Huang et al. 2004). The rhizosphere environment is found to be capable of bioremediating polluted soils; however, several microbes competent of degenerating definite types of organic contaminants are unable to stay alive and attain bioremediation in the soil surroundings, as they cannot compete with the local microbial population. In the meantime, various microbes which are strong in the rhizosphere do not demonstrate or are found to have only incomplete capability to degrade natural contaminants. Through the progress of molecular biology, the genetically engineered rhizospheric bacteria with the pollutant-degrading gene are created to accomplish the bioremediation in rhizosphere (Ojuederie and Babalola 2017).

6 PGPR Mediate Stress Tolerance on Crop Plants

So far, the development of stress-tolerant varieties has been the strategy employed to alleviate the harmful results of stress on crop plant and yield (Barrow et al. 2008; Eisenstein 2013). Conservative plant breeding methods have permitted the progress of high-yielding, stress- tolerant cultivars. The shortcoming of this strategy is that it is a lengthy, cumbersome process which is associated with the failure of losing some good characters of the host genome. Moreover, the conventional breeding transfer benefits to single crops which are nontransferable (Ashraf 2010; Eisenstein 2013; Philippot et al. 2013). Additionally, as soon as the genetically modified crops arrive in the market, their favorable outcome is not certain because the response of customer to these products varies country to country (Fedoroff et al. 2010).

Application of rhizobacteria has been acknowledged to adapt abiotic stress control through direct and indirect methods which stimulate systemic tolerance (Yang et al. 2009). Several rhizospheric microbes are continuously studied for their function in developing plant-water associations, ion homeostasis, and photosynthetic effectiveness within plants in stress conditions (Ilangumaran and Smith 2017). Their improvement methods are complex and not easy to comprehend. These methods are controlled by a set of intricate system of signals which takes place all through plant-bacteria communications and as a result help in mitigating stress (Smith et al. 2017). Several workers studied and demonstrated the function related to PGPR in stress alleviation.

Habib et al. (2016) reported that beneficial root-associated bacteria consisting of 1-aminocyclopropane-1-carboxylate (ACC) deaminase assist in high salt pressure immunity associated with okra by inducing ROS-scavenging enzymes. Beneficial plant root-associated bacteria applied in okra showed improved germination and development limits, along with chlorophyll percentage as compared to positive control. Improved antioxidant enzyme functions like SOD, APX, and CAT plus management of ROS pathway genes (CAT, APX, GR, and DHAR) were found among okra under high salt pressure. Amid a few exemptions, application of *Enterobacter* sp. UPMR18 had a considerable effect related to majority of tested parameters under high salt pressure, as contrast to additionally applied bacteria. Barnawal et al. (2017) revealed in their study that abiotic stresses like salt and drought symbolize unfavorable ecological circumstances which considerably harm plant development and farm production. His results advocated that rhizospheric strains, *Arthrobacter protophormiae* (SA3) and *Dietzia natronolimnaea* (STR1), can make possible wheat crop in tolerating salt stress, whereas *Bacillus subtilis* (LDR2) can give tolerance to wheat crop toward drought stress. These rhizospheric isolates improve photosynthetic effectiveness in salt and drought stress environments.

7 Climate Change, Crop Productivity, and PGPR Role

The randomness of the water cycle has created a grave face before agriculturists, in addition to the international population, regarding their consequence with gathering food requirements pertaining to humans as well as animals. Population of the world is rising by a regular speed and presents a huge gap between the food supply and demand to feed the population. Consecutively, to meet the ever-increasing water demand, the farmers increase the application of every available source of water, which results in salinization of soil. Nonetheless, the utilization of beneficial plant root-associated bacteria as bioinoculants can help in fighting abiotic climate-induced alteration which hinders the normal functioning of crop plant under stress conditions (Staudinger et al. 2016; Alori et al. 2017b) (Fig. 8.2).

Several PGPR with the specific enzymes essential for the degradation plant secretions are capable of protecting crop plants from drought and salt stress. These rhizobacteria secrete different compounds like enzyme (Saleem et al. 2015a, b), plant growth regulator indole acetic acid, siderophore, phosphate-solubilizing

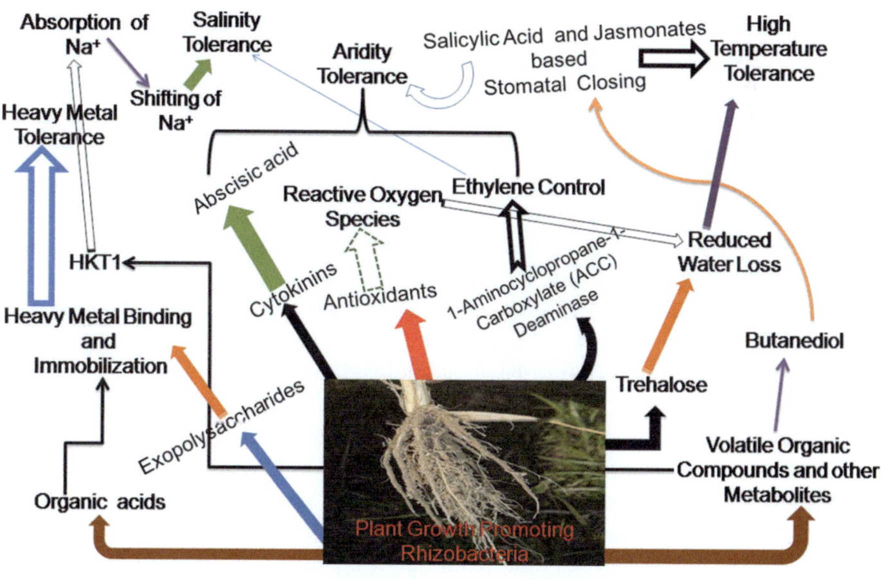

Fig. 8.2 Mechanism of different types of tolerance of stress in plants induced by beneficial plant root-associated bacteria. (Partly adapted from Shameer and Prasad 2018)

enzyme (Kumari and Khanna 2016), salicylic acid (Ekinci et al. 2014), as well as microbiocidal/biostatic enzyme (Moustaine et al. 2017).

Plant root-associated beneficial bacteria serve as the main character related to the struggle for viable growth under adverse environment starting after human actions as well as climatic changes. The types of microorganism associated with these genera are *Micrococcaceae* HW-2 (Hong et al. 2016), *Bradyrhizobium* (Masciarelli et al. 2014), *Bacillus* (Kasim et al. 2016), *Microbacterium*, *Pseudomonas*, *Curtobacterium* (Cardinale et al. 2015), *Variovorax*, *Paenibacillus* (Yolcu et al. 2011), *Pantoea* (Damam et al. 2014), and others.

8 Enhance Metabolic Activity and Persistence in Biofilm Mode Under Stress Conditions

Biofilms are microbial community connected through living and nonliving surface in addition to surrounded in personally formed means of extracellular polymeric substances (EPS). Chemically EPS is mixed mostly prepared from polysaccharides, extracellular proteins and enzymes, DNA, and supplementary materials. The main advantage of biofilm state intended for rhizospheric bacteria is defense toward aridity. The rhizosphere is a vibrant bacterial environment meant for biological microniches and endlessly examined to comprehend microbial multiplicity and its control. Polyphasic strategies are employed to know these habitats (Hinsinger et al. 2009; Bogino et al. 2013). The rhizosphere environment is explicitly effected by roots and their secretions. Movement of microbes from bulk soil to rhizospheric soil or rhizoplane influenced the habitation of roots by soil microbes. Successful colonization of beneficial plant root-associated bacteria is determined by attachment and microcolony creation. Biofilm developments participate in a vital function in bacterial existence and physiology. The bacterium has to cooperate by numerous microbes and must attach themselves as multispecies biofilm in the rhizosphere (Compant et al. 2010). The communications of biofilm bacteria with plants may be beneficial or harmful. Biofilm also results in source of nourishment yield and decrease in living and nonliving stress (Angus and Hirsch 2013).

The capability to develop biofilm via large number of bacteria such as nitrogen-fixing rhizobacteria both symbiotic (*Rhizobium alamii*, *R. leguminosarum* bv. *viciae* 3841, *R. leguminosarum*, *Rhizobium* sp. NGR234, *Rhizobium*, *Sinorhizobium*) and nonsymbiotic nitrogen fixers (*Azospirillum brasilense*, *Azorhizobium caulinodans*, *Azotobacter chroococcum*) has been acknowledged through numerous people (Shelud'ko et al. 2010; Krysciak et al. 2011; Robledo et al. 2012). Likewise biocontrol bacteria (species of *Bacillus*, *Pseudomonas*) were also found to create biofilm by several scientists (Beauregard et al. 2013; Yasmin et al. 2014) who defend host plant from disease-causing microbes by biocontrol method under biofilm mode. The aerobic endospore-forming bacteria (*Bacillus* and associated genera) are commonly found in soil, water, air, and plant, and they have the capability to develop resistant endospores and produce different antibiotics. Utilizing these potentials, the bacteria are capable of inhabiting various places in agroecosystems and can dislodge new microbes in rhizosphere/rhizoplane. As a result, the attachment spot of bacteria is more repeatedly secured, and these bacteria can be applied in accuracy organization of agroecosystems. Likely, endospore-forming species of *Paenibacillus polymyxa* inhabit as biofilms in rhizoplane (Timmusk et al. 2005). The bacterial biofilms can defend plants toward pathogens plus against abiotic stress conditions (Milošević et al. 2012; Timmusk et al. 2013). Other plant root-associated beneficial bacteria such as *Burkholderia cepacia* and *Klebsiella pneumoniae* were also studied related to biofilm development and their role in plant health (Ji et al. 2010; Liu et al. 2011). For that reason, current activities on plant root-linked biofilm specify a growing fashion of investigation in this area and culminate in their acceptance.

9 Nanoparticles and Performance of Plant Growth-Promoting Rhizobacteria

The relevance of current tools like nanotechnology has remarkable capacity to transform the agricultural sector. Nanoagriculture that at present targets on specific cultivation which entails the application of nanosized units, for example, nanofertilizer, presents selected technique meant for enhancing the output of the crops via competent nutrient absorption (Tarafdar et al. 2013). The exclusive characteristics related to nanosized particles in reference to their physical, chemical, and biological parameters in contrast to the above given at a bigger level offer the possibility in the direction of plant defense, identify infection of plant, check development, improve food value, augment food manufacture, as well as decrease waste. Nanofertilizers are more beneficial as compared to fertilizers beacuse they decrease nitrogen loss owing to leeching, emissions, and long-term amalgamation through soil microbes (Liu et al. 2006). Moreover, Suman et al. (2010) have established the benefits related to the use of nanofertilizers by presenting that restricted discharge of fertilizers might also develop the soil by lessening the poisonous outcome connected through the overuse of synthetic fertilizers. The application of beneficial plant root-associated bacteria as bioinoculants is not very successful because about 90% are vanished during usage. Moreover, they are not tolerant to adverse environmental conditions like heat, drought, etc. Nanoencapsulation technique might be applied as an adaptable means to guard rhizobacteria, improving their existence and distribution in fertilizer composition and permitting the restricted discharge of the rhizobacteria (Vejan et al. 2016).

Timmusk et al. (2018) during an investigation revealed the use of nanotitania (TNs) as career in the nanointerface communication among plants and attachment of PGPR. The efficiency of PGPR is associated with the efficacy of the technique applied for their composition. TNs created by the Captigel patented SolGel strategy, distinguished by the transmission and scanning electron microscopy, were employed for formulation of the adverse environment PGPR isolates. The study further revealed that alteration in the biomass of wheat seedlings and in the concentration of single and double inoculants with and without TNs was examined during 2 weeks of stress stimulated by drought, salt, and the pathogen *Fusarium culmorum*. The results demonstrated that double inoculants with TNs can bind firmly to plant roots.

10 Conclusion and Future Prospects

PGPR confers an ecologically stable and feasible strategy for increasing crop yield. In recent period, the knowledge related to intricate atmosphere of the rhizosphere, functional diversity among rhizobacteria, methods of their functions, mechanisms of bioinoculant development, and their release methods has improved significantly. On the other hand, there are a number of restrictions that are related to the commercial applications of PGPR. One of the main problems is their unpredictable functioning in field environment. One of the possible ways to mitigate the problem of inconsistency is the application of PGPR in biofilm mode. As the knowledge related to the understanding of different mechanisms of action of plant growth-promoting rhizobacteria develop, it will become easier to improve the capability to enhance the plant development by altering different characters of plant growth-promoting rhizobacteria by adding desired genes inside other microbes. These types of genetically modified microorganisms through multiple plant growth-promoting characters could improve colonization and growth effectiveness, leading to higher crop yield simultaneously maintaining soil fertility along with the removal of contaminants generated by the application of synthetic chemicals. Climate change at present is believed as one of the main dangers to agricultural productivity. The role of PGPR in reducing adverse effect of climate change on crop production through providing tolerance to heat, drought, etc. needs to be explored and exploited in the future.

References

Abhilash PC, Tripathi V, Edrisi SA et al (2016) Sustainability of crop production from polluted lands. Energy Ecol Environ 1(1):54–65

Adesemoye AO, Obini M, Ugoji EO (2008) Comparison of plant growth-promotion with *Pseudomonas aeruginosa* and *Bacillus subtilis* in three vegetables. Braz J Microbiol 39:423–426

Ahemad M, Kibret M (2014) Mechanisms and applications of plant growth promoting rhizobacteria: current perspective. J King Saud Univ Sci 26:1–20

Ahmad F, Ahmad I, Khan MS (2008) Screening of free-living rhizospheric bacteria for their multiple plant growth promoting activities. Microbiol Res 163:173–181

Ahmad F, Ahmad I, Altaf MM et al (2016) Characterization of *Paenibacillus durus* (PNF16) a new isolate and its synergistic interaction with other isolated rhizobacteria in promoting growth and yield of chickpea. J Microbiol Biotechnol Food Sci 5(4):345–350

Alexandre G, Greer SE, Zhulin IB (2000) Energy taxis is the dominant behavior in *Azospirillum brasilense*. J Bacteriol 182(21):6042–6048

Alori ET, Glick BR, Babalola OO (2017a) Microbial phosphorus solubilization and its potential for use in sustainable agriculture. Front Microbiol 8:971

Alori ET, Dare MO, Babalola OO (2017b) Microbial inoculants for soil quality and plant fitness. In: Lichtfouse E (ed) Sustainable agriculture review. Springer, Berlin, pp 181–308

Altaf MM (2016) *In vitro* screening for biofilm forming ability of rhizospheric bacteria and their plant growth promoting potential. Ph.D. thesis, Aligarh Muslim University

Altaf MM, Ahmad I (2017) *In vitro* and *in vivo* biofilm formation by *Azotobacter* isolates and its relevance to rhizosphere colonization. Rhizosphere 3:138–142

Amara U, Khalid R, Hayat R (2015) Soil bacteria and phytohormones for sustainable crop production. In: Maheshwari D (ed) Bacterial metabolites in sustainable agroecosystem. Sustainable development and biodiversity. Springer, Cham, pp 87–103

Anand K, Kumari B, Mallick MA (2016) Phosphate solubilizing microbes: an effectiveand alternative approach as bio-fertilizers. Int J Pharm Sci 8(2):37–40

Angus AA, Hirsch AM (2013) Biofilm Formation in the Rhizosphere: Multispecies Interactions and Implications for Plant Growth. In: De Bruijn FJ (ed) Molecular microbial ecology of the rhizosphere. Wiley, New York, pp 703–712

Arshad M, Frankenberger WT Jr (1993) Microbial production of plant growth regulators. In: Metting FB Jr (ed) Soil microbial ecology. Marcel Dekker Inc., New York, pp 307–347

Arshad M, Frankenberger WT Jr (1998) Plant growth regulating substances in the rhizosphere: microbial production and functions. Adv Agron 62:146–151

Ashraf M (2010) Inducing drought tolerance in plants: some recent advances. Biotechnol Adv 28:169–183

Bach E, dos Santos SGD, de Carvalho FG et al (2016) Evaluation of biological control and rhizosphere competence of plant growth promoting bacteria. Appl Soil Ecol 99:141–149

Barnawal D, Bharti N, Pandey SS et al (2017) Plant growth promoting rhizobacteria enhances wheat salt and drought stress tolerance by altering endogenous phytohormone levels and *TaCTR1/TaDREB2* expression. Physiol Plant 161(4):502–514. https://doi.org/10.1111/ppl.12614

Barrow JR, Lucero ME, Reyes-Vera I (2008) Do symbiotic microbes have a role in plant evolution, performance and response to stress? Commun Integr Biol 1:69–73

Beauregard PB, Chai Y, Vlamakis H, Losick R, Kolter R (2013) *Bacillus subtilis* biofilm induction by plant polysaccharides. Proc Natl Acad Sci USA 110:E1621–E1630

Beneduzi A, Peres D, da Costa PB et al (2008) Genetic and phenotypic diversity of plant-growth-promoting bacilli isolated from wheat fields in southern Brazil. Res Microbiol 159(4):244–250

Berger B, Wiesner M, Brock AK (2015) *K. radicincitans*: a beneficial bacteria that promotes radish growth under field conditions. Agron Sustain Dev 35:1521–1528

Bhardwaj D, Ansari MW, Saho RK (2014) Biofertilizers function as key player in sustainable agriculture by improving soil fertility, plant tolerance and crop productivity. Microbial Cell Fact 13(66):1–10

Bhattacharyya PN, Jha DK (2012) Plant growth-promoting rhizobacteria (PGPR): emergence in agriculture. World J Microbiol Biotechnol 28:1327–1350

Bogino P, Abod A, Nievas F et al (2013) Water-limiting conditions alter the structure and biofilm-forming ability of bacterial multispecies communities in the alfalfa rhizosphere. PLoS ONE 8(11):e79614

Cardinale M, Ratering S, Suarez C et al (2015) Paradox of plant growth promotion potential of rhizobacteria and their actual promotion effect on growth of barley (*Hordeum vulgare* L.) under salt stress. Microbiol Res 181:22–32

Castillo P, Molina R, Andrade A et al (2015) Phytohormones and other plant growth regulators produced by PGPR: the genus *Azospirillum*. Handbook for *Azospirillum*. Springer, Cham

Cheng Z, Park E, Glick BR (2007) 1-aminocyclopropane-1-carboxylate deaminase from *Pseudomonas putida* UW4 facilitates the growth of canola in the presence of salt. Can J Microbiol 53:912–918

Compant S, Duffy B, Nowak J (2005) Use of plant growth promoting bacteria for biocontrol of plant diseases: principles, mechanisms of action, and future prospects. Appl Environ Microbiol 71(9):4951–4959

Compant S, Clément C, Sessitsch A (2010) Plant growth-promoting bacteria in the rhizo- and endosphere of plants: their role, colonization, mechanisms involved and prospects for utilization. Soil Biol Biochem 42:669–678

Couillerot O, Prigent-Combaret C, Caballero-Mellado J et al (2009) *Pseudomonas fluorescens* and closely-related fluorescent pseudomonads as biocontrol agents of soil-borne phytopathogens. Lett Appl Microbiol 48(5):505–512

Damam M, Gaddam B, Kausar R (2014) Effect of plant growth promoting rhizobacteria (PGPR) on *Coleus forskohlii*. Int J Curr Microbiol Appl Sci 3(9):266–274

Dastager SG, Deepa CK, Puneet SC et al (2009) Isolation and characterization of plant growth-promoting strain *Pantoea* NII-186 from Western Ghat Forest soil, India. Lett Appl Microbiol 49:20–25

Dastager SG, Kumaran DC, Pandey A (2010) Characterization of plant growth promoting rhizobacterium *Exiguobacterium* NII-0906 for its growth promotion of cowpea *Vigna unguiculata*. Biologia 65:197–203

Datta C, Basu P (2000) Indole acetic acid production by a *Rhizobium* species from root nodules of a leguminous shrub *Cajanus cajan*. Microbiol Res 155:123–127

De Weert S, Vermeiren H, Mulders IH (2002) Flagella-driven chemotaxis towards exudate components is an important trait for tomato root colonization by *Pseudomonas fluorescens*. Mol Plant Microbe Interact 15(11):1173–1180

Ding Y, Wang J, Liu Y et al (2005) Isolation and identification of nitrogen fixing bacilli from plant rhizospheres in Beijing region. J Appl Microbiol 99:1271–1281

Dobbelaere S, Vanderleyden J, Okon Y (2003) Plant growth-promoting effects of diazotrophs in the rhizosphere. Crit Rev Plant Sci 22:107–149

Duffy B, Keel C, Defago G (2004) Potential role of pathogen signaling in multitrophic plant-microbe interactions involved in disease protection. Appl Environ Microbiol 70(3):1836–1842

Eisenstein M (2013) Discovery in a dry spell. Nature 501:S7–S9

Ekinci M, Turan M, Yildirim E et al (2014) Effect of plant growth promoting rhizobacteria on growth, nutrient, organic acid, amino acid and hormone content of cauliflower (*Brassica oleracea* L. var. *botrytis*) transplants. Acta Sci Pol Hortorum Cultus 13(6):71–85

Farajzadeh D, Yakhchali B, Aliasgharzad N (2012) Plant growth promoting characterization of indigenous Azotobacteria isolated from soils in Iran. Curr Microbiol 64:397–403

Fatnassi IC, Chiboub M, Saadani O et al (2015) Phytostabilization of moderate copper contaminated soils using co-inoculation of *Vicia faba* with plant growth promoting bacteria. J Basic Microbiol 55(3):303–311

Fedoroff NV, Battisti DS, Beachy RN et al (2010) Radically rethinking agriculture for the 21st century. Science 327:833–834

Fernandes GC, Trarbach LJ, Campos SB et al (2014) Alternative nitrogenase and pseudogenes: unique features of the *Paenibacillus riograndensis* nitrogen fixation system. Res Microbiol 165:571–580

Fernando WGD, Nakkeeran S, Zhang Y (2006) Biosynthesis of antibiotics by PGPR and its relation in biocontrol of plant diseases. In: Siddiqui ZA (ed) PGPR: biocontrol and biofertilization. Springer, Dordrecht, pp 67–109

Gabriela F, Casanovas EM, Quillehauquy V (2015) Azospirillum inoculation effects on growth, product quality and storage life of lettuce plants grown under salt stress. Sci Hortic 195:154–162

Gaur AC (1990) Physiological functions of phosphate solubilizing microorganisms. In: Gaur AC (ed) Phosphate solubilizing microorganisms as biofertilizers. Omega Scientific Publishers, New Delhi, pp 16–72

Glick BR (1995) The enhancement of plant growth by free-living bacteria. Can J Microbiol 41(2):109–117

Glick BR (2010) Using soil bacteria to facilitate phytoremediation. Biotechnol Adv 28:367–374

Glick BR (2012) Plant growth-promoting bacteria: mechanisms and applications. Scientifica 2012:1–15

Glick BR (2014) Bacteria with ACC deaminase can promote plant growth and help to feed the world. Microbiol Res 169:30–39

Glick BR, Cheng Z, Czarny J et al (2007) Promotion of plant growth by ACC deaminase containing soil bacteria. Eur J Plant Pathol 119:329–339

Gomiero T, Pimentel D, Paoletti MG (2011) Environmental impact of different agricultural management practices: conventional vs. organic agriculture. Crit Rev Plant Sci 30:95–124

Goswami D, Thakker JN, Dhandhukia PC (2016) Portraying mechanics of plant growth promoting rhizobacteria (PGPR): a review. Cogent Food Agric 2:1127500

Gouda S, Kerry RG, Das G (2018) Revitalization of plant growth promoting rhizobacteria for sustainable development in agriculture. Microbiol Res 206:131–140
Gupta G, Parihar SS, Ahirwar NK et al (2015) Plant growth promoting Rhizobacteria (PGPR): current and future prospects for development of sustainable agriculture. J Microbiol Biochem 7:96–102
Gyaneshwar P, Naresh KG, Parekh LJ et al (2002) Role of soil microorganisms in improving P nutrition of plants. Plant Soil 245:83–93
Haas D, Défago G (2005) Biological control of soil-borne pathogens by fluorescent pseudomonads. Nat Rev Microbiol 3:307–319
Haas D, Keel C (2003) Regulation of antibiotic production in root-colonizing *Pseudomonas* spp. and relevance for biological control of plant disease. Annu Rev Phytopathol 41:117–153
Habib SH, Kausar H, Saud HM (2016) Plant growth-promoting rhizobacteria enhance salinity stress tolerance in okra through ROS-scavenging enzymes. Biomed Res Int 2016:6284547
Hameeda B, Harinib G, Rupelab OP (2008) Growth promotion of maize by phosphate solubilizing bacteria isolated from composts and macrofauna. Microbiol Res 163:234–242
Hashem A, Abd-Allah EF, Alqarawi AA (2016) Comparing symbiotic performance and physiological responses of two soybean cultivars to arbuscular mycorrhizal fungi under salt stress. Saudi J Biol Sci 26(1):38–48. https://doi.org/10.1016/j.sjbs.2016.11.015
Hassen AI, Bopape FL, Sanger LK (2016) Microbial inoculants as agents of growth promotion and abiotic stress tolerance in plants. In: Singh D, Singh H, Prabha R (eds) Microbial inoculants in sustainable agricultural productivity. Springer, New Delhi, pp 23–36
Hinsinger P, Bengough AG, Vetterlein D et al (2009) Rhizosphere: biophysics, biogeochemistry, and ecological relevance. Plant Soil 321:117–152
Hong SH, Ham SY, Kim JS et al (2016) Application of sodium polyacrylate and plant growth-promoting bacterium, *Micrococcaceae* HW-2 on the growth of plants cultivated in the rooftop. Int Biodeterior Biodegrad 133:297–303
Honma M, Shimomura T (1978) Metabolism of 1-aminocyclopropane-1-carboxylic acid. Agric Biol Chem 43:1825–1831
Huang XD, El-Alawi Y, Penrose DM et al (2004) Responses of three grass species to creosote during phytoremediation. Environ Pollut 130:453–463
Ilangumaran G, Smith DL (2017) Plant growth promoting rhizobacteria in amelioration of salinity stress: a systems biology perspective. Front Plant Sci 8:1768
Jeon J, Lee S, Kim H et al (2003) Plant growth promotion in soil by some inoculated microorganisms. J Microbiol 41:271–276
Jha CK, Saraf M (2015) Plant growth promoting rhizobacteria (PGPR): a review. E3 J Agric Res Dev 5:108–119
Ji X, Lu G, Gai Y et al (2010) Colonization of *Morus alba* L. by the plant-growth-promoting and antagonist bacterium *Burkholderia cepacia* strain Lu10-1. BMC Microbiol 10:243–254
Kamkar B (2016) Sustainable development principles for agricultural activities. Adv Plant Agric Res 3(5):1–2
Kang BG, Kim WT, Yun HS et al (2010) Use of plant growth-promoting rhizobacteria to control stress responses of plant roots. Plant Biotechnol Rep 4:179–183
Kasa P, Modugapalem H, Battini K (2015) Isolation, screening, and molecular characterization of plant growth promoting rhizobacteria isolates of *Azotobacter* and *Trichoderma* and their beneficial activities. J Nat Sci Biol Med 6:360–363
Kasim WA, Gaafar RM, Abou-Ali RM et al (2016) Effect of biofilm forming plant growth promoting rhizobacteria on salinity tolerance in barley. Ann Agric Sci 61(2):217–227
Khan AL, Waqas M, Kang S-M, Al-Harrasi A, Hussain J, Al-Rawahi A, Al-Khiziri S, Ullah I, Ali L, Jung H-Y, Lee I-J (2014) Bacterial endophyte Sphingomonas sp. LK11 produces gibberellins and IAA and promotes tomato plant growth. J Microbiol 52(8):689–695
Kloepper JW, Schroth MN (1981) Relationship of in vitro antibiosis of plant growth promoting rhizobacteria to plant growth and the displacement of root microflora. Phytopathology 71:1020–1024

Kloepper JW, Lifshitz R, Zablotowicz RM (1989) Free-living bacterial inocula for enhancing crop productivity. Trends Biotechnol 7:39–44

Kong Z, Glick BR (2017) The role of plant growth-promoting bacteria in metal phytoremediation. Adv Microb Physiol 71:97–132

Krysciak D, Schmeisser C, Preuß S et al (2011) Involvement of multiple loci in quorum quenching of autoinducer I molecules in the nitrogen-fixing symbiont *Rhizobium* (*Sinorhizobium*) sp. strain NGR234. Appl Environ Microbiol 77:5089–5099

Kuan KB, Othman R, Abdul Rahim K et al (2016) Plant growth-promoting rhizobacteria inoculation to enhance vegetative growth, nitrogen fixation and nitrogen remobilisation of maize under greenhouse conditions. PLoS ONE 11(3):e0152478

Kumar A, Verma JP (2018) Does plant–microbe interaction confer stress tolerance in plants: a review? Microbiol Res 207:41–52

Kumar KV, Singh N, Behl HM et al (2008) Influence of plant growth promoting bacteria and its mutant on heavy metal toxicity in *Brassica juncea* grown in fly ash amended soil. Chemosphere 72:678–683

Kumar AS, Lakshmanan V, Caplan JL et al (2012) Rhizobacteria *Bacillus subtilis* restricts foliar pathogen entry through stomata. Plant J 72:694–706

Kumar M, Mishra S, Dixit V (2016) Synergistic effect of *Pseudomonas putida* and *Bacillus amyloliquefaciens* ameliorates drought stress in chickpea (*Cicer arietinum* L.). Plant Signal Behav 11(1):e1071004

Kumari P, Khanna V (2016) Biodiversity of *Pseudomonas* and *Bacillus* possessing both bioantagonistic and plant growth promoting traits in chickpea rhizosphere. Int J Sci Nat 7(1):153–158

Liu XM, Feng ZB, Zhang FD et al (2006) Preparation and testing of cementing and coating nano-subnanocomposites of slow/controlled-release fertilizer. Agric Sci China 5:700–706

Liu Y, Wang H, Sun X et al (2011) Study on mechanisms of colonization of nitrogen-fixing PGPB, *Klebsiella pneumoniae* NG14 on the root surface of rice and the formation of biofilm. Curr Microbiol 62(4):1113–1122

Liu D, Lian B, Dong H et al (2012) Isolation of *Paenibacillus* sp. and assessment of its potential for enhancing mineral weathering. J Geomicrobiol 29:413–421

Loper JE, Gross H (2007) Genomic analysis of antifungal metabolite production by *Pseudomonas fluorescens* Pf-5. Eur J Plant Pathol 119:265–278

Lugtenberg B, Kamilova F (2009) Plant-growth-promoting Rhizobacteria. Ann Rev Microbiol 63:541–556

Lynch JM (1985) Origin, nature and biological activity of aliphatic substances and growth hormones found in soil. In: Vaughan D, Malcom RE (eds) Soil organic matter and biological activity. Martinus Nijhoff/Dr. Junk W, Publishers, Dordrecht/Boston, pp 151–174

Mäder P, Kaiser F, Adholeya A et al (2011) Inoculation of root microorganisms for sustainable wheat-rice and wheat-black gram rotations in India. Soil Biol Biochem 43:609–619

Maheshwari DK, Dheeman S, Agarwal M (2015) Phytohormone-producing PGPR for sustainable agriculture. Bacterial metabolites in sustainable agroecosystem. Springer, Cham, pp 159–182

Mantelin S, Touraine B (2004) Plant growth-promoting bacteria and nitrate availability: impacts on root development and nitrate uptake. J Exp Bot 55:27–34

Masciarelli O, Llanes A, Luna V (2014) A new PGPR co-inoculated with *Bradyrhizobium japonicum* enhances soybean nodulation. Microbiol Res 169:609–615

Mazzola M (2004) Assessment and management of soil microbial community structure for disease suppression. Annu Rev Phytopathol 42:35–59

McGill WB, Cole CV (1981) Comparative aspects of cycling of organic C, N, S and P through soil organic matter. Geoderma 26:267–268

Miller JB, Oldroyd GE (2012) The role of diffusible signals in the establishment of rhizobial and mycorrhizal symbioses. Signaling and communication in plant symbiosis. Springer, Berlin, pp 1–30

Milošević NA, Marinković JB, Tintor BB (2012) Mitigating abiotic stress in crop plants by microorganisms. Proc Nat Sci Matica Srpska Novi Sad 123:17–26

Moustaine M, Elkahkahi R, Benbouazza A et al (2017) Effect of plant growth promoting rhizobacterial (PGPR) inoculation on growth in tomato (*Solanum lycopersicum* L) and characterization for direct PGP abilities in Morocco. Int J Environ Agric Biotechnol 2(2):590–596

Murray JD, Muni RR, Torres-Jerez I et al (2011) Vapyrin, a gene essential for intracellular progression of arbuscular mycorrhizal symbiosis, is also essential for infection by rhizobia in the nodule symbiosis of *Medicago truncatula*. Plant J 65:244–252

Mus F, Crook MB, Garcia K et al (2016) Symbiotic nitrogen fixation and the challenges to its extension to nonlegumes. Appl Environ Microbiol 82(13):3698–3710

Nannipieri P, Giagnoni L, Landi L et al (2011) Role of phosphatase enzymes in soil. In: Bunemann E, Oberson A, Frossard E (eds) Phosphorus in action: Biological processes in soil phosphorus cycling. Springer, Heidelberg, pp 251–244

Neilands JB (1995) Siderophores: structure and function of microbial iron transport compounds. J Biol Chem 270:26723–26726

Neubauer U, Furrer G, Kayser A et al (2000) Siderophores, NTA, and citrate: potential soil amendments to enhance heavy metal mobility in phytoremediation. Int J Phytorem 2:353–368

Ojuederie OB, Babalola OO (2017) Microbial and plant-assisted bioremediation of heavy metal polluted environments: a review. Int J Environ Res Public Health 14(12):1–26

Okon Y, Bloemberg GV, Lugtenberg BJJ (1998) Biotechnology of biofertilization and phytostimulation. In: Altman A (ed) Agricultural biotechnology. Marcel Dekker, New York, pp 327–349

Oteino N, Lally RD, Kiwanuka S et al (2015) Plant growth promotion induced by phosphate solubilizing endophytic *Pseudomonas* isolates. Front Microbiol 6:745

Parmar P, Sindhu SS (2013) Potassium solubilisation by rhizosphere bacteria: influence of nutritional and environmental conditions. J Microbial Res 3:25–31

Patten CL, Glick BR (1996) Bacterial biosynthesis of indole-3-acetic acid. Can J Microbiol 42:207–220

Paustian K, Lehmann J, Ogle S et al (2016) Climatesmart soils. Nature 532:49–57

Perez YM, Charest C, Dalpe Y et al (2016) Effect of inoculation with Arbuscular Mycorrhizal fungi on selected spring wheat lines. Sustain Agric Res 5(4):24–29

Pérez-Montaño F, Alías-Villegas C, Bellogín RA et al (2014) Plant growth promotion in cereal and leguminous agricultural important plants: from microorganism capacities to crop production. Microbiol Res 169:325–336

Philippot L, Raaijmakers JM, Lemanceau P et al (2013) Going back to the roots: the microbial ecology of the rhizosphere. Nat Rev Microbiol 11:789–799

Picard C, Di Cello F, Ventura M et al (2000) Frequency and biodiversity of 2, 4 diacetylphloroglucinol-producing bacteria isolated from the maize rhizosphere at different stages of plant growth. Appl Environ Microbiol 66(3):948–955

Plaxton WC, Tran HT (2011) Metabolic adaptations of phosphate-starved plants. Plant Physiol 156:1006–1015

Porcel R, Zamarreño ÁM, García-Mina JM et al (2014) Involvement of plant endogenous ABA in *Bacillus megaterium* PGPR activity in tomato plants. BMC Plant Biol 14(1):1

Poupin MJ, Timmermann T, Vega A et al (2013) Effects of the plant growth-promoting bacterium *Burkholderia phytofirmans* PsJN throughout the life cycle of Arabidopsis thaliana. PLoS One 8(7):e69435

Prapagdee B, Chanprasert M, Mongkolsuk S (2013) Bioaugmentation with cadmium-resistant plant growth-promoting rhizobacteria to assist cadmium phytoextraction by *Helianthus annuus*. Chemosphere 92:659–666

Raaijmakers JM, Weller DM (1998) Natural plant protection by 2,4-diacetylphloroglucinol producing *Pseudomonas* sp. in takeall decline soils. Mol Plant Microbe Interact 11:144–152

Raaijmakers JM, Vlami M, de Souza JT (2002) Antibiotic production by bacterial biocontrol agents. Anton Leeuw 81:537–547

Rajkumar M, Ae N, Prasad MNV et al (2010) Potential of siderophore producing bacteria for improving heavy metal phytoextraction. Trends Biotechnol 28(3):142–149

Ramaekers L, Remans R, Rao IM et al (2010) Strategies for improving phosphorus acquisition efficiency of crop plants. Field Crops Res 117:169–176

Raymond J, Siefert JL, Staples CR et al (2004) The natural history of nitrogen fixation. Mol Biol Evol 21:541–554

Rezzonico F, Zala M, Keel C et al (2007) Is the ability of biocontrol fluorescent pseudomonads to produce the antifungal metabolite 2, 4 diacetylphloroglucinol really synonymous with higher plant protection? New Phytol 173(4):861–872

Richardson AE (1994) Soil microorganisms and phosphorus availability. In: Ankhurst CE, Doube BM, Gupta VVSR et al (eds) Management of the soil biota in sustainable farming systems. CSIRO Publishing, Melbourne, pp 50–62

Richardson AE, Hocking PJ, Simpson RJ et al (2009) Plant mechanisms to optimize access to soil phosphorus. Crop Pasture Sci 60:124–143

Riedlinger J, Schrey SD, Tarkka MT et al (2006) Auxofuran, a novel metabolite that stimulates the growth of fly agaric, is produced by the mycorrhiza helper bacterium *Streptomyces* strain AcH 505. Appl Environ Microbiol 72:3550–3557

Robledo M, Rivera L, Jiménez-Zurdo JI et al (2012) Role of *Rhizobium* endoglucanase $CelC_2$ in cellulose biosynthesis and biofilm formation on plant roots and abiotic surfaces. Microb Cell Fact 11:125

Rodriguez H, Fraga R, Gonzalez T et al (2006) Genetics of phosphate solubilisation and its potential applications for improving plant growth promoting bacteria. Plant Soil 287:15–21

Sachdev DP, Chaudhari HG, Kasure VM et al (2009) Isolation and characterization of indole acetic acid (IAA) producing *Klebsiella pneumoniae* strains from rhizosphere of wheat (*Triticum aestivum*) and their effect on plant growth. Indian J Exp Biol 47:993–1000

Saleem AR, Bangash N, Mahmood T et al (2015a) Rhizobacteria capable of producing ACC deaminase promote growth of velvet bean (*Mucuna pruriens*) under water stress condition. Int J Agric Biol 17:663–667

Saleem M, Zamir MSI, Haq I et al (2015b) Yield and quality of forage oat (*Avena sativa* L.) cultivars as affected by seed inoculation with nitrogenous strains. Am J Plant Sci 6(19):3251

Saleh S, Huang XD, Greenberg BM et al (2004) Phytoremediation of persistent organic contaminants in the environment. In: Singh A, Ward O (eds) Soil biology: applied bioremediation and phytoremediation. Springer, Berlin, pp 115–134

Sapre S, Gontia-Mishra I, Tiwari S (2018) *Klebsiella* sp. confers enhanced tolerance to salinity and plant growth promotion in oat seedlings (*Avena sativa*). Microbiol Res 206:25–32

Seldin L, Van Elsas JD, Penids EGC (1984) *Bacillus azotofixans* sp. nov. A nitrogen fixing species from Brazilian soils and grass roots. Int J syst Bacteriol 34:451–456

Shahzad R, Waqas M, Khan AL et al (2016) Seed-borne endophytic *Bacillus amyloliquefaciens* RWL-1 produces gibberellins and regulates endogenous phytohormones of *Oryza sativa*. Plant Physiol Biochem 106:236–243

Shakeel M, Rais A, Hassan MN et al (2015) Root Associated *Bacillus* sp. improves growth, yield and zinc translocation for basmati rice (*Oryza sativa*) varieties. Front Microbiol 6:1286

Shameer S, Prasad TNVKV (2018) Plant growth promoting rhizobacteria for sustainable agricultural practices with special reference to biotic and abiotic stresses. J Plant Growth Regul 84:1–13

Sharma SB, Sayyed RZ, Trivedi MH et al (2013) Phosphate solubilizing microbes: sustainable approach for managing phosphorus deficiency in agricultural soils. SpringerPlus 2:587

Shelud'ko AV, Shirokov AA, Sokolova MK et al (2010) Wheat root colonization by *Azospirillum brasilense* strains with different motility. Microbiology 9(5):688–695

Smith DL, Gravel V, Yergeau E (2017) Editorial: signaling in the phytomicrobiome. Front Plant Sci 8:611

Souza R, Beneduzi A, Ambrosini A et al (2013) The effect of plant growth-promoting rhizobacteria on the growth of rice (*Oryza sativa* L.) cropped in southern Brazilian fields. Plant Soil 366:1–19

Spaepen S, Vanderleyden J (2011) Auxin and plant-microbe interactions. Cold Spring Harb Perspect Biol 3(4):a001438

Spaepen S, Vanderleyden J, Remans R (2007) Indole-3-acetic acid in microbial and microorganism-plant signaling. FEMS Microbiol Rev 31:425–448

Stacey G, Burris RH, Evans HJ (1992) Biological nitrogen fixation. Springer, Berlin

Staudinger C, Mehmeti-Tershani V, Gil-Quintana E et al (2016) Evidence for a rhizobia induced drought stress response strategy in *Medicago truncatula*. J Proteomics 136:202–213

Stephane C, Brion D, Jerzy N et al (2005) Use of plant growth bacteria for biocontrol of plant diseases: principles, mechanisms of action and future prospects. Appl Environ Microbiol 71(9):4951–4959

Suman PR, Jain VK, Varma A (2010) Role of nanomaterials in symbiotic fungus growth enhancement. Curr Sci 99:1189–1191

Sureshbabu K, Amaresan N, Kumar K (2016) Amazing multiple function properties of plant growth promoting rhizobacteria in the rhizosphere soil. Int J Curr Microbiol Appl Sci 5(2):661–683

Syed S (2016) Haloalkaliphilic *Bacillus* species from solar salterns: an ideal prokaryote for bioprospecting studies. Ann Microbiol 66(3):1315–1327

Syed S, Chinthala P (2015) Heavy metal detoxification by different *Bacillus* species Isolated from solar salterns. Scientifica (Cairo) 2015(2015):319760

Tabassum B, Khan A, Tariq M et al (2017) Bottlenecks in commercialisation and future prospects of PGPR. Appl Soil Ecol 121:102–117

Tarafdar A, Raliya R, Wang WN et al (2013) Green synthesis of TiO_2 nanoparticle using *Aspergillus tubingensis*. Adv Sci Eng Med 5:943–949

Timmusk S, Grantcharova N, Wagner EGH (2005) *Paenibacillus polymyxa* invades plant roots and forms biofilms. Appl Environ Microbiol 71:7292–7300

Timmusk S, Timmusk K, Behers L (2013) Rhizobacterial plant drought stress tolerance enhancement. J Food Security 1:10–16

Timmusk S, Seisenbaeva G, Behers L (2018) Titania (TiO_2) nanoparticles enhance the performance of growth-promoting rhizobacteria. Sci Rep 8:617

Trifi H, Ben Salem I, Kolsi Benzina N et al (2017) Effectiveness of the plant growth-promoting rhizobacterium *Pantoea* sp. BRM17 in enhancing *Brassica napus* growth in phosphogypsum-amended soil. Pedosphere. https://doi.org/10.1016/S1002-0160(17)60454-5

Vacheron J, Desbrosses G, Bouffaud ML et al (2014) Plant growth-promoting rhizobacteria and root system functioning. Ecophysiol Root Syst Environ Interact 166. https://doi.org/10.3389/fpls.2013.00356

Vejan P, Abdullah R, Khadiran T (2016) Role of plant growth promoting Rhizobacteria in agricultural sustainability- a review. Molecules 21(573):1–17

Verma JP, Yada J, Tiwari KN et al (2013) Effect of indigenous *Mesorhizobium* spp. and plant growth promoting rhizobacteria on yields and nutrients uptake of chickpea (*Cicer arietinum* L.) under sustainable agriculture. Ecol Eng 51:282–286

Viswanathan R, Samiyappan R (1999) Induction of systemic resistance by plant growth promoting rhizobacteria against red rot disease in sugarcane. Sugar Tech 1(3):67–76

Website used.: https://ourworldindata.org/world-population-growth. Accessed 15 Oct 2018

Yang J, Kloepper JW, Ryu CM (2009) Rhizosphere bacteria help plants tolerate abiotic stress. Trends Plant Sci 14:1–4

Yasmin S, Hafeez FY, Rasul G (2014) Evaluation of *Pseudomonas aeruginosa* Z5 for biocontrol of cotton seedling disease caused by *Fusarium oxysporum*. Biocontrol Sci Technol 24(11):1227–1242

Yolcu H, Turan M, Lithourgidis A et al (2011) Effects of plant growth-promoting rhizobacteria and manure on yield and quality characteristics of Italian ryegrass under semi-arid conditions. Aust J Crop Sci 5(13):1730–1736

Yu X, Liu X, Zhu TH et al (2011) Isolation and characterization of phosphate-solubilizing bacteria from walnut and their effect on growth and phosphorus mobilization. Biol Fert Soils 47:437–446

Zhao Y (2010) Auxin biosynthesis and its role in plant development. Annu Rev Plant Biol 61:49–64

Microbial Augmentation of Salt-Affected Soils: Emphasis on Haloalkalitolerant PGPR

9

M. Gavit Pavankumar, B. Chaudhari Ambalal,
D. Shelar Rajendra, and D. Dandi Navin

1 Introduction

The inordinate agriculture practices, human activities, and population growth have levitated soluble salt concentration in agriculture soils from arid and semiarid regions of the world notably, northern Africa, the Middle East, USSR, Central Asia, Australia, and the America (Rengasamy 2006). A plethora of soluble salts and eventual alkalinity build-up is collectively referred to as salt-affected soils (SAS) (Qadir et al. 2000). The SAS is mainly formed due to diminution of divalent cations such as Ca^{2+} and Mg^{2+}, concomitant accumulation of sodium and carbonates, and high alkalinity and pH (Sorokin et al. 2008), collectively responsible for salt and alkali stress to agricultural crops and microbial diversity in rhizosphere (Chang et al. 2014). Overall, the SAS may pose a serious menace to one-third of the world's food production (Munns 2002).

At a global scale, the SAS region is estimated to be 1307 million hectares (Mha) (et al. 2012) and represents more than 20% of the world's crop land with world economic losses of US$ 12 billion per year (Ghassemi et al. 1995). Of 1.53 billion hectares (Bha) cultivated cropland (Rengasamy 2006; Foley 2011), 23% (0.34×10^9 ha) and 37% (0.56×10^9 ha) are saline and alkaline soils, respectively (Tanji 2002). The cropland is anticipated to be lost up to 30–50% by 2050 (Singh et al. 2011).

In India, about 44% of estimated land area is degraded (Mythili and Goedecke 2016). Of the total 142.5 Mha net crop land in India, 4.12 Mha of alkali soil, 3.26 Mha of saline, and 4.62 Mha of saline-alkali soil (Bhadauria et al. 2010) are distributed in states of Andhra Pradesh, Gujarat, Karnataka, Madhya Pradesh, Maharashtra, Tamil

M. Gavit Pavankumar · B. Chaudhari Ambalal (✉) · D. Dandi Navin
School of Life Science, Kavayitri Bahinabai Chaudhari North Maharashtra University, Jalgaon, India

D. Shelar Rajendra
Department of Microbiology, Z. B. Patil College, Dhule, India

D. P. Singh et al. (eds.), *Microbial Interventions in Agriculture and Environment*,
https://doi.org/10.1007/978-981-13-8383-0_9

Nadu, West Bengal, and predominantly Indo-Gangetic Plains (Mandal et al. 2010). Nearly 50% of the canal-irrigated area is saline and/or sodic owing to inadequate drainage, maladroit water management, and distorted subsidized energy pricing (State of the Environment 2001). The cost of land degradation in India is estimated to be about US$ 5.35 billion in 2009 (Mythili and Goedecke 2016) with 6.2 million ton of production loss due to salinity alone (Young 1994). The estimated area under SAS in India may increase to 16.2 Mha by 2050 (CSSRI 2015).

At present, considerable research on the causes and effects of SAS have been envisaged, but performance of only few sustainable salt removal strategies at field scale are available. Hence, it is imperative to resurrect SAS to support the productive land use for cultivation of more crops in order to abate the upcoming challenges of global food shortage.

2 Genesis of Salt-Affected Soils

The soil is a dynamic, structured, heterogeneous, discontinuous system that exists as organic and inorganic matrix formed by the biotic and abiotic processes (Gobat et al. 2004). Organic soils contain 20% organic matter, but the major land area of inorganic (mineral) soils consists only of 1–6% organic matter (Brady and Weil 2002). The physical and chemical weathering of minerals gradually release the soluble connate salts which eventually form soil profile causing accumulation of salt over time in the soils once drainage and leaching fail (McBride 1994). In this regard, Fitzpatrick et al. (2001) suggested three determinants, namely, (1) hydrological status, (2) natural and induced status, and (3) soil chemical status of soluble salts. But few natural factors also contribute to soluble salts and alkalinity build-up, as (1) arid/semiarid or subhumid climate, (2) topology, (3) seawater intrusion onto land, (4) poor quality irrigation water or uncontrolled irrigation, (5) shallow saline groundwater rise through capillary action, (6) freeze-thaw action, and (7) inadequate drainage (Szabolcs et al. 1998). Some anthropogenic sources including (1) indiscriminate chloride/sulfate-based soluble fertilizers use, (2) inferior soil amendments, (3) devegetation, (4) waterlogging of the soil, (5) excessive technogenic pollution, etc. also elevate salinization in the soil layers (Bhadauria et al. 2010; Rengasamy 2010). Further, the inordinate human activities exacerbated the dissolved mineral salts mostly of the cations (sodium, calcium, magnesium, boron) and the anions (bicarbonate, chlorides, sulfate, nitrate) in the soil to promulgate sodicity in the soil profile, proving inhospitable for crop survival (Rengasamy 2010). The major soluble salts are usually neutral in its reaction such as chlorides and sulfates of sodium, calcium, and magnesium, but the soda salts such as carbonates and bicarbonates of sodium are capable of alkalinity build-up in the soil with pH between 8.5 and 10 (Rengasamy 2010). The alkalinity of soils is mainly due to hydrolysis of (1) exchangeable cations or (2) carbonates such as $CaCO_3$, $CaMg(CO_3)_2$, and Na_2CO_3.

Salt-impacted soils appear as white grayish ash-colored salt incrustations during the dry period (Rai et al. 2010). SAS can be categorized pedogenically into (1) saline soils with high quantity of soluble salts such as Ca^{2+}, Mg^{2+}, Cl^-, SO_4^{-2}, and NH_4^+ (Fitzpatrick 2002), (2) sodic soils with a higher concentration of Na^+ ions relative to Ca^{2+} and Mg^{2+}, and (3) alkali-saline or saline-sodic soils which contain an inexorable quantity of exchangeable sodium ions and soluble salts (Rengasamy 2010). These soils are recognized locally as Kallar/Thur (Punjab), Usar/ Reh (Uttar Pradesh), Luni (Rajasthan), Khar/Kshar (Gujrat, Maharashtra), Chouddu/Uippu (Andhra Pradesh), and Choppan (Karnataka) (Bhadauria et al. 2010). The SAS have been realized as a widespread global environmental problem (Yang et al. 2010) limiting agricultural crop productivity and sustainability in arid as well as semiarid areas of the world (Pitman and Läuchli 2002; Qadir et al. 2006).

3 Need for Increased Crop Productivity

An increase in agricultural yield over the past several decades has come with some costs (Gliessman 2004). Degradation of soil through accelerated erosion is a serious threat with consequent decline in ecosystem functions and services especially in tropic and subtropic regions of the developing countries which is expected to sustain during the twenty-first century (Lal 2001, 2015). In India, the average agriculture land holding area has progressively reduced from 2.30 to 1.16 ha during 1970 to 2010 because of increasing population pressure. About 60% rainfed land is low in productivity, causing high interannual changes in agricultural output. The present global climate change, intensive agricultural practices, and rise in salt-affected region may likely exacerbate the necessity for fertile soils to feed 9.3 billion (34%) populations by 2050 (Singh et al. 2011). The average worldwide food consumption is also raised by 54% that needs 60% rise in agriculture productivity (Godfrey et al. 2010).

Although the land area contributes around 99% of the world's food supply, the present per capita availability of world food grains (80%) is on the verge of decline (Pimentel and Giampietro 1994; Kendall and Pimentel 1994). Hence, the alternative strategy is to either increase the cropland area or boost the agricultural productivity. At present, fertilizer amendment to the soils has failed to increase agricultural productivity any further, and hence, increase in cropland area through cultivation of crops in the SAS using microbial augmentation is the most reasonable path owing to their genetic and metabolic diversity to cope saline and sodic microenvironment by haloalkalitolerant species. Hence, the assessment of alkalinity and salinity is indispensable for the success of microbial amendments to the SAS.

4 Parameters to Rate Salt-Affected Soils

Several parameters based on physical and chemical analysis are currently employed to characterize salinity and sodacity of the soils. These parameters analyze quantum of soluble salts in soil solution and exchangeable Na^+. The parameters summarized in Table 9.1 offer a realistic characterization of soils. Based on some key parameters, SAS have been broadly categorized (Table 9.2) into three types: (1) saline soils, (2) alkaline or sodic soils, and (3) saline-sodic soils.

Table 9.1 Assessment of physical and nutrient status of SAS

Sr. no.	Parameter	Property/function	Measurement	Assessment suggest
1.	pH (scale of 0–14)	Acidic, neutral, or alkaline	pH meter	Suitability of pH for a particular crop and indicate solubility of soil nutrients and minerals
2.	EC (dSm^{-1} or $mmhos.cm^{-1}$)	Capacity to conduct electricity	Conductivity meter	More dissolved mineral ions
3.	Exchangeable cations (meq per 100 g)	Cations (Ca^{2+}, Mg^{2+}, K^+, Na^+) associated with soil clay particles	Atomic adsorption spectroscopy (AAS) or voltammetry/ion chromatography	Sodium hazard and assist in calculating gypsum requirement for sodic soils
4.	Specific elements	B^+, Cl^-, Na^+ concentration	Ion chromatography	Elemental ion toxicity
5.	Sodium adsorption ratio (SAR)	Na^+, Mg^{2+}, Ca^{2+}	Ion chromatography, voltammetry $SAR = \frac{Na^+}{\sqrt{\frac{Ca^{2+} + Mg^{2+}}{2}}}$	Sodium toxicity
6.	Exchangeable sodium percentage (ESP)	Cation exchange sites covered by Na^+ ions	$ESP = \frac{(100 \times \text{Exchangeable } Na^+)}{\text{CEC}(\text{Cation Exchange Capacity})}$	Sodium toxicity and estimate GR for SAS
7.	Cation exchange capacity (CEC., meq/100 g)	Soil flocculating capacity based on the soil's ability to hold cations (Ca^{2+}, Mg^{2+})	Atomic adsorption spectroscopy (AAS)	Calculate GR to adjust ESP and SAR
8.	Calcium carbonate equivalent	Residual $CaCO_3$ (undissolved)	Titrimetric method	Elemental S or H_2SO_4 amendments for alkali soils in case pH<8
9.	Organic carbon	Slightly processed organic residues of plant and animal origin, humus, charcoal, fossil organic matter, microorganisms	Dichromate oxidation method (Kalembasa and Jenkinson 1973)	Soil fertility and soil physical conditions

10.	Total N	N compounds of the organic matter of the soil biomass and also inorganic forms of N	Kjeldahl method/titrimetric method/acetylene reduction assay (Bremner and Mulvaney 1982)	N availability as macronutrient to the plants
11.	Total K	Potassium for proper plant growth, reproduction, and overall health	Flame photometry (Allen et al. 1974)	K availability for plant growth
12.	Total P	Phosphorous for photosynthesis, energy storage and transfer, cell division, cell enlargement, etc.	Olsen's method (Olsen 1954)	P availability as a macronutrient and vital component in photosynthesis
13.	Microbial biomass C and biomass N	Organic C, total N and ammonia N, and ninhydrin-reactive N	Chloroform fumigation extraction method (Brookes et al. 1985; Vance et al. 1987)	Direct estimate of overall microbial presence in the soil
14.	Enzyme activity	Measures the enzyme activity of microbial populations	Fluorescein diacetate hydrolysis (FDH) assay (Schnürer and Rosswall 1982)	Overall microbial activity in soil
15.	Basal respiration	Measures CO_2 evolution	Titrimetric method (Anderson 1982)	Indirect estimate of overall microbial activity in soil

Table 9.2 Types and properties of salts affected soils

Soil type	EC (dSm^{-1})	SAR	ESP ($Cmolkg^{1-}$)	pH	Soil structure	Peculiarities	Ions	Mode of impact on biomass production	Possible location	Common reclamation strategies	References
Non-saline	<0.75	<5	<5	<7.5	Flocculated	Good soil tilth Nutrient rich	Optimal proportion of NH_4^+, K^+, Ca^{2+}, Mg^{2+}, Cl^-, NO_3^-, PO_4^{3-},SO_4^{2-}	–	–	–	Horneck et al. (2007)
Saline or white alkali or solonchak soil	>4	<13	<15	<8.5	Flocculated	White fluffy encrustation on surface during dry periods Good soil tilth Poor vegetation due to salinity	Na^+, K^+, Ca^{2+}, Mg^{2+}, Al^{3+} and Cl^-, NO_3^-, PO_4^{3-}, SO_4^{2-}, CO_3^{2-}-HCO_3^-	High osmotic pressure, ion specific toxicity, disproportionate cation nutrition, soil structure impacts	Dominate in arid and semiarid regions	Leaching with high-quality water, provision of drainage	
Alkaline or solonets or sodic soil	<4	>13	>15	>8.5	Dispersed	Poor soil tilth Low penetration Black surface crust due to dispersion of organic matter	Na^+ and alkali hydrolysis	Alkali pH stress, soil structure impacts, ions specific toxicity especially Na^+	Dominate in semiarid and subhumid regions	Use of chemical amendments (e.g., gypsum, sulfur, sulfuric acid, etc.), phytoremediation by salt-tolerant grasses (e.g., Karnal grass)	
Alkaline-saline	>4	>13	>15	≥8.5	Largely flocculated	Dispersed if pH exceeds 8.5 Poor soil structure	Na^+, K^+, Ca^{2+}, Mg^{2+}	High osmotic pressure, alkali pH stress, soil structure impacts	Arid and semiarid regions	Leaching with application of chemical amendments	

Alberta soil	>4	>13	NS	>8.5	Columnar/dispersed	Columnar structure Hard consistence when dry Dark organic staining in the B horizon	Na^+, Mg^{2+}, Ca^{2+}, SO_4^{2-}, HCO_3^-,Cl^-	Alkali pH, soil structure impacts	Alberta, Canada	Deep plowing, subsoiling, application of Ca^{2+} amendments	McNeil et al. (1994) and Miller and Brierley (2011)
Technogenic saline	>4	NS	NS	NS	Columnar	Sodium and chlorine compounds with deicing salt-sand mixtures Aggravated by the presence of heavy metals	Na^+, K^+, Cl^- with deicing salt-sand mixtures	Alkaline reaction of the environment, high osmotic pressure, toxic effect of readily soluble salts	Oil extraction sites, mining industries, near roads and man-made ore-bearing tailing dumps, snow banks, landfills, coal, and ash disposal areas	–	Charzyński et al. (2013) and Artamonova et al. (2010)

NS: Not specified

5 Impact of Salt-Affected Soil on Plant Growth

The impact of SAS on growth of plant has attracted more research efforts throughout the world for nearly a century (Fougere et al. 1991). SAS contribute specific structural problems due to (1) physical slaking, (2) swelling, (3) dispersion of clay, (4) surface crusting, and (5) hard setting (Qadir and Schubert 2002). Consequently, it influences (1) mobility of water and air, (2) available water holding capacity of plant, (3) root penetration, (4) seedling emergence, (5) photosynthesis, (6) microbial content, (7) plant-microbe interactions, (8) BNF, and (9) agriculture tillage and sowing operations (Fitzpatrick 2002).

Excess salt and alkali is accountable for (1) increased osmotic stress and (2) Na^+ and Cl^- ion-induced toxicity, (3) influence metal ions and phosphorous to precipitate, thereby affecting (a) ion excess toxicity; (b) photosynthetic dysfunction due to nitrate reductase inhibition; (c) absorption of Na^+, Cl^-, NO_3^-, and $H_2PO_4^-$ to disrupt ionic imbalance; (d) pH homeostasis; and (e) physical soil structure (Yang et al. 2007; Patel et al. 2012), and (4) eventually impact the choice of crops, crop growth, and ultimately yield, due to (a) osmotic and (b) ion-specific damage and (c) disproportionate cation nutrition leading to several nutrient deficiencies including Mg^{2+} and Ca^{2+} (Naidu and Rengasamy 1993; Singh and Charath 2001). Accordingly, Munns (2005) proposed a concept of biphasic growth response to describe salinity effect on the plant (Table 9.3). Of these, osmotic stress effect is instantaneous, but ionic stress impacts during later phase of plant growth with less effect (Patel et al. 2012). Any survival and plant growth in low or moderate SAS are the consequence of (1) ion transport and compartmentation, (2) osmolyte synthesis, and (3) turnover of protein for cellular repair (Hong et al. 2000). The established plants nullify the stress with more energy turnover to make organic and inorganic solutes (Fitzpatrick 2002; Brady and Weil 2002) and maintain a potential gradient against influx of water (Sohan et al. 1999).

Excess amount of soluble salts in the soil primarily target plant cells in seedling emergence to dehydrate and failure to extract water that significantly influence plant growth

Table 9.3 Summary of some observable impacts of various parameters on plant growth

	Risk impact		
Parameter	Negligible	Prominent	Observable impact on plant
EC (dSm^{-1})	<0.75	<4 or more	High-soluble salts impede seed germination and stunted growth of plant
Na^+ (%)	<10	10–40 or more	Leaf burn, scorch and dead tissue around the leaf periphery, necrosis of the leaves, accumulate in shoots and foliage, ion-specific damage, inactivate enzyme, and protein synthesis
Cl^- (ppm)	<175	175–700 or more	Wilting, browning of the leaf tips and leaf drop, inhibit photosynthesis, accumulate in leaves
B (ppm)	<0.5	0.5–4 or more	Yellow-colored older leaves and drying of the leaf at the periphery
CO^{3-} (ppm)	<0.5	16–18 or more	Yellow leaf veins due to Fe^{2+}/Zn^{2+} deficiency
ESP (%)	<5	5–15 or more	Impair structure of soil, impede water movement
SAR	<5	5–13 or more	Poor soil structure and restricted water movement
pH	<7.5	7.5–8.5 or more	Unavailability of Fe^{2+}, Mg^{2+}, micronutrients to the plant

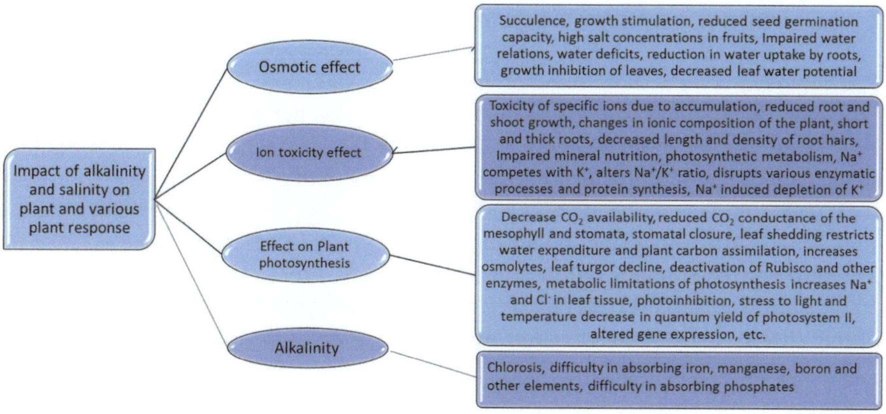

Fig. 9.1 Impact of salinity and alkalinity on plants

and briefly outlined in Fig. 9.1 (Cicek and Cakirlar 2002). The yield of wheat crop is expected to be lowered by 10, 25, and 50% at EC of 7.4, 9.5, and 13 dSm^{-1}, respectively, but crop yield remains unaffected at EC below 6 dSm^{-1} (Horneck et al. 2007).

6 Impact on Plant Developmental Stages

Seed emergence and its growth are two distinct plant growth stages. Salt stress delays germination in several plants, inadvertently affecting the emerging seedling (Grattan and Oster 2003) rendering susceptibility to hypocotyl and cotyledon injury as well as attack by the pathogens (Esechie et al. 2002). The high pH environment in the root rhizosphere can cause precipitation of metal ions and P, thereby affecting absorption of Na^+, K^+, Cl^-, NO_3^-, and $H_2PO_4^-$ to alter ionic balance and pH homeostasis in plant tissues (Yang et al. 2007, 2009).

Salinity stress causes inadequate Ca^{2+} supply which influences (1) membrane integrity, (2) root growth (Cramer 1992), (3) increased Na^+ deposition on the root tips, and (4) decreased selectivity for K^+ (Zhong and Lauchli 1994). Salt stress in the root region cumulatively inhibits plant growth due to (1) induced plant water deficits, (2) more absorption of an individual ion (mainly Na^+ and Cl^-), and (3) preferential absorption of specific ion which retards the absorption of other essential plant nutrients (Ashraf 2004). Kurth et al. (1986) observed that roots grown at high Na^+/Ca^{2+} are shorter and thicker, but Ca^{2+} amendment recoups root elongation and shoot growth and restores root hair ramification (Shabala et al. 2003). The plant shoot growth is commonly observed by a reduced leaf surface area and stunted shoots with spotty stands of crops in the SAS.

The excess concentration of soluble salts in the root rhizosphere can lead to decrease in leaf water potential and, eventually, interfere with the capacity to abstract water from soil and uphold turgor pressure (Sohan et al. 1999). Apse and Blumwald (2007) suggested that Na^+ transport is mainly unidirectional and, hence, causes progressive accumulation of Na^+ as the leaves age.

7 Impact on Physiological Processes

The photosynthetic activity of plant grown under salt stress is critically suppressed (Wei et al. 2006; Chaves et al. 2009) due to (1) weakened CO_2 diffusion (Ma et al. 1997; Sultana et al. 1999; Ouerghi et al. 2000), (2) impaired photosynthetic components such as chloroplast ultrastructure and intracellular space in mesophyll (Ma et al. 1997), and (3) reduced photochemical efficiency (Delfine et al. 1999; Fidalgo et al. 2004). The effects can be attributed to (1) reduced diffusion conductance to CO_2 through the stomata and mesophyll under salt stress conditions (Flexas et al. 2007; Chaves et al. 2009), (2) modification of photosynthetic reaction (Lawlor and Cornic 2002), and (3) inactivation of Rubisco (Meyer and Genty 1998). Munns (2005) found nonfunctional photosynthesis-related metabolism due to higher accumulation of Na^+ and Cl^- in the leaf parenchyma, while Chaves et al. (2009) suggested rapid alteration in gene expression pattern under salt stress conditions. Accumulation of Cl^- leads to a change in photosynthetic function through the inhibition of nitrate reductase (Singh and Charath 2001; Munns 2002), while accumulation of Na^+ out-competes K^+ which can disrupt various enzymatic processes in cytoplasm including protein synthesis (Blaha et al. 2000).

8 Effects on Soil Properties

A plethora of natural soluble salts as chlorides and sulfates of Na^+, Ca^{2+}, Mg^{2+}, and sometimes nitrate (NO_3^-) (Fitzpatrick et al. 2001) and their gradual accumulation make the upper soil layer sodic and subsoil to become saline (Fetter 2001; Rengasamy 2002). More accumulation of exchangeable Na^+ ions than Ca^{2+} and Mg^{2+} causes saline soil to become saline-sodic or sodic soil. A significant amount of Na^+ in the soil further reduces flocculation of soil particles and small clay particles that move downward through soil profile to obstruct the pore spaces, leading to poor water infiltration and waterlogging. It is estimated that even 6% Na^+ can affect the soil structure (Naidu and Rengasamy 1993). Excess Na^+ in the clay fraction of soil is also responsible for slaking and swelling upon hydration and causes upward movement of clay to form large mounds and gilgai (Fitzpatrick et al. 1994). More frequent wetting and drying causes compact soil with significant loss of soil structure (Warrence et al. 2002) and forms thin (about 10 mm), but dense, crust, which prevents seedling emergence and torn out roots after emergence (Fitzpatrick et al. 2001). High pH and Na^+ in sodic soils cause low level of organic matter with meager biological activity (Naidu and Rengasamy 1993; Peinemann et al. 2005). Hence, excess Na^+ must be removed or replaced to allow formation of linkages between clay particles by organic matter (Rengasamy and Olsson 1991).

9 Effects on Soil Microbiota

The peculiar structure of SAS restricts functional activity of plant roots and the rhizobiome. The meager organic matter content, high-soluble salts (Na^+), and high pH significantly impede microbial metabolic routes (Rietz and Haynes 2003; Sardinha et al.

2003) to mineralize available carbon sources (Rietz and Haynes 2003; Sardinha et al. 2003; Yuan et al. 2007; Setia et al. 2010; Li et al. 2012), since no (1) vegetation, (2) root biomass, (3) litter debris, and (4) accumulated decayed organic matter are available as energy sources to microbial communities (Sardinha et al. 2003; Wichern et al. 2006; Elmajdoub and Marschner 2013). The ratio of bacteria/fungi is generally more as revealed from low ergosterol concentration (Sardinha et al. 2003; Chowdhury et al. 2011). Low fungal abundance impacts nutrient cycling, carbon incorporation efficiency, and carbon sequestration because fungi have better enzymatic activity to decompose complex compounds. Besides, nitrogen mineralization is greatly reduced, but ammonification is unaffected compared to nitrification, and hence, ammonia is accumulated (Laura 1974; Kaushik and Sethi 2005). Low osmotic potential causes microbes to synthesize osmolytes similar to plants (Oren 2001) that demand excess energy, eventually restricting further growth of microbes (Zahran et al. 1995; Hagemann 2011). This is substantiated by Elmajdoub and Marschner (2013) who observed failure of cellulase induction in soil microbial community to utilize cellulose simply because of necessity of osmolyte synthesis in low osmotic condition and, thus, limited the growth of microbes in soil. But Dodd and Perez-Alfocea (2012) showed that the presence of soil microbiota relives the plant salinity stress through modification of hormonal root-shoot signaling and phytohormone milieu in salt-affected conditions.

Variation in the phytohormone milieu is implicated to control vegetative growth under low osmotic potential through biomass and leaf senescence (Pérez-Alfocea et al. 2010; Ghanem et al. 2011). Few soil microbes may modulate novel mechanisms to enhance plant salinity tolerance. For instance, mycorrhizae, namely, *Glomus intraradices*, accumulate silicon which helps to (1) limit Na^+ uptake, (2) improve photochemical efficiency, (3) protect cellular membrane integrity, (4) increase antioxidant enzyme activity, and (5) decrease water loss (Romero-Aranda et al. 2006). Several earlier reports have reviewed the impact of salt-affected condition on plant and soil properties, but very few reports realized the significance of microbial systems in SAS.

Hence, the study about microbial systems in SAS may provide bespoke microbe as inoculant(s) for crop cultivation under stressed conditions.

10 Remediation Strategies for Salt-Affected Soils

The problem of SAS due to dissolved salt and/or exchangeable Na^+ is growing at the rate of 10% every year (Shrivastava and Kumar 2015). Hence, the removal of surplus soluble salts or exchangeable Na^+ from the plant rhizosphere region is now considered a priority (Qadir et al. 2007). The criteria for selection of appropriate strategy to recoup SAS are primarily based on (1) economic benefit, (2) on-site drainage, (3) inherent lime/gypsum extent in the subsoil, (4) clay content, and (5) loading of additional soluble salts (Horneck et al. 2007). Several on-site approaches have been recommended for removal of excess amount of soluble salts or exchangeable Na^+ from SAS. In India, about 1 Mha of SAS has been recovered using some hydro-chemical technology and achieved food production of INR 10 billion per annum (Minhas and Sharma 2003). Table 9.4 summarizes benefits and drawbacks of various traditional physicochemical amelioration methods for SAS.

Table 9.4 Summary of various traditional approaches tried for reclamation of SAS

Sr.	Method	Mechanism	Soil properties improved	Cost of process (in INR, unless otherwise stated)	Advantages	Drawbacks	Efficiency	Reference
1.	Excavation	Replace affected soils with fertile soil	–	1,35,000–1,65,000 ha^{-1}	Simple Efficient on small scale	Requires heavy-duty machines for transportation Intensive Cost-intensive	–	Sorvari et al. (2009) and Wirthensohn et al. (2009)
2.	Leaching	Application of low salt water to the soil in order to dissolve soluble salts and move them below the root zone	Salinity	2.95 (for 10,000 L) (as of 1991)	Approach suits for low to moderately saline soils	Effective only when low moisture and deep ground water table exists Effective with gypsiferous soils and needs adequate drainage in gypsum layer Requires deionized water	Up to 80%	Qadir et al. (2001, 2007) and Khosla et al. (1979) and Sangal (1991)
3.	Electrokinetic	Electrical gradient applied across electrodes installed in soils and electrolysis, produces H^+ and OH^- ions, respectively	EC and Na^+	5400–7800 ton^{-1}	Suitable for removal of salts as charged ions May lower up to 60% EC in 96 h and Na^+ in 10 days	Failed at field level Generates strong oxidizing harmful compounds Removes essential ions along with undesirable ions Causes environmental pollution	Reduction in EC by 64–97% in 1–2 weeks	Cho et al. (2009), Trombly (1994) and Lee et al. (2012)

4.	Drainage	Prevents waterlogging and removal of excess water beyond rhizosphere with surface or sub-surface drainage systems	Salinity, hydraulic conductivity	1250–1500 ha^{-1}, surface drainage 65,000 ha^{-1}, sub-surface drainage	Efficiently removes excess water with reduction in salinity and alkalinity Increases hydraulic conductivity	Unaffordable to poor/marginal farmers Skill oriented and requires proper guidance	Reduction in EC by 29%	Pandey (2009) and Grismer (1990)
5.	Application of gypsum	Provides Ca^{2+} pool and prevents Na^+ exchange sites	Hydraulic conductivity, soil structure	474 ton^{-1} of gypsum + leaching cost	Improves soil properties Improves hydraulic conductivity Stops dispersion of clay fraction Low cost Readily available as bulk or sack Moderately soluble Easy broadcast application Sustained release Sulfate in gypsum nontoxic to plants	No clear guidelines on application rates Will not reduce pH in alkaline-saline soils	Reduction in (a) Na^+ by 38–46%; (b) EC by 35–52%; (c) SAR by 52–78%; (d) ESP by 29–45%	Hajkowicz and Young (2002), Horneck et al. (2007) and Makoi and Verplancke (2010)

(continued)

Table 9.4 (continued)

Sr.	Method	Mechanism	Soil properties improved	Cost of process (in INR, unless otherwise stated)	Advantages	Drawbacks	Efficiency	Reference
6.	Application of compost/ manure	Humic acid chelates soluble salts and recruits biological activity, increase organic C content	Soil structure, hydraulic conductivity, microbial status	15,000–18,000 ton^{-1}	Increases total C, N, P Improves soil structure and permeability Reduction in surface evaporation Inhibition of salt accumulation in surface soils Increases soil respiration rate Recruitment of beneficial microbes for nutrient cycling Low input, effective agro-technological approach	Anaerobic condition may lead to odors and/or the development of toxic compounds Bag swelling and bursting Impact on plant growth due to reduced oxygen in the soil root zone Presence of phototoxic compounds Inferior quality compost contain excess amount of heavy metals and salts such as Pb, Cd, Cu, and Zn Pesticides or industrial effluents in compost may be phototoxic, nitrogen deficient decreasing crop yield	Reduction in (a) EC by 21–32%; (b) pH by 0.3 units; (c) ESP by 80%; (d) SAR by up to 86%	Murillo et al. (1995),William (2000),Unsal and Ok (2001),Liang et al. (2005) and Lakhdar et al. (2009)

7.	Application of biochar	High carbon residence time, improves physical, chemical, and biological properties of soil	Improves soil CEC, increased soil organic C, improved availability of macro- and micronutrients, decreases ESP/SAR	US$ 120–180 ton^{-1}	Enhanced physical properties by balancing water content and air porosity, stabilization of soil structure, increased retention of polyvalent cations, and replacement of Na, improves soil microbial activity, may hasten salt leaching	Excess application leads to increase in salinity/sodicity High cost associated with production and transport Not all types applicable to particular soil likewise not a single type applicable to all soils Most studies undertaken in laboratory or greenhouse setting Need to undertake cost/benefit analysis in SAS		Blackwell et al. (2009) and Dahlawi et al. (2018)
8.	Cyanobacteria	Nitrogen-fixing cyanobacterial application reduces salinity by possible entrapment of Na^+ in extracellular mucopclysaccharide sheaths	Available C and N, Na^+	Less than 1 US$+ labor cost (for 35 kg N/ha) (US$ rate as of 1989)	Improves carbon and nitrogen status of soil Temporarily immobilizes the deleterious Na^+-reducing salinity	Most fixed nitrogen confined as biomass Permanent removal of salt requires removal of cyanobacterial mats Not feasible on a large scale nor rewarding since such a step removes much of newly fixed carbon and nitrogen Does not provide permanent solution	26–38% reduction in salinity	Apte and Thomas (1997) and Singh and Singh (1989)

(continued)

Table 9.4 (continued)

Sr.	Method	Mechanism	Soil properties improved	Cost of process (in INR, unless otherwise stated)	Advantages	Drawbacks	Efficiency	Reference
9.	Application of calcium chloride and calcium carbonate	Provides Ca^{2+} pool and prevents Na^+ exchange sites	Soil structure, hydraulic conductivity	7500–18,000 ton^{-1} $CaCl_2$ 15,000–15,500 ton^{-1} $CaCo_3$	Can be obtained as industrial waste product High solubility give high electrolyte levels and high water intake rates More efficient than gypsum in high ESP soils	Expensive than gypsum with time; however, slow dissolution of gypsum may be more significant than $CaCl_2$ $CaCO_3$ insoluble in alkaline soils Adding $CaCO_3$ adds nothing to calcareous soils in terms of increasing Ca^{2+} levels Application of excess $CaCl_2$ will increase salinity		Shainberg et al. (1982) and Walworth (2012)

10.	Application of sulfur and sulfuric acid	Oxidation of S to H_2SO_4 by soil microorganisms helps alkaline soils by reducing pH, supplies sulfates to plants, mends soil structure, and increases availability of certain plant nutrients like phosphorous, iron, manganese, zinc, etc.	Salinity, pH, hydraulic conductivity, availability of nutrients	11,400–23,400 ton^{-1} H_2SO_4 7200–30,000 ton^{-1} Elemental S	More efficient in correcting ESP than gypsum More effective in correcting pH than gypsum Helps in dissolution of $CaCO_3$ and reduction in salinity/sodicity Improves soil hydraulic conductivity Helps crop to take up more iron, manganese, zinc, and phosphates	Expensive than gypsum Elemental sulfur is insoluble in water Requires to undergo microbial oxidation, hence slow process May prove largely ineffective unless soil is leached periodically	Reduction in pH to 7.6 from 8.5	López-Aguirre et al. (2007) and Sadiq et al. (2007)

(continued)

Table 9.4 (continued)

Sr.	Method	Mechanism	Soil properties improved	Cost of process (in INR, unless otherwise stated)	Advantages	Drawbacks	Efficiency	Reference
11.	Development of transgenic varieties	Development of the transgenic cultivars by careful selection, insertion, and expression of desired genes			Has provided extensive number of tolerant varieties Probably best approach to obtain superior lines bearing abiotic stress tolerance	Requires understanding of cellular processes and genes Stability of insert in plants depends on many factors Needed more information to understand and manipulate complex quantitative traits Reproductive barrier and narrow genetic variations in food crops Majority of engineered salinity-tolerant plants tested in laboratory/greenhouse conditions, remain unproven in field Transformation of monocots other than rice still not routing Stress-related metabolic phenomenon not well understood Large number of transgenic lines obtained but none has been released as field-tested salt-tolerant variety except 1 or 2	–	Yamaguchi and Blumwald (2005), Ashraf and Akram (2009), Xoconostle-Cazares et al. (2010) and Turan et al. (2012)

In the last century, many conventional physicochemical approaches involving chemical amendments, irrigation and drainage, tillage operations, water-related approaches, and electrical currents are explored, but they proved to be uneconomic to redeem SAS. These conventional methods appeared to be (1) inefficient to remove exchangeable Na^+ from soils, (2) expensive, (3) unreliable, (4) non-sustainable, and (5) publicly unacceptable (Pandey et al. 2005). Whereas, biological methods using living or dead organic matter (1) improved soil structure, (2) facilitated salt leaching, (3) reduced surface evaporation, and (4) inhibited salt accumulation in the surface layers but are proved to be cost-intensive and offered temporary solution to recoup SAS. Although traditional breeding and use of transgenic plants appeared as a better alternative, the cost of genomic development is enormous and unaffordable to marginal farmers.

11 Why Alternative System for Salt-Affected Soils?

Besides the adoption of various strategies, the treatment of SAS in poor or developing countries is slow simply because (1) large areas of SAS are owned by subsistence marginal or resource poor farmers, (2) meager or slow efforts realized by the government, (3) chemical amendment cost to nearly 60%, (4) benefit/cost ratio ranged from 1.29 to 1.42 with payback period of 2–4 years, (5) no access to assured quality irrigation water, (6) lack of lab to land contingent program for landless or marginal farmers, (7) lack of technical know-how guidelines to the farmers, and (8) lack of optimization of chemical amendments to obtain sustained release of nutrient in synchrony with the plant needs. Hence, most acceptable environment-friendly strategy for the treatment of SAS is desirable to be put for effective crop cultivation purpose. At present, detailed information pertinent to effects of SAS on the physical and chemical properties of soil and plant growth is available (Levy 2000; Sardinha et al. 2003), but almost scarce effort about application of microbial systems to such problem soils is observed (Yuan et al. 2007). It is evident that soil health is highly dependent on organic matter and microbial nutrient cycling in the soil.

The comparative cost analysis for wheat crop regarding required nitrogen (46 kg) from urea (INR 4715 ton$-^1$), composted manure (INR 3000 ton$-^1$), and Twin N (Meritt 2006) diazotrophic bacteria (at US$ 15–60/per vial where 1 vial treats 5 ha) for target production of 4 ton per hectares containing 13% protein revealed some interesting findings such as (1) 100% N efficiency to plants using bacteria, 65% using compost, and 50% using urea, (2) higher cost effectiveness for N supply with bacteria than compost and urea, (3) low cost nitrogen application with bacteria (US$ 7 ha^{-1}; INR 350 ha^{-1}) compared to urea (INR 700 ha^{-1} for 100 kg) and compost (INR 3000 ha^{-1}), (4) long-term sustainability with microbial system but average in urea and compost application, and (5) no loss of nitrogen through leaching in bacteria case but high with compost and urea. The critical comparative analysis, thus, indicated the valuable potential of microbial system for sustainable agricultural productivity.

Hence, most pragmatic alternative is to select the microbial species with desirable attributes from natural habitats so that SAS can be made suitable for cultivation of agricultural crops. This may be probably because (1) certain microbes pursue unusual metabolic apparatus to operate in edaphic conditions, (2) microbial entity plays a pivotal role in nutrient cycling, (3) incredible diversity of microbes permits them to inhabit hostile ecological habitats, (4) microbial system offers eco-friendly and economically sustainable solution and (5) simple growth requirements, and (6) GRAS microbes enjoy wide public acceptance.

12 The Nature of the Problem and Solution to Salt-Affected Soils

Generally, alkaline-saline soils are deficient in nitrogen, and hence, N input to SAS is the prime necessity (Sprent and Sprent 1990). The better source of N input in alkali-saline soils can be achieved with the augmentation of N_2-fixing diazotrophic bacteria. Ehrlich et al. (2001) indicated that microbial interactions with crop plants are peremptory for nutrient cycling on this planet. But most potent legume *Rhizobium* symbiosis is highly sensitive under alkali-saline conditions (Shamseldin and Werner 2005) and shows decrease in nodule respiration, leghemoglobin secretion, and nitrogenase activity. Hence, bespoke microbes required to abate alkali-saline stressed condition must have the capacity to (1) perform efficiently, (2) display robustness, (3) effectively remove exchangeable Na^+, (4) exhibit haloalkalitolerant diazotroph, and (5) secrete array of metabolites to boost plant growth. Toward this end, limited ecological niches such as mangrove and soda solonchak soils have been investigated to isolate free-living diazotrophic bacteria such as *Bacillus arsenicoselenatis* (Blum et al. 1998), *Amphibacillus tropicus* (Zhilina et al. 2001), *Clostridium alkalicellum* (Zhilina et al. 2005), *Geoalkalibacter ferrihydriticus* (Zavarzina et al. 2006), and *Phyllobacterium* spp. (Thatoi et al. 2012). However, these microbial strains remain to be explored in relevance to plant growth promotion.

13 Microbial System for Bioaugmentation of Salt-Affected Soils

The significance of microbial system in biogeochemistry is known since long (Conrad 1996), and 80–90% of the reactions in the soil involve microbial systems which contribute to nutrient cycling, soil organic matter, soil structure, and plant growth promotion (Kennedy 1999). Interest in application of beneficial microbes as a biofertilizer is known for 60 years as a most pragmatic, cost-effective, eco-friendly, and sustainable approach to agricultural productivity (Tilak et al. 2005). Plant-microbe interactions in the rhizosphere is known since long for (1) nutrient acquisition, (2) metal tolerance, (3) phytopathogen suppression, and (4) ecological fitness of the plants (Glick 1995) through mucilage production to alter physical soil aggregate formation (Forster 1990).

Beneficial microbes in the rhizosphere area mainly contain PGPR and arbuscular mycorrhizal fungi (AMF). But SAS inflicts on microbial communities near the surface of the soil eventually and lowers biomass and subsequent carbon and nitrogen mineralization (Yuan et al. 2007; Paul and Nair 2008). On the contrary, several microbes are reported to exert physiological challenges in the extreme haloalkaline conditions, and hence, investigation of such haloalkalitolerant microbes from ecological niche may offer an alternative solution for crop cultivation in SAS. Toward this end, Nautiyal et al. (2000) notified that bacteria tolerant to edaphic conditions may enhance crop productivity in semiarid and arid regions of the world. Salt-impacted soils impose restrictions on plant growth through (1) reduced water potential, (2) smaller leaf size, (3) altered cellular metabolism, and (4) elevated stress hormone (ethylene) to inhibit plant growth. Most crop plants are sensitive and are unable to survive under salt-impacted stress (Siddiqui et al. 2009).

Hence, the major challenge is to improve the growth of plants under salt-stressed soil. The tolerance against salt-impacted stress among plants may be improved with salt-tolerant alkaliphilic microbial augmentation to withstand ion homeostasis, water deficit/homeostasis, and phytohormone and nitrogen requirement. The initial efforts of Mayak et al. (2004) insinuated application of halotolerant bacteria adapted to rhizospheric living conditions as a key to crop productivity in SAS. These tolerant halophytic bacteria can (1) withstand the high osmotic strength (low water potential) and ion toxicity, (2) facilitate sustained availability of plant nutrients, and (3) protect plants from phytopathogenic fungal challenges (Lugtenberg et al. 2001; Egamberdieva 2012). Zahran et al. (1995) demonstrated 20 bacterial strains out of 400 isolates from saline soils (5% NaCl) which exhibited nitrogenase, cellulolytic, amylolytic, and pectinolytic activity for maintaining soil fertility and productivity.

Some haloalkalitolerant PGPR are free-living diazotrophic microbes with attributes of persistence, survival, and competition under stressed salt and alkali conditions. Such microbes may cause desirable engineering for plant growth in three ways, viz., (1) enviable physicochemical environment of root zone, (2) supportive nutrient cycling in the root zone, and (3) PGP effect (Lugtenberg et al. 2001; Egamberdieva 2012).

The haloalkaliphilic microbial strains for cultivation of crops in SAS may bestow an array of metabolic apparatus including (1) nitrogen fixation, (2) phosphate solubilization, (3) ammonia production, (4) release of phytohormone (indole-3-acetic acid or IAA), (5) biosurfactant production (Mulligan 2009), (6) siderophore secretion (Deepa et al. 2010), (7) role in biogeochemical cycles for soil nutrients and micronutrients, (8) resistance against antibiotics (streptomycin 100 μgL^{-1}) and certain heavy metals for outcompetitiveness (Fernando et al. 2006), (9) removal of xenobiotics, and (10) denitrification. These halotolerant rhizobacteria can lower ethylene in plant tissue and can improve plant grown in SAS. The studies on halotolerant PGPR w.r.t. physiology as well as properties in SAS are scanty, and its field application is still a challenging task (Egamberdieva et al. 2008). A very few reviews recently focused contribution of microbial PGPR to recoup SAS condition for better crop productivity (Grover et al. 2011; Egamberdieva 2012; Dodd and Pérez-Alfocea 2012; de-Bashan et al. 2012; Kim et al. 2012). These reports have indicated a

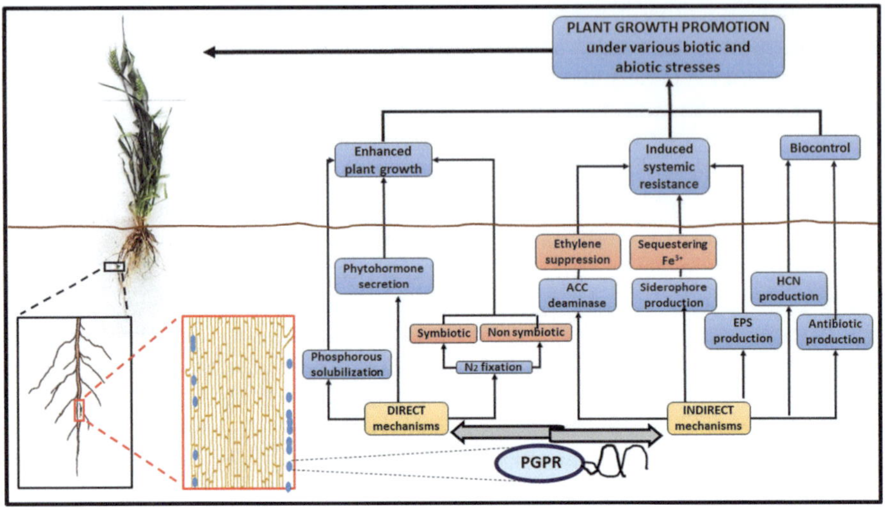

Fig. 9.2 Variety of mechanisms adopted by bacteria for promotion of plant growth

variety of mechanisms in halotolerant bacteria including production of plant growth regulators, siderophores, HCN, NH_3, antibiotics, and diazotrophy to promote plant growth in SAS (Fig. 9.2).

14 Activities and Significance of Haloalkalitolerant PGPR

Survival of beneficial halophytic microbes in SAS is dependent on (1) adaptation to nutrient deficiency, (2) efficient utilization of plant root exudates, (3) specific plant-microbe interactions, and (4) availability of empty niche to outcompete indigenous microflora (Bull et al. 1991; Rekha et al. 2007). Egamberdieva (2012) showed that bacterial endophytes survived in the roots and persist in salinized soil condition.

Shelar et al. (2010) observed haloalkalitolerant *Rheinheimera* spp. L6 to reduce acetylene to ethylene, suggesting nonsymbiotic nitrogenase activity in alkaline saline condition. Further, Shelar (2014) established successful PGP activity of the strain for wheat attributing to traits such as nitrogen fixation, phosphate solubilization, and production of siderophore, hydrolases (catalase, oxidase, protease), etc. Upadhyay et al. (2011) evaluated activity of 11 PGPR isolates activity in SAS (80 gL^{-1} NaCl) at pot level revealed production of auxin, proline, phosphate solubilization, and increased wheat biomass after inoculation.

14.1 Phytohormone Secretion

Microbial system promotes plant growth through mobilization of nutrients and secretion of phytohormones (auxin) such as gibberellin, abscisic acid (ABA), or

IAA (Lugtenberg et al. 2001; Patten and Glick 2002). Munns (2002) and Albacete et al. (2008) confirmed auxin activity during plant growth under salinity. About 80% of the rhizobacteria synthesize IAA, and most of the rhizosphere bacteria utilize tryptophan present in plant root exudate as a precursor (Spaepen et al. 2009; Dodd et al. 2010). These phytohormones modulate physiological processes such as root cell growth and tissue differentiation and boost rhizospheric microbes (Patten and Glick 2002; van Loon 2007). Of the phytohormones, IAA is implicated in (1) orientation of root and shoot growth, (2) apical dominance, (3) initiation of lateral and advantageous roots, (4) differentiation of plant xylem and phloem tissue, (5) elongation of stems and roots, and (6) seed and tuber germination (Glick 1995). Halotolerant *P. aurantiaca* TSAU22 and *P. extremorientalis* TSAU20 showed 79% germination through secretion of IAA (Egamberdieva 2009).

Patel et al. (2012) found that 10 out of 13 isolates produce IAA up to 10–26 $\mu g\ ml^{-1}$ after 72 h. Similarly, Hasnain and Sabri (1997) noticed that wheat inoculated with *Pseudomonas* spp. enhanced growth in salt-impacted condition by minimizing plant uptake of toxic compounds and secreting more phytohormone auxin. In several metal phytoremediation studies, PGP bacteria demonstrated the secretion of IAA (Glick 2010). These research findings substantiate better suitability of halotolerant PGPR to alleviate salt-impacted stress through phytohormone secretion or better yield of wheat in SAS (Egamberdieva et al. 2013). Yang et al. (2009) proposed PGPR for induced systemic tolerance (IST) through phytohormone secretion in salt-impacted condition for root stimulation to confer plant defense against phytopathogens, while Nabti et al. (2010) hypothesized that IAA production in *Azospirillum brasilense* NH contributes to increased salt tolerance in wheat plants.

14.2 ACC Deaminase to Alleviate Ethylene Stress

The plant growth in SAS is inhibited mainly due to toxic effects of Na^+, slows down the water uptake of plants, decreases germination, and delays seedling emergence (Jamil et al. 2006) and release of gaseous stress hormone known as ethylene (Mayak et al. 2004). The plant ethylene controls a spectrum of plant developmental processes including root hair formation, root growth, flower wilting, fruit ripening, leaf abscission, seed germination, and leaf senescence (Dugardeyn and van der Straeten 2008; Dodd et al. 2010). Several root-associated PGP bacteria express pyridoxal 5-phosphate (PLP)-dependent ACC deaminase to hydrolyze ethylene precursor 1-amino cyclopropane-1-carboxylate (ACC) to α-keto butyrate and ammonia (Glick et al. 1998) and eventually ensured that the ethylene level does not exceed the threshold limit to cause impairment of plant growth (Wang et al. 2000; Glick et al. 2007). Mayak et al. (2004) suggested that ACC deaminase of a PGPR, namely, *Achromobacter piechaudii* ARV8, conferred IST in tomato and pepper grown at 172 mM NaCl. Recently, Jha et al. (2012) demonstrated ACC deaminase activity in several PGPR strains, viz., *Zhihengliuella* spp., *Brevibacterium casei*, *Haererehalobacter* spp., and *Halomonas* spp. isolated from halophyte roots of *Salicornia* spp. grown in coastal saline region. ACC deaminase gene (*acd*S) of each strain showed 77–83% similarities with *acd*S from *Pseudomonas* spp. CH-GRS, *Acidovorax facilis*, and *Phyllobacterium brassicacearum*. Saravanakumar and Samiyappan (2007) reported ACC deaminase-positive *P. fluorescens* TDK1 for growth of groundnut sporophyte under salt stress vis-à-vis control treatments. Greenberg et al. (2008) confirmed the PGP activity of PGPR strains in salt-impacted soils.

Salt-impacted conditions can decrease shoot growth and root, xylem, and root cytokinins of tomato (Dodd and Perez-Alfocea 2012). However, augmenting root-to-shoot cytokinin signaling by PGPR *Bacillus subtilis* (Arkhipova et al. 2007) increases zeatin riboside with 30% rise in fruit yield. Although cytokinin production in PGPR is a common trait, careful selection of potential microbial system able to grow in salt-affected condition is highly required for crop productivity (Dodd et al. 2010).

14.3 Exopolysaccharide Secretion

Certain microbes modify physical barriers around the roots through secreting exopolysaccharides (EPS) under salinity-stressed conditions. EPS may bind to soil particles to form micro- or macro-aggregates and, thus, facilitate plant roots and fungal hyphal network to impregnate in the soil pores (Grover et al. 2011; Upadhyay et al. 2011). In this context, Sandhya et al. (2009) showed that plants treated with EPS-producing microbes increased plant resistance against water homeostasis possibly through improved conductance of soil structure. Besides, microbial EPS binds to cations including Na^+, thereby excluding the excess Na^+ ion toxicity of the plants (Grover et al. 2011; Upadhyay et al. 2011). Thus, PGPR articulate physical barrier around the plant roots through exopolysaccharide production which preferentially binds to Na^+ in roots and prevents plant Na^+ accumulation (Ashraf et al. 2004; Qurashi and Sabri 2012). Microbes can modify root uptake of toxic ions and nutrients through modification in host physiology or reduce foliar accumulation of Na^+ and Cl^- ions. *B. subtilis* mediate *in radix* processes in *Arabidopsis* to alter plant ionic status through the release of organic compound to exclude Na^+.

Root colonization by PGPR induces accumulation of osmolyte for osmotic adjustment under salinity stress (Chen et al. 2007). Enhanced accumulation of proline in plants against salinity stress was observed among *Burkholderia* spp. (Barka et al. 2006), *Arthrobacter* spp., and *Bacillus* spp. (Sziderics et al. 2007). Similarly, co-inoculation of *Rhizobium* and *Pseudomonas* showed increased tolerance toward salt in *Zea mays* due to (1) higher proline turnover, (2) decreased electrolyte leakage, and (3) selective uptake of K^+. But rhizobia secrete trehalose that imparts adaptation toward abiotic stress in leguminous plants (Suárez et al. 2008) and increases plant growth and N content (Figueiredo et al. 2008). Trehalose metabolism in certain microbes showed key role as an osmoprotectant and stabilizes enzyme and membrane and protects biological structure from desiccation damage (Dodd and Perez-Alfocea 2012).

14.4 Siderophore Secretion

Iron is the fourth major requirement to plants and microbes, but it is unavailable for direct assimilation because ferric ion is (1) predominant in nature, (2) sparingly soluble (10^{-18} M), and (3) scanty to support plant growth. Hence, plant and microbes sequester biologically available iron through secretion of siderophores in iron-stressed condition (Neilands 1995). Bacteria synthesize low molecular weight siderophores, hydroxamate or catecholate molecules, to bind with high affinity to ferric (Fe^{3+}) ion and make available soluble iron from soil to the plant. Plant siderophores have low affinity compared to bacterial siderophores, and hence, plant satisfies iron requirements through microbial colonization on root surfaces for uptake of iron-bacterial siderophore complexes (Wang et al. 1993). Kohler et al. (2006) reported that certain PGPR secrete siderophores that may be more important to the plant under salinity stress. *Pseudomonas* spp., a most promising potential PGPR strain, secrete a wide variety of siderophores that confer selective advantage such as rapid colonization in plant roots of several crops and biocontrol against several plant diseases. The siderophore-based suppression is based on competition for iron with fungal pathogen. Besides, siderophores are thought to induce ISR (Baker et al. 2003). Patel et al. (2012) demonstrated that 13 PGPR isolates secreted hydroxamate siderophores in the presence of NaCl (11.5 mM to 1 M).

14.5 Diazotrophic Interaction: An Essential Component to Outcompete in the Stressed Environment

Microbes may alter some of the biochemical pathways and biochemical activities like nitrogen cycle under stressed conditions of salt and alkaline pH. The rate of mineralization and immobilization of nitrogen as well as nitrification and ammonification (Wollenweber and Zechmeister-Boltenstern 1989) is decreased significantly in saline and alkaline soils. The nodulation in the clover inoculated with *Rhizobium trifolii* TA1 in SAS revealed that the native isolates outcompete the time-tested inoculant (Denton et al. 2003). For this purpose, Cummings (2005) proposed appropriate selection of "bespoke" inoculants with competence to edaphic conditions for sustainable activity in specific soil environments. The SAS habitats are N deficient (Sprent and Sprent 1990); therefore, the search for free-living diazotrophic haloalkalitolerant microbes is a prerequisite to enable cultivation of crops in SAS. Valiela et al. (1976) observed increased production of crops in salt-affected habitats due to increasing supply of N. Further, Barua et al. (2011) suggested bacterial nitrogen fixation as an alternative of N input in saline habitats. The higher rates of N_2 fixation may be favored in stressed condition possibly due to low oxygen tension in saline-alkaline soils (Wollenweber and Zechmeister-Boltenstern 1989). Several N_2-fixing diazotrophic haloalkalitolerant microbial species with efficient nitrogen fixation potential at higher levels of salt (15–20%) and at pH 9.0 have been delineated. Thus, haloalkalitolerant microbial species isolated from saline-alkaline habitats may serve as a bespoke bioinoculant for sustainable agriculture.

The reports on halotolerant diazotrophic PGPR are still limited, and a very few salt-tolerant PGPR strains such as *Azospirillum halopreferens* (Kallar grass) (Reinhold et al. 1987), *A. brasilense* strain NH (Nabti et al. 2007), and *Swaminathania salitolerans* (Loganathan and Nair 2004) are reported. Sorokin et al. (2008) systematically isolated Gram-positive spore-bearing *Amphibacillus tropicus* from soda solonchak soil of southwestern Siberia (Russia) and reported active diazotrophic activity even at pH 10 and 2–3 M total Na^+. In another study, Sorokin et al. (2008) isolated a novel Gram-positive anaerobic, diazotrophic, haloalkaliphile, and highly chloride-independent sodaphile, namely, *Nantronobacillus azotifigens* from soda-rich habitats capable of nitrogen fixation up to 16–80 nmol N (h^{-1}) at pH 9.5–10 and 0.5–1.5 M Na^+. Recently, Jha et al. (2012) isolated Gram-positive halotolerant diazotrophs (*Brachybacterium*, *Brevibacterium*, *Zhihengliuella*) and Gram-negative halotolerant diazotrophs (*Haererehalobacter*, *Halomonas*, *Rhizobium radiobacter*) from the roots of halophytic *Salicornia brachiata*. Each isolate showed NaCl salt tolerance (3–4% w/v), PGPR activity (IAA, ACC deaminase production), phosphate solubilization, and siderophore production.

Similarly, Ravikumar et al. (2004) isolated *Azotobacter* spp. from mangrove region from southern India and identified N_2 fixation and in vitro production of IAA at 3% NaCl salinity. Azobacterization of *Rhizophora* seedling showed (1) 98.2% root biomass, (2) 277.86% leaf area, and (3) 158.75% carotenoids, suggesting suitability for vigorous cultivation of mangroves in coastal wetlands.

Table 9.5 summarized certain microbes examined both at laboratory and in the field for performance under salt-impacted conditions.

Similarly, Sahin et al. (2011) systematically explored possibility of two bacterial strains (*B. subtilis* OSU-142 and *B. megaterium* M3) and three fungal isolates (*Aspergillus* spp. FS9, 11 and *Alternaria* spp. FS8) with PGPR activity in combination with gypsum and alone using column experiment for the treatment of four types of SAS (EC 5.3–42.8 dSm^{-1}). The saturated hydraulic conductivities (Ks) of soil columns increased significantly ($p < 0.01$) after leaching water treatments with microbes, and K values were 1.3, 3.4, 2.4, and 1.6 times more than treatments without microbes for experimental soils I, II, III, and IV, respectively. The increase in K values was possibly due to organic acids secreted by microorganisms and, thus, indicated the potential of microorganisms to improve water movement through saline-sodic soils.

15 Conclusion

Alkaline and saline soils are worldwide environmental problem with 30% loss of fertile soils expected globally within next 25 years. Saline and alkaline soils are known to (1) induce osmotic stress, (2) reduce plant growth, and (3) lower crop productivity in irrigated areas of arid and semiarid regions of the world including India. The high levels of Na^+ and pH along with low microbial activity are key parameters for lower or almost no crop productivity in SAS (Peinemann et al. 2005).

Table 9.5 Some bespcke microbes explored for plant growth-promoting properties in SAS

Sr. no.	Organism	Plant growth-promoting traits	Plant used	SAS used with composition	Pot assay	Field trial	Removal/reduction/consequence	Reference
1.	*Anabaena torulosa*	Nitrogen fixation, extracellular mucopolysaccharide	Rice	pH, 8.3; EC, >15 dSm^{-1}; organic C, 0.72%; available P, 84.22 $kgha^{-1}$; available K, 1064 $kgha^{-1}$	Yes	Yes	EC reduction, 21–34% (laboratory conditions) EC reduction 12–36% (field conditions)	Apte and Thomas (1997)
2.	*A. hydrophila*	EPS production	Wheat	pH, 8.2; EC, 8.0 dSm^{-1}, organic C, 0.29%; SAR, 11.7	Yes	Yes	Decreased Na^+ uptake by roots	Ashraf et al. (2004)
3.	*Bacillus insolitus*	EPS production	Wheat	pH, 8.2; EC, 8.0 dSm^{-1}, organic C, 0.29%; SAR, 11.7	Yes	Yes	Decreased Na^+ uptake by roots	
4.	*Bacillus* spp.	EPS production	Wheat	pH, 8.2; EC, 8.0 dSm^{-1}, organic C, 0.29%; SAR, 11.7	Yes	Yes	Decreased Na^+ uptake by roots	
5.	*Azospirillum brasilense* NH (*Ulva lactuca* extracts as osmoprotectants)	Nitrogen fixation, IAA production	*Triticum durum* var. *waha*	pH, 6.4; salinity, 0.85%; organic matter, 31%; total N, 19.48%; mineral N, 3.77%; P, 5.2%; K, 13%; water capacity, 25%.	Yes	Yes	Osmoprotection	Nabti et al. (2007)
6.	*Pseudomonas fluorescens*	ACC deaminase activity	Groundnut (*Arachis hypogaea* L.)	pH 9–11	Yes	Yes	ISR	Saravanakumar and Samiyappan (2007)

(continued)

Table 9.5 (continued)

Sr. no.	Organism	Plant growth-promoting traits	Plant used	SAS used with composition	Pot assay	Field trial	Removal/reduction/consequence	Reference
7.	*Pseudomonas* spp.	ACC deaminase	Barley, oats, tall fescue, tall wheatgrass	Site 1: EC 17.6 dSm^{-1} Site 2: EC 6.5 dSm^{-1} Site 3: EC 23.5 dSm^{-1}	Yes	Yes	Salinity reduction by 30% and 60% at Site 1 and 2, respectively. No change in Site 3	Wu (2009)
8.	*P. extremerientalis* TSAU6	Auxin production	Wheat	pH 8.0; EC: 6.2 dSm^{-1}; organic matter 0.79%	Yes	–	ISR	Egamberdieva (2009)
9.	*P. aurantiaca* TSAU22	ACC deaminase activity, auxin production	Wheat	pH 8.0; EC: 6.2 dSm^{-1}; organic matter 0.79%	Yes	–	ISR	
10.	*S. plymuthica*, *S. rhizophila*, *P. fluorescens*	HCN production, ACC deaminase activity, IAA production	Cucumber	EC, 6.59 dSm^{-1}, pH 8.0; organic matter 0.694%	Yes	–	ISR	Egamberdieva et al. (2011)
11.	*Azospirillum brasilense*	Nitrogen fixation	Sugarcane variety 8071	Salinity 4%	Yes	–	Increase in plant height, number of leaves, and biomass	Bapurao (2012)
12.	*Azotobacter chroococcum*	Nitrogen fixation	Sugarcane variety 8071	Salinity 4%	Yes	–	Increase in plant height, number of leaves, and biomass	Nakade et al. (2012)
13.	*Planococcus rifietoensis*	P solubilization, Zn solubilization, ACC deaminase activity, IAA production	Wheat	NS	Yes	Yes	ISR, increase in (a) plant height, (b) shoot fresh weight, (c) shoot dry weight, (d) root length, (e) root area, and (f) root dry weight	Rajput et al. (2013)

14.	*Halobacillus* spp.	IAA, HCN, NH_3 production, P solubilization, siderophore production, antimicrobial activity	*Sesuvium portulacastrum*	NS	Yes	–	Increased root length and shoot length	Desale et al. (2013)
15.	*Halomonas* spp.	IAA, HCN, NH_3 production, P solubilization, iron chelation	*Sesuvium portulacastrum*	NS	Yes	–	Increased root length and shoot length	
16.	*P. aurantiaca* TSAU20	Auxin production	Milk thistle (*Silybum marianum*)		Yes	–		Egamberdieva et al. (2013)
17.	*Halomonas* spp. (RAP3)	Nitrogen fixation, IAA production, ACC deaminase activity, siderophore production, EPS/biosurfactant production	Wheat (Tapovan variety)	pH 9.5; EC 5.15 dSm^{-1}; ESP, 20.2; SAR, 2.5	Yes	–	Reduction in (a) EC by 12–16%; (b) pH by 1.1–1.3 units; (c) ESP by up to 15%; (d) SAR by 28–36%	Shelar R. D (2014)
18.	*Rheinheimera* spp. (RAL6)	Nitrogen fixation, IAA production, ACC deaminase activity, Siderophore production, EPS/biosurfactant production	Wheat (Tapovan variety)	pH 9.5; EC, 5.15 dSm^{-1}; ESP, 20.2; SAR, 2.5	Yes	–	Reduction in (a) EC by 12–16%; (b) pH by 1.1–1.3 units; (c) ESP by up to 15%; (d) SAR by 28–36%	Shelar R.D. (2014)

(continued)

Table 9.5 (continued)

Sr. no.	Organism	Plant growth-promoting traits	Plant used	SAS used with composition	Pot assay	Field trial	Removal/reduction/ consequence	Reference
19.	*Bacillus licheniformis* SKU 3	EPS production, IAA production	*Triticum aestivum* L.	pH, 8.5 ± 1.03; EC (dSm^{-1}), 8.7 ± 1.12; Na^+ (mg kg^{-1}), 24.5 ± 2.21; K^+ (mg kg^{-1}), 2.3 ± 0.12; organic C (g kg^{-1}), 0.6 ± 0.2; total N (g kg^{-1}), 0.3 ± 0.1; mineral N (μg g^{-1}), 14.2 ± 1.23; silt (g kg^{-1}), 284 ± 9.5; clay (g kg^{-1}), 129 ± 10.3; sand (g kg^{-1}), 587 ± 71.2	Yes	–	Total dry weight increased by 50–230% (21d) and 73–269% (41d) under salinity stress compared to the control treatments The shoot weight increased by 12–202% (21d) and 7–226% (41d) under saline condition compared to the control treatments Plant K^+ uptake decreased by 1.2–11.8% and Na^+ uptake decreased by 2.5–6.3% over the control under both non-saline and saline conditions	Upadhyay et al. (2011)
	Bacillus pumilus SKU 4	EPS production, IAA production, P solubilization						
	Bacillus sp. SKU 5	EPS production, IAA production, P solubilization						
	Burkholderia cepacia SKU 6	EPS production, IAA production, P solubilization						
	Enterobacter sp. SKU 7	EPS production, IAA production, P solubilization						
	Enterobacter sp. SKU 8	EPS production, IAA production, P solubilization						
	Microbacterium sp. SKU 9	EPS production, IAA production						
	Paenibacillus macerans SKU 10	EPS production, IAA production						
	Paenibacillus sp. SKU 11	EPS production, IAA production, P solubilization						
	Bacillus coagulans SKU 12	EPS production, IAA production, P solubilization						
	Bacillus insolitus SKU 13	EPS production, IAA production, P solubilization						

20.	*Halomonas variabilis* (HT1)	EPS production, biofilm formation	*Cicer arietinum* var. CM-98	Varying salt concentration (up to 200 mM NaCl)	Yes	No	Increased soil aggregation around roots by 808% compared to non-inoculated control at 100 mM NaCl	Qurashi and Sabri (2012)
	Planococcus rifietoensis (RT4)	EPS production, biofilm formation					Increased soil aggregation around roots by 666% compared to non-inoculated control at 100 mM NaCl	
21.	*Bacillus licheniformis* strain A2	P solubilization, IAA production, siderophore production, ammonia production, antifungal activity	*Arachis hypogaea* (ground nut)	pH 6.8; sodium (as Na^+) 0.038%; potassium (as K^+) 0.051%; phosphate as P_2O_5 (mg kg^{-1} of soil) 14.6%; SO_4 (mg kg^{-1} of soil) 0.32%; magnesium% (as Mg^+) 0.006%; EC 0.21 mS cm^{-1}; original soil salinity, equivalent to 32 mM NaCl; EC after 50 mM NaCl supplementation, 0.64 mS cm^{-1}; total salinity after 50 mM NaCl supplementation, equivalent to 82 mM NaCl	Yes	–	31% total plant length and 43% increase in fresh biomass in additional 50 mM NaCl	Goswami et al. (2014)

(continued)

Table 9.5 (continued)

Sr. no.	Organism	Plant growth-promoting traits	Plant used	SAS used with composition	Pot assay	Field trial	Removal/reduction/consequence	Reference
22.	*Klebsiella* sp. D5A	ACC deaminase activity, IAA production, P solubilization, siderophore production	*Festuca arundinacea* L. (tall fescue)	pH 9.7 (1: 2.5 water), electrical conductivity (1: 5) 404 μS cm^{-1}; CEC 4.94 cmol kg^{-1}; organic matter 4.26 g kg^{-1}; $(NH_4)_2SO_4$ at 250 mg N kg^{-1}; K_2HPO_4 100 mg P kg^{-1} and NH_4OAc extractable K 176 mg kg^{-1}. TPH (total petroleum hydrocarbon) concentration 16,920 mg kg^{-1}	Yes	–	Improved (a) germination rate, (b) shoot height, (c) shoot dry weight, (d) root length, (e) root dry weight (73%), (f) chlorophyll content, (g) root activity (100%), (h) TPH removal efficiency (16%)	Liu et al. (2014)
23.	*Enterobacter asburiae*, S-24	IAA production, ACC deaminase activity	*Triticum aestivum* L.	pH, 7.2; EC, 0.45; dS m^{-1} and organic matter, 0.65% (salt amended gradually up to 200 mM)	Yes	No	Increased (a) root length, (b) shoot length, (c) fresh seedling biomass, (d) dry seedling biomass, (e) number of tillers (at 150–200 mM NaCl stress)	Raheem and Ali (2015)
	Bacillus thuringiensis, S-26							
	Bacillus thuringiensis, S-50							
	Moraxella pluranimalium, S-29							
	Pseudomonas stutzeri, S-80							

24.	*Bacillus subtilis* EY2	NS	*Raphanus sativus* L. cv "Cherry Belle" (radish)	pH, 7.36, EC, 1.12 dS cm^{-1}, nitrogen (N), 12.9 mg/kg, phosphorus (P), 14.55 mg/kg, exchangeable potassium (K), 1.55 meq/100 g soil, organic matter, 4.5%. Supplemented with base complete nutrient solution (SoFertig). EC of SoFertig solutions were 1.86 dS m^{-1} for 0 mM NaCl and 12.14 dS m^{-1} for 100 mM NaCl	Yes	No	Greater emergence percentage and decreased mean emergence time, decreased electrolyte leakage, increased (a) shoot and root fresh weight, (b) shoot and root dry weight, (c) chlorophyll content	Yildrim et al. (2008)
	Bacillus atrophaeus EY6							
	Bacillus sphaericus GC subgrup B EY30							

(continued)

Table 9.5 (continued)

Sr. no.	Organism	Plant growth-promoting traits	Plant used	SAS used with composition	Pot assay	Field trial	Removal/reduction/consequence	Reference
25.	*S. plymuthica* RR2-5-10	IAA production, antifungal activity	Cucumber cv. Simbal	Organic matter, 0.694%; total C, 2.506%; total N, 0.091%; Ca, 63.5 g/kg; Mg, 20.7 g/kg; K, 6.2 g/kg; P, 1.2 g/kg; Cl, 0.1 g/kg; Na, 0.7 g/kg, pH 8.0, EC, 659 mS m^{-1} (for pot experiment)	Yes	No	Increased biocontrol of cucumber foot and root rot, increased dry weight	Egamberdieva et al. (2011)
	S. rhizophila e-p10	IAA production, ACC deaminase activity, antifungal activity		EC, 560 mS m^{-1}; organic matter, 2.4%; N, 0.1%; P, 1.34%; K, 7.1%; pH 7.8 (for greenhouse experiment)			Increased plant height and fruit yield	
	P. fluorescens SPB 2145	IAA production, antifungal activity						
	P. extremorientalis TSAU20	IAA production, antifungal activity						
	P. fluorescens PCL1751	IAA production, antifungal activity						

26.	*Bacillus pumilus* (STR2)	Phosphate solubilization, ACC deaminase production	(*Bacopa monnieri*) brahmi	pH, 7.05; EC, 0.40 dS m^{-1}; organic carbon, 4.40 g kg^{-1}; available N 133 kg ha^{-1}, available P, 10.28 kg ha^{-1}, available K 96.25 kg ha^{-1}; potting mixture, soil: vermicompost (3:1 w/w). primary salinity by adding 4 g NaCl kg^{-1} soil and secondary salinization generated via irrigation with saline solution	Yes	No	Increased fresh weight and bacoside content, increased proline content (induce plant resistance), increased leaf catalase content	Bharti et al. (2013)
	Exiguobacterium oxidotolerans (STR36)	Phosphate solubilization, EPS production						
27.	*P. pseudoalcaligenes*	NS	(*Oryza sativa* L.) rice	Soil parameters not specified (saline condition maintained at different salinity levels by adding saline solution to the pots)	Yes	No	Induction of osmoprotectant and antioxidant proteins, increased dry weight and plant height	Jha et al. (2011)
	B. pumilus							

(continued)

Table 9.5 (continued)

Sr. no.	Organism	Plant growth-promoting traits	Plant used	SAS used with composition	Pot assay	Field trial	Removal/reduction/consequence	Reference
28.	*Bacillus subtilis* SU47	IAA production, gibberellin production, P solubilization, EPS production	(*Triticum aestivum* L. Raj 3077) wheat	ECe, 2 dS m^{-1}, and 6 dS m^{-1}	Yes	No	Increase in dry biomass, decreased accumulation of Na^{+} in plant tissue, increased TSS content, increased proline accumulation, decreased activity of antioxidant enzymes	Upadhyay et al. (2012)
	Arthrobacter sp. SU18	IAA production, gibberellin production, P solubilization, siderophore production						
29.	*Streptomyces*	IAA production, siderophore production	(*Triticum aestivum* cul. Chamran) wheat	160 mM NaCl salinity, pH 6.9	Yes	No	Increased shoot height, increased shoot fresh weight, increased uptake of nutrients	Sadeghi et al. (2012)
30.	*Pseudomonas chlororaphis* TSAU13	HCN production, IAA production, antifungal activity	(*Lycopersicon esculentum* Mill cv. Bella) tomato	EC, 659 mS m^{-1}; organic matter, 0.694%; total C, 2.506%; total N, 0.091%; Ca, 63.5 g/kg; Mg, 20.7 g/kg; K, 6.2 g/kg; P, 1.2 g/kg; Cl, 0.1 g/kg; Na, 0.7 g/kg; pH, 8.0.	Yes	No	Increased (a) shoot growth (up to 26%), (b) plant height (up to 12%), (c) fruit (up to 16%), biocontrol against *F. solani*	Egamberdieva (2012)
			(*Cucumis sativus*, cv. Simbal) cucumber				Increased (a) shoot growth (32%), (b) dry matter (43%), (c) plant height (8%), (d) fruit (up to 16%), biocontrol against *F. solani*	

31.	*Pseudomonas aurantiaca* SR1	Antibiotics and siderophore production, IAA production, antifungal activity	(*Triticum aestivum* L. cv. Klein Flecha) wheat (*Zea mays L. cv.* NKTDMAX 940) maize	pH (in water), 6.30; electrical conductivity, (dS m^{-1}), 0.28; organic matter, 2.56%; nitrate, 51 mg g^{-1}; available P, 19.7 mg g^{-1}.	Yes	No	Increased shoot and root dry weight, increased grain yield	Rosas et al. (2009)
32.	*Klebsiella oxytoca* Rs-5	IAA production, P solubilization	(*Gossypium hirsutum* L., variety Xinluzao13) cotton	C, 23.3 g Kg^{-1}; N, 1.20 g Kg^{-1}; P 1.03 g Kg^{-1}; K, 0.10 g Kg^{-1}, Mg 0.04 g Kg−1; Na, 1.10 g Kg^{-1}; Ca, 0.16 g Kg^{-1}; Total salt, 3.5 g Kg^{-1}; pH 8.85	Yes	No	Increased (a) plant height (14.9%), (b) dry weight (26.9%), (c) absorption of N, P, K, and Ca, decreased absorption of Na	Yue et al. (2007)

Knowledge of the origins and consequences of salinity, alkalinity (sodicity), and control is of paramount importance.

Several conventional strategies, viz., physical, chemical, and biological approaches, are available to combat SAS but proved inadequate to recoup SAS. Although the use of electro-reclamation, traditional breeding, and transgenic plants may offer better alternative, the cost of development is enormous and unaffordable to developing country like India. Salt- and alkali-tolerant PGP bacteria or fungi may be selected from affected habitats and/or tailor-made to protect a spectrum of different salt-stressed plants. It is easier to select and/or modify a few dozen soil bacteria than hundreds or even thousands of different plant cultivars. Microbial augmentation of SAS has emerged as the most effective alternative. The extensive use of PGP bacteria and/or fungi at periodic interval may decrease dependence on chemical fertilizers worldwide. Moreover, it is a technology that can be readily accessible to farmers in both developed and developing countries. Hence, the sustainable alternative to tackle the issue of undertaking agriculture in salinized-sodic soils is to select the microbial species with desirable attributes from natural extreme habitat(s) so that salt-affected problem soils can be made suitable for cultivation of agricultural crops.

Acknowledgments One of the authors (P.M.G.) acknowledges the fellowship from the University Grants Commission, New Delhi, under its UGC-RGNF-ST (Rajiv Gandhi National Fellowship for higher education of ST students) scheme. Financial support from University Grants Commission, New Delhi, and Department of Sciences and Technology, New Delhi, for strengthening the research facilities at the School under SAP–DRS (F.4-23/2015/DRS-II [SAPII]) and FIST (SR/FST/LSI-433/2010) programs, respectively, is gratefully acknowledged.

This research did not receive any specific grant from funding agencies in the public, commercial, or nonprofit sectors.

References

Albacete A, Ghanem ME, Martínez-Andújar C, Acosta M, Sánchez-Bravo J, Martínez V, Lutts S, Dodd IC, Pérez-Alfocea F (2008) Hormonal changes in relation to biomass partitioning and shoot growth impairment in salinized tomato (*Solanum lycopersicum* L.) plants. J Exp Bot 59:4119–4131

Allen SE, Grimshaw HM, Parkinson JA, Quarmby CL (1974) Chemical analysis of ecological materials. Blackwell Scientific Publications, Oxford

Anderson JPE (1982) Soil respiration. In: Page AL (ed) Methods of soil analysis part 2 chemical and microbiological properties. Soil Science Society of America, Madison, pp 837–871

Apse MP, Blumwald E (2007) Na^+ transport in plants. FEBS Lett 581:2247–2254

Apte SK, Thomas J (1997) Possible amelioration of coastal soil salinity using halotolerant nitrogen-fixing Cyanobacteria. Plant Soil 189:205–211

Arkhipova TN, Prinsen E, Veselov SU, Martinenko EV, Melentiev AI, Kudoyarova GR (2007) Cytokinin producing bacteria enhance plant growth in drying soil. Plant Soil 292:305–315

Artamonova VS, Dits LY, Elizarova TN, Lyutykh IV (2010) Technogenic salinization of soils and their microbiological characterization. Contemp Probl Ecol 3(3):323–330

Ashraf M (2004) Some important physiological selection criteria for salt tolerance in plants. Flora 199:361–376

Ashraf M, Akram NA (2009) Improving salinity tolerance of plants through conventional breeding and genetic engineering: an analytical Comparison. Biotechnol Adv 27:744–752

Ashraf M, Hasnain S, Berge O, Mahmood T (2004) Inoculating wheat seedlings with exopolysaccharide-producing bacteria restricts sodium uptake and stimulates plant growth under salt stress. Biol Fertil Soils 40:157–162

Baker HM, Anderson BF, Baker EN (2003) Dealing with iron: Common structural principles in proteins that transport iron and heme. Proc Natl Acad Sci USA 100:3579–3583

Bapurao ND (2012) Potential halotolerant *Azospirillum brasilense* N-30 as a biofertilizer for saline soils. Acta Biol Indica 1:198–206

Barka EA, Nowak J, Clement C (2006) Enhancement of chilling resistance of inoculated grapevine plantlets with a plant growth-promoting rhizobacterium, *Burkholderia phytofirmans* strain PsJN. Appl Environ Microbiol 72:7246–7252

Barua S, Tripathi S, Chakraborty A, Ghosh S, Chakrabarti K (2011) Studies on non-symbiotic diazotrophic bacterial populations of coastal arable saline soils of India. Indian J Microbiol 51:369–376

Bhadauria S, Sengar RMS, Mohan D, Dhanai CS, Kushwah BS (2010) Sustainable land use planning through utilization of alkaline wasteland by biotechnological intervention. Middle-East J Sci Res 6:580–592

Bharti N, Yadav D, Barnawal D, Maji D, Kalra A (2013) *Exiguobacterium oxidotolerans*, a halotolerant plant growth promoting rhizobacteria, improves yield and content of secondary metabolites in *Bacopa monnieri* (L.) Pennell under primary and secondary salt stress. World J Microbiol Biotechnol 29:379–387

Blackwell P, Reithmuller G, Collins M (2009) Biochar application to soil. In: Lehmann J, Joseph S (eds) Biochar for environmental management: science and technology, vol 1. Earthscan, London, pp 207–226

Blaha CAG, Revers LF, Passaglia LMP, Frazzon J, Schrank IS (2000) Characterization of an *Azospirillum brasilense* Tn5 mutant with enhanced nitrogen fixation. In: Pedrosa FO, Hungria M, Yates G, Newton WE (eds) Nitrogen fixation: from molecules to crop productivity, vol 38. Springer, Dordrecht, pp 93–94

Blum JS, Bindi AB, Buzzelli J, Stolz JF, Oremland RS (1998) *Bacillus arsenicoselenatis*, sp. nov., and *Bacillus selenitireducens*, sp. nov.: two haloalkaliphiles from Mono Lake, California that respire oxyanions of selenium and arsenic. Arch Microbiol 171:19–30

Brady N, Weil R (2002) The nature and properties of soils, 13th edn. Prentice Hall, Upper Saddle River

Bremner JM, Mulvaney CS (1982) Nitrogen-total. In: Page AL, Miller RH, Keeney DR (eds) Methods of soil analysis part 2. Chemical and microbiological properties, Madison, pp 595–624

Brookes PC, Landman A, Pruden G, Jenkinson DS (1985) Chloroform fumigation and the release of soil nitrogen: a rapid direct extraction method to measure microbial biomass nitrogen in soil. Soil Biol Biochem 17:837–842

Bull CT, Weller DM, Thomashow LS (1991) Relationship between root colonization and suppression of *Gaeumannomyces graminis* var. *tritici* by *Pseudomonas fluorescens* strain 2–79. Phytopathology 81:954–959

Chang P, Gerhardt KE, Huang XD, Yu DM, Glick BR, Gerwing PD, Greenber BM (2014) Plant growth promoting bacteria facilitate the growth of barley and oats in salt impacted soil: implications for phytoremediation of saline soils. Int J Phytoremediation 16:1133–1147

Charzyński P, Hulisz P, Bednarek RM (2013) Technogenic soils of Poland. Polish Society of Soil Science, Torun, pp 35–36

Chaves MM, Flexas J, Pinheiro C (2009) Photosynthesis under drought and salt stress: regulation mechanisms from whole plant to cell. Ann Bot 103:551–560

Chen Z, Cuin TA, Zhou M, Twomey A, Naidu BP, Shabala S (2007) Compatible solute accumulation and stress-mitigating effects in barley genotypes contrasting in their salt tolerance. J Exp Bot 58:4245–4255

Cho JM, Kim KJ, Chung KY, Hyun S, Baek K (2009) Restoration of saline soil in cultivated land using electrokinetic process. Sep Sci Technol 44:2371–2384

Chowdhury N, Marschner P, Burns R (2011) Response of microbial activity and community structure to decreasing soil osmotic and matric potential. Plant Soil 344:241–254

Cicek N, Cakirlar H (2002) The effect of salinity on some physiological parameters in two maize cultivars. Bulgarian J Plant Physiol 281:66–74

Conrad R (1996) Soil microorganisms as controllers of atmospheric trace gases (H_2, CO, CH_4, OCS, N_2O, and NO). Microbiol Rev 60:609–640

Cramer GR (1992) Kinetics of maize leaf elongation: II. Response of a Na-excluding cultivar and a Na-including cultivar to varying Na/Ca salinities. J Exp Bot 43:857–864

CSSRI (2015) Vision 2050. ICAR-Central Soil Salinity Research Institute, Karnal

Cummings SP (2005) The role and future potential of nitrogen fixing bacteria to boost productivity in organic and low-input sustainable farming systems. Environ Biotechnol 1:1–10

Dahlawi S, Naeem A, Rengel Z, Naidu R (2018) Biochar application for the remediation of salt-affected soils: Challenges and opportunities. Sci Total Environ 625:320–335

de-Bashan LE, Hernandez JP, Bashan Y (2012) The potential contribution of plant growth-promoting bacteria to reduce environmental degradation – A comprehensive evaluation. Appl Soil Ecol 61:171–189

Deepa CK, Syed GD, Pandey A (2010) Isolation and characterization of plant growth promoting bacteria from non-rhizospheric soil and their effect on cowpea (*Vigna unguiculata* (L.) Walp. seedling growth. World J Microbiol Biotechnol 26:1233–1240

Delfine S, Alvino A, Villani MC, Loreto F (1999) Restrictions to carbon dioxide conductance and photosynthesis in spinach leaves recovering from salt stress. Plant Physiol 119:1101–1106

Denton MD, Reeve WG, Howieson JG, Coventry DR (2003) Competitive abilities of common field isolates and a commercial strain of *Rhizobium leguminosarum* bv trifolii for clover nodule occupancy. Soil Biol Biochem 35:1039–1048

Desale P, Patel B, Singh S, Malhotra A, Nawani N (2013) Plant growth promoting properties of *Halobacillus* sp. and *Halomonas* sp. in the presence of salinity and heavy metals. J Basic Microbiol 54:781–791

Dodd IC, Pérez-Alfocea F (2012) Microbial amelioration of crop salinity stress. J Exp Bot 63:3415–3428

Dodd IC, Zinovkina NY, Safronova VI, Belimov AA (2010) Rhizobacterial mediation of plant hormone status. Ann Appl Biol 157:361–379

Dugardeyn J, van der Straeten D (2008) Ethylene: fine-tuning plant growth and development by stimulation and inhibition of elongation. Plant Sci 175:59–70

Egamberdieva D (2009) Alleviation of salt stress by plant growth regulators and IAA producing bacteria in wheat. Acta Physiol Plant 31(4):861–864

Egamberdieva D (2012) *Pseudomonas chlororaphis*: a salt-tolerant bacterial inoculant for plant growth stimulation under saline soil conditions. Acta Physiol Plant 34:751–756

Egamberdieva D, Kamilova F, Validov S, Gafurova L, Kucharova Z, Lugtenberg B (2008) High incidence of plant growth-stimulating bacteria associated with the rhizosphere of wheat grown on salinated soil in Uzbekistan. Environ Microbiol 10:1–9

Egamberdieva D, Kucharova Z, Davranov K, Berg G, Makarova N, Azarova T, Chebotar V, Tikhonovich I, Kamilova F, Validov SZ, Lugtenberg B (2011) Bacteria able to control foot and root rot and to promote growth of cucumber in salinated soils. Biol Fertil Soils 47:197–205

Egamberdieva D, Jabborova D, Wirth S (2013) Alleviation of salt stress in legumes by co-inoculation with *Pseudomonas* and *Rhizobium*. In: Plant microbe symbiosis: fundamentals and advances. Springer, New Delhi, pp 291–303

Ehrlich HL, Oremland RS, Zehr JP (2001) Biogeochemical cycles. eLS

Elmajdoub B, Marschner P (2013) Salinity reduces the ability of soil microbes to utilise cellulose. Biol Fertil Soils 49:379–386

Esechie HA, Al-Saidi A, Al-Khanjari S (2002) Effect of sodium chloride salinity on seedling emergence in chickpea. J Agron Crop Sci 188:155–160

Fernando WGD, Nalkeeran S, Yilan Z (2006) Biosynthesis of antibiotics by PGPR and its relation in biocontrol of plant diseases. In: Siddiqui ZA (ed) PGPR biocontrol and biofertilization. Springer, Dordrecht, pp 67–109

Fetter CW (ed) (2001) Applied hydrogeology. Prentice Hall, Upper Saddle River. Supplemental website http://www.appliedhydrogeology.info

Fidalgo F, Santos A, Santos I, Salema R (2004) Effects of long-term salt stress on antioxidant defence systems, leaf water relations and chloroplast ultrastructure of potato plants. Ann Appl Biol 145:185–192

Figueiredo MVB, Burity HA, Martinez CR, Chanway CP (2008) Alleviation of water stress effects in common bean (*Phaseolus vulgaris* L.) by co-inoculation *Paenibacillus* x *Rhizobium tropici*. Appl Soil Ecol 40:182–188

Fitzpatrick RW (2002) Land degradation processes. In: TR MV, Rui L, Walker J, Fitzpatrick RW, Liu C (eds) Regional water and soil assessment for managing sustainable agriculture in China and Australia, ACIAR Monograph, vol 84, pp 119–129

Fitzpatrick RW, Boucher SC, Naidu R, Fritsch E (1994) Environmental consequences of soil sodicity. Aust J Soil Res 32:1069–1093

Fitzpatrick RW, Rengasamy P, Merry RH, Cox JW (2001) Is dryland soil salinisation reversible? In: National dryland salinity programme (NDSP), Canberra

Flexas J, Ortuño MF, Ribas-Carbo M, Diaz-Espejo A, Flórez-Sarasa ID, Medrano H (2007) Mesophyll conductance to CO_2 in *Arabidopsis thaliana*. New Phytol 175:501–511

Foley JA (2011) Can we feed the world and sustain the planet? Sci Am 305:60–65

Forster SM (1990) The role of microorganisms in aggregate formation and soil stabilization: types of aggregation. Arid Land Res Manag 4:85–98

Fougere F, Ruduiler D, Streeter G (1991) Effect of salt stress on amino acid, organic acid and carbohydrate composition of roots, bacteroides and cytosol of Alfalfa (*Medicago sativa* L.). Plant Physiol 96:1228–1236

Ghanem M, Albacete A, Smigocki AC, Frébort I, Pospíšilová H, Martínez-Andújar C, Pérez-Alfocea F (2011) Root-synthesized cytokinins improve shoot growth and fruit yield in salinized tomato (*Solanum lycopersicum* L.) plants. J Exp Bot 62:125–140

Ghassemi F, Jakeman AJ, Nix HA (1995) Salinisation of land and water resources: Human causes, extent, management and case studies. CABI Publishing, Wallingford

Glick BR (1995) The enhancement of plant growth by free living bacteria. Can J Microbiol 41:109–117

Glick BR (2010) Using soil bacteria to facilitate phytoremediation. Adv Biotechnol 28:367–374

Glick BR, Penrose DM, Li J (1998) A model for the lowering of plant ethylene concentrations by plant growth-promoting bacteria. J Theor Biol 190:63–68

Glick BR, Cheng Z, Czarny J, Duan J (2007) Promotion of plant growth by ACC deaminase-producing soil bacteria. In: New perspectives and approaches in plant growth-promoting rhizobacteria research. Springer, Dordrecht, pp 329–339

Gliessman SR (2004) Integrating agroecological processes into cropping systems research. J Crop Improv 11:61–80

Gobat JM, Aragno M, Matthey W (2004) The living soil: fundamentals of soil science and soil biology. Science Publishers, Enfield

Godfrey ZE, Pax Blamey FC, Colin Asher L (2010) Some observations on the culture of attached groundnut (*Arachis hypogaea* L.) gynophores in simplified nutrient solutions. African J Plant Sci 4:296–300

Goswami D, Dhandhukia P, Patel P, Thakker JN (2014) Screening of PGPR from saline desert of Kutch: growth promotion in *Arachis hypogea* by *Bacillus licheniformis* A2. Microbiol Res 169:66–75

Grattan SR, Oster JD (2003) Use and reuse of saline-sodic waters for irrigation of crops. J Crop Prod 7:131–162

Greenberg BM, Huang XD, Gerwing P, Yu XM, Chang P, Wu SS, Glick B (2008) Phytoremediation of salt impacted soils: greenhouse and the field trials of plant growth promoting rhizobacteria (PGPR) to improve plant growth and salt phytoaccumulation. In: Proceeding of the 33rd AMOP Technical seminar on environmental contamination and response. Environment Canada, Ottawa, pp 627–637

Grismer M (1990) Leaching fraction, soil salinity, and drainage efficiency. Calif Agric 44:24–26

Grover M, Ali SZ, Sandhya V, Rasul A, Venkateswarlu B (2011) Role of microorganisms in adaptation of agriculture crops to abiotic stresses. World J Microbiol Biotechnol 27:1231–1240
Hagemann M (2011) Molecular biology of cyanobacterial salt acclimation. FEMS Microbiol Rev 35:87–123
Hajkowicz SA, Young MD (2002) Value of returns to land and water and costs of degradation. Final report to the National Land and Water Resources Audit, Policy and Economic Research Unit, CSIRO Land and Water, Folio Ref, 2, p 477
Hasnain S, Sabri AN (1997) Growth stimulation of *Triticum aestivum* seedlings under Cr-stresses by non-rhizospheric *Pseudomonas* strains. Environ Pollut 97:265–273
Hong Z, Lakkineni K, Zhang Z, Verma DP (2000) Removal of feedback inhibition of delta(1)-pyrroline-5-carboxylate synthetase results in increased proline accumulation and protection of plants from osmotic stress. Plant Physiol 122:1129–1136
Horneck DA, Ellsworth JW, Hopkins BG, Sullivan DM, Stevens RG (2007) Managing salt-affected soils for crop production. PNW 601-E, Oregon State University Extension Service, USA. http://ir.library.oregonstate.edu/concern/administrative_report_or_publications/pr76f371k
Jamil M, Lee DB, Jung KY, Ashraf M, Lee SC, Rha ES (2006) Effect of salt (NaCl) stress on germination and early seedling growth of four vegetables species. J Cent Eur Agric 7:273–282
Jha Y, Subramanian RB, Patel S (2011) Combination of endophytic and rhizospheric plant growth promoting rhizobacteria in *Oryza sativa* shows higher accumulation of osmoprotectant against saline stress. Acta Physiol Plant 33:797–802
Jha B, Gontia I, Hartmann A (2012) The roots of the halophyte *Salicornia brachiata* are a source of new halotolerant diazotrophic bacteria with plant growth-promoting potential. Plant Soil 356:265–277
Kalembasa SJ, Jenkinson DS (1973) A comparative study of titrimetric and gravimetric methods for the determination of organic carbon in soil. J Sci Food Agric 24:1085–1090
Kaushik A, Sethi V (2005) Salinity effects on nitrifying and free diazotrophic bacterial populations in the rhizosphere of rice. Bull Natl Inst Ecol 15:139–144
Kendall HW, Pimentel D (1994) Constraints on the expansion of the global food supply. Ambio 23:198–205
Kennedy AC (1999) Bacterial diversity in agroecosystems. Agric Ecosyst Environ 74(1):65–76
Khosla BK, Gupta RK, Abrol IP (1979) Salt leaching and the effect of gypsum application in a saline-sodic soil. Agric Water Manag 2:193–202
Kim S, Raybum AL, Voiqt T, Parrish A, Lee DK (2012) Salinity effects on germination and plant growth of prairie cord grass and switchgrass. Bioenergy Res 5:225–235
Kohler J, Caravaca F, Carrasco L, Roldán A (2006) Contribution of *Pseudomonas mendocina* and *Glomus intraradices* to aggregate stabilization and promotion of biological fertility in rhizosphere soil of lettuce plants under field conditions. Soil Use Manag 22:298–304
Kurth E, Cramer GR, Lauchli A, Epstein E (1986) Effects of NaCl and $CaC1_2$ on cell enlargement and cell production in cotton roots. Plant Physiol 82:1102–1106
Lakhdar A, Rabhi M, Ghnaya T, Montemurro F, Jedidi N, Abdelly C (2009) Effectiveness of compost use in salt-affected soil. J Hazard Mater 15(171):29–37
Lal R (2001) Soil degradation by erosion. Land Degrad Dev 12:519–539
Lal R (2015) Restoring soil quality to mitigate soil degradation. Sustainability 7:5875–5895
Laura RD (1974) Effects of neutral salts on carbon and nitrogen mineralisation of organic matter in soil. Plant Soil 41:113–127
Lawlor DW, Cornic G (2002) Photosynthetic carbon assimilation and associated metabolism in relation to water deficits in higher plants. Plant Cell Environ 25:275–294
Lee YJ, Choi JH, Lee HG, Ha TH, Bae JH (2012) Effect of electrode materials on electrokinetic reduction of soil salinity. Sep Sci Technol 47:22–29
Levy GJ (2000) Sodicity. In: Sumner ME (ed) Handbook of soil science. CRC Press, Boca Raton, pp 27–63
Li XG, Shi XM, Wang DJ, Zhou W (2012) Effect of alkalized magnesic salinity on soil respiration changes with substrate availability and incubation time. Biol Fertil Soils 48:597–602

Liang Y, Si J, Nikolic M, Peng Y, Chen W, Jiang Y (2005) Organic manure stimulates biological activity and barley growth in soil subject to secondary salinization. Soil Biol Biochem 37:1185–1195

Liu W, Hou J, Wang Q, Ding L, Luo Y (2014) Isolation and characterization of plant growth-promoting rhizobacteria and their effects on phytoremediation of petroleum-contaminated saline-alkali soil. Chemosphere 117:303–308

Loganathan P, Nair S (2004) *Swaminathania salitolerans* gen. nov., sp. nov., a salt-tolerant, nitrogen-fixing and phosphate-solubilizing bacterium from wild rice (*Porteresia coarctata Tateoka*). Int J Syst Evol Microbiol 54:1185–1190

López-Aguirre JG, Farias-Larios J, Molina-Ochoa J, Aguilar-Espinosa S, Flores-Bello M, González-Ramírez M (2007) Salt leaching process in an alkaline soil treated with elemental sulphur under dry tropic conditions. World J Agric Sci 3:356–362

Lugtenberg BJ, Dekkers L, Bloemberg GV (2001) Molecular determinants of rhizosphere colonization by Pseudomonas. Annu Rev Phytopathol 39:461–490

Ma HC, Fung L, Wang SS, Altman A, Hüttermann A (1997) Photosynthetic response of *Populus euphratica* to salt stress. For Ecol Manag 93:55–61

Makoi JH, Verplancke H (2010) Effect of gypsum placement on the physical chemical properties of a saline sandy loam soil. Asian J Crop Sci 4:556–556

Mandal AK, Sharma RL, Raingh G, Dagar JL (2010) Computerized database on SAS in India. http://krishikosh.egranth.ac.in/bitstream/1/2046527/1/CSSRI249.pdf

Mayak S, Tirosh T, Glick BR (2004) Plant growth-promoting bacteria that confer resistance to water stress in tomatoes and peppers. Plant Sci 166:525–530

McBride MB (1994) Environmental chemistry of soils. Oxford University Press/Wiley, New York

McNeil RL, Howitt RW, Whitson IR, Chartier AG (1994) Soil survey of the County of Forty Mile No. 8, Alberta. Alberta Research Council. Environmental Research and Engineering Department, Edmonton, AB. Alberta Soil Survey Report (54). http://www.agric.gov.ab.ca/soil/survey-reports/ab54/ab54_report.pdf

Meritt KA (2006) Method and system for promoting microbial fixation activity. Australian Patent, WO2006/005100, PCT/AU/2005/000785, pp 1–30

Meyer S, Genty B (1998) Mapping intercellular CO_2 mole fraction (Ci) in *Rosa rubiginosa* leaves fed with abscisic acid by using chlorophyll fluorescence imaging: significance of Ci estimated from leaf gas exchange. Plant Physiol 116:947–957

Miller JJ, Brierley JA (2011) Solonetzic soils of Canada: Genesis, distribution, and classification. Can J Soil Sci 91:889–902

Minhas PS, Sharma OP (2003) Management of soil salinity and alkalinity problems in India. J Crop Prod 7:181–230

Mulligan CN (2009) Recent advances in the environmental applications of biosurfactants. Curr Opin Colloid Interface Sci 14:372–378

Munns R (2002) Comparative physiology of salt and water stress. Plant Cell Environ 25:239–250

Munns R (2005) Genes and salt tolerance: bringing them together. New Phytol 167(3):645–663

Murillo JM, López R, Cabrera E, Martín-Olmedo P (1995) Testing a low-quality urban compost as a fertilizer for arable farming. Soil Use Manag 11:127–131

Mythili G, Goedecke J (2016) Economics of land degradation in India. In: Nkonya E, Mirzabaev A, Von Braun J (eds) Economics of land degradation and improvement-A global assessment for sustainable development. Springer, pp 431–469

Nabti E, Sahnoune M, Adjrad S, Van Dommelen A, Ghoul M, Schmid M, Hartmann A (2007) A halophilic and osmotolerant *Azospirillum brasilense* strain from Algerian soil restores wheat growth under saline conditions. Eng Life Sci 7:354–360

Nabti E, Sahnoune M, Ghoul M, Fischer D, Hofmann A, Rothballer M, Hartmann A (2010) Restoration of growth of durum wheat (*Triticum durum* var. *waha*) under saline conditions due to inoculation with the rhizosphere bacterium *Azospirillum brasilense* NH and extracts of the marine alga *Ulva lactuca*. J Plant Growth Regul 29:6–22

Nachtergaele F, Van Velthuizen H, Verelst L, Batjes N, van Engelen V, Petri M, Shi X (2012) Harmonized world soil database. FAO/IIASA/ISRIC/ISS-CAS/JRC, Rome/Italy/Laxenburg

Naidu R, Rengasamy P (1993) Ion interactions and constraints to plant nutrition in Australian sodic soils. Aust J Soil Res 31:801–819

Nakade DB, Chonde SG, Pallavi B (2012) Halophilic nitrogen fixing *Azotobacter chroococcum* N-21 and its use as a biofertilizer for saline soils. J Microbiol Biotechnol Res 2:1017–1023

Nautiyal CS, Bhadauria S, Kumar P, Lal H, Mondal R, Verma D (2000) Stress induced phosphate solubilization in bacteria isolated from alkaline soils. FEMS Microbiol Lett 182:291–296

Neilands JB (1995) Siderophores: structure and function of microbial iron transport compounds. J Biol Chem 270(45):26723–26726

Olsen SR (1954) Estimation of available phosphorus in soils by extraction with sodium bicarbonate, Circular no. 939. US Department of Agriculture, Washington, DC

Oren A (2001) The bioenergetic basis for the decrease in metabolic diversity at increasing salt concentrations: implications for the functioning of salt lake ecosystems. Hydrobiologia 466:61–72

Ouerghi Z, Cornic G, Roudani M, Ayadi A, Brulfert J (2000) Effect of NaCl on photosynthesis of two wheat species (*Triticum durum* and *T. aestivum*) differing in their sensitivity to salt stress. J Plant Physiol 156:335–340

Pandey MM (2009) Country report India-Indian agriculture an introduction. Country report presented in the fourth session of the Technical Committee of APCAEM, held during 1–3 December, 2008 at Chiang Rai, Thailand

Pandey KD, Shukla PN, Giri DD, Kashyap AK (2005) Cyanobacteria in alkaline soil and the effect of cyanobacteria inoculation with pyrite amendments on their reclamation. Biol Fertil Soils 41:451–457

Patel D, Jha CK, Tank N, Saraf M (2012) Growth enhancement of Chickpea in saline soils using plant growth-promoting *Rhizobacteria*. Plant Growth Regul 31:53–62

Patten CL, Glick BR (2002) Regulation of indoleacetic acid production in *Pseudomonas putida* GR12-2 by tryptophan and the stationary-phase sigma factor RpoS. Can J Microbiol 48:635–642

Paul D, Nair S (2008) Stress adaptations in a plant growth promoting rhizobacterium (PGPR) with increasing salinity in the coastal agricultural soils. J Basic Microbiol 48:378–384

Peinemann N, Guggenberger G, Zech W (2005) Soil organic matter and its lignin component in surface horizons of salt-affected soils of the *Argentinian Pampa*. Catena 60:113–128

Pérez-Alfocea F, Albacete A, Ghanem ME, Dodd IC (2010) Hormonal regulation of source–sink relations to maintain crop productivity under salinity: a case study of root-to-shoot signaling in tomato. Funct Plant Biol 37:592–603

Pimentel D, Giampietro M (1994) Current Trends Global population, food and the environment. Trends Ecol Evol 9:239

Pitman MG, Läuchli A (2002) Global impact of salinity and agricultural ecosystems. In: Läuchli A, Lüttge U (eds) Salinity: environment–plants–molecules. Kluwer, Dordrecht, pp 3–20

Qadir M, Schubert S (2002) Degradation processes and nutrient constraints in sodic soils. Land Degrad Dev 13:275–294

Qadir M, Ghafoor A, Murtaza G (2000) Amelioration strategies for saline soils: a review. Land Degrad Dev 11:501–521

Qadir M, Ghafoor A, Murtaza G (2001) Use of saline–sodic waters through phytoremediation of calcareous saline–sodic soils. Agric Water Manag 50:197–210

Qadir M, Noble AD, Schubert S, Thomas RJ, Arslan A (2006) Sodicity-induced land degradation and its sustainable management: problems and prospects. Land Degrad Dev 17:661–676

Qadir M, Oster JD, Schubert S, Noble AD, Sahrawat KL (2007) Phytoremediation of sodic and saline-sodic soils. Adv Agron 96:197–247

Qurashi AW, Sabri AN (2012) Bacterial exopolysaccharide and biofilm formation stimulate chickpea growth and soil aggregation under salt stress. Braz J Microbiol 43:1183–1191

Raheem A, Ali B (2015) Halotolerant rhizobacteria: beneficial plant metabolites and growth enhancement of *Triticum aestivum* L. in salt-amended soils. Arch Agron Soil Sci 61:1691–1705

Rai TN, Rai KN, Prasad S, Sharma C, Mishra S, Gupta B (2010) Effect of organic amendments, bioinoculants and gypsum on the reclamation and soil chemical properties in sodic soils of Etawah. J Soil Water Conserv 9:197–200

Rajput L, Imran A, Mubeen F, Hafeez FY (2013) Salt-tolerant PGPR strain *Planococcus rifietoensis* promotes the growth and yield of wheat (*Triticum aestivum* L.) cultivated in saline soil. Pak J Bot 45:1955–1962

Ravikumar S, Kathiresan K, Ignatiammal S, Babu Selvam M, Shanthy S (2004) Nitrogen-fixing *Azotobacter* from mangrove habitat and their utility as marine biofertilizers. J Exp Mar Biol Ecol 312:5–17

Reinhold B, Hurek T, Fendrik I, Pot B, Gillis M, Kersters K, De Ley J (1987) *Azospirillum halopraeferens* sp. nov., a nitrogen-fixing organism associated with roots of Kallar grass (*Leptochloa fusca* (L.) Kunth). Int J Syst Bacteriol 37:43–51

Rekha PD, Lai WA, Arun AB, Young CC (2007) Effect of free and encapsulated *Pseudomonas putida* CC-FR2-4 and *Bacillus subtilis* CC-pg104 on plant growth under gnotobiotic conditions. Bioresour Technol 98:447–451

Rengasamy P (2002) Transient salinity and subsoil constraints to dryland farming in Australian sodic soils: an overview. Anim Prod Sci 42:351–361

Rengasamy P (2006) World salinization with emphasis on Australia. J Exp Bot 57:1017–1023

Rengasamy P (2010) Soil processes affecting crop production in salt-affected soils. Funct Plant Biol 37:613–620

Rengasamy P, Olsson KA (1991) Sodicity and soil structure. Soil Res 29:935–952

Rietz DN, Haynes RJ (2003) Effects of irrigation-induced salinity and sodicity on soil microbial activity. Soil Biol Biochem 35:845–854

Romero-Aranda MR, Jurad O, Cuartero J (2006) Silicon alleviates the deleterious salt effect on tomato plant growth by improving plant water status. J Plant Physiol 163:847–855

Rosas SB, Avanzini G, Carlier E, Pasluosta C, Pastor N, Rovera M (2009) Root colonization and growth promotion of wheat and maize by *Pseudomonas aurantiaca* SR1. Soil Biol Biochem 41:1802–1806

Sadeghi A, Karimi E, Dahaji PA, Javid MG, Dalvand Y, Askari H (2012) Plant growth promoting activity of an auxin and siderophore producing isolate of *Streptomyces* under saline soil conditions. World J Microbiol Biotechnol 28:1503–1509

Sadiq M, Hassan G, Mehdi SM, Hussain N, Jamil M (2007) Amelioration of saline-sodic soils with tillage implements and sulfuric acid application. Pedosphere 17:182–190

Sahin U, Eroğlu S, Sahin F (2011) Microbial application with gypsum increases the saturated hydraulic conductivity of saline–sodic soils. Appl Soil Ecol 48(2):247–250

Sandhya V, Grover M, Reddy G, Venkateswarlu B (2009) Alleviation of drought stress effects in sunflower seedlings by the exopolysaccharides producing *Pseudomonas putida* strain GAP-P45. Biol Fertil Soils 46:17–26

Sangal SP (1991) Pricing of irrigation waters in India. Econ Polit Wkly 24:2645–2651

Saravanakumar D, Samiyappan R (2007) ACC deaminase from *Pseudomonas fluorescens* mediated saline resistance in groundnut (*Arachis hypogea*) plants. J Appl Microbiol 102:1283–1292

Sardinha M, Müller T, Schmeisky H, Joergensen RG (2003) Microbial performance in soils along a salinity gradient under acidic conditions. Appl Soil Ecol 23:237–244

Schnürer J, Rosswall T (1982) Fluorescein diacetate hydrolysis as a measure of total microbial activity in soil and litter. Appl Environ Microbiol 43:1256–1261

Setia R, Marschner P, Baldock J, Chittleborough D (2010) Is CO_2 evolution in saline soils affected by an osmotic effect and calcium carbonate? Biol Fertil Soils 46:781–792

Shabala S, Shabala L, Volkenburgh EV (2003) Effect of calcium on root development and root ion fluxes in salinised barley seedlings. Funct Plant Biol 30:507 514

Shainberg I, Keren R, Frenkel H (1982) Response of sodic soils to gypsum and calcium chloride application. Soil Sci Soc Am J 46:113–117

Shamseldin A, Werner D (2005) High salt and high pH tolerance of new isolated *Rhizobium etli* strains from Egyptian soils. Curr Microbiol 50:11–16

Shelar RD (2014) Studies on some aspects of nitrogen fixation by microbes from alkaline and saline soils. PhD Thesis, North Maharashtra University, Jalgaon, India

Shelar RD, Dandi ND, Dandi BN, Chaudhari AB (2010) Isolation and screening of di-nitrogen fixing microbes from extreme habitat. J Adv Sci Technol 13:1–7

Shrivastava P, Kumar R (2015) Soil salinity: a serious environmental issue and plant growth promoting bacteria as one of the tools for its alleviation. Saudi Journal of Biological Sciences 22:123–131

Siddiqui MH, Mohammad F, Khan MN (2009) Morphological and physio-biochemical characterization of *Brassica juncea* L. Czern. and Coss genotypes under salt stress. J Plant Interact 4:67–80

Singh KN, Charath R (2001) Salinity Tolerance. In: Reynolds MP, McNab A (eds) Application of physiology in wheat breeding. CIMMYT, Mexico, pp 101–110

Singh AL, Singh PL (1989) Nitrogen fixation in Indian rice fields (Azolla and blue-green algae). Agro-Botanical Publishers, Bikaner

Singh JS, Pandey VC, Singh DP (2011) Efficient soil microorganisms: a new dimension for sustainable agriculture and environmental development. Agric Ecosyst Environ 140:339–353

Sohan D, Jasoni R, Zajicek J (1999) Plant-water relations of NaCl and calcium-treated sunflower plants. Environ Exp Bot 42:105–111

Sorokin ID, Kravchenko IK, Doroshenko EV, Boulygina ES, Zadorina EV, Tourova TP, Sorokin DY (2008) Haloalkaliphilic diazotrophs in soda solonchak soils. FEMS Microbiol Ecol 65:425–433

Sorvari J, Antikainen R, Kosola ML, Hokkanen P, Haavisto T (2009) Eco-efficiency in contaminated land management in Finland–Barriers and development needs. J Environ Manag 90:1715–1727

Spaepen S, Vanderleyden J, Okon Y (2009) Plant growth-promoting actions of rhizobacteria. Adv Bot Res 51:283–320

Sprent JI, Sprent P (1990) Nitrogen fixing organisms: pure and applied aspects. Chapman and Hall, London

State of the Environment: India (2001) Land degradation, part III. http://envfor.nic.in/sites/default/files/soer/2001/soer.html

Suárez R, Wong A, Ramírez M, Barraza A, Orozco MDC, Cevallos MA, Iturriaga G (2008) Improvement of drought tolerance and grain yield in common bean by over expressing trehalose-6-phosphate synthase in rhizobia. Mol Plant-Microbe Interact 21:958–966

Sultana N, Ikeda T, Itoh R (1999) Effect of NaCl salinity on photosynthesis and dry matter accumulation in developing rice grains. Environ Exp Bot 42:211–220

Szabolcs I, Lal R, Blum WH (1998) Salt build up as a factor of soil degradation. In: Lal R, Blum WH, Valentine C, Stewart BA (eds) Methods for assessment of soil degradation. CRC press, Boca Raton, pp 253–264

Sziderics AH, Rasche F, Trognitz F, Sessitsch A, Wilhelm E (2007) Bacterial endophytes contribute to abiotic stress adaptation in pepper plants (*Capsicum annuum* L.). Can J Microbiol 53:1195–1202

Tanji KK (2002) Salinity in the soil environment. In: Läuchli A, Lüttge U (eds) Salinity: environment–plants–molecules. Kluwer, Dordrecht, pp 21–51

Thatoi H, Chandra BB, Kanti DT, Ranjan MR (2012) Microbial biodiversity in mangrove soils of Bhitarkanika, Odisha, India. Int J Environ Biol 2:50–58

Tilak KVBR, Ranganayaki N, Pal KK, De R, Saxena AK, Nautiyal CS, Johri BN (2005) Diversity of plant growth and soil health supporting bacteria. Curr Sci 89:136–150

Trombly J (1994) Electrochemical remediation takes to the field. Environ Sci Technol 28:289A–291A

Turan S, Cornish K, Kumar S (2012) Salinity tolerance in plants: breeding and genetic engineering. Aust J Crop Sci 6:1337

Unsal T, Ok S (2001) Description of characteristics of humic substances from different waste materials. Bioresour Technol 78:239–242

Upadhyay SK, Singh JS, Singh DP (2011) Exopolysaccharide-producing plant growth-promoting rhizobacteria under salinity condition. Pedosphere 21:214–222

Upadhyay SK, Singh JS, Saxena AK, Singh DP (2012) Impact of PGPR inoculation on growth and antioxidant status of wheat under saline conditions. Plant Biol 14:605–611

Valiela I, Teal JM, Persson NY (1976) Production and dynamics of experimentally enriched salt marsh vegetation: belowground biomass. Limnol Oceanogr 21:245–252

van Loon LC (2007) Plant responses to plant growth-promoting rhizobacteria. Eur J Plant Pathol 119:243–254

Vance ED, Brookes PC, Jenkinson DS (1987) An extraction method for measuring soil microbial biomass C. Soil Biol Biochem 19:703–707

Walworth J (2012) Using gypsum and other calcium amendments in southwestern soils. http://hdl.handle.net/10150/246053

Wang Y, Brown HN, Crowley DE, Szaniszlo PJ (1993) Evidence for direct utilization of a siderophore, ferrioxamine B, in axenically grown cucumber. Plant Cell Environ 16:579–585

Wang C, Knill E, Glick BR, Défago G (2000) Effect of transferring 1-aminocyclopropane-1-carboxylic acid (ACC) deaminase genes into *Pseudomonas fluorescens* strain CHA0 and its gacA derivative CHA96 on their growth-promoting and disease-suppressive capacities. Can J Microbiol 46:898–907

Warrence NJ, Bauder JW, Pearson KE (2002) Basics of salinity and sodicity effects on soil physical properties. Department of Land Resources and Environmental Sciences, Montana State University-Bozeman, Bozeman, pp 1–29

Wei Y, Xu X, Tao H, Wang P (2006) Growth performance and physiological response in the halophyte *Lycium barbarum* grown at salt-affected soil. Ann Appl Biol 149:263–269

Wichern J, Wichern F, Joergensen RG (2006) Impact of salinity on soil microbial communities and the decomposition of maize in acidic soils. Geoderma 137:100–108

William FB (2000) Compost quality standards and guidelines. Final report by Woods End Research Laboratories for the New York State Association of Recyclers. pp 6–10. http://citeseerx.ist.psu.edu/messages/downloadsexceeded.html

Wirthensohn T, Schoeberl P, Ghosh U, Fuchs W (2009) Pilot plant experiences using physical and biological treatment steps for the remediation of groundwater from a former MGP site. J Hazard Mater 163:43–52

Wollenweber B, Zechmeister-Boltenstern S (1989) Nitrogen fixation and nitrogen assimilation in a temperate saline ecosystem. Bot Acta 102:96–105

Wu SS (2009) Enhanced phytoremediation of salt-impacted soils using plant growth-promoting rhizobacteria (PGPR). PhD Thesis, University of Waterloo, Ontario, Canada

Xoconostle-Cazares B, Ramirez-Ortega FA, Flores-Elenes L, Ruiz-Medrano R (2010) Drought tolerance in crop plants. Am J Plant Physiol 5:241–256

Yamaguchi T, Blumwald E (2005) Developing salt-tolerant crop plants: challenges and opportunities. Trends Plant Sci 10(12):615–620

Yang C, Chong J, Li C, Kim C, Shi D, Wang D (2007) Osmotic adjustment and ion balance traits of an alkali resistant halophyte *Kochia sieversiana* during adaptation to salt and alkali conditions. Plant Soil 294:263–276

Yang J, Kloepper JW, Ryu CM (2009) Rhizosphere bacteria help plants tolerate abiotic stress. Trends Plant Sci 14:1–4

Yang C, Wang Z, Li Y, Niu Y, Du M, He X, Xu P (2010) Metabolic versatility of halotolerant and alkaliphilic strains of *Halomonas* isolated from alkaline black liquor. Bioresour Technol 101:6778–6784

Yildrim E, Donmez MF, Turan M (2008) Use of bioinoculants in ameliorative effects on radish plants under salinity stress. J Plant Nutr 31:2059–2074

Young A (1994) Land degradation in South Asia: its severity, causes and effects upon the people. FAO, Rome

Yuan BC, Li ZZ, Liu H, Gao M, Zhang YY (2007) Microbial biomass and activity in SAS under arid conditions. Appl Soil Ecol 35:319–328

Yue H, Mo W, Li C, Zheng Y, Li H (2007) The salt stress relief and growth promotion effect of Rs-5 on cotton. Plant Soil 297:139–145

Zahran HH, Ahmad MS, Afkar EA (1995) Isolation and characterization of nitrogen-fixing moderate halophilic bacteria from saline soils of Egypt. J Basic Microbiol 35:269–275

Zavarzina DG, Kolganova TV, Boulygina ES, Kostrikina NA, Tourova TP, Zavarzin GA (2006) *Geoalkalibacter ferrihydriticus* gen. nov. sp. nov., the first alkaliphilic representative of the family *Geobacteraceae*, isolated from a soda lake. Microbiology 75:673–682

Zhilina TN, Garnova ES, Tourova TP, Kostrikina NA, Zavarzin GA (2001) *Amphibacillus fermentum* sp. nov. and *Amphibacillus tropicus* sp. nov., new alkaliphilic, facultatively anaerobic, saccharolytic bacilli from lake Magadi. Microbiology 70:711–722

Zhilina TN, Kevbrin VV, Tourova TP, Lysenko AM, Kostrikina NA, Zavarzin GA (2005) *Clostridium alkalicellum* sp. nov., an obligately alkaliphilic cellulolytic bacterium from a soda lake in the Baikal region. Microbiology 74:557–566

Zhong HL, Lauchli A (1994) Spatial-distribution of solutes, K, Na, Ca and their deposition rates in the growth zone of primary cotton roots: effects of NaCl and $CaCl_2$. Planta 194:34–41

Impact of Plant-Associated Microbial Communities on Host Plants Under Abiotic Stresses

10

Saumya Arora and Prabhat Nath Jha

1 Introduction

Environmental impact resulting from the reduction of arable land, water resources, increased global warming, and various biotic and abiotic stressors may cause decline in the yield of major food crops. The major concern is abiotic stress, an environmental impact which reduces plant growth and yield below the optimal level. These abiotic factors include drought, salinity, high or low temperature, and various pollutants such as toxic metals. Moreover, abiotic stressors can alter the interaction of plants with pest, making plants more susceptible to pathogenic organisms. Also, a weakened immune system of plant reduces their competing ability with weeds.

Plant's response to these stressors depends upon which abiotic stress they are affected by and at what developmental stage. Plants can be affected by single stress at a time or by the combinatorial effect of various stressors. Plants' response to these stressors is a very complex phenomenon which includes both physiological and biochemical changes within the host plants. These changes include a variation in an aspect ratio of root-shoot length, leaf wilting, leaf abscission, generation of reactive oxygen species (ROS), altered relative water content, and accumulation of free radicals that can disrupt cellular homeostasis, thereby affecting cell viability. Plants also produce stress hormones, namely, abscisic acid and stress ethylene, which induce closure of leaf stomata leading to reduced water loss from the plants and withstand various molecular and metabolic changes.

At present, different strategies are practiced to overcome the problem of abiotic stress in plants which include the use of water-saving irrigation and traditional breeding. Also, genetic engineering approaches are applied to construct

S. Arora · P. N. Jha (✉)
Department of Biological Sciences, Birla Institute of Technology and Science, Pilani, Pilani, Rajasthan, India
e-mail: prabhatjha@pilani.bits-pilani.ac.in

D. P. Singh et al. (eds.), *Microbial Interventions in Agriculture and Environment*,
https://doi.org/10.1007/978-981-13-8383-0_10

stress-tolerant transgenic plants to increase the plant ability to combat particular stress. However, these approaches are highly technical, time-consuming, and labor intensive and, therefore, are challenging to practice. Application of plant growth-promoting microorganisms (PGPM) which include rhizobacteria (PGPR) and mycorrhizal fungi is one of the potential alternatives for improving plant growth under drought and salinity condition. Certain PGPM can produce a wide range of metabolites and enzymes that help plants to withstand various biotic and abiotic stresses (Ngumbi and Kloepper 2016).

The microbial community includes both endophytic and rhizospheric microorganism. These microorganisms have become a major tool guard to protect a plant's health in an eco-friendly manner. They have been considered as a potential remedy to improve crop productivity as they have the potential to improve various physiological processes in plants either directly or indirectly.

This chapter covers the impact of various abiotic stresses such as drought, salinity, high temperature, and toxic metals on plant growth, how plants impart tolerance against them, and finally how microorganisms associated with plants assist plants in improving their growth and productivity.

2 Abiotic Stress

Expanding industrialization and urbanization have a direct impact on the environment as well as on the loss of available land that can be used for different agriculture practices. Moreover, both biotic and abiotic stresses affect the health of plants, thus decreasing the yield of crops. Abiotic stress reduces plant growth by decreasing water uptake and modifying biochemical and physiological processes in plants (Hashem et al. 2016).

Stress is an environmental condition which has negative impacts on the growth and development of plants and decreases the crop yield below the optimum level. Major abiotic stress factors faced by plants are salinity, drought, and high or low temperature (Fig. 10.1). Plants respond to these abiotic stressors either dynamically or complexly, and these responses can be either reversible or irreversible. All plants induce the response against these stressors; these responses can be either to escape the stress condition, e.g., seeds undergo metabolically inactive dormant stage, or plants resist these stresses, i.e., plants respond actively to overcome stress treatment. Stress resistance includes both stress avoidance and stress tolerance. Stress avoidance means that plants try to maintain unstressed condition at cellular and tissue level, whereas in stress tolerance, plants respond to altered environmental conditions. The plant responses at a physiological level to abiotic stressors depend on the tissue or organ which are exposed or affected directly by the stress. The earliest metabolic responses to abiotic stress include growth inhibition, a decrease in protein synthesis, while an increase in protein folding and processing. The energy metabolic pathways get affected under severe stress condition.

The plant responses to environmental/abiotic stress can be divided into three phases:

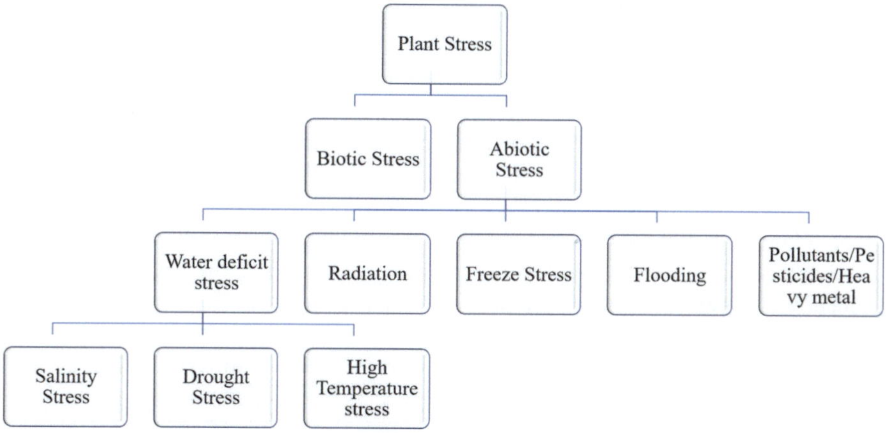

Fig. 10.1 Different types of stress challenged by the plants in its lifetime

(a) Stress stimulus sensing.
(b) Activation of a signal transduction mechanism into the cell.
(c) Activation of the suitable counter mechanism.

A plant produces various signals under stress condition. One of the earliest signals involves the generation of ROS and RNS (reactive nitrogen species), which can further alter gene regulation and activity of the various enzymes. Another major signal includes plant hormones. Abscisic acid (ABA) and ethylene are two important hormone regulators of plant response under various abiotic stresses. ABA is a phytohormone, which controls many developmental processes in plants such as stomatal closure, dormancy of seed and bud, and expression of genes in response to various biotic and abiotic stresses. ABA-mediated signalling controls plant responses to various abiotic stressors as well as the plant pathogen. ABA signalling involves three major components, ABA-soluble receptors (PYR/PYL/RCAR): 2C-type protein phosphatases (PP2C), which include ABI1 and ABI2; and protein kinases related to SNF1 family (SnRK2/OST1) (Hu et al. 2012). These complexes play an important role in ABA detection and signalling. Ethylene is a gaseous phytohormone which directs several developmental processes in plants such as promoting fruit ripening, leaf senescence and wilting, abscission, germination, root architecture, and flowering (Chandra and Sharma 2016; Abeles et al. 1992). At low concentration, ethylene impacts several benefits to the plants, but its higher concentration is detrimental as it results in root growth inhibition, abnormalities in the development of hypocotyl and its elongation, and defoliation and growth retardation; these phenomena are together termed as a triple response. Increased ethylene biosynthesis has been noticed during a stress condition, whether biotic or abiotic, which ultimately results in a decrease in the growth of root and shoot (Chandra and Sharma 2016).

Plant responses to stressors are mediated by various proteins which are directly involved in determining novel phenotype in an altered environment by adjustment of physiological traits. These proteins can be either structural proteins or regulatory proteins, and their function depends upon their cellular localization, post-translational modification, and their interactome along with their molecular structure. However, there are several other aspects, particular to different types of stress factors which may be responsible for the alleviation of stress conditions from the plant. Plant response to different abiotic stresses involves various interaction and cross-talk between multiple molecular pathways, as various signalling pathways get induced during environmental stress.

The responses to stressors activate with the perception of both external and internal signals through different interlinked or independent pathways which further regulate various responses for the development of tolerance (Hasanuzzaman et al. 2013). These external and internal signals can be certain stimuli, the interaction of various cofactors, or any signalling molecule. Signalling molecules along with transcriptional factors, through signal transduction mechanism, activate various stress-responsive genes depending upon plant type, development stage of the plants, and type of stress. Major signal transduction pathways involve mitogen-activated protein kinase (MAPK/MPKs), Ca-dependent protein kinases (CDPKs), nitrogen oxide (NO), phytohormones (Hasanuzzaman et al. 2013), and sugar (as signalling molecule). Upon activation, these stress-responsive genes activate detoxifying enzymes and free radical scavengers to detoxify ROS and redirect the synthesis of essential proteins and enzymes; all these processes will help to maintain the cellular homeostasis.

2.1 Drought and Salt Stress

Drought is a condition where soil water potential decreases which causes less water uptake by plant's root, whereas salinity stress is a condition where there is excess salt in the soil, inhibiting plant growth or can even lead to plant death. The decrease in water potential is a key effect in both drought and salt stress. Low water potential creates difficulties for plants to uptake water from the soil (Verslues et al. 2006). Plants face various physiological and biochemical changes during drought and salt stresses which hamper growth, maturity, and yield of the plants (Noman et al. 2017). Like other stressors, drought and salt (water stress) also induce production of stress ethylene which adversely affects root and shoot growth of the plants (Sharp and LeNoble 2002). Drought also has a significant influence on different processes including respiration, photosynthesis, ion uptake, translocation of nutrients, catabolism of carbohydrates, and the mobilization and re-mobilization of nutrients. Both drought and salinity stressors create water-deficit condition which inhibit plant growth as water uptake get declined into expanding cells. Also, rheological properties of cell wall get altered both enzymatically and non-enzymatically. Growth inhibition further leads to a decrease in photosynthetic and respiration rate (Forni et al. 2016).

The stress signals generated from drought and salt stress can be categorized as a primary and secondary stress signal. A hyperosmotic signal, a primary signal caused by both drought and salt, can be referred to as osmotic stress which decreases the water-absorbing ability of the root systems and accelerate water loss from the leaves. The secondary stress signals are more severe and much complex. They include oxidative damage to cellular components (oxidative stress) and dysfunction of various important metabolic pathways including respiration and photosynthesis, nutrient imbalance, and decrease in stomatal aperture. Hyperosmotic signals result in accumulation of stress hormone called abscisic acid in plants which activates adaptive responses in plants. As in osmotic stress, salinity stress is also hyper-ionic stress as it leads to the accumulation of Na^+ and Cl^- ions in tissues of plants grown in NaCl-rich soil. The resulting accumulated ions cause ion imbalance and other physiological interventions. For instance, the accumulation of Na^+ inhibits K^+ ion uptake, later is required for the growth and development of plants. Ion homeostasis, osmoprotectant protection, modulation in hormones, and activation of antioxidant pathways are few key physiological and biochemical changes that plants adapt to deal with drought and salinity stress.

2.2 High Temperature/Heat Stress

Numerous biochemical reactions involving plant growth and development are temperature sensitive. Any change in global temperature directly or indirectly affects the plant metabolism. Increasing global warming has resulted in increased Earth's temperature by about 0.2 °C in every 10 years, which may lead to about 2–4 °C rise in temperature by 2100 in comparison to the current level (Asthir 2015; Hasanuzzaman et al. 2013). The major cause of the rise in global temperature is the discharge of greenhouse gases such as methane, nitrous oxide, and carbon dioxide (Asthir 2015). These changes in global temperature have altered geographical distribution and climate change. Therefore, a high temperature beyond a threshold value limits plant growth, their metabolism, and ultimately the crop productivity (Wahid et al. 2007). High-temperature stress includes three major aspects, i.e., intensity, time duration, and frequency of increased temperature over a period (Wahid et al. 2007). High-temperature stress has thus become a key issue, which plants are facing and have to face in the coming future. High temperature leads to enzymes inactivation, denaturation of heat-sensitive proteins, and generation of ROS and can also lead to cell death in adverse conditions (Hu et al. 2018). Plant response to elevated temperature depends upon the degree and extent of the raised temperature as well as the type of plant. Species of plants which required less temperature for their growth, such as temperate crops, are under major risk to increasing global temperature (Asthir 2015).

Major effects of heat stress on plants include loss of water, oxidative damage by generation of ROS, reduction in plant growth, improper development, reduced plant yield, alteration in photosynthesis, and inhibition of seed germination. Heat stress creates a metabolic imbalance as it disturbs the stability of various proteins and enzymes and membrane integrity and reduced enzyme efficiency. For example, starch synthesis is one of the important factors in grain filling. Therefore, crop yield is affected by the endosperm enzyme and soluble starch synthase of the cereal crops. These enzymes are very much heat-liable, and even 30 °C temperature for 3 hour can reduce its activity, thereby decreasing crop yield as the major proportion of grain weight is due to starch alone (Hu et al. 2018). Morphological changes due to heat stress include leaves' damage, decoloration of leaves, leaf senescence and abscission, fruit discoloration, retarded shoot growth due to a reduction in the length of the first internode, inhibition of root growth, and even disturbance in reproduction and fertilization processes (Wahid et al. 2007). At sub-cellular level, photosynthesis, the most important phenomenon, is greatly affected by heat stress, especially in C_3 plants. Both carbon metabolism and photochemical reactions occurring in stroma and thylakoid membrane, respectively, of the chloroplast are majorly impaired by high temperature. Structural organization of thylakoid membrane is disturbed including swelling of grana and loss of grana stacking. The heat stress leads to several changes in mesophyll cells, which include rounding up of chloroplasts, swollen stromal lamella, aggregation and clumping of vacuole content, disruption of cristae, and altered mitochondrial function (Wahid et al. 2007; Rodríguez et al. 2005). All the abovementioned events result in reduced photosynthetic and respiratory activity as such changes result in malfunctioning of thermo-liable photosystem II (PSII) (Wahid et al. 2007). High-temperature stress also disturbs microtubule organization by spindle splitting and elongation, microtubule aster formation in mitotic cells, and phragmoplast microtubule elongation (Wahid et al. 2007).

Plants have to struggle for their survival under rapidly changing environmental conditions continuously. Plants cope with heat stress by possessing various mechanisms including adaptive, avoidance, tolerance, or acclimation. Plant's response to a change in temperature depends upon their ability to perceive the high-temperature signals, further transmitting those signals, and activating appropriate physiological and biochemical changes (Hasanuzzaman et al. 2013). To offset the stress-induced physiological and biochemical changes, plants activate stress tolerance mechanism through signalling cascades including induction of MAPK and CDPK, chaperone signalling, and transcriptional regulation involving ion transporters, different proteins and enzymes, osmoprotectants, antioxidants, transcription factors, and other aspects (Wahid et al. 2007). At permissive high temperature, many changes occur at gene as well as transcript level to synthesize and/or activate heat protection proteins which include heat shock proteins (HSPs), galactinol synthase, and ascorbate peroxidase, which in turn provide thermotolerance to them (Hu et al. 2018). Genes responsible for osmoprotectant synthesis, regulatory proteins, transport proteins, and detoxifying proteins also get activated (Hasanuzzaman et al. 2013).

2.3 Heavy Metal Stress

Human activities like mining, industrialization, sewage processing, automobile emissions, and use of pesticides and chemical/inorganic fertilizers have increased the number of heavy metals above the toxic level in the soil which is neither suitable for the plants nor the animals. Plants act as a major source for entering these heavy metal contaminants into the food cycle, threatening the human life as well (Etesami 2018; DalCorso et al. 2013). Generally, metals are classified as essential (Fe, Zn, Cu, Mn, Ni, etc.) and non-essential (Cr, Pb, Hg, Cd and As) as far as their biological role is concerned (DalCorso et al. 2013). Heavy metals are those who weigh more than 5.0 g cm^{-3} and are categorized into three groups: (1) toxic metals (e.g., lead (Pb), nickel (Ni), mercury (Hg), copper (Cu), cobalt (Co), chromium (Cr), zinc (Zn), tin (Sn) cadmium (Cd), arsenic (As), etc.), (2) ornamental metals (e.g., gold (Au), silver (Ag), palladium (Pd), platinum (Pt), ruthenium (Ru), etc.), and (3) radioactive metals (e.g., thorium (Th), radium (Ra), uranium (U), americium (Am), etc.) (Etesami 2018). However, beyond threshold limits all metals are toxic to the environment, plants, microbial communities of rhizosphere niche, as well as human beings as they are not chemically and biologically degradable (Etesami 2018). Moreover, they induce stress symptoms including necrosis, chlorosis, wilting and retarded growth, senescence, leaf rolling, and eventually death in the plants (Paredes-Páliz et al. 2018; DalCorso et al. 2013), thereby, having negative impact on photosynthetic activity, biomass production, and growth of plants (Etesami 2018). Excess metal concentration can affect ion homeostasis and enzyme activity leading to the altered plant metabolism including germination, photosynthesis, growth, and nutrient uptake. Metal stress includes the generation of ROS (Hossain et al. 2012). Also, it interferes with Ca^{2+}/calmodulin signalling (DalCorso et al. 2013). Metals such as cobalt, nickel, and cadmium disturb the Ca^{2+} flow by altering the conformation of calcium ion channels; they also replace calcium from calmodulin-binding sites, thereby inhibiting calcium-dependent signalling (DalCorso et al. 2013). Excess metal ion concentration also causes damage to DNA (Lin et al. 2012) and causes peroxidation of membrane lipids (Upadhyay and Panda 2010) along with interfering the uptake of crucial metal ions (Etesami 2018).

In general, following are the four major mechanisms by which excess metal ion concentration exerts a negative impact on plants (DalCorso et al. 2013):

1. They raise competition with nutrient species with similar properties for absorption by the root cell membrane.
2. Metals such as mercury, cadmium, lead, and arsenic react with sulfhydryl groups (–SH) of a cysteine residue in the polypeptide chain, disrupting the protein/enzyme's structure and function. Several metals also react with a carboxylic group of the protein.
3. Induction of oxidative stress by producing ROS.

4. They displace essential metal co-factor from various enzymes and signalling proteins. For example, cadmium can displace zinc, iron, and manganese from the enzyme superoxide dismutase, thus affecting the enzyme activity. Also, Cd displaces zinc from zinc finger transcription factors, which interferes with gene expression.

Higher plants usually uptake the metal ions from the soils through various membrane transporters located in the plasma membrane of the root hairs. There are three main families of ion transporters, namely, natural resistance-associated macrophage protein (NRAMP), copper transporter (CTR/COPT), and ZTR/IRT-related proteins (ZIP) which along with some aquaporins are required for uptake of metal ions (DalCorso et al. 2013). The NRAMP transporters are responsible for the transport of several divalent cations such as manganese, zinc, copper, iron, cadmium, nickel, and cobalt across the plasma membrane and tonoplast in plant's root. The ZIP transporters also transport divalent cations across membranes, but different members of this family can be expressed either in roots alone or roots and shoot both (DalCorso et al. 2013). From the roots these metal ions along with their chelator molecules are transported to shoots through the xylem, involving various transport proteins such as P-type ATPases, MATEs (multidrug and toxic compound extrusion), and OPT (oligopeptide transporter) protein families (DalCorso et al. 2013). Phloem is another route for the translocation of metal ions.

Metal chelation by root exudates, transport, and compartmentalization into vacuoles and cell wall are few strategies which the plants have acquired to overcome or to circumvent the damaging effects from accumulation of excess metal ions. Other methods include cell wall modification, secondary metabolite production, metal translocation to phloem, ethylene synthesis, and antioxidant production (Paredes-Páliz et al. 2018; DalCorso et al. 2013). Some of the stress pathways including expression of genes encoding enzyme cysteine synthase, glutathione reductase, and glutathione transferase are activated by metals. These enzymes are required for the synthesis of the glutathione and phytochelatins, which are involved in the detoxification mechanism (Paredes-Páliz et al. 2018).

3 Plant Adaption to Abiotic Stress

3.1 Physiological and Biochemical Response

Plants adapt to various physiological and biochemical mechanisms to generate tolerance against various abiotic stresses, which include ion homeostasis by ion transport and uptake, biosynthesis of osmoprotectant and compatible solutes and secondary metabolites, and phytohormone modulation (Fig. 10.2).

Ion Homeostasis Maintaining ion balance is not only crucial for normal plant growth but also essential for the development of the plants under salinity stress conditions. Plants achieve ion homeostasis by either ion transport, ion uptake, or ion

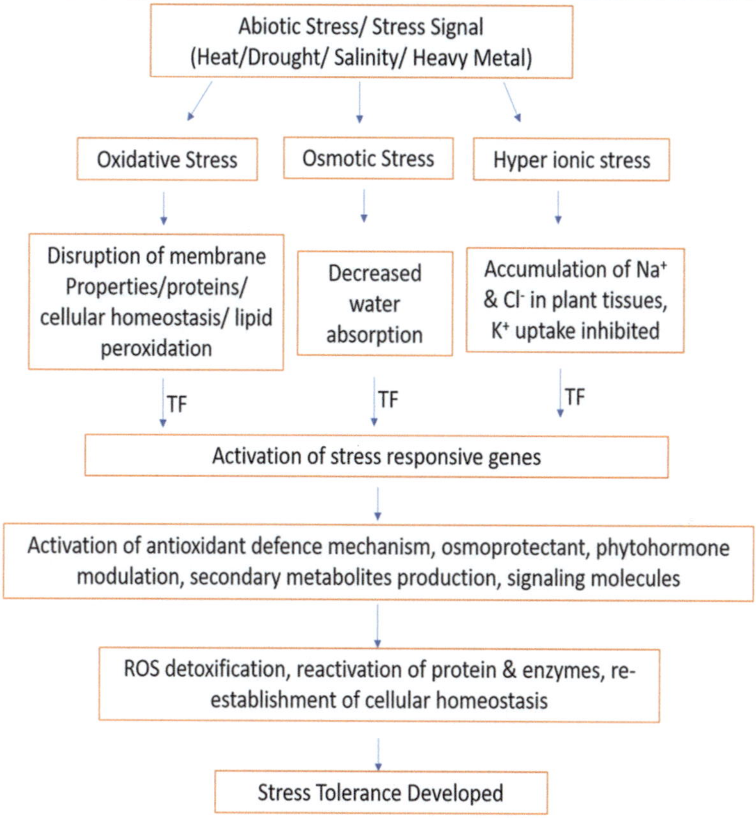

Fig. 10.2 Representation of abiotic stress-induced signal transduction mechanism and stress tolerance development in plants. In response to various stress signals, plants activate various stress-responsive genes. Their gene product helps plants to combat stress by re-establishing the cellular homeostasis and developing stress tolerance (Na^+ sodium ion, Cl^-, chloride ion, K^+ potassium ion, *TF* transcription factors, *ROS* reactive oxygen species)

compartmentalization. Both halophytes (salt tolerant) and glycophytes (salt sensitive) cannot tolerate high salt concentration in their cytoplasm. But halophytes can overcome high salt concentration as they can either transport excess salt to vacuoles or requisition to older tissues, which eventually die and protect younger tissues from high salt concentration.

Salt stress activates salt overlay sensitive (SOS), the Na^+ influx transporter HKT1, and the tonoplast Na^+/H^+ antiporter NHX1 stress signalling pathways to maintain ion homeostasis (Verslues et al. 2006). Most of the soils contain an excess amount of NaCl. Excess Na^+ that enters the cytoplasm is transported to the vacuoles by Na^+/H^+ antiporters (NHX1). V-ATPase is one of the two H^+ pumps present in the vacuole membrane, which under normal conditions maintains solute balance, provides energy for secondary transport, and facilitates vesicle fusion. Under salt stress, the activity of this pump increases, and it is important for plant survival during stress condition, but the activity of another H^+ pump, i.e., the vacuolar pyrophosphatase (V-PPase), is either inhibited or plays a minor role under the same conditions (Gupta and Huang 2014).

SOS signalling pathway also plays an important role in ion homeostasis. SOS1, SOS2, and SOS3 are three major proteins of this pathway. *SOS1* gene encodes plasma membrane Na^+/H^+ antiporter (SOS1 protein) which regulates Na^+ efflux and facilitates Na^+ transport from root to shoot. *SOS2* gene encodes serine/threonine kinase (SOS2 protein) and gets activated by salt stress-elicited Ca^{2+} signal, and SOS3 is a calcium-binding protein. Interaction of SOS2 and SOS3 activates protein kinase; this complex also controls the expression and activity of ion transporters (Gupta and Huang 2014; Rodríguez et al. 2005). Activated protein kinase then phosphorylates SOS1 protein, thereby increasing its activity, i.e., efflux of Na^+, and reducing Na^+ toxicity, maintaining ion homeostasis, and detoxifying the response. This protein also maintains pH balance, membrane vesicle trafficking, and vacuole functions.

Osmoprotectant and Compatible Solutes Compatible solutes can be defined as uncharged, polar, and highly soluble organic compounds/osmolytes, which do not interfere with the metabolism of the cell even at high concentration. They are low-molecular-weight compounds, which include sugars, proline, glycine betaine, and polyols. Their concentration within the cell is maintained either by their irreversible synthesis or by maintaining the balance between their synthesis and degradation. These osmolytes protect important biological structures and maintain cellular osmotic balance via constant water influx, reducing osmotic potential and thereby improving cell water retention ability, termed as the osmotic adjustment under various abiotic stresses including drought, salt, or temperature stress. Osmotic adjustment sustains cell structure and photosynthetic activity even at low water potential, therefore, delay leaf senescence and death. They also can reorganize proteins and other cellular structure to maintain turgidity of the cell (Hasanuzzaman et al. 2013).

During salt and drought stress, the concentration of amino acids such as cysteine, arginine, and methionine significantly decreases, whereas there is a marked increase in the concentration of amino acid proline. Glutamate acts as a precursor for proline biosynthesis, and major regulatory enzymes responsible for its synthesis are pyrroline carboxylic acid synthetase and pyrroline carboxylic acid reductase. Proline has very strong hydration ability as its hydrophobic part interacts with proteins and hydrophilic part interacts with the water molecule, enhancing protein solubility and preventing them from denaturation and degradation during osmotic stress. Proline maintains cell turgor pressure, ameliorates activity of certain antioxidative enzymes, increases photosynthetic activity, as well as acts as organic nitrogen reserve during stress condition, thereby helping in stress recovery. It also reduces cell acidity by adjusting the redox potential of the cell.

Further, γ-4-Aminobutyric acid (GABA), a non-protein amino acid, is widely distributed molecule in all living organism which also acts as a compatible solute. It is being synthesized from an amino acid, glutamic acid with the help of enzyme glutamate decarboxylase (GAD) under acidic pH. GAD activity is also regulated by calmodulin under heat stress (Wahid et al. 2007). GABA accumulates under various environmental stresses and helps in mitigating stress effects as it plays an important role in stress signal perception for the activation of physiological response to overcome the stress conditions. In addition to proline, several other osmoprotectants are known to ameliorate abiotic stressors (Box 10.1).

Box 10.1: Various Osmoprotectants Synthesized by Plants in Response to Stress

Glycine betaine: Glycine betaine is non-toxic, amphoteric quaternary ammonium compound and cellular osmolyte which protects the cell from osmotic stress by increasing osmolarity of the cell during the stress period. It is synthesized from choline or glycine. Along with osmotic adjustment, glycine betaine in combination with proline stabilizes proteins and protects the photosynthetic apparatus from ROS and other stress damage.

Polyols: Polyols are cyclic or non-cyclic compounds with many reactive hydroxyl groups in their structure. Sugar alcohols are the major class of polyols which act as compatible solutes and ROS scavengers. Mannitol is a non-cyclic sugar alcohol, produced by NADPH-dependent mannose-6-phosphate reductase enzyme during osmotic stress and protects enzymes and membrane structure from stress damage.

Sugars: Sugars (e.g., glucose, fructose, fructans, and trehalose) and starch also get accumulated during osmotic stress and help in stress mitigation either by acting as osmoprotectants, carbon storage, or ROS scavenger. Starch degradation in osmotically stressed tissues helps in carbon export and osmolyte accumulation in different tissues. Soluble sugars accumulate to adjust osmotic balance and to procure the needed turgor pressure under water scarcity. Trehalose, a reducing disaccharide, as the most affected sugar during osmotic stress, leads to cross-talk with ABA signalling for stomata closure as trehalose-6-P is required for ABA-dependent germination. Trehalose stabilizes the structure of proteins, nucleic acid, and other macromolecules by blocking transformation of phospholipid bilayer plasma membrane from the liquid crystal state to a solid state under drought condition.

Secondary Metabolites Secondary metabolites are the metabolites which do not have any functional importance in the maintenance of life processes in the plant but provide extra benefits, if synthesized, to plants such as protection, interaction among different species, and competition. They help plants to interact with their environment for adaption and defense (Akula and Ravishankar 2011). They are usually accumulated in plants under various abiotic stress conditions. They are synthesized from primary metabolites. Phenolic compounds are the most vital category of secondary metabolites produced by the plants which provide tolerance to them against stress condition. Abiotic stressors are also known to induce production of phenolic compounds, for instance, flavonoids and phenylpropanoids (Akula and Ravishankar 2011; Wahid et al. 2007) which can have a role in stress amelioration. Jasmonate and methyl jasmonate stimulate the biosynthesis of various secondary metabolites as a defense response. Anthocyanin, a flavonoid, is accumulated during salt and drought stress, but its concentration decreases during heat stress and metal stress. Plant tissues containing anthocyanin are usually resistant to drought stress.

Another important class of secondary metabolites is polyamines. They are small, low-molecular-weight and polycationic aliphatic compounds these polyamines are ubiquitous to plant kingdom and contribute in plant growth and development including regulation of cell proliferation, differentiation, and morphogenesis, somatic embryogenesis, breaking dormancy and seed germination, and development and senescence of fruit and flower. Some of the most common polyamines present in plant kingdom include di- (PUT), tri- (SPD), and tetra-amine spermine (SPM). PUT, the smallest polyamine, is synthesized from either of the two amino acids, ornithine and arginine, by enzymes ornithine decarboxylase (ODC) and arginine decarboxylase (ADC), respectively. PUT further acts as a prime substrate for the synthesis of high molecular weight polyamines such as SPM and SPD. The synthesis of SPD is catalyzed by spermidine synthase which adds an aminopropyl group to diamine putrescine, whereas spermine synthase catalyzes the addition of the aminopropyl group to tri-amine spermidine to synthesize tetra-amine spermidine (Vílchez et al. 2018).

Level of endogenous polyamines increases during cold, drought, and salinity stress, thereby maintaining membrane integrity, controlling the expression of genes involved in the synthesis of osmotically active solutes, reducing ROS production, and controlling ion accumulation. Polyamine level within the cell is regulated by the enzyme copper-binding diamine oxidases along with FAD-binding polyamine oxidases, hence playing a significant role in stress tolerance. The balance between polyamine biosynthesis and catabolism is regulated by drought, and salinity stress condition is thereby acting as a cellular signal in regulating abscisic acid (ABA) hormonal pathways in response to stress. SPM and SPD also induce another important small, volatile gaseous signalling molecule, nitric oxide (NO), under salinity stress. Nitric oxide regulates numerous growth and development processes in the plant, for example, the growth of roots, flowering, stomata opening, respiration, seed germination, and stress responses. NO triggers expression of many genes regulating redox stress either directly or indirectly and reacts with lipid radicals, thereby preventing their oxidation, acting as a scavenger of superoxide radicals, as well as activating antioxidant enzymes.

Phytohormone Regulation Phytohormones play a major role in sensing the adverse environmental conditions, making plants adaptable and tolerant against these conditions. Cross-talk between hormonal signalling generates different stress signals, integrate those inputs and make plants respond to those signals appropriately (Wahid et al. 2007). Abscisic acid and ethylene are major phytohormones which act as signal molecules under various stress conditions.

Abscisic acid (ABA) has many physiological functions in plants like seed development, control dormancy of seed and bud, embryo maturation, seed germination, and ripening of fruits. It is also considered as stress hormones as it is upregulated in plant roots and shoots during water-deficit conditions. ABA reduces water loss by controlling several stress-adaption responses such as activation of those genes which are responsible for osmotic balance, limiting the rate of transpiration and wilting of leaves, as well as it compartmentalizes ion and regulates the growth of root and shoot. ABA accumulation helps plant to mitigate stress condition as it maintains ion balance by amelioration of K^+ and Ca^{2+} accumulation and increased production of compatible solutes, for example, proline and sugars, thereby counterbalancing the Na^+ and Cl^- ion uptake. It also acts as a cellular signal to modulate many salt and water deficit-responsive gene expression. ABA also persuades closure of leaf stomata, thereby reducing water loss during transpiration, as well as decreases photosynthesis rate to improve plant efficiency to use water. ABA level also gets increased during heat stress, and it gets accumulated during the heat recovery period. ABA modifies the expression of certain genes including induction of several heat shock proteins (HSPs), which help plants in providing thermotolerance.

The activity of compounds like brassinosteroids (BR) as well as salicylic acid (SA) also increases during abiotic stress. Although these compounds are not phytohormones, they have hormonal properties. Salicylic acid restores membrane potential and prevents K^+ loss by rectifying GORK (guard cell outward-rectifying potassium ion) channels under salinity stress. They also upregulate the activity of H^+-ATPase, thereby improving retention of potassium ion, whereas brassinosteroids help to overcome oxidative stress by improving the activity of antioxidant enzymes and accumulation of other antioxidants. Salicylic acid (SA) is also involved in heat stress response. It acts as a signal molecule in various heat response pathways and stabilizes the heat-shock transcription factor trimers, which help them to associate with heat shock elements (HSE) to the promoters of genes related to heat shock (Wahid et al. 2007). SA provides thermotolerance by involving both Ca^{2+} homeostasis and antioxidant system. Salicylic acid derivatives like sulphosalicylic acid (SSA) and methyl salicylate (MeSA) act as signal molecules and have multiple functions including removal of ROS such as hydrogen peroxide (H_2O_2) or by elevating the concentration of antioxidants like ascorbic acid and α-tocopherol, thereby increasing thermotolerance (Wahid et al. 2007; Shi et al. 2006; Llusia et al. 2005).

3.2 Molecular Responses

Along with physiological and biochemical changes, abiotic stressors also induce changes at the molecular level regarding induction/alteration of gene expression, activation of regulatory proteins, and initiation of plant signalling to combat the effect of given stressors. Under abiotic stress, changes in plants at molecular level include activation of antioxidant mechanism, synthesis/activation of stress proteins, and production of metal chelators.

Antioxidant Enzymes and Other Compounds Disturbance in electron transport chain (ETC) in mitochondria and chloroplast during biotic and abiotic stresses causes molecular oxygen (O_2) to accept an electron, and this leads to the accumulation of reactive oxygen species (ROS) in the cell. These ROS (e.g., singlet oxygen ($^1O_2$), superoxide radicals ($^{\bullet}O^{-}{}_2$), and hydroxyl radicals ($^{\bullet}OH$)) act as oxidizing agents and damage cellular integrity. The overaccumulation of ROS can damage phospholipids and fatty acids present in the plasma membrane, resulting in membrane lipid peroxidation and protein carbonylation. Excess ROS can also damage cellular DNA. Also, ROS stress causes the disturbance in the spatial configuration of various enzymes and other membrane proteins, which leads to increased membrane permeability and ion leakage, chlorophyll annihilation, metabolism disquiets, and even death of the plant. Abiotic stress results in increased antioxidant metabolism due to the imbalance between electron generation and utilization to produce antioxidant enzymes and other antioxidant compounds to detoxify these ROS in chloroplasts, peroxisomes, mitochondria, endoplasmic reticulum, plasma membrane, and cell wall as a defense mechanism to maintain redox homeostasis of the cell. Antioxidant enzymes include catalase (CAT), glutathione peroxidase (GPX), superoxide dismutase (SOD), ascorbate peroxidase (APX), and glutathione reductase (GR), whereas non-enzymatic antioxidants are anthocyanin, ascorbate, glutathione, and α-tocopherol (vitamin E). In addition to antioxidative enzymes, silicon, ascorbic acid, α-tocopherol, and glutathione are also involved in antioxidative responses.

Silicon influences the hormonal and antioxidant response in rice under salinity stress and increases plant growth by reducing sodium accumulation and lipid peroxidation (Gupta and Huang 2014). Ascorbic acid, synthesized from D-glucose in mitochondria, is distributed to other organelles through diffusion and protects plants from harmful effects of hydrogen peroxide and other ROS. The α-tocopherol is a predominant antioxidant in chloroplast membrane and protects plants from photooxidative damage. It mainly prevents lipid peroxidation (Ahmad et al. 2010). Glutathione is a tripeptide of glutamic acid, cysteine, as well as glycine and exists in almost all the cellular compartments. It acts as both antioxidant and metal ion chelator. Also, it acts as a substrate molecule for the biosynthesis of phytochelatins (DalCorso et al. 2013; Ahmad et al. 2010).

Stress Proteins Stress proteins are highly expressed during abiotic stress. They are mostly water-soluble and provide stress tolerance to plants by hydrating cellular structures (Akula and Ravishankar 2011). Some of the stress-related proteins are described.

Late Embryogenesis Abundant (LEA) Proteins: LEA proteins have low molecular weight (10–30 kDa), and hyper hydrophilic proteins formed during seed development are abundant in basic amino acids like lysine, glycine, and serine but lack cysteine and tyrosine residues. They are accumulated at the time of seed desiccation (at a later stage of embryogenesis) as well as during water-deficit conditions persisting in cold, drought, or high salinity stress. Due to the biased amino acid composition in their sequence, they are categorized into nine distinct groups (Hundertmark and Hincha 2008). They protect biological macromolecules by redirecting water distribution within the cell, protect cells from ion damage as they bind with inorganic ions, and prevent dehydration of the cells. They also prevent conformational changes in enzymes induced due to water-deficit condition, maintaining their biological activity and preventing them from denaturation or aggregation (Olvera-Carrillo et al. 2011). The promoter region of many genes encoding LEA protein has an abscisic acid response (ABRE) along with low-temperature response elements (LTRE) and is activated by drought, cold, or abscisic acid according to the respective element present in their promoter.

Aquaporins Aquaporins are water transport channels in the plants. They are highly selective to regulate processes like seed germination and stomatal movements. They facilitate the passive transport of water along with small uncharged solutes from high concentration to low across the vacuolar and plasma membrane, thereby maintaining moisture balance during drought and salt stress. Aquaporins are transmembrane proteins, forming channels and maintaining cellular water homeostasis (Zargar et al. 2017). The number of aquaporin proteins present vary among plant genera, for example, *Arabidopsis*, maize, potato, cotton, tomato, and tea possess 35, 36, 41, 71, 47, and 20 aquaporin proteins, respectively (Feng et al. 201). These aquaporin proteins are categorized into five subclasses based on the sequence similarity in their amino acid composition and their membrane localization. These subclasses include plasma membrane intrinsic proteins (PIPs), NOD26-like intrinsic proteins (NIPs), small basic intrinsic proteins (SIPs), tonoplast intrinsic proteins (TIPs), and the unrecognized X intrinsic proteins (XIPs) (Feng et al. 2018; Zargar et al. 2017). The activity of the abovementioned proteins is regulated at both transcriptional and post-transcriptional level. At the transcriptional level, endogenous signals (such as gibberellin, abscisic acid, and brassinolide), in addition to environmental factors (such as drought, salinity, cold), modulate their activity, whereas post-transcriptional regulation includes phosphorylation, heteromerization, change in cytosolic pH, and calcium ion concentration (Zargar et al. 2017).

Heat Shock Proteins (HSPs) Heat Shock Proteins (HSP) are stress proteins with a varied molecular weight (about 10 to 200 kDa), which are ubiquitous in nature and found in the cells of all the organisms. They have molecular chaperone-like function and are associated with various signal transduction pathways. They are encoded by heat shock genes (HSGs), which are stimulated by heat stress. HSPs are dynamic protein family with extremely heterogeneous nature (Hasanuzzaman et al. 2013). HSPs are categorized into five different classes in plants: HSP100 (or ClpB), HSP90, HSP70 (or DnaK), HSP60 (or GroE), and HSP 20 (or small HSP, sHSP). Among various HSPs, small heat shock proteins (sHSPs), ranging from 15 to 42 kDa in molecular weight, are abundantly present in plants under heat stress and otherwise usually remain undetected. Under normal environmental conditions, their synthesis is constrained only to few developmental stages, for instance, fruit maturation, pollen development/

microsporogenesis, seed germination, and embryogenesis. The nuclear gene encodes these proteins in plants, which are further divided into six categories (Hasanuzzaman et al. 2013; Sun et al. 2002). They are localized into different compartments, for example, three of the six classes are confined in cytosol or nucleus, whereas the other three classes are localized in either plastids, endoplasmic reticulum, or mitochondria (Sun et al. 2002). They ensure correct folding of the newly synthesized polypeptide chain. They also stabilize thermo-labile proteins from heat denaturation or activate their degradation after plants have overcome stress or injury period (Hu et al. 2018; Whitley et al. 1999). These stress proteins are induced under heat shock and controlled by heat shock transcription factors. Various stress signals activate these transcription factors which further get associated with the heat shock-responsive element (HSE), a greatly conserved upstream response element, located in the promoter of heat shock genes. Heat shock-responsive element comprises of the palindromic nucleotide sequence (5-AGAANNTTCT-3) which has a dual role in recognizing and binding to heat shock transcription factors. Under the unstressed condition, these heat shock transcription factors are present as a monomeric unit, bound to HSP70 molecule in the cytoplasm. Whenever there is a stress signal, HSP70 dissociates from these transcription factors and gets activated. Upon activation they enter the nucleus; the monomeric unit trimerizes, binds to the promoter region, and prompts the expression of stress protein genes (Fig. 10.3) (Hasanuzzaman et al. 2013; Sun et al. 2002; Whitley et al. 1999). HSPs are synthesized both in HSE-dependent and HSE-independent manner. They help in improving various physiological phenomena under heat stress including membrane stability, water, and

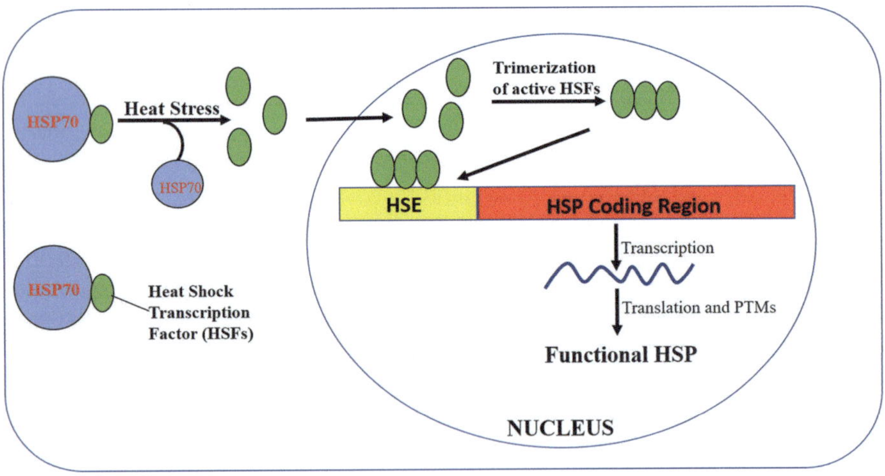

Fig. 10.3 Regulatory mechanism for activation and synthesis of heat shock protein. Under non-stress conditions, HSFs remain associated with HSP70. Upon induction of heat stress, HSFs get dissociated from HSP70 and enter the nucleus, where they trimerize and bind to heat shock-responsive element present in the promoter region of HSP genes. As soon as HSFs bind to HSE, expression of HSP genes initiates (*HSFs* heat shock transcription factors, *HSE* heat shock-responsive elements, *HSP* heat shock proteins, *PTMs* post-translational modifications)

nutrient efficiency along with photosynthesis (Wahid et al. 2007). At the molecular level, heat shock also alters gene expression to provide direct protection to plants against the stress condition.

HSP production is not only restricted to high-temperature stress. It has been stated that certain sHSP genes are also expressed under osmotic stress. HaHsp17.6 and HaHsp17.9 genes are expressed in sunflower under water stress. Also, homologous to these genes are found in resurrection plant *Craterostigma plantagineum*, and their level increases under drought stress via abscisic acid-dependent mechanisms (Sun et al. 2002). Several sHsp-encoding genes are also triggered by oxidative stress, low-temperature stress, UV radiation, heavy metal, ozone, and gamma irradiation, suggesting the role of HSP under different abiotic stresses (Sun et al. 2002).

Metal Chelation Metal chelation by root exudates is one of the stress tolerance mechanisms amended by the plants to prevent the accumulation of toxic metals within the cell, and it behaves as the first line of defense as plants come in contact to metal ions in soil matrix through roots. Plants also restrict metal ions into the apoplast, preventing their accumulation in root cells. The cell wall immobilizes metal ions as they are rich in pectic and histidyl groups along with extracellular carbohydrates, for example, callose and mucilage (DalCorso et al. 2013). Crossing the root barrier, metal ions are also chelated in the cytosol of the cell by binding to high-affinity ligands molecules, for instance, phytochelatins and metallothioneins. The metallothioneins are metal-binding peptides synthesized in plants during abiotic stresses such as metals, extreme temperatures, and nutrient deficit and development stimuli.

Moreover, amino acids (histidine and cysteine), as well as organic acids (citric acid and malic acid), act as metal ion chelator, thereby preventing tolerance to plants under metal stress (Wycisk et al. 2004). Phytochelatins are non-protein cysteine thiol group, produced from glutathione by the help of the enzyme phytochelatin synthase, which form complex with metals, which is further transported to different compartments such as vacuoles through transporter proteins present in tonoplast, reducing cytosolic concentration of free metal (DalCorso et al. 2013; Mohamed et al. 2012). Several heavy metals including zinc, silver, copper, cadmium, gold, mercury, and lead are known to activate phytochelatin synthase (DalCorso et al. 2013).

3.3 Transcriptional Regulation and Gene Expression

Expression of genes is significantly affected to combat the deleterious effect of abiotic stressors. Gene expression can be regulated at different stages of central dogma including transcription initiation, RNA processing, translation, or post-translational modification of proteins. These gene products can be categorized into two categories. One of these categories includes those genes which encode proteins that protect the cell from water stress. These genes include osmolyte biosynthesis genes, water channel, and membrane transporter genes, those genes which govern protection to the cell structure from oxidative damage, fatty acid metabolism enzymes, LEA proteins, and molecular chaperons to protect macromolecules other than proteins, protease inhibitors, and proteins involved in lipid transfer. Other categories include regulatory proteins and RNAs, which regulate stress signal transduction pathways and also modulate expression of other genes. Transcription factors included in this category are considered to be the most important regulator of gene expression. There are several families of transcription factors including WRKY, bZIP, NAC, C_2H_2 zinc finger, AP_2, and DREB that consist of a huge number of stress-responsive transcription factors. Other than transcription factors, various protein phosphatases, proteinases, as well as protein kinases (MAP kinase, receptor protein kinase, and CDP kinase) also play an instrumental role in signal transduction along with gene expression regulation. This second category of genes controls the expression of the first group (Lata and Prasad 2011).

Expression of various stress-related genes can be controlled by transcription factors as these TFs can bind to cis-acting elements in the protomer regions of these genes. These TF-controlled genes are further controlled either by the ABA-dependent pathway or ABA-independent pathway. ABA-dependent signalling system includes ABA-responsive element-binding protein/ABA-binding factor (AREB/ABF) regulon and myelocytomatosis oncogene/myeloblastosis oncogene (MYC/MYB) regulon, whereas ABA-independent signalling system includes the C-repeat binding factor/dehydration-responsive element-binding (CBF/DREB) regulon and the NAC and ZF-HD (zinc finger homeodomain) regulon (Lata and Prasad 2011). Based on the functions, genes and transcription factors that are upregulated under salt stress have been categorized as ion transport or homeostasis (e.g., SOS genes, AtNHX1, and H -ATPase), dehydration-related transcription factors (e.g., DREB), molecular chaperones (e.g., HSP genes), and senescence-associated genes (e.g., SAG).

One of the prime important transcription factors, dehydration-responsive element-binding (DREB) genes, activates the expression of several stress-related genes, mainly in an ABA-independent manner. Dehydration-responsive element (DRE) is a 9 bp conserved core sequence (5′-TACCGACAT-3′) present in the promoter region of many abiotic responsive genes, to which DREB interacts, for instance, late embryogenesis-abundant (LEA) and cold-responsive (COR) proteins. Another transcription factor which belongs to the subfamily of the bZIP transcription factor is ABA-responsive element-binding proteins/factors (AREBs/ABFs) which are ABA-dependent and upregulated during drought stress; their upregulation and activation require ABA. Transcription factor MYB2 gets upregulated under various abiotic stresses and impacts tolerance against them by regulating H_2O_2 and malondialdehyde accumulation. It also controls the expression of genes encoding various transporter proteins and proline synthases (Lata and Prasad 2011). Another important family of the transcription factor is a zinc finger. Cys_2/His_2 (C_2H_2) is a type of zinc finger protein which encodes ZPT2–3 gene; its constitutive overexpression impart tolerance to dehydration stress, whereas DST zinc finger protein functions as a negative controller of drought as well as salinity stress; it controls the H_2O_2-mediated stomatal closure genes (Fang and Xiong 2015). High-temperature stress involves phytochrome-interacting factor 4 (PIF4) transcription factor, which has a basic helix-loop-helix (bHLH) structure.

3.4 The Role of Regulatory RNAs in Stress Response in Plants

Transcription process in higher eukaryotes is much more complex than it seems. It has been observed that more than 90% of the genome is transcribed, but only 1–2% encodes for proteins. Most of the RNAs lack protein-coding capacity, and these RNAs are termed as non-coding RNA (ncRNA) (Shafiq et al. 2016; Liu et al. 2015; Zhu and Wang 2012). These ncRNAs are further grouped into small ncRNAs (18–30 nt.), medium-sized ncRNAs (31–200 nt.), and long ncRNAs (lncRNA) (>200 nt.).

Small ncRNAs are 20–24 nucleotide long and include small interfering RNAs (siRNAs) along with microRNAs (miRNAs). They suppress the target genes' expression either through degradation or inhibition of mRNA targets, thereby regulating gene expression by translational inhibition. They play a vital role in RNA-mediated gene silencing. They are synthesized from the primary (pri)-miRNA transcripts. Enzyme RNA polymerase II transcribes the nuclear-encoded miRNA (MIR) genes (Wang et al. 2017; Shriram et al. 2016; Liu et al. 2015). Under stress condition, plants assign these miRNAs as transcriptional and post-transcriptional gene regulators to mitigate plant's growth and development. Their expression is either downregulated or upregulated in response to various stressors, for example, the expression of miR-160 is upregulated, while miR-169 expression is downregulated in *Arabidopsis* under nitrogen deficiency. Upregulated miR-160 decreases the expression of genes coding for auxin response factors (ARFs), repressing plant growth and thus increasing stress tolerance, whereas downregulated miR-169 is involved in the increased accumulation of nitrate transporters (Wang et al. 2017).

Some miRNAs are conserved among plants species, displaying similar expression pattern, whereas some show opposing expression pattern in different species. For examples, miR-169 is upregulated under nitrogen deficiency in all plant species, but miR-156 shows differential expression pattern in different plant species under drought stress. Also, the expression pattern of miRNAs is expressed in a stress-dependent manner. For example, the expression of the miR-169 is upregulated under cold stress but decreased during a drought in *Arabidopsis* (Wang et al. 2017).

Long non-coding RNAs (lncRNAs) are those RNAs which are more than 200 bp in length. They act as riboregulators in complex organisms, including humans, animals, and plants, regulating many developmental processes or stress responses in the cell by controlling mRNA translation and stability (Shafiq et al. 2016; Amor et al. 2009). The lncRNA lacks open reading frame. In plants, they are transcribed by RNA polymerase II or III, also furthermore by polymerase IV/V. Once transcribed, these lncRNAs further undergo processing via splicing or non-splicing mechanism and also can be polyadenylation or non-polyadenylation (Liu et al.

2012). They are situated either in the nucleus or the cytoplasm of the cell. lncRNAs regulate many biological functions in plants such as affecting the function of coding RNAs by target mimicry and alternative splicing, histone modification, gene silencing, flowering time regulation, phosphate homeostasis, as well as acting as molecular cargo for protein re-localization (Shafiq et al. 2016; Zhu and Wang 2012). Exploration of ESTs, NGS RNA-sequencing, and tiling array data are used to identify the majority of plant lncRNAs till now and based on their genome origin; they are classified as sense, antisense, and intronic and intergenic lncRNAs. Long intergenic RNAs are encoded by the DNA segment located between the protein-coding genes, whereas intronic lncRNAs are the product of transcribed introns of the protein-coding genes. On the other hand, antisense and sense lncRNAs are the transcripts representing antisense and sense strand, respectively, of an open reading frame (Shafiq et al. 2016).

Recent studies have revealed that ncRNAs, particularly long ncRNAs (lncRNAs) along with microRNAs (miRNAs), are the important regulatory molecules which affect various stress-responsive genes' expression in tissue-dependent mode (Wang et al. 2017). A total of 1060 lncRNAs expressed under 17 different biotic as well as abiotic stresses in 43 plants are submitted in PLNlncRNA database. Analysis of this data suggests that lncRNAs react to stresses in five different manners: target mimicry, sRNA precursor, DNA methylation, histone modification, and antisense transcription. In target mimicry, some lncRNAs binds to specific miRNAs, thus acting as competing for endogenous RNAs (ceRNAs), and prevent interactions of miRNAs with their targets. This phenomenon is used for maintaining phosphate homeostasis in *Arabidopsis*. Under phosphate starvation, a member of the TPS1/Mt4 gene family called *phosphate starvation1* (IPS1) acts as a target mimic for miR-399. Owing to partial complementarity of a 23-nt motif of IPS1 to miR399, it counterbalances the effect of increased miR399 accumulation under phosphate starvation, which results in coordinating PHO2 expression as well as phosphate uptake (Wang et al. 2017; Zhu and Wang 2012). Certain plant lncRNAs are also the precursors of some sRNAs, for example, siRNAs and miRNAs. For instance, during powdery mildew infection in wheat crop, three lncRNAs act as the precursors for miR-2066 and miR-2004, whereas 16 lncRNAs are required to produce 97 siRNAs. Antisense lncRNAs interact with sense RNAs and regulate gene expression at various levels such as transcription process, transcription interference, RNA stability, genomic imprinting, chromatin modification, and alternative splicing. Various antisense lncRNAs have been detected in various stress responses such as under N deficiency, in maize under drought stress, wheat infected with stripe rust pathogen. Several lncRNAs, for example, Air, HOTAIR, Kcnq1ot1, and Xist, are involved in chromatin modification regulatory pathways. In *Arabidopsis*, under cold stress, lncRNAs COOLAIR and COLDAIR suppress FLOWERING LOCUS C (FLC) gene expression through lncRNAs mediated-chromatin modification regulation pathway. FLC acts as a floral repressor and inhibits flowering under cold stress. lncRNAs are also important components of the RNA-directed DNA methylation pathway in various biotic and abiotic stress responses (Wang et al. 2017; Zhu and Wang 2012).

4 Plant Growth-Promoting Rhizobacteria in Ameliorating Abiotic Stress

A slender region of soil surrounding roots of the plants is defined as rhizosphere; it acts as an ecological niche. Plants' roots besides providing nutrition, water, and anchorage in the soil also release certain chemical substances, called root exudates, containing organic acids, phytosiderophores, etc., that not only attract microbial population but also help in their proliferation (Vacheron et al. 2013). This rhizo-microbiome consists of diversified microorganism, both bacteria and fungi, and may or may not be establishing a symbiotic relationship with plants. In symbiotic relationship between plants and microbes, cost and benefits are shared by both, and this interaction is further categorized as mutualistic and associative interaction (Vacheron et al. 2013). Mutualistic interaction is obligatory for plants and is restricted to compatible host plant only.

Moreover, these microbes form special structures in plants for their interaction and influence mineral and metal bioavailability and mobility such as root nodule formation during nitrogen fixation. On the other hand, associative interaction, also called cooperative interaction, is less obligatory, and microbes colonize root surface to promote growth and health of the plant by releasing plant growth-promoting substances (Vacheron et al. 2013; DalCorso et al. 2013). These associative bacteria are named as plant growth-promoting rhizobacteria (PGPR), and they interact with a wide range of host plant species (Vacheron et al. 2013).

PGPR (plant growth-promoting rhizobacteria) can be defined as free-living soil bacteria that colonize the rhizosphere (plant roots), and when applied to crops or seeds, they promote the plant growth as well as yield (Kumar et al. 2014). They can be found either on root surface or in endophytic association with plant and facilitate the growth of the plants in optimal, abiotic, or biotic stress conditions either directly or indirectly. They promote the growth of the plants by a variety of mechanisms, for example, through nitrogen fixation, synthesizing phytohormones, solubilizing mineral nutrients, decomposing organic matter, recycling essential elements, degrading organic pollutants, as well as stimulating root growth. They also make various metals bioavailable to the plants by altering pH of the soil; producing chelator molecules such as siderophores, organic acids, and biosurfactants; adsorption of metal ion directly on root surface; or performing the various biochemical reactions. PGPR has shown promising results in the agricultural field by increasing soil fertility and plant growth, as well as they suppress the growth of phytopathogens and promote eco-friendly sustainable agriculture practice (Gupta et al. 2015). Application of PGPR can be one of the most suitable and sustainable approaches to overcoming the negative impact of the effect of various abiotic stresses. Some of the PGPR having ACC (1-aminocyclopropane-1-carboxylate) deaminase property can protect plants from various abiotic and biotic stressors by reducing the level of stress ethylene produced during stress (Singh et al. 2015). In addition to ACC deaminase-mediated stress amelioration, the PGPR can modulate plant physiology which favors the growth of plants under given stress. The ability of PGPR to combat abiotic stressors through ACC deaminase and other mechanisms is referred to as IST or induced systemic tolerance (Jha et al. 2013).

4.1 Induced Systemic Tolerance (IST)

Microbes mediate induction of chemical as well as physical changes in plants at a physical, biochemical, or molecular level under various abiotic stressors to alleviate their deleterious effect, a process called as IST or induced systemic tolerance (Yang et al. 2010). IST leads to the protection of plants against both abiotic and biotic stressors, and cumulatively, it leads to the foundation of eco-friendly stress management strategy (Arya et al. 2018). Induction of IST by bacteria involves the production of various compounds including lipopolysaccharides from bacterial outer membranes, volatile organic compounds (VOC), biosurfactants, siderophores, and antibiotics, along with some other metabolites, amendment of phytohormone profile, activation of antioxidant defense mechanism, production of osmoprotectants, and activation of stress-responsive proteins (Etesami 2018; Vílchez et al. 2018). This may also involve changes in gene expression not only at transcriptional level but also at the translational level. One of the most common mechanisms of IST is a synthesis of enzyme ACC deaminase by PGPRs to combat the deleterious effects of ethylene (Fig. 10.4, Chandra and Sharma 2016).

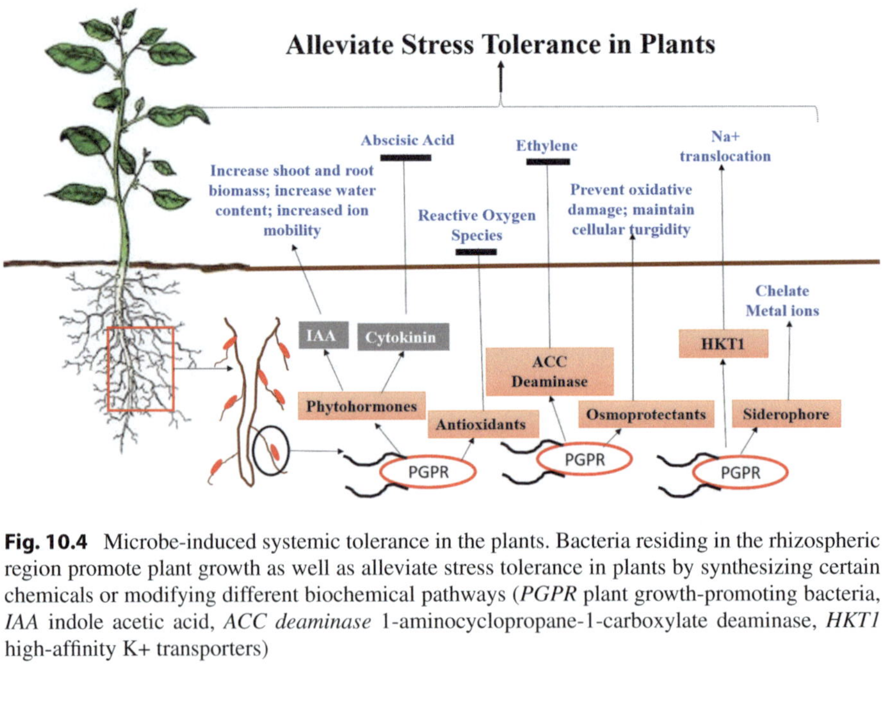

Fig. 10.4 Microbe-induced systemic tolerance in the plants. Bacteria residing in the rhizospheric region promote plant growth as well as alleviate stress tolerance in plants by synthesizing certain chemicals or modifying different biochemical pathways (*PGPR* plant growth-promoting bacteria, *IAA* indole acetic acid, *ACC deaminase* 1-aminocyclopropane-1-carboxylate deaminase, *HKT1* high-affinity K+ transporters)

4.2 PGPR- Mediated Drought and Salt Tolerance

The PGPR can ameliorate drought stress by altering the physiology of host plants directly by ACC deaminase activity or indirectly by inducing the differential activation or expression of various metabolites which can mitigate the harmful effects of stressors. These metabolites range from osmoprotectants to stress-related proteins, hormones, and ions (Ghatak et al. 2017). To ameliorate the ion-induced damage caused due to high salinity condition, plant growth-promoting bacteria improve the growth of the plants through the differential expression of high-affinity K^+ transporters (HKT1) (Singh et al. 2015). HKT1 transporter controls sodium ion (Na^+) import/uptake in the roots and the shoot; it is involved in Na^+ translocation from xylem to the phloem. It also adjusts the level of Na^+ and K^+ depending upon the tissue type (Vacheron et al. 2013; Yang et al. 2010). Under salt stress, PGPR promotes differential expression of HKT1 gene, which is upregulated in shoots but downregulated in roots. This differential expression reduces the accumulation of Na^+ but increases the aggregation of K^+ in shoot and roots (Singh et al. 2015).

Increase in the growth of plants in the presence of PGPR (plant growth-promoting bacteria) in different crops in response to water stress has been acknowledged in several reports all across the world. PGPR shows ACC deaminase activity which suppresses the level of stress-induced ethylene in plants and renders plant growth. It has been testified that plants growing in association with PGPR thrive better under drought stress. Many drought-tolerant bacterial species have been isolated from various plant crops by different research groups, for instance, Niu et al. (2018) have isolated *Enterobacter hormaechei*, *Pseudomonas fluorescens*, and *Pseudomonas migulae* from drought-tolerant crop foxtail millet which has shown potential in alleviating drought stress. Their plant inoculation studies revealed that these bacterial isolates show enhanced shoot and root length along with increased root and shoot biomass. Lin Chen and his co-researchers from China have investigated that rhizobacterium *Bacillus amyloliquefaciens* SQR9 induces salt tolerance in plants by the synthesis of spermidine. Spermidine produced by rhizobacteria increased the expression of those genes which were accountable for increased levels of glutathione, a scavenger of ROS. They have also reported that SQR9-produced spermidine also upregulates NHX1 and NHX7 gene expression, which confiscates Na^+ into vacuoles and expels Na^+ from the cell, thereby reducing ion toxicity.

Endophytes, *Phoma glomerata*, and *Penicillium* sp., associated with cucumber plant under salinity and drought stress, significantly increase plant biomass and other growth-related parameters along with assimilation of other essential nutrients (Lata et al. 2018; Waqas 2012). Malinowski and Belesky (2000) reported that endophytes impact drought tolerance to inoculated plants by enhancing accumulation of solutes or by reducing leaf conductance and reducing transpiration stream or by the formation of thicker cuticles. According to Vílchez et al. (2018), pepper plant can be protected from drought stress by *Microbacterium* sp. 3J1, an endophytic bacterium. Plant inoculated with *Microbacterium* sp. 3 J1 produces osmoprotectants as well as antioxidants which mediate higher resistance to water-deficit condition. The comparative metabolomic approach revealed that *Microbacterium* sp. 3J1 induces an increase in accumulation of sugars (e.g., trehalose, melibiose, glucose, and fructose) in host plants (pepper and tomato) which protected the plant as well as bacterium under drought conditions. Their major finding was that amino acid glutamine increases in the range of decrease in the content of α-ketoglutarate, suggesting a balance between carbon and nitrogen metabolism in the existence of *Microbacterium* sp. 3J1. They also found re-routing of the polyamine synthesis pathway for lignin modification as spermidine production increases which is accompanied by a decrease in ethylene and GABA concentration. Such alteration in lignin allows the release of sugars and facilitates water, nutrients, and Xero- and osmoprotectant transition by bacteria to offset oxidation damage caused due to stomatal closure. Mishra et al. (2016) have worked on the wheat crop to enhance drought tolerance with the help of plant growth-promoting bacteria. They have investigated the role of PGPR strains, IG 3 (*Klebsiella* sp.), IG 10 (*Enterobacter ludwigii*), and IG 15 (*Flavobacterium* sp.) in wheat in improving drought tolerance. They found that the plants grown in the presence of PGPR showed attenuation of certain stress-related genes (DREB2A and CAT1) which were otherwise upregulated in un-inoculated plants under drought stress. According to Kumar et al., PGPR *Bacillus amyloliquefaciens* NBRISN13 and *Pseudomonas putida* NBRIRA can ameliorate drought stress in chickpea (*Cicer arietinum* L.). They have reported that both these strains can work synergistically to enhance PGP attributes in chickpea to impact drought resistance. They have performed compatibility assay to check their compatibility with each other. Also, pre-formed different PGP traits individually and in the consortium and their finding revealed that inoculation with consortia increased plant biomass in drought-tolerant and drought-sensitive cultivators of chickpea significantly. They have also assayed different defense enzymes and found their down-regulation in PGPR-inoculated plants.

4.3 PGPR-Mediated Heat Stress Tolerance

Global warming has resulted in climate change, land degradation, an increase in drought frequency, and finally desertification. Further, global warming has resulted in increased Earth's temperature which leads to a rise in drought intensity as loss of water through transpiration pull increases. The high temperature may also affect the soil microbial diversity as optimum temperature needed by rhizospheric microbes for their growth is around 28–35 °C; a temperature beyond 37 °C is detrimental for their growth although bacteria isolated from the hot and dry environment can withstand temperature up to 45 °C and few of the heat-tolerant rhizobia actively form nodules and fix atmospheric nitrogen, thereby promoting plant growth. These microbes have adapted themselves through various complex regulatory processes. Under high-temperature stress, expression of a huge number of small heat shock proteins has been observed in these microorganisms along with variations in cell surface composition including proteins, extracellular polymeric substances (EPS), and lipopolysaccharides (LPS) (Alexandre and Oliveira 2010). Ali et al. (2011) have investigated the importance of thermotolerant PGPR *Pseudomonas putida* strain AKMP7 on the growth of wheat crop under high-temperature stress. This strain helped the host plant to endure under high temperature by significantly increasing the length of root and shoot along with increased tiller, dry biomass, grain formation, and spikelet. PGPR counterbalances the adverse effects of heat by lowering the ROS generation and by the production of various osmoprotectants. Also Alexandre and Oliveira in 2010 reported that many heat-tolerant rhizobia of chickpea plant show high upregulation of major chaperone genes, dnaKJ and groESL, upon stress. Chaperone systems GroEL-GroES and DnaK-DnaJ are important components of the heat stress response; they help denatured proteins to retain their native conformation.

4.4 PGPR-Mediated Metal Stress Tolerance

Human activities such as mining, agricultural practice, metal processing have increased the rate of metal accumulation in land and water. Moreover, inadequate waste clearance, inappropriate waste disposal, and the use of inappropriate materials have increased environmental pollution. All these activities cause leaching of toxic metals in the soil, decreasing growth as well as the yield of crops, contaminating waters bodies, and ultimately entering into the food chain. At present different physiochemical techniques are being used to decontaminate the soil with metal ions, but these are expensive procedures and also reduce soil fertility. The recently polluted soil is reclaimed by a green technology termed as phytoremediation. In this technique, plants are used to detoxify, translocate, as well as accumulate toxic metals from the soil to recover polluted areas. Besides phytoremediation, rhizospheric bacteria have shown promising results in detoxifying metal accumulation (Paredes-Páliz et al. 2018; Etesami 2018).

Besides promoting plant growth, PGPR also has a vital role in counterbalancing the soil from metal and organic pollutant by several mechanisms (Etesami 2018). PGPRs detoxify heavy metals from the soil by metal mobilization, phytoextraction, and phytoremediation. They efficiently mobilize the metals by making them more soluble, acidifying the rhizosphere region, enhancing the surface area of roots, and increasing the release of the root exudates, but this increases the metal content in the edible parts of the plants (Etesami 2018). An alternative to this is heavy metal-tolerant PGPR; they stimulate the growth of the plants even in the presence of toxic metals in the soil and reduce their bioavailability and uptake by the plants. Hence these crops can be regarded as safe for human consumption.

Mechanisms adapted by metal tolerant-microorganism to reduce metal toxicity include:

1. Metal exclusion by permeability barrier: bacteria make alterations in their cell wall, cell membrane, cellular envelope, or surface layer (S layer) while protecting their important cellular components to exclude metal ions. Some of the bacterial cell walls are efficient in adsorbing a high level of dissolved ion (Etesami 2018).

2. Active transport of metals: accumulated metals are effluxed outside the cell by various transporters. Ion-selective ATPase pumps are ATP driven and the most commonly used transporter protein; they pump metal ions from cytoplasm to periplasmic space, whereas CDF transporters expel metal ions using a chemiosmotic gradient (Etesami 2018).
3. Metal complexation: metal ions are trapped either extracellularly (extracellular space) or intracellularly (mainly cell vacuole) by metal-binding compounds or proteins. For example, metal cations can form complex with phosphate (PO4 3-), sulfur (S), and bicarbonate (HCO3-) or with metal-chelating proteins such as metallothioneins, metallochaperones, or low-molecular-mass cysteine-rich proteins (Etesami 2018).
4. Detoxification of metal ions by enzymes/biotransformation: enzymes convert toxic metals into a less toxic form either by methylation or through a metal reduction in the presence of electron donors. Methylation promotes diffusion of certain metals like Se, Sn, Te, and Pb as it converts them to their volatile form (Etesami 2018).
5. Metal biosorption and precipitation: bacteria secrete specific metabolites such as biosurfactants, siderophores, and exopolymer substances which help in absorption or precipitation of metal ion on the cell surface of microbial cells. Metals either bind to anionic functional groups of extracellular components such as sulfhydryl, hydroxyl, carboxyl, amine, or amide group, or they react with cell wall components and, thus, decrease the bioavailability of the metals (Etesami 2018).

Endophytic bacteria which reside in plant body improve mineral nutrition content to enhance plant growth as well as alleviate metal or metalloid toxicity. Endophytes, those are resistant to metal, help in providing tolerance to plants under metal stress by the production of metal chelators, organic acids, and siderophore or through acidification. For instance, *Brassica napus* have been inoculated with strains of lead-resistant bacteria *Pseudomonas fluorescens* and *Mycobacterium* sp., and they have shown potential to increase plant biomass production. Also, they have improved lead uptake into the shoots from a non-soluble phase in the soil. Pb-resistant bacteria also synthesize ACC deaminase, decreasing the ethylene level and enhancing the growth of the plants. They also enhance the activity of antioxidant enzymes, overcoming the negative impact of ROS (DalCorso et al. 2013).

In addition to PGPR, mycorrhizae, a symbiotic association between plant root and fungi, also improve plant growth due to an extensive hyphal system. This large hyphal system extends plant roots, increasing surface area for absorption of micronutrients and phosphate as well as water uptake (DalCorso et al. 2013). They improve plant tolerance to metal toxicity either by secreting organic acids which help plants to uptake metal via root-to-shoot transport or by immobilization and precipitation of metal ions. Metal ions are immobilized onto phosphate granules, or they are adsorbed by fungal cell wall containing chitin. Fungi secrete insoluble glycoproteins called glomalin which binds with metal ions and prevents accumulation in plants (DalCorso et al. 2013).

Metal resistance in certain microorganism may be due to the presence of some resistant genes located on transposon, plasmid, or chromosomal DNA. These resistances are mainly plasmid borne and are highly specific for particular ion and can be for both essential and non-essential metal ion (Etesami 2018). Bacteria promote plant growth under metal toxicity by producing metal-binding compounds such as organic acids, metallophores, biosurfactants, and siderophores, producing phytohormones, decreasing ethylene level by the production of ACC deaminase, facilitating mineral availability to the plants, modifying root and shoot growth, and competing with pathogenic microorganisms (Etesami 2018).

5 Mode of Action of PGPR in Alleviating Abiotic Stress Tolerance in Plant

Plants and microbes show a symbiotic relationship with each other; their interaction during stress condition plays a vital role for the adaption as well as the survival of both microbes and plant. Plant-associated microbial community can alleviate abiotic stress in plants with their metabolic abilities. Microbes facilitate plant growth by decreasing the burden of environmental stress through various mechanisms (Fig. 10.5).

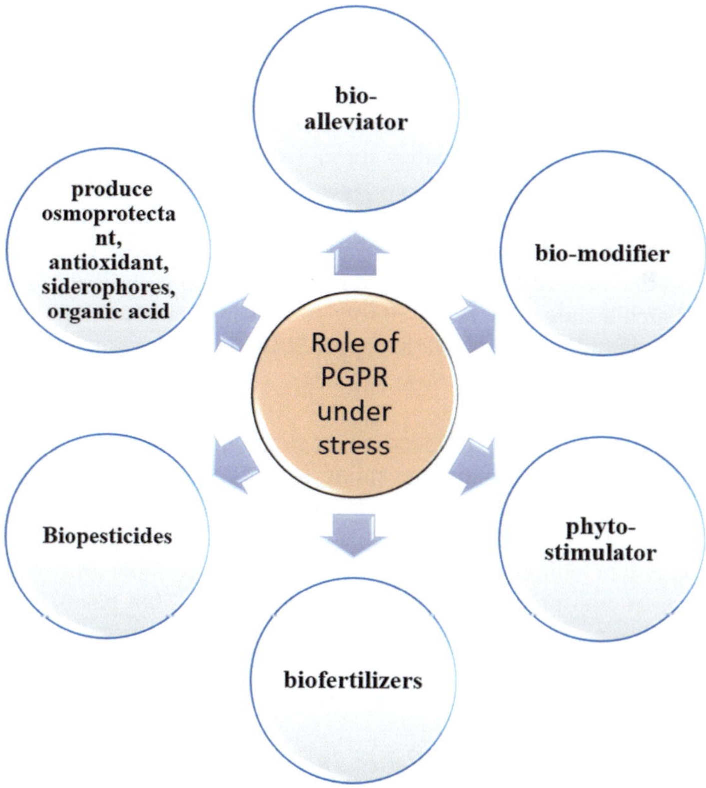

Fig. 10.5 Mode of action of PGPR in alleviating abiotic stress tolerance in plant

5.1 Microbes as Bio-alleviators

1-Aminocyclopropane-1-Carboxylic Acid (ACC) Deaminase Production Ethylene is a gaseous hormone whose rate of production depends upon the stage of development and the tissue type. Its production accelerates during abscission of leaves, flower senescence, ripening of fruits, mechanical wounding, and stress condition (Abeles et al. 1992). Methionine acts as a substrate for ethylene production, whereas its immediate precursor is 1-aminocyclopropane-1-carboxylate (ACC). ACC deaminase is a pyridoxal 5′-phosphate (PLP)-dependent multimeric enzyme. Some of the PGPR can have ACC deaminase which can hydrolyze ACC into α- ketoglutarate and ammonia by catalyzing the fragmentation of cyclopropane ring and deamination of ACC, thereby preventing ethylene biosynthesis (Chandra and Sharma 2016). This hydrolysis reaction decreases the level of endogenous ACC and ethylene, thereby eliminating the inhibitory effect of elevated ethylene concentrations. As a result, the plant can maintain their normal growth. Stress ethylene results in reduced root length, accumulation of ROS, and apoptosis (Etesami 2018).

5.2 Microbes as Bio-modifiers

Root Growth Characteristics Roots are the only source of water uptake by the plants from the soil. Roots have morphological plasticity which helps them respond better if there is any change in the physical condition of the soil. This property is particularly helpful to plants when they are subjected to an osmotic stress condition. Under osmotic stress, root number increases with a smaller diameter, and they even get deeper so that they can access the low water level of the soil. The resulting changes also increase hydraulic conductance as they are more in contact with soil water due to the increased surface area. Plants inoculated with PGPR have shown altered root architecture with increased root growth and better nutrients and uptake of water from the soil (Ngumbi and Kloepper 2016).

Shoot Growth Characteristic Under drought stress, plants limit their shoot growth as it decreases leaf area thereby preventing evaporative loss of limited available water. Now, instead of producing essential solutes required for growth, plants have diverted themselves to perform stress-related functions such as osmolyte adjustments, synthesis of antioxidant compounds, etc. Decreased shoot growth has a negative impact on crop productivity as plant size decreases, but plants treated with PGPR have shown normal to increased shoot growth, even under drought stress (Ngumbi and Kloepper 2016).

Relative Water Content (RWC)/Relative Turgidity RWC is the measure of leaf conductance, i.e., it is a ratio of actual water-holding capacity of leaf to its maximum water-holding capacity. It provides a water-deficit measurement of the leaf. It measures plant water stress status as it integrates both leaf water potential and the effect of the osmotic adjustment. It provides with the metabolic activity of the tissue. Under stress condition, RWC declines which reflects the loss of turgidity that results in diminished cell expansion which consequently results in reduced plant growth. It has been observed that PGPR-inoculated plants show higher RWC in comparison with non-inoculated plants, suggesting that PGPR ameliorates plant tolerance against water stress as higher RWC counterbalance the osmotic as well as oxidative stress (Ngumbi and Kloepper 2016).

5.3 Microbes as Phyto-stimulators

Phytohormones Production Plant growth and development is under the control of various plant growth regulators and phytohormones including auxin, cytokinin, ethylene, gibberellins, and abscisic acid. Auxin and cytokinin along with gibberellins are positive growth regulators, whereas ethylene and abscisic acid are considered as a negative regulator of growth. Abiotic stress leads to increase in the concentration of negative growth regulators to counterbalance the loss caused due to environmental stress. PGPR, when applied to plants, manipulates and modifies the content of phytohormones under stress condition. Bacterial phytohormones either directly complement plant hormones, or they indirectly increase the root surface area, which is followed by increased root exudates. Root exudates further alleviate stress condition including increasing plant resistivity to toxic heavy metals (Etesami 2018).

Indole-3-acetic acid (IAA) is naturally occurring auxin which is involved in every aspect of growth and development of plant including root growth including lateral and adventitious roots initiation, cell division, vascular differentiation, and wound healing, apical dominance, fruit growth, and development as well as leaf abscission. Plants under drought stress, when treated with PGPR, have shown increased shoot and root biomass along with increased water content, and these increased growth parameters are correlated with increased IAA production. Up to 80% of rhizobacteria when colonized on seed or root surface can synthesize IAA, which in conjugation with endogenous IAA promotes many developmental processes in the plant (Etesami 2018). Tryptophan acts as a precursor molecule for the synthesis of indole-3-acetic acid in bacteria, which is present in root exudates of plants. These phytohormones are continuously released by bacteria as slow rate, which provides benefit to the plants as higher concentration can be detrimental for growth.

IAA produced by the bacteria can loosen the plant cell walls, increasing the amount of root exudates, which further support growth of rhizosphere microorganism by adding more nutrition as it increases the mobility of nutrients such as sodium, phosphorous, potassium, iron, manganese, zinc, copper, etc. (Etesami 2018). Moreover, root exudates prevent heavy metals from entering the cell symplast as root secretions can chelate metals either in the rhizosphere or the apoplast.

Several PGPRs, for example, *Azotobacter* sp., *Bacillus subtilis*, *Pantoea agglomerans*, *Rhizobium* sp., etc., can produce cytokinin or gibberellins or both at low concentration, but several plant pathogens can also produce cytokinin at a higher concentration in comparison to PGPR. This indicates that the effect of cytokinin from PGPR is stimulatory on plant growth whereas from pathogens is inhibitory. *Azospirillum brasilense* produces auxin, IAA, which strengthens the growth of bean, sorghum, bean, wheat, and maize. Auxin produced by bacteria during plant-microbe interaction induces tumor growth in plants, which is further utilized by the symbiotic microorganism to promote plant growth. Also, auxin may act as a signalling molecule and activates amino acid catabolism, regulates biofilm formation, and helps bacteria to maintain stable copies of plasmid (DalCorso et al. 2013).

ABA (abscisic acid) is considered as a stress hormone as its concentration increases during a stress condition, which further induces physiological changes in plant and modulates their growth. PGPR-inoculated plants have also shown the elevated concentration of ABA which enhance plant ability to tolerate drought and salinity stress. ABA enhances drought tolerance of plants by regulating leaf transpiration rate and hydraulic conductivity of roots or by upregulating aquaporins.

Microbes facilitate a broad range of adaptive response through phytohormone production, for example, IAA, ABA, brassinosteroids, cytokinin, and gibberellin, which play a vital role in adapting plants to heavy metal toxicity stress. For instance, gibberellin can alleviate heavy metal toxicity by reducing uptake of cadmium, activate the antioxidant system to balance oxidative stress, alter the hormonal balance, and regulate the activity of various proteases and catalase.

5.4 Microbes as Biofertilizers

Nitrogen Fixation Nitrogen is the most important element for all form of life, an essential mineral nutrient necessary for the plants for their growth and productivity. It is the constituents of amino acids, nucleic acid, and vitamins. Although it is the most abundant gas (78%) present in the Earth's atmosphere, plants cannot utilize it directly. No plant species can fix free nitrogen into ammonia for their utilization. The atmospheric nitrogen is fixed to ammonia by nitrogen-fixing microorganism by a complex enzyme system using nitrogenase.

Plant growth-promoting nitrogen-fixing bacteria fix the nitrogen present in the atmosphere and make it available to the plants either symbiotically or non-symbiotically. In symbiotic nitrogen fixation, rhizobacteria colonize and form root nodules in leguminous plants, where they fix atmospheric nitrogen to ammonia and provide it for plant use. It is a mutualistic relationship between the plant and the microbes. Non-symbiotic nitrogen fixation is carried out by diazotrophs (free-living nitrogen-fixing bacteria), which stimulates growth in the non-leguminous plant. Both symbiotic and free-living nitrogen fixer contain nitrogenase (*nif*) genes, which include structural genes accountable for the synthesis of iron-molybdenum cofactor, activation of the iron protein, and electron donation, along with other regulatory genes required for biosynthesis and functioning of the enzyme (Gupta et al. 2015).

Phosphate Solubilization Phosphorous is another important element required by the plant. Although it is present abundantly in the soil, it cannot be utilized by the plants due to its insoluble, immobilized, and precipitated form. The monobasic (H_2PO_4) and the dibasic (HPO_4^{2-}) ions are only two soluble forms of phosphate which can be utilized by the plant. PGPR makes available the insoluble form of phosphate to plant useable form by different strategies, which include the release of compounds which form a complex with the insoluble phosphorous or dissolve them such as carbon dioxide, hydroxyl ion, or organic acid anions. Other mechanisms include the liberation of extracellular enzymes, or they may release phosphorous during degradation of the substrate. These bacteria are known as PSB (phosphate-solubilizing bacteria) and comprise around 1–50% of the rhizospheric bacterial population. Phosphorous released by the PSB can also immobilize the heavy metal from the metal contaminated soil, reducing metal toxicity through the formation of an insoluble precipitate complex with heavy metals (Gupta et al. 2015).

5.5 Microbes as Biopesticides

Prevention of Deleterious Effects of Phytopathogens Some strain of plant growth-promoting bacteria (PGPR) can promote the growth of the plants and biocontrol plant diseases. Biocontrol can be defined as a process by which microbes restrict the growth and multiplication of harmful microbes or pathogens (Gopalakrishnan et al. 2014). They can control plant pathogen such as *Xanthomonas* spp., *Pseudomonas* spp., *Pythium*, etc. (Liu et al. 2017; Gopalakrishnan et al. 2014). PGPRs mediate biocontrol activity by releasing certain secondary metabolites such as hydrogen cyanide (HCN), tensin, phenazines, etc. against phytopathogens (Liu et al. 2017), by competing with them for nutrients (Arora et al. 2001), by producing cell wall-degrading enzymes (Ozkoc and Deliveli 2001), and by producing antibiotic and siderophore (Gopalakrishnan et al. 2014). Plant resistance to pathogen mediated by bacteria is called ISR (induced systemic resistance); it is induced by jasmonate and ethylene, which act as signal molecules for the activation of a signal transduction pathway.

5.6 Other Mechanisms

Production of Osmoprotectant Accumulation of compatible solute, also known as the osmoprotectant, protects cellular membrane, organelles, protein, enzymes, and other macromolecules from oxidative damage. These compounds maintain cellular turgidity and help in decreasing plant water potential without reducing actual water content. Plants inoculated with PGPR have shown an increased concentration of the compatible solutes under abiotic stress.

Antioxidant Metabolism Abiotic stress leads to the production of ROS in plants which reduce normal metabolism in the plant through oxidative damage. They attack all types of biomolecules in the cells including lipids, nucleic acid, and proteins, causing irreplaceable damage, leading to cell death (Etesami 2018). The plant produces enzymatic (catalase, peroxidase, glutathione-S-transferase, glutathione peroxidase, ascorbate peroxidase, superoxide dismutase, and glutathione reductase) along with non-enzymatic (ascorbate, glutathione, proline, and alpha-tocopherol) antioxidants to avoid these damages. Treatment of plants with PGPR leads to elevated accumulation of these antioxidants in plants and minimizes oxidative injuries, thereby contributing to abiotic stress tolerance. PGPRs could enhance gene/ mRNA expression of plant antioxidant enzymes, increasing their activity.

Siderophore Production A predominant form of iron (Fe) present in soil is ferric ion (Fe^{3+}), which is not easily assimilated by plants and even bacteria. But bacteria have evolved a special mechanism for the utilization of Fe, which includes synthesis of iron-chelating compounds. These chelators such as siderophore are low-molecular-weight compounds (400–4000 kDa) which are synthesized under nutrient-depleted condition by the bacteria and help in transportation of iron into their cells. They act as an iron scavenger. The siderophore secreted by PGPR has enhanced plant growth either directly or indirectly. They can sequester Fe^{3+} from the root area, making unavailable for pathogen bacteria, thereby preventing their proliferation. Many plants can utilize siderophores as iron sources, but the uptake concentration is considered low. Siderophore-producing bacteria increase the plant growth and chlorophyll content under stress condition by selectively promoting the iron absorption from the complex of other heavy metals which are competing to enter the plant (Etesami 2018). Absorption of heavy metals by the bacteria depends upon the composition of bacteria and plant and also upon the type of heavy metal. Also, the rate of siderophore production depends upon the availability of iron, pH, and nutrition condition of the soil, concentration, and type of the heavy metal present as well as on plant status such as its nutrients absorption ability, translocation of nutrients from root to shoots, and amount of root exudates released.

PGPRs producing siderophores can overcome the metal-induced stress by the following mechanism:

- Increasing uptake of iron and reducing uptake of heavy metals, for example, lead, nickel, or cadmium. Amount of iron absorption depends directly on its concentration in the root environment.
- Counterbalancing the oxidative stress.
- By binding to heavy metals, making them unavailable for plant uptake.
- Regulating phytohormone production in plants and preventing microbial phytohormones from oxidative damage through metal chelation.

Siderophores also form complex with metal ions other than iron such as zinc, lead, aluminum, and cadmium and solubilize the inaccessible form of these ions, making their uptake possible by bacterial cell directly and plants indirectly (DalCorso et al. 2013). For instance, soil bacteria *Pseudomonas fluorescens* strain C7 release the iron-siderophore pyoverdine, this complex is taken up by the plants and iron concentration increase in their shoots, which further improve plant growth, whereas *Pseudomonas aeruginosa* produces siderophores which increase bioavailability of chromium and lead which are taken up by plant's root (DalCorso et al. 2013).

Organic Acid Production In the rhizosphere, microorganism secretes organic acids like glucuronate, citrate, and oxalate. They are of low-molecular-weight with one or more carboxylic groups and form complexes with metal. They are influenced by the pH of the soil. These acids increase the mobility as well as uptake of both essential and non-essential ions (DalCorso et al. 2013). Organic acid-producing PGPRs support growth of the plants by alleviating the metal-induced stress as they can form less phytotoxic metal ion complexes, which in turn diminish the cytological impacts of free/unbound metal ions (Etesami 2018). These bacteria also force the plants exposed to heavy metal to release organic compounds from their root exudates, which can solubilize the heavy metal of the soil, making them inaccessible for plant uptake (Etesami 2018). For instance, from the rhizosphere region of *Sedum alfredii*, Cd/Zn-resistant bacteria were isolated which produces organic acids such as formic, tartaric, oxalic, and succinic and thereby increase the solubility of zinc and cadmium in water (DalCorso et al. 2013).

Biosurfactant Production They are surface active, amphiphilic biomolecules which are produced on the bacterial cell surface (Etesami 2018; DalCorso et al. 2013). They are composed of one or more compounds, for example, phospholipids, glycolipids, fatty acids, lipoproteins, and mycolic acid (Etesami 2018). They increase heavy metal tolerance as they mobilize the metals by desorbing them from the soil matrix, form complex with them, and increase their bioavailability (DalCorso et al. 2013). Biosurfactants are also secreted extracellularly by the microbes, reducing surface and interfacial tension (Etesami 2018). For example, biosurfactants produced by *Bacillus* sp. J119 enhance cadmium uptake by the plants (DalCorso et al. 2013); biosurfactant di-rhamnolipid enhances the uptake of lead and cadmium (Etesami 2018). *Pseudomonas aeruginosa* and *Bacillus subtilis* produce biosurfactants surfactin and rhamnolipid which enhance metal resistance (DalCorso et al. 2013).

Exopolymer Production EPS (extracellular polymeric substance) are high-molecular weight natural polymers, secreted by the microbes in the environment. They are composed of homo- or hetero-polysaccharides. These exopolymers attach to the bacterial cell surface via capsule or slime layer and help them to form biofilms. Glucose, galactose, and mannose are essential monomers in the structure of EPS along with another monosaccharide like uronic acids, neutral sugars, amino sugars, organic ester-linked substituent, or pyruvate ketals, depending upon the bacterial strain. Acyl group provides anionic character to the EPS, which enhances the lipophilicity and their interaction with other compounds (Etesami 2018).

EPS helps in alleviating the metal stress as they bind strongly with the toxic heavy metals, reducing their mobility in the soil and therefore their availability for the plants (Etesami 2018; Rajkumar et al. 2012; Kunito et al. 2001).

6 Conclusion

Plant growth and yield are largely affected by environmental abiotic stressors such as temperature, salinity, drought, heavy metals, and xenobiotic pollutants. These stressors can have a deleterious effect which stunts plant's growth through the generation of ROS/RNS, leaf wilting, altered relative water content, leaf abscission, and accumulation of free radicals that damage cellular components. To respond these stressors, plants adapt to various physiological and biochemical mechanisms to generate tolerance against various abiotic stressors, which includes ion homeostasis by ion transport and uptake, biosynthesis of osmoprotectant and compatible solutes, secondary metabolites, and phytohormone modulation. At the molecular level, plant responses include changes at the level of protein functions, up- or downregulation of structural and regulatory proteins, and differential expression of non-coding RNAs. Various stress proteins including LEA proteins, aquaporin, and HSPs are highly expressed to provide stress tolerance. Gene expression can be regulated at different stages of central dogma including transcription initiation, RNA processing, translation or post-translational modification of proteins. These genes encode proteins that can protect the cell from water stress and cellular damage through enzymes of osmolyte biosynthesis, water channel and membrane transporter, LEA proteins, and molecular chaperones, and/or encode regulatory proteins (TFs) and RNAs, protein phosphatases, proteinases, as well as protein kinases which regulate stress signal transduction pathways and also modulate expression of other genes. Recently, the role of regulatory RNAs including miRNA and lncRNA in modulating gene expression in response to abiotic stressors has received significant attention which can be exploited to improve abiotic tolerance in plants.

The problem of abiotic stressors regarding plant productivity can be overcome by developing stress-tolerant crop varieties through conventional plant breeding and transgenic approaches. However, these methods are either time-consuming or expensive. The application of induced systemic tolerance mediated by plant-associated bacteria is one of the most reliable and sustainable approaches to ameliorate abiotic stressors in plants hence improving plant growth and yield. Certain plant-associated bacteria mitigate the deleterious effects of plants through various mechanisms including maintaining ionic homeostasis; production of lipopolysaccharides from bacterial outer membranes, volatile organic compounds (VOC), biosurfactants, siderophores, and antibiotics, along with some other metabolites; amendment of phytohormone profile, and activation of antioxidant defense mechanism; and production of osmoprotectants and activation of stress-responsive proteins including membrane transporters and molecular chaperones. One of the most common mechanisms of IST involves bacterial ACC deaminase which reduces the level of stress ethylene by degrading ACC, an immediate precursor of ethylene. Though the role of plant-associated bacteria in stress amelioration is evident in several studies, the mechanism of bacteria-mediated stress amelioration at molecular levels such as changes in proteomic profiles, regulatory RNAs, and proteins is still not well understood. The understanding of the plant-microbiome relationship for managing abiotic stressors will lead to significant improvement in plant productivity employing the sustainable and cost-effective strategy.

References

Abeles FB, Morgan PW, Saltveit ME Jr (1992) Ethylene in plant biology, 2nd edn. Academic, New York

Ahmad P, Jaleel CA, Salem MA, Nabi G, Sharma S (2010) Roles of enzymatic and nonenzymatic antioxidants in plants during abiotic stress. Crit Rev Biotechnol 30:161–175

Akula R, Ravishankar GA (2011) Influence of abiotic stress signals on secondary metabolites in plants. Plant Signal Behav 6:1720–1731

Alexandre A, Oliveira S (2010) Most heat-tolerant rhizobia show high induction of major chaperone genes upon stress. FEMS Microbiol Ecol 75:28–36

Ali SZ, Sandhya V, Grover M, Linga VR, Bandi V (2011) Effect of inoculation with a thermotolerant plant growth promoting *Pseudomonas putida* strain AKMP7 on growth of wheat (*Triticum* spp.) under heat stress. J Plant Interact 6:239–246

Amor BB, Wirth S, Merchan F, Laporte P, d'Aubenton-Carafa Y, Hirsch J et al (2009) Novel long non-protein coding RNAs involved in Arabidopsis differentiation and stress responses. Genome Res 19(1):57–69

Arora NK, Kang SC, Maheshwari DK (2001) Isolation of siderophore producing strains of rhizobium meliloti and their biocontrol potential against *Macrophomina phaseolina* that causes charcoal rot of groundnut. Curr Sci 81:673–677

Arya B, Komala BR, Sumalatha NT, Surendra GM, Gurumurthy PR (2018) PGPR induced systemic tolerance in plant. Int J Curr Microbiol App Sci 7:453–462

Asthir B (2015) Protective mechanisms of heat tolerance in crop plants. J Plant Interact 10:202–210

Chandra D, Sharma AK (2016) Isolation and characterization of plant growth promoting Bacteria containing ACC deaminase from soil collected from central Himalayan region of Uttarakhand, India. Int J Curr Microbiol App Sci 5:436–445

DalCorso G, Manara A, Furini A (2013) An overview of heavy metal challenge in plants: from roots to shoots. Metallomics 5:1117–1132

Etesami H (2018) Bacterial mediated alleviation of heavy metal stress and decreased accumulation of metals in plant tissues: mechanisms and future prospects. Ecotoxicol Environ Safety 147:175–191

Fang Y, Xiong L (2015) General mechanisms of drought response and their application in drought resistance improvement in plants. Cell Mol Life Sci 72(4):673–689

Feng ZJ, Xu SC, Liu N, Zhang GW, Hua QZ, Xu ZS, Gong YM (2018) Identification of the AQP members involved in abiotic stress responses from Arabidopsis. Gene 646:64–73

Forni C, Duca D, Glick BR (2016) Mechanism of plant response to salt and drought stress and their alteration by rhizobacteria. Plant Soil 410(1–2):335–356

Ghatak A, Chaturvedi P, Weckwerth W (2017) Cereal crop proteomics: systemic analysis of crop drought stress responses towards marker-assisted selection breeding. Front Plant Sci 8:1–25

Gopalakrishnan S, Sathya A, Vijayabharathi R, Varshney RK, Gowda CL, Krishnamurthy L (2014) Plant growth promoting rhizobia: challenges and opportunities. 3 Biotech 5:355–377
Gupta B, Huang B (2014) Mechanism of salinity tolerance in plants: physiological, biochemical, and molecular characterization. Intl J Genomics 701596:1–18
Gupta G, Parihar SS, Ahirwar NK, Snehi SK, Singh V (2015) Plant growth promoting rhizobacteria (PGPR): current and future prospects for development of sustainable agriculture. J Microb Biochem Technol 7:96–102
Hasanuzzaman M, Nahar K, Md Alam M, Roychowdhury R, Fujita M (2013) Physiological, biochemical, and molecular mechanisms of heat stress tolerance in plants. Int J Mol Sci 14:9643–9684
Hashem A, Abd-Allah EF, Alqarawi AA, Al-Huqail AA, Wirth S, Egamberdieva D (2016) The interaction between arbuscular mycorrhizal fungi and endophytic bacteria enhances plant growth of *Acacia gerrardii* under salt stress. Front Microbiol 7:1–15
Hossain MA, Piyatida P, da Silva JAT, Fujita M (2012) Molecular mechanism of heavy metal toxicity and tolerance in plants: central role of glutathione in detoxification of reactive oxygen species and methylglyoxal and in heavy metal chelation. J Bot 872875:1–37
Hu L, Li H, Pang H, Fu J (2012) Responses of antioxidant gene, protein and enzymes to salinity stress in two genotypes of perennial ryegrass (*Lolium perenne*) differing in salt tolerance. J Plant Physiol 169(2):146–156
Hu XJ, Chen D, Mclntyre CL, Dreccer MF, Zhang ZB, Drenth J, Sundaravelpandian K, Chang H, Xue GP (2018) Heat shock factor C2a serves as a proactive mechanism for heat protection in developing grains in wheat via an ABA-mediated regulatory pathway. Plant Cell Environ 41:79–98
Hundertmark M, Hincha DK (2008) LEA (late embryogenesis abundant) proteins and their encoding genes in *Arabidopsis thaliana*. BMC Genomics 9(1):118
Jha PN, Gupta G, Jha P, Mehrotra R (2013) Association of rhizospheric/endophytic bacteria with plants: a potential gateway to sustainable agriculture. Greener J Agri Sci 3:73–84
Kumar A, Maurya BR, Raghuwanshi R (2014) Isolation and characterization of PGPR and their effect on growth, yield and nutrient content in wheat (*Triticum aestivum* L.). Biocatal Agric Biotechnol 3:121–128
Kunito T, Saeki K, Nagaoka K, Oyaizu H, Matsumoto S (2001) Characterization of copper-resistant bacterial community in rhizosphere of highly copper-contaminated soil. Eur J Soil Biol 37:95–102
Lata C, Prasad M (2011) Role of DREBs in regulation of abiotic stress responses in plants. J Exp Bot 62:4731–4748
Lata R, Chowdhury S, Gond SK, White JF Jr (2018) Induction of abiotic stress tolerance in plants by endophytic microbes. Lett Appl Microbiol 66:268–276
Lin CC, Chen YJ, Chen CY, Oyang YJ, Juan HF, Huang HC (2012) Crosstalk between transcription factors and microRNAs in human protein interaction network. BMC Syst Biol 6(1):18
Liu J, Jung C, Xu J, Wang H, Deng S, Bernad L et al (2012) Genome-wide analysis uncovers regulation of long intergenic noncoding RNAs in Arabidopsis. Plant Cell 24(11):4333–4345
Liu H, Sultan MARF, Xl L, Zhang J, Yu F, Hx Z (2015) Physiological and comparative proteomic analysis reveals different drought responses in roots and leaves of drought tolerant wild wheat (*Triticum boeoticum*). PLoS One 10:e0121852
Liu K, Newman M, McInroy JA, Hu CH, Klopper JW (2017) Selection and assessment of plant growth promoting rhizobacteria for biological control of multiple plant disease. Phytopathology 107:928–936
Llusia J, Penuelas J, Munne-Bosch S (2005) Sustained accumulation of methyl salicylate alters antioxidant protection and reduces tolerance of holm oak to heat stress. Physiol Plant 124:353–361
Malinowski DP, Belesky DP (2000) Adaptations of endophyte-infected cool-season grasses to environmental stresses. Crop Sci 40(4):923
Mishra G, Sapre S, Sharma A, Tiwari S (2016) Amelioration of drought tolerance in wheat by the interaction of plant growth promoting rhizobacteria. Plant Biol (Stuttg) 18:992–1000

Mohamed AA, Castagna A, Ranieri A, Sanità di Toppi L (2012) Cadmium tolerance in *Brassica juncea* roots and shoots is affected by antioxidant status and phytochelatin biosynthesis. Plant Physiol Biochem 57:15–22

Ngumbi E, Kloepper J (2016) Bacterial-mediated drought tolerance: current and future prospects. Appl Soil Ecol 105:109–125

Niu X, Song L, Xiao Y, Ge W (2018) Drought-tolerant plant growth-promoting rhizobacteria associated with foxtail millet in a semi-arid agroecosystem and their potential in alleviating drought stress. Front Microbiol 8:1–11

Noman A, Ali Q, Maqsood J, Iqbal N, Javed MT, Rasool N, Naseem J (2017) Deciphering physio-biochemical, yield, and nutritional quality attributes of water-stressed radish (*Raphanus sativus* L.) plants grown from Zn-Lys primed seeds. Chemosphere 195:175–189

Olvera-Carrillo Y, Luis Reyes J, Covarrubias AA (2011) Late embryogenesis abundant proteins: versatile players in the plant adaptation to water limiting environments. Plant Signal Behav 6(4):586–589

Ozkoc I, Deliveli MH (2001) *In vitro* inhibition of the mycelial growth of some root rot fungi by *Rhizobium leguminosarum* biovar phaseoli isolates. Turk J Biol 25:435–445

Paredes-Páliz K, Rodríguez-Vázquez R, Duarte B, Caviedes MA, Mateos-Naranjo E, Redondo-Gómez S, Caçador MI, Rodríguez-Llorente ID, Pajuelo E (2018) Investigating the mechanisms underlying phytoprotection by plant growth-promoting rhizobacteria in *Spartina densiflora* under metal stress. Plant Biol (Stuttg) 20:497–506

Rajkumar M, Sandhya S, Prasad MNV, Freitas H (2012) Perspectives of plant-associated microbes in heavy metal phytoremediation. Biotechnol Adv 30:1562–1574

Rodríguez M, Canales E, Borrás-Hidalgo O (2005) Molecular aspects of abiotic stress in plants. Biotecnología Aplicada 22:1–10

Shafiq S, Li J, Sun Q (2016) Functions of plants long non-coding RNAs. Biochimica et Biophysica Acta (BBA)-Gene Regulatory Mechanisms 1859(1):155–162

Sharp RE, LeNoble ME (2002) ABA, ethylene and the control of shoot and root growth under water stress. J Exp Bot 53(366):33–37

Shi Q, Bao Z, Zhu Z, Ying Q, Qian Q (2006) Effects of different treatments of salicylic acid on heat tolerance, chlorophyll fluorescence, and antioxidant enzyme activity in seedlings of *Cucumis sativa* L. Plant Growth Regul 48:127–135

Shriram V, Kumar V, Devarumath RM, Khare TS, Wani SH (2016) MicroRNAs as potential targets for abiotic stress tolerance in plants. Front Plant Sci 7:817

Singh RP, Shelke GM, Kumar A, Jha PN (2015) Biochemistry and genetics of ACC deaminase: a weapon to 'stress ethylene' produced in plants. Front Microbiol 6:1–14

Sun W, Montagu MV, Verbruggen N (2002) Small heat shock proteins and stress tolerance in plants. Biochim Biophys Acta 1577:1–9

Upadhyay R, Panda SK (2010) Influence of chromium salts on increased lipid peroxidation and differential pattern in antioxidant metabolism in *Pistia stratiotes* L. Braz Arch Biol Technol 53:1137–1144

Vacheron J, Desbrosses G, Bouffaud ML, Touraine B, Moënne-Loccoz Y, Muller D, Legendre L, Wisniewski-Dyé F, Prigent-Combaret C (2013) Plant growth-promoting rhizobacteria and root system functioning. Front Plant Sci 4:1–19

Verslues PE, Agarwal M, Katiyar-Agarwal S, Zhu J, Zhu JK (2006) Methods and concepts in quantifying resistance to drought, salt and freezing, abiotic stresses that affect plant water status. Plant J 45(4):523–539

Vílchez JI, Niehaus K, Dowling DN, González-López J, Manzanera M (2018) Protection of pepper plants from drought by *Microbacterium* sp. 3J1 by modulation of the Plant's glutamine and a-ketoglutarate content: a comparative metabolomics approach. Front Microbiol 9:1–17

Wahid A, Gelani S, Ashraf M, Foolad MR (2007) Heat tolerance in plants: an overview. Environ Exp Bot 61:199–223

Wang J, Meng X, Dobrovolskaya OB, Orlov YL, Chen M (2017) Non-coding RNAs and their roles in stress response in plants. Genomics Proteomics Bioinformatics 15(5):301–312

Waqas M, Khan AL, Kamran M, Hamayun M, Kang SM, Kim YH, & Lee IJ (2012) Endophytic fungi produce gibberellins and indoleacetic acid and promotes host-plant growth during stress. *Molecules*, 17 (9): 10754–10773

Whitley D, Goldberg SP, Jordan WD (1999) Heat shock proteins: a review of the molecular chaperones. J Vasc Surg 29:748–751

Wycisk K, Kim EJ, Schroeder JI, Kramer U (2004) Enhancing the first enzymatic step in the histidine biosynthesis pathway increases the free histidine pool and nickel tolerance in *Arabidopsis thaliana*. FEBS Lett 578:128–134

Yang J, Kloepper JW, Ryu CM (2010) Rhizosphere bacteria help plants tolerate abiotic stress. Trend Plant Sci 14:1–4

Zargar SM, Nagar P, Deshmukh R, Nazir M, Wani AA, Masoodi KZ et al (2017) Aquaporins as potential drought tolerance inducing proteins: towards instigating stress tolerance. J Proteome 169:233–238

Zhu QH, Wang MB (2012) Molecular functions of long non-coding RNAs in plants. Genes 3(1):176–190

Alleviating Drought Stress of Crops Through PGPR: Mechanism and Application

11

Firoz Ahmad Ansari and Iqbal Ahmad

1 Introduction

Drought ranks among the most severe environmental stressors that diminished crop productivity throughout the world. Alterations in the temperature of the global air are leading to drought for longer period of time, and these changes create more drought situations that lead to hampering the production of food in many countries (Lau and Lennon 2012). To overcome the adverse changes in the environment and to attain the required food production, several molecular tools and breeding-related programs are being adapted by researchers to enhance environmental stress tolerance by different varieties of plants. A number of physiological and biochemical changes are induced under drought stress which may lead to stunted plant growth and loss in the normal functions of the organs (Anjum et al. 2011). Stress response creates a dynamic and complex development which aimed at novel physiological disturbance in plants under hostile growth situations. A variety of specific drought-regulating mechanisms are described which include scavenging of reactive oxygen species (ROS), signaling of kinase cascade, hormone imbalance, regulation of gene expression, osmolyte production, structural variations in cells, ion channel activation, metabolism of carbohydrates and energy, metabolism of amino acids and fatty acids, as well as assimilation of nitrogen (Johnova et al. 2016). These processes actively interfere and play a specific role in the metabolic system and cellular structure which may lead to the final determination of phenotype in plants. The development of plant biochemistry and molecular biology resulted in the investigation of proteomic and metabolic basis of susceptible and drought-tolerant plants to understand the basis of adaptation under stressed environment. There are many omics tools that are used by researchers to investigate the response in plants due to abiotic stresses to explore new understanding. More precisely, system biology is a more

F. A. Ansari (✉) · I. Ahmad
Biofilm Research Laboratory, Department of Agricultural Microbiology, Faculty of Agricultural Sciences, Aligarh Muslim University, Aligarh, India

D. P. Singh et al. (eds.), *Microbial Interventions in Agriculture and Environment*,
https://doi.org/10.1007/978-981-13-8383-0_11

comprehensive approach for the clarification of complex regulatory system that regulates several responses in the plants under both abiotic and biotic stress situations. Water-limiting condition seriously affects the major crop plants which are grown in the climate of temperate region including cereals such as wheat, durum barley, and rice. Limitations in the investigation on the metabolic changes are still unexplored. Proteomic-based study for the pathways which are modulated by drought in the leaves of several plants is reported (Hill et al. 2013; Ullah et al. 2017). The insufficient determination of the diverse gene systems among the various cultivars of the sole crop plant is another major issue in emerging resistant varieties against environmental stress by exploiting the gene technology attitude (Kim et al. 2012). Consequently, alternative environment-friendly approaches are considerably substantial at present time. Growth-stimulating bacteria with resistance against environmental stress could be a progressive approach for growth and development under stress situations. Plant growth-promoting bacteria (PGPB) compete with pathogenic bacteria in the soil for the establishment and colonization of the phyllosphere, rhizosphere, and cortex of the plants. They are previously being used as an effective inoculant to increase growth as well as development of the plants under normal and environmental stress conditions. PGPB have the capacity of producing multifunctional growth regulators including auxin, cytokinins, ethylene, and gibberellins and several other growth-enhancing secondary metabolites like hydrogen cyanides, siderophores, nitrogenase, and phosphatase which impact a potential role on the growth stimulation of the plants even under environmental stress (Saikia et al. 2018). Under normal environmental situation, ethylene plays a beneficial role on the health of plant; however, secretion of ethylene increases during biotic and abiotic stress, and thus these abrupt changes in the level of ethylene lead to the negative effect on the growth and development. Rhizobacteria with the ability of ACC deaminase enzyme could enhance the growth of plants by controlling the ethylene synthesis under challenging or stressed environment (Tao et al. 2015; Egamberdieva et al. 2017).

2 Drought Stress on Growth and Development of Plants

Several functions of the plants (biochemical as well as physiological) are affected by the imbalance of turgor pressure and water potential under drought condition (Hsiao 2000; Rahdari and Hoseini 2012). Drought stress has demonstrated adverse effect on productivity of various crop plants such as rice, barley, wheat, and maize reduction, reported in their growth and productivity due to drought stress (Kamara et al. 2003; Samarah 2005; Rampino et al. 2006; Lafitte et al. 2006). Drought affects general plant attributes such as water content and fresh and dry matter content of the plants (Jaleel et al. 2009). Moreover, drought stress also influenced the uptake of water and nutrients in the soil system by the roots and hampered the transport system in the plants. Diffusion of the nutrients and flow of soluble nutrients including sulfates, nitrates, Ca, Si, Mg, and other essential trace elements significantly decrease under water-limiting condition (Selvakumar et al. 2012). Under water-limiting condition, free radicals are induced, and plant defense systems such as antioxidant levels are progressively diminished which leads to generation of oxidative stress and cell death. Reactive oxygen species at high concentration could adversely influence the biochemical and physiological process at different steps of molecular and cellular organization at the time of growth of the plants (Sgherri et al. 2000; Hendry 1993; Nair et al. 2008). Photosynthesis was also decreased due to the dysfunction of photosystems under drought stress (Anjum et al. 2011; Rahdari et al. 2012). For example, reduction in the productivity of *Paulownia imperialis*, bean, and *Carthamus tinctorius* has been documented under water-limiting condition (Siddiqi et al. 2009; Ayala-Astroga and Alcaraz-Melendez 2010). Plants' biochemical activities such as the nitrate reductase activity are also interrupted due to limited uptake of nitrate from the soil under drought environment (Caravaca et al. 2005). Ethylene biosynthesis is also emphasized which leads to growth inhibition of the plants via several mechanisms. Multidimensional interference of drought in the plant system causes damage and imbalance at different cellular level as well as to the organs of the plants (Choluj et al. 2004; Rahdari et al. 2012). Both qualitative and quantitative plant growths are adversely affected due to drought stress condition. Therefore, for sustainable agriculture, amelioration of environmental stress, mainly drought, is urgently required to ensure crop production.

3 Microbial Ecology Influenced by Drought Stress

There are enormous number of microorganisms that survive and exist around the rhizosphere and the surface of roots forming complex microbial ecology that influences the growth of the plants through the production of metabolites and root-microbe associations (Berg 2009; Lugtenberg and Kamilova 2009; Schmidt et al. 2014). Structural and morphological changes in root-associated bacteria in the rhizosphere as well as in the root zone may lead to the colonization which are adapted to environmental stress and enhance tolerance against stressors to promote resistance to drought and protect plant health (Schmidt et al. 2014; Cherif et al. 2015). Distribution of bacteria in the rhizosphere, endosphere, and root-adhered soil particles was observed in comparison with uncultivated soil with profound drought pepper plant in the dryland farming condition representing a selection pressure by the metabolism activity of plant on the microbiome (Marasco et al. 2012). In another investigation, inoculation of pepper plants with drought-tolerant bacteria from dryland revealed more resistance against water-limiting situation in comparison to the control set of uninoculated plants. Plants inoculated with drought-tolerant bacteria increased the root growth up to 40% which enhances the capacity of water uptake of plants (Marasco et al. 2013). *Salicornia* plants cultivated in high salinity condition and their rhizobacterial microbiome are resistant to abiotic stresses and have the ability to perform multifarious PGP traits and more rhizosphere-colonizing capacity suggesting that halotolerant rhizobacteria isolated from *Salicornia* plants demonstrated an ability to enhance plant growth even under saline soil and water-limiting condition (Mapelli et al. 2013). Microorganisms in the soil ameliorate the environmental stress to the plants by modulating traits related to plant growth (Lau and Lennon 2011). Manipulation in the native microbial flora of *Brassica rapa* in the soil system resulted in smaller growth and physiology of plants in comparison to specific indigenous microbial flora which showed maximum growth (Lau and Lennon 2011). Inoculation of chamomile seedling with three different native gram-positive bacterial isolates including *S. subrutilus* Wbn2–11, *B. subtilis* Col-6, and *P. polymyxa* Mc5Re-14 from Egypt and three gram-negative bacterial isolates such as *P. fluorescens* LI3–6-12, *S. rhizophila* P69, and *S. plymuthica* 3RE4–18 from Europe region exhibited substantial changes in morphological and functional traits of the community of bacteria. These differences clearly revealed the microbial shift within the bacterial population structure. Furthermore, *B. subtilis* Col-6 and *P. polymyxa* Mc5Re-14 exhibited in the improvement of bioactive secondary metabolites apiginin-7-O-glucoside. Treatment with *Pseudomonas* sp. strain AKM-P6 and *P. putida* strain AKM-P7 increased the tolerance to high-temperature stress in the wheat and sorghum seedling mainly due to the production of high-molecular-weight proteins as well as enhanced the cellular metabolites. Hereafter, this study clearly indicated that bacterial inoculants perform a novel activity to cooperate with microbiome of rhizosphere and influence the metabolome of the plant (Schmidt et al. 2014). A comparative study on the diversity of microorganisms in the desert soil of native region in Egypt and the agricultural soil which was irrigated with organic manure continuously for 30 years indicated that there is a substantial difference in the microbiome. Communities of the soil bacteria in the agricultural fields revealed a maximum diversity and improved ecosystem utility for the growth and health of the plant (Köberl et al. 2011, 2013; Ding et al. 2013). Abiotic environmental factors including salinity, pH, water content, and concentration of nutrient in the soil bacterial diversity recovered from arid, Mediterranean, and semiarid localities receiving 400, 300, and 100 mm precipitation per annum, respectively, indicated that soil precipitation decreases the abundance of bacteria. Furthermore, the composition of bacterial population was found to be sole or unique to each environment and deviates from time to time and adapts to survive, creating favorable condition.

The synergistic interaction between roots and beneficial microbiome of the rhizosphere ameliorates stress to the plants through various mechanisms (Rolli et al. 2015). Among them, PGPR directly affect homeostasis of the phytohormones and increase uptake of micronutrients or directly boost the immune machinery of the plant to fight against phytopathogens and protect plant health (Balloi et al. 2010). ACC deaminase (ACCd) enzyme producing PGPB breaks down the precursor of ethylene ACC and depresses the level of ethylene in the plants and therefore defends the plant from the water-limiting condition (Glick 2004). Likewise, *Acinetobacter* and *Pseudomonas* enhanced growth in grapevine plant by increasing the metabolic rate as well as physiological attributes by lowering the level of ACC under drought-stressed environment (Rolli et al. 2014). In the *Stenotrophomonas rhizophila* strain DSM14405T, several genes responsible for the function in the energy production, cell motility, and stress protection were induced in response to salinity as well as extracts of the root. During the investigation in the experimental analysis, biosynthesis genes such as alginate and involvement of downregulating genes in flagella and upregulating genes in biofilm development were recognized as a solitary systematic mechanism toward salinity stress (Alavi et al. 2013). Endophytic PGPR colonize the leaves, stems, tubers, and other organelles of the plants and provide protection against environmental stress through different mechanisms (Gray and Smith 2005; Compant et al. 2008). The level of colonization by endophytic bacteria to the plant host and cortical tissues reflects PGPR that have the ability to adapt to these ecosystems and accommodate and make a favorable environment for their existence (Hallman et al. 1997; Gray and Smith 2005). Rhizosphere colonization by endophytic and epiphytic bacteria as putative inflorescence including *B. phytofirmans* strain PsJN in *Vitis vinifera* L. grown in natural soil system revealed the difference between plants for colorization of bacteria of inflorescences, while analysis at microscopic level exposed the PsJN as a flourishing endophyte found in inflorescence organs followed by the colonization process (Compant et al. 2008). Entry in the cortex of the root surface by *V. vinifera* and *Burkholderia* sp. strain PsJN was done through multistep process including attachment on the root surface and finally formation of internode on the whole leaf, and furthermore an enzyme such as cell wall-degrading endoglucanase and endopolygalacturonase was produced by the bacteria and degrades the cell wall following the entry into the leaves of the plants (Compant et al. 2005). Endophytic bacterial ecology of the date palm roots from oases in the Tunisian Sahara exposed that bacteria quickly cross and colonized the cortical region of root of different palm varieties and mustard plants. Under drought stress endophytes significantly enhanced the biomass indicating that the roots of date palm shape communities of endophytes and also have the ability to promote plant growth, thus contributing an important ecological role to the entire oasis ecosystem, and increase resistance against environmental stress in plants (Cherif et al. 2015; Theocharis et al. 2012). Transcription of stress-related genes and levels of metabolite production increased very fast at 4 °C in *Burkholderia phytofirmans* PsJN-inoculated grapevine plantlets compared to uninoculated plants. There are several changes associated with cold stress tolerance deliberated which was observed by the occurrence of PsJN (Theocharis et al. 2012). Correspondingly, enhancement in drought, cold, and salt tolerance in the rice plants through symbiosis with class 2 endophytes recovered from plants cultivating under salinity and moisture gradients. Treatment with bacteria to the plants under drought environment increased plant growth, biomass, and yield and also decreased the consumption of water. This symbiotic approach clearly indicates that this is useful in alleviating the effect of climate change and enhancing productivity of the agricultural crops on the marginal as well as global level (Redman et al. 2011). Mutualistic symbiosis interaction among the fungal endophytes, viruses, as well as plants in deliberating temperature tolerance revealed that when fungal isolates preserved of virus are incapable to confer temperature tolerance but temperature tolerance reinstated after the virus is restored. The fungus infected with virus deliberates temperature tolerance not merely to its local monocot plant but also to a dicot plant which proposes that the highlighted mechanism comprises regulatory pathways conserved among these plants (Márquez et al. 2007).

4 PGPR in Drought Stress

The bacterial populations which are commonly found in the rhizosphere act a crucial and beneficial role in the nutrient management, plant growth, as well as biocontrol properties. These bacteria colonize the rhizosphere and endorhizosphere successfully of the plants and enhanced growth through several indirect or direct routes (Grover et al. 2011). Moreover, the crucial role of the soil bacteria in the biotic and abiotic stress management is gaining prominence in sustainable agriculture in the environmental stress. There are various possible mechanisms of tolerance against drought in the plants induced by rhizobacteria such as phytohormone synthesis like indole-3-acetic acid (IAA), abscisic acid (ABA), cytokinins, gibberellic acid, and ACC deaminase to decrease ethylene level in the root and bacterial EPS and induced systemic resistance by bacterial volatile compounds (Timmusk and Nevo 2011; Kim et al. 2012; Timmusk et al. 2014) (Fig. 11.1). This chapter hence attempts to focus on pathways as well as regulatory mechanisms that involve in microbial especially bacterial species and retain environmental stress ameliorating ability and their application in sustainable agriculture. A summary of the present chapter reported on the role of beneficial bacteria in tolerance and protection to plants against environmental stress especially drought has been provided briefly in this chapter.

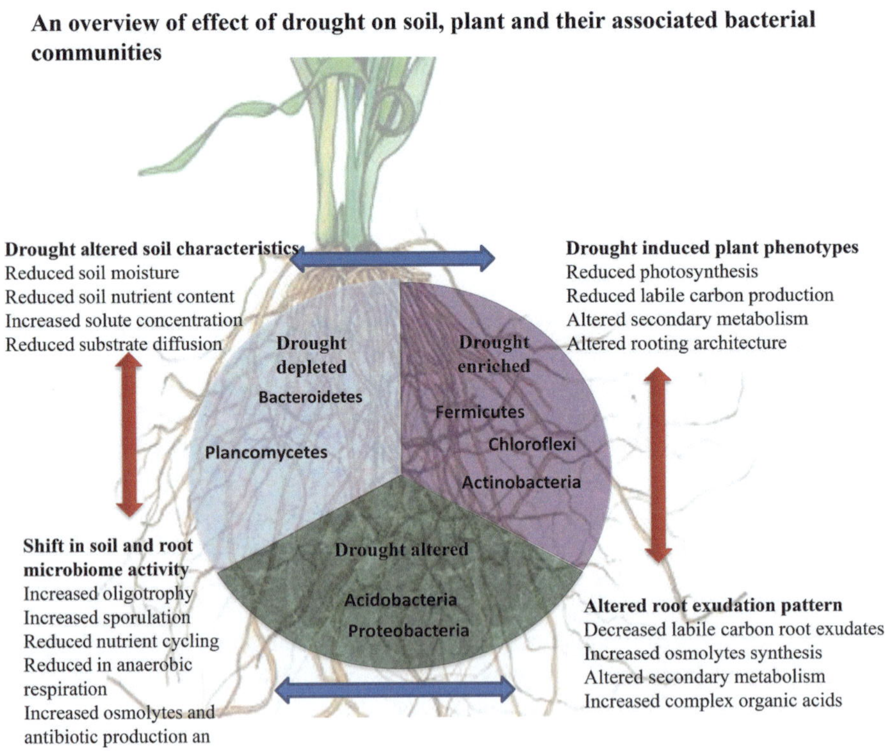

Fig. 11.1 An overview of effect of drought on soil, plant and their associated bacterial communities

5 PGPR-Mediated Mechanisms in Drought Stress Tolerance

There are various mechanisms proposed for the enhancement of drought tolerance of plants mediated by PGPR. It includes activity relating to production of phytohormones, EPS, and volatile compounds, accumulation of osmolytes, ACC deaminase activity, antioxidant enzyme defense system, and morphological changes in the root. The role of the induced systemic resistance (ISR) and induced systemic tolerance (IST) has been conceived to accommodate the induction in the chemical as well as physical variations due to microbial application in the plants resulting in the increased tolerance to hostile environment (Yang et al. 2009). The possible role of some of these biochemical mechanisms is briefly described.

5.1 Growth Regulators and Drought Tolerance in the Plants

Plant hormones including indole-3-acetic acid (IAA), gibberellin, ethylene, cytokinin, abscisic acid, and jasmonic acid are synthesized and secreted by the plants which contribute to a crucial role for their development and enhance stress tolerance (Teale et al. 2006; Egamberdieva 2013). The crucial role of the phytohormones involved alleviation of stress (Skirycz and Inzé 2010; Fahad et al. 2015). Auxin is the most active IAA physiologically as well as biochemically, and they are involved in the root differentiation and root generation in the plants. Inoculation with indole acetic acid-synthesizing bacteria enhanced growth and increased lateral root formation, root cap, as well as root hairs (Dimkpa et al. 2009). *Azospirillum* with IAA synthesizing capacity increased drought tolerance to the plants (Dimkpa et al. 2009). Nitric oxide is a tiny in size diffusible gas produced by *A. brasilense*, and this acts as a signal transduction molecule in the synthesis of IAA mechanism and acts a determinant role for adventitious root formation in tomato plant (Creus et al. 2005; Molina-Favero et al. 2008). Correspondingly, *A. brasilense* Cd inoculations enhanced all the regions of the root including the exact root area and exact root length as well as determined the distribution of the root in the soil system in the common bean plant in comparison to uninoculated plants in drought condition (German et al. 2000; Dimkpa et al. 2009).

5.2 Osmolyte Synthesis and Drought Tolerance in Plants

In the drought stress condition, plants are adapted and create a normal condition for their survival by the adjustment in the physiology and metabolism by regulating several compatible solutes and molecules such as osmolytes, polyamines, sugars, betaines, quaternary ammonium compounds, proline, amino acids, and polyhydric alcohols (Yancey et al. 1982; Close 1996). PGPR secreted a variety of osmolytes in response to drought, promote plant growth, maintain physiology of the cells, as well as alleviate drought stress to the plants (Paul et al. 2008). Enhancement in plant biomass due to adjustment in biochemical and physiological systems has been documented by the inoculation of *Pseudomonas putida* GAP-P45 under drought-stressed environment (Sandhya et al. 2010). The sizable quantity of the proline accumulation was increased on the application of PGPR, and therefore an enhancement of the growth parameters was recorded even in the presence of drought condition (Ansary et al. 2012). Treatment of *Lavandula dentata* with *B. thuringiensis* enhances the growth of the plants by increasing the concentration of proline in the shoot in comparison to control plants under drought condition (Armada et al. 2014). There are several biosynthesis genes that regulate the proline in response to stress condition. These genes are upregulated and increase the level of proline concentration by upholding the water status within the cell and give protection to membranes and cell proteins against environmental stress (Sandhya et al. 2010). Similarly, treatment with phosphate solubilizer *Bacillus polymyxa* to tomato plant produced excess proline content to overcome drought stress and maintain the cell protein level (Shintu and Jayaram 2015). Mixtures of the PGPR (*Pseudomonas jessenii* R62, *Pseudomonas synxantha* R81, and *Arthrobacter nitroguajacolicus* strain YB3 and strain YB5) are well known for their result on growth of plant and stress alleviation in comparison to the single inoculation under stressed environment in the different varieties of rice cultivars. Accumulation of higher proline content in the treated plants indicated maximum tolerance toward water-liming condition (Gusain et al. 2015). Treatment of the plants with consortium increases severalfold the proline in the plants, and this enhancement in the proline suggests a pivotal role of proline in the osmoregulation pathways which leads to amelioration and enhancement of drought tolerance to the plants (Gusain et al. 2015). Many free amino acids and water-soluble sugars also act a modulatory role in growth promotion on the application of PGPR isolates and PGPR *P. putida* GAP-P45 and *A. lipoferum* in the maize plants even in the presence of drought stress condition (Sandhya et al. 2010; Bano et al. 2013). Sucrose and trehalose are nonreducing sugars stabilized by many of the dehydrated enzymes and membrane biosynthesis and therefore act as osmoprotectants (Yang et al. 2010). Inoculation of *P. vulgaris* with *Rhizobium etli* enhances the drought tolerance by the overexpressing gene of trehalose-6-phosphate synthase in comparison to inoculated plants with wild isolates (Suarez et al. 2008). Treatment of nodule plants with overexpressing gene of trehalose-6-phosphate synthase isolates showed the upregulation of the genes involved in the metabolism of C and N, signaling metabolism of trehalose, and upregulation of stress tolerance gene indicating that trehalose plays a crucial role in the signaling system (Suarez et al. 2008). It is clearly well known that trehalose plays a substantial role as a signal molecule in balancing the metabolism of carbohydrates in the plant systems (Paul et al. 2008). Similar impacts were recorded as overexpression of biosynthesis gene of the trehalose regulating the metabolism of carbohydrates in the maize (Rodriguez-Salazar et al. 2009).

5.3 ACC Deaminase in the Drought Stress Tolerance

Environmental stressors play an influential role in the biosynthetic mechanism of ethylene as conversion of S-adenosylmethionine (S-AdoMet) to 1-aminocycloprop ane-1-carboxylate (ACC) by the action of 1-aminocyclopopane-1-carboxylate synthase (ACS). The phytohormone (ethylene) controlled the homeostasis endogenously which results in the reduction of plant vegetative growth. The level of ethylene regulates the activity of the plant, and ethylene biosynthesis is controlled by environmental as well as biotic stresses (Hardoim et al. 2008). Rhizospheric bacteria with ACC deaminase are degraded and sequestered ACC to supply nitrogen to the plants and diminished the deleterious effect of drought stress. Secretion of ACC deaminase enzymes by *Achromobacter piechaudii* ARV8 lowered the production of ethylene and therefore enhanced dry and fresh weight of pepper and tomato plants significantly under water-limiting condition (Mayak et al. 2004). It has been stated that rhizobacteria isolated from water stress field exhibit more ability to tolerate or alleviate drought stress in comparison to those bacterial population recovered from regular water-irrigated field (Mayak et al. 2004). For example, *A. piechaudii* ARV8 recovered from dryland or arid region perform better in growth parameters of the plant seedling as compared to *P. putida* GR12–2 which was recovered from grass rhizosphere where water is sufficient (Lifshitz et al. 1986). Similarly, pea plant inoculated with *Variovorax paradoxus* 5C-2 congaing ACC deaminase enzyme was more evident and consistent under dry soil environment (Dodd et al. 2004). Plants inoculated with ACC deaminase containing rhizobacteria increased the production of yield, seed number, nitrogen accumulation, and reestablishment of the nodule under drought-stressed soil system (Dodd et al. 2004). Soil bacteria with ACC deaminase producing ability eradicated drought effect on different stages of the growth and development of even under field as well as pot house condition (Arshad et al. 2008). Treatment of pea plant with *Pseudomonas fluorescens* biotype G (ACC-5) containing ACC deaminase enhances the size of the root which may lead to improved water uptake from the soil under drought environment (Zahir et al. 2008). Treatment of pea plant with *V. paradoxus 5C-2* enhanced plant growth as well as water use efficiency even under water-limiting condition (Belimov et al. 2009). In the healthy nodules, there is a large and sufficient number of nitrogen fixing bacteria, and these bacteria may prevent a deleterious effect of the drought in the reduction of nodules and content of nitrogen in the seeds. Successful colonization of the rhizosphere by rhizobacteria increased the yield and quality of nutritional value in the plant through both systems of signaling of hormone and local interference in the metabolic system under dry soil condition. Therefore, these rhizobacteria provide an effortlessly and economic worth of crop yield under drought land agriculture system (Belimov et al. 2009). Axenic study revealed that inoculation of wheat plants with ACC deaminase containing rhizobacteria enhances root numbers and shoot numbers in comparison to untreated control plants. Improvement in root numbers and root growth helps plants to uptake water and nutrients from the soil easily which therefore results in increase in plant growth parameters under water-limiting condition (Shakir et al. 2012). In another study, consortium inoculants with ACC deaminase-positive isolates (*Bacillus* isolates 23B and *Pseudomonas* 6-P with *Mesorhizobium ciceri*) cause reduction in the deleterious effect of drought, protect plants to the stress, and increased the growth in green gram. Treatment with the abovementioned PGPR to different varieties of plants was associated with higher content of proline in their roots and shoots under drought stress and provides protection against environmental stress to the plants. Consortium inoculation (23-B with

Mesorhizobium) was also found to be better effective under water-limiting environment (Sharma et al. 2013).

Choline is another water-soluble essential element found in plants and involves a signaling molecule in the stressed environment. Choline acts as a major part in the enhancement of stress to the plant mainly by the improvement of the glycinebetaine (GB) synthesis and its accumulation (Zeisel 2006; Zhang et al. 2010). There are several reports available highlighting the influential role of GB in mustard plant. The synthesis of GB is modulated by the rhizobacterial strain *B. subtilis* GB03 (Zhang et al. 2010) and, in maize plant choline, was enhanced and accumulated, indicating regulation of GB metabolism by the choline as precursor on the application of *Pseudomonas fluorescence* YX2, *Klebsiella variicola* F2, and *Raoultella planticola*. Accumulation of the dry matter content and increased content of water in the leaf directly relate to the accumulation of GB in the plants (Zhang et al. 2010; Gou et al. 2015). Choline plays a crucial part in water-limiting condition by providing nutrition also in the maize plant (Zeisel 2006; Zhang et al. 2010).

5.4 Rhizobacterial Exopolysaccharide (EPS) in the Alleviation of Drought Stress

Biological and physiological properties of the bacteria in the soil are disturbed due to unfavorable condition created by drought stress, and therefore soil bacterial activity and productivity of crop were ultimately diminished. Availability of water controls the production of extracellular polysaccharides and extracellular proteins by the soil bacteria. Soil microorganisms produced exopolysaccharides as slime and capsular material in the soil, and these EPSs are absorbed by the clay soil surfaces due to several chemical bondings (anion absorption, cation bridges, van der Waals forces, and hydrogen bonding) between them, thus creating a protective layer of capsule along with soil aggregates, and thus the structure and properties of the soil are influenced directly (Tisdall and Oades 1982; Roberson and Firestone 1992; Sandhya et al. 2009; Roberson and Firestone 1992). Microbial exopolysaccharides give a protection to them from unfavorable or inhospitable environmental condition and enable their existence and survival in the stress condition by forming meshwork or network-like structure of the EPS material. *A. brasilense* Sp245 produced carbohydrate complex of the high-molecular-weight (lipopolysaccharides-protein [LP] complex and polysaccharide-lipid [PL] complex) harboring to give a protection against stress condition like desiccation. The amendment of these sugar complexes to the liquid suspension of encapsulated *A. brasilense* cells leads to significant improvement under drought environment (Konnova et al. 2001). EPS creates a microenvironment that holds and maintains moisture as well as water level which may help to protect themselves and the root of the plants against dehydration, and those which does not produced EPS in their surrounding environment as well as the root does not protect themselves against harsh environmental condition (Hepper 1975). EPSs of the soil bacterial populations have been showed to enhance permeability by enhancing aggregation of soil and retaining maximum water potential along with the root, and thereby nutrient uptake by the plants was increased and provided protection against water-limiting condition and increased growth attributes of the plants (Alami et al. 2000; Selvakumar et al. 2012). Plants treated with EPS-secreting soil bacteria enhanced growth as well as resistance against water stress or limitation to the plants (Bensalim et al. 1998). Rhizobial isolate YAS34 produced EPS in significant amount, and this strain increased the ratio of root-adhering soil per root tissue (RAS/RT) which was found in rhizosphere of soil under water limitations (Alami et al. 2000). Rhizobacterial treatment could colonize the rhizosphere, root-adhering soil particles, as well as rhizoplane efficiently and improve the stability of soil. Improvement in the RAS aggregation increases the level of water uptake as well as nutrients from the rhizosphere, thus increasing growth and development of plants under drought environment (Sandhya et al. 2009). In another study, seed of maize plant treated with higher EPS-secreting strain *Proteus penneri* (AF3) has increased its length, dry matter content, area of the leaf, and plant biomass as well as has improved its moisture contents in the soil. Rhizobacterial-treated plants showed an increment in the proline content and carbohydrates, therefore lowering the level of stress markers like antioxidant enzymes and several osmolytes under water-limiting conditions (Naseem and Bano 2014). Catalase and EPS producing *Rhizobium phaseoli* (MR-2), *Rhizobium leguminosarum* (LR-30), and *Mesorhizobium ciceri* (CR-30 and CR-39) interact positively to wheat plant root even under drought condition and enhanced growth and protect the plant against drought stress. Inoculation of wheat plant with high EPS-producing rhizobacteria leads to enhanced growth and tolerance index under PEG-6000 (polyethylene glycol)-simulated drought stress (Hussain et al. 2014; Mishra et al. 2017).

5.5 Antioxidant Defense System in Stress Management

Antioxidative systems in the plants act a crucial role for the protection and maintenance of the physiological functions under environmental stress conditions. There are various oxidative pathways leading to the generation of reactive oxygen species (ROS), namely, hydrogen peroxide (H_2O_2), hydroxyl radicals (OH), superoxide anion radicals (O_2^-), and alkoxy radicals (RO), which undergo alteration in their activity when exposed to drought. Reactive oxygen species reacts with macromolecules (lipids, deoxyribonucleic acid, and protein) and hampered the functions of cells by oxidative imbalance as well as oxidative damage in the plant. In order to overwhelm these impacts, plant developed antioxidant defense system consisting of enzymatic and nonenzymatic mechanisms that reduce the accumulation of ROS and ameliorate oxidative imbalance under water-limiting condition (Miller et al. 2010). Antioxidant enzymatic mechanisms including superoxide dismutase (SOD), catalase (CAT), glutathione reductase (GR), and ascorbate peroxidase (APX) in the plants are activated under environmental stress situation (Zandalinas et al. 2017). There are also some nonenzymatic systems including glutathione, cysteine, and ascorbic acid which are also involved in the drought alleviation and provide protection to the plants against severe environmental stress (Kaushal and Wani 2016).

Inoculation of maize plant with five drought-tolerant plant growth-stimulating bacteria, namely, *P. entomophila*, *P. putida*, *P. syringae*, *P. montelli*, and *P. stutzeri*, under drought condition revealed lowering of the antioxidant enzyme activity in comparison to untreated plants under stress (Sandhya et al. 2010). Likewise, treatment of maize seedling with *Bacillus* spp. showed significant reduction in the antioxidant enzyme (GPX and APX) activity in the plants under drought stress (Vardharajula et al. 2011). Treatment of the basil plant with *Pseudomonas* sp. in the drought under field condition enhances the catalase enzyme activity significantly. In the same condition, microbial consortia (*Bacillus lentus*, *Pseudomonas* sp., and *A. brasilense*) revealed maximum activity of APX and GPX (Heidari and Golpayegani 2012). Treatment of maize plant with EPS-producing soil bacteria increased the growth of plant by decreasing the activity of GPX, APX, and CAT (Naseem and Bano 2014). PGPR colonize the rhizoplane as well as rhizosphere and reduce the drought stress by secreting EPS and maintain plant water relation for a certain period of time. Antioxidant systems are also modulated by soil microbes by lowering the stress to the plants.

6 Molecular Studies in Drought Stress Alleviation by the PGPR

Nowadays, there are several molecular strategies for environmental stress management. Likewise, genetic modification (addition or deletion) in the gene responsible for stress tolerance is coming in current scenario for stress alleviation. Gene expression studies are the regulating tool to compare and understand the universal response of microorganisms to their surroundings (Schlauch et al. 2010). The transcript of the transcriptome is expressed in the cells and organisms at specific stages of the development under different environmental situations. Several tools as well as technologies discovered nowadays for evaluation of transcriptome analysis are based on microarray to RNA sequencing (Trewavas 2006; Wang et al. 2009). Under drought condition, expression of gene linked with physiological functions was characterized using molecular approaches in respect to stress tolerance (Kandasamy et al. 2009; Yuwono et al. 2005). Drought tolerance of the plants enhances at the transcription level by PGRP to the water-limiting condition on the *Paenibacillus polymyxa* B2 which improved tolerance in *Arabidopsis thaliana* under drought-stressed soil system. RNA-exhibited mRNA transcription of the responsible gene to drought, EARLY RESPONSE TO DEHYDRATION 15 (ERD15), was amplified in the drought condition in the treated plant in comparison to untreated plant (Timmusk and Wagner 1999). With the help of 2D-PAGE (2D polyacrylamide gel electrophoresis) and DD-PCR (differential display polymerase chain reaction), expression of six different stress proteins was observed under drought condition which are recognized in the treated pepper plants with *B. licheniformis* K11. Among the proteins of stress markers, specific genes responsible to drought were sHSP, Cadhn, and VA which showed more than 1.5-fold increase in inoculated plants in comparison to uninoculated plants (Lim and Kim 2013). Using RT-PCR, genes related to stress in the wheat plant (APX1, SAMS1, and HSP17.8) were upregulated and enhance the activity of enzymes when priming with *Bacillus amyloliquefaciens* 5113 and *A. brasilense* NO40. These upregulated genes confer drought tolerance in wheat plant and ameliorate the negative impact of water-limiting condition (Kasim et al. 2013). A cluster of genes responsible for drought were downregulated in *P. chlororaphis* O6 colonization on the application of *P. chlororaphis* O6 colonization on the mustard plant in comparison to those without treatment of bacteria in the drought stress by using DNA microarray technology. PR-1 and VSP-1 are the genes which are responsible for transcription in jasmonic acid, and these genes regulate the salicylic acid, pdf-1.2, and ethylene responsible gene which were upregulated in PGPR-colonized plants but dissimilar in the response of drought stress (Cho et al. 2013). Interactive effect between the sugarcane cv. SP70–1143 and diazotroph *Gluconacetobacter diaazotrophicus* PAL5 under water-limiting condition concluded that bacterial treatment activates the signaling gene responsible for ABA deliberating drought tolerance in sugar cane cv. SP70–1143, and this demonstration was observed by using Illumina sequencing (Vargas et al. 2014; Zandalinas et al. 2017).

7 Conclusion

Crop productivity is severely affected by the drought stress considering climate change. The impact will be of higher magnitude in the future. Stress tolerance by plant growth-promoting rhizobacteria performs a crucial role in conferring adaptation and resistance to the drought condition and has a prospective role in stress management of plants. Under drought stress, interaction of PGPR to the plant usually affects the growth of the plants as well as soil properties. There are several mechanisms triggered by PGPR including osmotic response eliciting, and initiation of new genes plays a crucial effect in assuring survival of the plant under drought stress condition. Plant breeding and genetic engineering will perform a vital role in the improvement of drought tolerance crop varieties, but the application of these mechanisms is time-consuming, whereas drought amelioration to the plant by using PGPR inoculation opens a new dimension in stress management to plants and improving crop productivity. Considering the current available reports, identification of efficient microorganisms with desired traits for addressing the problems of plant specificity and their field assessment under stress condition will be mandatory for practical utility. An expectation was whispered that improved understanding of molecular mechanisms of plant-bacteria interaction under stress condition will be the key for successful exploitation of microbial based stress management to plants.

References

Alami Y, Achouak W, Marol C, Heulin T (2000) Rhizosphere soil aggregation and plant growth promotion of sunflowers by an exopolysaccharide-producing *Rhizobium* sp. strain isolated from sunflower roots. Appl Environ Microbiol 66(8):3393–3398

Alavi P, Starcher M, Zachow C, Müller H, Berg G (2013) Root-microbe systems: the effect and mode of interaction of stress protecting agent (SPA) *Stenotrophomonas rhizophila* DSM14405T. Front Plant Sci 4:141

Ali SZ, Sandhya V, Rao LV (2014) Isolation and characterization of drought-tolerant ACC deaminase and exopolysaccharide-producing fluorescent Pseudomonas sp. Ann Microbiol 64(2):493–502

Anjum SA, Xie XY, Wang LC, Saleem MF, Man C, Lei W (2011) Morphological, physiological and biochemical responses of plants to drought stress. Afr J Agric Res 6(9):2026–2032

Ansary MH, Rahmani HA, Ardakani MR, Paknejad F, Habibi D, Mafakheri S (2012) Effect of *Pseudomonas fluorescent* on proline and phytohormonal status of maize (*Zea mays* L.) under water deficit stress. Ann Biol Res 3(2):1054–1062

Armada E, Roldán A, Azcon R (2014) Differential activity of autochthonous bacteria in controlling drought stress in native Lavandula and Salvia plants species under drought conditions in natural arid soil. Microb Ecol 1 67(2):410–420

Arshad M, Shaharoona B, Mahmood T (2008) Inoculation with *Pseudomonas* spp. containing ACC-deaminase partially eliminates the effects of drought stress on growth, yield, and ripening of pea (*Pisum sativum* L.). Pedosphere 18(5):611–620

Ayala-Astorga GI, Alcaraz-Meléndez L (2010) Salinity effects on protein content, lipid peroxidation, pigments, and proline in *Paulownia imperialis* (Siebold & Zuccarini) and *Paulownia fortunei* (Seemann & Hemsley) grown in vitro. Electron J Biotechnol 13(5):13–14

Balloi A, Rolli E, Marasco R, Mapelli F, Tamagnini I, Cappitelli F, Borin S, Daffonchio D (2010) The role of microorganisms in bioremediation and phytoremediation of polluted and stressed soils. Agrochimica 54(6):353–369

Bano QU, Ilyas N, Bano A, Zafar NA, Akram AB, Hassan F (2013) Effect of Azospirillum inoculation on maize (*Zea mays* L.) under drought stress. Pak J Bot 45(S1):13–20

Belimov AA, Dodd IC, Hontzeas N, Theobald JC, Safronova VI, Davies WJ (2009) Rhizosphere bacteria containing 1-aminocyclopropane-1-carboxylate deaminase increase yield of plants grown in drying soil via both local and systemic hormone signalling. New Phytol 181(2):413–423

Bensalim S, Nowak J, Asiedu SK (1998) A plant growth promoting rhizobacterium and temperature effects on performance of 18 clones of potato. Am J Potato Res 75(3):145–152

Berg G (2009) Plant–microbe interactions promoting plant growth and health: perspectives for controlled use of microorganisms in agriculture. Appl Microbiol Biotechnol 84(1):11–18

Caravaca F, Alguacil MM, Hernández JA, Roldán A (2005) Involvement of antioxidant enzyme and nitrate reductase activities during water stress and recovery of mycorrhizal *Myrtus communis* and *Phillyrea angustifolia* plants. Plant Sci 169(1):191–197

Cherif H, Marasco R, Rolli E, Ferjani R, Fusi M, Soussi A, Mapelli F, Blilou I, Borin S, Boudabous A, Cherif A (2015) Oasis desert farming selects environment-specific date palm root endophytic communities and cultivable bacteria that promote resistance to drought. Environ Microbiol Rep 7(4):668–678

Cho SM, Kang BR, Kim YC (2013) Transcriptome analysis of induced systemic drought tolerance elicited by *Pseudomonas chlororaphis* O6 in *Arabidopsis thaliana*. Plant Pathol J 29(2):209

Choluj D, Karwowska R, Jasinska M, Haber G (2004) Growth and dry matter partitioning in sugar beet plants (*Beta vulgaris* L.) under moderate drought. Plant Soil Environ 50(6):265–272

Close TJ (1996) Dehydrins: emergence of a biochemical role of a family of plant dehydration proteins. Physiol Plant 97(4):795–803

Compant S, Reiter B, Sessitsch A, Nowak J, Clément C, Barka EA (2005) Endophytic colonization of Vitis vinifera L. by plant growth-promoting bacterium *Burkholderia* sp. strain PsJN. Appl Environ Microbiol 71(4):1685–1693

Compant S, Kaplan H, Sessitsch A, Nowak J, Ait Barka E, Clément C (2008) Endophytic colonization of *Vitis vinifera* L. by *Burkholderia phytofirmans* strain PsJN: from the rhizosphere to inflorescence tissues. FEMS Microbiol Ecol 63(1):84–93

Creus CM, Graziano M, Casanovas EM, Pereyra MA, Simontacchi M, Puntarulo S, Barassi CA, Lamattina L (2005) Nitric oxide is involved in the *Azospirillum brasilense*-induced lateral root formation in tomato. Planta 221(2):297–303

Dimkpa C, Weinand T, Asch F (2009) Plant–rhizobacteria interactions alleviate abiotic stress conditions. Plant Cell Environ 32(12):1682–1694

Ding GC, Piceno YM, Heuer H, Weinert N, Dohrmann AB, Carrillo A, Andersen GL, Castellanos T, Tebbe CC, Smalla K (2013) Changes of soil bacterial diversity as a consequence of agricultural land use in a semi-arid ecosystem. PLoS One 8(3):e59497

Dodd IC, Belimov AA, Sobeih WY, Safronova VI, Grierson D, Davies WJ (2004) Will modifying plant ethylene status improve plant productivity in water-limited environments. In: Proceedings for the 4th international crop science congress, vol 26, Brisbane, Australia

Egamberdieva D (2013) The role of phytohormone producing bacteria in alleviating salt stress in crop plants. Biotechnological techniques of stress tolerance in plants. Studium, Houston, pp 21–39

Egamberdieva D, Davranov K, Wirth S, Hashem A, Abd Allah EF (2017) Impact of soil salinity on the plant-growth–promoting and biological control abilities of root associated bacteria. Saudi J Biol Sci 24(7):1601–1608

Fahad S, Hussain S, Bano A, Saud S, Hassan S, Shan D, Khan FA, Khan F, Chen Y, Wu C, Tabassum MA (2015) Potential role of phytohormones and plant growth-promoting rhizobacteria in abiotic stresses: consequences for changing environment. Environ Sci Pollut Res 22(7):4907–4921

German MA, Burdman S, Okon Y, Kigel J (2000) Effects of *Azospirillum brasilense* on root morphology of common bean (*Phaseolus vulgaris* L.) under different water regimes. Biol Fertil Soils 32(3):259–264

Glick BR (2004) Bacterial ACC deaminase and the alleviation of plant stress. Adv Appl Microbiol 56:291–312

Gou W, Tian L, Ruan Z, Zheng PE, Chen FU, Zhang L, Cui Z, Zheng P, Li Z, Gao M, Shi W (2015) Accumulation of choline and glycinebetaine and drought stress tolerance induced in maize (*Zea mays*) by three plant growth promoting rhizobacteria (PGPR) strains. Pak J Bot 47(2):581–586

Gray EJ, Smith DL (2005) Intracellular and extracellular PGPR: commonalities and distinctions in the plant–bacterium signaling processes. Soil Biol Biochem 37(3):395–412

Grover M, Ali SZ, Sandhya V, Rasul A, Venkateswarlu B (2011) Role of microorganisms in adaptation of agriculture crops to abiotic stresses. World J Microbiol Biotechnol 27(5):1231–1240

Gusain YS, Singh US, Sharma AK (2015) Bacterial mediated amelioration of drought stress in drought tolerant and susceptible cultivars of rice (*Oryza sativa* L.). Afr J Biotechnol 14(9):764–773

Hallmann J, Quandt-Hallmann A, Mahaffee WF, Kloepper JW (1997) Bacterial endophytes in agricultural crops. Can J Microbiol 43(10):895–914

Hardoim PR, van Overbeek LS, van Elsas JD (2008) Properties of bacterial endophytes and their proposed role in plant growth. Trends Microbiol 16(10):463–471

Heidari M, Golpayegani A (2012) Effects of water stress and inoculation with plant growth promoting rhizobacteria (PGPR) on antioxidant status and photosynthetic pigments in basil (*Ocimum basilicum* L.). J Saudi Soc Agric Sci 11(1):57–61

Hendry GA (1993) Oxygen, free radical processes and seed longevity. Seed Sci Res 3(3):141–153

Hepper CM (1975) Extracellular polysaccharides of soil bacteria. In: Walker N (ed) Soil microbiology. Elsevier, Amsterdam

Hill CB, Taylor JD, Edwards J, Mather D, Bacic A, Langridge P, Roessner U (2013) Whole genome mapping of agronomic and metabolic traits to identify novel quantitative trait loci in bread wheat grown in a water-limited environment. Plant Physiol 162:1266–1281

Hsiao TC (2000) Leaf and root growth in relation to water status. Hort Sci 35:1051–1058

Hussain MB, Zahir ZA, Asghar HN, Asgher M (2014) Can catalase and exopolysaccharides producing rhizobia ameliorate drought stress in wheat? Int J Agric Biol 16(1)

Jaleel CA, Manivannan PA, Wahid A, Farooq M, Al-Juburi HJ, Somasundaram RA, Panneerselvam R (2009) Drought stress in plants: a review on morphological characteristics and pigments composition. Int J Agric Biol 11(1):100–105

Johnová P, Skalák J, Saiz-Fernández I, Brzobohatý B (2016) Plant responses to ambient temperature fluctuations and water-limiting conditions: a proteome-wide perspective. Biochimica et Biophysica Acta (BBA)-proteins and. Proteomics 1864(8):916–931

Kamara AY, Menkir A, Badu-Apraku B, Ibikunle O (2003) The influence of drought stress on growth, yield and yield components of selected maize genotypes. J Agric Sci 141(1):43–50

Kandasamy S, Loganathan K, Muthuraj R, Duraisamy S, Seetharaman S, Thiruvengadam R, Ponnusamy B, Ramasamy S (2009) Understanding the molecular basis of plant growth promotional effect of *Pseudomonas fluorescens* on rice through protein profiling. Proteome Sci 7(1):47

Kasim WA, Osman ME, Omar MN, El-Daim IA, Bejai S, Meijer J (2013) Control of drought stress in wheat using plant-growth-promoting bacteria. J Plant Growth Regul 32(1):122–130

Kaushal M, Wani SP (2016) Plant-growth-promoting rhizobacteria: drought stress alleviators to ameliorate crop production in drylands. Ann Microbiol 66(1):35–42

Kim YC, Glick BR, Bashan Y, Ryu CM (2012) Enhancement of plant drought tolerance by microbes. In: Plant responses to drought stress. Springer, Berlin/Heidelberg, pp 383–413

Köberl M, Müller H, Ramadan EM, Berg G (2011) Desert farming benefits from microbial potential in arid soils and promotes diversity and plant health. PLoS One 6:e24452. https://doi.org/10.1371/journal.pone.0024452

Köberl M, Schmidt R, Ramadan EM, Bauer R, Berg G (2013) The microbiome of medicinal plants: diversity and importance for plant growth, quality and health. Front Microbiol 20(4):400

Konnova SA, Brykova OS, Sachkova OA, Egorenkova IV, Ignatov VV (2001) Protective role of the polysaccharide-containing capsular components of *Azospirillum brasilense*. Microbiology 70(4):436–440

Lafitte HR, Yongsheng G, Yan S, Li ZK (2006) Whole plant responses, key processes, and adaptation to drought stress: the case of rice. J Exp Bot 58(2):169–175

Lau JA, Lennon JT (2011) Evolutionary ecology of plant–microbe interactions: soil microbial structure alters selection on plant traits. New Phytol 192(1):215–224

Lau JA, Lennon JT (2012) Rapid responses of soil microorganisms improve plant fitness in novel environments. Proc Natl Acad Sci 109(35):14058–14062

Lifshitz R, Kloepper JW, Scher FM, Tipping EM, Laliberté M (1986) Nitrogen-fixing pseudomonads isolated from roots of plants grown in the Canadian high Arctic. Appl Environ Microbiol 51(2):251–255

Lim JH, Kim SD (2013) Induction of drought stress resistance by multi-functional PGPR *Bacillus licheniformis* K11 in pepper. Plant Pathol J 29(2):201

Lugtenberg B, Kamilova F (2009) Plant-growth-promoting rhizobacteria. Annu Rev Microbiol 63:541–556

Mapelli F, Marasco R, Rolli E, Barbato M, Cherif H, Guesmi A, Ouzari I, Daffonchio D, Borin S (2013) Potential for plant growth promotion of rhizobacteria associated with Salicornia growing in Tunisian hypersaline soils. Biomed Res Int 2013:1

Marasco R, Rolli E, Ettoumi B, Vigani G, Mapelli F, Borin S, Abou-Hadid AF, El-Behairy UA, Sorlini C, Cherif A, Zocchi G (2012) A drought resistance-promoting microbiome is selected by root system under desert farming. PLoS One 7(10):e48479

Marasco R, Rolli E, Vigani G, Borin S, Sorlini C, Ouzari H, Zocchi G, Daffonchio D (2013) Are drought-resistance promoting bacteria cross-compatible with different plant models? Plant Signal Behav 8(10):e26741

Márquez LM, Redman RS, Rodriguez RJ, Roossinck MJ (2007) A virus in a fungus in a plant: three-way symbiosis required for thermal tolerance. Science 315(5811):513–515

Mayak S, Tirosh T, Glick BR (2004) Plant growth-promoting bacteria confer resistance in tomato plants to salt stress. Plant Physiol Biochem 42(6):565–572

Miller GA, Suzuki N, Ciftci-Yilmaz SU, Mittler RO (2010) Reactive oxygen species homeostasis and signalling during drought and salinity stresses. Plant Cell Environ 33(4):453–467

Mishra SK, Khan MH, Misra S, Dixit VK, Khare P, Srivastava S, Chauhan PS (2017 Feb 1) Characterization of *Pseudomonas* spp. and *Ochrobactrum* sp. isolated from volcanic soil. Antonie Van Leeuwenhoek 110(2):253–270

Molina-Favero C, Creus CM, Simontacchi M, Puntarulo S, Lamattina L (2008) Aerobic nitric oxide production by *Azospirillum brasilense* Sp245 and its influence on root architecture in tomato. Mol Plant-Microbe Interact 21(7):1001–1009

Nair AS, Abraham TK, Jaya DS (2008) Studies on the changes in lipid peroxidation and antioxidants in drought stress induced cowpea (*Vigna unguiculata* L.) varieties. J Environ Biol 29:689–691

Naseem H, Bano A (2014) Role of plant growth-promoting rhizobacteria and their exopolysaccharide in drought tolerance of maize. J Plant Interact 9(1):689–701

Paul MJ, Primavesi LF, Jhurreea D, Zhang Y (2008) Trehalose metabolism and signaling. Annu Rev Plant Biol 59:417–441

Rahdari P, Hoseini SM (2012) Drought stress: a review. Int J Agronomy Plant Production 3(10):443–446

Rahdari P, Tavakoli S, Hosseini SM (2012) Studying of salinity stress effect on germination, proline, sugar, protein, lipid and chlorophyll content in purslane (*Portulaca oleracea* L.) leaves. J Stress Physiol Biochem 8(1):182–193

Rampino P, Pataleo S, Gerardi C, Mita G, Perrotta C (2006) Drought stress response in wheat: physiological and molecular analysis of resistant and sensitive genotypes. Plant Cell Environ 29(12):2143–2152

Redman RS, Kim YO, Woodward CJ, Greer C, Espino L, Doty SL, Rodriguez RJ (2011) Increased fitness of rice plants to abiotic stress via habitat adapted symbiosis: a strategy for mitigating impacts of climate change. PLoS One 6(7):e14823

Roberson EB, Firestone MK (1992) Relationship between desiccation and exopolysaccharide production in a soil Pseudomonas sp. Appl Environ Microbiol 58(4):1284–1291

Rodríguez-Salazar J, Suárez R, Caballero-Mellado J, Iturriaga G (2009) Trehalose accumulation in Azospirillum brasilense improves drought tolerance and biomass in maize plants. FEMS Microbiol Lett 296(1):52–59

Rolli E, Marasco R, Vigani G, Ettoumi B, Mapelli F, Deangelis ML, Gandolfi C, Casati E, Previtali F, Gerbino R, Pierotti Cei F (2015) Improved plant resistance to drought is promoted by the root-associated microbiome as a water stress-dependent trait. Environ Microbiol 17(2):316–331

Saikia J, Sarma RK, Dhandia R, Yadav A, Bharali R, Gupta VK, Saikia R (2018) Alleviation of drought stress in pulse crops with ACC deaminase producing rhizobacteria isolated from acidic soil of Northeast India. Sci Rep 8(1):3560

Samarah NH (2005) Effects of drought stress on growth and yield of barley. Agron Sustain Dev 25(1):145–149

Sandhya VZ, Grover M, Reddy G, Venkateswarlu B (2009) Alleviation of drought stress effects in sunflower seedlings by the exopolysaccharides producing *Pseudomonas putida* strain GAP-P45. Biol Fertil Soils 46(1):17–26

Sandhya VS, Ali SZ, Grover M, Reddy G, Venkateswarlu B (2010) Effect of plant growth promoting *Pseudomonas* spp. on compatible solutes, antioxidant status and plant growth of maize under drought stress. Plant Growth Regul 62(1):21–30

Schlauch KA, Grimplet J, Cushman J, Cramer GR (2010) Transcriptomics analysis methods: microarray data processing, analysis and visualization using the Affymetrix Genechip® *Vitis Vinifera* genome Array. In: Methodologies and results in grapevine research. Springer, Dordrecht, pp 317–334

Schmidt R, Köberl M, Mostafa A, Ramadan EM, Monschein M, Jensen KB, Bauer R, Berg G (2014) Effects of bacterial inoculants on the indigenous microbiome and secondary metabolites of chamomile plants. Front Microbiol 5:64

Selvakumar G, Panneerselvam P, Ganeshamurthy AN (2012) Bacterial mediated alleviation of abiotic stress in crops. In: Bacteria in agrobiology: stress management. Springer, Berlin, Heidelberg, pp 205–224

Sgherri CL, Maffei M, Navari-Izzo F (2000) Antioxidative enzymes in wheat subjected to increasing water deficit and rewatering. Journal of Plant 157(3):273–279

Shakir MA, Bano A, Arshad M (2012) Rhizosphere bacteria containing ACC-deaminase conferred drought tolerance in wheat grown under semi-arid climate. Soil & Environment 31(1):108–112

Sharma P, Khanna V, Kumari P (2013) Efficacy of aminocyclopropane-1-carboxylic acid (ACC)-deaminase-producing rhizobacteria in ameliorating water stress in chickpea under axenic conditions. Afr J Microbiol Res 7(50):5749–5757

Shintu PV, Jayaram KM (2015) Phosphate solubilising bacteria (*Bacillus polymyxa*)-an effective approach to mitigate drought in tomato (*Lycopersicon esculentum* Mill). Tropic Plant Res 2:17–22

Siddiqi EH, Ashraf M, Hussain M, Jamil A (2009) Assessment of intercultivar variation for salt tolerance in safflower (*Carthamus tinctorius* L.) using gas exchange characteristics as selection criteria. Pak J Bot 41(5):2251–2259

Skirycz A, Inzé D (2010) More from less: plant growth under limited water. Curr Opin Biotechnol 21(2):197–203

Suárez R, Wong A, Ramírez M, Barraza A, Orozco MD, Cevallos MA, Lara M, Hernández G, Iturriaga G (2008) Improvement of drought tolerance and grain yield in common bean by overexpressing trehalose-6-phosphate synthase in rhizobia. Mol Plant-Microbe Interact 21(7):958–966

Tao JJ, Chen HW, Ma B, Zhang WK, Chen SY, Zhang JS (2015 Nov 27) The role of ethylene in plants under salinity stress. Front Plant Sci 6:1059

Teale WD, Paponov IA, Palme K (2006) Auxin in action: signalling, transport and the control of plant growth and development. Nat Rev Mol Cell Biol 7(11):847

Theocharis A, Bordiec S, Fernandez O, Paquis S, Dhondt-Cordelier S, Baillieul F, Clément C, Barka EA (2012) *Burkholderia phytofirmans* PsJN primes *Vitis vinifera* L. and confers a better tolerance to low nonfreezing temperatures. Mol Plant-Microbe Interact 25(2):241–249

Timmusk S, Nevo E (2011) Plant root associated biofilms. In: Maheshwari DK (ed) Bacteria in agrobiology (vol 3) plant nutrient management. Springer Verlag, Berlin. http://pub-epsilon.slu.se:8080/2534/

Timmusk S, Wagner EG (1999) The plant-growth-promoting rhizobacterium *Paenibacillus polymyxa* induces changes in *Arabidopsis thaliana* gene expression: a possible connection between biotic and abiotic stress responses. Mol Plant Microb Interact 12:951

Timmusk S, El-Daim IA, Copolovici L, Tanilas T, Kännaste A, Behers L, Nevo E, Seisenbaeva G, Stenström E, Niinemets Ü (2014) Drought-tolerance of wheat improved by rhizosphere bacteria from harsh environments: enhanced biomass production and reduced emissions of stress volatiles. PLoS One 9(5):e96086

Tisdall JM, Oades J (1982) Organic matter and water-stable aggregates in soils. J Soil Sci 33(2):141–163

Trewavas A (2006) A brief history of systems biology: "every object that biology studies is a system of systems." Francois Jacob (1974). Plant Cell 18(10):2420–2430

Ullah N, Yüce M, Gökçe ZN, Budak H (2017) Comparative metabolite profiling of drought stress in roots and leaves of seven Triticeae species. BMC Genomics 18(1):969

Vardharajula S, Zulfikar Ali S, Grover M, Reddy G, Bandi V (2011) Drought-tolerant plant growth promoting *Bacillus* spp.: effect on growth, osmolytes, and antioxidant status of maize under drought stress. J Plant Interact 6(1):1–4

Vargas L, Santa Brígida AB, Mota Filho JP, de Carvalho TG, Rojas CA, Vaneechoutte D, Van Bel M, Farrinelli L, Ferreira PC, Vandepoele K, Hemerly AS (2014) Drought tolerance conferred to sugarcane by association with *Gluconacetobacter diazotrophicus*: a transcriptomic view of hormone pathways. PLoS One 9(12):e114744

Wang Z, Gerstein M, Snyder M (2009) RNA-Seq: a revolutionary tool for transcriptomics. Nat Rev Genet 10(1):57

Yancey PH, Clark ME, Hand SC, Bowlus RD, Somero GN (1982) Living with water stress: evolution of osmolyte systems. Science 217(4566):1214–1222

Yang J, Kloepper JW, Ryu CM (2009) Rhizosphere bacteria help plants tolerate abiotic stress. Trends Plant Sci 14(1):1–4

Yang J, Benyamin B, McEvoy BP, Gordon S, Henders AK, Nyholt DR, Madden PA, Heath AC, Martin NG, Montgomery GW, Goddard ME (2010) Common SNPs explain a large proportion of the heritability for human height. Nat Genet 42(7):565

Yuwono T, Handayani D, Soedarsono J (2005) The role of osmotolerant rhizobacteria in rice growth under different drought conditions. Aust J Agric Res 56(7):715–721

Zahir ZA, Munir A, Asghar HN, Shaharoona B, Arshad M (2008) Effectiveness of rhizobacteria containing ACC deaminase for growth promotion of peas (*Pisum sativum*) under drought conditions. J Microbiol Biotechnol 18(5):958–963

Zandalinas SI, Balfagón D, Arbona V, Gómez-Cadenas A (2017) Modulation of antioxidant defense system is associated with combined drought and heat stress tolerance in citrus. Front Plant Sci 7(8):953

Zeisel SH (2006) Choline: critical role during fetal development and dietary requirements in adults. Annu Rev Nutr 26:229–250

Zhang H, Murzello C, Sun Y, Kim MS, Xie X, Jeter RM, Zak JC, Dowd SE, Paré PW (2010) Choline and osmotic-stress tolerance induced in Arabidopsis by the soil microbe *Bacillus subtilis* (GB03). Mol Plant-Microbe Interact 23(8):1097–1104

Fertilizer Nitrogen as a Significant Driver of Rhizosphere Microbiome in Rice Paddies

12

Balasubramanian Ramakrishnan, Prasanta Kumar Prusty, Swati Sagar, M. M. Elakkya, and Anjul Rana

1 Introduction

Plants are colonized by microorganisms abundantly, with more of their genes and population densities than a total number of host plant's genes and cells (Mendes et al. 2013). These plant-associated microorganisms with their genomes, referred to as the plant microbiome, have significant influences at all the developmental stages, from seed germination to maturity. The zone surrounding the plant roots, the rhizosphere, also supports microorganisms, protozoa, nematodes, and other organisms including viruses. The plants' rhizospheres, rich with diverse microorganisms, are one of the highly complex ecosystems (Raaijmakers et al. 2009). The nutrients released by plants through exudates, border cells, and mucilage (rhizodeposits) determine the assemblage and activities of microorganisms in this zone. Generally, the rhizosphere microbial communities require the plant photosynthates released from roots, between 20 and 50% for the carbon skeletons and energy generation (Kuzyakov and Domanski 2000). The structure and dynamic changes in the root-associated microorganisms (rhizosphere microbiome) determine the flow of C and energy in the complex food web that involves nematodes, protozoa, and other organisms. More importantly, the microbiome mainly provides nutrients and phytohormones to plants, suppresses pathogens, and decomposes the plant debris and dead matter in soils. Understanding the assemblage and activities of the rhizosphere microbiome is critical to improving the plant productivity and sustainability in agriculture.

Rice (*Oryza sativa* L.), which has been domesticated about 8000–9000 years ago, is cultivated in more than 100 countries. Their annual production is more than 700 million tons from a total harvested area of about 158 million ha. Their grain

B. Ramakrishnan (✉) · P. K. Prusty · S. Sagar · M. M. Elakkya · A. Rana
Division of Microbiology, Indian Agricultural Research Institute, New Delhi, India
e-mail: brama@iari.res.in

D. P. Singh et al. (eds.), *Microbial Interventions in Agriculture and Environment*,
https://doi.org/10.1007/978-981-13-8383-0_12

yields vary, and the productivity ranges from less than 1 t ha^{-1} to 10 t ha^{-1}, depending on the resource and energy utilization for cultivation (Zeigler and Barclay 2008). Among the nutrients of significance, carbon is essential for photosynthesis and for all the microbial activities in the rhizosphere; carbon is chiefly obtained from carbon dioxide, currently comprising 0.04% of the atmosphere. Besides, the decomposition processes mediated by different microbial guilds with competition for electron donors or syntrophic cooperation provide carbon dioxide, methane, and other C-containing compounds, which serve as the important sources of C for microorganisms. These microbe-mediated activities can occur rapidly due to intensive rice cultivation. In general, the respirational and fermentative activities (root and rhizo-microbial) contribute more than 40% of the total CO_2 flux (Raich and Schlesinger 1992). The microbiome of flooded rice paddies utilizes the photosynthates, not only for respiration but also for methanogenesis. Up to 52% of the total emissions of methane from rice paddy ecosystem is chiefly produced by methanogenic archaea in the soil (Minoda et al. 1996; Yuan et al. 2012).

The exudation by rice plants contributes more to the stable forms of soil carbon such as organo-mineral complexes, and these complexes enhance carbon sequestration more than the above-ground plant biomass (Schmidt et al. 2011). Like all higher plants, rice plants with the associated microbiome in the rhizosphere and soil sustain and support the growth and activities of several taxonomically and functionally different members of *Archaea, Bacteria,* and *Eucarya* (Ramakrishnan et al. 2001; Edwards et al. 2015; Priya et al. 2015). Recent reports on the metagenomic analysis of rice microbiomes show a large diversity of microorganisms, in plants cultivated in different geographical regions (Hernandez et al. 2015; Edwards et al. 2015). These organisms are considered to offer an extension of functional capabilities to the host plants. Nevertheless, the rhizosphere microbial community structure, specifically those on the surface of roots, is primarily determined by root exudates (Yang and Crowley 2000; Dennis et al. 2010). Hence, the root exudates play vital roles in nutrient acquisition; plant growth is thus regulated by root exudation, which in turn determines the cooperative and competitive interactions among the members of the microbial communities (Einhellig et al. 1993).

Rice plants acquire nitrogen from several chemical forms. The crop management practices such as transplanting enhance the uptake of N at early growth stages, relative to other methods of cultivation (Ooyama 1975). The biological transformations such as microbial nitrogen mineralization and fixation provide N in the available forms to rice plants in the flooded (wet-puddled) soils. Fertilizers have become an indispensable input of intensive rice cultivation in the modern days. The total nutrient capacity (N, P_2O_5, and K_2O) is estimated to be about 310 million tonnes globally in 2018, while the actual supply of ammonia, phosphoric acid, and potash are expected to be 267 million tonnes (FAO 2017). India is the third largest producer and consumer of fertilizers. The national average consumption of plant nutrients is 165.12 kg ha^{-1} in 2017, which is, nevertheless, below the average consumption of 565.25 kg ha^{-1} in China. Since the soils of India are innately poor in fertility status, the judicious application of fertilizers becomes necessary to correct the inadequate level of nutrient availability. The success of

Table 12.1 Common nitrogen fertilizers with acidity or basicity

Fertilizer (%N)	Acidity (−) or basicity (+) (For kg $CaCO_3$ kg^{-1} N)
Urea (46%)	−1.79
Sulfur-coated urea (37%)	−3.19
Ammonium nitrate (34%)	−1.74
Ammonium sulfate (21%)	−5.24
Calcium nitrate (15%)	+1.33
Sodium nitrate (16%)	+1.81

the "Green Revolution" in India is chiefly due to high input uses (nutrient, irrigation, and plant protection chemicals), along with genetic manipulations of crops. However, sustaining the higher productivity of rice now depends on achieving more from the available land, water, and labor resources, without ecological and social harm. In the production systems of rice which have become highly intensive, both the inputs of C and N through the manures and fertilizers and the output of C and N through harvesting and emission of greenhouse gases regulate the nutrient budgets. The synthetic fertilizers that are applied to enhance the productivity of rice, especially N fertilizers with acidic or basic in nature, may alter the pathways of microbial C and N cycling significantly (Table 12.1).

The microbial processes that facilitate the C and N cycling have been largely investigated by measuring the process rates and enzyme activities. Several investigations focused on the carbon or nitrogen turnover individually, in terms of methane emission or biological nitrogen fixation while the turnover of C and N is a closely interspersed network of different processes (Rath et al. 1999; Zhao et al. 2016). Similarly, the major monitoring programs for soil health and fertility rely profoundly on the chemical and physical measures with lesser emphasis on the biological components, which are inadequately measured regarding the (bio)mass of microbial components, and activities of microbial enzymes (Mele and Crowley 2008). Traditionally, the microbial contributions to the carbon or nitrogen cycling in the rhizosphere have been studied using pure cultures isolated from these environments. However, the microbial processes that facilitate the C or N transformations involve the guilds of diverse microorganisms. Though the approach of reductionism to understanding from a few model microorganisms as pure, exponentially growing cultures in nutrient-rich medium has provided important information about individual microorganisms, the structure, activities, stability, adaptive responses, and evolution over time of soil microbial communities due to the complexities in the interaction of biotic and abiotic factors that take place in soils or rhizosphere are at present largely unknown.

The composition of functional guilds can now be quantified using the taxonomically related and functional genes of the microorganisms from the nucleic acids extracted from soils (Wallenstein and Vilgalys 2005). The methods of metagenomic analysis and quantitative polymerase chain reaction (qPCR) provide newer opportunities to investigate and gain better insights on the complex microbial communities. Hayden et al. (2010) also proposed the use of N cycling gene copy estimates to determine the soil health. Zhang et al. (2013) showed the sensitivities of nearly all genes (*nifH* of fixation, *chiA* of mineralization, *amoA* of ammonia-oxidizing bacteria (AOB) or ammonia-oxidizing archaea, *nirS*, *nirK,* and *nosZ* of denitrification) to nitrogen enrichment. The microbial guilds that are involved in C and N cycling and their responses to the environmental factors and different agricultural practices including the application of synthetic fertilizers have not been fully understood because of the challenges associated with studying the structure and dynamics of microbial communities in the rice rhizosphere.

2 Ecology of Microorganisms in Rice Ecosystems

Native populations of bacteria, archaea, fungi, algae, and several other living organisms are commonly present in soils. On the Earth, there could be more than one trillion (10^{12}) microbial species (Locey and Lennon 2016). The members of *Bacteria* and *Archaea* are the most abundant in the terrestrial soils, especially in the rhizospheres due to the increased availability of carbon and energy. The bacterial population that is cultured using the laboratory media can be about 10^7 to 10^8 g^{-1} soil, whereas total populations of culturable and yet-to-be-cultured organisms can be more than 10^{10} g^{-1} soil (Roszak and Colwell 1987). The reasons for their abundance in soils are due to the microbial capabilities for decomposition and nutrient cycling. More than 50% of the energy fixed by plants goes to the soil in various forms including dead plant tissues and excreta of animals. They become a central energy source of what Macfadyen (1963) called the "decomposer industry." Heterotrophic microorganisms and animals control the rates and the utilization patterns of substrates during decomposition, and the ensuing nutrient cycling. Of the soil mass, the microorganisms constitute less than 0.5 percent (w/w), and the concentrations of carbon in microbial biomass comprise about 1–4% of total soil organic carbon. The laboratory cultivation of these microorganisms, sampled from the non-rhizosphere (bulk) and rhizosphere soils, and their identification involve morphological, physiological, and biochemical analyses, which are now considered to provide inadequate information (Prosser et al. 2007). Apart from these constraints, this traditional approach suffers largely from the biases resulting from the media and the conditions employed for cultivation.

The natural or evolutionary relationships within microbial communities do not correlate well with conventional classification. The traditional methods provide only "snapshots" of microbial members. The ribosomal RNAs are now employed to assess the phylogenetic relationships among microorganisms. The rRNA sequences, particularly 16S rRNA as molecular chronometers for archaea and bacteria, have high evolutionary information content. The simplification of methods for determining rRNA sequences provides larger gains in the information on the ecology of microbial communities, from a descriptive to a quantitative manner. The PCR-based amplification of nucleic acids from soils or plant tissues, independent of cultivation using media for microorganisms, allows profiling of microorganisms better. Thus, the recent developments in nucleic acid extraction methods, sequencing of nucleic acids and/or proteins, and labeling, microarray analysis (e.g., PhyloChip and GeoChip), and computational capabilities offer the new "lenses" of molecular sequences that provide better quantitative information on the structure and dynamic changes of the microbial communities (Valm et al. 2011; Marcus et al. 2013). They offer many advantages in identifying the members of different microbial taxa, more than 10,000 operational taxonomic units (OTUs) from each sample of soil, water, or tissues, and hence in the simultaneous and economic profiling of microbiomes.

Microorganisms do not exist as individuals, and they form communities depending on many factors, even in the production systems of rice. As the intricacy of the community increases, their simplicity is replaced by more complex positive or negative interactions. Because of their enormous phenotypic and genotypic diversity, the microbial communities continue to be challenging for characterization (Ovreas and Torsvik 1998; Mora et al. 2011). Certain techniques based on fatty acid or metabolic profiles can help to describe the structure of the microbial community to some extent. Bai et al. (2000) extracted and estimated phospholipid fatty acids (PLFA), hydroxy fatty acids of lipopolysaccharides (LPS-HYFA), and phospholipid ether lipids (PLEL) directly from the soil to determine the structure of microbial communities. The aerobic, facultative anaerobic, and archaea formed about 44%, 32%, and 24%, respectively, of the total PLFAs. The nucleic acid-based approaches such as fingerprinting or sequencing of 16S rRNA (rDNA) have become popular to understand the genomic variations and species differences within microorganisms (Torsvik 1980). Hengstmann et al. (1999) analyzed the bacterial rDNA sequences, obtained directly from soils, to examine the bacterial community structure of anoxic rice soil. Using the Italian rice soils, Chin et al. (1999) examined the culturable polysaccharolytic and saccharolytic fermenting bacteria by the most-probable-number (liquid serial dilution culture) methods; the use of xylan, pectin, or a mixture of seven mono- and disaccharides as the growth substrate led to highest MPN counts (up to 2.5×10^8 g^{-1} soil). The strains isolated from the positive tubes of higher dilutions belonged to *Verrucomicrobia*, the *Cytophaga-Flavobacterium-Bacteroides*, clostridial cluster XIVa, clostridial cluster IX, *Bacillus* spp., and the class *Actinobacteria*. The other molecular techniques such as cloning of soil DNA in bacteria and sequencing have made rapid advances in gaining information on the soil microorganisms (Handelsman et al. 2002; Tringe et al. 2005; Delmont et al. 2011).

The metaproteogenomic approach by Knief et al. (2012) provided better understanding of the bacterial and archaeal assemblages and their physiologies in the phyllosphere and rhizosphere of rice. The rhizosphere had those proteins related to methanogenesis and methanotrophy while those of methanol-based methylotrophy were in the phyllosphere. Though the rice phyllosphere bacteria had *nifH* genes, the dinitrogenase reductase was solely present in the rhizosphere. The application of N fertilizers at low (0 kg N), standard (30 kg N), and high (300 kg N per ha) levels altered the bacterial communities in the rice (Ikeda et al. 2014). The members of *Burkholderia*, *Bradyrhizobium*, and *Methylosinus* were abundant while those of methanogenic archaea decreased in the low (0 kg N) N-applied fields compared to the standard N-applied field plots. Evidently, the application of N altered the key microbial groups associated with plants, especially those guilds that mediate the biogeochemical processes.

High-throughput sequencing significantly advances the insights on the structure of the root-associated microbiomes (Edwards et al. 2015). The composition of microbiome was found to vary due to both the soil types and rice genotypes, under the controlled greenhouse conditions. On the contrary, the variations in the field conditions were mostly due to the geographic location and the cultivation practice such as the application of manures. About 119 out of 152 operational taxonomic units (OTUs) that were enriched in the rhizosphere were found to colonize the roots. Also, more than 271 OTUs were enriched in the rhizoplane and endosphere, suggesting their similarities. Breidnebach et al. (2016) observed that rice plants enriched the members of *Gemmatimonadetes*, *Proteobacteria,* and *Verrucomicrobia*. The rhizosphere had the potential iron reducers (*Geobacter* and *Anaeromyxobacter*) and fermenters (*Clostridiaceae* and *Opitutaceae*) more than the bulk soils. Edwards et al. (2018) monitored the archaeal and bacterial assemblies at different developmental stages and found that their assemblies dynamically changed at the vegetative stage and remained stable after, in rice. The nucleic acid-based methods, relative to other methods available, provide rapid information on the compositions and functions of microbial communities of the rice rhizospheres.

3 Carbon Flow in the Rhizosphere Microbiome

Flooding of soil for rice cultivation leads to the reduction processes in a sequential manner, resulting in the development of chemical gradients spatially and temporally, along with the dynamic changes in the microbial members that perform complimentary biogeochemical reactions (Fig. 12.1; Sethunathan et al. 1982; Kumaraswamy et al. 2000; Liesack et al. 2000). In the flooded fields, the bulk soils are predominantly anoxic while the rhizosphere soil is partially oxic. In the rhizosphere, the major one-carbon metabolism involves CH_4 and CO_2, with methanol and carbon monoxide to a lesser extent. Any reduction in the population and diversity of microbial guilds which compete for electron donors or cooperate syntrophically will affect not only their proliferation but also the uptake of nutrients by rice plants. The rice plants allocate about 30 to 60% of net photosynthesized carbon, relatively a high carbon cost, to roots. Interestingly, the rice root growth rates are found to be significantly correlated to the percentage distribution of photosynthates to the soil. Lu et al. (2004) found a strong relationship between microbial activity and rice plant photosynthesis. Fundamentally, the rice grain yields are associated with the efficiencies of photosynthesis and translocation of photosynthates to sinks. While the

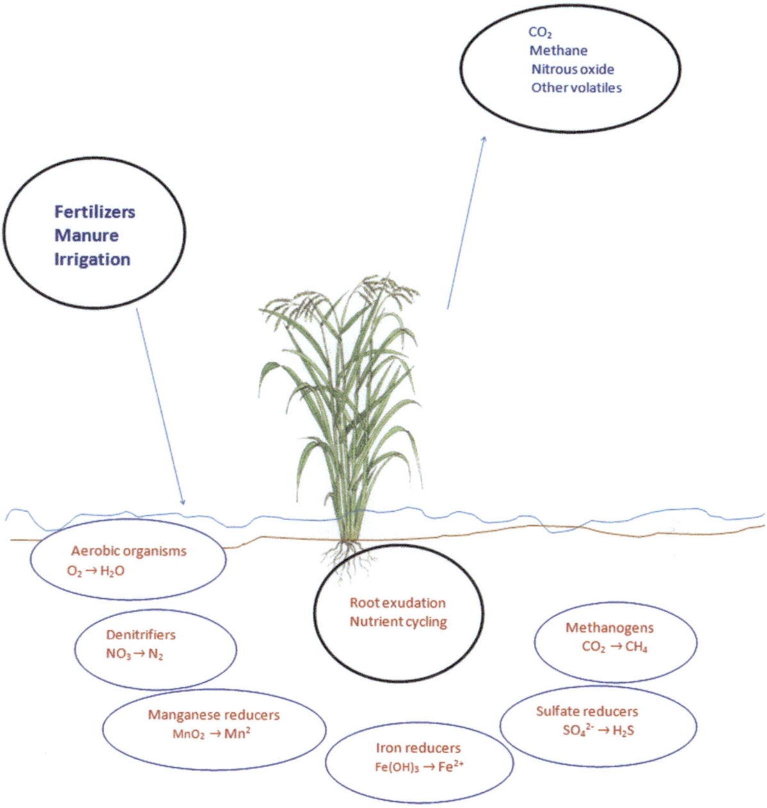

Fig. 12.1 Nature cycling in rhizosphere of rice with predominant microbial members (Adopted from Cao et al. 2003)

non-rhizosphere (bulk) soil is generally nutrient-limited, the root exudates can make or break microbial communities and provide niches for colonization and interactions with roots. High costs of carbon and energy involved in producing and releasing the root exudates strongly suggest the plant-mediated recruitment of microorganisms in the rhizosphere. The removal of spikelets, which in turn could impair grain development, was found to increase methane emissions (Denier van der Gon et al. 2002). The production of methane is chiefly by methanogenic archaea, which probably received more substrates from rice plants after the removal of spikelets.

In the rhizospheres, the concentrations of carbon can range from 10 to 100 μg C g^{-1} soil (Darrah 1991). The root exudation, both in terms of quality and quantity, will determine the enrichment of selective groups, thereby determining the assemblies and activities of microbial communities (Hartmann et al. 2009). Since methane production is primarily from the organic carbon of rice roots (Yuan et al. 2012), a significant proportion of host plant's photosynthates goes for methane production and oxidation in the rhizosphere. The ensuing methane emission by rice plants can

Table 12.2 Selected compounds for microbial utilization and their energetic values

Compound	Formula	Molecular weight	kJ mol^{-1}
Water	H_2O	18	−237.2
Carbon dioxide	CO_2	44	−394.4
Oxygen	O_2	16	0
Hydrogen	H_2	2	0
Glucose	$C_6H_{12}O_6$	180	−917.3
Acetate	C_2OH_4	52	−369.4
Methane	CH_4	16	−50.8
Nitrous oxide	N_2O	44	+104.2
Ammonium	NH_4^+	18	−79.4
Cellulose	$n(C_6H_{12}O_6)$	>400,000	–
Fats	$C_{57}H_{110}O_6(l)$	890	−75,520

Adapted from Dilly (2005)

influence the source-sink relationships considerably. On an annual basis, the tropical flooded rice ecosystem can serve as a net carbon sink, despite the loss of carbon through harvest, leachates of dissolved organic carbon, and the emissions of methane (Bhattacharyya et al. 2014). In the irrigated fields, the anoxic soil conditions favor the activities of several fastidious anaerobic microorganisms which are difficult for culturing and require compounds with different energetic values (Table 12.2). Despite the fundamental importance of carbon cycling in the agricultural production systems including that of rice, the taxonomic, genetic, and functional diversity of microorganisms on the roots and rhizosphere as influenced by various soil and plant factors, and the management practices remain a challenge to be characterized.

3.1 Methanogenic Archaea in Rice Paddies

The first field study of methane emission from rice was made in California, suggesting the soil-plant-atmosphere continuum of carbon flow (Cicerone and Shetter 1981). In 1988, Rajagopal and coworkers isolated and characterized methanogens from rice soils of Texas, USA. Early researches by the Indian workers showed that there were wide variations in the capacities of rice varieties for methane emission; these varieties were grouped based on their peak emissions occurring at vegetative, reproductive, or at both growth stages (Satpathy et al. 1998; Kumaraswamy et al. 2000). In a landmark report, Lu and Conrad (2005) showed the Rice Cluster I archaea as the key methanogenic group responsible for methane production, and these archaea utilized the photosynthates for about 85% of methane produced. The rice plants regulate methane emission primarily because they act as (i) a source of substrates for methanogenesis and (ii) conduits for methane for a well-developed system of aerenchyma, and (iii) they facilitate active oxidation of methane in the rhizosphere (Nouchi and Mariko 1993; Satpathy et al. 1997; Aulakh et al. 2000).

Archaea are one of the three major evolutionary lineages of life on the Earth and have several methanogenic microbial members. The molecular phylogenetic surveys of microorganisms in soils and oceans reveal an enormous diversity of archaea. Großkopf et al. (1998) identified hitherto unknown methanogens from Italian rice soils. An investigation using the method of terminal restriction fragment length polymorphism led to characterizing seven different archaeal ribotypes from the soil aggregates of Italian rice soils (Ramakrishnan et al. 2000). The methane production was strongly affected by the soil aggregate size and incubation time while the structure of the methanogenic microbial community was found to remain unchanged relatively. Nevertheless, the biogeographical distribution of methanogenic microbial communities can be different, and these differences may contribute to the observed variations in soil functioning regarding methanogenesis, and subsequently on the emission of methane. The patterns and dynamics of methanogenic archaeal communities were characteristically different in rice field soils of different geographical origin; these communities actively participated in energy and carbon flow (Ramakrishnan et al. 2001). Employing the propionate-oxidizing and hydrogen-producing syntroph, *Syntrophobacter fumaroxidans* in the coculture method, Sakai et al. (2007) isolated a methanogen (strain SANAE) belonging to RC-I from the Japanese rice paddies. The denser root colonization by RC-I methanogens, relative to that of *Methanomicrobiales*, led to higher methane emission (Conrad et al. 2008). The metaproteogenomic approach of Knief et al. (2012) provided stronger evidences that methanogenesis is the foremost one-carbon conversion process in rice roots. Hernandez et al. (2015) supported this observation and found that the root-associated bacterial populations could incorporate plant-derived carbon and a greater proportion was in the root compartment relative to those in the rhizosphere.

3.2 Methanotrophic Bacteria in the Rice Ecosystems

Half of the total organic carbon degraded by anaerobic microorganisms to methane is further oxidized by aerobic methane oxidizers into carbon dioxide, resulting in the atmospheric methane release of only about 0.5% of the total carbon turnover. Generally, the number of methanotrophs isolated as "colony-forming units" is a small fraction of the viable population (Hanson and Hanson 1996). Methanotrophic bacteria, a group of methylotrophs which can aerobically grow with CH_4 as their sole energy and carbon source, are mainly distinguished by their phylogenetic affiliation (γ- and α-*Proteobacteria*) and their carbon assimilation pathways (ribulose monophosphate and serine pathway, respectively). Type I methanotrophs affiliated to the genera *Methylomonas*, *Methylobacter*, and *Methylococcus*, and also to a novel type I methanotroph sublineage were abundant on rice roots than in the bulk soil, evidently from copies of *pmoA*, which encodes the subunit of the particulate methane monooxygenase (Horz et al. 2001). On the contrary, type II methanotrophs of the *Methylocystis-Methylosinus* group were more in both the soil and root compartments. Methanotroph diversity can also be analyzed by the recovery of genes coding for subunits of soluble methane monooxygenase (*mmoX*), methanol dehydrogenase (*mxaF*), and 16S ribosomal DNA (rDNA). In addition to aerobic/anaerobic methane oxidizers, bacteria that oxidize ammonium to nitrite (chemoautotrophic ammonium oxidizers) can oxidize methane to carbon dioxide (Bedard and Knowles 1989).

The contemporary knowledge about the microbial metabolism of carbon in the rice rhizosphere is largely derived from functional studies. There is a need to recognize the dynamic nature of microbial communities on rice roots, both on the surface and inside, that are performing various functions at different developmental growth stages. Like the microbial carbon pump in the oceans (Jiao and Zheng 2011), the C cycling mediated by microorganisms may play significant roles in soil fertility and carbon sequestration (Six et al. 2006; Liang and Balser 2011). The plant carbon inputs affect the assemblies and activities of microbial communities, and nutrient cycling significantly, which is evident from the varietal variations for root exudation (Wang and Adachi 2000; Aulakh et al. 2001). Besides, nitrogenous fertilizers are known to influence the production as well as oxidation of methane, and consequently its emission (Kumaraswamy et al. 1997; Kimura et al. 2004). Since the characterization of the microbial community structure or root exudation in situ are highly challenging, the functional metagenomic and metaproteogenomic analyses that allow community-wide analysis can facilitate the identification of the key members of microbiome and their functions in the C cycling. A complete understanding of the carbon flow in the rice rhizosphere is crucial since the balance between the carbon storage and loss is critical, not only for greenhouse gas mitigation but also for crop productivity.

4 Nitrogen Cycling in Rice Ecosystems

Rice provides about 20% and 31% of the total calories required by the global and Indian populations, respectively (Zeigler and Barclay 2008). For each ton of rough rice including straw, plants need about 16–17 kg N (De Datta 1981). The nitrogenous compounds in soils are found either in inorganic or in organic forms. More than 90% of total nitrogen present in organic matter composed primarily of amides (NH_2), in soils. The clay minerals may retain N in minor amounts, in the form of NH_4^- (Sahrawat and Narteh 2001), but not in higher amounts. During mineralization, nitrogen that is present in soil organic matter, crop residues, and manures is transformed from organic nitrogen to the ammonia (NH_3) and ammonium (NH_4^+) through ammonification and the ammonium to nitrate form (NO_3^-) through nitrification. The synthetic fertilizers are generally used to augment the rice production. Nearly one-third of global nitrogen fertilizers caters to the requirement of rice cultivation (Cassman et al. 1998). Since the efficiency of nitrogenous fertilizer is as low as 40%, different forms of N go as pollutants in the soil and water system (Chowdhury and Kennedy 2005).

4.1 Microbial Mediation of Rhizosphere Influences

In the rice production systems, the biological N transformations such as microbial nitrogen mineralization and fixation provide nitrogen in the available forms to plants. The biotic and abiotic nitrogen-transforming reactions that occur in the rice paddies are many, including ammonification, N immobilization, nitrification, denitrification, dissimilatory nitrate reduction to ammonium (DNRA), anaerobic ammonium oxidation (anammox), and nitrogen fixation (Ishii et al. 2011). Ammonium production is the significant process for N nutrition in the lowland rice paddy ecosystems because the mineralization is considered to proceed up to ammonium production (Narteh and Sahrawat 2000). Arth et al. (1998) adapted the N_2 flux method and demonstrated that the coupled nitrification-denitrification process was localized in the rhizosphere. However, the culturable, ammonium-oxidizing nitrifiers in soils under submergence were of low densities (Reichardt et al. 2001) although there were large population shifts in other biogeochemically relevant members of microbial communities. In the rice paddies, the oxygen gradient is often formed in the root regions due to the leakage of oxygen (Liesack et al. 2000). This radial oxygen loss can significantly influence the assemblies and activities of microbial communities in the rhizosphere and on the surface of roots. The synthetic chemical fertilizers, when applied to enhance the productivity of rice, may alter the microbial dynamics and the N pathways. Several investigations focused on individual processes of nitrogen turnover such as biological nitrogen fixation while the N turnover is a network of closely interlinked processes (Ollivier et al. 2011). Since the input N and the output N through harvesting and environmental losses regulate the N budgets, the chemical N fertilizers, both in terms of quantity and type, can significantly alter the N cycling in the intensive rice production systems.

Nitrification activity is supported by the roots of rice plants, and the root-associated nitrification activity is a main determinant of the ratio of NH_4^--N and NO_3-N in the ammonium-fertilized plants (Briones et al. 2003). When rice plants were provided with ammonium and nitrate together, relative to the source of nitrogen in either form, the growth and yields are found to improve (Kronzucker et al. 1999; Kirk and Kronzucker 2005). Generally, the rice roots are provided with both NH_4^--N and NO_3-N, due to ammonium oxidation by native microbial communities in the paddy fields (Briones et al. 2003; He et al. 2007). Kirk and Kronzucker (2005) and Ying-Hua et al. (2006) also reported that partial nitrate nutrition was efficient in assimilating nitrate in the rice rhizosphere. Ghosh and Kashyap (2003) found a strong correlation between the nitrifying bacterial population enumerated by the culture-dependent methods and the biomass and air space of rice roots. Briones et al. (2002) reported the microscale differences in the oxygen availability due to the varietal difference of rice, and as a result, the changes in the composition and activity of root-associated AOB population. The forms of nitrogen available in the environments such as mineral nitrogen and organic nitrogen and the availability of nutrients such as phosphorus and potassium have stronger influences on the activities of ammonium-oxidizing microbial communities (Jia and Conrad 2009). In general, the AOB communities are dominant under high nitrogen availability (Meyer et al. 2013; Strauss et al. 2014).

More importantly, the differences among the rice cultivars for N-use efficiencies can be attributed to the root-associated nitrification activity.

Nitrogen cycling in soils and sediments includes the well-studied processes such as nitrogen fixation, ammonification of nitrification, and denitrification as well as the recently discovered processes such as anammox (anaerobic ammonium oxidation) and dissimilatory nitrate reduction to ammonium (DNRA) (Rutting et al. 2011; van Niftrik and Jetten 2012). Information on the microbial players that mediate the recently discovered processes in the rice rhizosphere is very limited. The rice plants, depending on the growth stages, can influence the structures and abundances of ammonia-oxidizing bacterial (AOB) and denitrifying communities (Hussain et al. 2011). Ikeda et al. (2014) observed the differences in the relative abundances of *Burkholderia*, *Bradyrhizobium*, and *Methylosinus* between the low N- and high N-root microbiomes. The amounts of nitrogenous fertilizers may determine the diversity and functions of not only the nitrogen-cycling but also the carbon-cycling microorganisms. Since several of the carbon- or nitrogen-cycling microorganisms are fastidious in their growth, the application of modern molecular analyses such as the metagenomic and real-time PCR analyses offers advantages not only in their detection but also in their quantification rapidly.

Nitrate, nitrite, NO, and N_2O are inhibitory to certain processes of C cycling, especially methanogenesis (Kluber and Conrad 1998). Nitrate weakly inhibited methanogenesis by *Methanosarcina barkeri* and *Methanobacterium bryantii* while N_2O and nitrite did for *Ms. barkeri*, and nitrite and N_2O did for *Mb. bryantii*. NO strongly inhibited methanogenesis in both bacteria (Kluber and Conrad 1998). Thus, the products and intermediates of denitrification can inhibit CH_4 production both reversibly and irreversibly, depending on the type of methanogenic bacterium and the concentration of the N-compound. Globally, the biological N_2 fixation is the primary source of reactive N with an annual production of about 90–130 Tg N (Galloway et al. 2004). The researches at both national and international levels have been attempted to derive benefits from the association of free-living nitrogen-fixing bacteria (Rao et al. 1998; Yanni et al. 1997; Ladha and Reddy 2000;) or from the endophytic association between diazotrophic bacteria and rice plants (Barraquio et al. 1997; Mano and Morisaki 2008; Hardoim et al. 2011).

Higher plants including rice have microorganisms intracellularly as endophytes, as obligate, or in the facultative association (Harley 1989; Hinton and Bacon 1985). Hardoim et al. (2012) found that about 74% of the rice seed-borne bacterial endophytes were interrelated with those found in the maturing and matured seed tissues, endosphere of root and leaf tissues. Using four species of *gfp*-tagged rhizobia, Chi et al. (2005) demonstrated the ascending migration of these endophytes from roots to leaves; these endophytes influenced the physiology of rice plants, with an increased rate of photosynthetic. Hardoim et al. (2008) defined the term "competent endophytes" as those microorganisms that could colonize the plant tissues, stimulate the physiological activities, and get favored for maintenance. What is unknown now is how the dynamics of endophytes, especially those "competent endophytes," gets influenced by the application of chemical fertilizers. The utilization of microbial consortia that interact synergistically in the presence of fertilizers has not been attempted hitherto. Significant gains can be made from information on the "core microbiome" obtained from the global analyses of metagenomes and proteomes of rice under oxic (upland) and anoxic (irrigated) cultivation. Since the nitrogen use efficiency of rice plants is less than about 50% (Lassaletta et al. 2014), the loss of reactive nitrogen from synthetic N fertilizers is enormous, resulting in contamination of groundwater and other water bodies, soils, and even the atmosphere. Therefore, the challenges of managing the soil fertility through the application of chemical fertilizers judiciously during the intensive rice cultivation have assumed greater importance now than ever before.

5 Future Perspectives

The assemblies and activities of microbial communities of the globally important agronomic plant of rice which is cultivated ranging from flooded (irrigated) to dry (upland) conditions in diverse ecosystems have been inadequately described until now. Of the total soil microbiome, the rice plants may associate only with certain microbial groups (Huang et al. 2014). The flooded rice paddies which support the anaerobic metabolism of nitrate, iron, manganese, and sulfate reducers and methanogens are difficult to study and interpret, with the cultivation-based methods (Liesack et al. 2000). Traditional methods of cultivation-based microbial analyses and indicators are now being criticized because of their biases. Higher plants including rice are found to cultivate their microbiomes by providing carbon and energy sources, adjusting the soil pH and reducing the competition for the beneficial microorganisms. Kristin and Miranda (2013) suggested that plants may use roots exudation to make an extension of its phenotypic traits through microbiota. Thus, the root exudates help to form a distinctive environment for attracting and initiating interactions with microorganisms in the rhizosphere. Since plants can fix atmospheric carbon photosynthetically ranging from 1080 to 2100 kg ha^{-1}, the carbon flow in the continuum of atmosphere-plant-microbial communities-soil will be an important determinant of crop productivity. The research group at the University of Queensland, Australia even went on to suggest that plants engulf and digest microorganisms as a source of nutrients, a clear deviation from the chemical theories suggested for the nutrition of plants (Paungfoo-Lonhienne et al. 2010). An improved understanding of microbiomes and their interactions with plants and soils, obliterating from the traditional models of plant growth and development, and application of beneficial microbiome members for efficient colonization of plant tissues are vital for efficient utilization of resources and enhanced productivities.

A gamut of different agronomic practices is gaining acceptance among the farmers, to save the scarce resources like water, labor, nutrients, and energy. The nutrient management, especially N for the rice cultivation with higher yields and nitrogen use efficiency, includes several practices including the application of microbial inoculants, site-specific N management practices (SSNM), real-time N management, and fixed-time adjustable-dose N management (FTNM) (Dobermann et al. 2002 and Peng et al. 2006). Since the soil functioning is chiefly determined by microorganisms, detailed knowledge and information on the composition and activities of microbial communities are warranted. The phylogenetic richness and functional redundancy of the microbial communities in the rice rhizosphere have only now begun to be understood (Edwards et al. 2015, 2018). The rapid and cheaper DNA sequence-based approaches, combined with qPCR assays, offer great potential for gaining insights on the dynamics of microbial communities (Rondon et al. 2000; Daniel 2005 and Petersen et al. 2012). The stimulation of microbial growth in the rice rhizosphere was evident from the abundances of 16S rRNA genes, assayed by the qPCR methods (Breidenbach et al. 2016). The taxonomic markers such as 16S rRNAs provide a limited understanding on the microbial community structure associated with the C or N cycling, while the functional gene markers are useful, not only in understanding but also in predicting the nutrient cycling pathways (Poretsky et al. 2014; Altieri 2018).

6 Conclusions

Because of the challenges associated with studying the soil microbial communities, a more systematic application of both the culture-based and sequencing techniques becomes a necessity to gain better insights and identification of the "core microbiome" of rice and devising methods for manipulating beneficial microorganisms favorably. In the longer term, developments of rapid and reliable methods are necessary to assess the dynamics of microbial communities and develop "better management practices" for sustainability and environmental security of the rice production systems.

Acknowledgments The study was partly funded by the SERB Project on "Archaeal- and anaerobic ammonia oxidative processes of nitrogen cycling in oxic and anoxic soils," DST, Government of India, and the ICAR extramural research project on "Soil microbiome modulation strategies to enhance nitrogen acquisition efficiency in rice," granted to BR. We are thankful to the Division of Microbiology, ICAR-IARI, New Delhi, for providing necessary facilities.

References

Altieri MA (2018) Agroecology: the science of sustainable agriculture. CRC Press, Boca Raton

Arth I, Frenzel P, Conrad R (1998) Denitrification coupled to nitrification in the rhizosphere of rice. Soil Biol Biochem 30(4):509–515

Aulakh MS, Bodenbender J, Wassmann R, Rennenberg H (2000) Methane transport capacity of rice plants. II. Variations among different rice cultivars and relationship with morphological characteristics. Nutr Cycl Agroecosys 58(1–3):367–375

Aulakh MS, Wassmann R, Bueno C, Kreuzwieser J, Rennenberg H (2001) Characterization of root exudates at different growth stages of ten rice (Oryza sativa L.) cultivars. Plant Biol 3(2):139–148

Bai Q, Gattinger A, Zelles L (2000) Characterization of microbial consortia in paddy rice soil by phospholipid analysis. Microb Ecol 39(4):273–281

Barraquio WL, Revilla L, Ladha JK (1997) Isolation of endophytic diazotrophic bacteria from wetland rice. Plant Soil 194:15–24

Bédard C, Knowles R (1989) Physiology, biochemistry, and specific inhibitors of CH4, NH4+, and CO oxidation by methanotrophs and nitrifiers. Microbiol Rev 53(1):68–84

Bhattacharyya P, Neogi S, Roy KS, Dash PK, Nayak AK, Mohapatra T (2014) Tropical low land rice ecosystem is a net carbon sink. Agri Ecosyst Environ 189:127–135
Breidenbach B, Pump J, Dumont MG (2016) Microbial community structure in the rhizosphere of rice plants. Front Microbiol 6:1537
Briones AM, Okabe S, Umemiya Y, Ramsing NB, Reichardt W, Okuyama H (2002) Influence of different cultivars on populations of ammonia-oxidizing bacteria in the root environment of rice. Applied Environ Microbiol 68(6):3067–3075
Briones AM, Okabe S, Umemiya Y, Ramsing NB, Reichardt W, Okuyama H (2003) Ammonia-oxidizing bacteria on root biofilms and their possible contribution to N use efficiency of different rice cultivars. Plant Soil 250(2):335–348
Cao S, Tanji KK, Scardaci SC (2003) Incorporating straw may induce sulfide toxicity in paddy rice. Calif Agr 57(2):55–59
Cassman KG, Peng S, Olk DC, Ladha JK, Reichardt W, Dobermann A, Singh U (1998) Opportunities for increased nitrogen-use efficiency from improved resource management in irrigated rice systems. Field Crop Res 56(1–2):7–39
Chi F, Shen SH, Cheng HP, Jing YX, Yanni YG, Dazzo FB (2005) Ascending migration of endophytic rhizobia, from roots to leaves, inside rice plants and assessment of benefits to rice growth physiology. Applied Environ Microbiol 71(11):7271–7278
Chin KJ, Hahn D, Hengstmann U, Liesack W, Janssen PH (1999) Characterization and identification of numerically abundant culturable bacteria from the anoxic bulk soil of rice paddy microcosms. Applied Environ Microbiol 65(11):5042–5049
Chowdhury A, Kennedy IR (2005) Nitrogen fertilizer losses from rice soils and control of environmental pollution problems. Commun Soil Sci Plan 36:1625–1639
Cicerone RJ, Shetter JD (1981) Sources of atmospheric methane: measurements in rice paddies and a discussion. J Geophys Res 86(C8):7203–7209
Conrad R, Klose M, Noll M, Kemnitz D, Bodelier PL (2008) Soil type links microbial colonization of rice roots to methane emission. Glob Change Biol 14(3):657–669
Daniel R (2005) The metagenomics of soil. Nature Rev Microbiol 3(6):470–478
Darrah PR (1991) Models of the rhizosphere. Plant Soil 133(2):187–199
De Datta SK (1981) Principles and practices of rice production. International Rice Research Institute, The Philippines
Delmont TO, Robe P, Cecillon S, Clark IM, Constancias F, Simonet P, Hirsch PR, Vogel TM (2011) Accessing the soil metagenome for studies of microbial diversity. Appl Environ Microbiol 77(4):1315–1324
Dennis PG, Miller AJ, Hirsch PR (2010) Are root exudates more important than other sources of rhizodeposits in structuring rhizosphere bacterial communities? FEMS Microbiol Ecol 72(3):313–327
Dilly O (2005) Microbial energetics in soils. In: Varma A, Buscot F (eds) Microorganisms in soils: roles in genesis and functions. Soil biology, vol 3. Springer, Berlin/Heidelberg
Dobermann A, Witt C, Dawe D, Abdulrachman S, Gines HC, Nagarajan R, Satawathananont S, Son TT, Tan PS, Wang GH, Chien NV (2002) Site-specific nutrient management for intensive rice cropping systems in Asia. Field Crops Res 74(1):37–66
Edwards J, Johnson C, Santos-Medellín C, Lurie E, Podishetty NK, Bhatnagar S, Eisen JA, Sundaresan V (2015) Structure, variation, and assembly of the root-associated microbiomes of rice. Proc Natl Acad Sci U S A 112(8):E911–E920
Edwards JA, Santos-Medellín CM, Liechty ZS, Nguyen B, Lurie E, Eason S, Phillips G, Sundaresan V (2018) Compositional shifts in root-associated bacterial and archaeal microbiota track the plant life cycle in field-grown rice. PLoS Biol 16(2):e2003862
Einhellig FA, Rasmussen JA, Hejl AM, Souza IF (1993) Effects of root exudate sorgoleone on photosynthesis. J Chem Ecol 19(2):369–375
FAO (2017) World fertilizer trends and outlook to 2020. Food and Agriculture Organization of the United Nations, Rome
Galloway JN, Dentener FJ, Capone DG, Boyer EW, Howarth RW, Seitzinger SP, Asner GP, Cleveland CC, Green PA, Holland EA, Karl DM (2004) Nitrogen cycles: past, present, and future. Biogeochemistry 70(2):153–226

Ghosh P, Kashyap AK (2003) Effect of rice cultivars on rate of N-mineralization, nitrification and nitrifier population size in an irrigated rice ecosystem. Appl Soil Ecol 24(1):27–41
Großkopf R, Janssen PH, Liesack W (1998) Diversity and structure of the methanogenic community in anoxic rice paddy soil microcosms as examined by cultivation and direct 16S rRNA gene sequence retrieval. Appl Environ Microbiol 64(3):960–969
Handelsman J, Liles M, Mann D, Riesenfeld C, Goodman RM (2002) Cloning the metagenome: culture-independent access to the diversity and functions of the uncultivated microbial world. Methods Microbiol 33:241–255
Hanson RS, Hanson TE (1996) Methanotrophic bacteria. Microbiol Rev 60(2):439–471
Hardoim PR, van Overbeek LS, van Elsas JD (2008) Properties of bacterial endophytes and their proposed role in plant growth. Trends Microbiol 16(10):463–471
Hardoim PR, Andreote FD, Reinhold-Hurek B, Sessitsch A, van Overbeek LS, van Elsas JD (2011) Rice root-associated bacteria: insights into community structures across 10 cultivars. FEMS Micrbiol Ecol 77(1):154–164
Hardoim PR, Hardoim CC, Van Overbeek LS, Van Elsas JD (2012) Dynamics of seed-borne rice endophytes on early plant growth stages. PLoS One 7(2):e30438
Harley JL (1989) The significance of mycorrhiza. Mycol Res 92(2):129–139
Hartmann A, Schmid M, Van Tuinen D, Berg G (2009) Plant-driven selection of microbes. Plant Soil 321(1–2):235–257
Hayden HL, Drake J, Imhof M, Oxley APA, Norng S, Mele PM (2010) The abundance of nitrogen cycle genes *amoA* and *nifH* depends on land-uses and soil types in South-Eastern Australia. Soil Biol Biochem 42:1774–1783
He JZ, Shen JP, Zhang LM, Zhu YG, Zheng YM, Xu MG, Di H (2007) Quantitative analyses of the abundance and composition of ammonia-oxidizing bacteria and ammonia-oxidizing archaea of a Chinese upland red soil under long-term fertilization practices. Environ Microbiol 9(9):2364–2374
Hengstmann U, Chin KJ, Janssen PH, Liesack W (1999) Comparative phylogenetic assignment of environmental sequences of genes encoding 16S rRNA and numerically abundant culturable bacteria from an anoxic rice paddy soil. Appl Environ Microbiol 65(11):5050–5058
Hernández M, Dumont MG, Yuan Q, Conrad R (2015) Different bacterial populations associated with the roots and rhizosphere of rice incorporate plant-derived carbon. Appl Environ Microbiol 81:2244–2253
Hinton DM, Bacon CW (1985) The distribution and ultrastructure of the endophyte of toxic tall fescue. Can J Bot 63(1):36–42
Horz HP, Yimga MT, Liesack W (2001) Detection of methanotroph diversity on roots of submerged rice plants by molecular retrieval of *pmoA, mmoX, mxaF,* and 16S rRNA and ribosomal DNA, including pmoA-based terminal restriction fragment length polymorphism profiling. Appl Environ Microbiol 67(9):4177–4185
Huang XF, Chaparro JM, Reardon KF, Zhang R, Shen Q, Vivanco JM (2014) Rhizosphere interactions: root exudates, microbes, and microbial communities. Botany 92(4):267–275
Hussain Q, Liu Y, Jin Z, Zhang A, Pan G, Li L, Crowley D, Zhang X, Song X, Cui L (2011) Temporal dynamics of ammonia oxidizer (amoA) and denitrifier (nirK) communities in the rhizosphere of a rice ecosystem from tai Lake region, China. Appl Soil Ecol 48(2):210–218
Ikeda S, Sasaki K, Okubo T, Yamashita A, Terasawa K, Bao Z, Liu D, Watanabe T, Murase J, Asakawa S, Eda S (2014) Low nitrogen fertilization adapts rice root microbiome to low nutrient environment by changing biogeochemical functions. Microbes Environ 29(1):50–59
Ishii S, Ikeda S, Minamisawa K, Senoo K (2011) Nitrogen cycling in rice paddy environments: past achievements and future challenges. Microbes Environ 26(4):282–292
Jia Z, Conrad R (2009) Bacteria rather than Archaea dominate microbial ammonia oxidation in an agricultural soil. Environ Microbiol 11(7):1658–1671
Jiao N, Zheng Q (2011) The microbial carbon pump: from genes to ecosystems. Appl Environ Microbiol 77(21):7439–7444
Kimura M, Murase J, Lu Y (2004) Carbon cycling in rice field ecosystems in the context of input, decomposition and translocation of organic materials and the fates of their end products (CO_2 and CH_4). Soil Biol Biochem 36(9):1399–1416

Kirk GJD, Kronzucker HJ (2005) The potential for nitrification and nitrate uptake in the rhizosphere of wetland plants: a modelling study. Ann Bot-Lond 96(4):639–646

Klüber HD, Conrad R (1998) Effects of nitrate, nitrite, NO and N_2O on methanogenesis and other redox processes in anoxic rice field soil. FEMS Microbiol Ecol 25(3):301–318

Knief C, Delmotte N, Chaffron S, Stark M, Innerebner G, Wassmann R, Von Mering C, Vorholt JA (2012) Metaproteogenomic analysis of microbial communities in the phyllosphere and rhizosphere of rice. ISME J 6(7):1378–1390

Kristin A, Miranda H (2013) The root microbiota—a fingerprint in the soil? Plant Soil 370(1–2):671–686

Kronzucker HJ, Siddiqi MY, Glass AD, Kirk GJ (1999) Nitrate-ammonium synergism in rice. A subcellular flux analysis. Plant Physiol 119(3):1041–1046

Kumaraswamy S, Ramakrishnan B, Satpathy SN, Rath AK, Misra S, Rao VR, Sethunathan N (1997) Spatial distribution of methane-oxidizing activity in a flooded rice soil. Plant Soil 191(2):241–248

Kumaraswamy S, Rath AK, Ramakrishnan B, Sethunathan N (2000) Wetland rice soils as sources and sinks of methane: a review and prospects for research. Biol Fertil Soils 31(6):449–461

Kuzyakov Y, Domanski G (2000) Carbon input by plants into the soil. Review. J Plant Nutr Soil Sci 163:421–431

Ladha JK, Reddy PM (2000) The quest for nitrogen fixation in rice. International Rice Research Institute, The Philippines

Lassaletta L, Billen G, Grizzetti B, Anglade J, Garnier J (2014) 50 year trends in nitrogen use efficiency of world cropping systems: the relationship between yield and nitrogen input to cropland. Environ Res Lett 9(10):105011

Liang C, Balser TC (2011) Microbial production of recalcitrant organic matter in global soils: implications for productivity and climate policy. Nature Rev Microbiol 9(1):75

Liesack W, Schnell S, Revsbech NP (2000) Microbiology of flooded rice paddies. FEMS Microbiol Rev 24(5):625–645

Locey KJ, Lennon JT (2016) Scaling laws predict global microbial diversity. Proc Natl Acad Sci U S A 113(21):5970–5975

Lu Y, Conrad R (2005) In situ stable isotope probing of methanogenic archaea in the rice rhizosphere. Science 309(5737):1088–1090

Lu Y, Murase J, Watanabe A, Sugimoto A, Kimura M (2004) Linking microbial community dynamics to rhizosphere carbon flow in a wetland rice soil. FEMS Microbiol Ecol 48(2):179–186

Macfadyen A (1963) Contribution of the microfauna to total soil metabolism. In: Doeksen J, Van der Drift J (eds) Soil organisms, Amsterdam, pp 3–16

Mano H, Morisaki H (2008) Endophytic bacteria in the rice plant. Microbes Environ 23(2):109–117

Marcus IM, Wilder HA, Quazi SJ, Walker SL (2013) Linking microbial community structure to function in representative simulated systems. Appl Environ Microbiol 79:2552–2559

Mele PM, Crowley DE (2008) Application of self-organizing maps for assessing soil biological quality. Agri Ecosyst Environ 126:139–152

Mendes R, Garbeva P, Raaijmakers JM (2013) The rhizosphere microbiome: significance of plant beneficial, plant pathogenic, and human pathogenic microorganisms. FEMS Microbiol Rev 37:634–663

Meyer A, Focks A, Radl V, Keil D, Welzl G, Schöning I, Boch S, Marhan S, Kandeler E, Schloter M (2013) Different land use intensities in grassland ecosystems drive ecology of microbial communities involved in nitrogen turnover in soil. PLoS One 8(9):e73536

Minoda T, Kimura M, Wada E (1996) Photosynthates as dominant source of CH_4 and CO_2 in soil water and CH_4 emitted to the atmosphere from paddy fields. J Geophys Res 101(D15):21091–21097

Mora C, Tittensor DP, Adl S, Simpson AG, Worm B (2011) How many species are there on Earth and in the ocean? PLoS Biol 9(8):e1001127

Narteh LT, Sahrawat KL (2000) Ammonium in solution of flooded west African soils. Geoderma 95(3–4):205–214

Nouchi I, Mariko S (1993) Mechanism of methane transport by rice plants. In: Oremland RS (ed) Biogeochemistry of global change. Springer, Boston, MA, pp 336–352
Ollivier J, Töwe S, Bannert A, Hai B, Kastl EM, Meyer A, Su MX, Kleineidam K, Schloter M (2011) Nitrogen turnover in soil and global change. FEMS Microbiol Ecol 78(1):3–16
Ooyama N (1975) Nitrogen supplying patterns of paddy soils for rice plant in temperate area in Japan. J Sci Soil Manure 46:297–302
Øvreås L, Torsvik V (1998) Microbial diversity and community structure in two different agricultural soil communities. Microb Ecol 36(3–4):303–315
Paungfoo-Lonhienne C, Rentsch D, Robatzek S, Webb RI, Sagulenko E, Näsholm T, Schmidt S, Lonhienne TG (2010) Turning the table: plants consume microbes as a source of nutrients. PLoS One 5(7):e11915
Peng S, Bouman B, Visperas RM, Castañeda A, Nie L, Park HK (2006) Comparison between aerobic and flooded rice in the tropics: agronomic performance in an eight-season experiment. Field Crops Res 96(2–3):252–259
Petersen DG, Blazewicz SJ, Firestone M, Herman DJ, Turetsky M, Waldrop M (2012) Abundance of microbial genes associated with nitrogen cycling as indices of biogeochemical process rates across a vegetation gradient in Alaska. Environ Microbiol 14(4):993–1008
Poretsky R, Rodriguez-R LM, Luo C, Tsementzi D, Konstantinidis KT (2014) Strengths and limitations of 16S rRNA gene amplicon sequencing in revealing temporal microbial community dynamics. PLoS One 9(4):e93827
Priya H, Prasanna R, Ramakrishnan B, Bidyarani N, Babu S, Thapa S, Renuka N (2015) Influence of cyanobacterial inoculation on the culturable microbiome and growth of rice. Microbiol Res 171:78–89
Prosser JI, Bohannan BJ, Curtis TP, Ellis RJ, Firestone MK, Freckleton RP, Green JL, Green LE, Killham K, Lennon JJ, Osborn AM (2007) The role of ecological theory in microbial ecology. Nature Rev Microbiol 5(5):384–392
Raaijmakers JM, Paulitz TC, Steinberg C, Alabouvette C, Moënne-Loccoz Y (2009) The rhizosphere: a playground and battlefield for soilborne pathogens and beneficial microorganisms. Plant Soil 321:341–361
Raich JW, Schlesinger WH (1992) The global carbon dioxide flux in soil respiration and its relationship to vegetation and climate. Tellus B 44(2):81–99
Rajagopal BS, Belay N, Daniels L (1988) Isolation and characterization of methanogenic bacteria from rice paddies. FEMS Microbiol Lett 53(3–4):153–158
Ramakrishnan B, Lueders T, Conrad R, Friedrich M (2000) Effect of soil aggregate size on methanogenesis and archaeal community structure in anoxic rice field soil. FEMS Microbiol Ecol 32(3):261–270
Ramakrishnan B, Lueders T, Dunfield PF, Conrad R, Friedrich MW (2001) Archaeal community structures in rice soils from different geographical regions before and after initiation of methane production. FEMS Microbiol Ecol 37(2):175–186
Rao VR, Ramakrishnan B, Adhya TK, Kanungo PK, Nayak DN (1998) Review: current status and future prospects of associative nitrogen fixation in rice. World J Microbiol Biotechnol 14:621–633
Rath AK, Swain B, Ramakrishnan B, Panda D, Adhya TK, Rao VR, Sethunathan N (1999) Influence of fertilizer management and water regime on methane emission from rice fields. Agri Ecosyst Environ 76(2–3):99–107
Reichardt W, Briones A, De Jesus R, Padre B (2001) Microbial population shifts in experimental rice systems. Applied Soil Ecol 17(2):151–163
Rondon MR, August PR, Bettermann AD, Brady SF, Grossman TH, Liles MR, Loiacono KA, Lynch BA, MacNeil IA, Minor C, Tiong CL (2000) Cloning the soil metagenome: a strategy for accessing the genetic and functional diversity of uncultured microorganisms. Appl Environ Microbiol 66(6):2541–2547
Roszak DB, Colwell RR (1987) Survival strategies of bacteria in the natural environment. Microbiol Rev 51(3):365–379

Rütting T, Boeckx P, Müller C, Klemedtsson L (2011) Assessment of the importance of dissimilatory nitrate reduction to ammonium for the terrestrial nitrogen cycle. Biogeosciences 8(7):1779–1791

Sahrawat KL, Narteh LT (2001) Organic matter and reducible iron control of ammonium production in submerged soils. Commun Soil Sci Plan 32(9–10):1543–1550

Sakai S, Imachi H, Sekiguchi Y, Ohashi A, Harada H, Kamagata Y (2007) Isolation of key methanogens for global methane emission from rice paddy fields: a novel isolate affiliated with the clone cluster rice cluster I. Appl Environ Microbiol 73(13):4326–4331

Satpathy SN, Rath AK, Ramakrishnan B, Rao VR, Adhya TK, Sethunathan N (1997) Diurnal variation in methane efflux at different growth stages of tropical rice. Plant Soil 195(2):267–271

Satpathy SN, Mishra S, Adhya TK, Ramakrishnan B, Rao VR, Sethunathan N (1998) Cultivar variation in methane efflux from tropical rice. Plant Soil 202(2):223–229

Schmidt MW, Torn MS, Abiven S, Dittmar T, Guggenberger G, Janssens A, Kleber M, Kogel-Knabner I, Lehmann J, Manning DA, Nanipieri P (2011) Persistence of soil organic matter as an ecosystem property. Nature 478(7367):49

Sethunathan N, Rao VR, Adhya TK, Raghu K (1982) Microbiology of rice soils. CRC Crit Rev Microbiol 10(2):125–172

Six J, Frey SD, Thiet RK, Batten KM (2006) Bacterial and fungal contributions to carbon sequestration in agroecosystems. Soil Sci Soc Am J 70(2):555–569

Strauss SL, Reardon CL, Mazzola M (2014) The response of ammonia-oxidizer activity and community structure to fertilizer amendment of orchard soils. Soil Biol Biochem 68:410–418

Torsvik VL (1980) Isolation of bacterial DNA from soil. Soil Biol Biochem 12(1):15–21

Tringe SG, Von Mering C, Kobayashi A, Salamov AA, Chen K, Chang HW, Podar M, Short JM, Mathur EJ, Detter JC, Bork P (2005) Comparative metagenomics of microbial communities. Science 308(5721):554–557

Valm AM, Welch JLM, Rieken CW, Hasegawa Y, Sogin ML, Oldenbourg R, Dewhirst FE, Borisy GG (2011) Systems-level analysis of microbial community organization through combinatorial labeling and spectral imaging. Proc Natl Acad Sci U S A 108(10):4152–4157

van Der Gon HD, Kropff MJ, Van Breemen N, Wassmann R, Lantin RS, Aduna E, Corton TM, Van Laar HH (2002) Optimizing grain yields reduces CH4 emissions from rice paddy fields. Proc Natl Acad Sci U S A 99(19):12021–12024

van Niftrik L, Jetten MS (2012) Anaerobic ammonium-oxidizing bacteria: unique microorganisms with exceptional properties. Microbiol Mol Biol Rev 76(3):585–596

Wallenstein MD, Vilgalys RJ (2005) Quantitative analyses of nitrogen cycling genes in soils. Pedobiologia 49:665–672

Wang B, Adachi K (2000) Differences among rice cultivars in root exudation, methane oxidation, and populations of methanogenic and methanotrophic bacteria in relation to methane emission. Nutr Cycl Agroecosys 58(1–3):349–356

Yang CH, Crowley DE (2000) Rhizosphere microbial community structure in relation to root location and plant iron nutritional status. Appl Environ Microbiol 66:345–351

Yanni YG, Rizk RY, Corich V, Squartini A, Ninke K, Philip-Hollingsworth S, Orgambide G, De Bruijn F, Stoltzfus J, Buckley D, Schmidt TM (1997) Natural endophytic association between *Rhizobium leguminosarum* bv. *trifolii* and rice roots and assessment of its potential to promote rice growth. Plant Soil 194(1–2):99–114

Ying-Hua D, Ya-Li Z, Qi-Rong S, Song-Wei W (2006) Nitrate effect on Rice growth and nitrogen absorption and assimilation at different growth stages. Pedosphere 16(6):707–717

Yuan Q, Pump J, Conrad R (2012) Partitioning of CH_4 and CO_2 production originating from rice straw, soil and root organic carbon in rice microcosms. PLoS One 7(11):e49073

Zeigler RS, Barclay A (2008) The relevance of rice. Rice 1:3–10

Zhang X, Liu W, Schloter M, Zhang G, Chen Q, Huang J, Li L, Elser JJ, Han X (2013) Response of the abundance of key soil microbial nitrogen-cycling genes to multi-factorial global changes. PLoS One 8(10):e76500

Zhao X, Liu SL, Pu C, Zhang XQ, Xue JF, Zhang R, Wang YQ, Lal R, Zhang HL, Chen F (2016) Methane and nitrous oxide emissions under no-till farming in China: a meta-analysis. Glob Change Biol 22(4):1372–1384

Environmental Remediation: Microbial and Nonmicrobial Prospects

13

J. Godheja, D. R. Modi, V. Kolla, A. M. Pereira, R. Bajpai, M. Mishra, S. V. Sharma, K. Sinha, and S. K. Shekhar

1 Introduction

Remediation is a process of detoxifying the contaminants from an environment. The selection of a suitable and affordable remediation strategy is therefore very important. Many remediation strategies do not focus on complete biodegradation of the contaminants. However, they lower the limits of contamination below a regulatory toxic level which is considered comfortable and safe for the environment. The level of contaminants and their spread can be minimized either by removing them from the site or by immobilizing them in the site which prevents its migration from subsurface geoenvironment to external environment.

Both domestic and industrial activities result in the accumulation of contaminants that when deposited in the soil and water give rise to contaminated sediments containing many inorganic and organic pollutants which are recalcitrant and a major threat to ecosystem. They can also enter the food chain when absorbed by plants or ingested by aquatic organisms. Soil sediment contamination is very severe around the waterways of industrialized and industrializing countries. The scientific and regulatory bodies mainly consider five major types of contaminations. The first ones are the raw sewage consisting of phosphorus and nitrogenous compounds like ammonia and organic matter. The second ones are organic hydrocarbons comprising of oil and grease. The third category belongs to halogenated hydrocarbons and pesticides. The fourth ones are polycyclic aromatic hydrocarbons which are mostly petroleum products and by-products. The fifth category are heavy metals like iron,

J. Godheja (✉) · V. Kolla · A. M. Pereira · R. Bajpai · M. Mishra · S. V. Sharma · K. Sinha
School of Life and Allied Sciences, ITM University, Naya Raipur, Chhattisgarh, India

D. R. Modi · S. K. Shekhar
Department of Biotechnology, Babasaheb Bhimrao Ambedkar University, Lucknow, Uttar Pradesh, India

D. P. Singh et al. (eds.), *Microbial Interventions in Agriculture and Environment*,
https://doi.org/10.1007/978-981-13-8383-0_13

manganese, lead, cadmium, zinc, and mercury with metalloids, usually comprising arsenic and selenium which can pose serious threat to the flora and fauna.

2 Site Assessment

An important task before starting remediation is site assessment which helps in determining the project cost, as well as the most suitable and efficient technology. Two important assessments must be made before initiating the project like identification of any potential hazards to the workers and the effect of the pollutants on the overall local community. Some other tasks are to assess the types of contamination and the historical use of the site. Raw materials used on the site and the products obtained will determine the assessment strategy, type of sampling, and type of chemical analysis to be done. Soil consisting of topsoil and subsoil as well as surface and groundwater of nearby land should also be tested, both before and after any remediation. The government and the local people living either in the contaminated zone or nearby unaffected zone play a critical role in a successful remediation.

3 Types of Remediation

Remediation process can be done wherever a contamination is identified like soil, groundwater, sediment, and surface water. Water remediation includes contamination removal from both the groundwater and surface water. Likewise, contamination removal from topsoil, subsoil, and sediments comes under soil remediation. Both of these remediations can be done separately or simultaneously, depending on the contaminant type and the level of pollution. Contaminants leaching through the soil and sediment are the main reasons of groundwater pollution. Improper industrial practices based on mining and drilling for natural gas and oil isolation may also cause groundwater pollution. The cause of soil contamination is the same with the cause of groundwater contamination. There are many remediation technologies which can be categorized into in situ and ex situ methods. Excavation of the soil is not needed in in situ remediation methods whereas ex situ remediation processes engage soil or sediment excavation from the sites and treating the same somewhere else.

This chapter describes the remediation technologies based on microbial and non-microbial techniques. Various techniques described under both categories can be either in situ or ex situ in nature.

4 Biological Remediation Techniques (Bioremediation)

Bioremediation may be defined as a process of treating contaminated soils and subsurface materials and/or water bodies and sediments by changing their environmental conditions in such a way that can stimulate growth of microbial population to degrade the targeted pollutants. In various cases, bioremediation becomes less expensive, economic, and a sustainable tool in comparison to other remediation measures. A list of various microbial techniques is given in Table 13.1.

Table 13.1 Biological remediation techniques

Remediation techniques	Advantages	Disadvantages	References
Cometabolism	Able to degrade contaminants to trace concentrations	High costs of maintenance on growth substrates	Catia et al. 2010, Nzila (2013), and Li et al. (2016)
Biopiling	In situ technology No transportation cost	Requires control of abiotic loss Mass transfer problem Limiting bioavailability	Jorgensen et al. (2000), Filler et al. (2001), Mohn et al. (2001), Li et al. (2002), and Iturbe et al. (2004)
Windrows	Relatively low cost treatment process Saleable products Improved nutrient quality	Dependent on weather conditions Health issues related to aerosols generated High use of water	Barr (2002), Hobson et al. (2005), and Coulon et al. (2010)
Bioreactors/ bioslurry	Fast degradation by the use of microbial inoculants and surfactants	Expensive method, needs soil transport	Zhang et al. (2001a), Baptista et al. (2005), and Chikere et al. (2012)
Land farming	Simple design, implementable at short treatment times	Area requirement is high Dust and vapor create air pollution	Hejazi (2002), Paudyn et al. (2008), Katsivela et al. (2005), and Kuo et al. (2011)
Bioventing	Economic and easy to install, clubbed with other techniques	High content may become toxic for microorganisms Limiting factor is low soil permeability	Mihopoulos et al. (2002), Diele et al. (2002), Kao et al. (2008), Mau and Yue (2010)
Biosparging	Ready equipment, cost competitive, in situ method	Complex biochemical and physiological interactions Constituents migration is an issue	RAAG (2000), Kao et al. (2008)
Bioslurping	Applied at shallow as well as deep sites Recovers free product, thus speeding remediation	Limit low soil permeability, soil moisture, and oxygen content Low temperature reduces remediation process	Yen et al. (2003)

(continued)

Table 13.1 (continued)

Remediation techniques	Advantages	Disadvantages	References
Phytoremediation	Minimum ecological disturbance, solar energy-driven technology, large-scale use, cost-effective	Time-consuming	Meagher (2000), Zhang et al. (2001b), Wang et al. (2002), Rupassara et al. (2002), Burken (2004), Danika and Norman (2005), Kvesitadze et al. (2006), Subramanian et al. (2006), Marchiol et al. (2007), Mendez and Maier (2008), Sorek et al. (2008), and Burken et al. (2011)
Natural attenuation	Remediation waste use, can be clubbed with any other technology	Costly and complex	Khan and Husain (2002, 2003), Peter et al. (2006), Serrano et al. (2008), Agarry et al. (2010), and Liu et al. (2010)
Composting	Cheap having rapid reaction rate	More treatment time, requires nitrogen supplement	Atagana (2004, 2008)
Biofiltration	Biofilters are self-regenerating; thus, they maintain maximum adsorption capacity. Contaminants are destroyed, not just separated	Low temperatures may slow or stop degradation. Moisture levels, pH, temperature, and other filter conditions such as fungi growth should be monitored to maintain high removal efficiencies. Accumulation of excess bacteria may plug filters	Chaudhary et al. (2003) and Álvarez-Hornos et al. (2008)

4.1 Cometabolism

Cometabolism involves the simultaneous degradation of two compounds: one is the nutrient and the other is the contaminant. An injection of a dilute solution of nutrients such as methane and oxygen is introduced into the contaminated groundwater or soil. The microbes that utilize and metabolize these nutrients produce those enzymes also which have the ability to metabolize organic contaminant and degrade it to a nontoxic compound. Thus, this method has shown good results against most recalcitrant compounds like TCE, MTBE, PCE, dioxane, atrazine, and many more. Methanotrophs have been shown to produce methane monooxygense, an oxidase that can degrade over 300 organic compounds. The cometabolic bioremediation method also results in minimum residual concentrations of contaminants since the metabolic activities of microbes are not dependent on the contaminant for carbon or energy.

4.2 Biopiling

This technique is an ex situ simple remediation technique comprising of piling the excavated contaminated soil layer by layer following nutrient amendment and/or enhancing aeration to efficiently manage the process of bioremediation. This increases the microbial growth and their metabolic activities. The advantages of this technique have led to its popularity which includes cost-effectiveness, which enables effective bioremediation under optimum nutrition, temperature, and aeration (Whelan et al. 2015). Heating can also be associated with the biopile methodology with flexibility to introduce air and steam at random which increase the microbial activities and contaminant availability (Aislabie et al. 2006). Biopiling reduces the extent of volatilization and can also be used effectively in cold regions (Dias et al. 2015; Gomez and Sartaj 2014; Whelan et al. 2015). Biopiles have also shown good results on clay and sandy soil (Chemlal et al. 2013). Sanscartier et al. reported that biopiles with high moisture content had a very low residual hydrocarbon content in comparison to the heated and passive biopiles. Also, biopiles can be applied for treating huge volumes of polluted soils in a limited space. Finally, scaling up the biopile technique from laboratory to pilot scale is comparatively easy (Chemlal et al. 2013). The efficiency of biopile can be increased by sieving, aeration, and the addition of other agents like straw, saw dust, bark, or wood chips (Delille et al. 2008; Rodrı'guez-Rodrı'guez et al. 2010).

4.3 Windrows

An extension of the biopile technique, windrows operate by periodical turning and mixing of piled waste materials for enhancing air exchange for greater bioremediation/decomposition by activating the indigenous bacteria present in contaminated soil. Regular turning of polluted soil, together with water addition, may increase aeration and promote uniform distribution of pollutants, nutrients, and microbes, speeding up the degradation. Remediation can be achieved through assimilation, accumulation, biotransformation, and mineralization (Barr 2002). Windrow technique showed a higher rate of hydrocarbon removal when compared to biopile and this depends on the soil type (Coulon et al. 2010). However, this technique is not good when soil is polluted with toxic volatiles and has been shown to release CH_4 (greenhouse gas). Anaerobic zone within piled contaminated soil causes the release of methane gas (Hobson et al. 2005).

4.4 Bioreactor

This ex situ technique consists of a vessel in which contaminated and waste materials can be biologically converted into simple nontoxic product(s) following a series of microbiological reactions in batch, fed batch, or continuous modes. Bioreactors are designed to maintain the natural environment inside the vessel to support optimum growth conditions for microorganisms. Contaminated samples are fed into a bioreactor either as dried products or in the form of slurry, where they undergo treatment. Bioreactors offer many advantages in comparison to other existing ex situ bioremediation methods like excellent control of bioprocess parameters (agitation, aeration, temperature, pH, substrate, and inoculum concentrations) which leads to efficient bioremediation. Other problems like biostimulation, nutrient addition, bioavailability, and mass transfer can be a limiting factor for the process of bioremediation and thus can be resolved by using bioreactors. Volatile compounds can be effectively treated and the most important character is the flexible nature of bioreactor design. Tracking the metabolic changes in short- or long-term operations inside the reactor is another advantage (Chikere et al. 2012; Zangi-Kotler et al. 2015).

4.5 Land Farming

Land farming is a combination of both ex situ and in situ technique and is popular because of its low-cost investment and maintenance. The depth of pollutant is important in determining the nature of bioremediation and whether tilling (for <1 m) or excavation (for >1.7 m) of soil should be followed (Nikolopoulou et al. 2013). The soils having contaminants or polluted materials are to be kept on the solid support above the ground surface where aerobic biodegradation of pollutant can occur (Philp and Atlas 2005; Paudyn et al. 2008; Volpe et al. 2012; Silva-Castro et al. 2015). Tillage involves aeration and addition of nutrients, followed by irrigation which activates the indigenous microorganisms to enhance bioremediation.

4.6 Bioventing

Bioinventing is an in situ method involving controlled airflow stimulation by diffusing oxygen to unsaturated (vadose) zone and adding minerals and moisture content to enhance the bioremediation process (Philp and Atlas 2005). The main advantage of this method is restoration of sites which has led to its popularity among other in situ bioremediation methods (Hohener and Ponsin 2014). Sui and Li (2011) designed a model with the facility of air injection which helped in bioremediation of toluene contamination. Frutos et al. (2010) reported that 93% of phenanthrene contamination in soil was removed after 7 months of treatment. Bioventing is economical during long air injection intervals with low air injection rate in the clay soils which are polluted by diesel (Thome et al. 2014). Rayner et al. (2007) reported that for the removal of hydrocarbons, single-well bioventing was found to be ineffective in majority of the cold regions. The results were attributed to low water and thin soil cover. Applying a microbioventing system using nine small injection rods (0.5 m apart) on the same site increased the hydrocarbon removal due to optimal uniform distribution of oxygen. Bioventing design enhances aeration in unsaturated zone and at the same time may be utilized for anaerobic bioremediation.

4.7 Biosparging

Biosparging mimics bioventing and is a process of air injection into soil subsurface for stimulating microbial activity. Unlike bioventing, air injection is performed at the saturated zone. This initiates the movement of organic compounds of volatile nature in the unsaturated zone of soils for enhancing biodegradation. Efficiency of biosparging depends on pollutant biodegradability and soil permeability, which greatly determines the pollutant bioavailability (Philp and Atlas 2005). This method is effective to treat petroleum material-contaminated treated water, especially with the diesel and kerosene oils. Kao et al. (2008) reported the biosparging of BTEX-contaminated aquifer plume which was evident from increased heterotrophs, dissolved oxygen, nitrate, sulfate with a gradual decrease in dissolved sulfide, ferrous iron, methane, and total anaerobes. The only limitation is the direction of airflow while performing biosparging.

4.8 Bioslurping

Bioslurping is a unique method which follows other techniques like bioventing, soil vapor extraction, and vacuum-enhanced pumping to achieve maximum remediation in soil and groundwater-contaminated areas. The mechanism acts like a straw that slurps any liquid from a vessel, and the contaminants are brought upward to the surface. On the surface, contaminants are separated from water and air and then bioventing is used to degrade the contaminants (Kim et al. 2014). The site is provided by indirect oxygen supply which stimulates contaminant biodegradation (Gidarakos and Aivalioti 2007). The main disadvantage is the excessive moisture in the soils that ultimately limits air permeability and the rate of oxygen transfer and results in reducing microbial activities. Bioslurping is also not suitable for the remediation of the soils with low permeability. This sometimes needs establishment of vacuum in highly permeable sites, but it may be economical due to the requirement of less amount of groundwater, thus minimizing storage, treatment, and disposal costs.

4.9 Phytoremediation

This technique uses the interactions between plant and microbes at physical, biological, chemical, and microbiological level in the polluted sites to nullify the toxic impact of pollutants. There are different mechanisms of phytoremediation (accumulation, filtration, stabilization, and volatilization) involved in the removal of pollutant from the contaminated sites. Plants like willow and alfalfa also lead to mineralization (Meagher 2000; Kuiper et al. 2004). Selection of plants depends on some factors like extent of root system, feasibility of site monitoring, growth rate of plant, disease and pest resistance of plant, and, above all, the time required to achieve the degradation (Lee 2013). Miguel et al. (2013) reported that the mechanism of removal of contaminants by the plant involves uptake of contaminants; translocation of contaminants from roots to shoots, which is carried out by xylem flow; and finally their accumulation in shoots. Indigenous plants can act as potent phytoremediators to optimize remediation potentials of these plants growing in polluted sites, which can be done either by bioaugmentation or by biostimulation. Plant growth-promoting rhizobacteria (PGPR) also plays an important role in phytoremediation. It enhances the biomass production and increases the tolerance of plants against heavy metals (Yancheshmeh et al. 2011; de-Bashan et al. 2012). Grobelak et al. (2015) reported increased root and stem length and growth after exogenous inoculation of microbial inoculants to *Brassica napus* L. subsp. napus and *Festuca ovinia* L. at the time of seed germination. This protected seeds and plants from the inhibitory impacts of heavy metals in metal-polluted soils. Similarly, endogenous bacteria resulted in increased metal accumulation during phytoremediation of metal-contaminated water bodies with *Spartina maritime* (Mesa et al. 2015). Some other plant species like *Brachiaria mutica* and *Zea mays* have shown good results in terms of phytoremediators in heavy metal-contaminated sites (Ijaz et al. 2015; Tiecher et al. 2016). There are many reports about other plants used in phytoremediation (Kuiper et al. 2004; Wang et al. 2012; Ali et al. 2013; Yavari et al. 2015). Phytomining is an important outcome of phytoremediation because some metals can bioaccumulate in some various plants which can be recovered later. Wu et al. (2015) reported the recovery of selenium-enriched material from phytoremediation sites. Other advantages of phytoremediation are low-cost installation and maintenance, environmentally friendly, conservation of soils, erosion prevention, and metal leaching (Van Aken 2009; Ali et al. 2013). Soil fertility can be also improved due to organic matter (Mench et al. 2009). Phytoremediation has some disadvantages also like longer remediation time, slow growth rate, excess pollutant accumulation, toxicity, and bioavailability to plant and depth of plant roots (Kuiper et al. 2004; Vangronsveld et al. 2009; Ali et al. 2013).

4.10 Natural Attenuation

This is an in situ natural process in which passive remediation of the polluted sites occurs without any human intervention and this makes it an inexpensive method.

Aerobic and anaerobic pathways help in biodegradation of recalcitrant contaminants but continuous monitoring is essential. Criteria that must be met in natural bioremediation are amount of contaminants loss, ability of microorganisms to biodegrade or transform contaminants, and the evidence of feasibility of biodegradation potentials in the field. Hydrocarbon-degrading bacteria from the refinery oil-contaminated soils were isolated by Mrassi et al. (2015). The study demonstrated the biodegradation potentials of the isolates when they were grown in substrates saturated and unsaturated with hydrocarbon as the sole carbon sources. They have the capacity to reduce hydrocarbon concentrations. It has been reported that most of the contaminated sediment sites reflected the presence of higher bacterial diversity and were found abundant in culturable degraders of microbial population during monitoring of intrinsic bioremediation of polluted marine coastal areas. Studies therefore suggested that bacterial communities can be used as sensitive indicators of contamination in marine sediment (Catania et al. 2015). Attenuation when combined with other treatments like biostimulation and bioaugmentation can be very effective as reported by Adetutu et al. (2015) when he saw successful reduction or dechlorination of groundwater contaminated with trichloroethene (TCE).

4.11 Composting

Composting is an excellent method to increase the natural aerobic process of microbial organic waste degradation with an important end product called humus that is principally used for agricultural purposes. This method has been increasingly used as a remediation technology also for removing biodegradable contaminants. It also modulates heavy metal bioavailability in phytoremediation strategies.

4.12 Biofiltration

In this ex situ technology, the volatile organic contaminants like fuel hydrocarbons are simply passed through a soil bed and after adsorption are degraded by the indigenous microorganisms. The major advantage of the method is its application for bioaugmentation so that specific bacterial strains can be introduced for selective degradation. Other advantages of biofilters over conventional activated carbon adsorbers are self-regeneration and maximum adsorption capacity. Besides this, the contaminants are separated as well as degraded.

5 Nonbiological Thermal Remediation Strategies (NBTRS)

Another way to remediate pollutants is to use low-temperature and high-temperature treatments for effective removal of contaminants. A list of these methods is given in Table 13.2.

Table 13.2 Nonbiological thermal techniques

Remediation techniques	Advantages	Disadvantages	References
In situ thermal remediation (ISTR) Consists of a variety of technologies: Electrical resistance heating (ERH), three-phase heating (TPH), six-phase heating (SPH), dynamic underground stripping (DUS), hydrous pyrolysis oxidation (HPO), thermal enhanced vapor extraction system (TEVES), and vitrification	Contaminant toxicity is checked Commercially viable, usable, clean technology	Destroys metals	Beyke and Fleming 2002, Schmidt et al. (2002) Kaslusky and Udell (2002), Baker and Heron (2004), Wait and Thomas (2003), Dermatas and Meng (2003), and U.S. EPA (1995)
ERH/TPH/SPH	Enhanced removal of VOCs and SVOCs in unsaturated clay-rich soils	Extraction is the problem buried metals pose safety hazards	Heron and Nielsen (2005)
DUS and HPO	HPO and DUS offer faster and more complete remediation of DNAPLs cost effective	Microbial life is destroyed	Newmark (1994)
TEVES	Commercially available	Soil having high permeability is a problem	La Chance et al. (2006)
Vitrification	Ability to process both organic and inorganic waste Significant volume reduction of specific waste	Durability of the waste form is an issue	
Ex situ thermal remediation (ESTR) Involves the destruction or removal of contaminants through exposure to high temperature in treatment cells, combustion chambers, or other means used to contain the contaminated media during the remediation process	Results are accurate No soil transportation required Easy and flexible	More risk of material handling	Pavel and Gavrilescu (2008), Hutton (2009), Santoleri et al. (2000), Venderbosch et al. (2010), and Rofiqul Islam et al. (2008)
Hot gas decontamination	Disposal of waste is easy as they are stockpiled Permits reuse or disposal of scrap	Higher costs than open burning Slow rate of decontamination	Hyman and Dupont (2001)

(continued)

Table 13.2 (continued)

Remediation techniques	Advantages	Disadvantages	References
Incineration	Very effective against soils contaminated with explosives and hazardous wastes	Specific material feed is required Volatile heavy metals May cause air pollution	Sabbas et al. (2003)
Open burn (OB)	Very effective for agricultural plastic, tires, oil spills, and gas flares	Safety concerns are high	Lemieux et al. (2003), and Gullett (2003)
Pyrolysis	Effective for SVOCs and pesticides and other organic contaminants in soils and oily sludges	Requires specific feed size and material handling	Stals et al. (2010), Thuan and Chang (2012), and Nkansah et al. (2011)
Vitrification	Vitrification at broad range of solid media Proven commercial technology	Use, storage, or disposal of vitrified slag is a problem	Bingham and Hand (2006)

5.1 Thermal Strategies

These methods are based on both high temperature (>500 °C) and low temperature (<500 °C) and efficiently used for those contaminants which have a high volatilization potential. Methods followed in high-temperature processes are electric pyrolysis, incineration, and in situ vitrification, whereas those followed in low-temperature treatments include thermal aeration, infrared furnace treatment, and thermal stripping. In high temperatures, oxidation leads to degradation, whereas effective contamination separation occurs at low temperature. Low-temperature treatment may increase the phase transfer rate of contaminants from liquid to gaseous phase. For the soil samples contaminated with volatile and semi-volatile compounds, radio frequency (RF) heating and steam stripping are used for the in situ thermal decontamination. In both processes, hot air, water, or steam is injected into the ground resulting in increased volatilization of contaminants with the help of vacuum. Thermal strategies become more effective by adding chemicals which are able to increase the volatility of the contaminants. The disadvantages are high cost, ineffectiveness with some contaminants, and production of toxic gases.

5.2 In Situ Thermal Strategies (ISTR)

This is a useful and important technique for organic compound remediation. It depends on the application of heat (50 °C to 100 °C) to the subsurface by various methods. Contaminants are vaporized due to subsurface heating, and later on, extraction of the gas mixture is done from the subsurface by soil vapor extraction (SVE). In some cases, the contaminants are degraded at higher temperatures (above 120 °C). Air treatment systems like activated carbon filters or catalytic oxidation (CatOx) are used to cool the extracted contamination and partial hot soil gas mixture. One major advantage is in its application in urban areas but certain risks can occur, which should be monitored properly.

Another example of in situ thermal strategy is electrical resistance heating (ERH) that applies electricity into the ground through electrodes. In ERH, the electrodes are installed both vertically (100 feet) and horizontally below the soil. Soils contaminated with volatile and semi-volatile organic compounds (VOCs and SVOCs) are easily recovered by this technique. When combined, ERH assists SVE by heating the contaminants in the soil, thus raising the vapor pressure of VOCs and SVOCs, increasing volatilization and removal. Application of electrical resistance heating also dries the soil as well as creates a source of steam that strips contaminants from soils. Good variants of ERH are three-phase heating (TPH) and six-phase heating (SPH). SPH splits the normal three-phase system into six separate phases, with each phase delivered to single electrodes which are placed in a hexagonal pattern. The whole system has the vapor extraction well located in the center of the hexagon which efficiently removes the contaminants. SPH is very good for circular areas (< 65-ft diameter). For larger and rough areas, three-phase system is ideal.

Further, in improving the ERH technique, two methods called dynamic underground stripping (DUS) and hydrous pyrolysis oxidation (HPO) are used to remediate soil and groundwater contamination. In both techniques, injection and extraction wells are installed in both saturated and unsaturated zones. In less-permeable clays, ERH is used to vaporize contaminants and drive them into the steam zone. Sometimes DUS uses an underground imaging system called electrical resistance tomography (ERT) in combination with hydrous pyrolysis/oxidation (HPO) that delineates heated areas to monitor cleanup and process control. When injection is halted, the steam condenses and gets mixed with contaminated groundwater further enhancing natural attenuation of pollutants by providing nutrients to thermophiles.

One more advancement in thermal strategies is the introduction of thermal enhanced vapor extraction system (TEVES) which uses a combination of other soil-heating processes like electrical resistance, electromagnetic heating, fiber-optic/radio-frequency heating, and/or hot-air/steam injection to enhance soil vapor extraction (SVE). It can be used effectively for semi-volatile organic compounds and volatile organic compounds. Three rows of electrodes are placed under the soil till a depth of 25 feet which is sufficient to heat the soil and drive off the moisture. Water is added regularly to maintain the electrical conductivity. As the soil heats and dries, the current between electrodes stops flowing, and at this point, RFH method is applied which uses electromagnetic energy to heat soil to over 300 °C. Soil permeability is increased due to raised contaminant vapor pressure leading to enhanced SVE. The extracted vapor can then be treated by a variety of technologies, such as granular activated carbon (GAC).

In situ vitrification (ISV) technique uses electric power for creating heat needed to melt soil through the electrodes inserted in the polluted area, melting the soil between them as extremely high temperatures (160 to 200 °C) are reached near the ground surface. Electrodes sink further into the ground as the upper soil melts causing the electrodes to move deeper in the soil. Vitrification is carried out by turning the power off, and as the melted soil cools, it turns into a solid block of glass-like material and the electrodes become part of the block. Any contaminants that remain underground become trapped in the vitrified block, thus retarding their migration of compounds encapsulated in the glass. ISV is also useful for volatile contaminants in combination with pyrolysis. A vacuum hood is used to collect the flue gases which are treated before release. All of the radionuclides and heavy metals are retained within the molten soil. Planar ISV is more efficient than top-down melting as it melts the soil only in specific areas of the subsurface rather than substantial overmelting. Another simple technique is in situ thermal desorption (ISTD) which is based on simple heating by thermal conduction in combination with fluid extraction. Contaminants are easily mobilized as higher temperatures are reached in subsurface and finally extracted using a solvent. ISTD is very effective for organic source zones and can be combined with less aggressive methods for complete site remediation.

5.3 Ex Situ Thermal Strategies

Ex situ thermal techniques degrade or immobilize contaminants through high-temperature treatments in a specified combustion chamber where the excavated soil samples are kept for remediation. Regular monitoring, shorter time durations, and uniformity in the treatment process are some of the main advantages of ex situ treatments. However, it requires excavation of soils, which increases the running costs and needs to deal with purchase of large equipments, permission from authorities, materials handling, and worker safety issues. Thermal processes use heat to separate, destroy, or immobilize contaminants. Hot gas decontamination and plasma high-temperature recovery are the techniques for separation, but pyrolysis and incineration degrade the contaminants. Other techniques like ex situ vitrification both degrade and separate the organic contaminants and sometimes also immobilize the inorganic contaminants if present. Incineration is the most popular technique based on high temperatures. A very popular technique is hot gas decontamination in which the temperature of contaminated solid material is raised to 260 °C for a specified period of time which vaporizes the volatile compounds. The gas effluent from the contaminated soil is then treated in an exhaust system to degrade all volatilized contaminants. The most advantageous objective of this method is the reuse or disposal of scrap as nonhazardous material. Another technique specially used for heavy metal-contaminated soil is plasma high-temperature recover which uses a thermal treatment process that purges contaminants as metal fumes. The vapors are then burned as fuel, thus recovering the metals which can be recycled.

Open burning is an effective but not a realistic means to eliminate all types of contaminants. According to Gullett (2003), some precautions that may improve the technique are as follows: avoiding noncombustible materials (glass, metals, and wet waste); avoiding wastes containing halogens (Lemieux et al. 2003); and avoiding wastes or materials containing catalytic metals (copper, iron, chromium, and aluminum). Contaminated materials to be burnt should be dry, homogeneous, or well blended. Some of the precautions to be taken during the burning process are supply of sufficient air, maintenance of steady burning, and minimizing smoldering. Most of the persistent organic pollutants are produced in the smoldering phase of burning (Lemieux et al. 2003).

To overcome the disadvantages of open burn, a very confined process which became very popular was incineration at high temperatures (870–1200 °C). This technique is used to combust and volatilize halogenated and other organics in contaminated soil. Common fuels are used to initiate and sustain combustion. A good incinerator has a basic design that caters the need toward different waste streams and different end products, and also the operating temperatures vary with the different designs. The efficiency of incineration is almost 100% but the off gases and residual contaminants need to be treated before disposal.

Pyrolysis is a form of treatment that chemically decomposes organic materials by heat in the absence of oxygen. Pyrolysis typically occurs under pressure and at operating temperatures above 430 °C. Since it is not practically possible to achieve a completely oxygen-free environment, a small amount of oxidation occurs. If the contaminated soil contains volatile or semi-volatile materials, thermal desorption can also occur.

Organic materials are transformed into gases and small quantities of liquid, and after the whole process, a solid residue is left behind containing carbon and ash. The off gases and particulates are then treated in a secondary oxidation unit. Many different types of pyrolysis units are available, including the rotary kiln, rotary hearth furnace, and fluidized bed furnace which are similar to incinerators except that they operate at lower temperatures and with less air supply.

A molten salt process like molten-salt oxidation (MSO) can be used where combustible waste is oxidized in a bath of molten salts (at 500–950 °C) which overcomes the problems associated with incineration like high temperature. Solid waste is injected with air below the surface of a molten salt bath that generates hot gases which rise through the molten salt bath and, as they are alkaline in nature, scrubs acids from the gases. Contaminated soil degrades and melts under the heat of the molten salt and by-products are also retained in the melt. Gases exiting the salt bath are effectively treated before discharge to the atmosphere and the spent molten salt is cooled and placed in a landfill.

Ex situ vitrification (ESV) is similar to ISV, except that it is done inside a chamber where heating is done by plasma torches or electric arc furnaces. Removal of molten material is carried out by slowing down the rotation which causes the slag to flow through a bottom opening. Gaseous products or effluent gases are kept in a separate chamber for further oxidation at high temperatures. Another heating method is the arc furnace which contains carbon electrodes, cooled sidewalls, a continuous feed system, and an off-gas treatment system. Contaminated soil is heated to temperatures above 1500 °C where it melts and, as it exits the vitrification unit, cools to form a glassy solid.

Another interesting ex situ thermal technique is the in-pile thermal desorption (IPTD), by which the contaminated solids are piled one above the other. The piles are then heated (330 °C) and treated using electrical heaters for SVOCs. Temperature can be raised or lowered depending upon the type of contaminants. The heat generated volatilizes both organic contaminants and moisture present within the soil.

6 Nonbiological Nonthermal Remedial Strategies (NBNTRS)

These methods are used to degrade contaminants without taking the help of microbes or plants and also are independent of thermal heating. A list of such methods has been summarized in Table 13.3.

6.1 Soil Vapor Extraction (Soil Venting)

It is the most common and successful method since it is economical. The technique is very effective for extracting contaminated groundwater and soil from a limited soil depth and then can be combined with other chemical and biological techniques for complete degradation of contaminants. Volatile organic compounds can be removed by vacuum technique called air sparging. Volatilization of organic compounds is done by a vacuum extraction probe placed in the vadose zone, and then injecting medium consisting of either oxygen or nitrogen is used to extract soil-water and/or soil-air mixture. This process is highly dependent on the soil structure which influences a lot on the passage of extracted water and vapor and also on the delivery of injecting medium. Granular soils are much better, whereas the presence of clay and organic matter restricts the free passage of both fluid and vapor. Some other factors which limit this process are nonvolatile organic matter, high density, and water content of soil leading to minimum transmissivity. Properties of volatile organic compounds such as solubility, sorption, vapor pressure, and concentration also influence the extraction process.

6.2 Soil Washing

It is a very simple physical method for remediation done by removing the contaminated soil with clean soil. This technique is only effective over small contaminated regions. The soil structure plays an important role like granular soils with less clay

Table 13.3 Nonbiological nonthermal techniques for bioremediation

Remediation techniques	Advantages	Disadvantages	References
Soil vapor extraction (soil venting)	Readily available and easy to install	Low efficacy	Zhan and Park (2002) and Halmemies et al. (2003)
Soil washing	Effectively reduces the volume of contaminants	Contaminant toxicity remains unchanged, less effective	Feng et al. (2001), Chu and Chan (2003)
Soil flushing	Usable for all soil contaminants	Soils having less permeability and heterogeneity are difficult to treat	Otterpohl (2002), Logsdon et al. (2002), and Di Palma et al. (2003)
Encapsulation	Physical isolation and containment	Efficiency of encapsulation decreases with time	Anderson and Mitchell (2003) and Robertson et al. (2003)
Stabilization/ solidification (S/S)	Useful and established remediation technology	Lack of expertise Lesser residual liability	Swarnalatha et al. (2006)
Stable isotope probing	Establishes the identity of microorganisms	Weaknesses of molecular methods	Boschker et al. (1998), Morris et al. (2002), Manefield et al. (2002a, b), Orphan et al. (2001, 2002),and MacGregor et al. (2002)
Nanotech remediation	Used to stabilize and guard enzymes	Yet to be exploited commercially	Cloete et al. (2010) and Xiaolei et al. (2013)
Air stripping	Better than 95% removal efficacy	The presence of solids in wastewaters can foul steam strippers	Benner et al. (2002) and Adams and Reddy (2003)
Dehalogenation	Target compounds are halogenated SVOCs and pesticides	High clay and high moisture content increases treatment costs	Taniguchi et al. (1997)
Electrokinetic remediation (EKR)/ electrodialytic soil remediation (EDR)	Has small impact on environment (soil removal is not required)	Efficiency reduced by alkaline soils	U.S. DOE (2002), Saichek and Reddy (2005), and Ottosen et al. (2005)
Photo catalytic degradation	Complete mineralization, low cost, no waste disposal problem	Limited to surface contaminants	Burrows et al. (2002);Boreen et al. (2003), and Walter Simmler (2011)

(continued)

Table 13.3 (continued)

Remediation techniques	Advantages	Disadvantages	References
Ultraviolet oxidation	Chemicals used do not pollute the environment	Low turbidity and suspended solids are necessary for good light transmission	Brillas et al. (2003) and Liang et al. (2003)
Precipitation/ flocculation	Cost-effective	Yet to be explored	Natrajan (2008)
Microfiltration	Removes dissolved solids effectively	Yet to be explored	Darvishzadeh and Priezjev (2012)
Analytical biosensors	Used for nutrient monitoring	Bioelements and chemicals used in the biosensors need to be prevented	Liu et al. (2004) and Nouri et al. (2009, 2011)
Chemical decontamination	Target treatment group is inorganic	Incomplete oxidation may occur	Block et al. (2004), Brown (2003), Haselow et al. (2006), Hrapovic et al. (2005), and Huling and Pivetz (2006)

content, and inorganic pollutants are much suitable for soil washing. Clayey soils can also be used after adding a chemical agent to break the retention of contaminants on the clay surface. Certain solvents or surfactants act as washing agents for soils contaminated with organic contaminants. Soil washing is a method of choice as an in situ technique where solvent alone or water mixed with solvents is used to wash the contaminants from the saturated zone. The chemicals are used to enhance contaminant release and mobility resulting in increased recovery and hence decreased soil contamination.

6.3 Soil Flushing

An alternative process of soil washing is soil flushing which can be used for extracting contaminants from the soil. Flushing dissolves the contaminants which are leached into the groundwater. Leached contaminants are then extracted and treated. Flushing solution can directly be injected into the groundwater which raises the water table into the capillary fringe just above the surface of the water table. This area has high concentrations of contaminants which can be removed by adding surfactants or solvents. The effectiveness of this technique depends on the type of soil, moisture content, and the type of contaminant.

6.4 Stabilization/Solidification (S/S)

A simple in situ and ex situ process of immobilizing toxic contaminants is performed in a single step or in two steps. In the single step, the contaminated soil is mixed with a special binder so that soil is fixed and rendered insoluble. In the two-step process, the contaminated soil is first made insoluble and unreactive, and in the second step, solidification is done so that the contaminants are fixed and therefore not free to contaminate the surroundings. This method is very effective in treating highly toxic pollutants but is mostly influenced by the transmissivity characteristics of the soil, viscosity, and setting time of the binder which limits its use to certain contaminants and geographical area.

In ex situ methods, contaminated soil is grinded, dispersed, and then mixed with a binder material (cement, lime, fly ash, clays, zeolites) which results in stabilized and solidified material before disposing in a well containing landfill. Leaching should be avoided as it will contaminate the groundwater. Organic binders which include bitumen, polyethylene, epoxy, and resins are also used for soil contaminated with organic pollutants.

6.5 Chemical Decontamination

This method is applicable for soils containing high concentrations of inorganic heavy metals. For effective extraction of contaminants, it is critical to understand the nature of bonding between the contaminant and the soil surface. Extractants used are complexing agents, electrolytes, weak acids, oxidizing and reducing agents, strong acids, and other compounds. Concentration of the inorganic heavy metals and the nature of the soil determine whether to use these extractants in pure form or in complex form. A permeable reactive barrier (PRB) or treatment wall (TW) is used which is packed with adsorbents and will retain inorganic heavy metals by exchange (sorption), complexation, or precipitation reaction. The transmission and the reaction time determine the thickness of the reactive barrier, and knowledge of heavy metals helps in selection of the adsorbent. The success rate of PRB depends upon its location, and thus, it is essential to have a good knowledge on the hydrogeological conditions where such barriers can be placed. Hydraulic conductivity of the PRB is very important which should be greater than or equal to the surrounding soil for proper permeation to occur. Further, reaction kinetics and barrier permeability are important in determining the thickness of the wall to be provided such that most of the contaminants are removed.

6.6 Electrokinetic Remediation

The principles of conductivity are efficiently employed in this method and mostly used for decontaminating granular soils. Two metal electrodes inserted into the soil mass act as anode and cathode. An electric field is established across these electrodes that produces electronic conduction as well as charge transfer between electrodes and contaminants in the soil-water system. Contaminants move from one electrode to the other as a result of electroosmosis and depending upon their charges. Contaminants are finally collected by a recovery system at the respective electrodes. Migration of contaminants is increased by adding surfactants and complexing agents. This method is commercially used for the removal of heavy metals from the soil such as uranium, mercury, and other metals.

6.7 Air Stripping

This technique involves the removal of those contaminants which are volatile, by transferring them from extracted water to air. The process takes place in a packed tower which is called an air stripper or sometimes in an aeration tank. The unique design of countercurrent packed-tower type air stripper offers a large surface area for mass transfer of volatile compounds and therefore offers significant advantages in efficiency and overall cost.

The height of the air strippers varies which is correlated to the chemical concentration of the contaminated water. A recent innovation in air strippers is the use of low-height air stripper which has a setup of horizontally arranged trays. Contaminated water is flown over the trays to maximize air-water contact and allow maximum contaminant volatilization. Since they are not so visible, they are increasingly being used for groundwater treatment.

6.8 Dehalogenation

This process removes chlorine and other halogens from soil or water. Contaminated soil is screened, processed, and mixed with chemicals before heating in a reactor. Dechlorination can be achieved either by replacing the chlorine molecules or by partially decomposing and partially volatizing contaminants. This method is effective for removal of furans, polychlorinated biphenyls, dioxins, and other chlorinated hydrocarbons such as pesticides from the soils. Base-catalyzed decomposition (BCD) is an example of dehalogenation technique where a base like sodium bicarbonate is added to the contaminated soil after the soil is screened and crushed before being heated to 330 °C in a reactor. The chlorine molecules volatilize and are captured, condensed, and treated separately in a compartment. Very high concentrations of PCBs (45,000 ppm) have been treated using BCD. Other chemicals such as sodium orthosilicate, sodium borohydride, and sodium phosphate are also being tested to improve overall efficiency.

6.9 Solar Detoxification

This technique uses the ultraviolet radiations to oxidize the organic contaminants in water. Two oxidizing agents such as ozone (O_3) or hydrogen peroxide (H_2O_2) are added to the contaminated groundwater before it is passed through a chamber where it is exposed to intense UV radiation provided by UV light bulbs. Oxidation of target contaminants is caused by direct reaction with the oxidizers and through the action of UV light in combination with ozone and/or hydrogen peroxide. For contaminated soil, vacuum extraction is used to remove contaminants from soils and then they are condensed before being fed into the reactor. An advantage of this system over conventional treatment processes, such as those using granular activated carbon or air stripping, is that it destroys the toxic compounds.

6.10 Precipitation

Precipitation technique has become the most widely selected process for removing heavy metals from groundwater using the pump-and-treat technology. Some remediation technologies such as chemical oxidation or air stripping use it as a pretreatment process, where the presence of metals would interfere with treatment. Water is pumped to the surface, where the precipitants convert soluble heavy metals to insoluble metals that either settle down or are filtered out of the water.

Adjustment of pH is critical in this technique. Addition of chemicals that stimulate precipitation and coagulants is also important which are mixed together in a device called a flocculator which coalesces the particles. The chemical precipitants, coagulants, and flocculator are all used to increase particle size so that the precipitation occurs rapidly. Commonly used precipitants include carbonates, sulfates, sulfides, lime, and other hydroxides. Precipitants and coagulants have a different role to play; the former is used to generate very fine particles that are held in suspension and the latter is added to aggregate the suspended particles. Flocculator promotes contact among the particles, which in turn promotes particle aggregation and precipitation.

6.11 Biosensors

Biosensors are designed using a specific bioactive component for the desired conversion to yield a signal that can be monitored. There are various biological components that have been developed that may be useful as biosensors. An example of this may be the potentiometric microbial electrode designed for detection of organophosphorus pesticides [Mulchandani et al. 1998]. There could be another option where the inherent property of a cell or tissue is used to generate the signal [Wang et al. 1996]. Of the different types of biosensors, the enzyme-based biosensors have been extensively explored as they generate the signal either by product formation, the disappearance of substrate, or coenzyme conversion. Various enzyme-based biosensors have been developed for the detection of pesticides in contaminated samples. The enzymes used in one report are cholinesterase and choline oxidase for the detection of organophosphorus pesticides [Rekha et al. 2000]. Recently, luminescence property of some compounds has shown a trend where biosensors can be designed on enzymes linked with light reactive compounds (Billiar and Dubow 1998).

6.12 Encapsulation

This process is different from other techniques where encapsulation ensures that contaminants cannot spread any further in the surroundings. Contaminated soil sites are marked and then mixed with lime, cement, and concrete which makes the area inaccessible, thus preventing any other fresh soil from coming in contact with the contaminants contained inside. The only disadvantage is that the soil cannot be used again for any cultivation purpose. Therefore, encapsulation is done only when it is clear that the soil in question is never going to be used in any capacity for growing anything.

6.13 Stable Isotope Probing (SIP)

It is a technique used to track the contaminant fate and analyze whether biodegradation is occurring at a site. Stable isotope probing is used to conclusively determine whether in situ biodegradation of contaminants like petroleum hydrocarbons (e.g., BTEX, polycyclic aromatic hydrocarbons (PAHs)) and oxygenates (e.g., MTBE, TBA) is occurring. The "probe" is a synthesized version of the contaminant compound composed of the heavier stable isotope (e.g., ^{13}C, ^{15}N) rather than the more common light isotope of that element (e.g., ^{12}C, ^{14}N). The heavy isotope serves as the "label" to track the environmental fate of the contaminant and determine if biodegradation is occurring. Results are often used to evaluate the feasibility of monitored natural attenuation (MNA) as a site management strategy (Busch-Harris et al. 2008; Geyer et al. 2005; Key et al. 2014; Williams et al. 2014). SIP can also be used to evaluate the feasibility and performance of engineered bioremediation approaches. SIP can also be used with DNA-based analyses to help identify the organisms involved in specific biodegradation processes (Aslett et al. 2011; Hatzinger and Fuller 2014; Key et al. 2013).

6.14 Nanoremediation

This is the most advanced form of remediation and uses a variety of nanoscale materials with environmental applications. These nanoscale materials are used to remediate contaminated soil and groundwater which contain chlorinated solvents or oil spills. Nanoscale materials used for remediation are more reactive compared to same material at much larger sizes (U.S. EPA 2007). Nanomaterials can also be manipulated to create novel properties which may not be present in materials at the micro- or macroscale. Nanoscale materials exhibit altered reaction rates because of availability of many reaction sites. These properties help in rapid reduction of contaminant concentrations from the site. These fine nanoscale materials can easily enter the very small spaces in the subsurface and remain suspended if appropriate coatings are used which allows the particles to travel farther than macro-sized particles, achieve wider distribution, and therefore improve contaminant reduction. A lot of research is going on to scale up the technology from a pilot scale to full scale. Certain nanoscale materials are being researched (TiO_2, self-assembled monolayers on mesoporous supports, dendrimers, carbon nanotubes, metalloporphyrinogens) for their use in specific contaminated areas. Scientists are evaluating how to apply the unique chemical and physical properties of these nanoscale materials for use in full-scale environmental remediation (Sanchez et al. 2011; Crane and Scott 2012). A lot of questions are still to be answered like understanding the fate and transport of unused or free nanoscale materials in the environment, persistency, and toxicity on various biological systems. A lot of thinking is still to be done so that the theoretical benefits of nanoscale materials can be realized in broad commercial use (U.S. EPA 2008).

6.15 Photocatalytic Degradation

It is a self-cleaning process and has been suggested as a remediation technology mainly for nitrogen-containing aromatic compounds. The process uses the photocatalytic properties of a thin layer of metal oxide like TiO_2 at the surface of the matrix such as glass or embedded in paints. The application of TiO_2 photocatalysts as an emerging air pollution control technology is increasing worldwide, but its effectiveness is still under observation and needs to be worked on so that the people living in the urban areas are protected from the toxic effects of pollutants.

6.16 Electrodialytic Soil Remediation (EDR)

This method utilizes ion exchange membranes to separate soil and processing solutions. Remediation of suspended soil is possible because these membranes avoid direct mixing of ions from processing solutions with the soil suspension. EDR is used in two ways, first by treating the soil as a stationary and wet matrix and secondly by treating the soil in a suspension. For the latter method, EDR can be combined with soil washing. The choice depends on both contaminated site and soil characteristics. The second method is suited for calcareous soils because in these soils the acidic front progresses slowly, hampering fast remediation process (Ottosen et al., 2005). The distance for electro-migrating heavy metal ions is shorter. The option of placing the electrodes directly in the soil is very economical for a site where the upper half meter is polluted, the groundwater level is deep, and there is no specific request for fast remediation.

6.17 Site Characterization and Analysis Penetrometer System (SCAPS)

This rapid in-field soil and groundwater analysis method provides characterization of soil conditions and uses a cone penetrometer system and other instruments. The whole set up is loaded on a truck to directly insert a probe in soils for rapid characterization of their types including delineation of the presence. Laser-induced fluorescence (LIF) and X-ray fluorescence sensors are attached to the probe to detect different compounds like petroleum compounds and metals. The UVOST (Ultraviolet Optical Screening Tool) is a recently developed technique used in sites where leaks of gasoline, diesel, hydraulic fluids, and oils are suspected to have occurred. This technology works on the principle of fluorescence of polycyclic aromatic hydrocarbons (PAHs) located in soil and/or groundwater when irradiated by ultraviolet light. Thus, different types of PAHs will fluoresce at different wave lengths giving a characteristic fluorescent signature. Various other sensors and sampling tools mounted on the SCAPS have been tested which include the thermal desorption and Hydrosparge sensors/samplers used to detect volatile organic compounds (VOCs) in soil and groundwater.

6.18 Geotubes

Geotubes have been used in the past years as a low-cost and high-volume dewatering solution. The process was proven to be a simple and effective way to handle sludge, hazardous contaminated soils, or dredge waste materials. The process consists of the following phases: filling and dewatering. In the first phase, sludge is filled via pump into the geotubes container before adding polymers which binds the solid materials and thus separating water. In the second phase, clear effluent water simply drains from the container through the small pores of a specially engineered textile. As the dewatering process continues, it leads to an efficient volume reduction of the contained materials, giving space for the repeated filling of the geotubes container. Efficiency of geotubes is about 99% and the decanted water or the clear filtrate can be collected and recirculated through the system. After the final cycle of filling and dewatering, the solids remain in the bag and continue to densify as residual water vapor escapes through the fabric. When the geotubes container is full, the contents can be deposited at a landfill and remain on-site or the solids can be removed from the container for land application, if appropriate.

7 Conclusion

All the methods mentioned in this chapter have some pros and cons and can be effectively used alone or in combination of two or more methods. Undoubtedly, the microbial and phyto-based methods are more economical and ecofriendly in spite of being time-consuming. Thermal and nonthermal strategies are quick but costly, and some produces toxic products which need to be sequestered properly.

Acknowledgments My sincere thanks to Dr. D.R. Modi (Professor and Head of Department, Biotechnology), Babasaheb Bhimrao Ambedkar University, Lucknow, for his kind support, guidance, and providing valuable input to improve this chapter. Also, my sincere thanks to all the coauthors who gave me the necessary inputs for completing this book chapter.

References

Adams JA, Reddy KR (2003) Extent of benzene biodegradation in saturated soil column during air sparging. Ground Water Monit Remed 23(3):85–94

Adetutu EM, Gundry TD, Patil SS, Golneshin A, Adigun J, Bhaskarla V, Aleer S, Shahsavari E, Ross E, Ball AS (2015) Exploiting the intrinsic microbial degradative potential for field-based in situ dechlorination of trichloroethene contaminated groundwater. J Hazard Mater 300:48–57

Agarry S, Owabor C, Yusuf R (2010) Bioremediation of soil artificially contaminated with petroleum hydrocarbon oil mixtures: evaluation of the use of animal manure and chemical fertilizer. Biorem J 14:189–195

Aislabie J, Saul DJ, Foght JM (2006) Bioremediation of hydrocarbon contaminated polar soils. Extremophiles 10:171–179

Ali H, Khan E, Sajad MA (2013) Phytoremediation of heavy metals—concepts and applications. Chemosphere 91:869–881

Álvarez-Hornos FJ, Gabaldón C, Martínez-Soria V, Martín M, Marzal P, Penya-roja JP (2008) Biofiltration of ethylbenzene vapors: influence of the packing material. Bioresour Technol 99:269–276

Anderson A, Mitchell P (2003) Treatment of mercury-contaminated soil, mine waste and sludge using silica micro-encapsulation. TMS annual meeting, extraction and processing division, March 2–6 2003, San Diego, CA, pp 265–274

Aslett D, Haas J, Hyman M (2011) Identification of tertiary butyl alcohol(TBA)-utilizing organisms in BioGAC reactors using ^{13}C-DNA stable isotope probing. Biodegradation 22(5):961–972

Atagana HI (2004) Co-composting of PAH-contaminated soil with poultry manure. Lett Appl Microbiol 39:163–168

Atagana H (2008) Compost bioremediation of hydrocarbon-contaminated soil inoculated with organic manure. Afr J Biotechnol 7:1516–1525

Baker RS, Heron G (2004) In-situ delivery of heat by thermal conduction and steam injection for improved DNAPL remediation. In: Proceedings of the 4th international conference on remediation of chlorinated and recalcitrant compounds, Monterey, CA, May 24–27, 2004, Battelle Columbus, OH

Baptista JS, Cammarota CM, Carvalhofreire DD (2005) Production of CO_2 in crude oil bioremediation in clay soil. Braz Arch Biol Technol 48:249–255

Barr D (2002) Biological methods for assessment and remediation of contaminated land: case studies. Construction Industry Research and Information Association, London

Benner ML, Mohtar RH, Lee LS (2002) Factors affecting air sparging remediation systems using field data and numerical simulations. J Hazard Mater 95(3):305–329

Beyke G, Fleming D (2002) Enhanced removal of separate phase viscous fuel by electrical resistance heating and multi-phase extraction. 9th annual international petroleum environmental conference, October 22–25, 2002, Albuquerque, NM

Bingham PA, Hand RJ (2006) Vitrification of toxic wastes: a brief review. Adv Appl Ceram 105(1):21–31

Block P, Brown R, Robinson D (2004) Novel activation technologies for sodium persulfate *in situ* chemical oxidation. In: Proceedings of the fourth international conference on the remediation of chlorinated and recalcitrant compounds, Monterey, CA

Boreen AL, Arnold WA, McNeill K (2003) Photodegradation of pharmaceuticals in the aquatic environment: a review. Aquat Sci 65(4):320–341

Boschker HTS, Nold SC, Wellsbury P, Bos D, de Graaf W, Pel R, Parkes RJ, Cappenberg TE (1998) Direct linking of microbial populations to specific biogeochemical processes by 13C-labelling of biomarkers. Nature 392:801–804

Brillas E, Calpe JC, Cabot PL (2003) Degradation of the herbicide 2,4- dichlorophenoxyacetic acid by ozonation catalyzed with Fe2þ and UVA light. Appl Catal B Environ 46(2):381–391

Brown R (2003) In situ chemical oxidation: performance, practice, and pitfalls. 2003 AFCEE technology transfer workshop, February 25 2003, San Antonio, TX

Burken JG (2004) Uptake and metabolism of organic compounds: Green-liver model. In: McCutcheon SC, Schnoor JL (eds) Phytoremediation: transformation and control of contaminants, a Wiley-Interscience series of texts and monographs. John Wiley, Hoboken, p 59

Burken J, Vroblesky D, Balouet JC (2011) Phytoforensics, dendrochemistry, and phytoscreening: new green tools for delineating contaminants from past and present. Environ Sci Technol 45(15):6218–6226

Burrows HD, Canle L, Santaballa M, Steenken S (2002) Reaction pathways and mechanisms of photodegradation of pesticides. J Photochem Photobiol B Biol 67(2):71–108

Busch-Harris J, Sublette K, Roberts KP, Landrum C, Peacock AD, Davis G, Ogles D, Holmes WE, Harris D, Ota C, Yang X (2008) Bio-traps coupled with molecular biological methods and stable isotope probing demonstrate the *in situ* biodegradation potential of MTBE and TBA in gasoline-contaminated aquifers. Groundw Monit Remediat 28(4):47–62

Catania V, Santisi S, Signa G, Vizzini S, Mazzola A, Cappello S, Yakimov MM, Quatrini P (2015) Intrinsic bioremediation potential of a chronically polluted marine coastal area. Mar Pollut Bull 99:138–149

Catia TDP, Mariano M, Janaína F, Medeiros B, Susana JK, Carlos A, Veiga B (2010) Biodegradation of phenol by free and encapsulated cells of a new *Aspergillus* sp. isolated from a contaminated site in southern Brazil. Afri J Biotechnol 9:6716–6720

Chaudhary DS, Vigneswaran S, Ngo HH, Shim WG, Moon H (2003) Biofilter in water and wastewater treatment. Korean J Chem Eng 20(6):1054–1065

Chemlal R, Abdi N, Lounici H, Drouiche N, Pauss A, Mameri N (2013) Modeling and qualitative study of diesel biodegradation using biopile process in sandy soil. Int Biodeterior Biodegradation 78:43–48

Chikere CB, Chikere BO, Okpokwasili GC (2012) Bioreactor-based bioremediation of hydrocarbon-polluted Niger Delta marine sediment, Nigeria. 3 Biotech 2:53–66

Chu W, Chan KH (2003) The mechanism of the surfactant-aided soil washing system for hydrophobic and partial hydrophobic organics. Sci Total Environ 307(1–3):83–92

Cloete TE et al (eds) (2010) Nanotechnology in water treatment applications. Caister Academic Press, Norfolk

Coulon F, Al Awadi M, Cowie W, Mardlin D, Pollard S, Cunningham C, Risdon G, Arthur P, Semple KT, Paton GI (2010) When is a soil remediated? Comparison of biopiled and windrowed soils contaminated with bunker-fuel in a full-scale trial. Environ Pollut 158:3032–3040

Crane RA, Scott TB (2012) Nanoscale zero-valent iron: future prospects for an emerging water treatment technology. J Hazard Mater 211-212:112–125

Danika L, Norman T (2005) Phytoremediation of toxic trace elements in soil and water. J Ind Microbiol Biotechnol 32:514–520

Darvishzadeh T, Priezjev NV (2012) Effects of cross flow velocity and transmembrane pressure on microfiltration of oil-in-water emulsions. J Membr Sci 423–424:468–476
de-Bashan LE, Hernandez JP, Bashan Y (2012) The potential contribution of plant growth-promoting bacteria to reduce environmental degradation – a comprehensive evaluation. Appl Soil Ecol 61:171–189
Delille D, Duval A, Pelletier E (2008) Highly efficient pilot biopiles for on-site fertilization treatment of diesel oil-contaminated sub-Antarctic soil. Cold Reg Sci Technol 54:7–18
Dermatas D, Meng X (2003) Utilization of fly ash for stabilization/ solidification of heavy metal contaminated soils. Eng Geol 70(3–4):377–394
Di Palma L, Ferrantelli P, Merli C, Biancifiori F (2003) Recovery of EDTA and metal precipitation from soil flushing solutions. J Hazard Mater 103(1–2):153–168
Dias RL, Ruberto L, Calabró A, Balbo AL, Del Panno MT, Mac Cormack WP (2015) Hydrocarbon removal and bacterial community structure in on-site biostimulated biopile systems designed for bioremediation of diesel-contaminated Antarctic soil. Polar Biol 38:677–687
Diele F, Notarnicola F, Sgura I (2002) Uniform air velocity field for a bioventing system design: some numerical results. Int J Eng Sci 40(11):1199–1210
Feng D, Lorenzen L, Aldrich C, Mare PW (2001) Ex-situ diesel contaminated soil washing with mechanical methods. Miner Eng 14(9):1093–1100
Filler DM, Lindstrom JE, Braddock JF, Johnson RA, Nickalaski R (2001) Integral biopile components for successful bioremediation in the Arctic. Cold Reg Sci Technol 32(2–3):143–156
Frutos FJG, Escolano O, Garcia S, Mar Babin M, Fernandez MD (2010) Bioventing remediation and ecotoxicity evaluation of phenanthrene-contaminated soil. J Hazard Mater 183:806–813
Geyer R, Peacock AD, Miltner A, Richnow HH, White DC, Sublette KL, Kästner M (2005) In situ assessment of biodegradation potential using biotraps amended with ^{13}C-labeled benzene or toluene. Environ Sci Technol 39(13):4983–4989
Gidarakos E, Aivalioti M (2007) Large scale and long term application of bioslurping: the case of a Greek petroleum refinery site. J Hazard Mater 149:574–581
Gomez F, Sartaj M (2014) Optimization of field scale biopiles for bioremediation of petroleum hydrocarbon contaminated soil at low temperature conditions by response surface methodology (RSM). Int Biodeterior Biodegradation 89:103–109
Grobelak A, Napora A, Kacprzak M (2015) Using plant growthpromoting rhizobacteria (PGPR) to improve plant growth. Ecol Eng 84:22–28
Gullett BK (2003) PCDD/F from uncontrolled combustion. First Session, Expert Group on Best Available Techniques and Best Environmental Practices
Halmemies S, Grondahl S, Arffman M, Nenonen K, Tuhkanen T (2003) Vacuum extraction based response equipment for recovery of fresh fuel spills from soil. J Hazard Mater 97(1–4):127–143
Haselow J, Block P, Sessa F (2006) Pilot scale application of heat-activated persulfate at a former petroleum underground storage tank area. The 22nd international conference on soils, sediments, and water, University of Massachusetts, Amherst, MA, October 16–19, 2006
Hatzinger P, Fuller M (2014) New approaches to evaluate the biological degradation of RDX in groundwater. Project ER-1607. Strategic Environmental Research Development Program, Arlington
Hejazi RF (2002) Oily sludge degradation study under arid conditions using a combination of landfarm and bioreactor technologies. PhD thesis, Faculty of Engineering and Applied Science, Memorial University of Newfoundland, St John's, Canada
Heron GSC, Nielsen SJD (2005) Full-scale removal of DNAPL constituents using steam enhanced extraction and electrical resistance heating. Ground Water Monit Remediat 25(4, Winter):92–107
Hobson AM, Frederickson J, Dise NB (2005) CH_4 and N_2O from mechanically turned windrow and vermicomposting systems following in-vessel pre-treatment. Waste Manag 25:345–352
Hohener P, Ponsin V (2014) *In situ* vadose zone bioremediation. Curr Opin Biotechnol 27:1–7
Hrapovic L, Brent E, David J, Hood ED (2005) Laboratory study of treatment of trichloroethene by chemical oxidation followed by bioremediation. Environ Sci Technol 39(8):2888–2897
Huling S, Pivetz B (2006) Engineering issue: In Situ chemical oxidation, EPA 600/R-06/072

Hutton B (2009) Waste management options to control greenhouse gas emissions – landfill, compost or incineration? Paper for the international solid waste association (ISWA) conference, Portugal 12–15 October 2009, pp 1–10
Hyman M, Dupont RR (2001) Groundwater and soil remediation. Process design and cost estimating of proven technologies, ASCE Press, Reston
Ijaz A, Shabir G, Khan QM, Afzal M (2015) Enhanced remediation of sewage effluent by endophyte-assisted floating treatment wetlands. Ecol Eng 84:58–66
Iturbe R, Flores C, Chavez C, Bautista G, Torres L (2004) Remediation of contaminated soil using soil washing and biopile methodologies at a field level. J Soils Sediments 4:115–122
Jorgensen K, Puustinen J, Suortti A (2000) Bioremediation of petroleum hydrocarbon-contaminated soil by composting in biopiles. Environ Pollut 107:245–254
Kao CM, Chen CY, Chen SC, Chien HY, Chen YL (2008) Application of in situ biosparging to remediate a petroleum hydrocarbon spill site: field and microbial evaluation. Chemosphere 70:1492–1499
Kaslusky SF, Udell KS (2002) A theoretical model of air and steam co-injection to prevent the downward migration of DNAPLs during steam enhanced extraction. J Contam Hydrol 55:213–232
Katsivela E, Moore E, Maroukli D, Strompl C, Pieper D, Kalogerakis N (2005) Bacterial community dynamics during in situ bioremediation of petroleum waste sludge in landfarming sites. Biodegradation 16:169–180
Key KC, Sublette KL, Duncan K, Mackay DM, Scow KM, Ogles D (2013) Using DNA-stable isotope probing to identify MTBE-and TBA-degrading microorganisms in contaminated groundwater. Groundw Monitor Remed 33(4):57–68
Key KC, Sublette KL, Johannes TW, Ogles D, Baldwin B, Biernacki A (2014) Assessing BTEX biodegradation potential at a refinery using molecular biological tools. Groundw Monitor Remed 34(1):35–48
Khan FI, Husain T (2002) Evaluation of contaminated sites using risk based monitored natural attenuation, chemical engineering progress-AIChE, USA January 34–44
Khan FI, Husain T (2003) Evaluation of a petroleum hydrocarbon contaminated site for natural attenuation using 'RBMNA' methodology. Environ Model Softw 18:179–194
Kim S, Krajmalnik Brown R, Kim JO, Chung J (2014) Remediation of petroleum hydrocarbon-contaminated sites by DNA diagnosis- based bioslurping technology. Sci Total Environ 497:250–259
Kuiper I, Lagendijk EL, Bloemberg GV, Lugtenberg BJJ (2004) Rhizoremediation: a beneficial plant-microbe interaction. Mol Plant-Microbe Interact 7:6–15
Kuo Y, Wang S, Kao C, Chen C, Sung W (2011) Using enhanced landfarming system to remediate diesel oil contaminated soils. Appl Mech Mater 121–126:554–558
Kvesitadze G, Khatisashvili G, Sadunishvili T, Ramsden JJ (2006) Biochemical mechanisms of detoxification in higher plants. Springer, Berlin/Heidelberg
La Chance J, Heron G, Baker R (2006) Verification of an improved approach for implementing in-situ thermal desorption for the remediation of chlorinated solvents. Paper F-32 in: Bruce M. Sass (Conference Chair), Remediation of Chlorinated and Recalcitrant Compounds—2006. Proceedings of the fifth international conference on remediation of chlorinated and recalcitrant compounds, Monterey, CA; May 2006
Lee JH (2013) An overview of phytoremediation as a potentially promising technology for environmental pollution control. Biotechnol Bioprocess Eng 18:431–439
Lemieux PM, Gullett BK, Lutes CC, Winterrowd CK, Winters D (2003) Variables affecting emissions of PCDD/F from uncontrolled combustion of household waste in barrels. J Air Waste Manage Assoc 53:523–531
Li P, Sun T, Stagnitti F, Zhang C, Zhang H, Xiong X, Allinson G, Ma X, Allinson M (2002) Field-scale bioremediation of soil contaminated with crude oil. Environ Eng Sci 19(5):277–289
Li S, Wang S, Yan W (2016) Biodegradation of Methyl tert-Butyl Ether by Co-Metabolism with a *Pseudomonas* sp. Strain. Int J Environ Res Public Health 13(9)

Liang S, Min JH, Davis MK, Green JF, Remer DS (2003) Use of pulsed-UV processes to destroy NDMA. Am Water Works Assoc 95(9):121–131

Liu J, Olsson G, Mattiasson B (2004) Short-term BOD (BODst) as a parameter for on-line monitoring of biological treatment process Part I. A novel design of BOD biosensor for easy renewal of bio-receptor. Biosen Bioelect 20(3):562–570

Liu W, Luo Y, Teng Y, Li Z, Ma LQ (2010) Bioremediation of oily sludge-contaminated soil by biostimulating indigenous microbes. Environ Geochem Health 32:23–29

Logsdon SD, Keller KE, Moorman TB (2002) Measured and predicted solute leaching from multiple undisturbed soil columns. Soil Sci Soc Am J 66(3):686–695

MacGregor BJ, Bruchert V, Fleischer S, Amann R (2002) Isolation of small-subunit rRNA for stable isotopic characterization. Environ Microbiol 4:451–464

Manefield M, Whiteley AS, GriKths RI, Bailey M (2002a) RNA stable isotope probing, a novel means of linking microbial community function to phylogeny. Appl Environ Microbiol 68:5367–5373

Manefield M, Whiteley AS, Ostle N, Ineson P, Bailey MJ (2002b) Technical considerations for RNA-based stable isotope probing: an approach to associating microbial diversity with microbial community function. Rapid Commun Mass Spectrom 16:2179–2183

Marchiol L, Fellet G, Perosa D, Zerbi G (2007) Removal of trace metals by Sorghum bicolor and *Helianthus annuus* in a site polluted by industrial wastes: a field experience. Plant Physiol Biochem 45(5):379–387

Mau L, Yue Q (2010) Remediation of diesel contaminated soil by bioventing and composting technology. In: Proceedings of international conference on challenges in environmental science and computer engineering vol 2, pp 3–6

Meagher RB (2000) Phytoremediation of toxic elemental organic pollutants. Curr Opin Plant Biol 3(2):153–162

Mench M, Schwitzguebel JP, Schroeder P, Bert V, Gawronski S, Gupta S (2009) Assessment of successful experiments and limitations of phytotechnologies: contaminant uptake, detoxification and sequestration, and consequences for food safety. Environ Sci Pollut Res Int 16:876–900

Mendez MO, Maier RM (2008) Phytostabilization of mine tailings in arid and semiarid environments—an emerging remediation technology. Environ Health Perspect 116(3):278–283

Mesa J, Rodrı'guez-Llorente JD, Pajuelo E, Piedras JMB, Caviedes MA, Redondo-Go'mez S, Mateos-Naranjo E (2015) Moving closer towards restoration of contaminated estuaries: bioaugmentation with autochthonous rhizobacteria improves metal rhizoaccumulation in native *Spartina maritima*. J Hazard Mater 300:263–271

Miguel AS, Ravanel P, Raveton M (2013) A comparative study on the uptake and translocation of organochlorines by *Phragmites australis*. J Hazard Mater 244:60–69

Mihopoulos PG, Suidan MT, Sayles GD, Kaskassian S (2002) Numerical modeling of oxygen exclusion experiments of anaerobic bioventing. J Contam Hydrol 58(3–4):209–220

Mohn W, Radziminski C, Fortin M, Reimer K (2001) On site bioremediation of hydrocarbon-contaminated Arctic tundra soils in inoculated biopiles. Appl Microbiol Biotechnol 57:242–247

Morris SA, Radajewski S, Willison TW, Murrell JC (2002) Identification of the functionally active methanotroph population in a peat soil microcosm by stable-isotope probing. Appl Environ Microbiol 68:1446–1453

Mrassi AG, Bensalah F, Gury J, Duran R (2015) Isolation and characterization of different bacterial strains for bioremediation of n-alkanes and polycyclic aromatic hydrocarbons. Environ Sci Pollut Res Int 22:15332–15346

Mulchandani A, Mulchandani P, Kaneva I, Chen W (1998) Biosensor for direct determination of organophosphorous nerve agent using recombinant *E. coli* with surface-expressed organophosphorous hydrolase. Potentiometric microbial electrode. Anal Chem 70:4140–4145

Natrajan KA (2008) Microbial aspects of acid mine drainage and its bioremediation. Trans Nonferrous Metals Soc China 18:1352–1360

Newmark RL (1994) Demonstration of dynamic underground stripping at the Lawrence Livermore National Laboratory Gasoline Spill site. Final report UCRL-ID-116964, vol 1–4. Lawrence Livermore National Laboratory, Livermore, CA

Nikolopoulou M, Pasadakis N, Norf H, Kalogerakis N (2013) Enhanced ex situ bioremediation of crude oil contaminated beach sand by supplementation with nutrients and rhamnolipids. Mar Pollut Bull 77:37–44

Nkansah MA, Christy AA, Barth T (2011) The use of anthracene as a model compound in a comparative study of hydrous pyrolysis methods for industrial waste remediation. Chemosphere 84:403–408

Nouri J, Khorasani N, Lorestani B, Karami MH, Hassani AH, Yousefi N (2009) Accumulation of heavy metals in soil and uptake by plant species with phytoremediation potential. Environ Earth Sci 59(2):315–323

Nouri J, Lorestani B, Yousefi N, Khorasani N, Hasani AH, Seif S, Cheraghi M (2011) Phytoremediation potential of native plants grown in the vicinity of Ahangaran lead– zinc mine (Hamedan, Iran). Environ Earth Sci 62(3):639–644

Nzila A (2013) Update on the cometabolism of organic pollutants by bacteria. Environ Pollut 178:474–482

Orphan VJ, House CH, Hinrichs KU, McKeegan KD, DeLong EF (2001) Methane-consuming Archaea revealed by directly coupled isotopic and phylogenetic analysis. Science 293:484–486

Orphan VJ, House C, Hinrichs KU, McKeegan KD, DeLong EF (2002) Multiple archaeal groups mediate methane oxidation in anoxic cold seep sediments. Proc Natl Acad Sci USA 99:7663–7668

Otterpohl R (2002) Options for alternative types of sewerage and treatment systems directed to improvement of the overall performance. Water Sci Technol 45(3):149–158

Ottosen LM, Pedersen AJ, Ribeiro AB, Hansen HK (2005) Case study on the strategy and application of enhancement solutions to improve remediation of soils contaminated with cu, Pb and Zn by means of electrodialysis. Eng Geol 77:317–329

Paudyn K, Rutter A, Kerry Rowe R, Poland JS (2008) Remediation of hydrocarbon contaminated soils in the Canadian Arctic by landfarming. Cold Reg Sci Technol 53:102–114

Pavel LV, Gavrilescu M (2008) Overview of ex-situ decontamination techniques for soil clean-up. Environ Eng Manag J 7:815–834

Peter AB, Miles L Vigelahn G, Teutsch G (2006) Evaluation of natural attenuation at the contaminated site Niedergörsdorf, Tank farm 1 (in German). Fachinformationen des Landesumweltamtes Brandenburg, Heft 12

Philp JC, Atlas RM (2005) Bioremediation of contaminated soils and aquifers. In: Atlas RM, Philp JC (eds) Bioremediation: applied microbial solutions for real-world environmental cleanup. American Society for Microbiology (ASM) Press, Washington, DC, pp 139–236

RAAG (2000) Evaluation of risk based corrective action model. Remediation Alternative Assessment Group, Memorial University of Newfoundland, St John's, NF, Canada

Rayner JL, Snape I, Walworth JL, Harvey PM, Ferguson SH (2007) Petroleum–hydrocarbon contamination and remediation by microbioventing at sub-Antarctic Macquarie Island. Cold Reg Sci Technol 48:139–153

Rekha K, Thakur MS, Karanth NG (2000) Biosensors for the detection of organophosphorous pesticides. Crit Rev Microbiol 20:213–235

Robertson D, Burnley S, Barratt R (2003) The immobilization of flue gas treatment residues through the use of a single staged wash and crystalline matrix encapsulation (CME) treatment process. In: Proceedings of the 11th Annual North American Waste to Energy Conferences (NAWTEC11), April 28–30 2003 Tampa FL, pp 135–143

Rodrı'guez-Rodrı'guez CE, Marco-Urrea E, Caminal G (2010) Degradation of naproxen and carbamazepine in spiked sludge by slurry and solid-phase Trametes versicolor systems. Bioresour Technol 101:2259–2266

Rofiqul Islam M, Haniu H, Rafi qul Alam Beg M (2008) Liquid fuels and chemicals from pyrolysis of motorcycle tire waste: product yields, compositions and related properties. Fuel 87:3112–3122

Rupassara SI, Larson RA, Sims GK, Marley KA (2002) Degradation of atrazine by hornwort in aquatic systems. Biorem J 6(3):217–224

Sabbas T, Polettini A, Pomi R, Astrup T, Hjelmar O, Mostbauer P, Cappai G, Magel G, Salhofer S, Speiser C (2003) Management of municipal solid waste incineration residues. Waste Manag 23:61–88

Saichek R, Reddy K (2005) Electrokinetically enhanced remediation of hydrophobic organic compounds in soils: a review. Crit Rev Environ Sci Technol 35:115–192

Sánchez A, Recillas S, Font X, Casals E, González E, Puntes V (2011) Ecotoxicity of, and remediation with, engineered inorganic nanoparticles in the environment. TrAC Trends Anal Chem 30(3):507–516

Santoleri JJ, Theodore L, Reynolds J (2000) Introduction to hazardous waste incineration. WileyInterscience, New York, pp 133–170

Schmidt R, Gudbjerg J, Sonnenborg TO, Jensen KH (2002) Removal of NAPLs from the unsaturated zone using steam: prevention of downward migration by injecting mixtures of steam and air. J Contam Hydrol 55:233–260

Serrano A, Gallego M, González J, Tejada M (2008) Natural attenuation of diesel aliphatic hydrocarbons in contaminated agricultural soil. Environ Pollut 151:494–502

Silva-Castro GA, Uad I, Rodrı'guez-Calvo A, Gonza'lez-Lo'pez J, Calvo C (2015) Response of autochthonous microbiota of diesel polluted soils to land- farming treatments. Environ Res 137:49–58

Sorek A, Atzmon N, Dahan O, Gerstl Z, Kushisin L, Laor Y, Mingelgrin U, Nasser A, Ronen D, Tsechansky L, Weisbrod N, Graber ER (2008) Phytoscreening: the use of trees for discovering subsurface contamination by VOCs. Environ Sci Technol 42(2):536–542

Stals M, Carleer R, Reggers G, Schreurs S, Yperman J (2010) Flash pyrolysis of heavy metal contaminated hardwoods from phytoremediation: characterisation of biomass, pyrolysis oil and char/ash fraction. J Anal Appl Pyrol 89:22–29

Subramanian M, Oliver DJ, Shanks JV (2006) TNT phytotransformation pathway characteristics in arabidopsis: role of aromatic hydroxylamines. Biotechnol Prog 22(1):208–216

Sui H, Li X (2011) Modeling for volatilization and bioremediation of toluene-contaminated soil by bioventing. Chin J Chem Eng 19:340–348

Swarnalatha S, Arasakumari M, Gnanamani A, Sekaran G (2006) Solidification/stabilization of thermally treated toxic tannery sludge. J Chem Technol Biotechnol 81:1307–1315

Taniguchi S, Murakami A, Hosomi M, Miyamura A, Uchida R (1997) Chemical decontamination of PCB-contaminated soil. Chemosphere 34(5–7):1631–1637

Thome A, Reginatto C, Cecchin I, Colla LM (2014) Bioventing in a residual clayey soil contaminated with a blend of biodiesel and diesel oil. J Environ Eng 140:1–6

Thuan NT, Chang MB (2012) Investigation of the degradation of pentachlorophenol in sandy soil via low-temperature pyrolysis. J Hazard Mater 229–230:411–418

Tiecher TL, Ceretta CA, Ferreira PAA, Lourenzi CR, Tiecher T, Girotto E, Nicoloso FT, Soriani HH, De Conti L, Mimmo T, Cesco S, Brunetto G (2016) The potential of Zea mays L. in remediating copper and zinc contaminated soils for grapevine production. Geoderma 262:52–61

U.S. Environmental Protection Agency (U.S. EPA) (2008) Nanotechnology for site remediation fact sheet. Solid Waste and Emergency Response, EPA 542-F-08-009, October 2008

U.S. EPA (1995) Geosafe Corporation in situ Vitrification innovative technology evaluation report, EPA 540/R-94/520. Office of Research and Development pp 148

U.S. EPA. Science Policy Council (2007) Nanotechnology white paper. U.S. Environmental Protection Agency, February 2007

U.S.DOE (2002) Final remedial action report for lasagna™ phase IIb in-situ remediation of Solid Waste Management Unit 91 at the Paducah Gaseous Diffusion Plant, Paducah, Kentucky, OR/072037 & D1, p 80

Van Aken B (2009) Transgenic plants for enhanced phytoremediation of toxic explosives. Curr Opin Biotechnol 20:231–236

Vangronsveld J, Herzig R, Weyens N, Boulet J, Adriaensen K, Ruttens R, Thewys T, Vassilev A, Meers E, Nehnevajova E, van der Lelie D, Mench M (2009) Phytoremediation of contaminated soils and groundwater: lessons from the field. Environ Sci Pollut Res 16:765–794

Venderbosch R, Ardiyanti A, Wildschut J, Oasmaa A, Heeres H (2010) Stabilization of biomass-derived pyrolysis oils. J Chem Technol Biotechnol 85:674–686

Volpe A, D'Arpa S, Del Moro G, Rossetti S, Tandoi V, Uricchio VF (2012) Fingerprinting hydrocarbons in a contaminated soil from an Italian natural reserve and assessment of the performance of a low-impact bioremediation approach. Water Air Soil Pollut 223:1773–1782

Wait ST, Thomas D (2003) The characterization of base oil recovered from the low temperature thermal desorption of drill cuttings, SPE/EPA exploration and production environmental conference, March 10–12, San Antonio, TX, pp 151–158

Walter Simmler (2011) Air, 6. Photochemical degradation in Ullmann's encyclopedia of industrial chemistry. Wiley-VCH, Weinheim

Wang J, Kane SA, Liu J, Symth MR, Rogers K (1996) Mushroom tissue based biosensor for inhibitor monitoring. Food Technol Biotechnol 34:51–55

Wang J, Zhao FJ, Meharg AA, Raab A, Feldmann J, McGrath SP (2002) Mechanisms of Arsenic Hyperaccumulation in *Pteris vittata*. uptake kinetics, Interactions with Phosphate, and Arsenic Speciation. Plant Physiol 130(3):1552–1561

Wang J, Feng X, Anderson CWN, Xing Y, Shang L (2012) Remediation of mercury contaminated sites. J Hazard Mater 221:1–18

Whelan MJ, Coulon F, Hince G, Rayner J, McWatters R, Spedding T, bSnape I (2015) Fate and transport of petroleum hydrocarbons in engineered biopiles in polar regions. Chemosphere 131:232–240

Williams N, Hyland A, Mitchener R, Sublette K, Key KC, Davis G, Ogles D, Baldwin B, Biernacki A (2014) Demonstrating the in situ biodegradation potential of phenol using bio-Sep® bio-traps® and stable isotope probing. Remediat J 23(1):7–22

Wu Z, Banuelos GS, Lin Z-Q, Liu Y, Yuan L, Yin X, Li M (2015) Biofortification and phytoremediation of Selenum in China. Front Plant Sci 6:1–8

Xiaolei Q, Pedro AJJ, Qilin L (2013) Applications of nanotechnology in water and wastewater treatment. Water Res 47(12):3931–3946

Yancheshmeh JB, Khavazi K, Pazira E, Solhi M (2011) Evaluation of inoculation of plant growth-promoting rhizobacteria on cadmium uptake by canola and barley. Afr J Microbiol Res 5:1747–1754

Yavari S, Malakahmad A, Sapari NB (2015) A review on phytoremediation of crude oil spills. Water Air Soil Pollut 226:1–18

Yen HK, Chang NB, Lin TF (2003) Bioslurping model for assessing light hydrocarbon recovery in contaminated unconfined aquifer. I: simulation analysis. Pract Period Hazard Toxic Radioact Waste Manage 7(2):114–130

Zangi-Kotler M, Ben-Dov E, Tiehm A, Kushmaro A (2015) Microbial community structure and dynamics in a membrane bioreactor supplemented with the flame retardant dibromoneopentyl glycol. Environ Sci Pollut Res Int 22:17615–17624

Zhan H, Park E (2002) Vapor flow to horizontal wells in unsaturated zones. Soil Sci Soc Am J 66(3):710–721

Zhang C, Daprato RC, Nishino SF, Spain JC, Hughes JB (2001a) Remediation of dinitrotoluene contaminated soils from former ammunition plants: soil washing efficiency and effective process monitoring in bioslurry reactors. J Hazard Mater 87(1–3):139–154

Zhang Q, Davis LC, Erickson LE (2001b) Plant uptake of methyl-tert-butyl ether (MTBE) from groundwater. Pract Period Hazard Toxic Radioact Waste Manage 5:136–140

Tree Ecosystem: Microbial Dynamics and Functionality

14

Samiksha Joshi, Manvika Sahgal, Salil K. Tewari, and Bhavdish N. Johri

1 Introduction

Forests cover 40 million km^2 globally which is about 30% of the total area on the earth. They inhabit most of earth's biome. Tropical rain forests cover about 17 million km^2 of area which is almost 12% of earth's ice-free land situated mostly in the Amazon basin (South America), Congo basin (Central Africa), and Southeast Asia (Ashton et al. 2012). Trees are the dominant primary producers and supply the bulk of the carbon (C) that finally enters the earth's ecosystem. This is why forests are an important carbon (C) sinks in the form of recalcitrant organic matter in their soils. A small fraction of organic C remains as simple organic molecules and a significant proportion in the form of complex biomass composed of wood, litter, and roots. Trees that represent phyllosphere, rhizosphere, wood, litter, or roots affect soil ecosystem due to penetration of their roots, litter production, development of ground vegetation, and terrain changes during uprooting. Thus they constitute major custodians of microbial, plant, and animal biodiversity. Trees also prevent soil erosion and play important role in global climate change and biogeochemical cycling (Gibson et al. 2011). Forests especially in tropics are under great threat due to conversion to other uses (Bawa et al. 2004; Stork et al. 2009). This is why the forest and tree cropping systems are of great significance on the earth. Microbial communities from soils, rhizosphere, and root-associated organisms (collectively called as microbiota) are essential to sustain development and productivity of the forests and fitness

S. Joshi · M. Sahgal (✉)
Department of Microbiology, G. B. Pant University of Agriculture & Technology, Pantnagar, Uttarakhand, India

S. K. Tewari
Department of Genetics and Plant Breeding, G. B. Pant University of Agriculture & Technology, Pantnagar, Uttarakhand, India

B. N. Johri
Department of Biotechnology, Barkatullah University, Bhopal, India

D. P. Singh et al. (eds.), *Microbial Interventions in Agriculture and Environment*,
https://doi.org/10.1007/978-981-13-8383-0_14

of the trees throughout their lives. The study of the microbial life and their interactions with trees and forests at the belowground level is critical to understand their contributions in the process of nutrient cycling. Here, we summarize different habitats of forest ecosystem, associated microbiota, and drivers of microbial diversity and point out future perspectives concerned with the microbial-life form management in forests to sustain tree crops.

2 Forest Types

Broadly, forests are of three types, boreal, temperate, and tropical. Boreal forests, (also known as taiga) are spread over formerly glaciated and presently permafrost areas across North America and Eurasia specifically inland Canada and Alaska, most parts of the Sweden, Finland, inland Norway, major portion of Russia, and the northern parts of Kazakhastan, Mongolia, and Japan. It is found in the areas having characteristic long and severe winters (at temperatures below freezing points), short summers with 50–100 frost-free days, and 15–20 inches of mean annual precipitation. These forests account for 29% of the world's forest cover. The word "taiga" represents forests in barren northern locations, while boreal for forests intemperate south area. The dominant vegetation is coniferous evergreen trees like spruce (*Picea*), fir (*Abies*), and pine (*Pinus*) and the deciduous larch (*Larix*), alder (*Alnus*), birch (*Betula*), aspen (*Populus*), and jack pine (*Pinus banksiana*) with understories of shrubs, mosses, and lichens. Temperate forests characteristically have high levels of precipitation (20 and 60 inches annually in the form of rain or snow), humidity, and a wide range of temperatures. These are typically located in the Northern Hemisphere, for example, in Eastern Asia, Central and Western Europe, and the Eastern United States. In temperate forests, dominant vegetation is deciduous trees that exist in several layers: (a) forest canopy tier (maple trees, walnut trees, birch trees), (b) small tree tier (dogwoods, redbuds, shadbush), (c) shrub tier (azaleas, mountain laurel, huckleberries), (d) herb tier (blue bead lily, Indian cucumber, wild sarsaparilla), and (e) floor tier (lichens and mosses). Mosses are the small, dense, nonvascular plants that thrive in the moist areas and prevent soil erosion. They also act as insulation during cold months. Lichens constitute symbiotic relationship between algae or cyanobacteria and fungi. They are important decomposers to help recycle plant leaves and generate fertile soils. Tropical forests cover 12% of the earth's ice-free land and contain almost 50% of the world's plants and animals. They receive heavy to very heavy rainfall throughout the year. Temperature and humidity range between 25 and 30°C and 77 and 88%, respectively. They are home to gigantic trees with a lifespan of 50–100 years. Legume trees along with mahoganies, teaks, and rosewood are the dominant vegetation. They provide valuable timber, fruits, nuts, and products like rubber and turpentine oils. These forests are spread over Central and South America, West and Central Africa, Southeast Asia, Eastern Madagascar, New Guinea, Queensland, Mexico, and Pacific Islands (Table 14.1).

Table 14.1 Major tropical forests of the world

Forest	Area	Countries	Specific features including vegetation
Amazon basin forest	6–8 million km^2	South American countries of Brazil, Peru, Colombia, Venezuela, Ecuador, Bolivia, Guyana, Suriname, and French Guiana	Largest tropical forest and harbors 390 billion individual trees representing 16, 000 species
Congo rain forest	1.78 million km^2	Cameroon, the Central African Republic, Republic of Congo, the Democratic Republic of Congo, Equatorial Guinea, and Gabon	It harbors 600 tree species and 10,000 animal species
Daintree rain forest	2600 km^2	Northeast coast of Queensland in Australia	Drained by Daintree River and Daintree National Park
Southeast Asian rain forest	–	Asian countries like Indonesia, Laos, and Cambodia and the Malay Peninsula	–
Tongass National Forest	17 million acres	United States	Biggest natural forest
Santa Elena Cloud Forest Reserve	26,000 acres	Cordillera de Tilaran in Alajuela and Puntarenas provinces of Costa Rica	Divided into six ecological zones and holds over 2000 plant species. Almost all the trees of the forest are covered with epiphytes
Monteverde Forest	10,500 hectares	Costa Rica	Harbors almost 2500 plant species
Kinabalu National Park	–	West Coast of Sabah, Malaysian Borneo	Also known as Tama Kinabalu. It is home to more than 4,500 species of flora and fauna
Sinharaja Forest Reserve	8864 km^2	Sri Lanka	Holds about 830 endemic species.
Sundarban Reserve Forest	10,000 km^2	Bangladesh and India	Mangrove trees are the dominant vegetation
Sapo National Park rain forest	1804 km^2	Southwest Liberia	–

2.1 Legume Trees in Tropical Forests

The soils in modern tropical forests are poor in mineral nutrients and rich in nitrogen (Vitousek et al. 2002). The relatively high N status is probably due to presence of leguminous trees, phyllospheric cyanobacteria, and lichens (Pons et al. 2007). Leguminosae is more abundant in the tropics and neotropics than in boreal and temperate forests (Ter Steege et al. 2000; Hammond 2005). Legumes dominate in both total number of tree species and abundance. Tropical legume tree species belong to three subfamilies and together represent ~8000 woody species (out of 19,000 legume species) (Lewis et al. 2005). In all 5000 species of Papilionaceae, 2700 species of Mimosaceae are mostly woody and tropical in origin, whereas 1900 species of subfamily Caesalpiniaceae are mostly woody and distributed from subtropical to temperate zones (Sutherland and Sprent 1993). The legume trees in Western African rain forests are represented by diverse species. In these forests, 26% of all commercial timber species are legumes (Dupuy et al. 1997). In Amazon basin, 40% of the total phytomass is represented by leguminous woody species (Puig et al. 1990). Nodulation is a genetic trait (Sprent 2009) and is not present in all the legumes. Within legume subfamilies, 54% Mimosoideae, 62% Papilionoideae, and 5% Caesalpinioideae are capable of nodulation (Moreira and Franco 1994; Sprent 2005). Worldwide, the Amazonian region of Brazil has richest diversity of leguminous species with 1221 autochthonous species distributed among 141 genera (Da Silva et al. 1989; Moreira and Franco 1994) and 234 legume species including 61 Caesalpinioideae, 79 Mimosoideae, and 94 Papilionoideae have been identified in the Amazonian forest of Brazil. In French Guiana forests, 68 families and 1200 species of trees have been reported (Sabatier 1994). The native legume trees could have an important role in the eco-restoration of N-depleted soils and can be used as pioneer tree species for the rehabilitation of degraded and overexploited rain forests. However, despite their abundance, diversity, and economic importance, very few of them have been studied for their ability to nodulate and fix nitrogen. Only about 20% of the known species of Leguminosae worldwide have been examined for nodulation (Sprent 1995). There is difference in distribution of legume genera and their N_2-fixing ability across equatorial rain forests in South America, Africa, and Asia (Sprent 2009). The legumes in African rain forests belong to non-nitrogen-fixing taxa. For example, non-N_2-fixing caesalpinoid species are dominant in Congo River basin forests (Högberg 1986; Sprent 2009).

2.2 Forest Microbiome

The forests typically have layered vegetation, of which trees are dominant. They contribute 90% of the primary forest production (Peh et al. 2015). Of this quantity, almost 33–50% is fixed in the soils through plant roots (Högberg et al. 2001). Thus trees regulate aboveground and belowground ecosystem interactions (Wardle et al. 2004). They are grown over years after years up to centuries to end their life cycle and therefore are believed to have high impact on soil parameters through different processes such as deposition of litter, uptake of nutrients and minerals, and exudation of chemicals from roots. These attributes change physicochemical properties of the soils and determine temperature, oxygen consumption, soil porosity through root development, and water capacity in terms of root uptake (Augusto et al. 2002). However, these modifications vary according to the species of the trees and their standing (pure vs mixed population) (Augusto et al. 2015). Trees as well as soils in forest are associated with diverse assemblage of microbial species in the form of epiphytes or endophytes (Trivedi et al. 2016). The microbial interactions between plant and whole microbiota are usually dynamic and determine development, fitness, and health of the plants (Lederberg 2006; Kowalski et al. 2015). The belowground microbiota that have gained attention during the last decade are mostly bacterial and fungal population of secondary trophic level (decomposers, mutualists, pathogens, parasites, and root feeders) (Ingham 1999). Their structural and functional interactions have an impact on the aboveground ecosystems. Due to the perennial, long-living nature of trees with persistent and deeper root system, it is believed that belowground microbial communities could be managed by persistent changes. This is well demonstrated by the currently available and powerful metagenomic approaches (Colagiero et al. 2017). Microbial communities regulate nutrient cycling and thus enhance tree nutrition and plant health. These communities also directly mobilize and transfer nutrients bound to organic matter to the plants. In this context, we can wonder how the microbiomes affect the functioning and homeostasis of forest ecosystems (Hacquard and Schadt 2015). Other habitats provided by forest ecosystems include the ubiquitous soil, litter, atmosphere, ground vegetation, deadwood, invertebrates, wetlands, streams, rocks, and habitats associated with trees—foliage, wood, bark, roots and rhizosphere, and nodules (in case of legume trees). The habitats differ in functions like nutrient availability and environmental conditions that altogether affect microbial abundance and community composition (Table 14.2). These varied habitats constitute forest microbiome.

3 Drivers of Microbial Community in Forest Soil

The growth of trees in the forest over the long duration has allowed for the acquisition of a diverse set of habitats like soils, plant tissues, tree surfaces, streams, rocks, etc. as habitats for the microbial communities (Hardoim et al. 2015). Five phyla found abundantly in most forest soils are *Acidobacteria*, *Actinobacteria*, *Proteobacteria*, *Bacteroidetes*, and *Firmicutes* (Lauber et al. 2009). Factors playing key role in shaping microbial community structure are soil physicochemical properties, particle size distribution, organic matter content, nutrient availability, climate conditions, anthropogenic activities, and biotic interactions (impact of vegetation) (Prevost-Boure et al. 2011). The spatial variation of these parameters evolves the hot spots of differential microbial activity in the soils; on the plant debris, litter, and deadwood; or around the plant roots (Kuzyakov and Blagodatskaya 2015). These niches hold specific properties and thus are inhabited by specific bacterial community (Fig. 14.1). Both plant and soil types can affect the microbial community structure, so there is a need to study interactions between them. Various factors acting as drivers of the microbial community in forest soil are discussed below.

Table 14.2 Habitats within forest ecosystems

Habitat	Carbon sources	Drivers, processes, and dynamics	Microbial biomass	Fungi/bacteria ratio	Dominant taxa, groups, ecology	References
Rock surface	CO_2, external C sources	Physical support, temperature and moisture fluctuation; interactions with mosses; weathering by organic acid production, leaching of cations	High if lichens are present	High in case of ECM weathering	ECM fungi, lichens, bacteria	Hoffland et al. (2004), Uroz et al. (2009) and Schmalenberger et al. (2015)
Invertebrates	Plant biomass in invertebrate gut, external C sources when invertebrates serve as vectors	Wood and litter decomposition using symbiotic gut microbes; grazing on fungal mycelia or bacterial colonies; soil mixing, symbiosis, commensalism, vectors; dispersal, outbreaks, seasonal reproduction	High in specific organs (guts, bacteriosomes)	High or low, depending on niche (gut, bacteriosome, surface)	*Bacteroidetes*, *Actinobacteria*, *Proteobacteria*, wood-decomposing fungi	Schloss et al. (2006), Adams et al. (2011), and Stursova et al. (Stursova et al. 2014)
Streams and lakes	CO_2; decomposition of allochthonous plant litter	Movement of zoosporic taxa, sedimentation, biofilm formation; fluctuation of allochthonous C input	10^6–10^8 bacterial cells ml^{-1}, 40–60 mg g^{-1} detrital mass fungi, annual production 0.2–2 t ha^{-1}	Very high on decomposing litter	Aquatic *Ascomycota*, zoosporic *Chytridiomycetes*, *Actinobacteria*, *Bacteroidetes*, *Proteobacteria*, protist predators	Hieber and Gessner (2002), Newton et al. (2011), Barlocher and Boddy (2016)

Roots and rhizosphere	Host photosynthates, root exudates	Interaction with host, rhizodeposition, commensalism, gradients of nutrient availability and chemistry, interaction of/with root-symbiotic organisms, priority effect in mycorrhizal colonization, root development, seasonality of root production and activity	Higher than in soil, very high in mycorrhizal roots, 10^8–10^{10} bacterial cells g^{-1}	Intermediate in rhizosphere, high fungal dominance in mycorrhizal roots	ECM fungi, AM fungi, endophytes, occurrence of wood decomposers, rhizosphere-specific bacteria, mycorrhizal helper bacteria	Prescott and Grayston (2013), Hardoim et al. (2015); Kohler et al. (2015), Lukesova et al. (2015), and Marupakula et al. (2016)
Soil	Humic compounds, dissolved and particulate OM, dead microbial biomass, dead roots	Microbial interactions; limited niche size (pores); interaction with fungal hyphae of ECM fungi; vertical stratification; spatial heterogeneity; seasonality of ECM mycelia and rhizodeposition, moisture variation	10^7–10^9 bacterial cells g^{-1}, 0.1–0.6 t ha^{-1} of ECM biomass; 0.2–0.7 mg g^{-1} fungal mycelia; production of 2 t ha^{-1} y^{-1} ECM mycelium, 0.2–1 t $ha^{-1}y^{-1}$ fungal fruiting bodies	Intermediate to low, decreases with soil depth	Mycorrhizal fungi, saprotrophic fungi, and bacteria of multiple phyla	Lindahl et al. (2007), Baldrian et al. (2012), Clemmensen et al. (2013), and Žifčáková et al. (2016)
Litter	Plant biomass (cellulose, hemicelluloses, lignin, proteins); high C/N ratio	Successive decomposition; fluctuations of moisture and temperature	10^8–10^9 bacterial cells g^{-1}, 0.7–7 mg g^{-1} fungal biomass	High to intermediate, varies during decomposition	Dominance of fungi, bacteria of diverse phyla, diverse ecology, high abundance of saprotrophs, mycophagous bacteria and fungi	Lindahl et al. (2007), Voriskova and Baldrian (2013), Lopez-Mondejar et al. (2015), and Žifčáková et al. (2016)

(continued)

Table 14.2 (continued)

Habitat	Carbon sources	Drivers, processes, and dynamics	Microbial biomass	Fungi/bacteria ratio	Dominant taxa, groups, ecology	References
Deadwood	Wood components (cellulose, hemicelluloses, lignin), fungal biomass; very high C/N ratio	Stochastic community assembly, priority effect, non-random development; high effect of microfauna; slow, successive fungi-dominated decomposition, moisture variation; commensalism, mycophagy, N2 fixation and translocation	Very high, up to 0.15 g g^{-1} fungal biomass; several t ha^{-1} of fungal fruiting bodies	Very high	Saprotrophic *Basidiomycota* and *Ascomycota*; lichens; *Myxomycota*; abundant *Proteobacteria*, *Acidobacteria*, and *Actinobacteria*; saproparasites	Stokland et al. (2012), Clissmann et al. (2015), Hoppe et al. (2015), Johnston et al. (2016), and Svensson et al. (2016)
Atmosphere	CO_2	Air movement, dispersal, radiation, precipitation, deposition, circadian deposition, circadian and moisture content, climatic events, windstorms	10^5–10^6 bacterial cells m^{-3} and 10^4–10^5 fungal spores m^{-3}	Low	Bacteria, fungal spores	Bowers et al. (2011)
Foliage	Plant biomass; high C/N ratio	Interaction with host, co-metabolism, methylotrophy, N_2 fixation, high fluctuation of moisture and temperature, foliage development and activity, circadian changes in leaf ecophysiology	Low biomass and diversity, high level of temporal fluctuation	Very high	*Proteobacteria*, *Firmicutes*, *Actinobacteria*, *Acidobacteria*, several fungi, e.g., *Dothideomycetes* and *Leotiomycetes*	Vorholt (2012), Carrell and Frank (2015), Hardoim et al. (2015), and Peršoh (2015)

Wood	Host photosynthates, plant biomass	Interaction with host, photosynthate flow, entry through injuries, seasonal changes in photosynthate flow	Low	High	*Ascomycota* (*Dothideomycetes*, *Leotiomycetes*), *Basidiomycota*, *Zygomycota*; endophytes and latent pathogens	Rayner and Boddy (1988) and Giordano et al. (2009)
Bark surface	Plant biomass (suberin), CO_2	Exposure to light and the atmosphere, fluctuation of moisture and temperature, N_2 fixation by lichen cyanobacteria	High if lichens are present	Uncertain	Lichens, yeasts, bacteria as symbionts of lichens, free-living cyanobacteria	Bhadra et al. (2008), Beck et al. (2014), and Grube et al. (2015)
Ground vegetation	Plant biomass (cellulose, hemicelluloses, lignin, proteins)	Vegetation diversity, interaction with host, seasonality of host activity	Low aboveground, high in and on roots	Very high	AM and ERM fungi, endophytic fungi and bacteria; saprotrophic *Ascomycota* on mosses	Öpik et al. (2008) and Davey et al. (2012)
Wetlands	Organic matter decomposition	Temporary or permanent O_2 limitation due to changes in water level; slow decomposition; habitat-specific vegetation (bryophytes); seasonal changes in vegetation activity, bacterial methanogenesis, methanotrophy, N_2 fixation	10^8–10^9 bacterial cells ml^{-1} in peat extract	Very low	*Acidobacteria*, *Proteobacteria*, *Actinobacteria*, *Verrucomicrobia*, *Planctomycetes*; Metazoa; high share of Archaea	Serkebaeva et al. (2013), Bragina et al. (2015), Schmidt et al. (2015), Kostka et al. (2016)

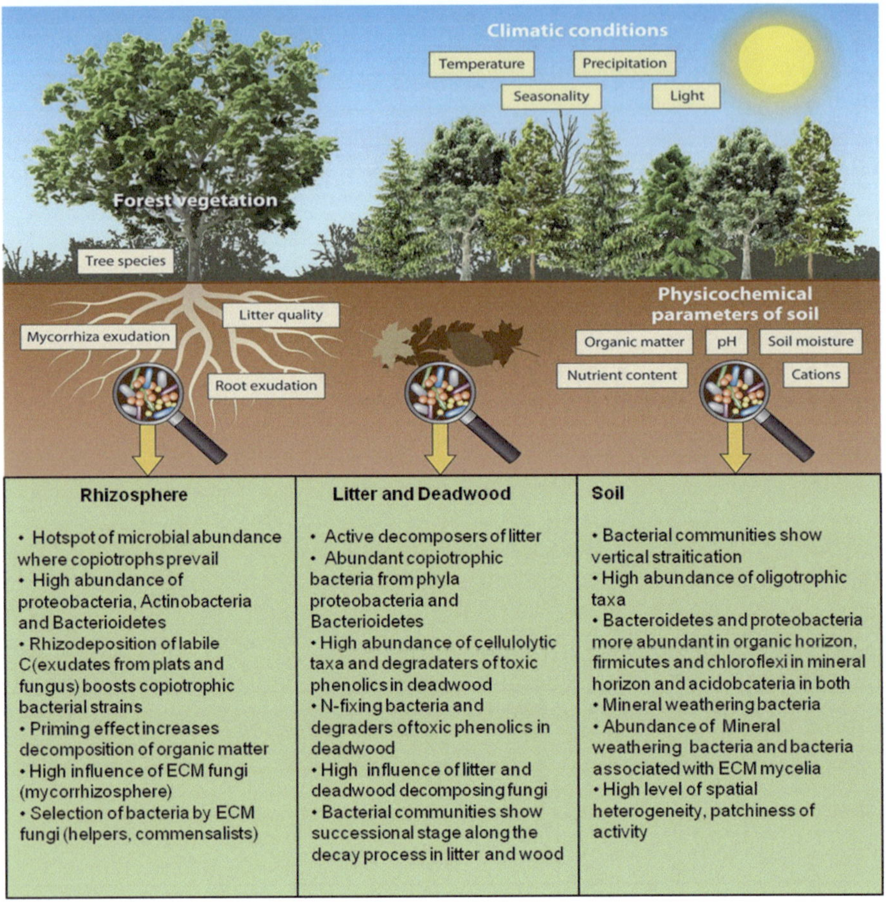

Fig. 14.1 Drivers of bacterial community composition in forest soils and their characteristics and roles (Lladó et al. 2017)

3.1 Tree Species

In forests, abundance and taxonomic identity of tree species strongly affect rhizospheric soil because of tree species-specific root exudation pattern (Augusto et al. 2015; Bonito et al. 2014; Lejon et al. 2005). A wide variety of compounds like sugars, amino acids, ethylene, organic acids, polysaccharides, vitamins, and enzymes are released as root exudates into the surrounding soil. These compounds create unique environments where interaction occurs between soil residing microbial taxa and specific root exudates. The exudation of labile C increases the chance of the availability of rhizosphere carbon that may result in the recruitment of rapidly growing bacterial strains (r-strategists). Tree identity also significantly affects the composition of microbial (fungal, bacterial, and protist) communities in the phyllosphere, litter, and bulk soil (Redford et al. 2010; Prescott and Grayston 2013;

Urbanova et al. 2015; Tedersoo et al. 2016). In bulk soil, communities of microbial forms are influenced by the tree identity and diversity. However, an impact is context dependent and partly by an effect of root exudation on soil chemistry (Tedersoo et al. 2016). The bacterial community structure and composition in the forest soils have relations with the tree species (Redford et al. 2010). However, the impact is not the same in monospecific and mixed forests. *Bacteroidetes* and *Betaproteobacteria* were more abundant on gymnosperms (evergreen), whereas *Actinobacteria* and *Gammaproteobacteria* on angiosperms (deciduous trees). Such an effect was also evidenced for crop plants such as maize, rape, wheat, and barrel clover (Haichar et al. 2008). Uroz et al. (2012) demonstrated that the impact of tree species on bacterial communities in monospecific and mixed forests varies. Marupakula et al. (2016) showed that mycorrhizal species strongly influenced bacterial community structure and composition. The community composition was shown to be highly dynamic in nature. Altogether, these observations reflected the shaping up of bacterial communities with the associated tree species through root exudation or indirectly via their mycorrhizal associations. In comparison to bacteria, fungal communities inhabiting forest soils have been extensively studied. Approximately 100,000 fungal species have been described so far (Hibbett et al. 2011). Tree host specificity plays an important role in the selection of symbiotic (Buée et al. 2009) and saprotrophic fungi (Ishida et al. 2007; Lang et al. 2011) in the rhizospheric region. It might affect mycorrhizal communities as well. For example, certain ectomycorrhizal (ECM) species (Ishida et al. 2007), sporocarp-forming fungi (Buée et al. 2011), and arbuscular mycorrhiza (AM) associated with tree species belonging to Fagaceae, Tiliaceae, Betulaceae, Oleaceae, and Aceraceae of boreal (Lang et al. 2011), tropical forests or temperate rain forests in the Southern Hemisphere (Tedersoo et al. 2008, 2010) depend on host tree species. These observations lead to conclude that the bacterial communities are shaped by their associated tree species in the forest ecosystem.

3.2 Dead Tree Material

Dead tree materials can affect the forest microbiome. ECM (Rosling et al. 2003; Buée et al. 2007) and saprotrophic fungi (Osono 2007; Stenlid et al. 2008) are dominant species on decomposing wooden logs, debris, and litter. Specific fungal communities (in terms of species richness, diversity, and community dynamics) growing on decaying wood and litter are determined by the tree species involved and variation in litter, and wood chemistry decides this selection (Küffer and Senn-Irlet 2005; Kulhánkova et al. 2006; Rajala et al. 2010; Renvall 1995; Yamashita et al. 2010). Furthermore, richness and assembly of fungal species in decaying wood have been shown to vary according to the availability of resources (Hottola et al. 2009), the volume of wood (Kubartova et al. 2012), and the stage of decay (i.e., primary or late) (Jönsson et al. 2008). Previous studies found that brown-rot fungi dominated during intermediate stages of decay, while white-rot fungi was detected across all decay phases excluding the final phase. Interestingly, it was also found that the

ECM fungi dominated during the last stage of decay. The study reflected that the ECM fungi can outcompete white- and brown-rot fungi when recalcitrant organic compounds are largely being degraded (Rajala et al. 2012).

3.3 Soil Parameters and Spatial Distribution of Microbial Diversity

Soil parameters and spatial distribution constitute the most significant drivers of the composition of soil microbial communities (Lauber et al. 2009; Talbot et al. 2014). In forest soils, different paedogenesis events lead to formation of various soil types and hence variation in edaphic parameters. In addition, tree species growing in the same soils may modify the edaphic properties of the soils (Augusto et al. 2002, 2015). Soil physicochemical properties and dominant plants affect community structure of microorganisms at the local level (scaling in centimeters to meters), whereas root exudates and soil heterogeneity at a fine scale (micrometers to centimeters) (Bardgett and van der Putten 2014). Soil pH and availability of organic (C, N, P) and inorganic (nutritive and toxic cations) nutrients (Fierer and Jackson 2006; Lauber et al. 2008; Rousk et al. 2010; Thomson et al. 2015) also affect composition of bacterial communities. Bacterial community composition is remarkably affected by the pH. In acidic soil environments, high abundance of *Acidobacteria* and *Alphaproteobacteria* is reflected (Baldrian et al. 2012; García-Fraile et al. 2015; Lladó et al. 2016), but *Bacteroidetes* and *Actinobacteria* are dominant in the soils with high pH range (Jeanbille et al. 2016; Lauber et al. 2008, 2009). Different soil horizons having distinct nutrition composition may contribute to the species abundance of the forest ecosystems and shape the structure of bacterial communities (Baldrian et al. 2012; Lopez-Mondejar et al. 2015). Uroz et al. (2016) reported that bacterial communities that inhabited the nutrient and mineral (apatite)-rich habitats became less diverse, and specific taxa like *Betaproteobacteria* were found dominant there compared to habitats rich in poorly weatherable minerals (e.g., phlogopite). Fungal communities are also strongly impacted by soil physicochemical characteristics. The strong impact of pH was also shown for ectomycorrhizal fungi inhabiting forest soils (Goldmann et al. 2015; Nacke et al. 2016). Furthermore, European beech forests have also shown diverse soil fungi at various soil pH conditions (Wubet et al. 2012; Coince et al. 2014).

3.3.1 *Dalbergia sissoo* Rhizosphere Microbiome

Legumes constitute a major proportion of the canopy trees in forests (Yahara et al. 2013). Several species of leguminous trees are useful for timber production, firewood, pharmacological products, ornamental usage, reforestation, and land reclamation activities. In addition to this, they are important in maintaining N balance of these forests through N_2 fixation ability (Pons et al. 2007). Soil microbial community associated with legumes is also important for nutrient biogeochemical cycling and soil fertility. Increased N availability (up to 15%) in legume tree rhizosphere and bulk soil leads to enhanced phosphatase activity (Treseder and Vitousek 2001).

Amongst various environmental factors plant species is the most important one responsible for shaping soil microbial community of legumes (McLaren and Turkington 2011). For example, Chen et al. (2014) showed that legume cover crops *Trifolium repens* greatly enriched alphaproteobacteria and reduced betaproteobacteria which also indicated improved soil fertility status. *Acacia holosericea*, a fast-growing leguminous tree, showed rhizobial symbiosis, as well as associations with arbuscular mycorrhiza (AM) or ectomycorrhiza (ECM (Founoune et al. 2002) to absorb mineral nutrients required for plant growth and efficient N_2 fixation. de Araujo Pereira et al. (2017) reported modification of the bacterial community composition in a monospecific cultivation system followed by introduction of a legume tree. Co-cultivation of *E.urophylla* or *E. grandis* with *A. mangium* (leguminous) leads to dominance of *Firmicutes* and *Proteobacteria* (Rachid et al. 2013) in soil. Conclusively, these studies reflect that legumes select specific microbial taxa, probably beneficial for themselves in their habitat (Bakker et al. 2013). *Dalbergia sissoo* Roxb., commonly known as shisham, is one of the important multipurpose leguminous tree species within the genus *Dalbergia* and family Fabaceae. It is a native of Indian subcontinent and grows naturally in Afghanistan, Pakistan, Nepal, Bangladesh, and India. Its timber is highly valued and important for a country's economy. This tree species is preferred for roadside plantations and agroforestry systems. It is also used for fuelwood, fodder, and shade. It is a pioneer tree species and is generally transplanted without considering the fertility status of soil (Bisht et al. 2009). Lately natural forests and plantations throughout the sub-Himalayan region, Indo-Gangetic Plains, and Indian plateau region are facing large-scale mortality. Diversity and composition of tree rhizomicrobiome are useful to ascertain the cause of large-scale mortality. Hence, *Dalbergia sissoo* Roxb. rhizosphere microbiome from three different provenances of Uttarakhand, Tanakpur, Lacchiwala, and Pantnagar was studied through Illumina high-throughput sequencing at Rhizosphere Microbiology Laboratory, G.B. Pant University of Agriculture and Technology, Pantnagar, India (Joshi 2018). A significant difference was observed in bacterial community composition between the provenances. There was abundance of specific taxa (phyla) like *Proteobacteria* and *Firmicutes* at Lachhiwala and Tanakpur, whereas *Acidobacteria* at Pantnagar. Bacterial genera such as *Pseudomonas*, *Flavobacterium*, *Bacillus*, *Paenibacillus*, *Sphingomonas*, *Nitrospira*, and *Massilia* within *Proteobacteria* and *Firmicutes* were reported in DS rhizosphere at Lacchiwala and Tanakpur, whereas *Williamsia*, *Opitutus*, *OM-43 clade*, and *uncultured Acidobacteria* at Pantnagar. Besides, differences in physicochemical properties (soil pH, nutrients, and C/N ratio) of native forest soils, tree genotype may also be responsible for selective recruitment of specific bacterial groups. Differences in the qualitative and quantitative composition of root exudates may be the key drivers to generate differences in the microbial community structure. The large numbers of unidentified OTUs indicated the presence of novel bacterial diversity in DS rhizosphere at all three provenances. This should be further studied with more high-resolution metagenomics techniques. Thus, the shift in forest characteristics with geographical location affected microbial community characteristics, with more apparent effects on bacterial diversity.

3.4 Seasonal Dynamics and Climatic Factors

The bacterial community structure and composition show seasonal variation (Lopez-Mondejar et al. 2015; Rasche et al. 2011) that impacts microbiome in different forest ecosystems (Kauserud et al. 2012). The metagenomic study of 126 forest soil samples from different latitudes in North America revealed that the temperature determines the bacterial and fungal community composition (Zhou et al. 2016). Globally, the precipitation may have a strong impact on the microbial community richness as per the specific edaphic conditions depending upon the pH and calcium or phosphorus soil content (Tedersoo et al. 2014). The effect of precipitation and temperature on mushroom fruiting is well established and illustrates seasonal effect on ecology of the fungal communities (Büntgen et al. 2011). The ECM community associated with the oak tree (*Quercus petraea*) revealed a fast turnover of their species across varied seasons where many species have shown adaptations toward environmental changes. For example, *Clavulina* sp. increased in the winters, while *Cenococcum geophilum* increased in the summers (Buée et al. 2005; Courty et al. 2008).

3.5 Impact of Forestry Practices and Anthropogenic Factors on Microbial Diversity

Soils in the forest ecosystem are less regularly managed than the agricultural soils under different cropping patterns. However, soil microbial communities are impacted by the forestry practices which are more natural (Frey et al. 2011; Purahong et al. 2014; Rakesh 2013). Nutrition harvesting in the forest soils is strongly associated with the removal of organic matter and minerals. Soil compaction and soil amendment (nitrogen) may lead to disturbances of fungal species richness, function, and community structure (Stenlid et al. 2008; Clemmensen et al. 2013; Rousk et al. 2011). Its pyrosequencing studies revealed decrease in abundance of ECM in conifer forests of British Colombia (Canada) after forest intensification (Hartmann et al. 2012, 2014). Diverse nutrient amendments done to increase soil fertility may change the structure of bacterial and fungal communities (Rakesh 2013). The heavy machinery used for tree harvesting results in soil compaction, modification in physicochemical properties, and sometimes formation of anoxic conditions. These resulted in changes in the structure, diversity, and abundance of bacterial communities and were observed over short term (one year post-harvest) as well as long term (15 years post-harvest) (Frey et al. 2011; Hartmann et al. 2012; Hartmann et al. 2014). For example, use of heavy machinery in forests where *Fagus sylvatica* and *P. abies* were dominant, transformed aerated soils into methane-producing soils leading to significant increase in the methanogenic bacterial communities (Frey et al. 2011). Various anthropogenic factors such as acid or nitrogen deposits or fires cause significant pressures on structure and composition of forest microbiome (Lilleskov et al. 2002; Rincón and Pueyo 2010).

4 Indices of Microbial Diversity and Abundance

Quantify the differences in microbial and other communities in the forests is creating a greater understanding toward how the biodiversity is distributed and how this functions. Biodiversity is measured or quantified in terms of "*species.*" The biodiversity is measured in terms of richness of species, relative abundance or proportion of a species in a community, and evenness of species. The scale at which diversity is measured is of critical importance. Diversity can be measured at different levels. The α diversity represents the local diversity (single site); β diversity, change in diversity between sites; γ diversity, regional diversity and ε diversity, diversity at a much larger scale. Various species-based diversity indices have been extensively applied to measure diversity of microbial communities (Hughes et al. 2001; Bohannan and Hughes 2003; Hill et al. 2003). These are species richness, Shannon-Wiener index, Simpson index, Sorensen index (also known as Bray Curtis), and Jaccard index.

4.1 Alpha Diversity

Alpha diversity is the species richness (number of taxa) within a single microbial ecosystem. It represents the number of different microbial species that could be detected in one sample, if one wants to know whether a definite habitat contains more diverse microbes than those habitats which are anthropogenically more active. What is the difference between the microbial communities associated with the disease-creating environment than those of healthy soils? Thus, a comparative status of the two distinct and contrasting habitats can define the species dominance in one or another. The measurements of α diversity include both qualitative and quantitative approaches. The qualitative species-based measurements, like Chao 1 (Chao 1984) or ACE (Chazdon et al. 1998), and quantitative species-based measurement, like the Shannon (Shannon and Weaver 1949) or Simpson (Simpson 1949) indices, were widely applicable.

4.2 Measures of β Diversity

In its simplest form, beta diversity represents the ratio of gamma (regional) and alpha (local) diversities (Whittaker 1960; Jost 2007). It informs about the degree of differentiation among biological communities in between two or more local assemblages (habitats) or between local and regional assemblages (habitats) (Koleff et al. 2003). It can also quantify the number of different communities in the regional or local habitat (Tuomisto 2010). For example, if both the local species richness (alpha diversity) and the regional species richness (gamma diversity) become equal, their ratio also becomes one, representing that their beta diversity equals unity. However, when the diversity of the local assemblages is completely different (maximum differentiation), gamma diversity will be equal to the multiplication of alpha diversity by the number of their sites (N). So beta diversity equals N, meaning that there are N different "communities." The β diversity measures can evaluate microbial community changes over time, biogeography, and different disease states. There are species-based and divergence-based methods for β diversity. Divergence-based methods evaluate the species to define whether similar environments contain the same species despite their distances and other geographic affiliations (Noguez et al. 2005). In this way, we can determine the spatial distribution of phylogenetic lineages, i.e., if microbes inhabiting physically separated and distinct environments are represented by any unique branch length, thus indicating unique evolution specific to any particular area. The divergence-based methods for β diversity are both qualitative and quantitative measures. The latter group has relations with the Jaccard and Bray-Curtis coefficients.

5 Microbial Activity in Forest Soils: Seasonal Variation

Soil microorganisms are crucial to the forest ecosystem. They play a critical role in fundamental ecological processes such as mineralization and decomposition (Wu et al. 2010; Corneo et al. 2013) and, hence, balance forest ecosystem. The environmental conditions like soil moisture, temperature, humidity, vegetation, and nutrient concentrations and seasonality of photosynthesis have strong impact on soil microbial community characteristics (Devi and Yadava 2006; Edwards et al. 2006; Hawkes et al. 2011; Frey et al. 2008). All these factors exhibit profound changes with changing seasons. In environments where such fluctuations are prevalent and majorly dominant, microbial communities may exhibit pronounced changes (Monson et al. 2006; McMahon et al. 2011). The microorganisms have physiological flexibility and react by adjusting their functions and compositions according to climatic and nutritional variations (Grayston et al. 2001; Wallenstein and Hall 2012). Impact of seasonal variation in the availability of nutritional resources is basically driven by the plants through their belowground organic C exudation and nutrient uptake. Such changes in the resource availability may induce shifts in microbial community composition (Kaiser et al. 2010, 2011). Microbial community dynamics of bacteria and fungi in relation to seasonal alterations have been shown by many workers (Lipson et al. 1999; Lipson et al. 2002). In subnival alpine regions, snow usually covers soils from November to May. The season in subnival alpine is distinguished by three major phases: (1) snow accumulation, (2) steady-state, and (3) snowmelt. The snow-covered soils are differently affected by UV radiation, light penetration, thermal insulation, water availability, and nutrient input (Edwards et al. 2007; Libois et al. 2013; Lazzaro et al. 2015). In these conditions, cold-adapted microbial life forms dominate below the winter snowpack (Schmidt and Lipson 2004; Buckeridge and Grogan 2010). At spring snowmelt, meltwater flowing down the snowpack causes nutrient flushing into the soils, affecting soil microbial composition (Schmidt and Lipson 2004; Edwards et al. 2007; Zinger et al. 2011). In temperate forest ecosystems, a seasonal pattern of microbial processes is observed. This relates to a seasonal variation in the substrate and variation in the temperature and moisture of the soils (Kaiser et al. 2010, 2011). There is low biomass turnover in winters, whereas high in summers (Vořiškova et al. 2014). In a deciduous temperate forest, saprotrophic taxa are maximum on freshly fallen litter in autumn, whereas ectomycorrhizal taxa are highly abundant during summer (Baldrian et al. 2013, Vořiškova et al. 2014). Within domain bacteria, only actinobacterial abundance exhibited seasonal variation in a forest soil (Kuffner et al. 2012).

Seasonal variations in the soil dynamics could be due to alterations in the microbial taxa abundance. Seasonal differences were markedly demonstrated in the soils. This corresponded to the observed changes in bacterial/fungal rRNA ratios in winter. Turner et al. (2013) reported that the fungi were more dominant in the rhizosphere than in the bulk soils. The activity of mycorrhizal fungi is reduced significantly in the winter season. The observed reduction in the ECM activity in the winter corresponds theoretically with the decrease in the inhibitory effects on decomposition of organic matter (Ekblad et al. 2013). The higher ECM activity in the summer corresponds with the increase in the abundance of *Planctomycetes* (Lindahl et al. 2010) and the bacterial taxa that harbor mycorrhiza helper bacteria such as *Burkholderia* spp., *Streptomyces* spp., or *Sphingomonas wittichii* (Churchland and Grayston 2014). Contrary to this, a few studies show that microbial enzyme activity in deciduous forests manifested seasonal variations in enzymatic processes (Kaiser et al. 2010;Vořiškova et al. 2014) and report no significant difference between seasons (Baldrian et al. 2013). In the future, the metatranscriptomics and metaproteomics may suggest and explore the dynamics of soil functions in deciduous forests.

6 Nutrient Cycling

Nutrient cycling maintains the cyclic exchange of nutrients and minerals in between the living and nonliving components of the forest soil ecosystem. Plants need huge amounts of nutrients like nitrogen (N), phosphorus (P), carbon (C), hydrogen (H), oxygen (O), potassium (K), calcium (Ca), and magnesium (Mg) but small amounts of boron (B), manganese (Mn), iron (Fe), copper (Cu), zinc (Zn), and chlorine (Cl) (micronutrients) continuously. Among these, carbon and nitrogen cycles are coupled to each other (Fig. 14.2). In forests, nutrient cycle also incorporates the uptake, storage, and transformations of nutrients in the perennial tissues; production and decomposition of litter, by the soil fauna and flora; addition of nutrient inputs from the atmosphere; weathering of primary minerals; and nutrient export from the soils by leaching or gaseous transfers. The trees put less demand on soil for nutrients than annual crops. Perhaps a major proportion of nutrient requirement of the trees is met through internal cycling. Nutrient recycling in the forest ecosystem is usually controlled by four key factors: climate, geography, abiotic properties (topography, parent material), and biotic communities.

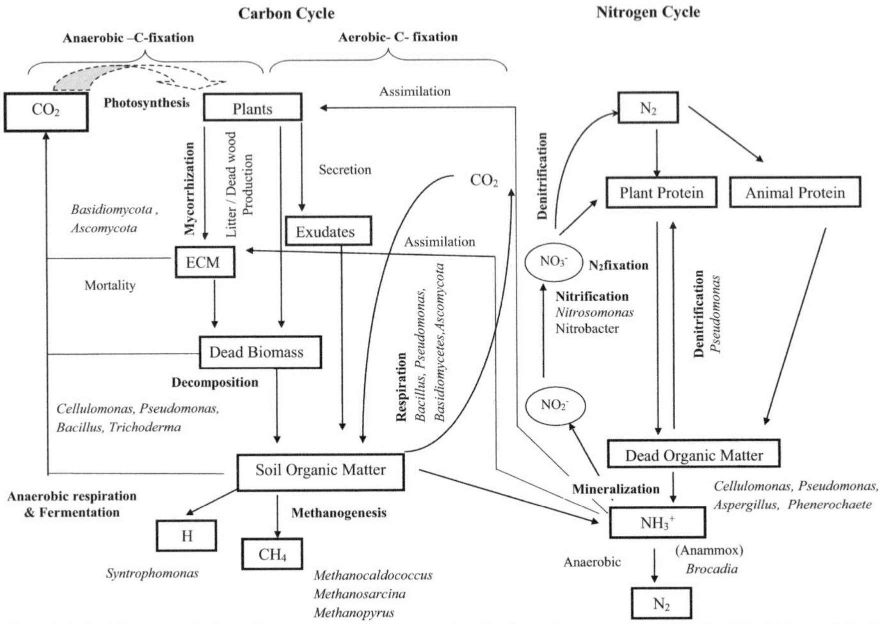

Fig. 14.2 The coupled carbon and nitrogen cycles in forest ecosystems. (Modified from Lladó et al. 2017)

6.1 Carbon Cycle

Forests store two-thirds of all terrestrial carbon deposits (Xia et al. 2015). The carbon flux is initiated by the carbon fixation from atmospheric CO_2 through photosynthesis. This is again mediated via the allocation of recalcitrant and general organic molecules into the soils. Assimilation of plant root exudates by soil and decomposition of the plant debris and biomass of microorganisms is highlighted as the key process regulating C flow in the soil systems that influences the ratio between C mineralization and immobilization (Xia et al. 2015; López-Mondéjar et al. 2016). Degradation of lignocellulosic plant biomass, a biopolymer of cellulose, hemicelluloses, and lignin, is a key step in the C cycle (Van Dyk and Pletschke 2012). The enzymes exocellulases, endocellulases, and glucosidases along with endoxylanases, xyloglucanases, xylosidases, mannosidases, endomannanases, fucosidases, pectinases, arabinosidases, and ligninolytic enzymes are necessary for the degradation of plant biomass (Rytioja et al. 2014). Several studies highlighted the important role of strains belonging to phyla *Acidobacteria* and *Actinobacteria* in the degradation of plant biomass under acidic soil conditions in temperate forests (Lladó et al. 2016). López-Mondejar et al. (2016) have shown that the cellulolytic bacteria *Mucilaginibacter*, *Luteibacter*, and *Pedobacter* are dominant from in deciduous forest topsoil. In the forests, fungal biomass represents a more readily decomposable substrate than lignocellulose, and bacteria are the major decomposers (Beier and Bertilsson 2013). The bacteria belonging to fungal biomass decomposition are *Ewingella*, *Pedobacter*, *Pseudomonas*, *Variovorax*, *Chitinophaga*, and *Stenotrophomonas* and all are known to produce chitinolytic enzymes especially β-1,3-glucanases (Brabcová et al. 2016; Berlemont and Martiny 2015; Bai et al. 2016).

Almost 1–5% of the photosynthetic carbon is released as root exudates from tree. It is mainly comprised of carbohydrates, amino acids, and organic acids (Wang et al. 2016). The largest fraction of C assimilated through photosynthesis is allocated to ECM fungi colonizing fine roots and root tips of forest trees (Korkama et al. 2007; Clemmensen et al. 2013). The rhizodeposition of C by plants is positively correlated with the rate of photosynthesis and shows seasonal variation (Högberg et al. 2010). Moreover, it constitutes a nutrient hot spot for other microorganisms (Marupakula et al. 2016). For example, soil bacteria grow abundantly and rapidly over *P. sylvestris* roots colonized by *Piloderma*. They utilize fungal sugars and organic acids (Fransson et al. 2016) more readily than plant sugars. It has been proposed that a specific bacterial community exists for hyphal exudates (Izumi et al. 2006). Besides using organic C, few soil bacteria also fix carbon dioxide. However, C input into the forest soils via CO_2 fixation is minimal (Žifčáková et al. 2016). These autotrophic bacteria fall within the genus *Bradyrhizobium* (Okubo et al. 2012).

Some soil bacteria from the forests can oxidize methane under aerobic conditions (Lau et al. 2015). Soils of boreal forests are rich in methanotrophic bacteria. These soils are a sink for the atmospheric methane. The forest soil methanotrophs belong to *Verrucomicrobia* (Nazaries et al. 2013), *Alphaproteobacteria*, and *Gammaproteobacteria* (Esson et al. 2016). These organisms can oxidize atmospheric methane/the methane generated by methanogenic archaea under waterlogged anaerobic conditions in some forest soils (Lau et al. 2015; Savi et al. 2016).

6.2 Nitrogen Cycle

Nitrogen, a limiting nutrient factor in the soils, enters the forest ecosystem majorly through the biological nitrogen fixation (BNF). The BNF process dominated by bacteria is responsible for >95% of the total required N input in the non-managed environments (Berthrong et al. 2014). Nitrogen-fixing bacterial diversity in the soils may rapidly change due to the exogenous application of N (Reed et al. 2011). The symbiotic as well as free-living N-fixing taxa have been observed in different temperate forest soils (Van Insberghe et al. 2015). These belong to *Alphaproteobacteria* (*Bradyrhizobium*, *Azospirillum*, *Hyphomicrobium*, and *Gluconacetobacter*) and *Deltaproteobacteria* (*Geobacter* spp.). Nitrification and denitrification lead to nitrogen loss from soils through NO, N_2O, and N_2 gas emissions as well as through the leaching of NO_3 (Pajares and Bohannan 2016). Among these, nitrous oxide is a powerful greenhouse gas with potential warming effect 300 times more than that of CO_2 (Jung et al. 2012). Nitrification is a multienzymatic biological oxidation of ammonia to nitrate. The *amoA* gene, encoding ammonia oxidase, which performs the first step of nitrification, is present in ammonia-oxidizing bacteria (AOB) as well as ammonia-oxidizing archaea (AOA), but AOB rather than AOA is rather more dominant in the soils (Long et al. 2012). However, *amoA* gene abundance appears to vary across forest soils (Cong et al. 2015). The reaction catalyzed by this enzyme is often limiting in temperate and boreal forest soils (Laverman et al. 2001) and is directly correlated with the AOB abundance in soil (Szukics et al. 2012). Most of the ammonia-oxidizing bacteria belong to phyla *Betaproteobacteria* and *Gammaproteobacteria* and represented by the genera *Nitrosomonas*, *Nitrosococcus*, and *Nitrosospira*. *Nitrosospira* seems to be the most abundant AOB in acidic forest soils with low NH_4^+ contents (Malchair and Carnol 2012). Nitrifying bacteria, *Nitrosomonas* and *Nitrosospira*, are affected by pH, acidity, and ammonia availability which in turn affect soil N availability which can result in dramatic shifts in the composition of the AOB community (Long et al. 2012). Denitrification is essential for maintaining global gaseous nitrogen (N_2 and nitrous oxide) pool. Ammonia-oxidizing bacteria are also involved in N losses from soil by producing N_2O during denitrification (Shaw et al. 2006). Denitrifying bacteria are abundant and widespread in forest soils. Denitrification genes have been found in bacterial strains belonging to the *Acidobacteria*, *Proteobacteria*, and *Firmicutes* (Priyanka and Koel 2015). The process can be addressed through marker genes involved in the enzymatic steps, such as the reduction of nitrate (*narG*), nitrite (*nirK*/*nirS*), nitric oxide(*norB*), and nitrous oxide (*nosZ*) (Jung et al. 2012). Incomplete denitrification is a powerful source of N_2O emissions from the soils, as denitrifying bacteria do not always perform all the steps of the dissimilatory pathway (Orellana et al. 2014).

6.3 Factors Affecting Nutrient Cycling

6.3.1 Climate

Nutrient storage in aboveground vegetation in forests generally increases in the order: boreal < temperate < tropical forests. However, due to slower decomposition in the cold conditions of higher latitudes, forest floor nutrient content increases from tropical to boreal forests. Nutrient cycling rates are low in subarctic woodland soils and Alaskan taiga forests due to extreme environment (Van Cleve et al. 1991). Because at low-temperature, microbial activity reduces accordingly, litter decomposition rates and nutrient availability also reduce. This increases C accumulation in the soils. Conditions in the tropical forest support good microbial activity throughout the year and results in fast decomposition (Foster and Bhatti 2006).

6.3.2 Anthropogenic Factors

Anthropogenic factors like fire, hurricanes, pest infestation, and harvesting are the key biogeochemical cycle regulators in almost 40% of the earth. Intensive wildfire leads to the volatilization and oxidation of live and decomposing plant materials, convection of ash under wind, percolation of solutes out of the soils, and water erosion of soil upper layer. The change in the nutrients (as a percentage of the amount present in the vegetation and litter pre- and post-fire condition) often varies in the order N > K > Mg >Ca> P. Revegetation may take a few months in the tropics, years in the temperate regions, and even a longer period of time in the boreal and tundra regions. In the tropics, negative impacts of the complete forest removal and slash burning can be observed on the microbial communities along with the site nutrients for the living biomass.

7 Specific Microbial Groups

7.1 Fungi

Fungi are generally represented by the organisms that include mildews, rusts, smuts, molds, yeasts, and mushrooms. Fungi vary from the unicellular life forms like yeasts to mycelial life forms that can cover hundreds of square meters and live for hundreds of years (McLaughlin et al. 2009). Biotic (plants and other organisms) and abiotic (soil pH, moisture, salinity, structure, and temperature) factors play an important role in regulating the diversity and activity of fungi (López-Bucio et al. 2015). The trees affect the composition of fungal communities, and fungal partner in return affects plant growth through mutualism, antagonism, nutrient cycling, and availability (Hannula et al. 2017). In addition, fungi also contribute to N fixation, hormone production, and antagonism against root pathogens (El Komy et al. 2015) and impart protection against drought stabilization of soil organic matter and decomposition of residues (Treseder and Lennon 2015). Fungal communities in leaves differ within the canopy of a single tree due to differences in stress intensity (such as irradiation). In tree leaves, the members of *Dothideomycetes* and *Leotiomycetes* fungi showed high level of dominance (Delhomme et al. 2015). In litter, fungal biomass exceeds 20-fold more than bacteria (Hieber and Gessner 2002). Fungi are relatively rich communities on the bark. Temporal variation in the fungal community composition on tree bark is shown in temperate forest, but seasonal development of the community pattern remains unclear (Beck et al. 2014). Fungi, due to their filamentous growth can penetrate wood easily and thus dominate this niche (de Boer et al. 2005a, b). Diverse communities of the fungi in *P. sylvestris* wood include endophytes and parasite species (Giordano et al. 2009), along with the saprotrophs that grow on deadwood (Parfitt et al. 2010). Also they dominate the eukaryotic transcript pool in spruce shoots (Delhomme et al. 2015). Fungi mostly found in association with wood decay are the filamentous species of *Basidiomycota* and *Ascomycota* (Arnstadt et al. 2016). The water mold *Phytophthora ramorum*, for example, plays role in the sudden oak death in the Pacific Northwest (Cobb et al. 2012). Exchange of plant nutrients within the forest soils is mainly mediated via root tissues of the plants and ECM or AM mycelia extending from the roots and rhizospheres into the bulk soil. ECM mycelia are characteristic of trees in temperate and boreal forests, majority of trees in tropical forests form AM associations. ECM represents almost one-third of the total microbial biomass and generate 50% of the organic carbon in these forests (Ekblad et al. 2013). Mineral weathering by *Paxillus involutus*, the ECM fungus, was generated by the carbon derived from the photosynthates of its *Pinus* host (Schmalenberger et al. 2015). Different fungi serve as a nutrient source for larvae, modifying the environment in a living tree to make it more favorable for the insects and responsible for tree mortality (Stokland et al. 2012).

7.2 Saprophytes

Saprophytes are the main decomposer organisms of wood debris and litter, acting as a substrate for energy and growth (Osono 2007; Stenlid et al. 2008). Nitrogen and phosphorus obtained through mycelial cords of the fungi act as promoters for wood decay into this nutrient-poor substrate (Wells and Boddy 1995; Bebber et al. 2011; Philpott et al. 2014). A nutritionally sound environment permits production of large quantity of spores of the fungi. These spores may enter wood through wounds leading to endophytic colonization. These may colonize the tissues of dead or develop into parasites when tree remains under stress (Parfitt et al. 2010). Several saprotrophic basidiomycetes, like *Phanerochaete velutina* or *Hypholoma fasciculare*, also act to penetrate their mycelia into the soil and decompose litter when they remain in the search of deadwood (Baldrian 2008). Likewise, root pathogen *Heterobasidion annosum* lives as a saprotroph on the deadwood. In optimum conditions, this further spreads itself through the soil to infect living trees (Piri 1996). Many fungi having capabilities of rotting the woods have also been isolated from the tree needles (Žifčáková et al. 2011).

7.3 N-Cycling Microbes

Biological N fixation (BNF) is the only key natural resource for the addition of new N to most of the terrestrial ecosystems (Galloway et al. 2004). Usually, the rate of BNF in tropical forests (15–36 kg N ha^{-1} $year^{-1}$) is the same or even higher than the estimations for temperate forests (7–27 kg N ha^{-1} $year^{-1}$) (Cleveland et al. 1999). Houlton et al. (2008) suggested that diazotrophs can be favorably used in the tropical forests due to their near-optimum temperature for natural N fixation. The *nifH* gene encoding the nitrogenase (reductase subunit) has widely been applied as a genetic marker to estimate genetic diversity and abundance of diazotrophic microorganisms (Zehr et al. 2003; Gaby and Buckley 2011). The low land tropical rain forest in Costa Rica were dominated by *nifH* clones from the genera *Heliobacterium* (*Firmicutes*), *Azospirillum*, *Gluconacetobacter*, *Methylobacterium*, and *Zymomonas* (α-proteobacteria) (Reed et al. 2010). Diazotrophic communities in primary and secondary rain forest soils are mainly comprised of the *Alphaproteobacteria* and *Betaproteobacteria* (the genera *Azospirillum*, *Azorhizobium*, *Bradyrhizobium*, *Methylobacterium*, *Burkholderia*), *Firmicutes* (*Paenibacillus*, *Heliobacterium*), and *Cyanobacteria* (*Anabaena, Nostoc*) (Mirza et al. 2014).

The process of nitrification in soils is autotrophic and heterotrophic in nature. The autotrophic nitrification is mainly represented by chemoautotrophic ammonium-oxidizing bacteria (AOB, represented by *Nitrosomonas*, *Nitrosospira*, and

Nitrosococcus species) and archaea (AOA, represented by *Nitrososphaera* and *Nitrosotalea* lineages from the phylum Thaumarchaeota) and nitrite-oxidizing bacteria (NOB, represented by *Nitrobacter*, *Nitrospira*, *Nitrococcus*, and *Nitrospira* species). Ammonia-oxidizing organisms can oxidize ammonia to nitrite through ammonia monooxygenase enzyme (Rotthauwe et al. 1997; Treusch et al. 2005). Microbial ammonia oxidation is a rate-limiting step of autotrophic nitrification. The chemolithotrophic *Gammaproteobacteria* and *Betaproteobacteria* play a primary role in this process in the soils (Kowalchuk and Stephen 2001). The belief that AOA acts as the ubiquitous constituent of the terrestrial life has changed the myth of aerobic nitrification (Leininger et al. 2006; Pester et al. 2012; Zhalnina et al. 2012; De Gannes et al. 2014). Establishing a relation in ammonia oxidation rates and the relative abundance of AOA and AOB is not easy (Nicol and Schleper 2006; Taylor et al. 2012). Heterotrophic nitrification is performed by heterotrophic bacteria and fungi having potentials to oxidize both organic and inorganic N compounds (Hayatsu et al. 2008). As an alternative pathway, nitrifier denitrification also involves the oxidation of ammonia to nitrite (Colliver and Stephenson 2000).

Denitrification is usually performed by taxonomically diverse microorganisms through diverse process of different evolutionary importance (Tiedje et al. 1983; Jones et al. 2008). The process that happens in the soils is majorly performed by the facultative aerobic heterotrophs (bacteria) like species of *Pseudomonas*, *Bacillus*, and *Paracoccus* and autotrophic bacteria like *Thiobacillus denitrificans* (Philippot et al. 2007; Demanèche et al. 2009). Denitrification is also reported from few archaea (Cabello et al. 2004; Bartossek et al. 2010) and fungi (Shoun et al. 1992; Hayatsu et al. 2008), including the members of *Ascomycota* (*Fusarium oxysporum*, *Fusarium solani*, *Cylindrocarpon tonkinense* and *Gibberella fujiuroii*) and *Basidiomycota* (*Trichosporon cutaneum*). Various diazotrophic bacteria like species of *Azospirillum* and *Bradyrhizobium* can also denitrify. Similarly, AOB belonging to either *Nitrosospira* or *Nitrosomonas* also have the capability to denitrify (Shaw et al. 2006). Fungi generally lack N_2O reductase enzyme (Shoun et al. 1992) and thus are responsible for N_2O emissions from the soil. Although microbial communities in the soils of tropical forests are usually dominated by the fungi (Hawksworth 2012), their contributions to N_2O emissions have not been addressed much.

7.4 Mycorrhiza

In addition to the biomass of the plants, fungal mycelial biomass also represent a major pool of organic matter in the forest litter and the soils (Baldrian et al. 2013). Ectomycorrhizal fungi (ECM) dominate the forest ecosystems having a biomass of up to 600 kg ha^{-1} (Tanaka and Nara 2009). Almost 90% of tree in the temperate and boreal forests take part in the ectomycorrhizal (ECM) symbiosis (Kluber et al. 2011), while AM associations are less common and generally dedicated to a limited number of tree species (Walker et al. 2014). ECM fungi form mantels and their mycelia extend around tree root tips up to the surrounding soils (Churchland and Grayston 2014). Large absorption areas and compound exudation by the ECM hyphae make mycorrhizosphere and mycosphere an important niche with definite features that differ from the nonmycorrhizal rhizosphere and bulk soil (Vik et al. 2013). High amount of labile C compounds in the tree root exudation or mycorrhizal hyphae enhances the availability of C to the inhabitant microbial communities in the rhizosphere. Consequently, microbial abundance and activity of extracellular enzymes in comparison to those in the bulk soils increase in the rhizosphere (Collignon et al. 2011; Brzostek et al. 2013). Trees are dependent on microbial symbiotic activities for growth-limiting nutrients (such as N). ECM mycelia may connect their host to soil or nutrient patches (Lindahl et al. 1999). Mycorrhizal networks transport soil, water, nitrogen, phosphorus, and carbon provided by the tree (Simard et al. 2012). ECM fungi may use organic N in decomposing deadwood or litter (Rajala et al. 2011) and mined rock surfaces for mineral nutrients and provide them to their plant hosts (Landeweert et al. 2001; Schmalenberger et al. 2015). Besides, ectomycorrhizal networks play a significant role in transfer of resource and stress signals from infected *Pseudotsuga menziesii* (Douglas-fir) to *Pine ponderosae* seedlings (Song et al. 2015). N-fixing bacteria and mycorrhiza account for the delivery of around 80% of all N and 75% of all P acquired by the plants in temperate and boreal forests (Van Der Heijden et al. 2008).

7.5 Endophytes

Bacteria and fungi can live as endophytes within plant tissues, without inducing symptoms. Endophytes are important players within the tree tissues and have the ability to tolerate extreme weather conditions like drought, heat, and cold and pathogen and herbivore attacks. Each tree in tropical forest harbors three to four specific endophytic fungi (Hawksworth 1991). Among all the nine different xylariaceous fungi, endophytically associated fungal organisms were found for healthy petioles of *Schefflera morototoni* in Puerto Rico (Laessøe and Lodge 1994), whereas 15 were with healthy green leaves of the palm *Euterpe oleracea* in Brazil (Rodrigues 1993). *Xylaria* species reported in both studies have wide host ranges. In temperate forest trees, species richness of endophytes ranged from a few taxa in *Betula* (Barengo et al. 2000), *Acer* spp. (Unterseher et al. 2007), *Quercus* spp. (Gennaro et al. 2003), *Abies* spp. (Carroll and Carroll 1978), and *Pinus* spp. (Sieber et al. 1999), *Picea abies* (Müller et al. 2001), *Abies alba* (Sieber-Canavesi and Sieber 1993), and *Carpinus caroliniana* (Bills and Polishook 1991). Foliar endophytes may be mutualists, latent plant pathogens, facultative entomopathogens, saprobes, and parasites, and a few have unknown functions (Carroll 1995). Foliar endophytes of deciduous trees usually are present as saprobes in litter or deadwood in winter, sporulate, and then reinvade living leaves, which is confirmed by their presence in both litter and living leaves (Unterseher et al. 2013). Several fungal foliar endophytes are asymptomatic inhabitants of living tree leaves or needles. However, a few can actively initiate litter decomposition (Müller et al. 2001; Szink et al. 2016). In woody tissues, they often appear as primary wood decay fungi (Boddy and Rayner 1983; Boddy 1992; Parfitt et al. 2010). Rhizosphere endophytes are similar to soil fungi, saprotrophic rhizoplane-inhabiting fungi, obligate and facultative pathogenic fungi, and mycorrhizal fungi (Jumpponen and Trappe 1998). These endophytes complete a part of their life cycle in the aquatic environment (Selosse et al. 2008). Spore dispersal via the atmosphere seems to be most likely the mode of reentry into the foliar structures of their hosts. Bacterial endophytes of the genus *Rhizobium*, which form nodules on the roots of *Alnus* and legume trees, take part in BNF (Boddey and Döbereiner 1995). *Paenibacillus polymyxa* provides up to 66% of the foliar N through BNF (Bal et al. 2012). *Pinus flexilis* (Moyes et al. 2016) and *P. contorta* show endophytic N fixation (Bal et al. 2012). Mycophyllas (mutualism between endophytic fungi and their host plants) are known for production of secondary metabolites that deter foraging organisms and pests (Clay 1988) in tropical forest trees. A tropical endophytic *Xylaria* species was found to be an aggressive antagonist of the fungus that causes witches' broom disease in *Theobroma* and *Herrania* species (Bravo-Velasquez and Hedger 1988). *P. fortinni* is a best known species of septate endophyte belonging to the PAC complex (*Phialocephala fortiniis – Acephalaap palanta).* It is mostly encountered in the roots of healthy coniferous trees in North American and European forests (Grünig et al. 2006, 2008). Endophytes belonging to *Trichoderma* induced the host defense against pathogens (Bailey et al. 2006) as well as mycoparasites (Bailey et al. 2008). A mycoparasite could sometimes work against a defense mutalist, for example, *Hydropisphaera fungicola* (Rossman and Palm-Hernández 2008) which feeds on an endophytic microbe *Ulocladium* that itself reduces the severity of leaf rust in *Populus* (Newcombe et al. 2010). *Beauveria bassiana*, an endophyte isolated from *Pinus monticola*, provided protection to pine trees against insects. According to Rodriguez et al. (2009), a few of the examined species belonging to this group had a positive effect on the growth of plants. The plants and endophytes growing individually could not survive the difficult environmental conditions. Technological advancements in the cultivation-independent techniques like metagenomics using next-generation sequencing (NGS) technology and association and network analysis using bioinformatics approaches will facilitate the identification of endophytic communities and their unraveling potential beneficial functions in the coming time.

7.6 Microbial Decomposers

In forest ecosystems, the major decomposers, symbionts, or pathogens are fungi. The activities of these fungi, especially saprophytic and symbiotic organisms, contribute largely to carbon recycling, inorganic nutrient cycling, and tree nutrition. The saprotrophic fungi are widely and densely distributed in forest soils. They are involved in degradation and decomposition of litter and wood debris to produce simple molecules (Lindahl et al. 2007; Rajala et al. 2012). These fungi exhibit considerable substrate specificity. For example, white-rot fungi decay both cellulose and lignin. In contrast, brown-rot fungi selectively degrade cellulose. The production and extracellular secretion of a large range of enzymes contributes to the degradation ability of white-rot fungi. For decay, two characteristic modes of the fungi were identified based on the presence or absence of ligninolytic peroxidase class II (Riley et al. 2014). The brown-rot fungi and ECM fungi utilize free radical-based non-specific Fenton reactions to degrade lignin in early stages of decay (Kersten and Cullen 2007; Rineau et al. 2012; Lindahl and Tunlid 2015). Saprophytic and ECM fungi complement each other in the process of organic matter decomposition as they coexist in similar habitat (Osono 2007; Rajala et al. 2015). The study on comparison of bacterial community in bulk soil and oak rhizosphere associated with ectomycorrhiza, *Scleroderma citrinum*, suggested that the ectomycorrhizosphere is enriched with bacteria capable of hydrolyzing glucopyranoside and chitin. In contrast, the surrounding soil appeared significantly enriched with bacteria capable of hydrolyzing cellobiose and N-acetyl glucosamine (Uroz et al. 2013). Certain strains of the genera *Collimonas* and *Pseudomonas* present in forest soils show laccase (polyphenol oxidase) activity, usually depicted by fungi. Microbial community in initial stages of litter decomposition is characterized by high fungal to bacterial biomass ratio (Purahong et al. 2015), whereas in later stages, bacteria are relatively abundant (Voriskova and Baldrian 2013).

The forest deadwood consists of fallen trees and large branches (Domke et al. 2016). It is characterized by physical and chemical characters like impermeability, high lignin, and low N content that make them resistant to the colonization by most of the bacteria (De Boer et al. 2005a, b). Thus, fungi, particularly saprotrophic cord-forming basidiomycetes, dominate deadwood (Eichlerová et al. 2015). The bacterial community-colonizing deadwood is composed of low pH-tolerant bacteria, including decomposers, commensals, and mycophages (Folman et al. 2008; Valaskova et al. 2009; Johnston et al. 2016). Decomposition of coarse wood such as logs or stumps typically takes about 10 years and is characterized by successive development of fungal communities with an initial dominance of decomposers and an increase of ECM fungi during late stages (Rajala et al. 2011; Baldrian et al. 2016). The taxonomy of primary colonizers of wood helps in the establishment of late-arriving species (Fukami et al. 2010; Lindner and Banik 2011; Hiscox et al. 2015). There are profound differences in the wood chemistry and decomposition rates (Baldrian 2008; Schilling et al. 2015; van der Wal et al. 2015). Climate, sun exposure, soil properties, and deadwood size and type (log, twig, or stump) are shown to be important determinants of fungal community composition (Seibold et al. 2015).

8 Conclusion and Future Research

Trees are the dominant component of the forest ecosystem. The forest ecosystem is dynamic. Microbial dynamics in a forest are driven by fires, insect outbreaks, climatic or seasonal variations, and anthropogenic activities occurring across tens to thousands of years. Microbial ecology studies have revealed that microorganisms are important drivers of various processes taking place in forests as well as responses to these changes. Microbial dynamic studies have been undertaken in temperate and boreal forests, whereas tropical forests are yet to be explored and characterized.

However, still, there are several knowledge gaps. For example, the description of the microbial community is incomplete or biased, and scanty information is available on ectomycorrhiza (ECM) and saprophytic wood-decomposing fungi. The microbial taxa actively participating in forest ecosystem processes have not yet been identified. Therefore, intensive studies on microbial activity and their interactions with trees and other forest biota are required, so that the impact of these processes at an ecosystem level can be analyzed.

Acknowledgments The authors would like to thank the financial assistance provided by ICAR and GOI. They also want to thank the State Forest Department of Tanakpur and Lachiwala range in Uttarakhand for allowing the sampling of natural sisso forests.

References

Adams AS, Jordan MS, Adams SM (2011) Cellulose-degrading bacteria associated with the invasive woodwasp *Sirex noctilio*. Int Soc Microb Ecol 5:1323–1331

Arnstadt T, Hoppe B, Kahl T et al (2016) Dynamics of fungal community composition, decomposition and resulting deadwood properties in logs of Fagus sylvatica, *Picea abies* and *Pinus sylvestris*. Forest Ecol and Manag 382:129–142

Ashton MS, Tyrrell ML, Spalding D, Gentry B (eds) (2012) Managing forest carbon in a changing climate. Springer Netherlands. https://www.springer.com/in/book/9789400722316. Accessed on 29 Jan 2019

Augusto L, Ranger J, Binkley D et al (2002) Impact of several common tree species of European temperate forests on soil fertility. Ann Forest Sci 59(3):233–253

Augusto L, De Schrijver A, Vesterdal L et al (2015) Influences of evergreen gymnosperm and deciduous angiosperm tree species on the functioning of temperate and boreal forests. Biol Rev 90(2):444–466

Bai Y, Eijsink VG, Kielak AM et al (2016) Genomic comparison of chitinolytic enzyme systems from terrestrial and aquatic bacteria. Environ Microbiol 18(1):38–49

Bailey BA, Bae H, Strem MD et al (2006) Fungal and plant gene expression during the colonization of cacao seedlings by endophytic isolates of four *Trichoderma* species. Planta 224(6):1449–1464

Bailey BA, Bae H, Strem MD et al (2008) Antibiosis, mycoparasitism, and colonization success for endophytic *Trichoderma* isolates with biological control potential in *Theobroma cacao*. Biol Control 46(1):24–35

Bakker MG, Otto-Hanson L, Lange AJ et al (2013) Plant monocultures produce more antagonistic soil *Streptomyces* communities than high-diversity plant communities. Soil Biol Biochem 65:304–312

Bal A, Anand R, Berge O et al (2012) Isolation and identification of diazotrophic bacteria from internal tissues of *Pinuscontorta* and *Thujaplicata*. Can J of For Res 42(4):807–813

Baldrian P (2008) Wood-inhabiting ligninolytic basidiomycetes in soils: ecology and constraints for applicability in bioremediation. Fungal Ecol 1(1):4–12

Baldrian P, Kolarik M, Stursova M et al (2012) Active and total microbial communities in forest soil are largely different and highly stratified during decomposition. Int Soci Microb Ecol J 6(2):248–258

Baldrian P, Snajdr J, Merhautova V et al (2013) Responses of the extracellular enzyme activities in hardwood forest to soil temperature and seasonality and the potential effects of climate change. Soil Biol Biochem 56:60–68

Baldrian P, Zrustova P, Tlaskal V et al (2016) Fungi associated with decomposing deadwood in a natural beech-dominated forest. Fungal Ecol 23:109–122

Bardgett RD, van der Putten WH (2014) Belowground biodiversity and ecosystem functioning. Nat 515(7528):505

Barengo N, Sieber TN, Holdenrieder O (2000) Diversity of endophytic mycobiota in leaves and twigs of pubescent birch (*Betula pubescens*). Sydowia 52(2):305–320

Barlocher F, Boddy L (2016) Aquatic fungal ecol—how does it differ from terrestrial? Fungal Ecol 19:5–13

Bartossek R, Nicol GW, Lanzen A et al (2010) Homologues of nitrite reductases in ammonia-oxidizing archaea: diversity and genomic context. Environ Microbiol 12(4):1075–1088

Bawa KS, Kress WJ, Nadkarni NM, Lele S, Raven PH, Janzen DH et al (2004) Tropical ecosystems into the 21st century. Science 306(5694):227–228

Bebber DP, Watkinson SC, Boddy L et al (2011) Simulated nitrogen deposition affects wood decomposition by cord-forming fungi. Oecol 167(4):1177–1184

Beck A, Persoh D, Rambold G (2014) First evidence for seasonal fluctuations in lichen-and bark-colonising fungal communities. Folia Microbiol 59(2):155–157

Beier S, Bertilsson S (2013) Bacterial chitin degradation—mechanisms and ecophysiological strategies. Front Microbiol 4:149

Berlemont R, Martiny AC (2015) Genomic potential for polysaccharide deconstruction in bacteria. Appl Environ Microbiol 81(4):1513–1519

Berthrong S, Yeager CM, Gallegos-Graves L et al (2014) Nitrogen fertilization has a stronger effect on soil N-fixing bacterial communities than elevated atmospheric CO2. Appl Environ Microbiol 80(10):3103–3112

Bhadra B, Rao RS, Singh PK (2008) Yeasts and yeast-like fungi associated with tree bark: diversity and identification of yeasts producing extracellular endoxylanases. Curr Microbiol 56:489–494

Bills GF, Polishook JD (1991) Microfungi from *Carpinus caroliniana*. Can J Bot 69(7):1477–1482

Bisht R, Chaturvedi S, Srivastava R et al (2009) Effect of arbuscular mycorrhizal fungi, *Pseudomonas fluorescens* and *Rhizobium leguminosarum* on the growth and nutrient status of *Dalbergia sissoo* Roxb. Tropical Ecol 50(2):231

Boddey RM, Döbereiner J (1995) Nitrogen fixation associated with grasses and cereals: recent progress and perspectives for the future. In: Nitrogen economy in tropical soils. Springer, Dordrecht, pp 241–250

Boddy L (1992) Microenvironmental aspects of xylem defenses to wood decay fungi. In: Defense mechanisms of woody plants against fungi. Springer, Berlin/Heidelberg, pp 96–132

Boddy L, Rayner ADM (1983) Ecological roles of basidiomycetes forming decay communities in attached oak branches. New Phytol 93(1):77–88

Bohannan BJ, Hughes J (2003) New approaches to analyzing microbial biodiversity data. Curr Opin Microbiol 6(3):282–287

Bonito G, Reynolds H, Robeson MS (2014) Plant host and soil origin influence fungal and bacterial assemblages in the roots of woody plants. Mol Ecol 23(13):3356–3370

Bowers RM, McLetchie S, Knight R (2011) Spatial variability in airborne bacterial communities across land-use types and their relationship to the bacterial communities of potential source environments. Int Soc Microb Ecol J 5:601–612

Brabcová V, Nováková M, Davidová A, Baldrian P (2016) Dead fungal mycelium in forest soil represents a decomposition hotspot and a habitat for a specific microbial community. New Phytologist 210(4):1369–1381

Bragina A, Berg C, Berg G (2015) The core microbiome bonds the Alpine bog vegetation to a transkingdom metacommunity. Mol Ecol 24:4795–4807

Bravo-Velasquez E, Hedger J (1988) The effect of ecological disturbance on competition between *Crinipellis perniciosa* and other tropical fungi. Proc R Soc Edinburgh, Sect B Biol Sci 94:159–166

Brzostek ER, Greco A, Drake JE et al (2013) Root carbon inputs to the rhizosphere stimulate extracellular enzyme activity and increase nitrogen availability in temperate forest soils. Biogeochem 115(1-3):65–76

Buckeridge KM, Grogan P (2010) Deepened snow increases late thaw biogeochemical pulses in mesic low arctic tundra. Biogeochem 101(1-3):105–121

Buée M, Vairelles D, Garbaye J (2005) Year-round monitoring of diversity and potential metabolic activity of the ectomycorrhizal community in a beech (Fagus silvatica) forest subjected to two thinning regimes. Mycorrhiza 15(4):235–245

Buée M, Courty PE, Mignot D et al (2007) Soil niche effect on species diversity and catabolic activities in an ectomycorrhizal fungal community. Soil Biol Biochem 39:1947–1955

Buée M, De Boer W, Martin F et al (2009) The rhizosphere zoo: an overview of plant-associated communities of microorganisms, including phages, bacteria, archaea, and fungi, and of some of their structuring factors. Plant Soil 321(1-2):189–212

Buée M, Maurice JP, Zeller B et al (2011) Influence of tree species on richness and diversity of epigeous fungal communities in a French temperate forest stand. Fungal Ecol 4:22–31

Büntgen U, Tegel W, Egli S et al (2011) Truffles and climate change. Front Ecol Env 9(3):150–151

Cabello P, Roldan MD, Moreno-Vivian C (2004) Nitrate reduction and the nitrogen cycle in archaea. Microbiology 150(11):3527–3546

Carrell AA, Frank C (2015) Bacterial endophyte communities in the foliage of coast redwood and giant sequoia Front Microbiol;6:1008

Carroll G (1995) Forest endophytes: pattern and process. Can J Bot 73(S1):1316–1324

Carroll GC, Carroll FE (1978) Studies on the incidence of coniferous needle endophytes in the Pacific Northwest. Can J Bot 56(24):3034–3043

Chao A (1984) Nonparametric estimation of the number of classes in a population Scandinavian. J Stat:265–270

Chazdon RL, Colwell RK, Denslow JS et al (1998) Statistical methods for estimating species richness of woody regeneration in primary and secondary rain forests of northeastern Costa Rica. No Man Biosph 20:285–309

Chen Y, Wen X, Sun Y et al (2014) Mulching practices altered soil bacterial community structure and improved orchard productivity and apple quality after five growing seasons. Sci Hort 172:248–257

Churchland C, Grayston SJ (2014) Specificity of plant-microbe interactions in the tree mycorrhizosphere biome and consequences for soil C cycling. Front Microbiol 5:261

Clay K (1988) Fungal endophytes of grasses: a defensive mutualism between plants and fungi. Ecol 69(1):10–16

Clemmensen KE, Bahr A, Ovaskainen O et al (2013) Roots and associated fungi drive long-term carbon sequestration in boreal forest. Sci 339(6127):1615–1618

Cleveland CC, Townsend AR, Schimel DS (1999) Global patterns of terrestrial biological nitrogen (N2) fixation in natural ecosystems. Global Biogeochem Cy 13(2):623–645

Clissmann F, Fiore-Donno AM, Hoppe B, Krüger D, Kahl T, Unterseher M, Schnittler M (2015) First insight into dead wood protistan diversity: a molecular sampling of bright-spored Myxomycetes (Amoebozoa, slime-moulds) in decaying beech logs. FEMS Microbiol Ecol 91(6)

Cobb RC, Filipe JA, Meentemeyer RK et al (2012) Ecosystem transformation by emerging infectious disease: loss of large tanoak from California forests. J Ecol 100(3):712–722

Coince A, Cordier T, Lengellé J (2014) Leaf and root-associated fungal assemblages do not follow similar elevational diversity patterns. PloS One 9(6):e100668

Colagiero M, Pentimone I, Rosso LC et al (2017) A metagenomic study on the effect of aboveground plant cover on soil bacterial diversity. In: Soil biological communities and ecosystem resilience. Springer, Cham, pp 97–106

Collignon C, Calvaruso C, Turpault MP (2011) Temporal dynamics of exchangeable K, Ca and Mg in acidic bulk soil and rhizosphere under Norway spruce (*Piceaabies abies Karst*) and beech (*Fagus sylvatica* L) stands. Plant and Soil 349(1-2):355–366

Colliver BB, Stephenson T (2000) Production of nitrogen oxide and dinitrogen oxide by autotrophic nitrifiers. Biotechnol Adv 18(3):219–232

Cong J, Yang Y, Liu X et al (2015) Analyses of soil microbial community compositions and functional genes reveal potential consequences of natural forest succession. Sci Rep 5:10007

Corneo PE, Pellegrini A, Cappellin L et al (2013) Microbial community structure in vineyard soils across altitudinal gradients and in different seasons. FEMS Microbiol Ecol 84(3):588–602

Courty PE, Franc A, Pierrat JC et al (2008) Temporal changes in the ectomycorrhizal community in two soil horizons of a temperate oak forest. Appl Environ Microbiol 74(18):5792–5801

Da Silva MF, Carreira LMM, Tavares AS, Ribeiro IC, Jardim MAG, Lobo MDAGA, Oliveira J (1989) As leguminosas da Amazo#nia Brasileira-Lista Pre$via Acta Bota [nica]. Anais do XXXIX Congresso Nacional de Botanica 2:193–237

Davey ML, Heegaard E, Halvorsen R (2012) Seasonal trends in the biomass and structure of bryophyte-associated fungal communities explored by 454 pyrosequencing. New Phytol 195:844–846

de Araujo Pereira AP, de Andrade PA, Bini D, Durrer A, Robin A, Bouillet JP, Andreote FD, Cardoso EJ (2017) Shifts in the bacterial community composition along deep soil profiles in monospecific and mixed stands of Eucalyptus grandis and *Acacia mangium*. PLoS One 12(7):e0180371

De Boer HJ, Kool A, Broberg A et al (2005a) Anti-fungal and anti-bacterial activity of some herbal remedies from Tanzania. J Ethnopharmacol 96(3):461–469

De Boer W, Folman LB, Summerbell RC et al (2005b) Living in a fungal world: impact of fungi on soil bacterial niche development. FEMS Microbiol Rev 29(4):795–811

De Gannes V, Eudoxie G, Hickey WJ (2014) Impacts of edaphic factors on communities of ammonia-oxidizing archaea, ammonia-oxidizing bacteria and nitrification in tropical soils. PLoS One 9(2):e89568

Delhomme N, Sundstrom G, Zamani N (2015) Serendipitous meta-transcriptomics: the fungal community of Norway spruce (*Picea abies*). PLoS One 10(9):e0139080

Demanèche S, Philippot L, David MM et al (2009) Characterization of denitrification gene clusters of soil bacteria via a metagenomic approach. Appl Environ Microbiol 75(2):534–537

Devi NB, Yadava PS (2006) Seasonal dynamics in soil microbial biomass C, N and P in a mixed-oak forest ecosystem of Manipur, North-east India. App Soil Ecol 31(3):220–227

Domke GM, Perry CH, Walters BF et al (2016) Estimating litter carbon stocks on forest land in the United States. Sci Total Environ 557:469–478

Dupuy B, Diahuissie A, Doumbia F et al (1997) Effet de deux types d'éclaircie en foret dense ivoirienne. Bois Forets Des Tropiques 253:5–19

Edwards M, Johns DG, Licandro P et al (2006) Ecological status report: results from the CPR survey 2004/2005. SAHFOS Tech Rep (3):1–8

Edwards AC, Scalenghe R, Freppaz M (2007) Changes in the seasonal snow cover of alpine regions and its effect on soil processes. Quat Int 162:172–181

Eichlerová I, Homolka L, Žifčáková L, Lisá L, Dobiášová P, Baldrian P (2015) Enzymatic systems involved in decomposition reflects the ecology and taxonomy of saprotrophic fungi. Fungal Ecol 13:10–22

Ekblad A, Wallander H, Godbold DL et al (2013) The production and turnover of extramatrical mycelium of ectomycorrhizal fungi in forest soils: role in carbon cycling. Plant Soil 366(1-2):1–27

El Komy MH, Saleh AA, Eranthodi A, Molan YY (2015) Characterization of novel *Trichoderma asperellum* isolates to select effective biocontrol agents against tomato Fusarium wilt. Plant Pathol J 31:50–60

Esson KC, Lin X, Kumaresan D et al (2016) Alpha and gamma proteobacterial methanotrophs co-dominate the active methane oxidizing communities in an acidic boreal peat bog. Appl Environ Microbiol. 82:2363–2371

Fierer N, Jackson RB (2006) The diversity and Biogeogr of soil bacterial communities. Proc Natl Acad Sci 103(3):626–631

Folman LB, Klein Gunnewiek PJ, Boddy L et al (2008) Impact of white-rot fungi on numbers and community composition of bacteria colonizing beech wood from forest soil. FEMS Microbiol Ecol 63(2):181–191

Foster NW, Bhatti JS (2006) Forest ecosystems: nutrient cycling Encyclopedia of soil science. Taylor Francis Group, New York, pp 718–721

Founoune H, Duponnois R, Ba AM et al (2002) Influence of the dual arbuscular endomycorrhizal/ ectomycorrhizal symbiosis on the growth of *Acacia holosericea* (A. Cunn. ex G. Don) in glasshouse conditions. Ann Forest Sci 59(1):93–98

Fransson P, Andersson A, Norström S et al (2016) Ectomycorrhizal exudates and pre-exposure to elevated CO2 affects soil bacterial growth and community structure. Fungal Ecol 20:211–224

Frey SD, Drijber R, Smith H et al (2008) Microbial biomass, functional capacity, and community structure after 12 years of soil warming. Soil Biol Biochem 40(11):2904–2907

Frey B, Niklaus PA, Kremer J et al (2011) Heavy machinery traffic impacts methane emission, abundance of methanogens and community structure in oxic forest soils. Appl Environ Microbiol 77:6060–6068

Fukami T, Dickie IA, Paula W et al (2010) Assembly history dictates ecosystem functioning: evidence from wood decomposer communities. Ecol Lett 13(6):675–684

Gaby JC, Buckley DH (2011) A global census of nitrogenase diversity. Environ Microbiol 13(7):1790–1799

Galloway JN, Dentener FJ, Capone DG (2004) Nitrogen cycles: past, present, and future. Biogeochemistry 70(2):153–226

García-Fraile P, Menéndez E, Rivas R (2015) Role of bacterial biofertilizers in agriculture and forestry. Bioengineering 2:183–205

Gennaro M, Gonthier P, Nicolotti G (2003) Fungal endophytic communities in healthy and declining *Quercus robur* L and *Q. cerris* L trees in northern Italy. J Phytopathol 151(10):529–534

Gibson L, Lee TM, Koh LP et al (2011) Primary forests are irreplaceable for sustaining tropical biodiversity. Nat 478:378–381

Giordano L, Gonthier P, Varese GC (2009) Mycobiota inhabiting sapwood of healthy and declining Scots pine (*Pinus sylvestris* L) trees in the Alps. Fungal Divers 38:69–83

Goldmann K, Schöning I, Buscot F (2015) Forest management type influences diversity and community composition of soil fungi across temperate forest ecosystems. Front Microbiol 6:1300

Grayston SJ, Griffith GS, Mawdsley JL et al (2001) Accounting for variability in soil microbial communities of temperate upland grassland ecosystems. Soil Biol Biochem 33(4-5):533–551

Grube M, Cernava T, Soh J et al (2015) Exploring functional contexts of symbiotic sustain within lichen-associated bacteria by comparative omics. Int Soci Microb Ecol J 9:412–424

Grünig CR, Duò A, Sieber TN (2006) Population genetic analysis of *Phialocephala fortinii sl* and *Acephala applanata* in two undisturbed forests in Switzerland and evidence for new cryptic species. Fungal Genet Biol 43(6):410–421

Grünig CR, Queloz V, Sieber TN et al (2008) Dark septate endophytes (DSE) of the *Phialocephala fortinii sl–Acephala applanata* species complex in tree roots: classification, population biology, and ecology. Botany 86(12):1355–1369

Hacquard S, Schadt CW (2015) Towards a holistic understanding of the beneficial interactions across the Populus microbiome. New Phytol 205(4):1424–1430

Haichar ZF, Marol C, Berge O et al (2008) Plant host habitat and root exudates shape soil bacterial community structure. Int Soci Microb Ecol 2(12):1221

Hammond DS (2005) Guianan forest dynamics: geomorphographic control and tropical forest change across diverging landscapes In: Hammond DS, ed. Tropical forests of the Guiana shield; ancient forests in a modern world. CABI Publishing, Wallingford, pp 343–379

Hannula SE, Morrien E, de Hollander M et al (2017) Shifts in rhizosphere fungal community during secondary succession following abandonment from agriculture. Int Soci Microb Ecol J 11(10):2294

Hardoim PR, van Overbeek LS, Berg G (2015) The hidden world within plants: ecological and evolutionary considerations for defining functioning of microbial endophytes. Microbiol Mol Biol R 79:293–320

Hartmann M, Howes CG, VanInsberghe D et al (2012) Significant and persistent impact of timber harvesting on soil microbial communities in northern coniferous forests. Int Soci Microb Ecol J 6(12):2199–2218

Hartmann M, Niklaus PA, Zimmermann S et al (2014) Resistance and resilience of the forest soil microbiome to logging-associated compaction. Int Soci Microb Ecol J 8(1):226

Hawkes CV, Kivlin SN, Rocca JD (2011) Fungal community responses to precipitation Global Change. Biol 17(4):1637–1645

Hawksworth DL (1991) The fungal dimension of biodiversity: magnitude, significance, and conservation. Mycol Res 95(6):641–655

Hawksworth DL (2012) Global species numbers of fungi: are tropical studies and molecular approaches contributing to a more robust estimate? Biodivers Conserv 21(9):2425–2433

Hayatsu M, Tago K, Saito M (2008) Various players in the nitrogen cycle: diversity and functions of the microorganisms involved in nitrification and denitrification. Soil Sci Plant Nutr 54(1):33–45

Hibbett DS, Ohman A, Glotzer D et al (2011) Progress in molecular and morphological taxon discovery in Fungi and options for formal classification of environmental sequences. Fungal Biol Rev 25(1):38–47

Hieber M, Gessner MO (2002) Contribution of stream detrivores, fungi, and bacteria to leaf breakdown based on biomass estimates. Ecol 83(4):1026–1038

Hill TC, Walsh KA, Harris JA, Moffett BF (2003) Using ecological diversity measures with bacterial communities. FEMS Microbiol Ecol 43(1):1–11

Hiscox J, Savoury M, Muller CT (2015) Priority effects during fungal community establishment in beech wood. Int Soci Microb Ecol J 9(10):2246

Hoffland E, Kuyper TW, Wallander H (2004) The role of fungi in weathering. Front Ecol Environ 2(5):258–264

Högberg P (1986) Soil nutrient availability, root symbioses and tree species composition in tropical Africa. J Trop Ecol 2:359–372

Högberg P, Nordgren A, Buchmann N et al (2001) Large-scale forest girdling shows that current photosynthesis drives soil respiration. *Nature* 411:789–792

Högberg MN, Briones MJ, Keel SG (2010) Quantification of effects of season and nitrogen supply on tree below-ground carbon transfer to ectomycorrhizal fungi and other soil organisms in a boreal pine forest. New Phytol 187(2):485–493

Hoppe B, Kruger D, Kahl T (2015) A pyrosequencing insight into sprawling bacterial diversity and community dynamics in decaying deadwood logs of *Fagus sylvatica* and *Picea abies*. Sci Rep 5:9456

Hottola J, Ovaskainen O, Hanski I (2009) A unified measure of the number, volume and diversity of dead trees and the response of fungal communities. J Ecol 97(6):1320–1328

Houlton BZ, Wang YP, Vitousek PM et al (2008) A unifying framework for dinitrogen fixation in the terrestrial biosphere. Nature 454(7202):327

Hughes JB, Hellmann JJ, Ricketts TH et al (2001) Counting the uncountable: statistical approaches to estimating microbial diversity. Appl Environ Microbiol 67(10):4399–4406

Ingham ER (1999) The soil biology primer chapter 1. The soil foodweb. NRCS Soil Quality Institute, USDA, p 48

Ishida TA, Nara K, Hogetsu T (2007) Host effects on ectomycorrhizal fungal communities: insight from eight host species in mixed conifer–broadleaf forests. New Phytol 174(2):430–440

Izumi H, Anderson IC, Alexander IJ et al (2006) Endobacteria in some ectomycorrhiza of Scots pine (*Pinus sylvestris*). FEMS Microbiol Ecol 56(1):34–43

Jeanbille M, Buée M, Bach C (2016) Soil parameters drive the structure, diversity and metabolic potentials of the bacterial communities across temperate beech forest soil sequences. Microb Ecol 71(2):482–493

Johnston SR, Boddy L, Weightman AJ (2016) Bacteria in decomposing wood and their interactions with wood-decay fungi. FEMS Microbiol Ecol:92–179

Jones CM, Stres B, Rosenquist M et al (2008) Phylogenetic analysis of nitrite, nitric oxide, and nitrous oxide respiratory enzymes reveal a complex evolutionary history for denitrification. Mol Biol Evol 25(9):1955–1966

Jönsson MT, Edman M, Jonsson BG (2008) Colonization and extinction patterns of wood-decaying fungi in a boreal old-growth *Picea abies* forest. J Ecol 96(5):1065–1075

Joshi S (2018) Rhizospheric bacterial diversity in different *Dalbergia sissoo* Roxb provenances PhD thesis, G.B. Pant University of Agriculture & Technology, UK, India

Jost L (2007) Partitioning diversity into independent alpha and beta components. Ecology 88:2427–2439

Jumpponen ARI, Trappe JM (1998) Dark septate endophytes: a review of facultative biotrophic root-colonizing fungi. New Phytol 140(2):295–310

Jung J, Yeom J, Han J et al (2012) Seasonal changes in nitrogen-cycle gene abundances and in bacterial communities in acidic forest soils. J Microbiol 50(3):365–373

Kaiser C, Koranda M, Kitzler B et al (2010) Belowground carbon allocation by trees drives seasonal patterns of extracellular enzyme activities by altering microbial community composition in a beech forest soil. New Phytol 187(3):843–858

Kaiser C, Fuchslueger L, Koranda M et al (2011) Plants control the seasonal dynamics of microbial N cycling in a beech forest soil by belowground C allocation. Ecology 92(5):1036–1051

Kauserud H, Heegaard E, Buntgen U et al (2012) Warming-induced shift in European mushroom fruiting phenology. Proc Natl Acad Sci 109:14488–14493

Kersten P, Cullen D (2007) Extracellular oxidative systems of the lignin-degrading Basidiomycete *Phanerochaete chrysosporium*. Fungal Genet Biol 44(2):77–87

Kluber LA, Smith JE, Myrold DD (2011) Distinctive fungal and bacterial communities are associated with mats formed by ectomycorrhizal fungi. Soil Biol Biochem 43(5):1042–1050

Kohler A, Kuo A, Nagy LG (2015) Convergent losses of decay mechanisms and rapid turnover of symbiosis genes in mycorrhizal mutualists. Nat Genet 47:410–U176

Koleff P, Gaston KJ, Lennon JJ (2003) Measuring beta diversity for presence–absence data. J Anim Ecol 72(3):367–382

Korkama T, Fritze H, Pakkanen A et al (2007) Interactions between extraradical ectomycorrhizal mycelia, microbes associated with the mycelia and growth rate of Norway spruce (*Picea abies*) clones. New Phytol 173(4):798–807

Kostka JE, Weston DJ, Glass JB (2016) The Sphagnum microbiome: new insights from an ancient plant lineage. New Phytol 211:57–64

Kowalchuk GA, Stephen JR (2001) Ammonia-oxidizing bacteria: a model for molecular microbial ecology. Annu Rev Microbiol 55(1):485–529

Kowalski KP, Bacon C, Bickford W et al (2015) Advancing the science of microbial symbiosis to support invasive species management: a case study on Phragmites in the Great Lakes. Front Microbiol 6:95

Kubartova A, Ottosson E, Dahlberg A et al (2012) Patterns of fungal communities among and within decaying logs, revealed by 454 sequencing. Mol Ecol 21(18):4514–4532

Küffer N, Senn-Irlet B (2005) Influence of forest management on the species richness and composition of wood-inhabiting basidiomycetes in Swiss forests. Biodivers Conserv 14(10):2419–2435

Kuffner M, Hai B, Rattei T et al (2012) Effects of season and experimental warming on the bacterial community in a temperate mountain forest soil assessed by 16S rRNA gene pyrosequencing. FEMS Microbiol Ecol 82(3):551–562

Kulhánkova A, Béguiristain T, Moukoumi J et al (2006) Spatial and temporal diversity of wood decomposer communities in different forest stands, determined by ITS rDNA targeted TGGE. Annals Forest Sci 63(5):547–556

Kuzyakov Y, Blagodatskaya E (2015) Microbial hotspots and hot moments in soil: concept & review. Soil Biol Biochem 83:184–199

Laessøe T, Lodge DJ (1994) Three host-specific *Xylaria* species. Mycol:436–446

Landeweert R, Hoffland E, Finlay RD et al (2001) Linking plants to rocks: ectomycorrhizal fungi mobilize nutrients from minerals. Trends Ecol Evol 16(5):248–254

Lang C, Seven J, Polle A (2011) Host preferences and differential contributions of deciduous tree species shape mycorrhizal species richness in a mixed Central European forest. Mycorrhiza 21(4):297–308

Lau MC, Stackhouse BT, Layton AC et al (2015) An active atmospheric methane sink in high Arctic mineral cryosols. Int Soci Microb Ecol J 9(8):1880

Lauber CL, Strickland MS, Bradford MA et al (2008) The influence of soil properties on the structure of bacterial and fungal communities across land-use types. Soil Biol Biochem 40(9):2407–2415

Lauber CL, Hamady M, Knight R et al (2009) Pyrosequencing-based assessment of soil pH as a predictor of soil bacterial community structure at the continental scale. Appl Environ Microbiol 75(15):5111–5120

Laverman AM, Speksnijder AGCL, Braster M et al (2001) Spatiotemporal stability of an ammonia-oxidizing community in a nitrogen-saturated forest soil. Microb Ecol 42(1):35–45

Lazzaro A, Hilfiker D, Zeyer J (2015) Structures of microbial communities in alpine soils: seasonal and elevational effects. Front Microbiol 6:1330

Lederberg J (2006) The microbe's contribution to biology-50 years after. Int Microbiol 9(3):155

Leininger S, Urich T, Schloter M et al (2006) Archaea predominate among ammonia-oxidizing prokaryotes in soils. Nat 442(7104):806

Lejon DP, Chaussod R, Ranger J et al (2005) Microbial community structure and density under different tree species in an acid forest soil (Morvan, France). Microb Ecol 50(4):614–625

Lewis G, Schrire B, Mackinder B, Lock M (2005) Legumes of the World, Royal Botanic Gardens, Kew:p592

Libois Q, PicardG FJL, Arnaud L et al (2013) Influence of grain shape on light penetration in snow. The Cryosphere 7(6):1803–1818

Lilleskov EA, Fahey TJ, Horton TR et al (2002) Belowground ectomycorrhizal fungal community change over a nitrogen deposition gradient in Alaska. Ecology 83(1):104–115

Lindahl BD, Tunlid A (2015) Ectomycorrhizal fungi–potential organic matter decomposers, yet not saprotrophs. New Phytol 205(4):1443–1447

Lindahl B, Stenlid JAN, Olsson S et al (1999) Translocation of ^{32}P between interacting mycelia of a wood-decomposing fungus and ectomycorrhizal fungi in microcosm systems. New Phytol 144(1):183–193

Lindahl BD, Ihrmark K, Boberg J et al (2007) Spatial separation of litter decomposition and mycorrhizal nitrogen uptake in a boreal forest. New Phytol 173(3):611–620

Lindahl BD, De Boer W, Finlay RD (2010) Disruption of root carbon transport into forest humus stimulates fungal opportunists at the expense of mycorrhizal fungi. Int Soci Microb Ecol J 4(7):872

Lindner DL, Banik MT (2011) Intragenomic variation in the ITS rDNA region obscures phylogenetic relationships and inflates estimates of operational taxonomic units in genus *Laetiporus*. Mycology 103(4):731–740

Lipson DA, Schmidt SK, Monson RK (1999) Links between microbial population dynamics and nitrogen availability in an alpine ecosystem. Ecol 80(5):1623–1631

Lipson DA, Schadt CW, Schmidt SK (2002) Changes in soil microbial community structure and function in an alpine dry meadow following spring snow melt. Microb Ecol 43(3):307–314

Lladó S, Zifcakova L, Vetrovsky T et al (2016) Functional screening of abundant bacteria from acidic forest soil indicates the metabolic potential of *Acidobacteria* subdivision 1 for polysaccharide decomposition. Biol Fert Soils 52(2):251–260

Lladó S, López-Mondéjar R, Baldrian P (2017) Forest soil Bacteria: diversity, involvement in ecosystem processes, and response to global change. Microbiol Mol Biol Rev 81(2)

Long X, Chen C, Xu Z et al (2012) Abundance and community structure of ammonia-oxidizing bacteria and archaea in a temperate forest ecosystem under ten-years elevated CO2. Soil Biol Biochem 46:163–171

López-Bucio J, Pelagio-Flores R, Herrera-Estrell A (2015) Trichoderma as biostimulant: exploiting the multilevel properties of a plant beneficial fungus. Sci. Hortic. 196:109–123

Lopez-Mondejar R, Voriskova J, Vetrovsky T et al (2015) The bacterial community inhabiting temperate deciduous forests is vertically stratified and undergoes seasonal dynamics. Soil Biol Biochem 87:43–50

López-Mondejar R, Zühlke D, Becher D et al (2016) Cellulose and hemicellulose decomposition by forest soil bacteria proceeds by the action of structurally variable enzymatic systems. Sci Rep 6:25279

Lukesova T, Kohout P, Vetrovsky T (2015) The potential of dark septate endophytes to form root symbioses with ectomycorrhizal and ericoid mycorrhizal middle European forest plants. PLoS One 10:e124752
Malchair S, Carnol M (2012) AOB community structure and richness under European beech, sessile oak, Norway spruce and Douglas fir at three temperate forest sites. Plant Soil 366:521–535
Marupakula S, Mahmood S, Finlay RD (2016) Analysis of single root tip microbiomes suggests that distinctive bacterial communities are selected by *Pinus sylvestris* roots colonized by different ectomycorrhizal fungi. Environ Microbiol 18(5):1470–1483
McLaren JR, Turkington R (2011) Plant identity influences decomposition through more than one mechanism. PLoS One 6(8)
McLaughlin DJ, Hibbett DS, Lutzoni F et al (2009) The search for the fungal tree of life. Trends Microbiol 17(11):488–497
McMahon SM, Harrison SP, Armbruster WS et al (2011) Improving assessment and modelling of climate change impacts on global terrestrial biodiversity. Trends Ecol Evol 26(5):249–259
Mirza BS, Potisap C, Nüsslein K et al (2014) Response of free-living nitrogen-fixing microorganisms to land use change in the Amazon rainforest. Appl Environ Microbiol 80(1):281–288
Monson RK, Lipson DL, Burns SP et al (2006) Winter forest soil respiration controlled by climate and microbial community composition. Nat 439(7077):711
Moreira FMS, Franco AA (1994) Rhizobia-host interactions in tropical ecosystems in Brazil. In: Sprent JI, McKey D (eds) Advances in legume systematics, Part 5: the nitrogen factor, vol 5. R BotGard, Kew, pp 63–74
Moyes AB, Kueppers LM, Pett-Ridge J et al (2016) Evidence for foliar endophytic nitrogen fixation in a widely distributed subalpine conifer. New Phytol 210(2):657–668
Müller MM, Valjakka R, Suokko A et al (2001) Diversity of endophytic fungi of single Norway spruce needles and their role as pioneer decomposers. Mol Ecol 10(7):1801–1810
Nacke H, Goldmann K, Schöning I et al (2016) Fine spatial scale variation of soil microbial communities under European beech and Norway spruce. Front Microbiol 7:2067
Nazaries L, Murrell JC, Millard P et al (2013) Methane, microbes and models: fundamental understanding of the soil methane cycle for future predictions. Environ Microbiol 15(9):2395–2417
Newcombe G, Martin F, Kohler A (2010) Defense and nutrient mutualisms in Populus. In: Genetics and genomics of populus. Springer, New York, pp 247–277
Newton RJ, Jones SE, Eiler A et al (2011) A guide to the natural history of freshwater lake bacteria. Microbiol Mol Biol R 75:14–49
Nicol GW, Schleper C (2006) Ammonia-oxidising Crenarchaeota: important players in the nitrogen cycle? Trends Microbiol 14(5):207–212
Nitrogen Cycling Down: https://organicsoiltechnologycom/nitrogen-cycling-downhtml
Noguez AM, Arita HT, Escalante AE et al (2005) Microbial macroecology: highly structured prokaryotic soil assemblages in a tropical deciduous forest. Global Ecol Biogeogr 14(3):241–248
Okubo T, Tsukui T, Maita H et al (2012) Complete genome sequence of *Bradyrhizobium* sp S23321: insights into symbiosis evolution in soil oligotrophs. Microb Environ 27(3):306–315
Öpik M, Moora M, Zobel M (2008) High diversity of arbuscular mycorrhizal fungi in a boreal herb-rich coniferous forest. New Phytol 179:867–876
Orellana LH, Rodriguez-R LM, Higgins S et al (2014) Detecting nitrous oxide reductase (nosZ) genes in soil metagenomes: method development and implications for the nitrogen cycle. Mol Biol 5(3):e01193–e01114
Osono T (2007) Ecology of ligninolytic fungi associated with leaf litter decomposition. Ecol Res 22(6):955–974
Pajares S, Bohannan BJ (2016) Ecology of nitrogen fixing, nitrifying, and denitrifying microorganisms in tropical forest soils. Front Microbiol 7:1045
Parfitt D, Hunt J, Dockrell D et al (2010) Do all trees carry the seeds of their own destruction? PCR reveals numerous wood decay fungi latently present in sapwood of a wide range of angiosperm trees. Fungal Ecol 3(4):338–346

Peh KSH, Corlett RT, Bergeron Y (2015) Routledge Handbook of Forest Ecology. Routledge
Peršoh D (2015) Plant-associated fungal communities in the light of meta'omics. Fungal Divers 75:1–25
Pester M, Rattei T, Flechl S et al (2012) amoA-based consensus phylogeny of ammonia-oxidizing archaea and deep sequencing of amoA genes from soils of four different geographic regions. Environ Microbiol 14(2):525–539
Philippot L, Hallin S, Schloter M (2007) Ecology of denitrifying prokaryotes in agricultural soil. Adv Agro 96:249–305
Philpott TJ, Prescott CE, Chapman WK et al (2014) Nitrogen translocation and accumulation by a cord-forming fungus (*Hypholomaf asciculare*) into simulated woody debris. Forest Ecol Manag 315:121–128
Piri T (1996) The spreading of the S type of *Heterobasidion annosum* from Norway spruce stumps to the subsequent tree stand. Eur J Forest Pathol 26(4):193–204
Pons TL, Perreijn K, Van Kessel C et al (2007) Symbiotic nitrogen fixation in a tropical rainforest: ^{15}N natural abundance measurements supported by experimental isotopic enrichment. New Phytol 173(1):154–167
Prescott CE, Grayston SJ (2013) Tree species influence on microbial communities in litter and soil: current knowledge and research needs. Forest Ecol Manag 309:19–27
Prevost-Boure NC, Maron PA, Ranjard L et al (2011) Seasonal dynamics of the bacterial community in forest soils under different quantities of leaf litter. Appl Soil Ecol 47(1):14–23
Priyanka J, Koel M (2015) Diversity study of nitrate reducing bacteria from soil samples-a metagenomics approach. J Comp Sci Syst Biol *8*(4):191
Puig H, Riera B, Lescure JP (1990) Phytomasse et productivite. Bois Forets Des Tropiques 220:25–32
Purahong W, Hoppe B, Kahl T et al (2014) Changes within a single land-use category alter microbial diversity and community structure: Molecular evidence from wood-inhabiting fungi in forest ecosystems. J Environ Manag 139:109–119
Purahong W, Kapturska D, Pecyna MJ et al (2015) Effects of forest management practices in temperate beech forests on bacterial and fungal communities involved in leaf litter degradation. Microb Ecol 69(4):905–913
Rachid CTCC, Balieiro FC, Peixoto RS et al (2013) Mixed plantations can promote microbial integration and soil nitrate increases with changes in the N cycling genes. Soil Biol Biochem 66:146–153
Rajala T, Peltoniemi M, Pennanen T et al (2010) Relationship between wood-inhabiting fungi determined by molecular analysis (denaturing gradient gel electrophoresis) and quality of decaying logs. Can J Forest Res 40(12):2384–2397
Rajala T, Peltoniemi M, Hantula J et al (2011) RNA reveals a succession of active fungi during the decay of Norway spruce logs. Fungal Ecol 4(6):437–448
Rajala T, Peltoniemi M, Pennanen T et al (2012) Fungal community dynamics in relation to substrate quality of decaying Norway spruce (*Picea abies* [L] Karst) logs in boreal forests. FEMS Microbiol Ecol 81(2):494–505
Rajala T, Tuomivirta T, Pennanen T et al (2015) Habitat models of wood inhabiting fungi along a decay gradient of Norway spruce logs. Fungal Ecol 18:48–55
Rakesh M (2013) Chronic N-amended soils exhibit an altered bacterial community structure in Harvard Forest, MA, USA. FEMS Microbiol Ecol 83(2):478–493
Rasche F, Knapp D, Kaiser C et al (2011) Seasonality and resource availability control bacterial and archaeal communities in soils of a temperate beech forest. Int Soci Microb Ecol J 5(3):389
Rayner AD, Boddy L (1988) Fungal decomposition of wood. Its biology and ecology. Wiley, New York
Redford AJ, Bowers RM, Knight R et al (2010) The ecology of the phyllosphere: geographic and phylogenetic variability in the distribution of bacteria on tree leaves. Environ Microbiol 12(11):2885–2893
Reed SC, Townsend AR, Cleveland CC et al (2010) Microbial community shifts influence patterns in tropical forest nitrogen fixation. Oecol 164(2):521–531

Reed BM, Sarasan V, Kane M et al (2011) Biodiversity conservation and conservation biotechnology tools. In vitro Cell Dev Biol-Pl 47(1):1–4
Renvall P (1995) Community structure and dynamics of wood-rotting Basidiomycetes on decomposing conifer trunks in northern Finland. Karstenia 35:1–51
Riley R, Salamov AA, Brown DW et al (2014) Extensive sampling of basidiomycete genomes demonstrates inadequacy of the white-rot/brown-rot paradigm for wood decay fungi. Proc Natl Acad Sci111(27):9923-9928
Rincón A, Pueyo JJ (2010) Effect of fire severity and site slope on diversity and structure of the ectomycorrhizal fungal community associated with post-fire regenerated *Pinus pinaster* Ait seedlings. Forest Ecol Manag 260(3):361–369
Rineau F, Roth D, Shah F et al (2012) The ectomycorrhizal fungus *Paxillus involutus* converts organic matter in plant litter using a trimmed brown-rot mechanism involving Fenton chemistry. Environ Microbiol 14(6):1477–1487
Rodrigues KF (1993) Endophytic species of *Xylaria*: cultural and isozymic studies. Sydowia 45:116–138
Rodriguez RJ, White JF Jr, Arnold AE et al (2009) Fungal endophytes: diversity and functional roles. New Phytol 182(2):314–330
Rosling A, Landeweert R, Lindahl BD et al (2003) Vertical distribution of ectomycorrhizal fungal taxa in a podzol soil profile. New Phytol 159(3):775–783
Rossman AY, Palm-Hernández ME (2008) Systematics of plant pathogenic fungi: why it matters. Plant Dis 92(10):1376–1386
Rotthauwe JH, Witzel KP, Liesack W (1997) The ammonia monooxygenase structural gene amoA as a functional marker: molecular fine-scale analysis of natural ammonia-oxidizing populations. Appl Environ Microbiol 63(12):4704–4712
Rousk J, Baath E, Brookes PC et al (2010) Soil bacterial and fungal communities across a pH gradient in an arable soil. Int Soci Microb Ecol J 4(10):1340
Rousk J, Brookes PC, Baath E (2011) Fungal and bacterial growth responses to N fertilization and pH in the 150-year 'Park Grass' UK grassland experiment. FEMS Microbiol Ecol 76(1):89–99
Rytioja J, Hilden K, Yuzon J et al (2014) Plant-polysaccharide-degrading enzymes from basidiomycetes. Microbiol Molr Biol Rev 78(4):614–649
Sabatier D (1994) Diversité des arbres et du peuplement forestier en Guyane. Gestion de l'écosysteme forestier et aménagement de l'espace régional, pp 41–47
Savi F, Di Bene C, Canfora L et al (2016) Environmental and biological controls on CH4 exchange over an evergreen Mediterranean forest. Agri Forest Meteorol 226:67–79
Schilling JS, Kaffenberger JT, Liew FJ et al (2015) SigNat wood modifications reveal decomposer community history. PloS one 10(3):e0120679
Schloss PD, Delalibera I, Handelsman J (2006) Bacteria associated with the guts of two wood-boring beetles: *Anoplophora glabripennis* and *Saperda vestita* (Cerambycidae). Environ Entomol 35:625–629
Schmalenberger A, Duran AL, Bray AW et al (2015) Oxalate secretion by ectomycorrhizal *Paxillus involutus* is mineral-specific and controls calcium weathering from minerals. Sci Rep 5:12187
Schmidt SK, Lipson DA (2004) Microbial growth under the snow: implications for nutrient and allelochemical availability in temperate soils. Plant Soil 259(1–2):1–7
Schmidt O, Horn MA, Kolb S (2015) Temperature impacts differentially on the methanogenic food web of cellulose supplemented peatland soil. Environ Microbiol 17:720–734
Seibold S, Bassler C, Brandl R et al (2015) Experimental studies of dead-wood biodiversity—a review identifying global gaps in knowledge. Biol Conserv 191:139–149
Selosse MA, Vohník M, Chauvet E (2008) Out of the rivers: are some aquatic hyphomycetes plant endophytes? New Phytol 178(1):3–7
Serkebaeva YM, Kim Y, Liesack W (2013) Pyrosequencing-based assessment of the bacteria diversity in surface and subsurface peat layers of a Northern Wetland, with focus on poorly studied phyla and candidate divisions. PLoS One 8:e63394
Shannon CE, Weaver W (1949) The mathematical theory of communication. Urbana: University of Illinois Press, Champaign

Shaw LJ, Nicol GW, Smith Z et al (2006) *Nitrosospira* spp can produce nitrous oxide via a nitrifier denitrification pathway. Environ Microbiol 8(2):214–222
Shoun H, Kim DH, Uchiyama H et al (1992) Denitrification by fungi. FEMS Microbiol Lett 94(3):277–281
Sieber TN, Rys J, Holdenrieder O (1999) Mycobiota in symptomless needles of *Pinusmugo* ssp *uncinata*. Mycol Res 103(3):306–310
Sieber-Canavesi F, Sieber TN (1993) Successional patterns of fungal communities in needles of European silver fir (*Abies alba* Mill). New Phytol 125(1):149–161
Simard SW, Beiler KJ, Bingham MA et al (2012) Mycorrhizal networks: mechanisms, ecology and modeling. Fungal Biol Rev 26(1):39–60
Simpson EH (1949) Measurement of diversity. Nature 163:688
Song Y, Chen D, Lu K et al (2015) Enhanced tomato disease resistance primed by arbuscular mycorrhizal fungus. Front Plant Sci 6:786
Sprent JL (1995) Legume trees and shrubs in the tropics N2-fixation in perspective. Soil Biol Biochem 27:401–407
Sprent JI (2005) Nodulated legume trees. In: Werner D, Newton WE (eds) Nitrogen fixation in agriculture, forestry, ecology and the environment. Springer, New York, pp 113–141
Sprent JI (2009) Legume nodulation: global perspective. Wiley-Blackwell, Oxford
Stenlid J, Penttila R, Dahlberg A (2008) Wood-decay basidiomycetes in boreal forests: distribution and community development. In: British mycological society symposia series, vol 28. Academic, pp 239–262
Stokland JN, Siitonen J, Jonsson BG (2012) Biodiversity in dead wood Cambridge University Press
Stork NE, Coddington JA, Colwell RK et al (2009) Vulnerability and resilience of tropical forest species to land-use change. Conserv Biol 23:1438–1447
Stursova M, Snajdr J, Cajthaml T (2014) When the forest dies: the response of forest soil fungi to a bark beetle-induced tree dieback. Int Soci for Microb Ecol J 8:1920–1931
Sutherland JM, and Sprent JI (1993) Nitrogen fixation by legume trees Symbiosis in Nitrogen fixing trees. 33-63
Svensson M, Johansson V, Dahlberg A (2016) The relative importance of stand and dead wood types for wood dependent lichens in managed boreal forests. Fungal Ecol 20:166–174
Szink I, Davis EL, Ricks KD et al (2016) New evidence for broad trophic status of leaf endophytic fungi of *Quercus gambelii*. Fungal Ecol 22:2–9
Szukics U, Hackl E, Zechmeister-Boltenstern S et al (2012) Rapid and dissimilar response of ammonia oxidizing archaea and bacteria to nitrogen and water amendment in two temperate forest soils. Microbiol Res 167(2):103–109
Talbot JM, Bruns TD, Taylor JW et al (2014) Endemism and functional convergence across the North American soil mycobiome. Proc Natl Acad Sci 111:6341–6346
Tanaka M, Nara K (2009) Phylogenetic diversity of non-nodulating *Rhizobium* associated with pine ectomycorrhizae. FEMS Microbiol Ecol 69(3):329–343
Taylor AE, Zeglin LH, Wanzek TA et al (2012) Dynamics of ammonia-oxidizing archaea and bacteria populations and contributions to soil nitrification potentials. Int Soci Microb Ecol J 6(11):2024
Tedersoo L, Suvi T, Jairus T (2008) Forest microsite effects on community composition of ectomycorrhizal fungi on seedlings of *Picea abies* and *Betula pendula*. Environ Microbiol 10(5):1189–1201
Tedersoo L, May TW, Smith ME (2010) Ectomycorrhizal lifestyle in fungi: global diversity, distribution, and evolution of phylogenetic lineages. Mycorrhiza 20:217–263
Tedersoo L, Bahram M, Polme S et al (2014) Global diversity and geography of soil fungi. Sci 346:1078–1087
Tedersoo L, Bahram M, Cajthaml T et al (2016) Tree diversity and species identity effects on soil fungi, protists and animals are context dependent. Int Soci Microb Ecol J 10(2):346–362

Ter Steege H, Sabatier D, Castellanos H et al (2000) An analysis of the floristic composition and diversity of Amazonian forests including those of the Guiana Shield. J Tropical Ecol 16:801–828

Thomson BC, Tisserant E, Plassart P et al (2015) Soil conditions and land use intensification effects on soil microbial communities across a range of European field sites. Soil Biol Biochem 88:403–413

Tiedje JM, Sexstone AJ, Myrold DD et al (1983) Denitrification: ecological niches, competition and survival. Anton Leeuw 48(6):569–583

Treseder KK, Lennon JT (2015) Fungal traits that drive ecosystem dynamics on land. Microbiol Mol Biol Rev 79(2):243–262

Treseder KK, Vitousek PM (2001) Effects of soil nutrient availability on investment in acquisition of N and P in Hawaiian rain forests. Ecol 82(4):946–954

Treusch AH, Leininger S, Kletzin A et al (2005) Novel genes for nitrite reductase and Amo-related proteins indicate a role of uncultivated mesophilic crenarchaeota in nitrogen cycling. Environ Microbiol 7(12):1985–1995

Trivedi P, Trivedi C, Grinyer J (2016) Harnessing host-vector microbiome for sustainable plant disease management of phloem-limited bacteria. Front Plant Sci 7:1423

Tuomisto H (2010) A diversity of beta diversities: straightening up a concept gone awry Part 1 Defining beta diversity as a function of alpha and gamma diversity. Ecography **33**:2–22

Turner BL, Yavitt JB, Harms KE et al (2013) Seasonal changes and treatment effects on soil inorganic nutrients following a decade of fertilizer addition in a lowland tropical forest. Soil Sci Soci Am J 77(4):1357–1369

Unterseher M, Reiher A, Finstermeier K et al (2007) Species richness and distribution patterns of leaf-inhabiting endophytic fungi in a temperate forest canopy. Mycol Prog 6(3):201–212

Unterseher M, Persoh D, Schnittler M (2013) Leaf-inhabiting endophytic fungi of European Beech *(Fagus sylvatica L)* co-occur in leaf litter but are rare on decaying wood of the same host. Fungal Divers 60(1):43–54

Urbanova M, Snajdr J, Baldrian P (2015) Composition of fungal and bacterial communities in forest litter and soil is largely determined by dominant trees. Soil Biol Biochem 84:53–64

Uroz S, Calvaruso C, Turpault MP et al (2009) Mineral weathering by bacteria: ecology, actors and mechanisms. Trends Microbiol 17(8):378–387

Uroz S, Oger P, Morin E et al (2012) Distinct ectomycorrhizospheres share similar bacterial communities as revealed by pyrosequencing-based analysis of 16S rRNA genes. Appl Environ Microbiol 78:3020–3024

Uroz S, Ioannidis P, Lengelle J et al (2013) Functional assays and metagenomic analyses reveals differences between the microbial communities inhabiting the soil horizons of a Norway spruce plantation. PLoS One 8(2):e55929

Uroz S, Oger P, Tisserand E et al (2016) Specific impacts of beech and Norway spruce on the structure and diversity of the rhizosphere and soil microbial communities. Sci Rep 6:27756

Valaskova V, De Boer W, Gunnewiek PJK et al (2009) Phylogenetic composition and properties of bacteria coexisting with the fungus *Hypholoma fasciculare* in decaying wood. Int Soci Microb Ecol J 3(10):1218

Van Cleve K, Chapin FS, Dyrness CT et al (1991) Element cycling in taiga forests: state-factor control. BioSci 41(2):78–88

Van Der Heijden MG, Bardgett RD, Van Straalen NM (2008) The unseen majority: soil microbes as drivers of plant diversity and productivity in terrestrial ecosystems. Ecol Lett 11(3):296–310

Van der Wal A, Ottosson E, de Boer W (2015) Neglected role of fungal community composition in explaining variation in wood decay rates. Ecol 96(1):124–133

Van Dyk JS, Pletschke BI (2012) A review of lignocellulose bioconversion using enzymatic hydrolysis and synergistic cooperation between enzymes factors affecting enzymes, conversion and synergy. Biotechnol Adv 30(6):1458–1480

Van Insberghe D, Maas KR, Cardenas E et al (2015) Non-symbiotic *Bradyrhizobium* ecotypes dominate North American forest soils. Int Soci Microb Ecol J 9(11):2435

Vik U, Logares R, Blaalid R et al (2013) Different bacterial communities in ectomycorrhizae and surrounding soil. Sci Rep 3:3471
Vitousek PM, Cassman K, Cleveland C et al (2002) Towards an ecological understanding of biological nitrogen fixation. Biogeochem 57:1–45
Vorholt JA (2012) Microbial life in the phyllosphere. Nat Rev Microbiol 10:828–840
Voriskova J, Baldrian P (2013) Fungal community on decomposing leaf litter undergoes rapid successional changes. Int Soci Microb Ecol J 7(3):477
Vořiškova J, Brabcova V, Cajthaml T et al (2014) Seasonal dynamics of fungal communities in a temperate oak forest soil. New Phytol 201(1):269–278
Walker JK, Cohen H, Higgins LM et al (2014) Testing the link between community structure and function for ectomycorrhizal fungi involved in a global tripartite symbiosis. New Phytol 202(1):287–296
Wallenstein MD, Hall EK (2012) A trait-based framework for predicting when and where microbial adaptation to climate change will affect ecosystem functioning. Biogeochem 109(1-3):35–47
Wang Y, Li C, Wang Q et al (2016) Environmental behaviors of phenolic acids dominated their rhizodeposition in boreal poplar plantation forest soils. J Soils Sediments 16:1858–1870
Wardle DA, Walker LR, Bardgett RD (2004) Ecosystem properties and forest decline in contrasting long-term chronosequences. Science 305(5683):509–513
Wells JM, Boddy L (1995) Phosphorus translocation by saprotrophic basidiomycete mycelial cord systems on the floor of mixed deciduous woodland. Mycol Res 99(8):977–980
Whittaker RH (1960) Vegetation of the Siskiyou mountains, Oregon and California. Ecol Monogr 30(3):279–338
Wu F, Yang W, Zhang J et al (2010) Litter decomposition in two subalpine forests during the freeze–thaw season. Acta Oecol 36(1):135–140
Wubet T, Christ S, Schoning I et al (2012) Differences in soil fungal communities between European beech (*Fagus sylvatica* L) dominated forests are related to soil and understory vegetation. PloS One 7(10):e47500
Xia M, Talhelm AF, Pregitzer KS (2015) Fine roots are the dominant source of recalcitrant plant litter in sugar mapledominated northern hardwood forests. New Phytol 208:715–726
Yahara T, Javadi F, Onoda Y et al (2013) Global legume diversity assessment: concepts, key indicators, and strategies. Taxon 62(2):249–266
Yamashita S, Hattori T, Abe H (2010) Host preference and species richness of wood-inhabiting aphyllophoraceous fungi in a cool temperate area of Japan. Mycol 102(1):11–19
Zehr JP, Jenkins BD, Short SM et al (2003) Nitrogenase gene diversity and microbial community structure: a cross-system comparison. Environ Microbiol 5(7):539–554
Zhalnina K, de Quadros PD, Camargo FA et al (2012) Drivers of archaeal ammonia-oxidizing communities in soil. Front Microbiol 3:210
Zhou J, Deng Y, Shen L et al (2016) Temperature mediates continental-scale diversity of microbes in forest soils. Nat Commun 7:12083
Žifčáková L, Dobiasova P, Kolarova Z et al (2011) Enzyme activities of fungi associated with *Picea abies* needles. Fungal Ecol 4(6):427–436
Žifčáková L, Vetrovsky T, Howe A (2016) Microbial activity in forest soil reflects the changes in ecosystem properties between summer and winter. Environ Microbiol 18:288–301
Zinger L, Lejon DP, Baptist F et al (2011) Contrasting diversity patterns of crenarchaeal, bacterial and fungal soil communities in an alpine landscape. PLoS One 75:5863–5870

Engineering Rhizobacterial Functions for the Improvement of Plant Growth and Disease Resistance

15

P. Jishma, A. Remakanthan, and E. K. Radhakrishnan

1 Rhizobacteria: Growth Boosters and Defense Elicitors of Plants

Plants interact with microorganisms of diverse species and the same at the rhizosphere region is highly remarkable. The rhizosphere region forms an interface between plant root and soil and has strong microbial activity which is due to the direct or indirect influence by the root. The term rhizosphere comes etymologically from the words "rhiza" which means root and "sphera" which means surroundings. This region has also been categorized into exorhizosphere which contains soil associated with the root after vigorous shaking, rhizoplane that corresponds to interface between soil and root and the endorhizosphere forming the intercellular space between the root tissues occupied by bacteria (Bowen and Rovira 1999).

Rhizosphere region is the zone of vital and powerful interactions among plant, soil, and associated microorganisms. Many previous studies have reported various biochemical communications between plants and microorganisms at the rhizosphere (Watt 2009). The biological and physicochemical characters of the rhizosphere mainly depend on the liberation of root exudates known as rhizodeposition. These exudates are mobilized through the root cell membrane and oozed out to the rhizosphere. The composition and nature of these depend on the species of plant, stage of development, nutritional status, soil type, and environmental conditions, such as temperature, light intensity, and soil water potential (Kochian et al. 2004; Noumavo et al. 2016).

P. Jishma · E. K. Radhakrishnan (✉)
School of Biosciences, Mahatma Gandhi University, Kottayam, Kerala, India
e-mail: radhakrishnanek@mgu.ac.in

A. Remakanthan
Department of Botany, University College, Thiruvananthapuram, Kerala, India

D. P. Singh et al. (eds.), *Microbial Interventions in Agriculture and Environment*,
https://doi.org/10.1007/978-981-13-8383-0_15

The rhizosphere is biochemically significant as it forms a distinct resource of nutrients like sugars, amino acids, organic acids, hormones (Badri et al. 2009), enzymes, nucleotides, fatty acids, sterols, growth factors, flavonoids, and other micromolecules released from the roots of plants. These rhizodeposits function as chemoattractants and nutrients for the rhizobacteria and mediate their colonization and biofilm formation. However, the difference in chemical profile of root exudome can have determining effect on diversity and density of rhizobacteria. As a result of the synergistic and favorable interactions of rhizobacteria and its impact on overall growth enhancement with associated protection to plants, they have been given the name plant growth-promoting rhizobacteria (PGPR) (Rout and Callaway 2012). Because of their extremely adapted plant beneficial mechanisms, PGPR also have molding effect on soil characteristics which thereby can convert infertile land into cultivable land (Fig. 15.1). Many previous reports have illustrated the augmented plant beneficial property of biofilm-forming PGPR when compared to non-biofilm candidates (Singh et al. 2014). The resistance mechanisms provided by biofilm-forming organisms were also reported to be remarkable (Lee et al. 2013). Due to wide array of plant beneficial mechanisms present in PGPR, they have also been considered as growth boosters and defense elicitors of plants.

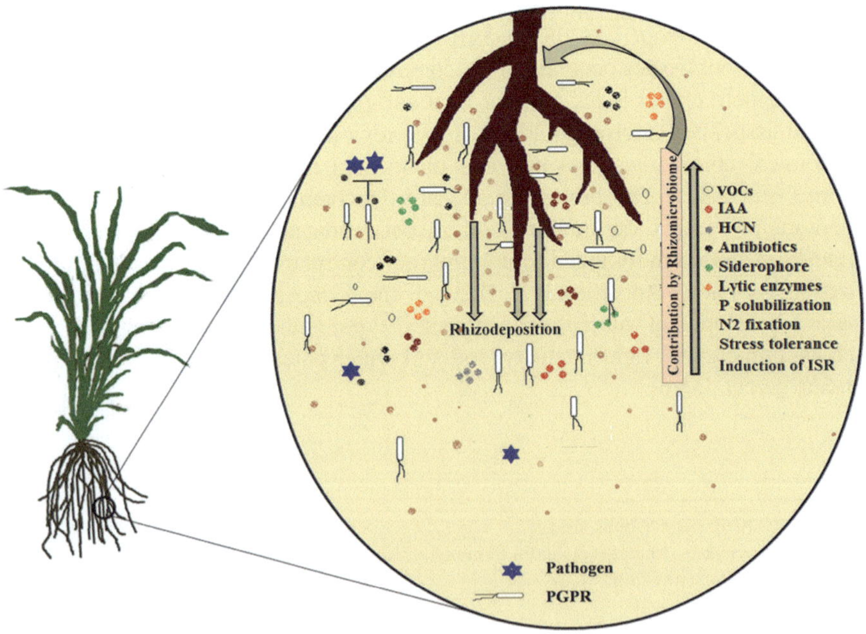

Fig. 15.1 Rhizobacterial functions to promote plant growth and disease resistance

2 Designing the Architecture of Rhizosphere by Plants and Rhizobacteria

Plants design the spatial construction of the soil through the growth and expansion of their roots. This also can design the chemical composition of the rhizodeposition and thereby the PGPR association (Lynch and Whipps 1990; Philippot et al. 2013). According to a previous study, 20–50% of photosynthate is considered to get transported from plant to belowground (Jones et al. 2009). In response to this chemical exudation, rhizomicrobiome provides support to plants through mobilizing soil nutrients, producing phytohormone, and providing tolerance to stress components like heat, high salt, drought, and phytopathogens (Yang et al. 2009; Haney et al. 2015). The rhizomicrobiome has also been established to perform decaying of organic matters in soil through biogeochemical cycling, soil formation, and augmentation of soil fertility (Breidenbach et al. 2016).

Various factors affect the microbial communities associated with plants (Jasim et al. 2016). A detailed study on same from maize showed difference between the rhizosphere and bulk soil microbial community, but more similar trend of rhizosphere communities was observed for plants grown in diverse soils (Peiffer et al. 2013). However, soil geochemistry including the soil nutrients can have role on the structure of rhizosphere community as per specific plant (Berg et al. 2014; Edwards et al. 2015).

Species and age of the plants, soil nutrients, soil type, and rhizodeposition are the general factors demonstrated to influence rhizobacteria. Rhizodeposition include the organic materials secreted and discharged by roots of plants to their surroundings. These can have the components like carbohydrates, amino acids, plant hormones, organic acids, vitamins, phenolics, sugar phosphate esters, ions, and many other carbon-containing secondary metabolites. The high-molecular-weight exudates include enzymes, proteins, and mucilage (polysaccharides) (Bais et al. 2006). These exudome components can serve as signaling molecules (Prashar et al. 2013) and have a central role in enhancement of chemotaxis in the soil by attracting more rhizobacteria and favoring their colonization on the plant roots. Hence, exudates from plant roots can have determining effect on rhizobacterial species which are get colonized on root.

Plants also prefer specific microorganisms to colonize within its tissues. In a study with *Arabidopsis thaliana*, microbial community associated with the plant has low diversity than bulk soil but was comparable in plants grown in geochemically different soils (Lundberg et al. 2012). Variation in metabolic activity and production of metabolites by plants at different stages of development or age can also determine the species of bacteria symbiotically associated with it. The alteration from seedling, flowering plant, to a plant in senescence may influence the microbial community composition (Houlden et al. 2008). Hence, all these determinants have significant role in transforming the architecture of the rhizosphere. The role and function of the chemical exudates also can be determined by properties of soil. The chemical secretions from recruited rhizobacteria may also manipulate the chemical architecture of rhizosphere.

3 Unculturable Rhizobacteria

As per the data available from the previously published reports, main fraction of the microbial community is unculturable which may have several beneficial traits, including the plant growth-promoting activities for sustainable agricultural production. Exploitation of such unculturable microorganisms can boost the understanding on currently used organic agricultural practices (Goel et al. 2018). As the ecological and functional roles of such unculturable bacteria are not completely known, information about natural composition of rhizosphere microorganisms is primitive. Eventhough the causes of their unresponsiveness to culture methods can be diverse, some may need other organisms for critical processes involved in survival (Tringe and Rubin 2005). This further adds complexity to rhizosphere.

Even though many microbes are unculturable in the laboratory, they can be active in rhizosphere functions (Rappe et al. 2002). Culturing of these microbes in artificial environment requires detailed information about their nutrient demand. Although methods have been used to convert the unculturable to cultivable (Sait et al. 2002; Zwirglmaier et al. 2004), these techniques are laborious and time-consuming. However, metagenomic approaches can characterize the unknown rhizobacterial genome to explore their agricultural promises. The advanced analyses of the microbial composition of rhizosphere soil can provide in-depth knowledge on the rhizomicrobiome specificity of plants and their applications (Lagos et al. 2015).

4 Rhizomicrobiome Diversity Analysis by Next-Generation Sequencing (NGS)

Soil is an unexplored wealth of metabolically active microbiome where microorganisms are expected to consist of less than 5% of the whole space occupied. Most of the microbial activity observed in the rhizosphere makes it to be a biochemical hot spot which is favored by the occurrence of high amount of organic matter (Pinton et al. 2001). Methods for studying diversity of rhizomicrobiome are cultivation-dependent or cultivation-independent. Traditional means to study microbial diversity are based on isolation and cultivation of microbes. With the emergence of functional genomics, it became a powerful platform for discovering novel genomic functions (Ahmad et al. 2011).

Techniques like PCR, quantitative RT-PCR, and DNA microarray have also been explored for identification and quantification of bacteria present in the rhizosphere (Shaw et al. 2015). However, these molecular-based methods rely on specific oligonucleotide probes, and the known sequences of the microbes are detected (Ye and Zhang 2011). In addition, these molecular techniques have limitations to identify diversity and abundance of all culturable and unculturable rhizobacteria. Recent technological advances in next-generation sequencing (NGS) offer better outlook of rhizosphere microorganisms and their diversity (Orlando et al. 2015). Currently, new sequencing methods have became important step for metagenomics analysis. NGS can quickly generate large amounts of DNA reads, and the technique is with reasonable cost (De Mandal and Panda 2015). Analysis of environmental genetic resources by this has resulted in the observation of numerous genes having promises for agricultural use, for industrial application, and also for pharmaceutical studies. The results can also indicate diversity, gene transfer, novel metabolic fractions, and genetic basis of enzymes such as chitinases and other compounds (Handelsman 2004; Elend et al. 2006; Vakhlu et al. 2008).

NGS provides rapid sequencing of millions of DNA sequences parallely which thereby reduce the cost compared to method of Sanger. This does not involve cloning step, as the NGS library synthesis is being done using DNA. So the output sequence data obtained can be used for analysis (Kircher and Kelso 2010). Any NGS study pursues several general steps like DNA purification, construction of library, the process of sequencing, analysis, and functional categorization (Field et al. 2008; Metzker 2010). Two types of NGS approaches used for analysis of microbial community are the deep amplicon-based sequencing and total analysis of metagenome or metatranscriptome. Shotgun sequencing method has also been used for the identification of microbial community. The common analysis process mainly involves assembly, binning, and prediction of function. However, amplicon sequencing has demonstrated to have use to profile community by using signature genes such as 16S rRNA from various sources (De Mandal and Panda 2015).

Proper knowledge from the transcriptome and proteome analysis is critical for the functional prediction of genes. Such efforts undeniably contribute to major progresses in agrigenomics (Papajorgji 2009; Van Emon 2016). Metagenomic analysis in agriculture has also demonstrated to have application to describe complex communications among the microorganisms (Carbonetto et al. 2014) in the rhizosphere (Mendes et al. 2014). Metagenomics also shown to have use to predict changes in composition and function of rhizosphere microorganisms with environmental changes and agricultural management (Bevivino et al. 2014; Souza et al. 2015). The array of applications of metagenomics also involve interpretation of functions of soil bacteria in plant sustenance (Lavecchia et al. 2015; Pii et al. 2016), in the cycle of the elements (Stempfhuber et al. 2015), and in the detection of new mechanisms for growth promotion in plants (Pandey et al. 2014).

5 Species-Specific Variation in Plant Growth-Enhancing Mechanisms

Exploration of interactions between plant and beneficial microorganisms can have applications to develop eco-friendly approach to reduce adverse plant stress caused by biotic and abiotic factors. Application of PGPR is a useful natural method to reduce the plant stress and is now widely used. Plants inoculated with rhizobacteria develop morphological and biochemical modifications to manage stress by induced systemic tolerance (Etesami and Maheshwari 2018). Usually, plant growth-promoting rhizobacteria make possible the enhanced plant growth by supporting the resource acquisition, altering the plant hormone levels, and declining the inhibitory effects of diverse pathogens as biocontrol agents (Ahemad and Kibret 2014). For this, PGPR produce ACC (1-aminocyclopropane-1-carboxylate) deaminase that can minimize the production of stress hormone ethylene, modify phytohormonal content, induce the production of plant antioxidative enzymes, enhance the uptake of necessary mineral elements, produce extracellular polymeric substance (EPS), decrease the absorption of excess nutrients or heavy metals, and induce expression of abiotic stress resistance genes (Etesami and Maheshwari 2018). However, all these properties are not exhibited by all the strains.

Viruses, bacteria, fungi, oomycetes, algae, archaea, nematodes, protozoa, and arthropods are the microbial communities inhabited in the rhizosphere. Rhizodeposits are the important driving force that regulates the diversity and activity of microorganisms in the rhizosphere. Plants can transform the rhizomicrobiome by exciting them selectively with traits which are favorable to plant health and growth. Various symbiotic (*Rhizobium* spp., *Bradyrhizobium* spp., and *Mesorhizobium* spp.) and nonsymbiotic (*Pseudomonas* spp., *Serratia* spp., *Bacillus* spp., *Azotobacter* spp., *Azospirillum* spp., and *Azomonas* spp.) rhizobacteria are now used worldwide as bioinoculants to augment growth and development of plants under stressed conditions (Ahemad and Khan 2010). These species support growth of plants and protect them from phytopathogens through various mechanisms.

5.1 Symbiotic Rhizobacteria

The symbiotic rhizobacteria including *Rhizobium* spp., *Bradyrhizobium* spp., and *Mesorhizobium* spp. augment the plant growth and soil fertility mainly by biological nitrogen fixation (BNF). This group of organisms colonize plant root and resides within the root nodule to convert inert atmospheric nitrogen (N_2) to utilizable form of ammonia (NH_3). Also, the bacteria can develop malate in plants as source for food and energy for the fixation of nitrogen (Spaink 2000). Biological nitrogen fixation actually occurs in the nodule symbiosome by the bacteroids using the extremely oxygen-sensitive enzyme complex, namely, nitrogenase (Caetano-Anolles and Gresshoff 1991). When ammonia is formed, the glutamate is acted to form glutamine, the common nitrogen source by glutamine synthase. This is either transformed to asparagine or exported to the nearby uninfected cell to get changed to the ureides, allantoin, and allantoic acid in its peroxisome. These four nitrogenous compounds can act as nitrogen source for the plants. The importance of biological nitrogen fixation by rhizobacteria is that it reduces the dependence on nitrogen-containing fertilizers (Graham and Vance 2003; Stacey et al. 2006).

Additionally, these symbiotic rhizobacteria are also involved in the production of indole-3-acetic acid (IAA), cytokinin, hydrogen cyanide (HCN), siderophore, and phosphate solubilization (Deshwal et al. 2003). IAA and cytokinins are growth-regulating phytohormones that promote division, elongation, differentiation, and extension of cells in plant roots and shoots. Their main function is on cell growth and differentiation and also they help the plant by delaying the senescence or aging of tissues (Sipes and Einset 1983; Spaepen et al. 2007). HCN is expected to inhibit electron transport chain and energy supply to cell, directing to the destruction of pathogenic cells (Dilkes et al. 2007). In the case of stress mediated by heavy metal, siderophores provide support to the plants to manage the stress. Besides scavenging iron from the surrounding and making the mineral available to the microbe, siderophores also have precious role in generating pathogen resistance in plants by limiting iron supply to pathogens. The form of phosphorus that the plant is needed for most of the growth and development is made available by the action of phosphate-solubilizing bacteria (PSB). Hence, introduction of PSB into the soil can increase the growth of plants and crop yield directly (Rivas et al. 2007). PSB lower the rhizospheric pH and solubilize the soil phosphate by producing low-molecular-weight organic acids such as gluconic and ketogluconic acids. Phosphorus solubilization capability of PSB has been directly related with the soil pH (Goldstein 1995).

5.2 Nonsymbiotic Rhizobacteria

5.2.1 *Pseudomonas* spp.

Pseudomonas spp. are ubiquitous organisms present in agricultural soil and are well adapted to exist in the rhizosphere. They have many traits to augment the plant growth and to function as biocontrol agent. They make use of root exudates and nutrients in the soil and colonize and multiply in the rhizosphere. At the same time, they generate range of bioactive metabolites, such as antibiotics, siderophores, volatile organic compounds, and growth-promoting substances in the rhizosphere. By this, they compete aggressively with other microorganisms. Also, Pseudomonads have a key role in the natural suppressiveness of many soilborne pathogens (Jimtha John et al. 2017). They also produce β-1,3-glucanase and chitinases like hydrolytic enzymes and through other metabolites such as phytoalexins they stimulate systemic resistance (ISR) (David et al. 2018). Important bioactive metabolites specifically produced by *Pseudomonas* spp. are phenazine derivatives such as phenazine-1-carboxylic acid (PCA), phenazine-1-carboxamide (PCN), 5-methyl-phenazine-1-carboxylic acid (5MPCA), 1-hydroxy-phenazine (1OHPZ), and pyocyanin and pyoluteorin, 2,4-diacetylphloroglucinol (Chen et al. 2015), and volatile organic compounds (Jishma et al. 2017). These compounds can inhibit the growth and establishment of phytopathogens by diverse mechanisms. Also, *Pseudomonas* spp. promote plant growth and development by producing IAA, siderophore, HCN, and solubilizing phosphate (Dey et al. 2004). *Pseudomonas fluorescens*, *Pseudomonas putida*, *Pseudomonas rhizophila*, *Pseudomonas taiwanensis*, *Pseudomonas monteilii*, and *Pseudomonas rhodesiae* are some of the PGPR which control plant diseases and augment plant growth (Jishma et al. 2017; David et al. 2018; Hassen et al. 2018).

5.2.2 *Bacillus* spp.

Bacillus spp. are widely used PGPR and were one of the first commercialized bacterial fertilizers. This has been successfully applied to amplify the yield of crop plants (Kang et al. 2015). Biocontrol property is the predominant characteristics of *Bacillus* spp. and is due to its ability to synthesize and release hydrolytic enzymes affecting the cell wall (Chernin and Chet 2002; Jasim et al. 2016b), iron-chelating siderophores, and several cyclic lipopeptides such as surfactin, iturin, fengycin (Jasim et al. 2016a; Jasim et al. 2016b), and volatile organic compounds (Yuan et al. 2012). These metabolites have antibacterial and antifungal activities against phytopathogenic microorganisms such as *Rhizoctonia* spp., *Pythium* spp., *Aspergillus* spp., *Botrytis cinerea*, and *Sclerotinia sclerotiorum* (Saidi et al. 2009). *Bacillus* spp. act as boosters of plant growth and activate plant defense mechanism by triggering the induced systemic resistance (ISR) (Compant et al. 2005). Some of the potential *Bacillus* species include *B. subtilis*, *B. amyloliquefaciens*, *B. polymyxa*, *B. mojavensis*, *B. vallismortis*, and *B. pumilus*. These generally have the potential to produce plant growth-promoting IAA, siderophore, and HCN. They can also solubilize phosphate and produce various antifungal compounds (Rajkumar et al. 2008; Shafi et al. 2017). Not all the strains may have the ability to produce all these compounds simultaneously, but certain strains can produce two or more compounds, and such strains are more potent to control plant diseases. All these properties of *Bacillus* spp. make it a suitable candidate for biocontrol formulation development that can minimize the application of chemical fertilizers and fungicides.

5.2.3 *Serratia* spp.

The rhizobacteria *Serratia* spp. commonly found in soil and they elicit ISR to inhibit phytopathogens and promote plant growth under both greenhouse and field conditions. They produce salicylic acid, volatile compounds, quorum-sensing autoinducers, and siderophores as determinants toward plant health (John and Radhakrishnan 2017). The capability to synthesize chitinase extracellularly has been described to be essential for *Serratia spp.* to function as biocontrol agent to suppress *Sclerotium rolfsii* (Compant et al. 2005). Among the PGP traits, *Serratia* spp. have members with phosphate solubilization and production of indole acetic acid, ACC deaminase, siderophore, and ammonia (Geros et al. 2016). Pigmented *Serratia* spp. produce the colored compound prodigiosin which can also inhibit wide range of fungal phytopathogens (Lapenda et al. 2015; John Jimtha et al. 2017). They also produce other important bioactive metabolites such as rhamnolipid and pyrrolnitrin. All these compounds have major role in controlling plant diseases by destroying the fungal phytopathogens (Chakraborty et al. 2010; John and Radhakrishnan 2017). *S. plymuthica*, *S. marcescens*, and *S. rubidaea* are some of the important rhizobacterial species of *Serratia* with multiplant growth-promoting traits (Zaheer et al. 2016).

5.2.4 *Azotobacter* spp., *Azomonas* spp., and *Azospirillum* spp.

Azotobacter spp. and *Azomonas* spp. belong to Azotobacteriaceae and are highly flexible in utilizing diverse carbon sources. Hence, the application of carbon-containing materials to the soil improves the nitrogen-fixing ability of diazotrophs. Azotobacters can form cysts which contain novel lipids. They have more than one type of nitrogenase enzymes and are very tolerant to oxygen due to the respiration protection of nitrogenase. *Azotobacter* spp. also have the ability to produce vitamins and growth substances which improve seed germination and phytohormone production resulting in enhanced root growth and nutrient absorption. Their phytopathogen inhibition is mediated by production of antifungal compounds and siderophores (Wani et al. 2016).

Azospirillum spp. are able to colonize hundreds of plant species and encourage their growth, development, and productivity. Free nitrogen fixation and additive mechanisms related to phytohormone production and other related molecules are considered to be the basis of its plant growth promotion effects on inoculated plants (Okon 1985). The biologically active plant regulators synthesized by *Azospirillum* sp. enable enhanced plant growth and plant disease management.

6 Manipulation of Rhizosphere for Improved Plant Growth and Health

The rhizosphere, an integrated network of soil, plant roots, and soil microbiome, can be manipulated or engineered for enhancing plant growth-promoting and disease resistance functions and thereby sustainable agriculture. This approach can be possible with the use of engineered rhizobacteria, plant selection, and soil inoculation (Ahkami et al. 2017). Biotechnological methods can generate rhizobacteria with novel phenotypes that help the plants to tolerate stressed conditions. Plants and rhizobacteria can be genetically engineered to synthesize and release exogenous compounds that advance plant nutrition, limit pathogenic microbes, and diminish the consequences of abiotic or biotic stresses (Ryan et al. 2009). These rhizobiotechnology can have utility as low-cost, sustainable, and environment-friendly technology for management of plant stress and disease (Gouda et al. 2018).

6.1 Engineering of Rhizobacteria

Bioengineering of rhizobacterial communities for plant growth promotion, disease resistance, and stress tolerance have significant applications. Even though countless bacterial strains have been identified to have many plant beneficial traits, all the strains may not exhibit expected outcome when introduced into the agricultural field. Two important factors that can influence the outcome of rhizobacterial application are their root colonization activity and mode of interaction with other microbial populations present in the applied field (Grosskopf and Soyer 2014). Engineering of advanced or novel features to PGPR strains for improved plant beneficial traits can be possible by genetic manipulations. By this, beneficial or targeted gene from a rhizobacteria can be inserted into most suitable and adaptable indigenous rhizobacteria. The genetically engineered rhizobacteria can further be used for multiplication and elevated expression of desired genes, which make it to be suitable for plant application.

The earlier studies with genetically engineered strains showed enhanced plant growth-promoting properties and ability to remove toxic heavy metals from the contaminated areas. A genetically modified *Pseudomonas* sp. with *chiA* gene from *Serratia marcescens* encoding chitinase production has been demonstrated for its acquired ability to suppress *Gaeumannomyces graminis* var. *tritici* and *Fusarium oxysporum* var. *redolens* more efficiently (Sundheim et al. 1988). Chitinase is known as a fungal cell wall lytic enzyme, which can degrade chitin of fungal phytopathogens. Thus, its activity results in the destruction of pathogen and thereby control of plant disease. Also *Pseudomonas fluorescens* CHA0 altered with gene *acdS* responsible for ACC deaminase from *P. putida* UW4 has been proved to augment the root length in canola seedlings. It has also proved to provide protection to cucumber seedlings against *Pythium* sp. infection (Wang et al. 2000). ACC deaminase catalyzes the cleavage of precursor of ACC. By decreasing ethylene concentration to very low level, plants can be protected from severe tissue damage due to the abiotic stress. Even though ethylene has role in plant physiology, its increase related with stress can damage physiological processes of plants and exaggerate disease symptoms (Glick et al. 2007). A cadmium-resistant *Pseudomonas aeruginosa* engineered with metallothionein gene has been demonstrated to exhibit the ability to adsorb Cd^{2+} through extracellular deposition and was found to have an increased ability for immobilizing Cd^{2+} ions from the external source. Application of genetically modified strain in cadmium-polluted soil was found to significantly enhance the biomass of plant and chlorophyll content in leaf (Huang et al. 2016). These reports clearly exposed the promises of genetic engineering of rhizobacteria.

Azospirilla, *Agrobacteria*, and *Rhizobia* have been conventionally applied in fields by seed priming which lead to enhanced yield of different crops, while biopriming with other inoculants such as Pseudomonads fail to produce expected yield under certain environmental conditions (Stewman and Lincoln 1981). To overcome this limitation, bacteria genetically modified with stress-tolerant and plant growth-promoting traits can be constructed and released. Here improved expression of beneficial traits under any adverse environmental condition can significantly improve plant yield (Fig. 15.2) (Amarger 2002).

The genomic alteration of naturally inhabiting rhizobacterial strains with genes responsible for plant growth-promoting traits will diminish the attack of pathogens and there by plant diseases. This will make possible the application of a single inoculant with multimodes of action to augment crop yield by minimizing the alternative inputs for protection of plants from diseases. Hence, consideration of the secure application of such favorable rhizobacteria will be an advantage to the environment and farming community (Nakkeeran et al. 2006).

7 Rhizobacteria-Based Products

Environment-friendly agricultural practices highlight the necessity to retain rhizospheric microbial balance by the application of PGPR (Jambhulkar and Sharma 2014). Bioformulations prepared from PGPR should have improved stability and shelf life to protect the bacteria against diverse environmental conditions. Biopriming of PGPR to augment crop yield and control disease is dependent on the commercial formulation development with appropriate carriers which maintain significant survival period for candidate bacteria.

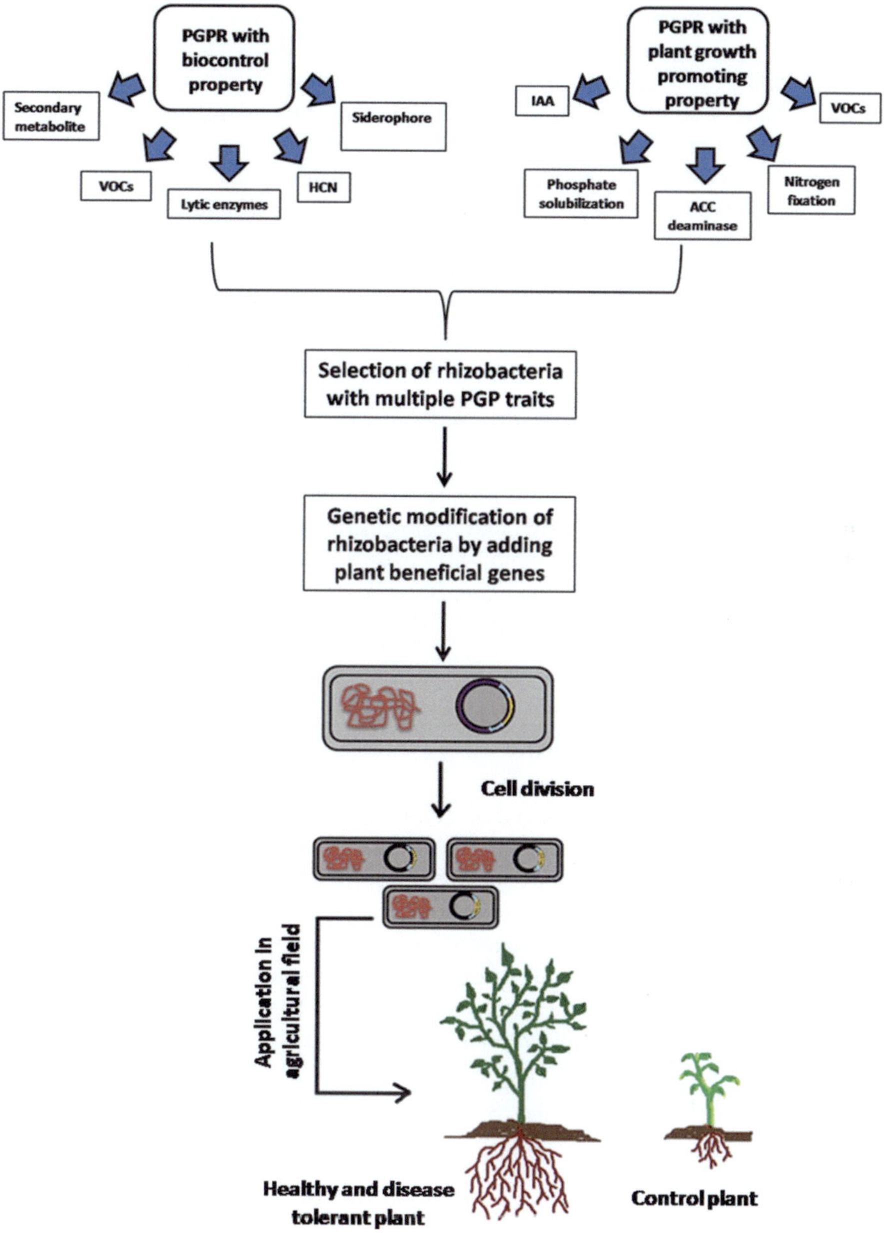

Fig. 15.2 Manipulation of rhizobacteria by recombinant DNA technology

Based on the potential biocontrol ability of *Pseudomonas* and *Bacillus* strains, numerous products are presently accessible in the international markets. Some of these products include Blightban A5061 (*P. fluoresens* A506), Cedomon (*P. chlororaphis* strain), Companion (*B. subtilis* GB03), Conquer (*P. fluoresens*), Epic (*B. subtilis*), Intercept (*P. cepacia*), and Serenade (*B. subtilis*). These PGPR strains can elicit resistance in plants by elevating the levels of phenylalanine ammonia lyase (PAL), peroxidase (PO), and polyphenol oxidase (PPO) (Kavitha et al. 2005). In India, around hundreds of public and private companies are involved in biofertilizer production with microbial consortia. Currently, biofertilizer production and consumption have gained importance in India, and there is an increase in organic agriculture practice in the country with around 1,000,000 ha (Garcia-Fraile et al. 2015). PGPR-based products available in Indian market include Gmax PGPR (*Azotobacter* sp., phosphobacteria, *Trichoderma viride*, and *Pseudomonas fluorescens*), P-SOL (phosphate-solubilizing bacteria), Mn-Sol (*Penicillium citrinum*), VAM (vesicular arbuscular mycorrhiza) biofertilizer, phospho-Bio-Com (phosphate-solubilizing bacteria), Bio-K and K-Sol (potash-mobilizing bacteria), Life, Biomix, Biozink, Biodine, Jet 9, Premium EMC and Calosphere (PGPR consortia), Calspiral (*Azospirillum* and PGPR), Symbion-N (*Azospirillum*, *Rhizobium*, *Acetobacter*, *Azotobacter*), Bio Power (*Azospirillum*, *Azotobacter*, VAM, PSB), and Bio Super (*Pseudomonas* sp., *Cellulomonas* sp., *Bacillus* sp., *Rhodococcus* sp.) (Sekar et al. 2016).

7.1 Selection of Bacteria

Though plant growth promotion and biocontrol using PGPR is safer and acceptable approach, the percentage of exploration of such beneficial strains as biocontrol agents for commercial use is extremely low. Bioformulation development with enhanced shelf life and multimode of action associated with reliable performance in agricultural field can favor commercialization of more PGPR-based products (Sendhilvel et al. 2007). Selection of PGPR for the preparation of bioformulation is a significant factor which determines the outcome of the product. Hence, selected PGPR should have certain characteristics like high saprophytic competency, ease for mass multiplication, high rhizosphere fitness, increased ability to promote plant growth, multiple range of action, excellent and consistent control, safety to ecosystem and high compatibility with other microorganisms present in the rhizosphere, etc. (Upadhyay et al. 2001).

7.2 Formulation Development

Though PGPR have an extremely superior ability to control pests and diseases, they could not be applied as culture suspension directly to the agricultural fields because of the unpredictable nature and environment of soil. Hence, sometimes expected outcome is hard to attain. Climatic fluctuations can also have a central role on the effectiveness of rhizobacteria, and unfavorable growth conditions in the agricultural field decline the functioning of PGPR (Zaidi et al. 2009). Hence, the rhizobacterial cells should be immobilized in suitable carriers and that should be developed as formulations for proper storage, field application, and commercialization. The carriers used for the preparation of bioformulation should satisfy certain properties such as longer shelf life, nontoxic to the plants, water solubility and easy release of cells, cost-effectiveness, etc.

The carrier is the major portion of the bioformulation which helps to release a suitable number of PGPR in better physiological condition. Hence, carrier is the vehicle that deliver live bacteria from the industry into the field. Also carrier should provide a consistent and suitable environment for the survival of PGPR that determine the shelf life of the formulation at room temperature. In addition to this, good carrier should have the capacity to absorb moisture, should be autoclavable or sterilizable by gamma irradiation, should be not lump-forming materials, should have pH buffering capacity, should be easily processable, and should be at low cost (Kakvan et al. 2013). Smaller-sized carrier with enhanced surface area can resist desiccation of rhizobacteria in the formulation by maximum coverage of cells. Organic and inorganic easily available material can be used as carrier which include talc, peat, zeolite, lignite, alginate, kaolinite, pyrophyllite, turf, montmorillonite, sawdust, press mud, and vermiculite (Mahalingam et al. 2017).

Talc, also known as steatite or soapstone, is composed of different minerals with chloride and carbonate. It is chemically magnesium silicate ($Mg_3Si_4O_{10}(OH)_2$) and is produced in industries in powder form which is suitable for many applications. Talc is used as a potential carrier due to its inert nature and accessibility from soapstone industries. A recent study reported the application of talc-based bioformulation of *Bacillus* sp. to have the potential to enhance germination of seeds and growth of *Vigna unguiculata* and *Abelmoschus esculentus*. In this study, bioformulation was developed by the mixing of biomass collected from 48-h-old bacterial culture suspension into 1 kg talc, 1 g carboxy methyl cellulose (CMC), and 15 g calcium carbonate ($CaCO_3$) (Basheer et al. 2018). In this method, CMC was used as the binder and $CaCO_3$ as the buffer (Mahalingam et al. 2017).

7.3 Application of Bioformulation

Bioformulations can maintain the viability of potential rhizobacterial strains for months by protecting them from adverse environmental conditions. There are different modes of applications of bioformulations in the field such as biopriming of seeds, foliar spray, seedling dip, and soil drenching.

Seed priming involves the coating of seeds with bioformulations before planting. This treatment helps to suppress the growth and establishment of phytopathogens through seeds. Sutruedee et al. (2013) reported the efficacy of bioformulation of *Pseudomonas fluorescens* to control fungal pathogens *Cercospora oryzae*, *Curvularia lunata*, *Alternaria padwickii*, *Helminthosporium oryzae*, *Fusarium semitectum*, and *Sarocladium oryzae* by applying as seed treatment (Sutruedee et al.

2013). A successful biocontrol agent should possess the ability to colonize the rhizosphere region during germination stage of seeds. Also, biopriming with rhizobacteria can enhance the seed germination rate and improve establishment of seedlings in the soil. Biopriming of seeds with biocontrol agents can prevent pathogens. In many crops, phytopathogens enter into the plant through seed, root, or aerial parts. The soilborne pathogens enter through the root and establish host-pathogen relationships. Hence, rhizosphere region can be protected by prior release and colonization of PGPR that can inhibit the establishment of pathogens.

Application of PGPR by dipping the seedlings in talc-based formulation suspension has been demonstrated to suppress disease incidence (Matthysse and McMahan 1998). As soil has the collection of both beneficial and harmful microbes, release of PGPR strains directly to the soil can enhance the population density of bacterial antagonists and thereby inhibit the survival of pathogens. So bioformulation directly added to the soil as drenching can inhibit the survival of pathogens in the soil.

The efficiency of biocontrol agents to suppress foliar diseases is significantly varied by instability of climate. Phyllosphere is subjected to deviation in temperature, rain, relative humidity, dew, radiation, and wind. Hence, water potential of microbes present in the phylloplane will be changeable frequently. Also the concentration of organic acids, amino acids, and sugars discharged through different pores on the leaves like stomata, lenticels, hydathodes, and wounds is highly variable. This affects the survival and efficiency of antagonist in the phylloplane. Hence, release of PGPR to the leaves can compete for nutrients actively on the surface of the leaf and thereby inhibit the germination of spores and growth of pathogens (Matthysse and McMahan 1998). As spores of pathogens mainly enter into the plant through stomata, the foliar spray of PGPR can inhibit its entry and establishment into the internal tissues.

Though PGPR perform well in the plant growth promotion and plant disease management, they are extremely susceptible to the variations in the environmental circumstances and are unpredictable in their efficiency. The performance of biocontrol agents can be boosted through simple approaches such as preparation of compatible consortia, using strains that elevate synergistic expression of biocontrol genes; use of spreaders, adjuvants, and stickers; application of genetically engineered PGPR strains; etc. For achieving agricultural sustainability, it is significant to understand the basics of rhizobacterial functions so that engineering the factors can also be done to maintain field competency of PGPR.

8 Conclusion

The utilization of PGPR inoculants to progress agricultural production has been established widely as an approach for sustainable agriculture. PGPR based on their means of action can be grouped as biofertilizers, biopesticides, and phytostimulators in which certain bacteria can be used for multiple purposes. But certain adverse environmental circumstances can affect the potential of PGPR adversely in agriculture fields. By using advanced techniques to manipulate the PGPR and develop bioformulations, such limitations can be managed to enhance plant nutrition, crop yields, and disease management.

References

Ahemad M, Khan MS (2010) *Pseudomonas aeruginosa* strain PS1 enhances growth parameters of greengram [*Vigna radiata* (L.) Wilczek] in insecticide-stressed soils. J Pest Sci 84:123–131

Ahemad M, Kibret M (2014) Mechanisms and applications of plant growth promoting rhizobacteria: current perspective. J King Saud Univ Sci 26:1–20

Ahkami AH et al (2017) Rhizosphere engineering: enhancing sustainable plant ecosystem productivity. Rhizosphere 3:233–243

Ahmad I et al (2011) Microbial applications in agriculture and the environment: a broad perspective. In: Microbes and microbial technology: agricultural and environmental applications. Springer, New York, pp 1–27. https://doi.org/10.1007/978-1-4419-7931-5

Amarger N (2002) Genetically modified bacteria in agriculture. Biochimie 84:1061–1072

Badri DV et al (2009) Rhizosphere chemical dialogues: plant–microbe interactions. Curr Opin Biotechnol 20:642–650

Bais HP et al (2006) The role of root exudates in rhizosphere interactions with plants and other organisms. Annu Rev Plant Biol 57:233–266

Basheer J et al (2018) Assessment of plant-probiotic performance of novel endophytic *Bacillus* sp. in talc-based formulation. Probiotics Antimicrob Proteins:1–8. https://doi.org/10.1007/s12602-018-9386-y

Berg G et al (2014) Illumina amplicon sequencing of 16S rRNA tag reveals bacterial community development in the rhizosphere of apple nurseries at a replant disease site and a new planting site. PLoS One 9:e111744

Bevivino A et al (2014) Soil bacterial community response to differences in agricultural management along with seasonal changes in a Mediterranean region. PLoS One 9:e105515

Bowen GD, Rovira AD (1999) The rhizosphere and its management to improve plant growth. Adv Agron 66:1–102

Breidenbach B et al (2016) Microbial community structure in the rhizosphere of rice plants. Front Microbiol 6:1537

Caetano-Anolles G, Gresshoff PM (1991) Plant genetic control of nodulation. Annu Rev Microbiol 45:345–382

Carbonetto B et al (2014) Structure, composition and metagenomic profile of soil microbiomes associated to agricultural land use and tillage systems in *Argentine Pampas*. PLoS One 9:e99949

Chakraborty U et al (2010) Influence of *Serratia marcescens* TRS-1 on growth promotion and induction of resistance in *Camellia sinensis* against *Fomes lamaoensis*. J Plant Interact 5:261–272

Chen Y et al (2015) Comparative genomic analysis and phenazine production of *Pseudomonas chlororaphis*, a plant growth-promoting rhizobacterium. Genomics Data 4:33–42

Chernin L, Chet I (2002) Microbial enzymes in the biocontrol of plant pathogens and pests 84. https://doi.org/10.1201/9780203904039.ch7

Compant S et al (2005) Use of plant growth-promoting bacteria for biocontrol of plant diseases: principles, mechanisms of action, and future prospects. Appl Environ Microbiol 71:4951–4959

David BV et al (2018) *Pseudomonas fluorescens*: a plant growth promoting rhizobacterium (PGPR) with potential role in biocontrol of pests of crops. In: Prasad R, Gill SS, Tuteja N (eds) Crop improvement through microbial biotechnology. Elsevier, Cambridge, MA, pp 221–243

De Mandal S, Panda A (2015) Microbial ecology in the era of next generation sequencing. J Next Gener Seq Appl 1(S1). https://doi.org/10.4172/2469-9853.S1-001

Deshwal VK et al (2003) Rhizobia as a biological control agent against soil borne plant pathogenic fungi. Indian J Exp Biol 41:1160–1164

Dey R et al (2004) Growth promotion and yield enhancement of peanut (*Arachis hypogaea* L.) by application of plant growth-promoting rhizobacteria. Microbiol Res 159:371–394
Dilkes B et al (2007) Host-plant selectivity of rhizobacteria in a crop/weed model system. PLoS One 2:e846
Edwards J et al (2015) Structure, variation, and assembly of the root-associated microbiomes of rice. Proc Natl Acad Sci 112:E911–E920
Elend C et al (2006) Isolation and biochemical characterization of two novel metagenome-derived esterases. Appl Environ Microbiol 72:3637–3645
Etesami H, Maheshwari DK (2018) Use of plant growth promoting rhizobacteria (PGPRs) with multiple plant growth promoting traits in stress agriculture: action mechanisms and future prospects. Ecotoxicol Environ Saf 156:225–246
Field D et al (2008) MetaSim—a sequencing simulator for genomics and metagenomics. PLoS One 3:e3373
Garcia-Fraile P et al (2015) Role of bacterial biofertilizers in agriculture and forestry. AIMS Bioeng 2:183–205
Geros H et al (2016) The multifarious PGPR *Serratia marcescens* CDP-13 augments induced systemic resistance and enhanced salinity tolerance of wheat (*Triticum aestivum* L.). PLoS One 11:e0155026
Glick BR et al (2007) Promotion of plant growth by ACC deaminase-producing soil bacteria. Eur J Plant Pathol 119:329–339
Goel R et al (2018) Toward the unculturable microbes for sustainable agricultural production. In: Role of rhizospheric microbes in soil: volume 1: stress management and agricultural sustainability. Springer, Singapore, pp 107–123
Goldstein AH (1995) Recent progress in understanding the molecular genetics and biochemistry of calcium phosphate solubilization by Gramnegative bacteria. Biol Agric Hortic 12:185–193
Gouda S et al (2018) Revitalization of plant growth promoting rhizobacteria for sustainable development in agriculture. Microbiol Res 206:131–140
Graham PH, Vance CP (2003) Legumes: importance and constraints to greater use. Plant Physiol 131:872–877
Grosskopf T, Soyer OS (2014) Synthetic microbial communities. Curr Opin Microbiol 18:72–77
Handelsman J (2004) Metagenomics: application of genomics to uncultured microorganisms. Microbiol Mol Biol Rev 68:669–685
Haney CH et al (2015) Associations with rhizosphere bacteria can confer an adaptive advantage to plants. Nat Plants 1:15051
Hassen W et al (2018) *Pseudomonas rhizophila* S211, a new plant growth-promoting rhizobacterium with potential in pesticide-bioremediation. Front Microbiol 9:34
Houlden A et al (2008) Influence of plant developmental stage on microbial community structure and activity in the rhizosphere of three field crops. FEMS Microbiol Ecol 65:193–201
Huang J et al (2016) Isolation and engineering of plant growth promoting rhizobacteria *Pseudomonas aeruginosa* for enhanced cadmium bioremediation. J Gen Appl Microbiol 62:258–265
Jambhulkar PP, Sharma P (2014) Development of bioformulation and delivery system of *Pseudomonas fluorescens* against bacterial leaf blight of rice (*Xanthomonas oryzae* pv. *oryzae*). J Environ Biol 35:843–849
Jasim B et al (2016) Identification of endophytic *Bacillus mojavensis* with highly specialized broad spectrum antibacterial activity. 3 Biotech 6:187
Jasim B et al (2016a) Identification of a novel endophytic *Bacillus* sp. from *Capsicum annuum* with highly efficient and broad spectrum plant probiotic effect. J Appl Microbiol 121:1079–1094
Jasim B et al (2016b) Surfactin, iturin, and fengycin biosynthesis by endophytic *Bacillus* sp. from *Bacopa monnieri*. Microb Ecol 72:106–119
Jimtha John C et al (2017) *Pseudomonas fluorescens* R68 assisted enhancement in growth and fertilizer utilization of *Amaranthus tricolor* (L.). 3 Biotech 7:256

Jishma P et al (2017) Strain-specific variation in plant growth promoting volatile organic compounds production by five different *Pseudomonas* spp. as confirmed by response of *Vigna radiata* seedlings. J Appl Microbiol 123:204–216

John Jimtha C et al (2017) Antifungal properties of prodigiosin producing rhizospheric *Serratia* sp. Rhizosphere 3:105–108

John CJ, Radhakrishnan EK (2017) Chemicobiological insight into anti-phytopathogenic properties of rhizospheric *Serratia plymuthica* R51. Proc Natl Acad Sci India Sect B Biol Sci

Jones DL et al (2009) Carbon flow in the rhizosphere: carbon trading at the soil–root interface. Plant Soil 321:5–33

Kakvan N et al (2013) Development of new bioformulations using *Trichoderma* and Talaromyces fungal antagonists for biological control of sugar beet damping-off disease. Crop Prot 53:80–84

Kang S-M et al (2015) Mechanism of plant growth promotion elicited by *Bacillus* sp. LKE15 in oriental melon. Acta Agric Scand Sect B Soil Plant Sci 65:637–647

Kavitha K et al (2005) Development of bioformulations of antagonistic bacteria for the management of damping off of Chilli (*Capsicum annuum* L). Arch Phytopathol Plant Protect 38:19–30

Kircher M, Kelso J (2010) High-throughput DNA sequencing–concepts and limitations. BioEssays 32:524–536

Kochian L et al (2004) The physiology, genetics and molecular biology of plant aluminum resistance and toxicity. Plant Soil 274:175–195

Lagos L et al (2015) Current overview on the study of bacteria in the rhizosphere by modern molecular techniques: a mini review. J Soil Sci Plant Nutr 15:504–523

Lapenda JC et al (2015) Antimicrobial activity of prodigiosin isolated from *Serratia marcescens* UFPEDA 398. World J Microbiol Biotechnol 31:399–406

Lavecchia A et al (2015) Microbial 16S gene-based composition of a sorghum cropped rhizosphere soil under different fertilization managements. Biol Fertil Soils 51:661–672

Lee KWK et al (2013) Biofilm development and enhanced stress resistance of a model, mixed-species community biofilm. ISME J 8:894–907

Lundberg DS et al (2012) Defining the core *Arabidopsis thaliana* root microbiome. Nature 488:86–90

Lynch JM, Whipps JM (1990) Substrate flow in the rhizosphere. Plant Soil 129:1–10

Mahalingam CA et al (2017) Development of bioformulation and its application against management of thrips and root rot disease of mulberry, *Morus alba*. Int J Curr Microbiol App Sci 6:2438–2449

Matthysse AG, McMahan S (1998) Root colonization by *Agrobacterium tumefaciens* is reduced in cel, attB, attD, and attR mutants. Appl Environ Microbiol 64:2341–2345

Mendes LW et al (2014) Taxonomical and functional microbial community selection in soybean rhizosphere. ISME J 8:1577–1587

Metzker ML (2010) Sequencing technologies – the next generation. Nat Rev Genet 11:31–46

Nakkeeran S et al (2006) Plant growth promoting rhizobacteria formulations and its scope in commercialization for the management of pests and diseases. In: Biocontrol and biofertilization. Springer Netherlands, Dordrecht, pp 257–296

Noumavo PA et al (2016) Plant growth promoting rhizobacteria: beneficial effects for healthy and sustainable agriculture. Afr J Biotechnol 15:1452–1463

Okon Y (1985) Azospirillum as a potential inoculant for agriculture. Trends Biotechnol 3:223–228

Orlando L et al (2015) Microbial community composition and diversity via 16S rRNA gene amplicons: evaluating the illumina platform. PLoS One 10:e0116955

Pandey GK et al (2014) Drought-tolerance of wheat improved by rhizosphere bacteria from harsh environments: enhanced biomass production and reduced emissions of stress volatiles. PLoS One 9:e96086

Papajorgji P (2009) Advances in modeling agricultural systems. Springer US, Boston

Peiffer JA et al (2013) Diversity and heritability of the maize rhizosphere microbiome under field conditions. Proc Natl Acad Sci 110:6548–6553

Philippot L et al (2013) Going back to the roots: the microbial ecology of the rhizosphere. Nat Rev Microbiol 11:789–799

Pii Y et al (2016) The interaction between iron nutrition, plant species and soil type shapes the rhizosphere microbiome. Plant Physiol Biochem 99:39–48
Pinton R et al (2001) The rhizosphere: biochemistry and organic substances at the soil-plant interface. Marcel Dekker, New York
Prashar P et al (2013) Rhizosphere: its structure, bacterial diversity and significance. Rev Environ Sci Biotechnol 13:63–77
Rajkumar M et al (2008) Characterization of metal-resistant plant-growth promoting *Bacillus weihenstephanensis* isolated from serpentine soil in Portugal. J Basic Microbiol 48:500–508
Rappe MS et al (2002) Cultivation of the ubiquitous SAR11 marine bacterioplankton clade. Nature 418:630–633
Rivas R et al (2007) Biodiversity of populations of phosphate solubilizing rhizobia that nodulates chickpea in different Spanish soils. Springer Netherlands, Dordrecht
Rout ME, Callaway RM (2012) Interactions between exotic invasive plants and soil microbes in the rhizosphere suggest that 'everything is not everywhere'. Ann Bot 110:213–222
Ryan PR et al (2009) Rhizosphere engineering and management for sustainable agriculture. Plant Soil 321:363–383
Saidi N et al (2009) Characterization and selection of *Bacillus* sp. strains, effective biocontrol agents against *Fusarium oxysporum* f. sp. *radicis-lycopersici*, the causal agent of Fusarium crown and root rot in tomato. Ann Microbiol 59:191
Sait M et al (2002) Cultivation of globally distributed soil bacteria from phylogenetic lineages previously only detected in cultivation-independent surveys. Environ Microbiol 4:654–666
Sekar J et al (2016) Microbial consortial products for sustainable agriculture: commercialization and regulatory issues in India. In: Singh HB, Sarma BK, Keswani C (eds) Agriculturally important microorganisms: commercialization and regulatory requirements in Asia. Springer, Singapore, pp 107–132
Sendhilvel V et al (2007) Talc-based formulation of *Pseudomonas fluorescens*-induced defense genes against powdery mildew of grapevine. Arch Phytopathol Plant Protect 40:81–89
Shafi J et al (2017) Bacillus species as versatile weapons for plant pathogens: a review. Biotechnol Biotechnol Equip 31:446–459
Shaw JLA et al (2015) Using amplicon sequencing to characterize and monitor bacterial diversity in drinking water distribution systems. Appl Environ Microbiol 81:6463–6473
Singh A et al (2014) Rhizosphere competent microbial consortium mediates rapid changes in phenolic profiles in chickpea during *Sclerotium rolfsii* infection. Microbiol Res 169:353–360
Sipes DL, Einset JW (1983) Cytokinin stimulation of abscission in lemon pistil explants. J Plant Growth Regul 2:73–80
Souza RC et al (2015) Metagenomic analysis reveals microbial functional redundancies and specificities in a soil under different tillage and crop-management regimes. Appl Soil Ecol 86:106–112
Spaepen S et al (2007) Indole-3-acetic acid in microbial and microorganism-plant signaling. FEMS Microbiol Rev 31:425–448
Spaink HP (2000) Root nodulation and infection factors produced by rhizobial bacteria. Annu Rev Microbiol 54:257–288
Stacey G et al (2006) Genetics and functional genomics of legume nodulation. Curr Opin Plant Biol 9:110–121
Stempfhuber B et al (2015) Spatial interaction of archaeal ammonia-oxidizers and nitrite-oxidizing bacteria in an unfertilized grassland soil. Front Microbiol 6:1567
Stewman S, Lincoln D (1981) Recombinant DNA breakthroughs in agriculture, industry, and medicine. Futures 13:128–140
Sundheim L et al (1988) Molecular cloning of two chitinase genes from *Serratia marcescens* and their expression in *Pseudomonas* species. Physiol Mol Plant Pathol 33:483–491
Sutruedee P et al (2013) Bioformulation *Pseudomonas fluorescens* SP007s against dirty panicle disease of rice. Afr J Microbiol Res 7:5274–5283
Tringe SG, Rubin EM (2005) Metagenomics: DNA sequencing of environmental samples. Nat Rev Genet 6:805–814

Upadhyay RK et al (2001) Biocontrol potential and its exploitation in sustainable agriculture. Springer US, Boston

Vakhlu J et al (2008) Metagenomics: future of microbial gene mining. Indian J Microbiol 48:202–215

Van Emon JM (2016) The omics revolution in agricultural research. J Agric Food Chem 64:36–44

Wang C et al (2000) Effect of transferring 1-aminocyclopropane-1-carboxylic acid (ACC) deaminase genes into *Pseudomonas fluorescens* strain CHA0 and its gacA derivative CHA96 on their growth-promoting and disease-suppressive capacities. Can J Microbiol 46:898–907

Wani SA et al (2016) *Azotobacter chroococcum* – a potential biofertilizer in agriculture: an overview. In: Hakeem KR, Akhtar J, Sabir M (eds) Soil science: agricultural and environmental prospectives. Springer International Publishing, Cham, pp 333–348

Watt M (2009) The rhizosphere: biochemistry and organic substances at the soil–plant interface. 2nd edn. Ann Bot 104:ix–ix

Yang J et al (2009) Rhizosphere bacteria help plants tolerate abiotic stress. Trends Plant Sci 14:1–4

Ye L, Zhang T (2011) Pathogenic bacteria in sewage treatment plants as revealed by 454 pyrosequencing. Environ Sci Technol 45:7173–7179

Yuan J et al (2012) Antifungal activity of *Bacillus amyloliquefaciens* NJN-6 volatile compounds against *Fusarium oxysporum* f. sp. *cubense*. Appl Environ Microbiol 78:5942–5944

Zaheer A et al (2016) Association of plant growth-promoting *Serratia* spp. with the root nodules of chickpea. Res Microbiol 167:510–520

Zaidi A et al (2009) Plant growth promotion by phosphate solubilizing bacteria. Acta Microbiol Immunol Hung 56:263–284

Zwirglmaier K et al (2004) Recognition of individual genes in a single bacterial cell by fluorescence *in situ* hybridization-ring-fish. Mol Microbiol 51:89–96

Impact Assessment of Microbial Formulations in Agricultural Soil

16

Rachana Jain and Jyoti Saxena

1 Introduction

Soil is the loose surface material that consists of inorganic particles and organic matter covering the land surface. The soil fit for plant growth is called fertile or agricultural soil. Fertile soils have various mineral components in balance (sand 0.05–2 mm, silt 0.002–0.05 mm, clay <0.002 mm), soil organic matter, water and air that allow optimum water retention, drainage, oxygen for roots, nutrients to plant growth and physical support for plants. However, the physico-chemical properties of soils vary from region to region due to factors such as age, parent material, climate, topography and vegetation under which they were formed. Microbial activity along with leaching and weathering processes is responsible for a range of different soil types, having particular strengths and weaknesses for agricultural production (Bronick and Lal 2005).

Seventeen essential elements are generally required for plant growth (Troeh and Thompson 2005) and lack of any one of them can cause adverse effect on plants. Of these, the primary macronutrients (N, P, and K) are required in the largest quantities but are mostly in short supply and the deficiency of these macronutrients shows limiting effect on plant growth and yield. Micronutrients (trace nutrients), on the other hand, are required in minute amounts and, if in excess, can be toxic to plants (Epstein 1994, Subbarao et al. 2003). Natural cycling of these nutrients occurs in soil in which nutrients move from soil to plants and animals and then back to the soil. It is a very important and complex phenomenon involving many biological, chemical and physical processes. Present agricultural practices are affecting the

R. Jain
Amity Food and Agriculture Foundation, Amity University, Noida, Uttar Pradesh, India

J. Saxena (✉)
Biochemical Engineering Department, B.T. Kumaon Institute of Technology, Dwarahat, Uttarakhand, India

D. P. Singh et al. (eds.), *Microbial Interventions in Agriculture and Environment*,
https://doi.org/10.1007/978-981-13-8383-0_16

natural cycling of nutrients, hence causing soil degradation (Parikh and James 2012).

In India, in the past 60 years, agriculture has become completely dependent on chemical fertilizers and pesticides. No doubt, chemical inputs played a vital part in the Green Revolution but their excessive use has generated many environmental issues, viz., global warming, eutrophication, soil degradation, water pollution and many other human health-related problems (Aggani 2013). Since our future agricultural growth is still dependent on chemical fertilizers, further loss in soil quality and water contamination are likely and will also cause unsustainable burden on agriculture (Rajasekaran et al. 2012). Therefore, in the present scenario, a sustainable solution is required and that is possible through the use of microbial formulations and organic inputs.

Microbial formulations are artificially multiplied cultures of certain soil microorganisms that can improve the nutrient input and growth, and control the plant pathogens. On the basis of their functions, these microbial formulations can be divided into three categories:

(i) Biofertilizers: they assist plant nutrient uptake by providing fixed nitrogen through nitrogen-fixing bacteria and other nutrients such as P and K through phosphate and potassium solubilizing bacteria, e.g. species of *Pseudomonas*, *Bacillus*, *Rhizobium leguminosarum bv. trifolii, Aspergillus*, *Penicillium*, etc. (Jain et al. 2010).

(ii) Phyto-stimulators: they promote plants growth, usually by producing plant hormones, e.g. *Azospirillum* sp., *Pseudomonas putida, Pseudomonas aeruginosa, Klebsiella* sp., *Enterobacter asburiae*, *Mesorhizobium* sp., etc. (Spaepen et al. 2007; Glick 2012).

(iii) Biocontrol agents: they protect plants from plant pathogens. The examples are *Trichoderma*, *Pseudomonas*, *Bacillus*, etc. (Dawar et al. 2010; Mohiddin et al. 2010).

Together these microorganisms are called plant growth promoting microorganisms (PGPMs). They are listed in Table 16.1. For decades, a variety of PGPMs have been studied and commercialized, which include the species of *Azobacter*, *Azosprillum*, *Enterobacter*, *Bacillus*, *Klebsiella*, *Pseudomonas*, *Serratia, Tricoderma, Variovorax*, etc. (Glick 2012). However, their utilization in agriculture is very limited worldwide (Bashan et al. 2014) because of inconsistent effects on crop growth and yield. The success of these formulations depends on their survival rate in the field, compatibility with plant on which they are inoculated, interaction with indigenous microflora, response to other agronomic practices and prevailing environmental factors (Martinez-Viveros et al. 2010). They cannot be used universally in all regions for all crops; this fact also limits their utilization, commercialization and production (Vejan et al. 2016). Therefore, there is a need to understand not only the rhizosphere and the interactions that are taking place there but also to understand the impact of the agronomic practices, soil carbon content and organic

Table 16.1 List of plant growth promoting microorganisms tested against different crops

PGPMs	Plant/crop	Condition	Target property	References
Biofertilizer				
Bradyrhizobium sp. 750, *Ochrobactrum cytisi, Pseudomonas* sp.	*Lupinus luteus*	Field	Nitrogen fixation	Dary et al. (2010)
Rhizobium sp.	Pea	Pot	Nitrogen fixation	Wani et al. (2007)
R. phaseoli	*Vigna radiata*	Pot	Nitrogen fixation	Zahir et al. (2010)
Rhizobium MRPI	Pea	Pot	Nitrogen and phosphorus nutrition	Ahemad and Khan (2011)
Burkholderia	Rice	Greenhouse	Nitrogen fixation	Kao et al. (2003)
A4, C1 and H6	Maize	Pot	Phosphate solubilization	Pande et al. (2017)
Serratia sp. J260 and *Pantoea*sp. J49	Maize and peanut	Pot	Phosphate solubilization and abiotic stress	Anzuay et al. (2017)
Azotobacter	Legume and non-legume crop	Field	Nitrogen fertilization	Gomare et al. (2013)
Azospirillum + *Azotobacter* + *Mesorhizobium* + *Pseudomonas*	Chickpea	Field	Nutrient enhancement	Rokhzadi et al. (2008)
Bacillus megaterium PSB12	Wheat	Pot and field	Phosphate solubilization	Mukhtar et al. (2017)
G. intraradices + *Rhizobium sp.* + *P. straita*	*C. arietinum*		Increased the biomass of plants and NPK uptake	Akhtar and Siddiqui (2008)
Trichoderma asperellum Q1	Cucumber	Hydroponic	Phosphate solubilization	Zhao and Zhang (2015)
Aneurinibacillus aneurinilyticus CKMV1	Tomato	Pot	Phosphate solubilization, nitrogen fixation	Chauhan et al. (2017)
Bacillus subtilis	*Cotton*	*Field*	*Enhance NPK*	Yao et al. (2006)
Serratia sp. J260 and *Pantoea*sp. J49	Maize and peanut	Pot	Phosphate solubilization	Anzuay et al. (2017)
Glomus intraradices and *Rhizobium tropici* CIAT899,	*Bean*	*Greenhouse*	*Enhance nitrogen phosphorus*	Tajini et al. (2012)
Pseudomonas jessenii, R62; *P. synxantha*, R81 and AM	*Wheat*	*Field*	*Enhance mineral content*	Mäder et al. (2011)

(continued)

Table 16.1 (continued)

PGPMs	Plant/crop	Condition	Target property	References
Bacillus amyloliquefaciens IN937a, *B. pumilus* T4 and *Glomus intraradices* (arbuscular mycorrhizal fungus)	*Tomato*	*Greenhouse*	*Enhance mineral content*	Adesemoye et al. (2009)
R. phaseoli	*Vigna radiata*	Pot	Nitrogen fixation	Zahir et al. (2010)
Aspergillus awamori S19	Mung	Pot	Phosphate solubilization	Jain et al. (2010)
A. awamori S33 A. tubingensis S36	Mung	Pot	Phosphate solubilization	Jain et al. (2014)
Aspergillus awamori S29	Mung	Pot	Phosphate solubilization	Jain et al. (2012)
Burkholderia vietnamensis	Rice	Greenhouse	Nitrogen fixation	Govindarajan et al. (2007)
Burkholderia cepacia BAM-12 and *Pseudomonas fluorescens BAM-4*	Wheat	Pot	Phosphate solubilization	Minaxi and Saxena (2011)
Biocontrol				
Sphingomonas	Tomato		Antibiotic production	Mazzola et al. (1995)
Streptomyces strain AzR-051	–	–	*Alternaria* alternate growth inhibition; IAA production	Verma et al. (2011)
Aneurinibacillus aneurinilyticus CKMV1	Tomato	Pot	Antifungal activity, Indole-3-acetic acid (IAA), HCN and siderophores production	Chauhan et al. (2017)
Trichoderma harzianum Tr6 + *Pseudomonas* sp. Ps14	*Cucumis sativus, A. thaliana*	–	Defence against *Fusarium oxysporum f. sp. radicis cucumerinum*	Alizadeh et al. (2013)
G. intraradices + Rhizobium sp. + P. straita	C. arietinum	–	Nematode galling and multiplication reduced	Akhtar and Siddiqui (2008)
Pseudomonas + T. harzianum + G. intraradices	*S. lycopersicum*	–	Increased seed germination	Srivastava et al. (2010)
Mixture of *Pseudomonads*	*Triticum aestivum*	–	Antibiosis	Duffy and Weller (1995)

(continued)

Table 16.1 (continued)

PGPMs	Plant/crop	Condition	Target property	References
Azospirillum; Pseudomonas fluorescens	Cotton	–	Against *Rhizoctonia bataticola*	Marimuthu et al. (2013)
Burkholderia cepacia BAM-6, B. cepacia BAM-12 and *Pseudomonas fluorescens BAM-4*	Mung Bean	Pot	Biocontrol activity against *Macrophomina phaseolina*; IAA production, siderophore and chitinase	Minaxi and Saxena (2010)
Stress reliever				
Trichoderma asperellum Q1	Cucumber	Hydroponic	Suppression of salt stress by phytohormone levels	Zhao and Zhang (2015)
Azospirillum brasilense	Tomato	Pot	Increase salicylic acid	Bashan and de-Bashan (2002)
Azospirillum sp., *Bacillus megaterium M3 Bacillus OSU-142, B. licheniformis* RC08, *Burkholderia cepacia* FS Tur, *Paenibacillus polymyxa* RC05, *Raoutella terrigena*	Wheat barley	Field	Biostimulant	Çakmakçi et al. (2014)
Pseudomonas aeruginosa	*Wheat*	*Greenhouse*	*Enhanced antioxidative enzyme, ascorbic acids and total phenolics*	Islam et al. (2014)
Aneurinibacillus aneurinilyticus CKMV1	Tomato	Pot	Production of siderophore, Indole-3-acetic acid (IAA) and HCN and antifungal activity	Chauhan et al. (2017)
R. phaseoli	*Vigna radiata*	Pot	Abiotic stress	Zahir et al. (2010)
Curtobacterium albidum SRV4	Rice	–	Salt stress	Vimal et al. (2018)
Serratia sp. J260 and *Pantoea* sp. J49	Maize and peanut	Pot	abiotic stress	Anzuay et al. (2017)
Azospirillum sp. AZ204 + *P. fluorescens Pf1*	*Gossypium hirsutum*	Pot	Plant growth promotion	Marimuthu et al. (2013)
Arthrobacter sp. *SU18; B. subtilis SU47*	*Wheat*	*Greenhouse*	*Increase salinity tolerance*	Upadhyay and Singh (2015)

(continued)

Table 16.1 (continued)

PGPMs	Plant/crop	Condition	Target property	References
Burkholderia vietnamiensis AR 1122	*Rice*	*Field*	*IAA production*	Araujo et al. (2013)
Bacillus sp. MML2551 + *B. licheniformis* MML2501 + *P. aeruginosa* MML2212 + *Streptomyces fradiae* MML1042	*Helianthus annuus*	–	Promoted plant growth	Srinivasan and Mathivanan (2009)
Bradyrhizobium diazoefficiens USDA 110(T) (BD1) and *Rhizobium tropici* CIAT 899(T) (RT1)	*Soybean*	*Greenhouse and field*	*Phytohormone production*	Marks et al. (2013)

amendments. In view of the above, an attempt has been made to discuss all these issues in detail in this chapter.

2 Agronomic Practices

There are several broadly accepted agronomic practices that may be useful. Tillage is one such practice that in the typical conventional system is used to prepare a good seed bed for crop establishment. Tillage involves the mechanical agitation of the soil by various ways such as digging, stirring and overturning. It disturbs the soil structure and helps in redistribution of the organic matter and microorganisms, reduces soil bulk density in the topsoil (Unger 1992; Tebrügge et al. 1999; Dam et al. 2005; Wang et al. 2015) and increases soil porosity, drainage and soil temperature (Pelegrin et al. 1990). It also helps in the improvement of soil hydraulic property (Unger 1992; Lampurlanés and Cantero-Martínez 2006; Schwen et al. 2011). Depending on the soil conditions and crop requirements, there are different ways of tillage (Morris et al. 2010) but it is roughly divided into two types, viz., conventional and conservational tillage. Conventional tillage is the most commonly used practice in India which involves the complete inversion of soil through ploughing. In contrast, conservational tillage (minimum tillage) includes zero tillage (no-tillage, NT), reduced (minimum) tillage (MT), ridge tillage, contour tillage and mulch tillage (El Titi 2003). In NT little or no soil surface disturbance is noticed; the only disturbance is during sowing, while MT involves lower soil manipulation through ploughing using primary tillage implements. Mulch tillage ensures that the plant residues or organic materials covering the surface are not disturbed as far as possible, whereas ridge tillage involves the preparation of rows of planting crops at the starting of the cropping season either along both sides or on top of the ridges. When tillage is at right angles to the direction of the slope, it is called contour tillage. Peters et al. (2003) performed 2- (barley and potato) and 3-year long (barley, red clover and potato) experiments with conservation tillage and observed the significant decrease in severity of dry rot (*Fusarium* spp.) and silver scurf (*Helminthosporium solani*) diseases.

As compared to conventional tillage, the conservational tillage increased more soil bulk density (Chassot et al. 2001; Deubel et al. 2011; Villamil and Nafziger 2015; Spiegel et al. 2015), water holding capacity (Alvarez and Steinbach 2009; Deubel et al. 2011; Villamil and Nafziger 2015), cation exchange capacity, microbial activity (Derpsch et al. 2010), resistance against pathogen penetration and side-by-side reduced soil erosion (Uri et al. 1998; Pittelkow et al. 2014). Recent research shows that conventional tillage often has adverse effects on soil biota but helps more in mitigation of weed and increases the decomposition and mineralization of organically bound nutrients (Bronick and Lal 2005). In this way, conservation tillage practices were often reported more useful in combination with other cover crops and mulches (Pittelkow et al. 2015). Cookson et al. (2008) observed that tillage decreased the fungal biomass but bacterial biomass was increased. A significantly higher earthworm population under no-till soil was observed during a 6-year study by Anderson (1987) over under-ploughed soil. In the same year, Kemper et al. (1987) reported increased activities of surface-feeding earthworms in case of less intense tillage. There are no studies available related to impact of tillage on microbial inoculation, thus emphasizing the need for work on this aspect.

Crop rotation, an excessively common agronomic practice, is the growing of different crops in succession on a particular field in a specified order. It improves the structure of the soil, plant nutrition and soil quality by increasing the carbon sequestration and nutrient flow (especially leguminous crops) (Dendoncker et al. 2004; Hamza and Anderson 2005), promotes biodiversity (Paoletti 1999) and prevents groundwater pollution (Albrecht 2003) and soil erosion (Buick et al. 1992). It also plays a very important role in weed and pest management (Marcroft et al. 2004; Beckie 2006). There are reports that quote enhancement in the soil organic matter and yield (Bullock 1992; West and Post 2002; Jarecki and Lal 2003) due to crop rotation. Legumes get special attention in organic crop rotation as they can fix nitrogen to the system through *Rhizobium* root nodules. They can be used in rotation as a harvested crop (e.g. alfalfa), green manure or cover crop (e.g. vetch or clover). It is well established that different roots produce different root exudates and have different structures that in turn affect the soil structure, soil quality and biodiversity (Balota et al. 2003; Govaerts et al. 2008). Earlier, Moore et al. (2000) found that crop rotation and plant cover at the time of sampling significantly affected the microbial biomass carbon. Alternate crops also diminish the incidence of pathogens and pests.

There are reports where rotation of beans with grain crops (wheat, barley, corn, rye, oat) has been helpful in decreasing the severity and incidence of root rot diseases caused by several fungal and nematodal pathogens (Snyder et al. 1959; Patrick et al. 1964; Glynne 1965; Cook and Baker 1983; Trivedi and Barker 1986; Abawi 1989). Rotation of wheat and maize (Govaerts et al. 2008) decreased nematodal incidences. Chen and Zhu (2011) discussed the benefits of planting diverse mixtures of rice cultivars in a field study; they reported significant yield increase and also very low levels of pathogen incidence so that farmers did not have to apply fungicides. The reason behind the reduction of different diseases was increase in organic matter that enhanced microbial diversity and number as a result of enhancement of the antibiosis activity in the root zone. Inoculation of *Phaseolus lunatus*, commonly known as lima bean, along with rhizobia have been found to provide more N than was allocated for plant growth and defence against herbivores (i.e. production of N-containing cyanogenic defence compounds). As a result, herbivory decreased significantly by the Mexican bean beetle in the rhizobia-inoculated plants (Thamer et al. 2011). Bakhshandeh et al. (2017) reported that the crop rotation of wheat and chickpea promoted the colonization of AMF in wheat root. Liu et al. (2018) found that crop rotation in combination with microbial inoculant reduced greenhouse gas emission.

3 Rhizosphere Engineering

The rhizosphere can be defined as the soil adjacent to plant roots that is influenced by root and its secretions, while the rest of the soil is called bulk soil (Prashar et al. 2013). The rhizosphere region as compared to bulk soil is more nutrient rich with high microbial density and biologically more active (Raaijmakers et al. 2009). The bulk soil, referred to as microbial seed bank, is a reservoir of microorganisms and contributes towards the microbial community of the rhizosphere, while the types of microorganisms that flourish and thrive in the rhizosphere is determined by plants. Hiltner (1904) first time proposed the term "rhizosphere," which was followed by many research papers highlighting the diversity of this region. Thereafter, the term plant growth promoting microorganisms (PGPMs) came into the picture, defining the microbes that were capable of modulating the root as well as soil system, which have a significant positive effect on crop and soil.

There are multiple mechanisms for the growth-promoting effects of these microbes and these could be classified under direct or indirect categories (Tailor and Joshi 2014; Chauhan et al. 2015). Direct effects include nutrient enrichment (Pii et al. 2015) and enhancing the plant growth by changing the auxin and/or cytokinins plant hormone synthesis and concentration of the hormone ethylene (Belimov et al. 2015). Indirect effects include biopesticide and biocontrol activities by the production of antibacterial, antifungal, nematicidal compounds (Saraf et al. 2014) and stimulation of the plant defence machinery such as induction of systemic resistance and increasing the intensity of pathogen-triggered immunity (Huang and Zimmerli 2014).

Basically, all three components, namely soil, plant and microbes of the rhizosphere can be manipulated. The soil can be amended not only to change its physicochemical properties but also to improve its overall quality. The plants can be transformed by introducing a novel trait of interest or designed in such a way that promoted the growth of a particular plant-beneficial microbe. Similarly, microorganisms can be engineered to support overall plant growth. This technology can prove to be good in replacing agrochemicals with beneficial microorganisms in an environmentally sustainable way.

Different types of soils nurture specific microbial communities. The plant species also affects the indigenous microflora as it influences the specific microbial flora. Type of microbial communities in rhizosphere depends on the plant type. Each plant type secretes a unique kind of root exudates that play a very important role in root-root and root–microbe interaction. It consists of enormous spectrum of organic compound. But researchers are only beginning to understand the role of single compound in mediating the interactions but the combined effect of different compound in single process is yet to understood (Bais et al. 2006; Haichar et al. 2008). The literature has several contrasting reports pointing towards the type of plant or soil as dominant factor (Grayston et al. 1998; Girvan et al. 2003; Nunan et al. 2005). With the help of microbial engineering, there is a possibility to enhance the desired microorganisms. Raaijmakers et al. (1995) introduced a siderophore receptor gene into *Pseudomonas flourescens* to make it more competitive. van Dillewijn et al. (2001) and Ma et al. (2004) used microbial engineering to increase nodule formation efficiency in *Sinorhizobium meliloti* by expressing gene proline dehydrogenase and 1-aminocyclopropane-1-carboxylate deaminase, respectively, whereas Suárez et al. (2008) increased nodulation by *Rhizobium etli* through overexpression of a trehalose-6-phosphate synthase gene. Likewise, Zhang et al. (2012) introduced chitinase gene into *Burkholderia vietnamiensis* P418 that increased its disease suppression efficiency against wheat sheath blight, tomato grey mould and cotton *Fusarium* wilt.

Most of the time, the success rate of microbial formulations is decreased when these are applied in field, the reason being biotic and abiotic factors. Biotic factors include antagonistic interactions and competition mainly from indigenous microbes for nutrients, whereas abiotic stress consists of temperature, heavy metals, water content and pollutants (Jain et al. 2010). This problem can be overcome by inoculating a consortium over a monoculture. There are reports where some microbes help the others in increasing their activity and in better establishment. Three strains of *Pseudomonas* biocontrol agents, viz., *Pseudomonas putida* PA14H7 and *P. fluorescens* PA3G8 and PA4C2, were isolated which showed biocontrol activity against *Dickeya* sp. that caused blackleg disease in potato. It was observed that these strains are more effective when applied together in comparison to individual inoculations (Liu et al. 2008; Raoul des Essarts 2015).

4 Soil Organic Carbon

Organic carbon is an essential element of life. Soil organic carbon (SOC) is the measurable amount of carbon present in soil as a component of soil organic matter (SOM). SOM comes from plants and animals that are in various stages of decay (Beare et al. 1992). It is the largest reservoir of organic carbon, even greater than plant biomass. It plays a very important role in soil structure, acts like 'glue' and helps in the aggregation of mineral particles and provides food supply for organisms (microorganisms, earthworms). Thus, organic matter contributes directly as well as indirectly to soil structure, which benefits both agriculture and the environment

(Sollins et al. 2007). It has also been found to increase soil fertility and improve the properties of the soil, both physical and biological, through the reduction of bulk density, improving soil structure and water-holding capacity and enhancing microbial activity.

Guo and Gifford (2002) performed a meta-analysis of data obtained from 74 research papers and concluded that soil carbon reserves declined after the changes taken place in land-use from pasture to plantation (−10%), native forest to cropland (−42%), native forest to plantation forest (−13%), and pasture to cropland (−59%). They also noticed the increase in soil C reserves after the change of native forest to pasture (+8%), cropland to plantation forest (+18%), cropland to pasture (+19%) and cropland to secondary forest (+53%). However, the factors such as plant species, annual precipitation and the length of study periods were responsible for varied results.

To increase organic carbon in soil there is need to increase plant production so that organic matter in soil can be increased. Increase in plant growth supports production of more root exudate that in turn increases soil organic matter. Another way of increasing soil C may be by growing cover crops such as annual ryegrass, cereal rye, oats, crimson clover, oilseed radishes, etc., between two seasons (Harrwing and Ulrich Ammon 2002; Fageria et al. 2005; Poeplau and Don 2015). Although there are reports where cover crops showed negative effects as they increased infiltration and decreased soil penetration resistance (Steele et al. 2012; Chen et al. 2014), but with regard to SOC, these crops were always beneficial. SOM contained nitrogen as well as C, so efforts to retain SOC were coupled with the retention of soil organic N (Powlson et al. 2011). It increased fixed nitrogen in soil that in reverse limits the C sequestration. Decomposition of SOM is influenced by structure of soil aspects such as gaseous exchange, moisture, porosity and physical location of C.

Bacteria, actinomycetes and fungi made the first line of decomposers and organic material is the food for them. Therefore, this organic carbon helps in deciding the microbial diversity, and number which is important for overall balance of the environment. The number of decomposer organisms depends on humus levels and organic material. Pathogenic microorganisms, once present, do not compete with other microorganisms and have fewer natural predators and hence are liable for more damage than if they were present in a more stable community. Many researchers are of the opinion that higher the diversity of species more will be the stability of microbial community (Dindal 1978). In other words, it can be related to the success of microbial formulations in soil. If soil is low in humus and organic carbon, then automatically microbial formulation will not be able to establish in it.

5 Soil Amendment

Soil amendment is the addition of organic and inorganic substances to soil that improves its physical and chemical properties, which include water retention and infiltration, permeability, aeration, drainage and structure leading to better plant productivity, microbial diversity and soil health. Some commonly used amendments are compost, organic manure, biochar, cover crop, mulches, etc., because they have low production costs and high organic matter content that ensure greater microbial activity and soil nitrogen supplying power. In addition to this, organic matter has a capacity to sink atmospheric CO_2. In recent years, improvement of soil quality and crop yield was observed through the use of soil amendments due to their sustainable nature (Chan et al. 2007; Dorraji et al. 2010; Belyaeva and Haynes 2012).

In India, it is not a new concept as organic amendments were in use since Vedic times. The description of organic amendments viz., Vrkshayurveda, Panchagavya, Kunapajala, Beejamrit, Jeevamirit, Compost tea, Matka khad, Vermiwash and Amrutpaniin are found in ancient Indian manuscripts. The amendments mainly consist of cow manure, cow urine, milk, yogurt and ghee; they are further combined with coconut water, jaggery, ripe banana and neem products and their preparation takes around 1 month time (Naresh et al. 2018). Nowadays, the most common soil organic amendments are animal manure, green manure and compost, though peat moss, straw, sawdust sewage sludge and wood chips are also used (Goss et al. 2013).

Compost addition to soil not only improves soil structure and lowers bulk density but also increases macro-aggregation and rhizosphere stability (de Leon-Gonzalez et al. 2000; Caravaca et al. 2002). It is dependent upon environmental conditions, for instance, the effectiveness of compost in aggregation is limited by drought (de Leon-Gonzalez et al. 2000). The addition of compost affects soil structure usually for a short period of time but has long-lasting positive effects on structural properties (Debosz et al. 2002).

There are also many reports observing that organic amendments help in plant disease management (Brown and Tworkoski 2004). Conn and Lazarovits (1999) and Lazarovits et al. (1999) reported that nitrogen-rich organic amendments (meat and bone meal, soymeal, and poultry manure) @ 37 t ha^{-1} reduced the occurrence of *Verticillium* wilt, and common scab of potato significantly in field trials. Xiao et al. (2016) reported that vermicompost significantly suppressed root pest *Meioidogyne incognita* via modulating soil properties, viz., pH, IAA concentration and soil microbial activity in susceptible cultivar of tomato. There are many more reports that are summarized in Table 16.2.

Use of biochar as soil amendment is not a new concept. There are evidences that it was used as soil amendment in Terra preta soils in Amazonia. It is a porous carbonaceous product produced by the pyrolysis of the biomass in an oxygen-depleted environment. It is a safe mode to store the carbon in the environment for long periods of time (Lehmann and Joseph 2009; Shackley et al. 2012). Addition of biochar in soil increased the water retention capacity of soil by 18% (Glaser et al. 2002), reduced leaching of the nutrients, helped in controlling the eutrophication (Sohi et al. 2009), reduced soil acidity, and increased soil electrical conductivity and cation exchange capacity (Liang et al. 2006; Gundale and DeLuca 2007; Warnock et al. 2007; Amonette and Joseph 2009; Laird et al. 2010) and increased soil organic carbon (Lugato et al. 2014). There are reports where it was found to lessen the bioaccumulation of heavy metals in plants. Namgay et al. (2010) reported that concentrations of heavy metals, viz., Cd, Cu and As in maize shoots significantly decreased

Table 16.2 Biocontrol by application of organic amendments

Pathogen	Plant	Organic amendment	References
Verticillium dahlia	Cotton	crab shell (chitin), soybean stalk and alfalfa	Huang et al. (2006)
Fusarium oxysporum	Cucumber	Sludge and manure compost	Huang et al. (2012)
Fusarium oxysporum f. sp. Cucumerinum	Cucumber	Vinegar residue compost	Shi et al. (2016)
Meioidogyne incognita	Tomato	Vermicompost	Xiao et al. (2016)
Rhizoctoniasolani AG 2-2IIIB	sugar beet	Inexpensive protein-rich waste products from food industry, i.e. hoof, meat, feather, blood and fish meal	Postma and Schilder (2015)
Fusarium wilt	Watermelon	Biogas slurry	Cao et al. (2016)
Cylindrocladium spathiphylli *Fusarium oxysporum* *Phytophthora nicotianae* *Phytophthora cinnamomi* *Rhizoctonia solani* *Verticillium dahlia*	Eggplant Cauliflower Tomato Lupin *Spathiphyllum* sp. *Pinus sylvestris* Flax	Green waste, domestic biowaste, manure	Termorshuizena et al. (2006)
R. solani	*Brassica*	Leaf residues from *Brassica rapa*, *B. napus* and *B.juncea*	Ascencion et al. (2015)
Aphanomyces euteiches	Pea	Compost from sewage sludge, wood chips	Lumsden et al. (1983)
Colletotrichum coccodes	Tomato	Compost from cannery waste	Abbasi et al. (2002)
Fusarium oxysporum f. sp. Dianthi	*Dianthus caryophyllus*	Compost from grass, ditch plants	Postma et al. (2003)
Fusarium oxysporum f. sp. Lycopersici	Tomato	Compost from vegetable, yard and animal wastes, sewage sludge	Cotxarrera et al. (2002)
Fusarium oxysporum f. sp. Lycopersici	Tomato	Compost from vermicompost	Szczech (1999)
Fusarium oxysporum f. sp. radicis-cucumerinum	Cucumber	Compost from dairy cattle manure	Kannangara et al. (2000)
Fusarium solani f. sp. Phaseoli	Bean	Compost from brewery waste	Abawi and Widmer (2000)
Macrophomina phaseolina	Clusterbean	Compost from pearl millet, cow dung	Lodha and Burman (2000)
Macrophomina phaseolina	Clusterbean	Compost from weeds, cow dung	Lodha and Burman (2000)
Macrophomina phaseolina	Clusterbean	Compost from cauliflower leaf residue and cow dung	Lodha et al. (2002)
Macrophomina phaseolina	Clusterbean	Compost from pearl millet, neem, cow dung, soil weeds, soil	Lodha and Burman (2000)

(continued)

Table 16.2 (continued)

Pathogen	Plant	Organic amendment	References
Pythium macrosporum	*Iris xyphium*	Compost from vegetable, fruit and garden waste	van Os and van Ginkel (2001)
Pythium ultimum	Bean	Compost from brewery waste	Abawi and Widmer (2000)
Pythium ultimum	Beetroot	Compost from cattle manure	Berner et al. (2000)
Pythium ultimum	Cucumber	Compost from shrimp waste, peat moss, sawdust	Labrie et al. (2001)
Rhizoctonia solani	Sugar beet	Compost from grass, ditch plants	Postma et al. (2003)
Rhizoctonia solani	Potato	Compost from cattle manure	Tsror et al. (2001)
Thielaviopsis basicola	Bean	Compost from brewery waste	Abawi and Widmer (2000)
Macrophomina phaseolina	Mung bean	Vermicompost, banana, NADEP, and Calotropis	Saxena et al. (2015)
Botrytis cinerea and *Leveillula taurica*	Pepper and tomato	Biochar	Elad et al. (2010)
Fusarium proliferatum and *F. oxysporum f. sp. asparagi*	Asparagus	Biochar	Elmer and Pignatello (2011)
Botrytis cinerea, Collatotrichum acutatum and *Podosphaera aphanis*	Strawberry	Biochar	Harel et al. (2012)

after application of biochar in soil (Freddo et al. 2012; Chan et al. 2012). Also, the biochar amendment significantly increased the rate of germination, growth and yield of plant (Glaser et al. 2002).There are reports where biochar was introduced in combination to PGPR and showed positive effects on soil structure, plant growth and microbial population (Saxena et al. 2013; Saxena et al. 2016; Tripti et al. 2017). Biochar has a potential to alleviate plant diseases and support plants' efficient defence (Burketova et al. 2015); some examples are mentioned in Table 16.2.

Biochar was used as carrier for inoculation of microbial formulation by Viveganandan and Jauhri (2000). The porous structure of biochar provided a safe niche for the microbial inoculants, protected them from predation and desiccation, provided carbon (C), energy and mineral nutrients and helped in establishment of the microbial formulation (Saito and Murumoto 2002). Asai et al. 2009 reported that biochar increased microbial activity in dried soil by hydrating the microbes with the help of high water-holding capacity. Saranya et al. (2011) showed that acacia wood and coconut shell based biochar could be used as a superior alternative carrier for the preparation of microbial formulations for commercialization.

6 Inter-species' Interaction

Plant–microbe, plant–plant and microbe–microbe interactions are examples of inter-species' interactions that take place in soil and also play a big role in soil health and shaping of the communities (Rey et al. 2013). These interactions could be positive, negative or neutral, and significantly influence the overall development of plants (Adesemoye et al. 2009; Ahmad et al. 2011; Lau and Lennon 2011). However, plant–microbe interaction is key to developing the rhizosphere effect. It plays a significant function in carbon sequestration, nutrient cycling, plant growth, soil structure and disease management.

The plant–microbe interaction is basically through the plant exudate secreted by plant roots, which can be ions, free oxygen, water, enzymes, mucilage, or a diverse array of primary and secondary metabolites (Gleba et al. 1999; Bais et al. 2001). These are divided into two categories on the basis of their function (Uren 2000; Bertin et al. 2003). The first category contains compounds that are waste products of plants, their secretion is gradient dependent and their function is unspecified, while the second category has those compounds that are purposefully prepared by the plants for secretion and their function is already defined such as defence, colonization, lubrication, etc. (Uren 2000; Bais et al. 2004). Plant–microbe interaction in soil also helps in carbon sequestration and nutrient cycling. Plant roots secrete huge amounts of secondary metabolites, thus encouraging the growth of microorganisms and also controlling the growth of pathogens.

Fons et al. (2003) reported that *Gypsophila paniculata* secreted saponin that promoted the growth of *Aquaspirillium* sp. in its rhizosphere, but this bacterium did not grow in the *Trifolium subterraneum* rhizosphere. From these results it can be concluded that saponin is responsible for the growth of *Aquaspirillium* sp. and adding saponin to the *Trifolium subterraneum* rhizosphere can help the bacteria to establish there. Legume–rhizobia association is another example that helps in nitrogen availability and soil fertility. It is a species-specific interaction, which means that specific rhizobial strains reside in specific host legumes, for which iso-flavanoids are responsible (Peters et al. 1986). Plant growth promoting microbes and mycorrhizae are also the examples of plant–microbe interaction.

Plant–microbe interaction is also responsible for abiotic stress tolerance. *Azospirillum brasilense* enhanced the abscisic acid (ABA) level and helped the *Arabidopsis thaliana* in drought tolerance (Cohen et al. 2015). Similarly, drought-tolerant bacterial strains, viz., *Alcaligenes faecalis* (AF3), *Proteus penneri* (Pp1) and *Pseudomonas aeruginosa* (Pa2) inoculation in maize plants enhanced the proline content, which in turn increased relative water content, protein and sugar (Naseem and Bano 2014).

There is one more interaction that occurs between microbes, i.e. microbe–microbe interaction, centred again on root exudates. Root exudates promote the growth of specific microbes, thereby influencing the other microflora of the rhizosphere. Dominant microbes also secrete signaling molecules that promote the growth of some microbes while inhibiting others. This microbe–microbe interaction also affects soil fertility and plant growth and helps in the development of pathogen-free environment. A very versatile example is *Trichoderma*, which parasitizes and kills other fungi. Presently we are using it as biofungicides in the field. In addition, some *Trichoderma* spp. can kill nematodes and hence have the potential for applications as bio-nematicides (Sharon et al. 2011). Understanding these interactions has led to the development of consortium-based microbial formulations, which in recent years have proved better than monocultures.

Plant roots and their rhizosphere affect the soil aggregation. Roots not only release exudates but also re-align and enmesh soil particles, which in turn influence cluster formation by altering the physical, chemical and biological properties. Aggregation was likely to increase significantly with increasing microbial associations, root length, density, glomalin and percent cover (Rillig et al. 2001). Caravaca et al. (2002) found that aggregate stability is greater in rhizosphere soil than in non-rhizosphere soil due to rhizodeposition, hyphal growth and mass, length, density, size and distribution of roots (Haynes and Beare 1997). The rhizosphere with a large population of micro- and macro-organisms also contributed to SOC and aggregation.

7 Conclusion

Understanding the impact of agronomic practices, organic carbon content and inter-species interaction on introduced microorganisms will help in success of bio-formulation. This will help in increase the number of beneficial microbial community in soil that will help naturally in improving soil health and plant growth. Therefore, there is a need to further study the developed bio-formulations in respect to other factors so that by their use we will get more benefited. Therefore, there is a need to further study the different aspects of microbial formulations in respect to abiotic and biotic factors so that their use can be encouraged for the production of organic crops.

References

Abawi GS (1989) Root rot. In: Schwartz HF, Pastor-Corrales MA (eds) Bean production problems: disease, insect, soil and climatic constraints of *Phaseolus vulgaris*. CIAT, Cali, Colombia, pp 105–157

Abawi GS, Widmer TL (2000) Impact of soil health management practices on soil borne pathogens, nematodes and root diseases of vegetable crops. Appl Soil Ecol 15:37–47

Abbasi PA, Al-Dahmani J, Sahin F, Hoitink HAJ, Miller SA (2002) Effect of compost amendments on disease severity and yield of tomato in conventional and organic production systems. Plant Dis 86(2):156–161

Adesemoye A, Torbert H, Kloepper J (2009) Plant growth-promoting rhizobacteria allow reduced application rates of chemical fertilizers. Microb Ecol 58(4):921–929

Aggani SL (2013) Development of biofertilizers and its future perspective. Sch Acad J Pharm 2(4):327–332

Ahemad M, Khan MS (2011) Effect of tebuconazole-tolerant and plant growth promoting *Rhizobium* isolate MRP1 on pea-Rhizobium symbiosis. Sci Hortic 129:266–272

Ahmad M, Zahir ZA, Asghar HN, Asghar M (2011) Inducing salt tolerance in mung bean through co-inoculation with rhizobia and plant-growth-promoting rhizobacteria containing 1-aminocyclopropane- 1-carboxylate deaminase. Can J Microbiol 57:578–589

Akhtar MS, Siddiqui ZA (2008) Biocontrol of a root-rot disease complex of chickpea by *Glomus intraradices*, *Rhizobium* sp. and *Pseudomonas straita*. Crop Prot 27:410–417

Albrecht H (2003) Suitability of arable weeds as indicator organisms to evaluate species conservation effects of management in agricultural ecosystems. Agric Ecosyst Environ 98:201–211

Alizadeh H, Behboudi K, Ahmadzadeh M, Javan-Nikkhah M, Zamioudis C, Pieterse CMJ, Bakker PAHM (2013) Induced systemic resistance in cucumber and *Arabidopsis thaliana* by the combination of *Trichoderma harzianum* Tr6 and *Pseudomonas* sp. Ps14. Biol Control 65:14–23

Alvarez R, Steinbach HS (2009) A review of the effects of tillage systems on some soil physical properties, water content, nitrate availability and crops yield in the Argentine Pampas. Soil Tillage Res 104:1–15

Amonette JE, Joseph S (2009) Characteristics of biochar: microchemical properties. Chapter 3. In: Lehmann J, Joseph S (eds) Biochar for environmental management science and technology. Earthscan, London, pp 33–52

Anderson EL (1987) Corn root growth and distribution as influenced by tillage and nitrogen fertilization. Agron J 79:544–549

Anzuay MS, Ciancio MGR, Luduena LM, Angelini JG, Barros G, Pastor N, Taurian T (2017) Growth promotion of peanut (*Arachis hypogaea* L.) and maize (*Zea mays* L.) plants by single and mixed cultures of efficient phosphate solubilizing bacteria that are tolerant to abiotic stress and pesticides. Microbiol Res 199:98–109

Araujo AES, Baldani VCD, Galisa PS, Pereira JA, Baldani JI (2013) Response of traditional upland rice varieties to inoculation with selected diazotrophic bacteria isolated from rice cropped at the northeast region of Brazil. Appl Soil Ecol 64:49–55

Asai H, Samson BK, Stephan HM, Songyikhangsuthor K, Homma K, Kiyono Y, Inoue Y, Shiraiwa T, Horie T (2009) Biochar amendment techniques for upland rice production in Northern Laos 1. Soil physical properties, leaf SPAD and grain yield. Field Crop Res 111:81–84

Ascencion LC, Liang W, Yen T (2015) Control of *Rhizoctonia solani* damping-off disease after soil amendment with dry tissues of *Brassica* results from increase in Actinomycetes population. Biol Control 82:21–30

Bais HP, Loyola-Vargas VM, Flores HE, Vivanco JM (2001) Root-specific metabolism: the biology and biochemistry of underground organs. In Vitro Plant 37:730–741

Bais HP, Park SW, Weir TL, Callaway RM, Vivanco M (2004) How plants communicate using the underground information superhighway. Trends Plant Sci 9:26–32

Bais HP, Weir TL, Perry LG, Gilroy S, Vivanco JM (2006) Role of root exudates in rhizosphere interactions with plants and other organisms. Annu Rev Plant Biol 57(233):266

Balota EL, Colozzi-Filho A, Andrade DS, Dick RP (2003) Microbial biomass in soils under different tillage in crop rotation systems. Biol Fertil Soils 38:15–20

Bakhshandeh S, Corneo PE, Mariotte P, Kertesz MA, Dijkstra FA (2017) Effect of crop rotation on mycorrhizal colonization and wheat yield under different fertilizer treatments. Agric Ecosyst Environ 247:130–136

Bashan Y, De-Bashan LE, Prabhu SR, Hernandez JP (2014) Advances in plant growth-promoting bacterial inoculant technology: Formulations and practical perspectives (1998–2013) (A Marschner Review). Plant Soil 378:1–33

Bashan Y, de-Bashan LE (2002) Reduction of bacterial speck (*Pseudomonas syringae* pv. tomato) of tomato by combined treatments of plant growth-promoting bacterium, *Azospirillum brasilense*, streptomycin sulfate, and chemothermal seed treatment. Eur J Plant Pathol 108:821–829

Beckie HJ (2006) Herbicide-resistant weeds: management tactics and practices. Weed Technol 20:793–814

Beare MH, Parmelee RW, Hendrix PF, Cheng W, Coleman DC, Crossley DA (1992) Microbial and faunal interactions and effects on litter nitrogen and decomposition in agroecosystem. Ecol Monogr 62:569–591

Belimov AA, Dodd IC, Safronova VI, Shaposhnikov AI, Azarova TS, Makarova NM, Davies WJ, Tikhonovich IA (2015) Rhizobacteria that produce auxins and contain 1-amino-cyclopropane-1-carboxylic acid deaminase decrease amino acid concentrations in the rhizosphere and improve growth and yield of well-watered and water-limited potato (*Solanum tuberosum*). Ann Appl Biol 167:11–25

Belyaeva ON, Haynes RJ (2012) Comparison of the effects of conventional organic amendments and biochar on the chemical, physical and microbial properties of coal fly ash as a plant growth medium. Environ Earth Sci 66:1987–1997

Berner A, Sousa H, Koller M, Mader P (2000) Suppressivity of garden waste and manure compost depending on the turning frequency and compost age. In: Proceedings 13th International IFOAM Scientific Conference Basel 5, p 55

Bertin C, Yang XH, Weston LA (2003) The role of root exudates and allelochemicals in the rhizosphere. Plant Soil 256:67–83

Bronick CJ, Lal R (2005) Soil structure and management: a review. Geoderma 124(1–2):3–22

Brown MW, Tworkoski T (2004) Pest management benefits of compost mulch in apple orchards. Agric Ecosyst Environ 103:465–472

Buick RD, Stone ND, Scheckler RK, Roach JW (1992) CROPS: a whole farm crop-rotation planning system to implement sustainable agriculture. AI Appl 6:29–50

Bullock DG (1992) Crop rotation. Crit Rev Plant Sci 11:309–326

Burketova L, Trda L, Ott PG, Valentova O (2015) Bio-based resistance inducers for sustainable plant protection against pathogens. Biotechnol Adv 33:994–1004

Çakmakçi R, Turan M, Güllüce M, Sahin F (2014) Rhizobacteria for reduced fertilizer inputs in wheat (*Triticum aestivum* spp. *vulgare*) and barley (*Hordeum vulgare*) on Aridisols in Turkey. Int J Plant Prod 8(2):163–181

Cao PB, Wang J, Wu H, Yan S, Guo D, Wang G, Ma Y (2016) Soil chemical and microbial responses to biogas slurry amendment and its effect on Fusarium wilt suppression. Appl Soil Ecol 107:116–123

Caravaca F, Hernandez T, Garcia C, Roldan A (2002) Improvement of rhizosphere aggregate stability of afforested semiarid plant species subjected to mycorrhizal inoculation and compost addition. Geoderma 108:133–144

Chan KY, Van ZL, Meszaros I, Downie A, Joseph S (2007) Agronomic values of biochar as a soil amendment. Aust J Soil Res 45:629–634

Chassot A, Stamp P, Richner W (2001) Root distribution and morphology of maize seedlings as affected by tillager and fertilizer placement. Plant Soil 231:123–135

Chauhan H, Bagyaraj DJ, Selvakumar G, Sundaram SP (2015) Novel plant growth promoting rhizobacteria – prospects and potential. Appl Soil Ecol 95:38–53

Chauhan A, Guleria S, Balgir PB, Walia A, Mahajan R, Mehta P, Shirkot CK (2017) Tricalcium phosphate solubilization and nitrogen fixation by newly isolated *Aneurinibacillus aneurinilyticus* CKMV1 from rhizosphere of *Valeriana jatamansi* and its growth promotional effect. Braz J Microbiol 48(2):294–304

Chan B, Yuan M, Qian L (2012) Enhanced bioremediation of PAHs-contaminated soil by immobilized bacteria with plant residue and biochar as carriers. J Soil Sedim 12:1350–1359

Chen G, Weil RR, Hill L (2014) Effects of compaction and cover crops on soil least limiting water range and air permeability. Soil Tillage Res 136:61–69

Chen H, Zhu Y (2011) Genetic diversity and disease control in rice. Nature 406:718–722

Cohen AC, Bottini R, Pontin M, Berli FJ, Moreno D, Boccanlandro H, Travaglia CN, Piccoli PN (2015) *Azospirillum brasilense* ameliorates the response of Arabidopsis thaliana to drought mainly via enhancement of ABA levels. Physiol Plant 153(1):79–90

Conn KL, Lazarovits G (1999) Impact of animal manures on verticillium wilt, potato scab, and soil microbial populations. Can J Plant Pathol 21:81–92

Cook RJ, Baker KF (1983) The nature and practice of biological control of plant pathogens. The American Phytopathological Society, St. Paul, p 539

Cookson WR, Murphy DV, Roper MM (2008) Characterizing the relationships between soil organic matter components and microbial function and composition along a tillage disturbance gradient. Soil Biol Biochem 40:763–777

Cotxarrera L, Trillas-Gay MI, Steinberg C, Alabouvette C (2002) Use of sewage sludge compost and *Trichoderma asperellum* isolates to suppress Fusarium wilt of tomato. Soil Biol Biochem 34:467–476

Debosz K, Vognsen L, Labouriau R (2002) Carbohydrates in hot water extracts of soil aggregates as influenced by long-term management. Commun Soil Sci Plant Anal 33:623–634

Dam RF, Mehdi BB, Burgess MSE, Madramootoo C, Mehuys GR, Callum IR (2005) Soil bulk density and crop yield under eleven consecutive years of corn with different tillage and residue practices in a sandy loam soil in central Canada. Soil Tillage Res 84:41–53

Dary M, Chamber-Pérez MA, Palomares AJ, Pajuelo E (2010) "In situ" phytostabilisation of heavy metal polluted soils using *Lupinus luteus* inoculated with metal resistant plant-growth promoting rhizobacteria. J Hazard Mater 177(1–3):323–330

Dawar S, Wahab S, Tariq M, Zaki MJ (2010) Application of *Bacillus* species in the control of root rot diseases of crop plants. Arch Phytopathol Plant Protect 43(4):412–418

de Leon-Gonzalez F, Hernandez-Serrano MM, Etchevers JD, Payan-Zelaya F, Ordaz-Chaparro V (2000) Short-term compost effect on macroaggregation in a sandy soil under low rainfall in the valley of Mexico. Soil Tillage Res 56:213–217

Dendoncker N, Van Wesemael B, Rounsevell MDA, Roelandt C, Lettens S (2004) Belgium's CO_2 mitigation potential under improved cropland management. Agric Ecosyst Environ 103:101–116

Derpsch R, Friedrich T, Kassam A, Hongwen L (2010) Current status of adoption of no-till farming in the world and some of its main benefits. Int J Agric Biol Eng 3:1–25

Deubel A, Hofmann B, Orzessek D (2011) Long-term effects of tillage on stratification and plant availability of phosphate and potassium in a loess chernozem. Soil Tillage Res 117:85–92

Dindal D (1978) Soil organisms and stabilizing wastes. Compost Science/ Land Utilization 19:8–11

Dorraji SS, Golchin A, Ahmadi S (2010) The effects of hydrophilic polymer and soil salinity on corn growth in sandy and loamy soils clean-Soil. Air Water 38:584–591

Duffy BK, Weller DM (1995) Use of *Gaeumannomyces graminis var. graminis* alone and in combination with fluorescent *Pseudomonas* spp. to suppress take-all of wheat. Plant Dis 79:907–911

El Titi A (ed) (2003) Soil tillage in agroecosystems. CRC Press, Boca Raton

Elad Y, David DR, Harel YM, Borenshtein M, Ben Kalifa H, Silber A (2010) Induction of systemic resistance in plants by biochar, a soil-applied carbon sequestering agent. Phytopathology 100:913–921

Elmer WH, Pignatello JJ (2011) Effect of biochar amendments on mycorrhizal associations and fusarium crown and root rot of asparagus in replant soils. Plant Dis 95:960–966

Epstein E (1994) The anomaly of silicon in plant biology. Proc Natl Acad Sci U S A 91:11–17

Fageria NK, Baligar VC, Bailey BA (2005) Role of cover crops in improving soil and row crop productivity. Commun Soil Sci Plant Anal 36:2733–2757

Freddo A, Cai C, Reid B (2012) Environmental contextualisation of potential toxic elements and polycyclic aromatic hydrocarbons in biochar. Environ Pollut 171:18–24

Fons F, Amellal N, Leyval C, Saint-Martin N, Henry M (2003) Effects of *Gypsophila* saponins on bacterial growth kinetics and on selection of subterranean clover rhizosphere bacteria. Can J Microbiol 49(6):367–373

Girvan MS, Bullimore J, Pretty JN, Osborn AM, Ball AS (2003) Soil type is the primary determinant of the composition of total and active bacterial communities in arable soils. Appl Environ Microbiol 69:1800–1809

Glaser B, Lehmann J, Zech W (2002) Ameliorating physical and chemical properties of highly weathered soils in the tropics with charcoal – a review. Biol Fertil Soils 35:219–230

Gleba D, Borisjuk NV, Borisjuk LG, Kneer R, Poulev A, Skarzhinskaya M (1999) Use of plant roots for phytoremediation and molecular farming. Proc Natl Acad Sci U S A 25:5973–5977

Glick BR, Todorovic B, Czarny J, Cheng Z, Duan J, McConkey B (2007) Promotion of plant growth by bacterial ACC deaminase. Crit Rev Plant Sci 26(5-6):227–242

Glick BR (2012) Plant growth-promoting bacteria: mechanisms and applications. Hindawi Publishing Corporation, Scientifica, Waterloo

Glynne MD (1965) Crop sequence in relation to soil-borne pathogens. In: Baker KF, Snyder WC (eds) Ecology of soil-borne plant pathogens, prelude to biological control. University of California Press, Berkeley, pp 423–433

Gomare KS, Mese M, Shetkar Y (2013) Isolation of *Azotobacter* and cost effective production of biofertilizer. Indian. Appl Res 3:54–56

Goss MJ, Tubeileh A, Goorahoo D (2013) A review of the use of organic amendments and the risk to human health. Adv Agron 120:275–379

Govaerts B, Mezzalama M, Sayre KD, Crossa J, Lichter K, Troch V, Vanherck K, De Corte P, Deckers J (2008) Long-term consequences of tillage, residue management, and crop rotation on selected soil microflora groups in the subtropical highlands. Appl Soil Ecol 38:197–210

Govindarajan M, Balandreau J, Kwon SW (2007) Effects of the inoculation of *Burkholderia vietnamensis* and related endophytic diazotrophic bacteria on grain yield of rice. Microb Ecol 55:21–37

Grayston SJ, Wang S, Campbell CD, Edwards AC (1998) Selective influence of plant species on microbial diversity in the rhizosphere. Soil Biol Biochem 30:369–378

Gundale MJ, DeLuca TH (2007) Charcoal effects on soil solution chemistry and growth of Koeleria macrantha in the ponderosa pine/Douglas-fir ecosystem. Biol Fertil Soils 43:303–311

Guo LB, Gifford RM (2002) Soil carbon stocks and land use change: a meta analysis. Glob Chang Biol 8:345–360

Haichar FZ, Marol C, Berge O, Rangel-Castro JI, Prosser JI, Balesdent J, Heulin T, Achouak W (2008) Plant host habitat and root exudates shape soil bacterial community structure. ISME J 2:1221–1230

Hamza MA, Anderson WK (2005) Soil compaction in cropping systems: a review of the nature, causes and possible solutions. Soil Tillage Res 82:121–145

Harel YM, Elad Y, Rav-David D, Borenstein M, Shulchani R, Lew B (2012) Biochar mediates systemic response of strawberry to foliar fungal pathogens. Plant Soil 357:245–257

Harrwing NL, Ulrich Ammon H (2002) Cover crops and living mulches. Weed Sci 50:688–699

Haynes RJ, Beare MH (1997) Influence of six crop species on aggregate stability and some labile organic matter fractions. Soil Biol Biochem 29(11):1647–1653

Hiltner L (1904) Über neuere Erfahrungen und Probleme auf dem Gebiet der Bodenbackteriologie und unter besonderer Berücksichtigung der Gründüngung und Brache. Arb Deut Landwirsch Ges 98:59–78. (in German)

Huang PY, Zimmerli L (2014) Enhancing crop innate immunity: new promising trends. Front Plant Sci 5:624

Huang J, Li H, Yuan H (2006) Effect of organic amendments on Verticillium wilt of cotton. Crop Prot 25(11):1167–1173

Huang X, Shi D, Sun F, Lu H, Liu J, Wu W (2012) Efficacy of sludge and manure compost amendments against Fusarium wilt of cucumber. Environ Sci Pollut Res Int 19(9):3895–3905

Islam F, Yasmeen T, Ali Q, Ali S, Arif MS, Hussain S, Rizvi H (2014) Influence of *Pseudomonas aeruginosa* as PGPR on oxidative stress tolerance in wheat under Zn stress. Ecotoxicol Environ Saf 104(1):285–293

Jain R, Saxena J, Sharma V (2010) The evaluation of free and encapsulated *Aspergillus awamori* for phosphate solubilization in fermentation and soil–plant system. Appl Soil Ecol 46(1):90–94

Jain R, Saxena J, Sharma V (2014) Differential effects of immobilized and free forms of phosphate-solubilizing fungal strains on the growth and phosphorus uptake of mung bean plants. Ann Microbiol 64(4):1523–1534

Jain R, Saxena J, Sharma V (2012) Effect of phosphate-solubilizing fungi *Aspergillus awamori* S29 on mungbean (*Vigna radiata* cv. RMG 492) growth. Folia Microbiol 57(6):533–541

Jarecki MK, Lal R (2003) Crop management for soil carbon sequestration. Crit Rev Plant Sci 22(6):471–502

Kannangara T, Utkhede RS, Paul JW, Punja ZK (2000) Effects of mesophilic and thermophilic composts on suppression of Fusarium root and stem rot of greenhouse cucumber. Can J Microbiol 46:1021–1028

Kao CM, Chen SC, Chen YS (2003) Detection of *Burkholderia pseudomallei* in rice fields with PCR-based technique. Folia Microbiol 48:521–524

Kemper WD, Trout TJ, Segeren A, Bullock M (1987) Worms and water. J Soil Water Conserv 42:401–404

Labrie C, Leclerc P, Cote N, Roy S, Brzezinski R, Hogue R, Beaulieu C (2001) Effect of chitin waste-based composts produced by two phase composting on two oomycete plant pathogens. Plant Soil 235:27–34

Lampurlanés J, Cantero-Martínez C (2006) Hydraulic conductivity, residue cover and soil surface roughness under different tillage systems in semiarid conditions. Soil Tillage Res 85:13–26. https://doi.org/10.1016/j.still.2004.11.006

Lau JA, Lennon JT (2011) Evolutionary ecology of plant–microbe interactions: soil microbial structure alters selection on plant traits. New Phytol 192:215–224

Lazarovits G, Conn KL, Potter J (1999) Reduction of potato scab, verticillium wilt, and nematodes by soymeal and meat and bone meal in two Ontario potato fields. Can J Plant Pathol 21:345–353

Lehmann J, Joseph S (2009) Biochar for environmental management: an introduction. In: Lehmann J, Joseph S (eds) Biochar for environmental management science and technology. Earthscans, London, pp 1–12

Liang B, Lehmann J, Solomon D, Kinyangi J, Grossman J, O'Neill B, Skjemstad JO, Thies J, Luizao FJ, Peterson J, Neves EG (2006) Black carbon increases cation exchange capacity in soils. Soil Sci Soc Am J 70:1719–1730

Laird D, Flaming P, Wang BQ, Horton R, Karlen D (2010) Biochar impact on nutrient leaching from Midwest agricultural soil. Geoderama 158(3–4):436–442

Liu H, Sarah J, Coulthurst SJ, Pritchard L, Hedley PE, Ravensdale M, Humphris S, Burr T, Takle G, Brurberg MB, Birch PRJ, Salmond GPC, Toth IK (2008) Quorum sensing coordinates brute force and stealth modes of infection in the plant pathogen *Pectobacterium atrosepticum*. PLoS Pathol 4:e1000093

Liu G, Ma J, Yang Y, Yu H, Zhang G, Xu H (2018) Effects of straw incorporation with a microbial inoculant on nitrous oxide and methane emissions from a wheat–rice rotation system. Pedosphere. In press

Lodha S, Burman U (2000) Efficacy of composts on nitrogen fixation, dry root-rot (*Macrophomina phaseolina*) intensity and yield of legumes. Indian J Agric Sci 70:846–849

Lodha S, Sharma SK, Aggarwal RK (2002) Inactivation of *Macrophomina phaseolina* propagules during composting and effect of composts on dry root-rot severity and on seed yield of clusterbean. Eur J Plant Pathol 108:253–261

Lugato E, Bampa F, Panagos P, Montanarella L, Jones A (2014) Potential carbon sequestration of European arable soils estimated by modelling a comprehensive set of management practices. Glob Chang Biol 20:3557–3567

Lumsden RD, Lewis JA, Millner PD (1983) Effect of composted sewage sludge on several soil borne pathogens and diseases. Phytopathology 73:1543–1548

Ma W, Charles TC, Glick BR (2004) Expression of an exogenous 1-aminocyclopropane-1-carboxylate deaminase gene in *Sinorhizobium meliloti* increases its ability to nodulate alfalfa. Appl Environ Microbiol 70:5891–5897

Mäder P, Kaiser F, Adholeya A, Singh R, Uppal HS, Sharma AK, Srivastava R, Sahai V, Aragno M, Wiemken A, Johri BN, Fried PA (2011) Inoculation of root microorganisms for sustainable wheat–rice and wheat–black gram rotations in India. Soil Biol Biochem 43:609–619

Marcroft SJ, Sprague SJ, Pymer SJ, Salisbury PA, Howlett BJ (2004) Crop isolation, not extended rotation length, reduces blackleg (*Leptosphaeria maculans*) severity of canola (*Brassica napus*) in south-eastern Australia. Aust J Exp Agric 44:601–606

Marks BB, Megías M, Nogueira MA, Hungria M (2013) Biotechnological potential of rhizobial metabolites to enhance the performance of *Bradyrhizobium* spp. and *Azospirillum brasilense*inoculants with soybean and maize. AMB Express 3:1–10

Marimuthu S, Ramamoorthy V, Samiyappan R, Subbian P (2013) Intercropping system with combined application of *Azospirillum* and *Pseudomonas fluorescens* reduces root rot incidence caused by *Rhizoctonia bataticola* and increases seed cotton yield. J Phytopathol 161:405–411

Martinez-Viveros O, Jorquera MA, Crowley DE, Gajardo G, Mora ML (2010) Mechanisms and practical considerations involved in plant growth promotion by rhizobacteria. J Soil Sci Plant Nutr 10:293–319

Mazzola M, Fujimoto DK, Thomashow LS (1995) Variation in sensitivity of *Gaeumannomyces graminis* to antibiotics produced by fluorescent *Pseudomonas* spp. and effect on biological control of take-all of wheat. Appl Environ Microbiol 61:2554–2559

Minaxi, Saxena J (2010) Disease suppression and crop improvement in moong beans (*Vigna radiata*) through *Pseudomonas* and *Burkholderia* strains isolated from semi arid region of Rajasthan, India. BioControl 55(66):799–810

Minaxi, Saxena J (2011) Efficacy of rhizobacterial strains encapsulated in nontoxic biodegradable gel matrices to promote growth and yield of wheat plants. Appl Soil Ecol 48(3):301–308

Mohiddin FA, Khan MR, Khan SM (2010) Why Trichoderma is considered super hero (super fungus) against the evil parasites? *Plant Pathol J* 9:1–11

Moore JM, Klose S, Tabatabai MA (2000) Soil microbial biomass carbon and nitrogen as affected by cropping systems. Biol Fertil Soils 31:200–210

Morris NL, Miller PCH, Froud-Williams RJ (2010) The adoption of non-inversion tillage systems in the United Kingdom and the agronomic impact on soil, crops and the environment—a review. Soil Tillage Res 108:1–15

Mukhtar S, Shahid I, Mehnaz S, Malik KA (2017) Assessment of two carrier materials for phosphate solubilizing biofertilizers and their effect on growth of wheat (*Triticum aestivum L.*). Microbiol Res 205:107–117

Namgay T, Singh B, Singh BP (2010) Influence of biochar application to soil on the availability of As, Cd, Cu, Pb, and Zn to maize (*Zea mays L.*). Aust J Soil Res 48:638–647

Naresh RK, Shukla AK, Kumar M, Kumar A, Gupta RK, Vivek SSP, Purushattom SPK, Kumar Y, Singh SP, Tomar SS, Singh V, Rathi RC, Mahajan NC, Kumar S, Singh S (2018) Cowpathy and vedic krishi to empower food and nutritional security and improve soil health: a review. J Pharmacog Phytochem 7(1):560–575

Naseem H, Bano A (2014) Role of plant growth-promoting rhizobacteria and their exopolysaccharide in drought tolerance of maize. J Plant Interact 9(1):689–701

Nunan N, Daniell TJ, Singh BK, Papert A, Mc Nicol JW, Prosser JI (2005) Links between plant and rhizoplane bacterial communities in grassland soils, characterized using molecular techniques. Appl Environ Microbiol 71:6784–6792

Rillig MC, Wright SF, Nichols KA, Schmidt WF, Torn MS (2001) Large contribution of arbuscular mycorrhizal fungi to soil carbon pools in tropical forest soils. Plant Soil 233:167–177

Pande A, Pandey P, Mehra S, Singh M, Kaushik S (2017) Phenotypic and genotypic characterization of phosphate solubilizing bacteria and their efficiency on the growth of maize. J Genet Eng Biotechnol 15(2):379–391

Paoletti MG (1999) Using bioindicators based on biodiversity to assess landscape sustainability. Agric Ecosyst Environ 74:1–18

Parikh SJ, James BR (2012) Soil: the foundation of agriculture. Nat Educ Knowl 3(10):2

Patrick ZA, Tousson TA, Koch LW (1964) Effect of crop residue decompostion products on plant roots. Annu Rev Phytopathol 2:267–292

Peters RD, Sturz AV, Carter MR, Sanderson JB (2003) Developing disease-suppressive soils through crop rotation and tillage management practices. Soil Tillage Res 72:181–192

Peters NK, Frost JW, Long SR (1986) A plant flavone, luteolin, induces expression of Rhizobium meliloti nodulation genes. Science 233:977–980

Pelegrin F, Moreno F, Martin-Aranda J, Camps M (1990) The influence of tillage methods on soil physical properties and water balance for a typical crop rotation in SW Spain. Soil Tillage Res 16:345–358

Pii Y, Mimmo T, Tomasi N, Terzano R, Cesco S, Crecchio C (2015) Microbial interactions in the rhizosphere: beneficial influences of plant growth-promoting rhizobacteria on nutrient acquisition process: a review. Biol Fertil Soils 51:403–415

Pittelkow CM, Linquist BA, Lundy ME, Liang X, van Groenigen KJ, Lee J, van Kessel C (2015) When does no-till yield more? A global meta-analysis. Field Crop Res 183:156–168

Pittelkow CM, Liang X, Linquist BA, van Groenigen KJ, Lee J, Lundy ME, van Gestel N, Six J, Venterea RT, van Kessel C (2014) Productivity limits and potentials of the principles of conservation agriculture. Nature 517:365–368

Poeplau C, Don A (2015) Carbon sequestration in agricultural soils via cultivation of cover crops – a meta-analysis. Agric Ecosyst Environ 200:33–41

Postma J, Schilder MT (2015) Enhancement of soil suppressiveness against *Rhizoctonia solani* in sugar beet by organic amendment. Appl Soil Ecol 94:72–79

Postma J, Montanari M, van den Boogert PHJF (2003) Microbial enrichment to enhance the disease suppressive activity of compost. Eur J Soil Biol 39(3):157–163

Powlson DS, Whitmore AP, Goulding KWT (2011) Soil carbon sequestration to mitigate climate change: a critical re-examination to identify the true and the false. Eur J Soil Sci 62:42–55

Prashar P, Kapoor N, Sachdeva S (2013) Rhizosphere: its structure, bacterial diversity and significance. Rev Environ Sci Biotechnol 13:63–77

Raaijmakers JM, Sluis LV, Bakker PAHM, Schippers B, Koster M, Weisbeek PJ (1995) Utilization of heterologous siderophores and rhizosphere competence of fluorescent *Pseudomonas* spp. Can J Microbiol 41:126–135

Raaijmakers JM, Paulitz TC, Steinberg C, Alabouvette C, Moënne-Loccoz Y (2009) The rhizosphere: a playground and battlefield for soilborne pathogens and beneficial microorganisms. Plant Soil 321(1–2):341–361

Rajasekaran S, Ganesh Shankar K, Jayakumar K, Rajesh M, Bhaaskaran C, Sundaramoorthy P (2012) Biofertilizers-current status of Indian agriculture. J Environ Bioenergy 4(3):176

Raoul des Essarts Y (2015) Pathogénie de Dickeya dianthicola et Dickeya solani chez *Solanum tuberosum*, Développement et Evaluation de Stratégie de Lutte Biologique. PhD thesis, Université Paris Sud

Rey FE, Gonzalez MD, Cheng J, Wu M, Ahern PP, Gordon JI (2013) Metabolic niche of a prominent sulfate-reducing human gut bacterium. Proc Natl Acad Sci USA 110

Rokhzadi A, Asgharzadeh A, Darvish F, Nourmohammadi G, Majidi E (2008) Influence of plant growth-promoting rhizobacteria on dry matter accumulation and yield of chickpea (*Cicer arietinum L.*) under field condition. Am-Eur J Agric Environ Sci 3:253–257

Saito M, Murumoto T (2002) Inoculation with arbuscular mycorrhizal fungi: the status quo in Japan and the future prospects. Plant Soil 244:273–279

Saraf M, Pandey U, Thakkar A (2014) Role of allelochemicals in plant growth promoting rhizobacteria for biocontrol of phytopathogens. Microbiol Res 169:18–29

Saranya K, Krishnan PS, Kumutha K, French J (2011) Potential for biochar as an alternate carrier to lignite for the preparation of biofertilizers in India. Int J Agric Environ Biotechnol 4(2):167–172

Saxena J, Rana G, Pandey M (2013) Impact of addition of biochar along with *Bacillus* sp. on growth and yield of French beans. Sci Hortic 162(23):351–356

Saxena J, Choudhary S, Pareek S, Choudhary AK, Equebal MA (2015) Recycling of organic waste through four different composts for disease suppression and growth enhancement in mung beans. Clean Soil Air Water 43(9999):1–6

Saxena J, Rawat J, Sanwal P (2016) Enhancement of growth and yield of *Glycine max* plants with inoculation of phosphate solubilizing fungus *Aspergillus niger* K7 and biochar amendment in soil. Commun Soil Sci Plant Anal 47(20)

Schwen A, Bodner G, Scholl P, Buchan GD, Loiskandl W (2011) Temporal dynamics of soil hydraulic properties and the water-conducting porosity under different tillage. Soil Tillage Res 113:89–98. https://doi.org/10.1016/j.still.2011.02.005

Shi L, Du N, Yuan Y, Shu S, Sun J, Guo S (2016) Vinegar residue compost as a growth substrate enhances cucumber resistance against the Fusarium wilt pathogen *Fusarium oxysporum* by regulating physiological and biochemical responses. Environ Sci Pollut Res Int 23(18):18277–18287

Shackley S, Carter S, Knowles T, Middelink E, Haefele S, Sohi S, Cross A, Haszeldine S (2012) Sustainable gasification-biochar systems? A case-study of rice-husk gasification in Cambodia, Part 1: Context, chemical properties, environmental and health and safety issues. Energy Policy 42:49–58

Sharon T, Moscovitch M, Gilboa A (2011) Rapid neocortical acquisition of long-term arbitrary associations independent of the hippocampus. Proc Natl Acad Sci 108:1146–1151

Snyder WC, Schroth MN, Christou T (1959) Effect of plant residues on root rot of beans. Phytopathology 49:755–756

Sohi S, Loez-Capel S, Krull E, Bol R (2009) Biochar's roles in soil and climate change: a review of research needs. CSIRO Land Water Sci Rep 05(09):64

Sollins P, Swanston C, Kramer M (2007) Stabilization and destabilization of soil organic matter-a new focus. Biogeochemistry 85:1–7

Spaepen S, Vanderleyden J, Remans R (2007) Indole-3-acetic acid in microbial and microorganism-plant signaling. FEMS Microbiol Rev 31(4):425–448

Spiegel H, Zavattaro L, Guzmán G, D'Hose T, Pecio A, Lehtinen T, Schlatter N, ten Berge H, Grignani C (2015) Compatibility of agricultural management practices and mitigation and soil health: impacts of soil management practices on crop productivity, on indicators for climate change mitigation, and on the chemical, physical and biological quality of soil deliverable. reference number: D3.371, CATCH-C Project

Srivastava R, Khalid A, Singh US, Sharma AK (2010) Evaluation of arbuscular mycorrhizal fungus, fluorescent *Pseudomonas* and *Trichoderma harzianum* formulation against *Fusarium oxysporum* f. sp. *lycopersici* for the management of tomato wilt. Biol Control 53:24–31

Srinivasan K, Mathivanan N (2009) Biological control of sunflower necrosis virus disease with powder and liquid formulations of plant growth promoting microbial consortia under field conditions. Biol Control 51:395–402

Steele MK, Coale FJ, Hill RL (2012) Winter annual cover crop impacts on no-till soil physical properties and organic matter. Soil Sci Soc Am J 76:2164–2173

Suárez R, Wong A, Ramírez M, Barraza A, Orozco Mdel C, Cevallos MA, Lara M, Hernández G, Iturriaga G (2008) Improvement of drought tolerance and grain yield in common bean by overexpressing trehalose-6- phosphate synthase in rhizobia. Mol Plant-Microbe Interact 21:958–966

Subbarao GV, Ito O, Berry WL, Wheeler RM (2003) Sodium – a functional plant nutrient. Crit Rev Plant Sci 22:391–416

Szczech MM (1999) Suppressiveness of vermicompost against Fusarium wilt of tomato. J Phytopathol 147:155–161

Tailor AJ, Joshi BH (2014) Harnessing plant growth promoting rhizobacteria beyond nature: a review. J Plant Nutr 37:1534–1571

Tajini F, Trabelsi M, Drevon JJ (2012) Combined inoculation with *Glomus intraradices* and *Rhizobium tropici* CIAT899 increases phosphorus use efficiency for symbiotic nitrogen fixation in common bean (*Phaseolus vulgaris* L.). Saudi J Biol Sci 19(2):157–163

Tebrügge F, Düring R, Du R, Tebru F (1999) Reducing tillage intensity–a review of results from a long-term study in Germany. Soil Tillage Res 53:15–28

Termorshuizena AJ, Rijna E, van der Gaagb DJ, Alabouvettec C, Chend Y, Lagerlofe J, Malandrakisf AA, Paplomatasf EJ, Ramerte B, Ryckeboerg J, Steinbergc C, Zmora-Nahum S (2006) Suppressiveness of 18 composts against 7 pathosystems: variability in pathogen response. Soil Biol Biochem 38:2461–2477

Thamer S, Schaedler M, Bonte D, Ballhorn DJ (2011) Dual benefit from a belowground symbiosis: nitrogen fixing rhizobia promote growth and defense against a specialist herbivore in a cyanogenic plant. Plant Soil 341:209–219

Tripti KA, Usmani Z, Kumar V, Anshumali (2017) Biochar and flyash inoculated with plant growth promoting rhizobacteria act as potential biofertilizer for luxuriant growth and yield of tomato plant. J Environ Manag 190:20–27

Trivedi PC, Barker KR (1986) Management of nematodes by cultural practices. Nematropica 16:213–236

Troeh FR, Thompson LM (2005) Soils and soil fertility, 5th edn. Blackwell, Iowa

Tsror L, Barak R, Sneh B (2001) Biological control of black scurf on potato under organic management. Crop Prot 20:145–150

Unger PW (1992) Infiltration of simulated rainfall: tillage system and crop residue effects. Soil Sci Soc Am J 56:283

Upadhyay SK, Singh DP (2015) Effect of salt-tolerant plant growth-promoting rhizobacteria on wheat plants and soil health in a saline environment. Plant Biol 17(1):288–293

Uren NC (2000) Types, amounts and possible functions of compounds released into the rhizosphere by soil grown plants. In: Pinton R, Varanini Z, Nannipieri P (eds) The rhizosphere: biochemistry and organic substances at the soil interface. Marcel Dekker, New York, pp 19–40

Uri ND, Atwood JD, Sanabria J (1998) The environmental benefits and costs of conservation tillage. Sci Total Environ 216:13–32

van Dillewijn P, Soto MJ, Villadas PJ, Toro N (2001) Construction and environmental release of a *Sinorhizobium meliloti* strain genetically modified to be more competitive for alfalfa nodulation. Appl Environ Microbiol 67:3860–3865

van Os GJ, van Ginkel JH (2001) Suppression of Pythium root rot in bulbous Iris in relation to biomass and activity of the soil microflora. Soil Biol Biochem 33:1447–1454

Vejan P, Abdullah R, Khadiran T, Ismail S, Boyce AN (2016) Role of plant growth promoting rhizobacteria in agricultural sustainability – a review. Molecules 21(5):573

Verma VC, Singh SK, Prakash S (2011) Bio-control and plant growth promotion potential of siderophore producing endophytic Streptomyces from *Azadirachta indica* A Juss. J Basic Microb 51:550–556

Vimal SR, Patel VK, Singh JS (2018) Plant growth promoting Curtobacterium albidum strain SRV4: an agriculturally important microbe to alleviate salinity stress in paddy plants. Ecol Indic. https://doi.org/10.1016/j.ecolind.2018.05.014

Viveganandan G, Jauhri KS (2000) Growth and survival of phosphate-solubilizing bacteria in calcium alginate. Microbiol Res 155:205–207

Villamil MB, Nafziger ED (2015) Corn residue, tillage, and nitrogen rate effects on soil carbon and nutrient stocks in Illinois. Geoderma 253–254:61–66

Wang X, Zhou B, Sun X, Yue Y, Ma W, Zhao M (2015) Soil tillage management affects maize grain yield by regulating spatial distribution coordination of roots, soil moisture and nitrogen status. PLoS One 10:e0129231. https://doi.org/10.1371/journal.pone.0129231

Wani PA, Khan MA, Zaidi A (2007) Synergistic effects of the inoculation with nitrogen-fixing and phosphate-solubilizing rhizobacteria on the performance of field-grown chickpea. J Plant Nutr Soil Sci 170(2):283–287

Warnock DD, Lehmann J, Kuyper TW, Rillig MC (2007) Mycorrhizal responses to biochar in soil – concepts and mechanisms. Plant Soil 300:9–20

West TO, Post WM (2002) Soil carbon sequestration by tillage and crop rotation: a global data analysis. Soil Sci Soc Am J 66:1930–1946

Xiao Z, Liu M, Jiang L, Chen X, Grffiths BS, Li H, Hu F (2016) Vermicompost increases defense against root-knot nematode (*Meloidogyne incognita*) in tomato plants. Appl Soil Ecol 105:177–186

Yao AV, Bochow DH, Kaimov S, Botourv U, Sanginboy S, Sharipov AK (2006) Effect of FZB 24® *Bacillus subtilis* as a biofertilizer on cotton yields in field tests. Arch Phytopathol Plant Protect 39(4):323–328

Zahir ZA, Shah MK, Naveed M, Akhter MJ (2010) Substrate-dependent auxin production by *Rhizobium phaseoli* improves the growth and yield of *Vigna radiata* L. under salt stress conditions. J Microbiol Biotechnol 20(9):1288–1294

Zhang X, Huang Y, Harvey PR, Ren Y, Zhang G, Zhou H, Yang H (2012) Enhancing plant disease suppression by *Burkholderia vietnamiensis* through chromosomal integration of *Bacillus subtilis* chitinase gene chi113. Biotechnol Lett 34:287–293

Zhao L, Zhang Y (2015) Effects of phosphate solubilization and phytohormone production of *Trichoderma asperellum* Q1 on promoting cucumber growth under salt stress. J Integr Agric 14(8):1588–1597

Harnessing the Microbial Interactions in Rhizosphere and Microbiome for Sustainable Agriculture

17

Anushree Suresh and Jayanthi Abraham

1 Introduction

Agriculture has been stated to be the backbone of income in most developing countries. However, the usage of intensive amounts of synthetic fertilisers, herbicides, and pesticides causes hazardous environmental conditions, thereby leading to health problems. Farmers use large amounts of synthetic nitrogen fertilisers, available in the form of ammonium nitrate, as fertiliser for better yield of crops. An influx in the concentration of ammonia present in the soil diminishes ammonia production, leading to a lower concentration of microorganisms present in the soil. Sometimes, the high level of ammonia deposited in the soil is used up by nitrogen-fixing bacteria, thereby producing nitrate in the soil. The nitrate so produced is utilised by nitrogen-fixing bacteria to produce N_2O, thereby leaching out the excess nitrate to the groundwater (Galloway et al. 2008). Hence, production of N_2O is increased by the processes of microbial denitrification and nitrification. Denitrification reduces nitrogen oxides to other gaseous products with the help of microorganisms and releasing them again into the environment, whereas nitrification is a double-step process wherein soil microbial flora convert atmospheric ammonia to nitrate (Butterbach et al. 2013).

For agricultural sustainability, yield of crops should be dependent on various parameters such as salt tolerance, disease resistance, drought tolerance, better nutritional value, and heavy metal stress tolerance. Microorganisms (bacteria, fungi, algae, etc.) can exploit crop yield as they have the capability to increase nutrient and water uptake by plants (Armanda et al. 2014). Diverse microorganisms are beneficial for plant growth; among these are those potential types called plant growth-promoting rhizobacteria (PGPR). PGPR help increase crop yield and also enhance the health of the plant without damaging the soil microbiota.

A. Suresh · J. Abraham (✉)
Microbial Biotechnology Laboratory, School of Biosciences and Technology, VIT University, Vellore, Tamil Nadu, India

D. P. Singh et al. (eds.), *Microbial Interventions in Agriculture and Environment*,
https://doi.org/10.1007/978-981-13-8383-0_17

For years, many organisms with PGPR activity have been reported and commercialized, including species of the genera *Pseudomonas*, *Bacillus*, *Enterobacter*, *Klebsiella*, *Azotobacter*, *Variovorax*, *Azospirillum*, and *Serratia* (Glick 2012). However, the usage of PGPR in the agriculture industry represents only a small fraction of agricultural practice worldwide (Bashan et al. 2014) because of the differing practices of the inoculated PGPR, which influence crop yield. The mechanism of action and its successful application in plants depend on its survival in the microbial soil biota, its compatibility with the plant onto which it is growing or inoculated, and interaction of these PGPR with the other microorganisms in the soil and various environmental factors (Martinez et al. 2010). Another challenge is that PGPR have different modes of action possessing different mechanisms (Dey et al. 2004; Choudhary et al. 2011). These disadvantages limit PGPR application in the agricultural field. Comparing chemical input with PGPRs as biofertilisers is becoming a new face of the global agricultural production system which needs to be available for the population boom. This new technique will be used by 8 billion people by 2025 to as many as 9 billion by 2050.

Among soil microbiota, PGPR have a vital role in sustainable agriculture for their plant growth promotional capability as well as their biocontrol potential. PGPR have emerged as beneficial soil microorganisms, involved in controlling a number of plant diseases and pests by synthesizing a wide range of antagonistic secondary metabolites. These ubiquitously found microorganisms can be used as an important part of management practices to attain sustainable yields.

2 Plant Growth-Promoting Rhizobacteria

The first report on promotion of plant growth and biological control of root pathogens by seed imbibition with the bacteria present in the soil was that by Kloepper et al. (1989), who described the plant growth-promoting activity of *Pseudomonas* strains that had antagonistic properties to a wide range of plant pathogens in vitro. This work also confirms that the rhizosphere microorganisms could be modified knowingly with bacteria into the plant material. Kloepper was the first scientist to coin the term PGPR to comprise bacteria residing in the root and rhizosphere soil having the ability to increase plant growth.

2.1 PGPR Background

Microorganisms have been used in earlier times to encourage plant growth. Theophrastus (372–287 BC) reported the assimilation of various soil samples for preventing various deficiencies and making the soil fertile (Tisdale and Nelson 1975). Hellriegel and Wilfarth (1888) have observed that plant roots such as those of grasses and legumes are inhabited by rhizospheric microorganisms that can convert atmospheric N_2 into other nitrogen forms available for plant growth.

PGPR are generally divided into three functional groups (Fig. 17.1).

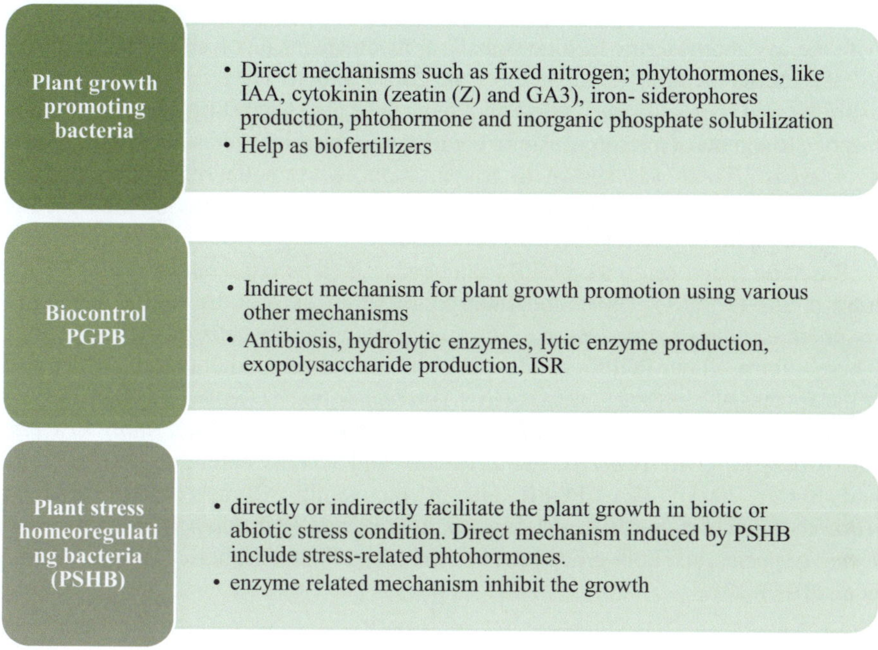

Fig. 17.1 Flowchart describing the three functional groups of plant growth-promoting rhizobacteria (PGPR)

2.2 Plants and Soil Microbiome Root Exudation

The rhizosphere of actively growing plants produces root exudates of which the chemical composition is of high importance for plant–microbe interactions (Badri and Vivanco 2009). The composition of the exudates produced by roots differs by plant species (Badri and Vivanco 2009) and within soil microbial communities. Differences in root exudates among various plant species and genotypes suggest the possibility of handling root exudation in agricultural crops to create selective effects on rhizosphere microbiota. Root exudates are composed of sugars (dextrose, sucrose, lactose), proteins, flavonoids, fatty acids, and amino acids (Badri and Vivanco 2009). These complex compounds can aid as growth substrates or signals for suitable microbial partners, and as antimicrobials or growth deterrents for other microbes (Badri and Vivanco 2009). In several classic examples of symbiosis, a very sophisticated interplay of chemical signalling mediates plant–microbe interactions. For example, legumes release flavonoids that regulate gene expression pattern in the rhizobia, initiating a series of complex and specific interactions that ultimately lead to atmospheric nitrogen fixation inside the nodules (Oldroyd and Downie 2008). An important upcoming task is to determine whether chemical signalling also has a major role in plant microbiome interactions in the rhizosphere. Evidence supports the idea that cross-kingdom interactions may be commonly mediated by chemical signalling. For instance, plants have been shown to interact with the acyl-homoserine lactone signalling mechanisms involved in bacterial cell-to-cell communication (Delalande et al. 2005) and to enhance antifungal gene expression in root-associated bacteria. Plants have also been shown to perceive and respond to signals of microbial origin. For instance, the rhizobacterium *Pseudomonas fluorescens* CHA0 was shown to trigger induced systemic resistance (ISR) in *Arabidopsis thaliana* through the production of the compound having the structure 2,4-diacetylphloroglucinol (Iavicoli et al. 2003).

Based on their origin, the PGPRs can be classified as extracellular (ePGPR) and intracellular (iPGPR). The extracellular PGPRs may exist in the habitat of the rhizosphere, on the rhizoplane, or in the spaces between the cells of the root cortex whereas intracellular PGPRs reside inside the specialized nodular structures of root cells. Species of bacterial genera such as *Agrobacterium*, *Arthrobacter*, *Azotobacter*, *Azospirillum*, *Bacillus*, *Burkholderia*, *Caulobacter*, *Chromobacterium*, *Erwinia*, *Flavobacterium*, *Micrococcus*, *Pseudomonas*, and *Serratia* belong to ePGPR (Gray and Smith 2005). The iPGPR are in the family Rhizobiaceae, including *Allorhizobium*, *Bradyrhizobium*, *Mesorhizobium*, and *Rhizobium*, endophytes, and *Frankia* species, and both groups can symbiotically fix atmospheric nitrogen (Verma et al. 2010).

3 Plant Growth Promotion: Mode of Action

PGPRs have a well-known phenomenon of plant growth promotion. They have a number of modes of action for promoting plant growth and development in diverse environmental conditions. According to Kloepper and Schroth (1981), plant growth promotion by PGPR results from changes in the whole microbial community composition in the rhizosphere because of the production of various substances. Generally, PGPRs promote plant growth directly either by supplying nutrients (N, P, K, Zn, Fe, and other essential minerals) or by modification of plant hormone levels. Indirectly, PGPRs can decrease the inhibitory impact of various pathogens on plant growth and act as microbial control agents and environmental protectors (Kloepper and Schroth 1981) (Fig. 17.2).

4 Direct Mechanisms

Plant growth-promoting rhizobacteria (PGPR) provide a direct mechanism for processes such as nitrogen fixation, solubilisation of mineral nutrients, organic compound mineralisation, and production of phytohormones (Arora et al. 2012; Bharadwaj et al. 2014).

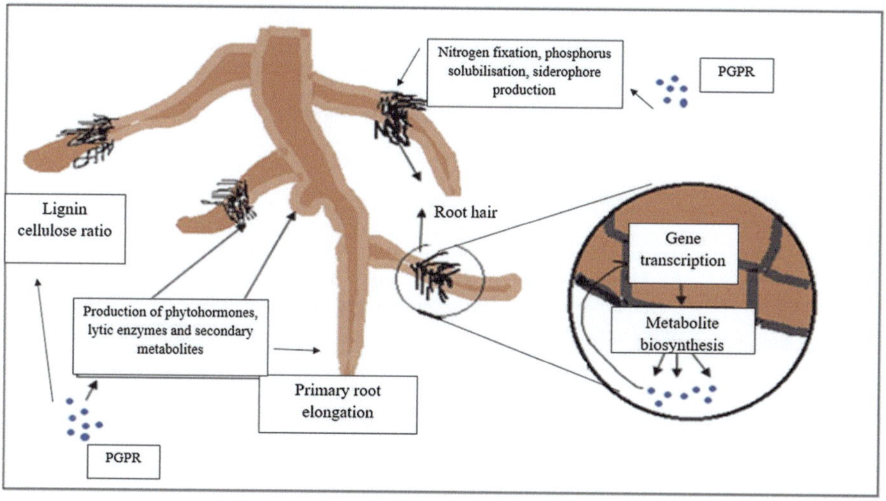

Fig. 17.2 Schematic representation of the PGPR mode of action for plant growth promotion in plants depicting the location and action of nitrogen fixation, phosphorus solubilisation, and siderophore production activity

4.1 Nitrogen Fixation Ability

Nitrogen has a pivotal function in plant growth and is also a requisite nutritional element for all life forms. In the atmosphere, nitrogen is present in a concentration of approximately 78% and is usually unavailable to plants. Unfortunately, plants are incapable of fixing atmospheric nitrogen to the usable form, ammonia, which can be directly used for the growth of the plant. Microorganisms become beneficial to plants by converting free nitrogen to ammonia. These nitrogen-fixing microorganisms are classified as biological nitrogen fixers (BNF) that convert atmospheric nitrogen to ammonia by the enzyme nitrogenase (Gaby and Buckley 2012). PGPR follow two mechanisms to fix atmospheric nitrogen to the usable plant form: symbiotic and nonsymbiotic. Symbiotic nitrogen fixation is a process of beneficial/mutual relationship between the PGPR and the plant. The PGPR microorganisms (listed in Table 17.1) primarily inhabit the root, forming root nodules where nitrogen fixation occurs. Rhizobacteria are microorganisms with a symbiotic relationship with the leguminous plant root by forming nodules that convert atmospheric nitrogen to ammonia, thus making it available for plants (Ahemad and Kibret 2014). Nitrogen fixation genes, namely, the *nif* genes (listed in Table 17.2), are usually found in both symbiotic and free-living systems such as *Klebsiella pneumoniae* (Reed et al. 2011). Nitrogenase (*nif*) genes consist of structural genes that function in electron donation, activation of Fe protein, iron molybdenum cofactor biosynthesis, and regulatory genes necessary for the synthesis and function of the enzyme.

Table 17.1 List of microorganisms involved in the nitrogen fixation activity

Nitrogen fixation steps	Microorganisms involved	References
As symbionts with leguminous plants	*Rhizobium, Bradyrhizobium, Sinorhizobium, Mesorhizobium*	Gupta et al. (2015)
Non-symbiotic nitrogen fixers	Free living diazotrophs	
N-fixing rhizospheric bacteria	*Azoarcus, Azotobacter, Acetobacter, Azospirillum, Burkholderia, Diazotrophicus, Enterobacter, Gluconacetobacter, Pseudomonas* and cyanobacteria (*Anabaena, Nostoc*)	

Table 17.2 *nif* genes and their role in nitrogen fixation activity

nif genes	Role in nitrogen fixation activity	References
nifH	Dinitrogenase reductase. Obligate electron donor to dinitrogenase during dinitrogenase turnover.	Klipp et al. (2004)
	Required for FeMo-Co biosynthesis and apodinitrogenase maturation.	
nifD	α-Subunit of dinitrogenase. Forms ß-tetramer with ß-subunit interface	
	FeMo-Co, the site substrate reduction is present	
nifK	ß-subunits of dinitrogenase.	
	ß-clusters are present at -subunit interface	
nifY	Aids in the insertion of FeMo-Co into apodinitrogenase in *K. pneumoniae* a	
nifE	Forms a ß-tetramer with nifN. Required for FeMo-Co synthesis	
nifN	Required for FeMo-Co synthesis	
nifW	Involved in the stability of dinitrogenase	
	Protects dinitrogenase from oxygen inactivation	
nifM	Required for the maturation of nifH	
nifB	Required FeMo-co synthesis *Nif*B-Co is the specific Fe and S donor to FeMo-Co complex	

Nitrogen fixation bacteria are inoculated onto the plant to promote plant growth, and also to combat the diseases targeting the crop and maintain the level of nitrogen in the agricultural field.

4.2 Phosphate Solubilisation

Phosphorus is an abundant element useful in the uptake of nutrients for plants. Phosphorus is crucial in many metabolic pathways, such as macromolecular biosynthesis, signal transduction, energy transfer, photosynthesis, and respiration (Khan et al. 2010). In the soil, phosphorus is usually found in two forms: organic and inorganic. Phosphorus is not taken up by plants because 95–99% of the element is present in immobilized, insoluble, and precipitated forms (Pandey and Maheshwari 2007). The absorbable versions of phosphorus are monobasic (HPO_4) and the

Table 17.3 List of microorganisms involved with phosphate solubilisation activity

Type of Phosphate solubilisation organisms	Microorganisms involved
Bacteria	*Arthrobacter, Bacillus, Beijerinckia, Burkholderia, Enterobacter, Erwinia, Flavobacterium, Microbacterium, Pseudomonas, Rhizobium, Rhodococcus*, and *Serratia*
Fungus	*Penicillium and Aspergillus, Rhizoctonia solani, Trichoderma.*
Yeasts	*Yarrowia lipolytica*

Table 17.4 Genes involved in phosphate solubilisation activity

Genes involved	Microorganisms (source)	Reference
napA	*Morganella morganii*	Fraga et al. (2001)
ushA	*Escherichia coli*	Burns and Beacham (1986)
Agp	*Escherichia coli*	Pradel and Boquet (1990)
cpdB	*Escherichia coli*	Beacham and Garrett (1980)

HPO_4^{2-} ions (Bhattacharyya and Jha 2012). PGPR present in the flora of the soil take up the unavailable forms of phosphate and through their enzymatic systems (genes tabulated in Table 17.4) break these down into simpler compounds that are later imbibed by the plants for their growth. The mechanisms attributed to phosphorus solubilisation by the PGPRs include (1) complex compounds or dissolved mineral compounds released, such as organic acid anions, protons, hydroxyl ions, and CO_2, (2) the release of the extracellular enzymes (biochemical phosphate mineralization), and (3) liberation of phosphate during degradation of substrate (biological phosphate mineralization) (Sharma et al. 2013). Various useful effects of PGPRs are reported (Table 17.3) during the process of phosphorus solubilisation (Zaidi et al. 2009).

4.3 Potassium Solubilisation

Potassium (K) is the third major essential nutrient required for plant growth promotion. The concentrations of soluble potassium present in the soil are usually very low: 90% and greater amounts of potassium present in the soil exist in the form of silicate minerals and insoluble rocks (Parmar and Sindhu 2013). However, usage of unbalanced content of fertiliser, and decline in potassium content in the soil, has resulted in a major problem for plant growth. Deficiency of potassium or inadequate potassium causes poorly developed roots, retarded growth with small seeds, and low yield. This concern has focused attention towards the search for an alternative indigenous source of potassium for plant uptake and to maintain potassium status in soils for sustainable crop production (Kumar and Dubey 2012). PGPR solubilize potassium rock through the production and secretion of organic acids (Han and Lee 2006). Thus, application of potassium-solubilising PGPR (Table 17.5) as biofertiliser for the improvement of agriculture can reduce use of chemical input and promote ecologically responsible crop production.

4.4 Siderophore Production

Iron is among the important micronutrients required for crop plants. Iron is not voluntarily assimilated by either plants or bacteria because the ferric ion, Fe^{3+}, the predominant form in nature, is only parsimoniously soluble so that the amount of iron available for assimilation by living organisms is extremely low (Ma 2005). However, microorganisms have developed a specialized mode of action for iron assimilation, including the low molecular weight production of iron-chelating compounds known as siderophores, which transport this element into their cells (Arora et al. 2013). Siderophores have basically three main categories depending on the characteristics of functional groups present in their structural composition: hydroxamates, catecholates, and carboxylates. At present approximately 500 different kinds of siderophores are known, of which 270 have been structurally modified (Cornelis 2010). The genes involved in siderophore formation include *asbD*, *asbG*, *asbC*, *asbF*, *asbB*, and *asbI*, produced by *Aeromonas salmonicida*. Siderophores are related to both the direct and indirect enhancement of plant growth by PGPR (Table 17.6). The direct assistance of siderophores produced by bacteria on the growth of plants has been established by using radiolabelled ferric siderophores as the sole source of iron, which confirmed that plants are able to take up the labelled iron by a large number of PGPR (as listed in Table 17.6).

Table 17.6 Microorganisms with siderophore activity

Mechanisms	Microorganisms involved
Bacterial reduction of ferric to ferrous ion	*Aeromonas*, *Azadirachta*, *Azotobacter*, *Pseudomonas* sp., *Rhizobium* sp., *Bacillus* sp., *Burkholderia* sp., *Serratia* and *Streptomyces* sp.
Reduction of ferric to ferrous iron	*Candida albicans, Cryptococcus neoformans, Geotrichum candidum, saccharomyces cerevisiae*
Siderophore procurement of ferric iron hydrooxamates (Rhodotorulic acids)	*Blastomyces dermatitidis, Epicoccum purpurescens, Histoplasma capsulatum, Stemphilium botryosum*
Siderophore procurement of ferric iron hydrooxamates (Ferrichromes)	*Aspergillus* sp., *Microsporum* sp., *Neurospora crassa, Trichophyton* spp., *Ustilago maydis*
Siderophore procurement of ferric iron hydrooxamates (Coprogens)	*Blastomyces dermatitidis, Curvularia lunata, Epicoccum purpurescens, Fusarium dimerum, Histoplasma capsulatum, Neurospora crassa*

Table 17.5 List of microorganisms involved with potash solubilisation activity

Type of microorgansism	Microorganisms involved
Bacteria	*Acidothiobacillus ferrooxidans, Bacillus edaphicus, Bacillus mucilaginosus, Burkholderia, Paenibacillus* sp., and *Pseudomonas*
Fungus	*Fomitopsis meliae, Aspergillus tubingensis, Penicillium citrinum*

4.5 Phytohormone Production

A great variety of microorganisms found in the habitat of the rhizosphere are able to synthesise substances that regulate growth and development of the plant. PGPR synthesises certain phytohormones such as gibberellins, auxins, cytokinins, and ethylene which can affect the proliferation of cells in the lateral roots and increase root hairs because of enhanced nutrient and water uptake (Arora et al. 2013).

4.6 Indole Acetic Acid (IAA)

Of all growth regulators, indole acetic acid (IAA) is a naturally found auxin with much biological importance relevant to the promotion of plant growth (Miransari and Smith 2014). As much as 80% of the rhizobacteria that colonize on the seed or root surfaces and can biosynthesize IAA or act in combination with endogenous IAA to stimulate proliferation of cells in the plants and enhance the host uptake of minerals and nutrients from the soil (Vessey 2003). The mode of action of IAA is through plant cell division, extension, and differentiation; stimulating seed and germination of tubers; increasing the xylem rate and development of roots; controlling the processes of vegetative growth; initiating the formation of lateral and adventitious roots; mediating light responses, gravity, and florescence; affecting the process of photosynthesis, formation of pigment, biosynthesis of various metabolites, and resistance activity to stressful conditions (Spaepen and Vanderleyden 2011). Tryptophan is an amino acid commonly found in the exudates of roots and has been identified as the precursor molecule for IAA biosynthesis in bacteria (Etesami et al. 2009). IAA synthesis by PGPR (Table 17.7) involves formation via indole-3-acetic aldehyde and indole-3-pyruvic acid, this being the most common mechanism in bacteria. The genes responsible for IAA production are *lao*, *ccc*A, *rid*A, and *rpo*D produced by *Gluconacetobacter diazotrophicus* (Rodrigues et al. 2016). The microorganisms involved with IAA formation are listed in Table 17.7.

Table 17.7 Microorganisms involved with IAA formation

Type of microorganisms	Microorganisms
Bacteria	*Pseudomonas, Rhizobium, Bradyrhizobium, Enterobacter, Acetobacter dizotrophicous, Klebsiella, Alkaligenes faecalis, Enterobacter cloacae,* species of *Azospirillum, Agrobacterium, Pseudomonas, Xanthomonas* sp.
Fungus	*Aspergillus* sp., *Microsporum* sp., *Neurospora crassa, Trichophyton* spp., *Ustilago maydis*

Table 17.8 Microorganisms involved with cytokinins and gibberellins production

Type of microorganisms	Microorganisms
Bacteria	*Pseudomonas*, *Rhizobium*, *Bradyrhizobium*, *Enterobacter*, *Acetobacter dizotrophicous*, *Klebsiella*, *Alkaligenes faecalis*, *Enterobacter cloacae*, species of *Azospirillum*, *Agrobacterium*, *Pseudomonas* and *Xanthomonas* sp.
Fungus	*Aspergillus* sp., *Microsporum* sp., *Neurospora crassa*, *Trichophyton* spp., *Ustilago maydis*

4.7 Cytokinins and Gibberellins

It has been observed that phytopathogens synthesize cytokinins for the plant. PGPR-producing cytokinins are of much lower level compared to the phytopathogen-producing cytokinins and thus promote plant growth, thereby reducing the phytopathogens. Ethylene is one of the most significant phytohormones for its various biological activities such as initiation of plant roots, stimulation of seed germination, ripening of fruits, lesser extent of wilting, promotion of leaf abscission, and production of various other plant hormones (Glick et al. 2007). The high concentration of ethylene production in the plant induces defoliation and inhibits the other cellular processes that may lead to reduced crop performance (Bhattacharyya and Jha 2012). One of the prerequisites for the production of ethylene is ACC carboxylic acid (1-aminocyclopropane-1), which is catalysed by the enzyme ACC oxidase. The microorganisms involved with cytokinins and gibberellin production are listed in Table 17.8. Another experiment conducted by Iqbal et al. (2012) reported the inoculation of the PGPR strain of *Pseudomonas* sp. into *Rhizobium leguminosarum* showed lower production of ethylene, thus improving other plant growth factors such as increase in the number of root nodules, increase in nodule dry weight, greater yield of straw and grains, and increased nitrogen content.

5 Biocontrol Plant Growth-Promoting Bacteria (Biocontrol PGPB)

The biocontrol PGPB provide a defensive mechanism for the plants by the production of microbial allelochemicals, such as biocidal volatiles, iron-chelating siderophores, hydrolytic enzymes, antibiotics, lytic enzymes, detoxification enzymes, and exopolysaccharides.

5.1 Antibiosis

Antibiotic production using plant growth-promoting bacteria (PGPB) is one of the defensive mechanisms acquired by plants. A large group of antibiotics have been discovered and studied, such as phenazine, pyoluteorin, pyrrolnitrin, tensin, tropolone, amphisin, 2,4-diacetylphloroglucinol (DAPG), hydrogen cyanide, oomycin A, and cyclic lipopeptides, produced by *Pseudomonas*, and oligomycin A, kanosamine, zwittermicin A, and xanthobaccin, produced by *Bacillus*, *Streptomyces*, and *Stenotrophomonas* spp. Antibiotics produced from PGPB can be used in experimental pharmaceuticals and hence provide a pathway to fight against multidrug-resistant (MDR) human pathogenic bacteria. The quantity of trace elements such as zinc and carbon present in the soil influences the stability or instability of microorganisms, thereby affecting the production of secondary metabolites. The genotype of the plant host is also important in suppressing the disease ability of microorganisms in interactions with plants.

5.2 Production of Lytic Enzymes

Most microorganisms have hyperparasitic ability, that is, they are able to attack the pathogens with the help of intracellular or extracellular enzymes that hydrolase the cell wall (Chernin and Chet 2002). One such example is *Serratia plymuthica*, which produces the enzyme chitinase that inhibits spore germination and germ tube elongation in *Botrytis cinerea* (Frankowski et al. 2001). Another organism, *Serratia marcescens*, has the ability to produce extracellular chitinases acting as antagonists against *Sclerotium rolfsii* (Ordentlich et al. 1988). Lim et al. (1991) demonstrated the synthesis of extracellular laminarinase and chitinase produced by *Pseudomonas stutzeri* to digest and lyse the mycelia of *Fusarium solani* (Lim et al. 1991). The β-1,3-glucanase synthesized by *Burkholderia cepacia* has the ability to destroy the cell wall integrity of certain microorganisms such as *Rhizobacterium solani*, *Sclerotium rolfsii*, and *Phythium ultimum*. Production of lytic enzyme (chitinase and protease enzyme) is regulated by various regulatory systems such as GrrA/GrrS or GacA/GacS (Ovadis et al. 2004).

5.3 Detoxification and Degradation of Virulence Factors by PGPR

Another main mechanism of biocontrol PGPB is detoxification of certain virulence factors produced by the pathogenic bacteria. For example, microbial control agents detoxify albicidin toxin, which is produced by *Xanthomonas albilineans* (Zhang and Birch 1997). Recent studies showed that some of the PGPB inhibit the quorum-sensing capability by degrading the autoinducer (AI) signals, which blocks the expression of several virulence genes (Dong et al. 2004). The mode of action of this auto-inducer-mediated quorum sensing is to turn on the cascade of genes responsible for virulence key factors such as phytotoxins and cell-degrading enzymes that will hold a vital potential for alleviating disease post infection, in a curative manner (Von Bodman et al. 2003). The toxins produced by the pathogens have broad-spectrum activity in inhibiting the development of competitive microorganisms or detoxifying the antibiotics synthesised by some biocontrol microorganisms as a self-defense mechanism against microbial control agents (Duffy et al. 2003, Schouten et al. 2013).

5.4 Production of Enzymes

The protection enzymes released by PGPR strains have biopesticidal properties. PGPR promote plant growth by controlling the production of phytopathogenic agents, primarily for metabolite production, contributing to the antibiosis and antifungal properties used as defense systems. The mechanism involves the production of hydrolytic enzymes, two examples of which are chitinase and glucanase. Fungal cell wall components are primarily composed of chitin and β-glucan; thus, chitinases and β-glucanases-producing bacteria would help in inhibition of fungal growth. Apart from unveiling the production of chitinase and beta-glucanases, *Pseudomonas* spp. inhibit *Rhizoctonia solani* and *Phytophthora capsici*, two of the most detrimental known crop pathogens.

5.5 Understanding Induced Systemic Resistance (ISR) and Systemic Acquired Resistance in Plants

Recent research has been going on in the field of conventional use of techniques in agriculture. Inhibition of pathogens has an important role in plant growth promotion. Understanding the mechanism of action of various plant pathogens allows preconditioning of plants from the harmful effects of diseases. PGPR itself cannot act as a direct antimicrobial agent as can the synthetic pesticides. PGPR act on the plants using two major mechanisms that can be distinguished on the basis of elicitor and regulatory systems used: (1) systemic acquired resistance (SAR) and (2) induced systemic resistance (ISR).

5.5.1 Systemic Acquired Resistance

This mode of mechanism is triggered or attained by exposing the plant to avirulent, virulent, and nonpathogenic microbes or by using artificial chemicals such as salicylic acid, 2,6-dichloro-isonicotinic acid (INA), or benzo-(1,2,3)-thiadiazole-7-carbothioic acid *S*-methyl ester (BTH) (Sticher et al. 1997). It has been documented that infection of plants with "necrotizing" pathogens (causing a hypersensitive reaction) results in augmented resistance to various infections by a variety of bacterial, fungal, and viral pathogens. This physiological immunity phenomenon is termed systemic acquired resistance (SAR). SAR in harmony with the induction of a set of SAR genes are encoded in proteins known as pathogenesis-related proteins (PR proteins) (Fig. 17.3).

5.5.2 Induced Systemic Resistance (ISR)

The ISR mechanism in plants involves the action of microbial flora present in the soil or plants that allows the plant to resist diseases. The activity of pathogenic

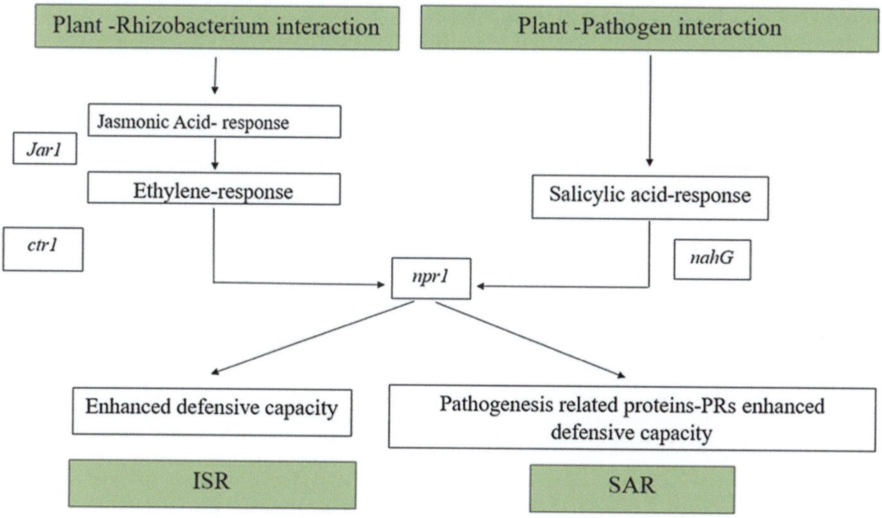

Fig. 17.3 Schematic representation of induced systemic resistance (ISR) and systemic acquired resistance (SAR) systems induced by PGPR

resistance is not the result of the PGPR antimicrobial activity: it happens through the production of antimicrobial agents. There are approximately six diverse types of plant disease resistance mechanisms: (1) parasite-specific, (2) cultivar-specific, (3) non-host (basic), (4) organ-specific, (5) plant age-related, and (6) induced (systemic and localized).

Some of the agents capable of eliciting ISR mechanisms have been tabulated in Table 17.9. Most of these agents are the secondary metabolites of PGPR organisms. They act onto the plants using two major mechanisms which can be distinguished on the basis of elicitor and regulatory systems used (Maleck et al. 2000).

6 Role of PGPRs as Biofertilisers in Sustainable Agriculture

Biofertilisers are vital in organic farming and in general agricultural practices on a global scale. These biofertilisers are synthesised using microorganisms that benefit the seed, plant surfaces, and rhizosphere and induce plant growth by making available the primary nutrients present in the soil (Vessey 2003). Mishra et al. (2013) reported the use of biofertiliser as a mixture of microorganisms promoting nitrogen fixing, cellulolytic, or P-solubilizers applied to the seed, roots, or composting areas for increasing microbial flora of PGPR and augmenting the availability of nutrients for plant growth promotion. Biofertiliser products are usually composed of plant growth-promoting microorganisms (PGPM). The PGPM can be categorized into three prominent groups of microorganisms based on an interdependent relationship with the plant: arbuscular mycorrhizal fungi (AMF) (Jeffries et al. 2003), PGPR

Table 17.9 Organisms/agents conferring ISR activity in plants

S.no	Agents	References
1	Fungi, bacteria, viruses, nematodes, insects	Averyanov et al. (2000)
2	Potassium and sodium phosphates, ferric chloride and silica	Kuc et al. (1959)
3	Glycine, β-aminobutyric acid, glutamic acid, α-aminobutyric acid, α-aminoisobutyric acid, D-alanine DL tryptophan and D-phenylalanine	Brasher (2000)
4	Salicylic acid, m-hydroxybenzoic acid, p-phydroxybenzoic acid, phloroglucinol, gallic acid, isovanillic acid, vanillic acid, protocatecheic acid, syringic acid and 1,3,5 benzene tricarboxylic acid	Averyanov et al. (2000)
5	Elaidoic acid, linoleic acid, linolenic acid, arachidonic acid and eicosapentaenoic acid	
6	Benzo(1,2,3) thiadiazolel-7-carbothioic acid s-methyl ester, 2,6-dichloroisonictonic acid	Kuc et al. (1959)
7	Riboflavin	
8	Probenazole and 2,2-dichloro-3,3-di-methyl cyclopropane carboxylic acid	
9	Penanthroline and phthalocyanine metal complexes (cobalt, iron and copper)	Brasher (2000)
10	Dodecyl L-valine and dodecyl DL-alanine	

(Podile and Kishore 2006), and nitrogen-fixing rhizobia (Franche et al. 2009). Of late, there have been reports that PGPR have been an aid in agricultural facilities, thereby showing rise in yield of crops and fertility of the soil. Thus, use of PGPR may lead to a conventional and sustainable agriculture (Khalid et al. 2009).

Previous studies have reported the practice of PGPR with compost to augment plant growth and also as a biocontrol agent for plants (Chen et al. 2011). *Bacillus* spp. (Gong et al. 2006) and *Pseudomonas* spp. (Leonardo et al. 2006) are two PGPR strains that have effective microbial control agents against plant pathogens. Among these bacterial PGPR species, *Bacillus amyloliquefaciens*, *Bacillus subtilis*, and *Bacillus cereus* were reported as most effective species at controlling plant diseases through various plant-induced mechanisms such as ISR and SAR (Francis et al. 2010). Endospore formation by *Bacillus* spp. and *Pseudomonas* spp. has the ability to sustain hostile environmental conditions and thus act as an effective biofertiliser formulation (Perez-Garcia et al. 2011). PGPR inoculants as biofertilizers are important in the uptake of nutritional elements such as N, P, and K for plant growth promotion and for combating plant pathogens (Vessey 2003; Perez-Garcia et al. 2011). The high release of N, P, and K enhances soil fertility, improves antagonistic biocontrol effects, and increases microorganism survival rates in soil (Yang et al. 2011). PGPR can be called effective biofertilisers as they help impart plant nourishment and maintain ecosystem balance by providing a better nutrient cycle relationship among the plant roots, soil, and microorganisms present.

7 Role of Nanotechnology for Agricultural Sustainability

Imparting our knowledge by using modern biotechnological methods such as nanotechnology has an immense capability to reinvent and lead conventional techniques in the agricultural industry. Nano-agriculture involves the practice of nanosized particles such as nano-biofertilizers and offers techniques to advance crop productivity and the availability of nutritional uptake for the plants (Tarafdar et al. 2013). Nanosized biofertilizers have certain physical, biological, and chemical properties that impart plant protection, detection of plant diseases, monitoring plant growth, inhibiting plant diseases, improving the quality of food, increasing the production of food, and reduction in food wastage.

The great advantage of nano-fertilizers compared to traditional fertilisers has been proven as they reduce nitrogen loss from emissions and leaching and allow lasting assimilation by soil microorganisms (Liu et al. 2006). Furthermore, Suman et al. (2010) have proven the advantage of using nano-fertilizers by showing that controlled release of fertilisers may also improve the soil by decreasing the toxic effects associated with the over-application of traditional synthetic fertilisers (Suman et al. 2010). Nano-encapsulation technology can be a multipurpose tool to protect PGPR, enhancing their service life and diffusion in fertiliser formulations, thus allowing the precise release of the PGPR.

8 Conclusion

With the constant increase in the human population, the demand for food is also always increasing. Approximately 70 years ago, the Green Revolution help to increase world agricultural production, thereby saving about 1 billion people from starvation and undernourishment. However, this process also triggered the development and production of chemical fertilisers along with the other technical advances. Thus, increase in the human demand for food has resulted in overexploitation of the soil ecosystem. Practice of the conventional crop approach cannot continue because these methods have increased anthropogenic activities such as intensive agriculture, crop monocultures, and the use of agrochemicals that are now causing great concern regarding the balance of all ecosystems.

The use of PGPR in biofertilisation, biocontrol, and bioremediation activities has exerted a positive influence on crop productivity and ecosystem functioning. For the improvement of modern technology in the field of agriculture and development, PGPR will be a boon for the constancy and efficiency of agro-ecosystems, thus leading towards an ideal agricultural system compatible with the environment.

Nanobiotechnology should be included in the agricultural field to reduce ecosystem damage and meet the global crop demand. Over the past decades, promising results and applications have been reported in the field of fertilisers, pesticides, and genetic modification for plant transformation. Based on this progress, a newer aspect of nanotechnology has to be developed with plant growth-promoting bacteria. Thus, nanotechnology has the techniques to expand the existing biofertilisers to support and improve agricultural sustainability globally.

References

Ahemad M, Kibret M (2014) Mechanisms and applications of plant growth promoting rhizobacteria: current perspective. J King Saud Univ Sci 26:1–20

Armanda E, Portela G, Roldan A, Azcon R (2014) Combined use of beneficial soil microorganism and agrowaste residue to cope with plant water limitation under semiarid conditions. Geoderma 232:640–648

Arora NK, Tewari S, Singh S, Lal N, Maheshwari DK (2012) PGPR for protection of plant health under saline conditions. In: Maheshwari DK (ed) Bacteria in agrobiology: stress management. Springer, Berlin/Heidelberg, pp 239–258

Arora NK, Tewari S, Singh S, Lal N, Maheshwari DK (2013) PGPR for protection of plant health under saline conditions. In: Maheshwari DK (ed) Bacteria in agrobiology: stress management, pp 239–258

Averyanov AA, Lapikova VP, Gaivornsky LM, Lebrun MH (2000) Two step oxidative burst associated with induced resistance to rice blast. In: First International Symposium Induced Resistance to Plant Diseases, Island of Corfu, Greece, May 22–27, Abstract pp 125–126

Badri DV, Vivanco GM (2009) Regulation and function of root exudates. Plant Cell Environ 32(6):666–681

Bashan Y, de-Bashan LE, Prabhu SR, Hernandez J-P (2014) Advances in plant growth-promoting bacterial inoculant technology: formulations and practical perspectives (1998–2013) (a Marschner review). Plant Soil 378:1–33

Beacham IR, Garrett S (1980) Isolation of *Escherichia coli* mutants (cpdB) deficient in periplasmic 2-cyclic phosphodiesterase and genetic mapping of the cpdB locus. J Gen Microbiol 119:31–34

Bhardwaj D, Ansari MW, Sahoo RK, Tuteja N (2014) Biofertilizers function as key player in sustainable agriculture by improving soil fertility, plant tolerance and crop productivity. Microb Cell Fact 13:66

Bhattacharyya NP, Jha DK (2012) Plant growth promoting rhizobacteria (PGPR): emergence in agriculture. World J Microbiol Biotechnol 28:1327–1350

Brasher P (2000) Protein for replacing pesticides approved. San Diego Union Tribune, April 29, p. A7

Burns DM, Beacham IR (1986) Nucleotide sequence and transcriptional analysis of the *Escherichia coli* ushA gene, encoding periplasmic UDP-sugar hydrolase (5′-nucleotidase): regulation of the ushA gene, and thesignal sequence of its encoded protein product. Nucleic Acids Res 14:4325–4342

Butterbach Bahl K, Baggs EM, Dannenmann M, Kiese R, Zechmeister-Boltenstern S (2013) Nitrous oxide emissions from soils: how well do we understand the processes and their controls? Philos Trans R Soc B 368(1621):20130122

Chen LH, Tang XM, Raze W, Li JH, Liu YX, Qiu MH, Zhang FG, Shen QR (2011) *Trichoderma harzianum* SQR-T037 rapidly degrades allelochemicals in rhizospheres continuously cropped cucumbers. Appl Microbiol Biotechnol 89:1653–1663

Chernin L, Chet I (2002) Microbial enzymes in biocontrol of plant pathogens and pests. In: Burns RG, Dick RP (eds) Enzymes in the environment: activity, ecology, and applications. Dekker, New York, pp 171–225

Choudhary DK, Sharma KP, Gaur RK (2011) Biotechnological perspectives of microbes in agroecosystems. Biotechnol Lett 33:1905–1910

Cornelis P (2010) Iron uptake and metabolism in pseudomonads. Appl Microbiol Biotechnol 86:1637–1645

Delalande L, Faure D, Raffoux A, Uroz S, D'Angelo-Picard C, Elasri M e a (2005) *N*-Hexanoyl-L-homoserine lactone, a mediator of bacterial quorum-sensing regulation, exhibits plant-dependent stability and may be inactivated by germinating *Lotus corniculatus* seedlings. FEMS Microbiol Ecol 52:13–20

Dey R, Pal KK, Bhatt DM, Chauhan SM (2004) Growth promotion and yield enhancement of peanut (*Arachis hypogea* L.) by application of plant growth-promoting rhizobacteria. Microbiol Res 159:371–394

Dong YH, Zhang XF, Xu JL, Zhang LH (2004) Insecticidal *Bacillus thuringiensis* silences *Erwinia carotovora* virulence by a new form of microbial antagonism, signal interference. Appl Environ Microbiol 70:954–960

Duffy CS, Jerzy N, Clement C, Barka EA (2003) Use of plant growth-promoting bacteria for biocontrol of plant diseases: principles, mechanisms of action, and future prospects. Appl Environ Microbiol 71(9):4951–4959

Etesami HA, Alikhani HA, Akbari A (2009) Evaluation of plant growth hormones production (IAA) ability by Iranian soils rhizobial strains and effects of superior strains application on wheat growth indexes. World Appl Sci J 6:1576–1584

Fraga R, Rodriguez H, Gonzalez T (2001) Transfer of the gene encoding the NapA acid phosphatase from *Morganella morganii* to a *Burkholderia cepacia* strain. Acta Biotechnol 21:359–369

Franche C, Lindström K, Elmerich C (2009) Nitrogen-fixing bacteria associated with leguminous and non-leguminous plants. Plant Soil 321:35–59

Francis I, Holsters M, Vereecke D (2010) The gram-positive side of plant–microbe interaction. Environ Microbiol 12:1–12

Frankowski J, Lorito M, Scala F, Schmidt R, Berg G, Bahl H (2001) Purification and properties of two chitinolytic enzymes of *Serratia plymuthica* HRO-C48. Arch Microbiol 176:421–426

Gaby JC, Buckley DH (2012) A comprehensive evaluation of PCR primers to amplify the *nifH* gene of nitrogenase. PLoS One 7. e42149 2004

Galloway JN, Townsend AR, Erisman JW, Bekunda M, Cai Z, Freney JR, Martinelli LA, Seitzinger SP, Sutton MA (2008) Transformation of the nitrogen cycle: recent trends, questions, and potential solutions. Science 320:889–892

Glick BR (2012) Plant growth-promoting bacteria: mechanisms and applications. Hindawi Publishing Scientifica, Waterloo

Glick BR, Todorovic B, Czarny J, Cheng Z, Duan J et al (2007) Promotion of plant growth by bacterial ACC deaminase. Crit Rev Plant Sci 26:227–242

Gong M, Wang JD, Zhang J, Yang H (2006) Study of the antifungal ability of *Bacillus subtilis* strain PY-1 in vitro and identification of its antifungal substance (Iturin A). Acta Biochim Biophys Sin 38:233–240

Gray EJ, Smith DL (2005) Intracellular and extracellular PGPR: commonalities and distinctions in the plant-bacterium signalling processes. Soil Biol Biochem 37:395–412

Gupta G, Parihar SS, Ahirwar NK, Snehi SK, Singh V (2015) Plant growth promoting rhizobacteria (PGPR): current and future prospects for development of sustainable agriculture. J Microb Biochem Technol 7:96–102. https://doi.org/10.4172/1948-5948.1000188

Han HS, Lee KD (2006) Effect of co-inoculation with phosphate and potassium solubilizing bacteria on mineral uptake and growth of pepper and cucumber. Plant Soil Environ 52:130–136

Hellriegel H, Wilfarth H (1888) Untersuchungen uber die Stickstoff-nahrung der Gramineen und Leguminosen. Beilageheft zu der Zeitschrift des Vereins fur Rubenzucker-Industrie Deutschen Reichs, p 234

Iavicoli A, Boutet E, Buchala A, Métraux JP (2003) Induced systemic resistance in *Arabidopsis thaliana* in response to root inoculation with *Pseudomonas fluorescens* CHA0. Mol Plant Microbe Interact 16:851–858

Iqbal MA, Khalid M, Shahzad SM, Ahmad M, Soleman N et al (2012) Integrated use of *Rhizobium leguminosarum*, plant growth promoting rhizobacteria and enriched compost for improving growth, nodulation and yield of lentil (*Lens culinaris* Medik). Chilean J Agric Res 72:104–110

Jeffries P, Gianinazzi S, Perotto S, Turnau K, Barea JM (2003) The contribution of arbuscular mycorrhizal fungi in sustainable maintenance of plant health and soil fertility. Biol Fertil Soils 37:1–16

Khalid A, Arshad M, Shaharoona B, Mahmood T (2009) Plant growth promoting rhizobacteria and sustainable agriculture. In: Microbial strategies for crop improvement. Springer, Berlin/Heidelberg, pp 133–160

Khan MS, Zaidi A, Ahemad M, Oves M, Wani PA (2010) Plant growth promotion by phosphate solubilizing fungi: current perspective. Arch Agron Soil Sci 56:73–98

Klipp W, Masepohl B, Gallon OR, Newton WE (eds) (2004) Genetics and regulation of nitrogen. Springer, Heidelberg

Kloepper JW, Schroth MN (1981) Relationship of in vitro antibiosis of plant growth promoting rhizobacteria to plant growth and the displacement of root microflora Phytopathology 71:1020–1024

Kloepper JW, Lifshitz R, Zablotowicz RM (1989) Free-living bacterial inocula for enhancing crop productity. Trends Biotechnol 7:39–43

Kuc J, Barnes E, Daftsios A, Williams E (1959) The effect of amino acids on susceptibility of apple varieties to scab. Phytopathology 49:313–315

Kumar P, Dubey RC (2012) Plant growth promoting rhizobacteria for biocontrol of phytopathogens and yield enhancement of *Phaseolus vulgaris*. J Curr Pers Appl Microbiol 1:6–38

Leonardo D, Blanca LF, Landa B, Weller DM (2006) Host crop affects rhizosphere colonization and competitiveness of 2,4-diacetylphloroglucinol-producing *Pseudomonas fluorescens*. Phytopathology 96:751–762

Lim H, Kim Y, Kim S (1991) *Pseudomonas stutzeri* YLP-1 genetic transformation and antifungal mechanism against *Fusarium solani*, an agent of plant root rot. Appl Environ Microbiol 57:510–516

Liu XM, Feng ZB, Zhang FD, Zhang SQ, He XS (2006) Preparation and testing of cementing and coating nano-subnanocomposites of slow/controlled-release fertilizer. Agric Sci China 5:700–706

Ma JF (2005) Plant root responses to three abundant soil minerals: silicon, aluminum and iron. Crit Rev Plant Sci 24:267–281

Maleck K, Levine A, Eulgem T., Morgan A, Schmid J, Lawton KA, Dangl JL Dietrich RA (2000) The transcriptome of *Arabidopsis thaliana* during systemic acquired resistance. Nat Genet 26:403–410

Martinez-Viveros O, Jorquera MA, Crowley DE, Gajardo G, Mora ML (2010) Mechanisms and practical considerations involved in plant growth promotion by rhizobacteria. J Soil Sci Plant Nutr 10:293–319

Miransari M, Smith DL (2014) Plant hormones and seed germination. Environ Exp Bot 99:110–121

Mishra D, Rajvir S, Mishra U, Kumar SS (2013) Role of bio-fertilizer in organic agriculture: a review. Res J Recent Sci 2:39–41

Mol Plant-Microbe Interact 17:1201–1211

Nitrogen-fixing *R. leguminosarum* sv. *trifolii*–white clover fixation in free living bacteria. Kluwer, Dordrecht

Oldroyd GE, Downie JA (2008) Coordinating nodule morphogenesis with rhizobial infection in legumes. Annu Rev Plant Biol 59:519–546

Ordentlich A, Elad Y, Chet I (1988) The role of chitinase of *Serratia marcescens* in biocontrol of *Sclerotium rolfsii*. Phytopathology 78:84–88

Ovadis M, Liu X, Gavriel S, Ismailov Z, Chet I, Chernin L (2004) The global regulator genes from biocontrol strain *Serratia plymuthica* IC1270: cloning, sequencing, and functional studies. J Bacteriol 186:4986–4993

Pandey P, Maheshwari DK (2007) Two species microbial consortium for growth promotion of *Cajanus cajan*. Curr Sci 92:1137–1142

Parmar P, Sindhu SS (2013) Potassium solubilization by rhizosphere bacteria: influence of nutritional and environmental conditions. J Microbiol Res 3:25–31

Perez-Garcia A, Romero D, de Vicente A (2011) Plant protection and growth simulation by micro-organism: biotechnological applications of *Bacillus* in agriculture. Curr Open Biotechnol 22:187–193

Podile AR, Kishore GK (2006) Plant growth-promoting rhizobacteria. In: Plant-associated bacteria. Springer, Dordrecht, pp 195–230

Pradel E, Boquet PL (1990) Nucleotide sequence and transcriptional analysis of the *Escherichia coli* agp gene encoding periplasmic acid glucose-1-phosphatase. J Bacteriol 172:802–807

Reed SC, Cleveland CC, Townsend AR (2011) Functional ecology of free-living nitrogen fixation: a contemporary perspective. Annu Rev Ecol Evol Syst 42:489–512

Rodrigues EP, Soares CP, Galvão PG, Imada EL, Simões-Araújo JL, Rouws LFM, ALM O, Vidal MS, Baldani JI (2016) Identification of genes involved in indole-3-acetic acid biosynthesis by *Gluconacetobacter diazotrophicus* PAL5 strain using transposon mutagenesis. Front Microbiol 7:1572. https://doi.org/10.3389/fmicb.2016.01572

Schouten A, van der Berg G, Edel-Hermann V, Steinberg C, Gautheron N, Sharma SB, Sayyed RZ, Trivedi MH, Gobi TA (2013) Phosphate solubilizing microbes: sustainable approach for managing phosphorus deficiency in agricultural soils. Springerplus 2:587

Sharma SB, Sayyed RZ, Trivedi MH, Gobi TA (2013) Phosphate solubilizing microbes: sustainable approach for managing phosphorus deficiency in agricultural soils. Springerplus 2:587

Spaepen S, Vanderleyden J (2011) Auxin and plant-microbe interactions. Cold Spring Harbor Perspect Biol 3:a001438

Sticher L, Mauch-Mani B, Metraux JP (1997) Systemic acquired resistance. Annu Rev Phytopathol 35:235–270

Suman PR, Jain VK, Varman A (2010) Role of nanomaterials in symbiotic fungus growth enhancement. Curr Sci 99:1189–1191

Tarafdar A, Raliya R, Wang WN, Biswas P, Tarafdar JC (2013) Green synthesis of TiO_2 nanoparticle using *Aspergillus tubingensis*. Adv Sci Eng Med 5:943–949

Tisdale SL, Nelson WL (1975) Soil fertility and fertilizers, 3rd edn. Macmillan, New York

Verma JP et al (2010) Impact of plant growth promoting rhizobacteria on crop production. Int J Agric Res 5:954–983

Vessey JK (2003) Plant growth promoting rhizobacteria as biofertilizers. Plant Soil 255:571–586

Von Bodman SB, Bauer WD, Coplin DL (2003) Quorum-sensing in plant-pathogenic bacteria. Annu Rev Phytopathol 41:455–482

Yang X, Chen L, Yong X, Shen Q (2011) Formulations can affect rhizosphere colonization and biocontrol efficiency of *Trichoderma harzianum* SQR-T037 against *Fusarium* wilt of cucumbers. Biol Fertil Soils 47:239–248

Zaidi A, Khan MS, Ahemad M, Oves M (2009) Plant growth promotion by phosphate solubilizing bacteria. Acta Microbiol Immunol Hung 56:263–284

Zhang L, Birch RG (1997) The gene for albicidin detoxification from *Pantoea dispersa* encodes an esterase and attenuates pathogenicity of *Xanthomonas albilineans* to sugarcane. Proc Natl Acad Sci USA 94:9984–9989

Plant-Microbiome Interaction and the Effects of Biotic and Abiotic Components in Agroecosystem

18

Indramani Kumar, Moumita Mondal, Raman Gurusamy, Sundarakrishnan Balakrishnan, and Sakthivel Natarajan

1 Introduction

The major goal in agriculture is to increase the food production through cost-effective strategies, and furthermore, the world economy is depending upon agricultural activities. As the world population is growing, the major priority is the eradication of poverty through the achievements of high crop yield (Gaba et al. 2014). Advent of technologies and understanding of fundamental knowledge on molecular genetics, molecular biology, ecology, and symbiotic relationships between microbes and plants are required for sustainable agriculture (Umesha et al. 2018). Research studies on plant rhizospheric microbial population provide new opportunities for agricultural developments that reduce the application of chemicals and also minimize the cost of agriculture (Umesha et al. 2018). The function of soil microbes is a prerequisite to manage the diverse species of plant-associated microbes and also remains a challenge (Toju et al. 2018). Plant microbiomes are the major components in the biogeochemical cycles and drive nutrient cycles with feedback on the functioning of the ecosystem. As the interactions of microbiomes toward plants and other soil microbes are highly complex, an array of tools, methodologies, and strategies have been employed by researchers (van der Heijden and Hartmann 2016). A diverse group of microbes has been associated with different parts of the plants. Bacteria and fungi are capable of living inside the tissue or colonize on the plant surfaces. Thousands of bacterial and fungal taxa colonize plant roots and leaves (Muegge et al. 2011; Berendsen et al. 2012; Hacquard et al. 2015).

I. Kumar · M. Mondal · S. Balakrishnan · S. Natarajan (✉)
Department of Biotechnology, School of Life Sciences, Pondicherry University, Puducherry, India

R. Gurusamy
Department of Life Sciences, Yeungnam University, Gyeongsan, Gyeongbuk, Republic of Korea

D. P. Singh et al. (eds.), *Microbial Interventions in Agriculture and Environment*,
https://doi.org/10.1007/978-981-13-8383-0_18

As ecological intensification of agriculture is a paradigm shift, there is a need to meet the challenges by manipulating biotic interactions. However, the multiple constraints and the complexities of agroecosystems result in difficulties to implement the changes in biotic interactions (Gaba et al. 2014). The nature of agricultural field soils is much complex with diverse ecosystem. It was reported that the hidden biodiversity of microbes in soil ranged up to 10^7 prokaryotes/gram of soil with approximately 10^5 different species of bacteria (Gans 2005).

Management of plant microbiomes promotes the functions of agroecosystems and subsequently increases crop yield and reduces the pest population through biological control (Bommarco et al. 2013). The soil microbial activities and plant communities influence ecosystems and their functions (Bardgett et al. 2005). Direct enhancement of biotic interactions through the microbial biofertilizers promotes plant growth and suppresses agricultural pests and pathogens (Doré et al. 2011; Ekström and Ekbom 2011). The functioning of terrestrial ecosystem is regulated by the complex interactions between three major biological components such as plants, soil, and microbes. Biological interactions also decide plant diversity and different ecophysiological traits and soil properties. Variations between biotic and abiotic factors further affect the soil microbiota (De Wit and Bouvier 2006). Studies have confirmed the biodiversity of soil microorganisms, their interaction with the host plants, and also the intervention of animals and human.

2 Diversity of Plant Microbiome

Highly diversified groups of microbes comprise the plant microbiome. The plant-associated microbes are the major players in determining plant health, physiology, development, growth, and yield (Mendes et al. 2013; Hartmann et al. 2014). Each part of plant hosts a different microbiome in the ecological niches, such as in the rhizosphere (Podile and Kishore 2007; Berendsen et al. 2012; Philippot et al. 2013), spermosphere (Schiltz et al. 2015), and phyllosphere (Vorholt 2012). Soil is considered to be the largest reservoir for the biological diversity. Almost 10^{11} microbial cells/gram root are harbored in the rhizospheric soil which accounts for more than 30,000 prokaryotic species (Egamberdieva et al. 2008; Berendsen et al. 2012). The root exudates and the nature of rhizodeposits influence the population dynamics and microbial diversity (Compant et al. 2010). Microbes such as bacteria, viruses, archaea, fungi, nematodes, algae, protozoa, and arthropods are the inhabitants of plant rhizosphere (Lynch 1990; Raaijmakers et al. 2002; Bonkowski et al. 2009; Buée et al. 2009; Mendes et al. 2011, 2013). A report from the rhizosphere of 14 species confirmed the presence of more than 1200 distinct bacterial taxa, where the group of *Proteobacteria* was identified as the predominant phylum (Hawkes et al. 2007). Roesch et al. (2007) reported 52,000 operational taxonomic units (OTUs) with *Bacteroidetes*, *Betaproteobacteria,* and *Alphaproteobacteria* as dominant species in four different soil types (Mendes et al. 2013). The vascular plants of Antarctic soil showed *Firmicutes* as the most widespread phylum (Teixeira et al. 2010) unlike *Proteobacteria* and *Actinobacteria* which are dominant in other rhizospheric soils such as mangrove wetland, grassland, and cucumber fields (Wang et al. 2018). Weinert et al. (2011) used PhyloChip analysis and elucidated different phylum populations of *Proteobacteria*, *Firmicutes*, *Actinobacteria*, *Bacteroidetes,* and *Acidobacteria*. A total of 37 bacterial genera from trace element-contaminated field and 29 from of the *Brassica napus* control field has been reported. The rhizospheric bacterial population is comprised of *Bacillus*, *Pseudomonas, Variovorax,* and *Staphylococcus*. From potato rhizosphere, *Ascomycota* was identified as a predominant fungus, and the rest included *Zygomycota, Chytridiomycota,* and *Basidiomycota* as unidentified communities (Qin et al. 2017). Recently, the effect of farming system on fungal diversity that comprises *Microsporidetes*, *Dothideomycetes* and *Eurotiomycetes* was reported (Kazerooni et al. 2017).

Seeds have been found to host microbial population not only on the surface but also in the embryo (Nelson 2004). The seed genotype, germination phase, and soil type greatly affect the spermospheric microbiome. In a study with peanut seeds, bacterial diversity was found to be higher in saline alkali soil as compared to nonsaline alkali soil. Out of the five classes, such as *Proteobacteria, Actinobacteria, Bacteroidetes, Acidobacteria* and *Firmicutes*, *Proteobacteria* and *Actinobacteria* were found to be dominant (Zhang et al. 2017). In another study, maximum microbial abundancy was found in organic-enriched soil (Buyer et al. 1999). In a study with turnip seeds after 72 h of germination, *Fusarium* and *Phythium* were identified as the most predominant fungi. Also, oomycetes *Achyla* and *Thraustotheca* were obtained in high numbers. Species of *Penicillium, Cephalosporium, Trichoderma, Cunninghamella*, *Mucor*, *Helicocephalum*, *Cylindrocarpon*, *Gliocladium* and *Rhizoctonia solani* were detected in the spermospheres of tomato, cabbage, onion, mustard, melon, and bean (Watson 1966a, b). In a study with spinach seed, more than 250 OTUs were noted having three major bacterial phyla of *Proteobacteria, Firmicutes* and *Actinobacteria*. Interestingly, both seed and cotyledon showed similar bacterial species and richness (Lopez-Velasco et al. 2013).

Metagenomic study in phyllosphere depicted microbial diversity that is comprised of almost 900 species which is comparatively lower than the rhizospheric microbial diversity (Berg et al. 2015). Leaf area can comprise of 10^6–10^7 bacterial cells/cm^2 that are epiphytic and endophytic (Lindow and Brandl 2003; Humphrey et al. 2014). About 6.4×10^8 km^2 terrestrial leaf surface area was inhabited by microbiome. These metaproteomic and metagenomic analyses revealed the presence of *Bacteroidetes, Actinobacteria,* and *Proteobacteria* as common bacterial phyla (Quiñones et al. 2005; Delmotte et al. 2009; Redford et al. 2010; Kim et al. 2012; Rastogi et al. 2012; Bringel and Couée 2015) and *Ascomycota* as the common yeast phylum from oak leaves (Jumpponen and Jones 2010; Voříšková and Baldrian 2013; Bringel and Couée 2015). In a study with five different vegetables such as spinach, rape, celery, broccoli, and cauliflower, *γ-Proteobacteria, Bacteroidetes, Firmicutes,* and *Cyanobacteria* species were identified (Zhang et al. 2010). Flavobacteria was found to be abundant in both the rhizosphere and phyllosphere of *A. thaliana* (Bodenhausen et al. 2013). In a study with 144 leaf samples from a tropical forest in Thailand, 1524 fungal OTUs were detected having 24 orders of *Ascomycota* and 21 orders of *Basidiomycota* and *Glomeromycota* (Izuno et al.

2016). *Xylariales, Hypocreales,* and *Tremellales* were also found in lower numbers. In another study, the authors reported a high level of fungal diversity in leaves in contrast to fruits and flowers. Different species of *Colletotrichum* genus including *C. clavatum, C. Acutatum,* and *C. karstii* have been reported as the predominant group on ripe fruits. *Pseudocercospora cladosporioides* was found to be dominant in all leaf samples (Abdelfattah et al. 2015). Increased microbial diversity in young leaves as compared to old matured leaves of sugar beets during the growing season was reported, and sugar beet exhibited a total of 78 bacterial species (Thompson et al. 1993). Interestingly, *Ercolani* form unique colonization pattern at different times of the year, particularly, peak diversity during cool and rainy periods (Ercolani 1991). Some distinct leaf colonizers that fail to sustain in roots, such as pigmented bacteria, *Azospirillum* (Jurkevitch and Shapira 2000) and *Rhizobia* (O'brien and Lindow 1989) were reported. Studies revealed that solar radiation affects the physiology and ecology of the plant microbiome (Jacobs and Sundin 2001).

3 Plant-Microbiome Interactions

The role of biotic and abiotic factors that determines the function of microbial communities and their interactions with biotic and abiotic factors is poorly understood. The evolution and ecology of a microbiome are important to understand the plant-microbiome interactions (Fitzpatrick et al. 2018). Recent studies confirmed the occurrence of high-level diversity of plant microbiomes. On the basis of specific habitats, plant microbiomes are classified as rhizosphere microbes, phyllosphere microbes, spermosphere microbes, epiphytic microbes, and endophytic microbes. All these microbiomes specifically interact with plants for essential functions (Berg et al. 2015).

After the advent of molecular tools, omics technologies, new microscopic and analytical analyses, and significant milestones on plant microbiome have been reported (Caporaso et al. 2012; Jansson et al. 2012). Insights into the population structure and functional understanding of the plant microbiome of *Arabidopsis* were reported (Bulgarelli et al. 2012; Lundberg et al. 2012). Detailed studies on the disease-suppressive plant microbiome in sugar beet have been documented (Mendes et al. 2011). Advancement of plant science research can be achieved through the realization of microbial communities and their primary roles in plant health and disease suppression throughout the life cycles of plants. In addition, describing microbial communities and their positive and negative interactions is essential to elucidate their assembly, dynamics, and functions (Waldor et al. 2015). Microbiomes in all plant ecological niches establish major microbial categories such as beneficial, neutral, and detrimental. Therefore, it is evident that plant microbiomes are influenced by plant genomes, and the microbes are considered as second genome and collectively described as pan-genome. An array of phytochemicals of root exudates from different genotypes of plants determines the microbial diversity in the area of roots and root-associated soil regions (Table 18.1). The plant, soil, and microbiome interactions are highly complex (Fig. 18.1).

Table 18.1 Phytochemicals present in the root exudates

Class of compounds	Components	References
Amino acids	α-Alanine, leucine, β-alanine, cysteine, glycine, cystine, lysine, isoleucine, asparagine, aspartate, methionine, glutamate, tryptophan, threonine, ornithine, arginine, histidine, valine, proline, L-hydroxyproline, α-aminoadipic acid, serine, homoserine, phenylalanine, γ-aminobutyric acid, mugineic acid	Dakora and Phillips (2002) and Badri and Vivanco (2009)
Organic acids and phenolic acids	Acetic acid, malic acid, butyric acid, malonic acid, valeric acid, citric acid, isocitric acid, fumaric acid, formic acid, tetronic acid, succinic acid, oxalic acid, aldonic acid, erythronic acid, glutaric acid, lactic acid, glyconic acid, piscidic acid, acotinic acid, pyruvic acid, L-aspartic acid, sinapic acid, shikimic acid, chorismic acid, salicylic acid, gallic acid and its methyl ester, caffeic acid, cinnamic acid, *p*-coumaric acid, ferulic acid, 3,4-dihyroxybenzoic acid, gentisic acid, tannic acid, vanillic acid, vanillyl alcohol, syringic acid, protocatechuic acid, 4-hydroxybenzaldehyde, syringaldehyde, benzoquinone, hydroquinone, 4-methoxycinnamic acid, *p*-hydroxybenzoic acid, tartaric acid, quercetin, rutin, naringenin, naringin, myricetin, kaempferol, strigolactone, genistein	Dakora and Phillips (2002), Kamilova et al. (2006), Makoi and Ndakidemi (2007), Wu et al. (2008a, b), Badri and Vivanco (2009), Mandal et al. (2009) and Tan et al. (2013)
Fatty acids and sugars	Palmitic acid, linoleic acid, stearic acid, oleic acid, glucose, galactose, fructose, rhamnose, xylose, arabinose, ribose, deoxyribose, raffinose, maltose, oligosaccharide, sucrose, mannitol	Dakora and Phillips (2002), Windstam and Nelson (2008) and Badri and Vivanco (2009)
Enzymes and proteins	Amylase, protease, acid/alkaline phosphatase, invertase, lipase, hydrolase, peroxidase, PR proteins, lectins, caffeoyl-CoA O-methyltransferase, L-phenylalanineammonia-lyase, cinnamate, 4-hydroxylase, 1-aminocyclopropane-1-carboxylate deaminase	Chen et al. (2000), Dakora and Phillips (2002), Achnine et al. (2004) and Croes et al. (2013)
Volatiles	2-Butanone, 2-aminoacetophenone, 2-pentylfuran, 2-methyl-n-1-tridecene, 2,3-butanediol, acetoin, 13-tetradecandien-1-ol, hydrogen cyanide	Bakker and Schippers (1987) and Velmourougane et al. (2017)

3.1 Rhizosphere Microbiome, Interaction, and Mechanism

The rhizosphere is rich in microbes by 10–100 folds than the neighboring "bulk" soil due to the carbon secreted by the roots, and this phenomenon is called the "Rhizosphere effect." Soil is mostly mesotrophic or oligotrophic that encompasses predominantly heterotrophic microbial communities and enriches microbes for their proliferation, persistence, and physiology (Hiltner 1904). All types of microbial population cannot harbor the rhizosphere due to the competition for carbon, nitrogen, and energy metabolism as per the availability of the rhizodeposits. Most of the rhizosphere microbes include *Archaea*, *Eubacteria*, *Actinobacteria*, viruses,

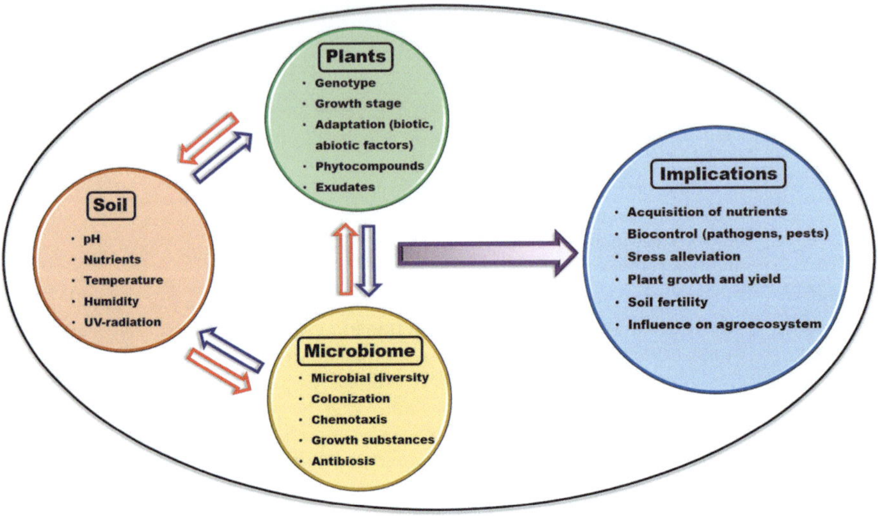

Fig. 18.1 Complex interactions between plant, soil, microbiome, and their implications on agroecosystem

and fungi (Philippot et al. 2013). Rhizospheric soil is a dynamic platform for various plant-microbe interactions which are in turn stimulated by the biotic and abiotic components along with the rhizodeposits. Such interactions can be either beneficial or antagonistic to the plants. Different microbes perform important biological functions and influence the agroecosystem (Giri et al. 2005). Three distinct phases are observed in the rhizosphere, namely, rhizosphere, rhizoplane, and root. The rhizosphere is the soil portion where the released substrates influence the microbial activities, while the rhizoplane is the root surface having adhered soil particles. An array of microbes, particularly endophytes, harbor in the roots (Barea et al. 2005).

3.1.1 Beneficial Plant-Microbe Interactions

Rhizosphere microbes such as bacteria and fungi that exhibit the detrimental or beneficial saprophytic or symbiotic relationships have been reported (Kobayashi and Crouch 2009). In beneficial plant-microbe interactions, both the parties get access to unavailable deep nutrients due to the microbial solubilization and mobilization processes, thus promoting plant growth and resistance against pathogens and also enhance the abiotic stress tolerance (Velmourougane et al. 2017).

Plant Growth Promotion

The beneficial microbial inoculants that enhance stimulation of plant growth and productivity are termed as plant growth-promoting rhizobacteria (PGPR) (Kloepper and Schroth 1981; Vessey 2003). The PGPRs are differentiated into extracellular PGPR and intracellular PGPR (Martínez-Viveros et al. 2010). Extracellular PGPRs are found in the rhizosphere or on the rhizoplane and sometimes in between the root cortex cells, while intracellular PGPRs reside in specialized nodular structures of root cells (Figueiredo et al. 2010). Common PGPRs include *Azospirillum*, *Bacillus*, *Agrobacterium*, *Azotobacter*, *Arthrobacter*, *Caulobacter*, *Micrococcus*, *Chromobacterium*, *Burkholderia*, *Erwinia*, *Flavobacterium*, *Serratia*, and *Pseudomonas*. The genera such as *Frankia*, *Allorhizobium*, *Azorhizobium*, *Bradyrhizobium*, *Mesorhizobium*, *Rhizobium*, and some endophytes comprise intracellular PGPRs (Bhattacharyya and Jha 2012). Such rhizospheric microbes aid in plant growth directly by facilitating the solubilization of phosphorus and potassium, nitrogen fixation, iron sequestration, and modulating phytohormone levels and secretions such as cytokinin, gibberellin, and indole-3-acetic acid or indirectly through their antagonistic role against plant pests and phytopathogens (Sakthivel and Gnanamanickam 1987; Naik and Sakthivel 2006; Naik et al. 2008b; Glick 2012; Velmourougane et al. 2017). Certain endophytes are reported for their role in seedling development, bioremediation, improved nutrient cycling and increased yield, and abiotic stress tolerance (Arnold 2007).

Nitrogen Fixation

Nitrogen (N) is a vital nutrient essential for plant growth and productivity, and it is not accessible by growing plants. Atmospheric nitrogen is transformed to utilizable forms for plants via biological nitrogen fixation. Biological nitrogen-fixing organisms such as rhizobia are usually symbiotic with leguminous plants (Ahemad and Khan 2012) and nonleguminous crop plants (Gauthier et al. 2000). The free-living *Azotobacter*, *Beijerinckia*, *Klebsiella*, and *Bacillus* are nonsymbiotic. The nitrogen-fixing endophytes include cyanobacteria (*Anabaena*, *Nostoc*, *Calothrix*), *Azotobacter*, *Azospirillum*, and *Gluconacetobacter* (Prasanna et al. 2009).

Phosphate Solubilization

Phosphorus (P) is a vital plant growth-limiting nutrient which is available in two forms such as organic and inorganic (Khan et al. 2009). Plants absorb phosphorus in monobasic and dibasic soluble forms (Bhattacharyya and Jha 2012). The insoluble form includes apatite, phospho-mono and triesters, and inositol phosphate (Glick 2012). Phosphate-solubilizing microorganisms (PSMs) are now introduced into soil as an alternative to harmful chemical fertilizers to maintain environmental friendliness and cost-effectiveness. Interaction between arbuscular micorrhizal fungi (AMF) and phosphate-solubilizing bacteria (PSB) facilitates the availability of phosphorus available to plants. Phosphate-solubilizing bacteria solubilize phosphorus ions, whereas AMF translocate them to the plants. Synergistic microbial interactions between AMF and phosphate-solubilizing bacteria have been reported (Azcón and Barea 2010). Some of the phosphate-solubilizing bacteria are *Azotobacter*, *Microbacterium*, *Erwinia*, *Bacillus*, *Beijerinckia*, *Serratia*, *Burkholderia*, *Enterobacter*, *Flavobacterium*, *Pseudomonas*, and *Rhizobium* (Naik et al. 2008a; Jha et al. 2009; Bhattacharyya and Jha 2012).

Iron Sequestration

Microbial siderophores are the metal-chelating agents that sequester insoluble ferric ion (Fe^{3+}) molecules. Iron (Fe) is vital to most life forms. Generally, Fe is found in insoluble hydroxide and oxyhydroxide forms unavailable to both plant and microbes. The microbial hydroxamate siderophores help plants to absorb iron from soil (Rajkumar et al. 2010).

3.1.2 Harmful Plant-Microbe Interactions

Some plant-microbe interactions are harmful to plants. These detrimental microbes are saprophytic or necrotrophic in nature. Higher colonization of phytopathogens present in rhizosphere affects the survival of host plants and also perturbs the diversity of native rhizomicroflora. In rhizosphere, the microbial changes in population density and community structure may cause the harmful effects toward both the host plants and associated microbes due to the change in nutrient availability or any metabolic activities (Velmourougane et al. 2017). The harmful plant pathogens include bacteria or fungi. Common bacterial pathogenic genera include *Agrobacterium*, *Pseudomonas*, *Xanthomonas*, *Ralstonia*, and *Erwinia* (Lacombe et al. 2010). The common phytopathogenic fungi are *Phytophthora*, *Mangnaporthe*, *Puccinia*, *Pythium*, *Rhizoctonia*, *Fusarium*, *Ustilago,* and *Alternaria* (Stukenbrock and McDonald 2008).

3.2 Spermosphere Microbiome, Interaction, and Mechanism

Slykhuis (1947) and Onorato Verona (1958) defined spermosphere as the region around seeds where high microbial activities prevail due to seed exudates and mucilage. Both beneficial and detrimental interactions occur between the soil, microbial communities, and germinating seeds. Beneficial microorganisms are naturally present and sometimes introduced by microbial inoculation in the spermosphere. Beneficial microbes enhance the germination of seeds and seedling vigor. Also, these beneficial microorganisms release phytohormones that promote physiological and morphological changes in seed tissues (Dodd et al. 2010). Moreover, during stress period or suboptimal temperature, these potential microorganisms also protect from oxidative damage and enhance seed germination (Mastouri et al. 2010). In order to explore the type and effect of interactions between seeds, microorganisms, and environment, two major experimental approaches are required. First step is the identification, quantification, and elucidation of the exudates from seeds at the time of germination of seed. Second step is the characterization of the microbiome composition. Combining the above two approaches gives an integrated functioning of spermosphere constituents (Buyer et al. 1999; Schiltz et al. 2015). Seeds are sexually produced from spermophytes that determine the genetic material transmission from generation to generation. Seeds vertically transmit the microorganisms and influence the plant growth and pathogenesis (Darrasse et al. 2010; Quesada-Moraga et al. 2014; Truyens et al. 2015). Microorganisms represent distinct microhabitats of individual seed compartments such as embryo, storage tissues, and seed coat (Lemanceau et al. 2017). Seed-associated microorganisms are commonly classified into seed-borne and seed-transmitted microorganisms. The former are ephemeral colonizers and the latter are involved in distinct plant developmental stages. Depending on the plant species and abiotic factors, during seed imbibition, various exudates such as organic acids, amino acids, fatty acids, and carbohydrates are released to the soil environment (Table 18.2). These exudates create a densely populated microbial zone competing for resources and space (Nelson 2004; Schiltz et al. 2015).

3.2.1 Chemotaxis in Spermosphere

Various spermosphere bacterial taxa such as *Bacillus, Pseudomonas,* and *Rhizobium* have shown chemotaxis-based motility toward seed exudates such as organic acids

Table 18.2 Phytocompounds released during seed germination

Class of compounds	Components	References
Amino acids	Alanine, glutamic acid, glutamine, homoserine, α-alanine, α-aminoadipic acid, β-alanine, β-pyrazolylalanine, uracil- alanine, α-γ- glutamylalanine, aspartate, arginine, asparagine, α-aminobutyric acid, γ-aminobutyric acid, glycine, serine, methionine, phenylalanine, leucine/ isoleucine, isoxazolin- 5-one, pyroglutamic acid, valine, tyrosine, threonine, tryptophan, aspartic acid, cysteic acid, cysteine, cystathionine, citrulline, proline, ornithine, pipecolic acid, lycine, histidine, homocysteic acid, homocysteine, glycine, α, ε-diaminopimelic acid, dihydroxyphenylalanine	Nelson (2004), Kamilova et al. (2006) and da Silva Lima et al. (2014)
Organic acids	Citric acid, pyruvic acid, malic acid, succinic acid, pyroglutamic acid, fumaric caid, acotinic acid	Kamilova et al. (2006) and da Silva Lima et al. (2014)
Fatty acids	Tridecanoic acid, hexadecenoic acid, octadecanoic acid isomers, palmitic acid, linoleic acid, oleic acid, myristic acid, azaleic acid, 4-(2,2,4-trimethylphenyl)- phenol, 5-(12-heptadecenyl)-resorcinol	Ruttledge and Nelson (1997) and Miche et al. (2003)
Phenolic compounds	Rosmarinic acid, paprazine, N-P-cis-coumaroyltyramine, caffeic acid, esculetin, quercetin, luteolin, Delphinidin, chrysoeriol, cyanidin, daidzein, apigenin, catechin, malvidin, myricetin, flavones, flavonols, genistein, kaempferol, stachydrine	Kelley et al. (1975), Kelley et al. (1976), Bhatti et al. (1992), Tian-Shung et al. (1994), Youssef and Frahm (1995), Zhang and Bao (2000) and Nelson (2004)
Volatiles	Acetone, ethane, methane, ethylene, formic acid, aldehyde, hydrogen cyanide, propylene, propionaldehyde, alcohol, carbon dioxide, volatile carboxylic acid	Stotzky et al. (1976), Gorecki et al. (1985, 1991), Nelson and Craft (1989) and Paulitz et al. (1992)
Sugars	Glucose, galactose, lactose, maltose, fructose, raffinose, mannose, rhamnose, sorbose, ribose, xylose, sucrose	Lugtenberg et al. (1999), Kamilova et al. (2006) and da Silva Lima et al. (2014)

and amino acids (Zheng and Sinclair 2000). Most work has been focused on *Bacillus* and *Pseudomonas* species. *Bacillus megaterium* strain B153-2-2 has exhibited positive chemotaxis toward soybean seed exudates that is largely comprised of alanine, threonine, malate, glutamine, serine, and asparagine (Zheng and Sinclair 2000). Chemotaxis is also reported with *B. megaterium* in response to malate, malonate, pyruvate, and succinate with no response to sugars. This chemotactic response is correlated to both seed colonization and suppression of *Rhizoctonia solani* that causes root rot of soybean (Zheng and Sinclair 2000). Another report of chemotactic response of *Pseudomonas fluorescens* and *P. putida* correlates with *B. megaterium* where mostly amino acids are present in exudates but not sugars and also positively chemotactic to soybean (Scher et al. 1985) and tomato seed exudates. Few *Rhizobium* species have also been reported to have chemotactic potential in response to amino acids present in soybean seed exudates (Barbour et al. 1991).

3.2.2 Regulation of Gene Expression in Sphermosphere

Regulation of gene expression in spermosphere has been mostly studied in *Pseudomonas* species and *B. cereus*. It has been proved that amino acid and sugar metabolic genes are basically induced by the components of seed exudates. Bayliss et al. (1997) have reported the expression of an ABC sugar transporter in *P. putida* GR12-2R3 due to canola seed exudates. In another report, corn seed exudate has been documented to induce the expression of aminotransferase gene that is part of lysine catabolism (Manuel and Ramos 2001). Koch et al. (2002) reported that sugar beet seed exudate can induce the GacS/GasA regulatory system in a *Pseudomonas* species that takes part in the fungal amphisin biosynthesis. GacS/GasA system plays a major role in the secondary metabolites biosynthesis and plant colonization in gram-negative bacteria (Heeb and Haas 2001; Haas and Keel 2003). Thus, evidences show that seed exudates in spermosphere can regulate complex microbial behavior at a molecular level.

3.3 Phylloshpere Microbiome, Interaction, and Mechanism

Plant microorganisms inhabit both below and above the ground, where above the ground plant parts also contain a dynamic microbial life (Vorholt 2012). Interestingly, profiles of leaf metabolites of *Arabidopsis thaliana* have been regulated by applications of soil microbes to root surfaces. Several amino acids' concentration was increased in the leaf metabolome which was attributed to increased herbivory by insects, probably due to cross talk between the above- and below-ground parts of the plant (Badri et al. 2013). The phyllosphere is defined as the microenvironment extending from the leaf surface outward to the outer edge of the boundary layer surrounding the leaf and inward into the leaf tissue. In short, it is the aerial habitat colonized by several microbes. Colonization of microbes is nonhomogeneous, and it is modulated by leaf structures such as veins, hairs, and stomata (Lindow and Brandl 2003). It is assumed that an average of 10^6–10^7 cm^{-2} bacteria inhabit the leaf surface. Also, phyllosphere bacterial population could comprise up to 10^{26} cells in the tropical plants (Morris 2001), but the total size of the fungal population of the phyllosphere has not yet been determined and is expected to be lower (Lindow and Brandl 2003; Baldotto and Olivares 2008). The composition of phyllosphere microbiome comprises several bacteria, fungi, *Archaea*, and nematodes (Lindow and Brandl 2003; Vorholt 2012) and interacts with both the host plant (Melotto et al. 2008; Baker et al. 2010) and with each other (Kemen 2014; Agler et al. 2016; Jakuschkin et al. 2016). Their interaction with mycoviruses (Marzano and Domier 2016), bacteriophages (Koskella 2013; Koskella and Parr 2015), and herbivorous arthropods (Crawford et al. 2010; Humphrey et al. 2014) has been reported. Albeit few investigations and reports on the microbial population of buds and flowers (Andrews and Harris 2000), most of the research on phyllosphere microbiology has been focused on leaves. Microbial growth starts with single cell population targeting different spots on the leaves for multiplication, resulting in microbial colonies. The bacterial cluster formation is an outcome of the conduciveness of specific spots and the type of offsprings produced (Mohanty et al. 2016).

3.3.1 Microbial Adaptation

At the onset of the vegetative season, using airborne dispersal mechanism, many microbial cells and spores reach the leaf and try to colonize. Microorganisms, after reaching the leaf cuticle, that are able to utilize the available nutrients in the phyllosphere can survive and reproduce in it. The expectation of microbial communities in this zone may be influenced by the structure and composition of the cuticle (Lemanceau et al. 2017). Nutritional heterogeneity on the leaf surface such as carbon sources of glucose, fructose, and sucrose leads to distinct microbial assemblages of varying community composition on leaf veins (Lindow and Brandl 2003; Vorholt 2012). Carbon sources such as amino acids, carbohydrates, and sugar alcohols have been identified as well on the leaf surface (Tukey Jr. 1970; Fiala et al. 1990). Such nutritional heterogenicity has been studied using bioreporters (Leveau and Lindow 2001; Miller et al. 2001; Remus-Emsermann and Leveau 2010). Besides, plant hormones such as indole-3-acetic acid (IAA) are also produced by bacteria and fungi (Brandi et al. 1996; Brandl and Lindow 1998; Limtong and Koowadjanakul 2012). Some reports stated the induction of cell wall loosening and release of saccharides due to the presence of IAA that results in an increase of nutrient availability (Fry 1989; Lindow and Brandl 2003). Foliar recognition of pathogens, stomatal aperture regulation, cell wall integrity, and foliar production of antimicrobial metabolites are considered to be important for phyllosphere microbes (Lemanceau et al. 2017). Horton et al. (2014) reported that plant loci are mainly responsible for defense and cell wall integrity of both the fungal and bacterial communities of phyllosphere of *A. thaliana*. Moreover, loci are responsible for the production of polysaccharides, such as callose that are used by plants for wound sealing during fungal infection, which drive the variation in the phyllosphere fungi (Horton et al. 2014). However, the signaling between plants and microbes is also interfered by the phyllospheric bacteria. For example, lipo-chito-oligosaccharides produced by microorganisms are cleaved due to chitinases produced by certain bacteria (Mohanty et al. 2016). Signaling molecules such as ethylene and gamma-amino benzoic acid mediate interactions between plants and microbes and thus influence the structure of phyllosphere microbial communities (Balint-Kurti et al. 2010; Bodenhausen et al. 2014). Additionally, there are also other plant-derived volatile metabolic substrates such as isoprenes and C_1 compounds (Fall and Benson 1996). A well-known C_1 source is methanol for phyllosphere microorganisms which is basically formed as a by-product of cell wall metabolism by pectin methyl esterase and is released during diurnal cycles (Fall and Benson 1996). Methanol has been reported to promote growth in situ (Sy et al. 2005; Abanda-Nkpwatt et al. 2006; Kawaguchi et al. 2011) by acting as a substrate for methylotrophic epiphytic bacteria and methylotrophic yeasts (Galbally and Kirstine 2002).

3.3.2 Phyllospheric Activities

Plant-microbe interaction is also influenced by quorum sensing signaling molecules. These signal molecules help plants to either stimulate immune responses against the harmful pathogens and boost the present beneficial microbes or facilitate their entry (Hartmann et al. 2014). One of the best studied examples is the influence of phyllospheric bacteria *Pseudomonas syringae* pv. *syringae* str. B728a toward a quorum sensing molecule *N*-acyl homoserine lactone (AHL). AHL synthase (AhlI) is responsible for 3-oxohexanoyl-homoserine lactone production which is regulated by AhlR and others. Production of extracellular polymeric substances that imparted resistance to desiccation and hydrogen peroxide was observed in *P. syringae* pv. *syringae* str. B728a. It was also shown to suppress motility (Quiñones et al. 2004, 2005). Thus, the quorum sensing enhances microbial epiphytic fitness. In 2008, Dulla and Lindow reported that *P. syringae*-AHL-dependent signaling in the phyllosphere is mainly influenced by the size of bacterial aggregates as well as the availability of water on leaves. Fast AHL accumulation is postulated to promote epiphyte survival during desiccation. Motility is a negative regulation of AHL signaling, and it is a kind of adaptation to dry conditions that allow bacteria to preserve energy and acquire resources which otherwise get used up in flagella production during movement restrictions. In plants, AHL stimulates phytohormone changes and boosts immunity. Plants also utilize their defense mechanism to combat signals from pathogens (Mohanty et al. 2016).

4 Effect of the Biotic and Abiotic Components in Agroecosystem

Biotic and abiotic factors are living and nonliving parts of an ecosystem, respectively. The abiotic factors are temperature, air, water, soil sunlight, and anything physical or chemical, and the biotic factors include plants and animals, insects, bacteria, fungi, birds, and other living organisms in an ecosystem.

4.1 Effect of Biotic Components

4.1.1 Autotrophs

This group of organisms synthesizes their own food through the conversion of chemical energy of sun into nutrients. Autotrophs such as photosynthetic bacteria use CO_2 as the sole carbon source. Autotrophs use different pigments to absorb light. However, some other autotrophs use chemosynthesis to create food by utilizing the organic compounds available in their surrounding area without sunlight. Bacteria of thermal vents in the sea absorb H_2S of seawater to produce food (Gupta and Malhotra 2009). Depending on their nutritional requirements, autotrophs are classified into photolithotrophic autotrophs and chemolithotrophic autotrophs. Photolithotrophic autotrophs use light as the energy source and inorganic hydrogen/electron (H/e^-) donor (e.g., green sulfur bacteria), and the chemolithotrophic autotrophs use organic chemical energy and organic H/e^- donor (e.g., denitrifying bacteria).

4.1.2 Heterotrophs

Most heterotrophs use organic nutrients as a source of both carbon and energy. Heterotrophs are extremely flexible with respect to carbon sources and make up about 95% of the diversity on earth (Chapin et al. 2002). On the basis of their nutritional requirements, they have been grouped into photo-organotrophic heterotrophs and chemo-organoheterotrophic heterotrophs. Photo-organotrophic heterotrophs use light as the energy source and organic H/e^- donor (e.g., green nonsulfur bacteria), and chemo-organoheterotrophic heterotrophs use organic chemicals as the energy source and organic H/e^- donor (e.g., nonphotosynthetic bacteria).

4.1.3 Detritivores

This group of organisms does not produce their nutrients and does not consume other living organisms. Detritivores depend on dead organisms, breaking down the dead animals and plants to avail energy (e.g., fungi, earthworms). Heterotrophs consume detritivores and derive energy to complete the nutrient cycle flow through the food web.

4.2 Effect of Abiotic Components

Abiotic components such as physical and chemical factors influence the ecosystem. Humidity, temperature, sunlight, and pH are the examples of abiotic components. These factors influence organisms in the ecosystem for their reproduction, survival, and adaptation, and are classified into edaphic and climatic factors.

4.2.1 Edaphic Factors

The edaphic factors include topography and soil characteristics such as pH. The structure and composition of soil affect the growth and development of plants that determine the animal and microbial population.

Soil

Soil is described as a complex environment that is composed of a mixture of nutrients such as minerals, gases, liquids, and organic matter that support the growth of plants and associated microbiome (Bronick and Lal 2005). The mineral constituents such as sand, silt, and clay decide the porosity and soil moisture which determine the soil fertility. The gradients of nutrients, pH, and gases create microenvironments for unique ecological niches. Soil organisms, bacteria, fungi, and viruses prefer distinct ecological niches (Hättenschwiler et al. 2005) and participate in the cycling of nutrients through the microbial degradation of organic matters and minerals that favor plant growth.

Physiological effects of plants are equally important to the abiotic factors. Plants promote the growth of the microbiome as well as influence abiotic properties which influence growth (Singh et al. 2009). It is well documented that soil characterization and geographic factors are the deciding factors for the structure of specific microbial communities (Girvan et al. 2003). Microbes play an important role in soil aggregates formation (Tisdall 1996; Bronick and Lal 2005). The soil moisture content is also important in determining the microbial community structure (Singh et al. 2009) as evidenced by research on the effects on extreme environments such as Antarctic soils, Canadian Arctic (Chu et al. 2011), and the Tibetan permafrost soil (Zhang et al. 2013). *Proteobacteria*, the dominant soil bacteria, was associated with the soil moisture content (Zhang et al. 2013). It was reported that in polar desert soils, the moisture content was closely associated with the abundance of specific bacterial communities (Geyer et al. 2014).

Soil pH

The hydrogen ion concentration that determines the acidity and alkalinity of the environment is pH. Soil pH is the indication of soil property as acid soil or alkaline soil on the basis of soil pH measurement that influences the microbiome composition (Lauber et al. 2009; Andrew et al. 2012; Zhalnina et al. 2014). Varying degrees of soil pH and a wide range of pH tolerance of microbes have been reported. Studies have indicated the specific association between soil pH and microbial communities present in the soil. Based on rRNA fingerprinting, the bacterial diversity has been correlated with soil pH as microbial diversity and richness of microbial communities showed variation in different soil types (Fierer and Jackson 2006; Lauber et al. 2009). An array of studies has been reported to confirm the soil pH as a key modulating the variations in the microbial community composition (Rousk et al. 2010). Soil pH is also considered as the predictor of microbial diversity. *Acidobacteria* is the predominant soil bacteria that show metabolic plasticity. The presence of such acidobacterial group of bacteria along with an elevational gradient pH was reported (Zhang et al. 2014). The effect of abiotic factors on nitrogen-fixing *Sinorhizobium meliloti* population indicated that soil pH and other geographical factors play a key role in shaping their microbial diversity (Donnarumma et al. 2014). Multiple factors such as climate, soil types and soil pH, nutrient availability, temperature, and altitude decide the composition of soil microbes (Angel et al. 2010; Singh et al. 2013).

Soil Nutrients

The role of nutrients in agricultural productivity was reported worldwide (Ryan and Sommer 2012). The three major nutritional factors such as nitrogen, carbon and phosphorus (Ryan and Sommer 2012) and the soil iron content affect the plant rhizosphere microbiome. Soil nutrient limitations are managed through the external application of chemical fertilizers. However, extensive use of chemicals affect soil environment (Geiger et al. 2010). Soil fertility is the result of complex biotic and abiotic interactions where microbiome plays a key role in decomposing the organic matters. Microbial decomposition of organic matters generates nutrients for plants, and in turn, the improved plant health permits the exploration of the roots for nutrient requisition and, thereby, facilitates the attachment of microbes to roots. Therefore, bioavailability of soil nutrients offers direct and indirect effects in determining the diversity of microbiome in the rhizosphere (Joergensen and Emmerling 2006). Nitrogen is the dominant factor in particular types of soil as it affects plant productivity and the composition of soil microbial communities (Clark et al. 2007). Enhanced plant productivity is observed as a result of enrichment of nitrogen content in soil. This also increases the richness of the bacterial community species and their diversity (Suding et al. 2005). When chronic nitrogen levels are low, they possess threat to the conservation of grassland ecosystem (Clark and Tilman 2008). Carbon forms the backbone of soil microbial communities (Degens et al. 2000; Drenovsky et al. 2004; Ahmed et al. 2008). Microbial catabolic evenness was used as a tool to evaluate the soil microbial diversity in soils that contain different organic carbon pools (Degens et al. 2000).

Phosphorus is an essential modulating factor of the plant microbiome. The effect of phosphorus applications has been extensively analyzed on the soil microbial diversity by employing the phospholipid, fatty acid profiling and denatured gradient gel electrophoresis techniques (Beauregard et al. 2010). Burning of grasslands increases the bioavailability of nitrogen and phosphorus, and the enrichment of nutrients enhances the bacterial populations (Coolon et al. 2013). Thus, the enrichment of soil nitrogen and phosphorus significantly altered the bacterial communities and their structure and diversity in ecosystems.

4.2.2 Climatic Factors

Climatic factors mainly comprise temperature, humidity, sunlight, wind, and water.

Temperature

Temperature and humidity are important factors for the survival and persistence of plant microbiome. Phytopathogenic bacteria prefer high humidity and temperature range of 25–30 °C (Smirnova et al. 2001) and fail to survive in cold temperature (Burdon et al. 1996). Furthermore, high moisture and humidity in tropical habitats favor the multiplicity of bacterial communities. Kinetics of microbial respiration and community structure are varied across a wide range of temperature due to global climate change.

UV Radiation

Selection of bacterial communities depends on plant rhizosphere, altered metabolism of roots, and release of nutrients. Since the chemical composition of root exudates influences the major energy and carbon sources for microorganisms, plants with different genotypes regulate the plant rhizobiome. However, UV radiation stratifies the phyllosphere microbiome. High UV-radiation-damaged bacterial DNA restricts phyllosphere bacteria (Beattie and Lindow 1999). Pigmented bacteria have the ability to withstand high UV exposure due to absorption of radiation and free radical quenching (Poplawsky et al. 2000; Jacobs et al. 2005; Gunasekera and Sundin 2006). In the absence of xanthomonadin pigment, the phytopathogenic bacteria, *Xanthomonas campestris,* showed 100-fold decrease in survival. Light penetration in tropical forests decides the foliar bacterial population and their tolerance to UV radiation (Chazdon and Fetcher 1984). Plant rhizosphere bacteria differ in their sensitivity to UV-B radiation-induced damage (Arrage et al. 1993). UV-radiation tolerance in phytopathogenic bacteria, *Pseudomonas syringae,* is conferred by a plasmid-coded *rulAB* operon (Cazorla et al. 2008). The extracellular polysaccharides (EPS) of *Xanthomonas campestris* confer the UV-radiation tolerance. The carotenoid pigments of *Erwinia herbicola* play a major role in their UV-A radiation protection (Whipps et al. 2008). Owing to the synthesis of melanin, the nonmotile, gram-positive Antarctic bacteria are shown to exhibit UV-protective mechanism (Bhattacharyya and Jha 2012). The *Archaea* bacteria increase their species diversity due to UV-B exposure (Robson et al. 2005). The Antarctica vascular plant rhizobacteria modify the growth of plants and their secondary metabolites due to UV radiation. The phyllosphere bacterial populations of *Arachis hypogaea* show UV tolerance, and they are predominantly spore-forming *Bacilli* (Sundin and Jacobs 1999). The UV radiation reduces the biomass of roots, thus resulting in the decrease in root-adhering microbial population (Caldwell et al. 2007).

5 Elucidation of Disease Resistance in Plants

Suppression of plant pathogens in soil relies upon the presence of specific group of microbes. Pathogen suppression is due to the complex interaction of antagonistic microbes. Enhanced suppression in the organic fields is due to higher microbial activity and lower soil nitrate level activity (Drinkwater et al. 1995). Specific groups of plant growth-promoting pseudomonads and arbuscular mycorrhizal fungi (AMF) play a role in disease suppression and plant growth promotion due to their biocontrol and plant growth promoting potential. Reduction of root necrosis symptoms of *Phytophthora* in strawberry was reported due to AMF. Antagonistic fluorescent pseudomonads inhabited in wheat rhizosphere enhance plant resistance because of their ability to produce multiple antibiotic compounds (Gu and Mazzola 2003). Specific pseudomonad groups of rhizobacteria and phyllosphere bacteria stimulate induced resistance (IR) in plants (Jetiyanon and Kloepper 2002; Jetiyanon et al. 2003). Specific groups of *Bacillus* sp. also induce the resistance and cause disease suppression. Induction of IR has been a strategy to suppress foliar pathogens. Biofumigation of crop residues controls cauliflower wilt and pathogenic nematodes.

5.1 Resistance Against Rhizosphere Diseases

In rhizosphere, it is known that root exudates act as signaling molecules between roots and microbial partners, and these may act as beneficial or detrimental to the microbes. Isoflavones secreted by soybean roots attract both mutualist (*Bradyrhizobium japonicum*) and pathogen (*Phytophthora sojae*) at the same time (Bais et al. 2006). Root cells contain defense proteins (phytoalexins) and several unidentified components that protect them from pathogenic bacteria (Flores et al. 1999). In some cases, degradation of plant and microbial compounds produces other allelopathic or toxic components that inhibit pathogenic microorganisms (Yang et al. 2001). Plants can be modulated to express AHL lactonase that can decrease the pathogenicity of plant pathogenic bacteria by disrupting quorum sensing, thus, in a way reducing the severity of the disease (Kalia 2013; Helman and Chernin 2015). Depending on plant species and their variation, bacteria differ in inducing resistance (Van Loon 1997). Biocontrol microbes, metabolites, and antimicrobial proteins have been reported to suppress plant pathogens (Sakthivel and Gnanamanickam 1987; Torres-Rubio et al. 2000; Notz et al. 2002; Naik and Sakthivel 2006; Kirubakaran and Sakthivel 2007; Singh et al. 2007; Kirubakaran et al. 2008).

Biocontrol activity through antibiotic production by fungi is mostly studied in *Trichoderma/Gliocladium* (Smith et al. 1990; Howell 1998) and *Talaromyces flavus* (Kim et al. 1990; Fravel and Roberts 1991). The strains of *Trichoderma virens* are classified into two groups of P (gliovirin-producing) (Howell and Stipanovic 1983) and Q (gliotoxin-producing) that differ in their physiology for antibiotic production and sensitivity toward different soilborne diseases (Howell 1991; Howell et al. 1993). P group strains are active against *Pythium ultimum* and not against the *Rhizoctonia solani* strain AG-4, whereas Q group strains are those which are quite potent against *R. solani* and less against *P. ultimum* (Howell et al. 1993). Thus, strains of the P group of *Trichoderma* act as biocontrol agents in controlling cotton seedling disease (Howell 1991), and the strains of Q group *Trichoderma* act as biocontrol agents in controlling cotton damping off disease (Howell et al. 1993). It has been also reported that rhizospheric hydrogen peroxide from the roots of *Talaromyces flavus* (Fravel and Roberts 1991) suppresses *Verticillium,* the wilt pathogen of eggplant (*Solanum tuberosum* L.) (Stosz et al. 1996; Melotto et al. 2008). *Chryseobacterium aquaticum* strain PUPC1 was reported to produce a novel antifungal protease from rice rhizosphere soil that showed broad-spectrum antifungal activity against plant pathogenic fungi and also enhanced plant growth (Gandhi Pragash et al. 2009).

5.2 Resistance Against Spermosphere Diseases

Suppression of pathogens of seed and seedling depends on the types, quantities, and time of release of inhibitors by biocontrol organisms or seed exudates (Raaijmakers et al. 2002). Antibiotic biosynthesis in spermosphere has been mostly concentrated on antibiotics produced by *P. fluorescens* and *P. aureofaciens* which suppress the infection by *P. ultimum.* Such antibiotics include oomycin A, pyoluteorin, 2,4-diacetylphloroglucinol (DAPG), and phenazine. It has been demonstrated that owing to the release of glucose from cotton seed, *afuE,* an oomycin A biosynthetic gene, was expressed in the spermosphere within 24 h after sowing (Howie and Suslow 1991). Similar observation was noted for cotton and cucumber seed spermospheres. Within 72 h of seed germination, *plt*, a pyoluteorin biosynthetic gene of *P. fluorescens* Pf-5, was expressed (Kraus and Loper 1995). In comparison to bulk soil, spermosphere and rhizosphere have been reported to harbor strains of *P. fluorescens* that synthesize vicosinamide, an antifungal compound (Nielsen and Sørensen 2003). *P. aureofaciens* PGS12 has been exhibited to produce phenazine in various plant species spermospheres. Zwittermicin A and kanosamine produced by *B. cereus* UW85 play a major role in *Phythium* species suppression (Silo-Suh et al. 1994; Stabb et al. 1994; Milner et al. 1995; Shang et al. 1999). Zwittermicin A is a broad-spectrum antibiotic (Silo-Suh et al. 1994) for the species of bacteria, fungi, and oomycetes (Silo-Suh et al. 1998). Kanosamine is the most effective against oomycetes and has little effect against bacteria and fungi (Milner et al. 1996).

Phytomicrobiome must prevent pathogen development and spread of infection. This is achieved either by production of inhibitors such as antibiotics or by elimination of essential nutritional energy and carbon sources within 12–24 h of sowing of seeds. Van Dijk and Nelson (2000) have demonstrated the competition for long-chain unsaturated fatty acids between *Enterobacter cloacae* and *P. ultimum* sporangia. *E. cloacae* can competitively limit the availability of fatty acids and thereby suppress *P. ultimum* sporangia infection to the plant. There are evidences to suggest that *Pseudomonas*, *Trichoderma*, *E. cloacae*, *Burkholderia cepacia*, and indigenous seed-colonizing microorganisms play a key role in suppressing seed and root infections (Dandurand and Knudsen 1993; Nelson 2004).

5.3 Resistance Against Phyllosphere Diseases

Plants produce a diverse group of secondary metabolites with antimicrobial activities. In addition, antimicrobial compounds are also produced by microorganisms. In general, plants are prone to attack by pests, pathogens, and herbivorous insects. The ratio between jasmonic acid induced by herbivorous insects and salicylic acid induced by various phyllospheric bacteria determines the sensitivity of the plant toward pathogens. In an experiment, prior to damage, the plant *Cardamine cordifolia* was administered with jasmonic acid or salicylic acid. It was monitored for changes in phyllospheric bacterial population and the effect on chewing of leaves by herbivores. Jasmonic acid-treated plants showed reduced herbivory, while the salicylic acid-treated ones showed increased herbivory. As compared to untreated plants, phyllospheric bacteria were found abundant in herbivore-damaged plants. Apart from the biocontrol through phytohormone production, plants have strengthened their innate defense system by controlling stomatal openings that act as ports for microbial entry (Mohanty et al. 2016).

6 Conclusions

Ecosystem is largely modulated by diverse interactions between soil and plant microbiomes. The interactions in all three plant compartments, namely, rhizosphere, spermosphere, and phyllosphere, together influence the plant nutrition and hosted microbial populations. The released exudates help in suppressing several plant diseases and tuning specific microbial diversity. Rhizospheric and spermospheric microbes are mostly dependent on the wise use of resources, while phyllospheric microbes thrive by efficient adaptation to biotic and abiotic stresses. Plants provide niches and act as nutritional sources to microbes, whereas microbes contribute to plant growth, nutrition, and suppression of phytopathogens. Insights into diverse plant traits and beneficial microbes will help in better plant breeding applications and decreased chemical inputs. Further, exploration and delineation of biomolecular networking between plants and microbes and the progress in metabolomics will help in discovering new signaling molecules that mediate the control of phytopathogens and substantiate the productivity and health of agroecosystem.

Acknowledgments The authors thank the University Grant Commission, New Delhi, for financial support through University Research Fellowship to Indramani Kumar and Rajiv Gandhi National Fellowship to Moumita Mondal. The authors also thank UGC-SAP and DST-FIST programs coordinated by Prof. N. Sakthivel for providing infrastructure facilities.

References

Abanda-Nkpwatt D, Musch M, Tschiersch J et al (2006) Molecular interaction between *Methylobacterium extorquens* and seedlings: growth promotion, methanol consumption, and localization of the methanol emission site. J Exp Bot. https://doi.org/10.1093/jxb/erl173

Abdelfattah A, Nicosia M, Cacciola SO et al (2015) Metabarcoding analysis of fungal diversity in the phyllosphere and carposphere of olive (*Olea europaea*). PLoS One 10:e0131069

Achnine L, Blancaflor EB, Rasmussen S, Dixon RA (2004) Colocalization of L-phenylalanine ammonia-lyase and cinnamate 4-hydroxylase for metabolic channeling in phenylpropanoid biosynthesis. Plant Cell 16:3098–3109

Agler MT, Ruhe J, Kroll S et al (2016) Microbial hub taxa link host and abiotic factors to plant microbiome variation. PLoS Biol. https://doi.org/10.1371/journal.pbio.1002352

Ahemad M, Khan MS (2012) Alleviation of fungicide-induced phytotoxicity in greengram [*Vigna radiata* (L.) Wilczek] using fungicide-tolerant and plant growth promoting *Pseudomonas* strain. Saudi J Biol Sci. https://doi.org/10.1016/j.sjbs.2012.06.003

Ahmed Z, Lim B-R, Cho J et al (2008) Biological nitrogen and phosphorus removal and changes in microbial community structure in a membrane bioreactor: effect of different carbon sources. Water Res. https://doi.org/10.1016/j.watres.2007.06.062

Andrew DR, Fitak RR, Munguia-Vega A et al (2012) Abiotic factors shape microbial diversity in Sonoran desert soils. Appl Environ Microbiol. https://doi.org/10.1128/AEM.01459-12

Andrews JH, Harris RF (2000) The ecology and biogeography of microorganisms on plant surfaces. Annu Rev Phytopathol 38:145–180

Angel R, Soares MIM, Ungar ED, Gillor O (2010) Biogeography of soil archaea and bacteria along a steep precipitation gradient. ISME J. https://doi.org/10.1038/ismej.2009.136

Arnold AE (2007) Understanding the diversity of foliar endophytic fungi: progress, challenges, and frontiers. Fungal Biol Rev 21:51–66

Arrage AA, Phelps TJ, Benoit RE, White DC (1993) Survival of subsurface microorganisms exposed to UV radiation and hydrogen peroxide. Appl Environ Microbiol 59(11):3545–3550

Azcón R, Barea J-M (2010) Mycorrhizosphere interactions for legume improvement. In: Microbes for legume improvement. Springer, pp 237–271

Badri DV, Vivanco JM (2009) Regulation and function of root exudates. Plant Cell Environ 32:666–681

Badri DV, Zolla G, Bakker MG et al (2013) Potential impact of soil microbiomes on the leaf metabolome and on herbivore feeding behavior. New Phytol. https://doi.org/10.1111/nph.12124

Bais HP, Weir TL, Perry LG et al (2006) The role of root exudates in rhizosphere interactions with plants and other organisms. Annu Rev Plant Biol. https://doi.org/10.1146/annurev.arplant.57.032905.105159

Baker CM, Chitrakar R, Obulareddy N et al (2010) Molecular battles between plant and pathogenic bacteria in the phyllosphere. Braz J Med Biol Res. https://doi.org/10.1590/S0100-879X2010007500060

Bakker AW, Schippers B (1987) Microbial cyanide production in the rhizosphere in relation to potato yield reduction and *Pseudomonas* spp-mediated plant growth-stimulation. Soil Biol Biochem 19:451–457

Baldotto LEB, Olivares FL (2008) Phylloepiphytic interaction between bacteria and different plant species in a tropical agricultural system. Can J Microbiol. https://doi.org/10.1139/W08-087

Balint-Kurti P, Simmons SJ, Blum JE et al (2010) Maize leaf epiphytic bacteria diversity patterns are genetically correlated with resistance to fungal pathogen infection. Mol Plant-Microbe Interact. https://doi.org/10.1094/MPMI-23-4-0473

Barbour WM, Hattermann DR, Stacey G (1991) Chemotaxis of Bradyrhizobium japonicum to soybean exudates. Appl Environ Microbiol 57(9):2635–2639

Bardgett RD, Bowman WD, Kaufmann R, Schmidt SK (2005) A temporal approach to linking aboveground and belowground ecology. Trends Ecol Evol 20:634–641

Barea JM, Pozo MJ, Azcón R, Azcón-Aguilar C (2005) Microbial co-operation in the rhizosphere. J Exp Bot 56(417):1761–1778

Bayliss C, Bent E, Culham DE et al (1997) Bacterial genetic loci implicated in the *Pseudomonas putida* GR12-2R3–canola mutualism: identification of an exudate-inducible sugar transporter. Can J Microbiol. https://doi.org/10.1139/m97-118

Beattie GA, Lindow SE (1999) Bacterial colonization of leaves: a spectrum of strategies. Phytopathology. https://doi.org/10.1094/PHYTO.1999.89.5.353

Beauregard MS, Hamel C, Atul-Nayyar, St-Arnaud M (2010) Long-term phosphorus fertilization impacts soil fungal and bacterial diversity but not AM fungal community in alfalfa. Microb Ecol. https://doi.org/10.1007/s00248-009-9583-z

Berendsen RL, Pieterse CMJ, Bakker PAHM (2012) The rhizosphere microbiome and plant health. Trends Plant Sci 17(8):478–486

Berg G, Rybakova D, Grube M, Köberl M (2015) The plant microbiome explored: implications for experimental botany. J Exp Bot 67:995–1002

Bhattacharyya PN, Jha DK (2012) Plant growth-promoting rhizobacteria (PGPR): emergence in agriculture. World J Microbiol Biotechnol 28(4):1327–1350

Bhatti MK, Akhtar F, Choudhary MI (1992) Alkaloids of *Fumaria indica*. Phytochemistry 31:2869–2872

Bodenhausen N, Horton MW, Bergelson J (2013) Bacterial communities associated with the leaves and the roots of *Arabidopsis thaliana*. PLoS One 8:e56329

Bodenhausen N, Bortfeld-Miller M, Ackermann M, Vorholt JA (2014) A aynthetic community approach reveals plant genotypes affecting the phyllosphere microbiota. PLoS Genet. https://doi.org/10.1371/journal.pgen.1004283

Bommarco R, Kleijn D, Potts SG (2013) Ecological intensification: harnessing ecosystem services for food security. Trends Ecol Evol 28(4):230–238

Bonkowski M, Villenave C, Griffiths B (2009) Rhizosphere fauna: the functional and structural diversity of intimate interactions of soil fauna with plant roots. Plant Soil 321:213–233

Brandi M, Clark EM, Lindow SE (1996) Characterization of the indole-3-acetic acid (IAA) biosynthetic pathway in an epiphytic strain of *Erwinia herbicola* and IAA production in vitro. Can J Microbiol. https://doi.org/10.1139/m96-079

Brandl MT, Lindow SE (1998) Contribution of indole-3-acetic acid production to the epiphytic fitness of Erwinia herbicola. Appl Environ Microbiol 64(9):3256–3263

Bringel F, Couée I (2015) Pivotal roles of phyllosphere microorganisms at the interface between plant functioning and atmospheric trace gas dynamics. Front Microbiol 6:486

Bronick CJ, Lal R (2005) Soil structure and management: a review. Geoderma. https://doi.org/10.1016/j.geoderma.2004.03.005

Buée M, De Boer W, Martin F et al (2009) The rhizosphere zoo: an overview of plant-associated communities of microorganisms, including phages, bacteria, archaea, and fungi, and of some of their structuring factors. Plant Soil 321:189–212

Bulgarelli D, Rott M, Schlaeppi K et al (2012) Revealing structure and assembly cues for Arabidopsis root-inhabiting bacterial microbiota. Nature 488(7409):91–95

Burdon JJ, Wennström A, Elmqvist T, Kirby GC (1996) The role of race specific resistance in natural plant populations. Oikos 76:411–416

Buyer JS, Roberts DP, Russek-Cohen E (1999) Microbial community structure and function in the spermosphere as affected by soil and seed type. Can J Microbiol. https://doi.org/10.1139/w98-227

Caldwell MM, Bornman JF, Ballaré CL et al (2007) Terrestrial ecosystems, increased solar ultraviolet radiation, and interactions with other climate change factors. Photochem Photobiol Sci. https://doi.org/10.1039/b700019g

Caporaso JG, Lauber CL, Walters WA et al (2012) Ultra-high-throughput microbial community analysis on the Illumina HiSeq and MiSeq platforms. ISME J. https://doi.org/10.1038/ismej.2012.8

Cazorla FM, Codina JC, Abad C et al (2008) 62-kb plasmids harboring rulAB homologues confer UV-tolerance and epiphytic fitness to *Pseudomonas syringae* pv. syringae mango isolates. Microb Ecol. https://doi.org/10.1007/s00248-007-9346-7

Chapin FS, Matson PA, Mooney HA, Chapin MC (2002) Principles of terrestrial ecosystem ecology. Springer, New York

Chazdon RL, Fetcher N (1984) Light environments of tropical forests. In: Medina E, Mooney HA, Vázquez-Yánes C (eds) Physiological ecology of plants of the wet tropics: proceedings of an international symposium held in Oxatepec and Los Tuxtlas, Mexico, June 29 to July 6, 1983. Springer Netherlands, Dordrecht, pp 27–36

Chen C, Meyermans H, Burggraeve B et al (2000) Cell-specific and conditional expression of caffeoyl-coenzyme A-3-O-methyltransferase in poplar. Plant Physiol 123:853–868

Chu H, Neufeld JD, Walker VK, Grogan P (2011) The influence of vegetation type on the dominant soil bacteria, archaea, and fungi in a low Arctic Tundra landscape. Soil Sci Soc Am J. https://doi.org/10.2136/sssaj2011.0057

Clark CM, Tilman D (2008) Loss of plant species after chronic low-level nitrogen deposition to prairie grasslands. Nature. https://doi.org/10.1038/nature06503

Clark CM, Cleland EE, Collins SL et al (2007) Environmental and plant community determinants of species loss following nitrogen enrichment. Ecol Lett. https://doi.org/10.1111/j.1461-0248.2007.01053.x

Compant S, Clément C, Sessitsch A (2010) Plant growth-promoting bacteria in the rhizo-and endosphere of plants: their role, colonization, mechanisms involved and prospects for utilization. Soil Biol Biochem 42:669–678

Coolon JD, Jones KL, Todd TC et al (2013) Long-term nitrogen amendment alters the diversity and assemblage of soil bacterial communities in Tallgrass Prairie. PLoS One. https://doi.org/10.1371/journal.pone.0067884

Crawford KM, Land JM, Rudgers JA (2010) Fungal endophytes of native grasses decrease insect herbivore preference and performance. Oecologia. https://doi.org/10.1007/s00442-010-1685-2

Croes S, Weyens N, Janssen J et al (2013) Bacterial communities associated with B rassica napus L. grown on trace element-contaminated and non-contaminated fields: a genotypic and phenotypic comparison. Microb Biotechnol 6:371–384

da Silva LL, Olivares FL, Rodrigues de Oliveira R et al (2014) Root exudate profiling of maize seedlings inoculated with *Herbaspirillum seropedicae* and humic acids. Chem Biol Technol Agric. https://doi.org/10.1186/s40538-014-0023-z

Dakora FD, Phillips DA (2002) Root exudates as mediators of mineral acquisition in low-nutrient environments. Plant Soil 245:35–47

Dandurand LM, Knudsen GR (1993) Influence of *Pseudomonas fluorescens* on hyphal growth and biocontrol activity of *Trichoderma harzianum* in the spermosphere and rhizosphere of pea. Phytopathology 83:265–270

Darrasse A, Darsonval A, Boureau T et al (2010) Transmission of plant-pathogenic bacteria by nonhost seeds without induction of an associated defense reaction at emergence. Appl Environ Microbiol. https://doi.org/10.1128/AEM.01098-10

De Wit R, Bouvier T (2006) "Everything is everywhere, but, the environment selects"; what did Baas Becking and Beijerinck really say? Environ Microbiol 8:755–758

Degens BP, Schipper LA, Sparling GP, Vojvodic-Vukovic M (2000) Decreases in organic C reserves in soils can reduce the catabolic diversity of soil microbial communities. Soil Biol Biochem. https://doi.org/10.1016/S0038-0717(99)00141-8

Delmotte N, Knief C, Chaffron S et al (2009) Community proteogenomics reveals insights into the physiology of phyllosphere bacteria. Proc Natl Acad Sci 106:16428–16433

Dodd IC, Zinovkina NY, Safronova VI, Belimov AA (2010) Rhizobacterial mediation of plant hormone status. Ann Appl Biol. https://doi.org/10.1111/j.1744-7348.2010.00439.x

Donnarumma F, Bazzicalupo M, Blažinkov M et al (2014) Biogeography of *Sinorhizobium meliloti* nodulating alfalfa in different Croatian regions. Res Microbiol. https://doi.org/10.1016/j.resmic.2014.06.001

Doré T, Makowski D, Malézieux E et al (2011) Facing up to the paradigm of ecological intensification in agronomy: revisiting methods, concepts and knowledge. Eur J Agron 34:197–210

Drenovsky RE, Vo D, Graham KJ, Scow KM (2004) Soil water content and organic carbon availability are major determinants of soil microbial community composition. Microb Ecol. https://doi.org/10.1007/s00248-003-1063-2

Drinkwater LE, Letourneau DK, Workneh F et al (1995) Fundamental differences between conventional and organic tomato agroecosystems in California. Ecol Appl. https://doi.org/10.2307/2269357

Dulla G, Lindow SE (2008) Quorum size of *Pseudomonas syringae* is small and dictated by water availability on the leaf surface. Proc Natl Acad Sci. https://doi.org/10.1073/pnas.0711723105

Egamberdieva D, Kamilova F, Validov S et al (2008) High incidence of plant growth-stimulating bacteria associated with the rhizosphere of wheat grown on salinated soil in Uzbekistan. Environ Microbiol 10:1–9

Ekström G, Ekbom B (2011) Pest control in agro-ecosystems: an ecological approach. CRC Crit Rev Plant Sci. https://doi.org/10.1080/07352689.2011.554354

Ercolani GL (1991) Distribution of epiphytic bacteria on olive leaves and the influence of leaf age and sampling time. Microb Ecol 21:35–48

Fall R, Benson AA (1996) Leaf methanol – The simplest natural product from plants. Trends Plant Sci 1:296–301

Fiala V, Glad C, Martin M et al (1990) Occurrence of soluble carbohydrates on the phylloplane of maize (Zea mays L.): variations in relation to leaf heterogeneity and position on the plant. New Phytol. https://doi.org/10.1111/j.1469-8137.1990.tb00492.x

Fierer N, Jackson RB (2006) The diversity and biogeography of soil bacterial communities. Proc Natl Acad Sci USA. https://doi.org/10.1073/pnas.0507535103

Figueiredo M do VB, Seldin L, de Araujo FF, Mariano R de LR (2010) Plant growth promoting rhizobacteria: fundamentals and applications. In: Plant growth and health promoting bacteria. Springer, pp 21–43

Fitzpatrick CR, Copeland J, Wang PW et al (2018) Assembly and ecological function of the root microbiome across angiosperm plant species. Proc Natl Acad Sci. https://doi.org/10.1073/pnas.1717617115

Flores HE, Vicanco JM, Loyola-Vargas VM (1999) "Radicle" biochemistry: the biology of root-specific metabolism. Trends Plant Sci 4:220–226

Fravel DR, Roberts DP (1991) *In situ* Evidence for the role of Glucose oxidase in the bio-control of verticillium wilt by *Talaromyces flavus*. Biocontrol Sci Tech. https://doi.org/10.1080/09583159109355189

Fry SC (1989) Cellulases, hemicelluloses and auxin-stimulated growth: a possible relationship. Physiol Plant. https://doi.org/10.1111/j.1399-3054.1989.tb05620.x

Gaba S, Bretagnolle F, Rigaud T, Philippot L (2014) Managing biotic interactions for ecological intensification of agroecosystems. Front Ecol Evol. https://doi.org/10.3389/fevo.2014.00029

Galbally IE, Kirstine W (2002) The production of methanol by flowering plants and the global cycle of methanol. J Atmos Chem. https://doi.org/10.1023/A:1020684815474

Gandhi Pragash M, Narayanan KB, Naik PR, Sakthivel N (2009) Characterization of Chryseobacterium aquaticum strain PUPC1 producing a novel antifungal protease from rice rhizosphere soil. J Microbiol Biotechnol 19:99–107

Gans C (2005) Checklist and bibliography of the amphisbaenia of the world. Bull Am Mus Nat Hist. https://doi.org/10.1206/0003-0090(2005)289<0001:CABOTA>2.0.CO;2

Gauthier D, Jaffré T, Prin Y (2000) Abundance of *Frankia* from *Gymnostoma* spp. in the rhizosphere of *Alphitonia neocaledonica*, a non-nodulated Rhamnaceae endemic to New Caledonia. Eur J Soil Biol. https://doi.org/10.1016/S1164-5563(00)01061-X

Geiger F, Bengtsson J, Berendse F et al (2010) Persistent negative effects of pesticides on biodiversity and biological control potential on European farmland. Basic Appl Ecol. https://doi.org/10.1016/j.baae.2009.12.001

Geyer KM, Altrichter AE, Takacs-Vesbach CD et al (2014) Bacterial community composition of divergent soil habitats in a polar desert. FEMS Microbiol Ecol. https://doi.org/10.1111/1574-6941.12306

Giri B, Giang PH, Kumari R et al (2005) Microbial diversity in soils. In: Microorganisms in soils: roles in genesis and functions. Springer, pp 19–55

Girvan MS, Bullimore J, Pretty JN et al (2003) Soil type is the primary determinant of the composition of the total and active bacterial communities in arable soils. Appl Environ Microbiol. https://doi.org/10.1128/AEM.69.3.1800-1809.2003

Glick BR (2012) Plant growth-promoting bacteria: mechanisms and applications. Scientifica (Cairo). https://doi.org/10.6064/2012/963401

Gorecki RJ, Harman GE, Mattick LR (1985) The volatile exudates from germinating pea seeds of different viability and vigor. Can J Bot 63:1035–1039

Gorecki RJ, Ashino H, Satoh S, Esashi Y (1991) Ethylene production in pea and cocklebur seeds of differing vigour. J Exp Bot. https://doi.org/10.1093/jxb/42.3.407

Gu YH, Mazzola M (2003) Modification of fluorescent pseudomonad community and control of apple replant disease induced in a wheat cultivar-specific manner. Appl Soil Ecol. https://doi.org/10.1016/S0929-1393(03)00066-0

Gunasekera TS, Sundin GW (2006) Role of nucleotide excision repair and photoreactivation in the solar UVB radiation survival of *Pseudomonas syringae* pv. syringae B728a. J Appl Microbiol. https://doi.org/10.1111/j.1365-2672.2006.02841.x

Gupta, KK, Malhotra S (2009) Everyday science. Frank Brothers

Haas D, Keel C (2003) Regulation of antibiotic production in root-colonizing *Peudomonas* spp. and relevance for biological control of plant disease. Annu Rev Phytopathol. https://doi.org/10.1146/annurev.phyto.41.052002.095656

Hacquard S, Garrido-Oter R, González A et al (2015) Microbiota and host nutrition across plant and animal kingdoms. Cell Host Microbe 17:603–616

Hartmann A, Rothballer M, Hense BA, Schröder P (2014) Bacterial quorum sensing compounds are important modulators of microbe-plant interactions. Front Plant Sci 5:131

Hättenschwiler S, Tiunov AV, Scheu S (2005) Biodiversity and litter decomposition in terrestrial ecosystems. Annu Rev Ecol Evol Syst. https://doi.org/10.1146/annurev.ecolsys.36.112904.151932

Hawkes CV, DeAngelis KM, Firestone MK (2007) Root interactions with soil microbial communities and processes. In: The rhizosphere. Elsevier, pp 1–29

Heeb S, Haas D (2001) Regulatory roles of the GacS/GacA two-component system in plant-associated and other gram-negative bacteria. Mol Plant-Microbe Interact 14:1351–1363

Helman Y, Chernin L (2015) Silencing the mob: disrupting quorum sensing as a means to fight plant disease. Mol Plant Pathol 16(3):316–329

Hiltner L (1904) Uber nevere Erfahrungen und Probleme auf dem Gebiet der Boden Bakteriologie und unter besonderer Beurchsichtigung der Grundungung und Broche. Arbeit Deut Landw Ges Berlin 98:59–78

Horton MW, Bodenhausen N, Beilsmith K et al (2014) Genome-wide association study of *Arabidopsis thaliana* leaf microbial community. Nat Commun. https://doi.org/10.1038/ncomms6320

Howell CR (1991) Biological control of Pythium damping-off of cotton with seed-coating preparations of *Gliocladium virens*. Phytopathology 81:738–741

Howell CR (1998) The role of antibiosis in biocontrol. Trichoderma Gliocladium 2:173–184

Howell CR, Stipanovic RD (1983) Gliovirin, a new antibiotic from *Gliocladium virens*, and its role in the biological control of *Pythium ultimum*. Can J Microbiol 29:321–324

Howell CR, Stipanovic RD, Lumsden RD (1993) Antibiotic production by strains of Gliocladium virens and its relation to the biocontrol of cotton seedling diseases. Biocontrol Sci Tech 3:435–441

Howie WJ, Suslow TV (1991) Role of antibiotic biosynthesis in the inhibition of *Pythium ultimum* in the cotton spermosphere and rhizosphere by *Pseudomonas fluorescens*. Mol Plant-Microbe Interact 4:393–399. https://doi.org/10.1094/MPMI-4-393

Humphrey PT, Nguyen TT, Villalobos MM, Whiteman NK (2014) Diversity and abundance of phyllosphere bacteria are linked to insect herbivory. Mol Ecol. https://doi.org/10.1111/mec.12657

Izuno A, Kanzaki M, Artchawakom T et al (2016) Vertical structure of phyllosphere fungal communities in a tropical forest in Thailand uncovered by high-throughput sequencing. PLoS One 11:e0166669

Jacobs JL, Sundin GW (2001) Effect of solar UV-B radiation on a phyllosphere bacterial community. Appl Environ Microbiol 67:5488–5496

Jacobs JL, Carroll TL, Sundin GW (2005) The role of pigmentation, ultraviolet radiation tolerance, and leaf colonization strategies in the epiphytic survival of phyllosphere bacteria. Microb Ecol. https://doi.org/10.1007/s00248-003-1061-4

Jakuschkin B, Fievet V, Schwaller L et al (2016) Deciphering the pathobiome: intra- and inter-kingdom interactions involving the pathogen *Erysiphe alphitoides*. Microb Ecol. https://doi.org/10.1007/s00248-016-0777-x

Jansson JK, Neufeld JD, Moran MA, Gilbert JA (2012) Omics for understanding microbial functional dynamics. Environ Microbiol 14(1):1–3

Jetiyanon K, Kloepper JW (2002) Mixtures of plant growth-promoting rhizobacteria for induction of systemic resistance against multiple plant diseases. Biol Control. https://doi.org/10.1016/S1049-9644(02)00022-1

Jetiyanon K, Fowler WD, Kloepper JW (2003) Broad-spectrum protection against several pathogens by PGPR mixtures under field conditions in Thailand. Plant Dis. https://doi.org/10.1094/PDIS.2003.87.11.1390

Jha BK, Pragash MG, Cletus J et al (2009) Simultaneous phosphate solubilization potential and antifungal activity of new fluorescent pseudomonad strains, *Pseudomonas aeruginosa*, *P. plecoglossicida* and *P. mosselii*. World J Microbiol Biotechnol 25:573–581

Joergensen RG, Emmerling C (2006) Methods for evaluating human impact on soil microorganisms based on their activity, biomass, and diversity in agricultural soils. J Plant Nutr Soil Sci 169:295–309

Jumpponen A, Jones KL (2010) Seasonally dynamic fungal communities in the *Quercus macrocarpa* phyllosphere differ between urban and nonurban environments. New Phytol 186:496–513
Jurkevitch EJ, Shapira G (2000) Structure and colonization dynamics of epiphytic bacterial communities and of selected component strains on tomato (*Lycopersicon esculentum*) leaves. Microb Ecol 40:300–308
Kalia VC (2013) Quorum sensing inhibitors: an overview. Biotechnol Adv 31(2):224–245
Kamilova F, Kravchenko LV, Shaposhnikov AI et al (2006) Organic Acids, Sugars, and L-tryptophane in exudates of vegetables growing on stonewool and their effects on activities of rhizosphere bacteria. Mol Plant-Microbe Interact. https://doi.org/10.1094/MPMI-19-0250
Kawaguchi K, Yurimoto H, Oku M, Sakai Y (2011) Yeast methylotrophy and autophagy in a methanol-oscillating environment on growing *Arabidopsis thaliana* leaves. PLoS One. https://doi.org/10.1371/journal.pone.0025257
Kazerooni EA, Maharachchikumbura SSN, Rethinasamy V et al (2017) Fungal diversity in tomato rhizosphere soil under conventional and desert farming systems. Front Microbiol 8:1462
Kelley CJ, Mahajan JR, Brooks LC et al (1975) Polyphenolic acids of *Lithospermum ruderale* (Boraginaceae). I Isolation and structure determination of lithospermic acid. J Org Chem 40:1804–1815
Kelley CJ, Harruff RC, Carmack M (1976) Polyphenolic acids of *Lithospermum ruderale*. II. Carbon-13 nuclear magnetic resonance of lithospermic and rosmarinic acids. J Org Chem 41:449–455
Kemen E (2014) Microbe-microbe interactions determine oomycete and fungal host colonization. Curr Opin Plant Biol 20:75–81
Khan MS, Zaidi A, Wani PA, Oves M (2009) Role of plant growth promoting rhizobacteria in the remediation of metal contaminated soils. Environ Chem Lett 7:1–19
Kim KK, Fravel DR, Papavizas GC (1990) Production, purification, and properties of glucose oxidase from the biocontrol fungus *Talaromyces flavus*. Can J Microbiol 36:199–205
Kim M, Singh D, Lai-Hoe A et al (2012) Distinctive phyllosphere bacterial communities in tropical trees. Microb Ecol 63:674–681
Kirubakaran SI, Sakthivel N (2007) Cloning and overexpression of antifungal barley chitinase gene in *Escherichia coli*. Protein Expr Purif 52:159–166
Kirubakaran SI, Begum SM, Ulaganathan K, Sakthivel N (2008) Characterization of a new antifungal lipid transfer protein from wheat. Plant Physiol Biochem 46:918–927
Kloepper JW, Schroth MN (1981) Relationship of in vitro antibiosis of plant growth-promoting rhizobacteria to plant growth and the displacement of root microflora. Phytopathology 71:1020-1024
Kobayashi DY, Crouch JA (2009) Bacterial/fungal interactions: from pathogens to mutualistic endosymbionts. Annu Rev Phytopathol. https://doi.org/10.1146/annurev-phyto-080508-081729
Koch B, Nielsen TH, Sørensen D et al (2002) Lipopeptide production in Pseudomonas sp. strain DSS73 is regulated by components of sugar beet seed exudate via the Gac two-component regulatory system. Appl Environ Microbiol 68(9):4509–4516
Koskella B (2013) Phage-mediated selection on microbiota of a long-lived host. Curr Biol. https://doi.org/10.1016/j.cub.2013.05.038
Koskella B, Parr N (2015) The evolution of bacterial resistance against bacteriophages in the horse chestnut phyllosphere is general across both space and time. Philos Trans R Soc B Biol Sci. https://doi.org/10.1098/rstb.2014.0297
Kraus J, Loper JE (1995) Characterization of a genomic region required for production of the antibiotic pyoluteorin by the biological control agent *Pseudomonas fluorescens* Pf-5. Appl Environ Microbiol 61(3):849–854
Lacombe S, Rougon-Cardoso A, Sherwood E et al (2010) Interfamily transfer of a plant pattern-recognition receptor confers broad-spectrum bacterial resistance. Nat Biotechnol. https://doi.org/10.1038/nbt.1613
Lauber CL, Hamady M, Knight R, Fierer N (2009) Pyrosequencing-based assessment of soil pH as a predictor of soil bacterial community structure at the continental scale. Appl Environ Microbiol. https://doi.org/10.1128/AEM.00335-09

Lemanceau P, Barret M, Mazurier S et al (2017) Plant communication with associated microbiota in the spermosphere, rhizosphere and phyllosphere. In: Advances in botanical research. Elsevier, pp 101–133

Leveau JHJ, Lindow SE (2001) Appetite of an epiphyte: quantitative monitoring of bacterial sugar consumption in the phyllosphere. Proc Natl Acad Sci. https://doi.org/10.1073/pnas.061629598

Limtong S, Koowadjanakul N (2012) Yeasts from phylloplane and their capability to produce indole-3-acetic acid. World J Microbiol Biotechnol. https://doi.org/10.1007/s11274-012-1144-9

Lindow SE, Brandl MT (2003) Microbiology of the phyllosphere. Appl Environ Microbiol 69:1875–1883

Lopez-Velasco G, Carder PA, Welbaum GE, Ponder MA (2013) Diversity of the spinach (*Spinacia oleracea*) spermosphere and phyllosphere bacterial communities. FEMS Microbiol Lett 346:146–154

Lugtenberg BJ, Kravchenko LV, Simons M (1999) Tomato seed and root exudate sugars: composition, utilization by *Pseudomonas* biocontrol strains and role in rhizosphere colonization. Environ Microbiol. https://doi.org/10.1046/j.1462-2920.1999.00054.x

Lundberg DS, Lebeis SL, Paredes SH et al (2012) Defining the core *Arabidopsis thaliana* root microbiome. Nature. https://doi.org/10.1038/nature11237

Lynch JM (1990) The rhizosphere. Wiley, New York. https://doi.org/10.1017/S0014479700019700

Makoi JHJR, Ndakidemi PA (2007) Biological, ecological and agronomic significance of plant phenolic compounds in rhizosphere of the symbiotic legumes. Afr J Biotechnol 6:1358–1368

Mandal S, Mandal M, Das A et al (2009) Stimulation of indoleacetic acid production in a Rhizobium isolate of *Vigna mungo* by root nodule phenolic acids. Arch Microbiol. https://doi.org/10.1007/s00203-008-0455-6

Manuel EU, Ramos JL (2001) Expression of a *Pseudomonas putida* aminotransferase involved in lysine catabolism is induced in the rhizosphere. Appl Environ Microbiol. https://doi.org/10.1128/AEM.67.11.5219-5224.2001

Martínez-Viveros O, Jorquera MA, Crowley DE et al (2010) Mechanisms and practical considerations involved in plant growth promotion by rhizobacteria. J Soil Sci Plant Nutr. https://doi.org/10.4067/S0718-95162010000100006

Marzano SYL, Domier LL (2016) Novel mycoviruses discovered from metatranscriptomics survey of soybean phyllosphere phytobiomes. Virus Res. https://doi.org/10.1016/j.virusres.2016.05.012

Mastouri F, Björkman T, Harman GE (2010) Seed treatment with *Trichoderma harzianum* alleviates biotic, abiotic, and physiological stresses in germinating seeds and seedlings. Phytopathology 100(11):1213–1221

Melotto M, Underwood W, He SY (2008) Role of stomata in plant innate immunity and foliar bacterial diseases. Annu Rev Phytopathol. https://doi.org/10.1146/annurev.phyto.121107.104959

Mendes R, Kruijt M, De Bruijn I et al (2011) Deciphering the rhizosphere microbiome for disease-suppressive bacteria. Science (80). https://doi.org/10.1126/science.1203980

Mendes R, Garbeva P, Raaijmakers JM (2013) The rhizosphere microbiome: significance of plant beneficial, plant pathogenic, and human pathogenic microorganisms. FEMS Microbiol Rev 37:634–663

Miche L, Belkin S, Rozen R, Balandreau J (2003) Rice seedling whole exudates and extracted alkylresorcinols induce stress-response in *Escherichia coli* biosensors. Environ Microbiol 5:403–411

Miller WG, Brandl MT, Quiñones B, Lindow SE (2001) Biological sensor for sucrose availability: relative sensitivities of various reporter genes. Appl Environ Microbiol. https://doi.org/10.1128/AEM.67.3.1308-1317.2001

Milner JL, Raffel SJ, Lethbridge BJ, Handelsman J (1995) Culture conditions that influence accumulation of zwittermicin A by *Bacillus cereus* UW85. Appl Microbiol Biotechnol. https://doi.org/10.1007/BF00164774

Milner JL, Silo-Suh L, Lee JC et al (1996) Production of kanosamine by *Bacillus cereus* UW85. Appl Environ Microbiol 62(8):3061–3065

Mohanty SR, Dubey G, Ahirwar U et al (2016) Prospect of phyllosphere microbiota: a case study on bioenergy crop *Jatropha Curcas*. In: Plant-microbe interaction: an approach to sustainable agriculture, pp 453–462

Morris CE (2001) Phyllosphere. In: Encyclopedia of life sciences. Nature Publishing Group, London. https://doi.org/10.1038/npg.els.0000400

Muegge BD, Kuczynski J, Knights D et al (2011) Diet drives convergence in gut microbiome functions across mammalian phylogeny and within humans. Science (80). https://doi.org/10.1126/science.1198719

Naik PR, Sakthivel N (2006) Functional characterization of a novel hydrocarbonoclastic Pseudomonas sp. strain PUP6 with plant-growth-promoting traits and antifungal potential. Res Microbiol 157:538–546

Naik PR, Raman G, Narayanan KB, Sakthivel N (2008a) Assessment of genetic and functional diversity of phosphate solubilizing fluorescent pseudomonads isolated from rhizospheric soil. BMC Microbiol 8:230

Naik PR, Sahoo N, Goswami D et al (2008b) Genetic and functional diversity among fluorescent pseudomonads isolated from the rhizosphere of banana. Microb Ecol 56:492–504

Nelson EB (2004) Microbial dynamics and interactions in the spermosphere. Annu Rev Phytopathol. https://doi.org/10.1146/annurev.phyto.42.121603.131041

Nelson EB, Craft CM (1989) Comparative germination of culture-produced and plant-produced sporangia of *Pythium ultimum* in response to soluble seed exudates and exudate components. Phytopathology. https://doi.org/10.1094/Phyto-79-1009

Nielsen TH, Sørensen J (2003) Production of cyclic lipopeptides by *Pseudomonas fluorescens* strains in bulk soil and in the sugar beet rhizosphere. Appl Environ Microbiol. https://doi.org/10.1128/AEM.69.2.861-868.2003

Notz R, Maurhofer M, Dubach H et al (2002) Fusaric acid-producing strains of *Fusarium oxysporum* alter 2,4-diacetylphloroglucinol biosynthetic gene expression in *Pseudomonas fluorescens* CHA0 in vitro and in the rhizosphere of wheat. Appl Environ Microbiol 68:2229–2235

O'brien RD, Lindow SE (1989) Effect of plant species and environmental conditions on epiphytic population sizes of *Pseudomonas syringae* and other bacteria. Phytopathology 79:619–627

Paulitz TC, Anas O, Fernando DG (1992) Biological control of Pythium damping-off by seed-treatment with *Pseudomonas putida*: relationship with ethanol production by pea and soybean seeds. Biocontrol Sci Tech. https://doi.org/10.1080/09583159209355233

Philippot L, Raaijmakers JM, Lemanceau P, Van Der Putten WH (2013) Going back to the roots: the microbial ecology of the rhizosphere. Nat Rev Microbiol 11(11):789–799

Podile AR, Kishore GK (2007) Plant growth-promoting rhizobacteria. In: Plant-associated bacteria. Springer, pp 195–230

Poplawsky AR, Urban SC, Chun W (2000) Biological role of xanthomonadin pigments in *Xanthomonas campestris* pv. campestris. Appl Environ Microbiol. https://doi.org/10.1128/AEM.66.12.5123-5127.2000

Prasanna R, Jaiswal P, Nayak S et al (2009) Cyanobacterial diversity in the rhizosphere of rice and its ecological significance. Indian J Microbiol. https://doi.org/10.1007/s12088-009-0009-x

Qin S, Yeboah S, Xu X et al (2017) Analysis on fungal diversity in rhizosphere soil of continuous cropping potato subjected to different furrow-ridge mulching managements. Front Microbiol 8:845

Quesada-Moraga E, López-Díaz C, Landa BB (2014) The hidden habit of the entomopathogenic fungus *Beauveria bassiana*: first demonstration of vertical plant transmission. PLoS One. https://doi.org/10.1371/journal.pone.0089278

Quiñones B, Pujol CJ, Lindow SE (2004) Regulation of AHL production and its contribution to epiphytic fitness in *Pseudomonas syringae*. Mol Plant-Microbe Interact. https://doi.org/10.1094/MPMI.2004.17.5.521

Quiñones B, Dulla G, Lindow SE (2005) Quorum sensing regulates exopolysaccharide production, motility, and virulence in *Pseudomonas syringae*. Mol Plant-Microbe Interact. https://doi.org/10.1094/MPMI-18-0682

Raaijmakers JM, Vlami M, de Souza JT (2002) Antibiotic production by bacterial biocontrol agents. Antonie Van Leeuwenhoek 81(1–4):537–547. https://doi.org/10.1023/A:1020501420831

Rajkumar M, Ae N, Prasad MNV, Freitas H (2010) Potential of siderophore-producing bacteria for improving heavy metal phytoextraction. Trends Biotechnol 28(3):142–149

Rastogi G, Sbodio A, Tech JJ et al (2012) Leaf microbiota in an agroecosystem: spatiotemporal variation in bacterial community composition on field-grown lettuce. ISME J 6:1812

Redford AJ, Bowers RM, Knight R et al (2010) The ecology of the phyllosphere: geographic and phylogenetic variability in the distribution of bacteria on tree leaves. Environ Microbiol 12:2885–2893

Remus-Emsermann MNP, Leveau JHJ (2010) Linking environmental heterogeneity and reproductive success at single-cell resolution. ISME J. https://doi.org/10.1038/ismej.2009.110

Robson TM, Pancotto VA, Scopel AL et al (2005) Solar UV-B influences microfaunal community composition in a Tierra del Fuego peatland. Soil Biol Biochem. https://doi.org/10.1016/j.soilbio.2005.04.002

Roesch LFW, Fulthorpe RR, Riva A et al (2007) Pyrosequencing enumerates and contrasts soil microbial diversity. ISME J 1:283

Rousk J, Bååth E, Brookes PC et al (2010) Soil bacterial and fungal communities across a pH gradient in an arable soil. ISME J. https://doi.org/10.1038/ismej.2010.58

Ruttledge TR, Nelson EB (1997) Extracted fatty acids from *Gossypium hirsutum* stimulatory to the seed-rotting fungus, *Pythium ultimum*. Phytochemistry 46:77–82

Ryan J, Sommer R (2012) Soil fertility and crop nutrition research at an international center in the Mediterranean region: achievements and future perspective. Arch Agron Soil Sci 58:s41–s54

Sakthivel N, Gnanamanickam SS (1987) Evaluation of *Pseudomonas fluorescens* for suppression of sheath rot disease and for enhancement of grain yields in rice (*Oryza sativa* L.). Appl Environ Microbiol 53:2056–2059

Scher FM, Kloepper JW, Singleton CA (1985) Chemotaxis of fluorescent Pseudomonas spp to soybean seed exudates in vitro and in soil. Can J Microbiol 31(6):570–574

Schiltz S, Gaillard I, Pawlicki-Jullian N et al (2015) A review: what is the spermosphere and how can it be studied? J Appl Microbiol 119(6):1467–1481

Shang H, Chen J, Handelsman J, Goodman RM (1999) Behavior of *Pythium torulosum* zoospores during their interaction with tobacco roots and *Bacillus cereus*. Curr Microbiol. https://doi.org/10.1007/PL00006787

Silo-Suh LA, Lethbridge BJ, Raffel SJ et al (1994) Biological activities of two fungistatic antibiotics produced by *Bacillus cereus* UW85. Appl Environ Microbiol. https://doi.org/10.1093/ecam/neq025

Silo-Suh LA, Stabb EV, Raffel SJ, Handelsman J (1998) Target range of Zwittermicin A, an aminopolyol antibiotic from *Bacillus cereus*. Curr Microbiol. https://doi.org/10.1007/s002849900328

Singh A, Kirubakaran SI, Sakthivel N (2007) Heterologous expression of new antifungal chitinase from wheat. Protein Expr Purif 56:100–109

Singh BK, Dawson LA, Macdonald CA, Buckland SM (2009) Impact of biotic and abiotic interaction on soil microbial communities and functions: a field study. Appl Soil Ecol. https://doi.org/10.1016/j.apsoil.2008.10.003

Singh D, Shi L, Adams JM (2013) Bacterial diversity in the mountains of South-West China: climate dominates over soil parameters. J Microbiol. https://doi.org/10.1007/s12275-013-2446-9

Slykhuis JT (1947) Studies on *Fusarium culmorum* blight of crested wheat and brome grass seedlings. Can J Res 25:155–180

Smirnova A, Li H, Weingart H et al (2001) Thermoregulated expression of virulence factors in plant-associated bacteria. Arch Microbiol. https://doi.org/10.1007/s002030100344

Smith VL, Wilcox WF, Harman GE (1990) Potential for biological control of *Phytophthora* root and crown rots of apple by *Trichoderma* and *Gliocladium* spp. Phytopathology 80:880–885

Stabb EV, Jacobson LM, Handelsman J (1994) Zwittermicin A-producing strains of *Bacillus cereus* from diverse soils. Appl Environ Microbiol 60(12):4404–4412

Stosz SK, Fravel DR, Roberts DP (1996) In vitro analysis of the role of glucose oxidase from Talaromyces flavus in biocontrol of the plant pathogen *Verticillium dahliae*. Appl Environ Microbiol 62(9):3183–3186

Stotzky G, Schenck S, Papavizas GC (1976) Volatile organic compounds and microorganisms. CRC Crit Rev Microbiol 4:333–382

Stukenbrock EH, McDonald BA (2008) The origins of plant pathogens in agro-ecosystems. Annu Rev Phytopathol. https://doi.org/10.1146/annurev.phyto.010708.154114

Suding KN, Collins SL, Gough L et al (2005) Functional- and abundance-based mechanisms explain diversity loss due to N fertilization. Proc Natl Acad Sci. https://doi.org/10.1073/pnas.0408648102

Sundin GW, Jacobs JL (1999) Ultraviolet radiation (UVR) sensitivity analysis and UVR survival strategies of a bacterial community from the phyllosphere of field-grown peanut (*Arachis hypogeae* L.). Microb Ecol. https://doi.org/10.1007/s002489900152

Sy A, Timmers ACJ, Knief C, Vorholt JA (2005) Methylotrophic metabolism is advantageous for *Methylobacterium extorquens* during colonization of *Medicago truncatula* under competitive conditions. Appl Environ Microbiol. https://doi.org/10.1128/AEM.71.11.7245-7252.2005

Tan S, Yang C, Mei X et al (2013) The effect of organic acids from tomato root exudates on rhizosphere colonization of *Bacillus amyloliquefaciens* T-5. Appl Soil Ecol 64:15–22

Teixeira LCRS, Peixoto RS, Cury JC et al (2010) Bacterial diversity in rhizosphere soil from Antarctic vascular plants of Admiralty Bay, maritime Antarctica. ISME J 4:989

Thompson IP, Bailey MJ, Fenlon JS et al (1993) Quantitative and qualitative seasonal changes in the microbial community from the phyllosphere of sugar beet (*Beta vulgaris*). Plant Soil 150:177–191

Tian-Shung W, Li-Fei O, Che-Ming T (1994) Aristolochic acids, aristolactam alkaloids and amides from *Aristolochia kankauensis*. Phytochemistry 36:1063–1068

Tisdall JM (1996) Formation of soil aggregates and accumulation of soil organic matter. Struct Org Matter Storage Agric soils:57–96

Toju H, Peay KG, Yamamichi M et al (2018) Core microbiomes for sustainable agroecosystems. Nat Plants 4(5):247–257

Torres-Rubio MG, Valencia-Plata SA, Bernal-Castillo J, Martínez-Nieto P (2000) Isolation of *Enterobacteria*, *Azotobacter* sp. and *Pseudomonas* sp., producers of indole-3-acetic acid and siderophores, from Colombian rice rhizosphere. Rev Latinoam Microbiol 42:171–176

Truyens S, Weyens N, Cuypers A, Vangronsveld J (2015) Bacterial seed endophytes: genera, vertical transmission and interaction with plants. Environ Microbiol Rep. https://doi.org/10.1111/1758-2229.12181

Tukey HB Jr (1970) The leaching of substances from plants. Annu Rev Plant Physiol 21:305–324

Umesha S, Singh KP, Singh RP (2018) Microbial biotechnology and sustainable agriculture. In: Biotechnology for sustainable agriculture. Elsevier. pp 185–205. ISBN: 9780128121603

van der Heijden MGA, Hartmann M (2016) Networking in the plant microbiome. PLoS Biol. https://doi.org/10.1371/journal.pbio.1002378

Van Dijk K, Nelson EB (2000) Fatty acid competition as a mechanism by which *Enterobacter cloacae* suppresses *Pythium ultimum* sporangium germination and damping-off. Appl Environ Microbiol. https://doi.org/10.1128/AEM.66.12.5340-5347.2000

Van Loon LC (1997) Induced resistance in plants and the role of pathogenesis-related proteins. Eur J Plant Pathol 103:753–765

Velmourougane K, Saxena G, Prasanna R (2017) Plant-microbe interactions in the rhizosphere: mechanisms and their ecological benefits. In: Plant-microbe interactions in agro-ecological perspectives. Springer, pp 193–219

Verona O (1958) The spermosphere. In: Annales de l'Institut Pasteur, pp 795–798

Vessey JK (2003) Plant growth promoting rhizobacteria as biofertilizers. Plant Soil 225:571–576

Vorholt JA (2012) Microbial life in the phyllosphere. Nat Rev Microbiol 10(12):828–840

Voříšková J, Baldrian P (2013) Fungal community on decomposing leaf litter undergoes rapid successional changes. ISME J 7:477

Waldor MK, Tyson G, Borenstein E et al (2015) Where next for microbiome research? PLoS Biol. https://doi.org/10.1371/journal.pbio.1002050

Wang X, Wang Z, Jiang P et al (2018) Bacterial diversity and community structure in the rhizosphere of four Ferula species. Sci Rep 8:5345

Watson AG (1966a) The effect of soil fungicide treatments on the inoculum potentials of spermosphere fungi and damping-off. New Zeal J Agric Res 9:931–955

Watson AG (1966b) Seasonal variation in the inoculum potentials of spermosphere fungi. New Zeal J Agric Res 9:956–963

Weinert N, Piceno Y, Ding G-C et al (2011) PhyloChip hybridization uncovered an enormous bacterial diversity in the rhizosphere of different potato cultivars: many common and few cultivar-dependent taxa. FEMS Microbiol Ecol 75:497–506

Whipps JM, Hand P, Pink D, Bending GD (2008) Phyllosphere microbiology with special reference to diversity and plant genotype. J Appl Microbiol 105(6):1744–1755

Windstam S, Nelson EB (2008) Temporal release of fatty acids and sugars in the spermosphere: impacts on *Enterobacter cloacae*-induced biological control. Appl Environ Microbiol 74:4292–4299

Wu H-S, Liu D-Y, Ning-Ling et al (2008a) Effects of vanillic acid on the growth and development of *Fusarium oxysporum* f. sp niveum. Allelopath J 22:111–121

Wu H-S, Raza W, Fan J-Q et al (2008b) Antibiotic effect of exogenously applied salicylic acid on in vitro soilborne pathogen, *Fusarium oxysporum* f. sp. niveum. Chemosphere 74:45–50

Yang CH, Crowley DE, Menge JA (2001) 16S rDNA fingerprinting of rhizosphere bacterial communities associated with healthy and Phytophthora infected avocado roots. FEMS Microbiol Ecol 35:129–135

Youssef D, Frahm AW (1995) Constituents of the Egyptian *Centaurea scoparia*; III. Phenolic constituents of the aerial parts. Planta Med 61:570–573

Zhalnina K, Dias R, de Quadros PD et al (2014) Soil pH Determines Microbial Diversity and Composition in the Park Grass Experiment. Microb Ecol 69(2):395–406

Zhang W, Bao W (2000) Studies on the chemical constituents of *Xanthoceras sorbifolia* Bunge. Acta Pharm Sin 35:124–127

Zhang B, Bai Z, Hoefel D et al (2010) Microbial diversity within the phyllosphere of different vegetable species. Curr Res Technol Educ Top Appl Microbiol Microb Biotechnol 2:1067–1077

Zhang XF, Zhao L, Xu SJ et al (2013) Soil moisture effect on bacterial and fungal community in Beilu River (Tibetan Plateau) permafrost soils with different vegetation types. J Appl Microbiol 114(4):1054–1065

Zhang Y, Cong J, Lu H et al (2014) Community structure and elevational diversity patterns of soil *Acidobacteria*. J Environ Sci 26:1717–1724

Zhang Z, Ci D, Zhang G et al (2017) Diversity of microbial community structure in the spermosphere of saline-alkali soil in shandong area. Acta Microbiol Sin 57:582–596

Zheng XY, Sinclair JB (2000) The effects of traits of *Bacillus megaterium* on seed and root colonization and their correlation with the suppression of *Rhizoctonia* root rot of soybean. BioControl 45:223–243

Plant-Microbe Communication: New Facets for Sustainable Agriculture

19

Purnima Bhandari and Neera Garg

1 Introduction

Agriculture, the supplier of human nutrition, is a profitable sector in the world that is being impended by a rapidly growing population, usage of non-sustainable agricultural practices, and changing climate (Etesami and Beattie 2017). Globally, population has been anticipated to touch 9.77 billion by 2050 (UN 2016; Meena and Lal 2018), which will necessitate a substantial surge in the productivity of major food commodities like wheat, rice, maize, and legumes. In addition, stresses of biotic and abiotic origin inhibit plant growth and development, among which several abiotic factors have been found responsible for the regression in crop productivity (Shelden and Roessner 2013; Filho et al. 2017). For mitigating stress effects, several strategies have been developed worldwide that include development of stress-tolerant varieties, resource management, etc. (Venkateswarlu and Shanker 2009; Chakraborty et al. 2018). However, due to their time-consuming and cost-intensive nature, nowadays, usage of multifaceted traits of several beneficial microorganisms has been professed as a cost-effective and environment-affable approach that can enhance plant stress tolerance, thereby managing ecosystem vigor (Meena et al. 2015; Talaat and Shawky 2017).

In nature, rhizosphere microbiome represents the core segment of plant-microbe symbiosis (Coats and Rumpho 2014; Filho et al. 2017), where beneficial microbes not only interact with the roots of host plants but also with each other via their signal cross talks, thus modulating plant morphophysiology and restoring soil fertility and leading to sustainable agriculture. Among the beneficial rhizospheric microbes, plant growth-promoting rhizobacteria (PGPRs) and mycorrhizal fungi represent important constituents of soil communities that ensure plant fitness and soil health

P. Bhandari
Mehr Chand Mahajan DAV College for Women, Chandigarh, India

N. Garg (✉)
Department of Botany, Panjab University, Chandigarh, India

D. P. Singh et al. (eds.), *Microbial Interventions in Agriculture and Environment*,
https://doi.org/10.1007/978-981-13-8383-0_19

both under stressed and unstressed conditions (Vimal et al. 2017). PGPRs have been reported to promote plant vigor by (1) improving plant nutrition, (2) modulating plant hormonal pathways, and (3) protecting plants against pathogens and parasites (Vacheron et al. 2013; Besset-Manzoni et al. 2018). Besides, use of arbuscular mycorrhizal (AM) inoculations has been validated to exert a constructive effect on the mobility of plant nutrients including phosphorus (P), nitrogen (N), potassium (K), sulfur (S), etc. Moreover, a growing evidence exists that indicates the crucial role of AM fungi in restoring soil fertility and maintaining membrane stability, thus imparting stress tolerance in plants (Garg and Bhandari 2016a, b; Garg and Kashyap 2017; Garg and Singh 2017). Recently, scientists have started adopting the usage of multi-strain microbial inoculant as a potent biotechnological approach to alleviate the lethal effects of different stresses on plant vigor (Garg and Pandey 2015; Talaat and Shawky 2017). In one of such studies, co-inoculation with *Dietzia natronolimnaea* and *Glomus intraradices* not only influenced growth characteristics of salt-stressed *Ocimum basilicum* but also modulated the buildup of soil microbial flora in salt-affected low-fertility soils (Bharti et al. 2016). The present review highlights (1) the interplay of signals and chemicals that leads to the establishment of plant-microbe symbiosis in the complex domain of the rhizosphere and (2) the role of multifaceted microbes (PGPRs and AM fungi) and the mechanisms therein in modulating the stress effects in crop plants.

2 Microbial Communication in the Rhizosphere

Soil represents one of the fullest microbial biomes (Gans et al. 2005; Singh et al. 2017) that harbors a variety of microorganisms, which live in close proximity with the plant root zone. Such soil-root interface where interactions among a myriad of organisms occur has been defined as *rhizosphere* or *microbial hotspot* (Hartmann et al. 2008; Nadarajah 2016) that not only modulates biogeochemical cycles but also plant growth and imparts stress tolerance. Microbial biota that resides within the rhizosphere have been reported to affect plant well-being by having overall impact on the natural ecosystem of the plants (Schnitzer et al. 2011; Mendes et al. 2011; Nadarajah 2016). Plants liberate as much as 20% of their photosynthates as root exudates such as amino acids, phenolic compounds, sugars, flavonoids, vitamins and hydrated polysaccharides, and proteins (Chandra and Singh 2016; Etesami and Beattie 2017; Yuan et al. 2018). Few of them act as chemical attractants for dynamically metabolizing soil microbial populations, e.g., flavonoids, while others such as glucose are linked with the establishment of the microbial community (Sengupta et al. 2017). These metabolites are released in the rhizosphere with the assistance of several membrane-bound proteins (Weston et al. 2012) such as ATP-binding cassette (ABC) transporters (Badri et al. 2008; Sugiyama et al. 2008) and aluminum-activated malate transporters (Weston et al. 2012; reviewed by Sengupta et al. 2017) and act as a nutritional source for microbes, thus helping in their sustenance (Nadeem et al. 2015). In response, microbes secrete an array of compounds such as secondary metabolites, enzymes, waste products, and phytohormones in the rhizosphere that modulate plant growth and defense (Ortíz-Castro et al. 2009; Etesami and Beattie 2017). Almost 7–15% of the total root surface is occupied by rhizosphere microbial communities (Nadeem et al. 2015) indicating that plants select only those microbes which provide greater benefit to them. In addition, some microbes have been observed to chiefly occur outside the roots (ectophytes), while those which inhabit the intercellular sites within roots (endophytes) have been suggested to impart greater benefit to the host due to their close proximity with the plant tissues, thus providing greater opportunities for chemical exchange (Etesami and Beattie 2017). Among the agriculturally beneficial microbial populations, a breadth of rhizospheric microorganisms have been observed to promote plant growth, among which bacteria and fungi are well documented. In one of the studies, the ratio of the microbial population in the rhizosphere (R) to that in the bulk soil (S), i.e., R/S value, has been estimated >=20 for bacteria and 10 for fungi (Bagyaraj and Rangaswami 2005; Haldar and Sengupta 2016). In the rhizosphere, bacteria, which are beneficial to the host plants, are collectively termed as plant growth-promoting rhizobacteria (PGPRs), precisely plant growth-promoting microorganisms (PGPMs) which include genera *Rhizobium*, *Agrobacterium*, *Azospirillum*, *Cellulomonas*, *Azotobacter*, *Micrococcus*, *Bacillus*, and *Pseudomonas*. Depending upon their localization, PGPRs have been classified as (1) intracellular PGPRs (iPGPRs) that reside within root cells or in specialized nodular structures and (2) extracellular PGPRs (ePGPRs) that exist in the rhizosphere (Sundaramoorthy and Balabaskar 2012; Sharma et al. 2016).

Among iPGPRs, members of the *Rhizobiaceae* family including *Rhizobium*, *Mesorhizobium*, and *Bradyrhizobium* species are able to form nodules on the roots of legumes, thus assisting in the nitrogen enrichment process. A growing evidence indicates that the process of root nodule symbiosis (RNS) has been restricted to the FaFaCuRo (Fabales, Fagales, Cucurbitales, and Rosales) clade, belonging to Eurosid I plants (Diédhiou and Diouf 2018). During the symbiotic nitrogen fixation process that occurs under N-limiting conditions, legumes secrete chemoattractants such as iso(flavonoids) and lectins to attract compatible N-fixing rhizobacteria present in the vicinity (Garg and Geetanjali 2007; Skorupska et al. 2010). On the recognition of compatible rhizobial species (i.e., host specificity), activation of a group of bacterial nodulation (*nod*) genes occurs, which leads to the synthesis of Nod factor (NF), a lipochitooligosaccharide (LCO) signal that initiates the formation of root nodules (Oldroyd et al. 2011; Wang et al. 2018). In general, nod genes have been categorized into common nod genes and host-specific nod genes (Kondorosi et al. 1984; Skorupska et al. 2010; Kaur and Kaur 2018). The common nod genes (*nodABC*) containing operon are conserved across all the rhizobia except *Bradyrhizobium* (Giraud et al. 2007) and are found to be critical for nodule progression. *nodC* gene encodes enzyme NodC which is a chitin-oligosaccharide synthase that links N-acetyl-D-glucosamine monomers; *nodB* gene encodes NodB enzyme (a chitin-oligosaccharide deacetylase) that eradicates the acetyl group from the terminal nonreducing sugar; and *nodA* gene encodes NodA (an N-acyltransferase) that adds fatty acyl chain (Spaink 2000; D'Haeze and Holsters 2002; Skorupska et al. 2010). Host-specific nod genes include *nodFE*, *nodH*, *nodG*, and *nodPQ* that differ among rhizobial species and are considered important in determining the host specificity and influence the rate of nodule formation. In addition, a third class of *nod* genes has been documented to exist that includes regulatory *nodD* gene (Spaink 2000) which encodes NodD protein that, in association with a flavonoid, binds conservative sequences upstream of nod operons and activates their expression (Kaur and Kaur 2018).

NFs, which are essential for nodule formation, were first identified from *Rhizobium meliloti* (Lerouge et al. 1990) and from *R. leguminosarum* bv. *viciae* (Spaink et al. 1991; Werner 2008). They have been reported to act at nanomolar concentrations, thus inducing many early molecular and physiological changes in the root hair (Via et al. 2015). At molecular level, Nod factors are sensed by Nod factor receptors (NFRs) (for instance, NFR1 and NFR5 in *Lotus japonicus*), which are lysine motif (LysM) extracellular domains (Radutoiu et al. 2003; Wang et al. 2018), that further leads to the activation of downstream nodulation signaling pathways (Broghammer et al. 2012). It has been observed that a single mutation in the NFR is enough to alter the specificity of the interaction at the species level (Radutoiu et al. 2007). Besides Nod factors, rhizobial surface polysaccharides comprising exopolysaccharides (EPS), cyclic glucans, lipopolysaccharides (LPS), and capsular polysaccharides (KPS) have been identified to play a crucial role in establishing symbiotic relationships (Jones et al. 2007; Gibson et al. 2008; Wang et al. 2018). Among them, EPS has been reported to act as a signaling molecule in various processes ranging from rhizobial infection to defense response (Parniske et al. 1994; Pellock et al. 2000; Via et al. 2015). In one of the studies, during *Sinorhizobium-Medicago* symbiosis, a major EPS – succinoglycan – secreted by rhizobia has been documented to enhance nodulation capacity (Jones 2012). However, Kelly et al. (2013) demonstrated the complicated role of EPS in the *Mesorhizobium-Lotus* interaction where a subset of EPS mutants of *M. loti* R7A exhibited severe nodulation deficiencies on *L. japonicus* and *L. corniculatus*, while other mutants formed effective nodules. Recently, Kawaharada et al. (2015, 2017) identified an EPS receptor (EPR3) in *L. japonicus*, which binds rhizobial EPS in a structurally specific manner, and suggested that the entry of bacteria to the host plant is controlled by two successive steps of receptor-arbitrated recognition of Nod factor and EPS signals (Wang et al. 2018).

LCOs thus produced have been documented to induce many events in the plant root cell including (1) deformation of root hairs (Lerouge et al. 1990), (2) root hair plasma membrane depolarization (Ehrhardt et al. 1996), and (3) nuclear calcium spiking (where a rapid increase in the levels of intracellular free Ca^{2+} in the root hairs occurs). Ca^{2+} oscillations observed are further decoded by a calcium- and calmodulin-dependent protein kinase (CCaMK/DMI3) which phosphorylates the CYCLOPS transcription factor to promote gene expression and activates the nodulation process (Yano et al. 2008; Singh et al. 2014; Diédhiou and Diouf 2018). In response to rhizobial genes, induction of plant genes – nodulin genes – occurs which is considered essential for the nodule development. These genes are specific and structurally as well as functionally conserved in several species of *Azorhizobium*, *Bradyrhizobium*, and *Rhizobium* (Dobert et al. 1994; Kaur and Kaur 2018). In general, rhizobia-induced genes have been categorized into (i) early nodulins (ENOD) which are induced within the first few days of infection, e.g., ENOD12, ENOD40, and RIP1, and (ii) late nodulins that have been documented to be induced several days after bacterial invasion and during nodule maturation (Garg and Geetanjali 2007).

Additionally, in response to Nod factors, change in the host cell pH occurs (Felle et al. 1996) which is trailed by root hair curling that further entraps rhizobia and initiates the process of infection (Downie et al. 1985). It is reported that root cortical cells dedifferentiate to establish nodule primordium as infection thread grows toward the pericycle and cortex (Oldroyd and Downie 2008; Kaur and Kaur 2018). In case of *Medicago truncatula*, the orthologue of *CYCLOPS*, *IPD3*, has been found to control infection thread formation as well as release of bacteroids in the nodule (Ovchinnikova et al. 2011; Diédhiou and Diouf 2018). Similarly, Xie et al. (2012) found that *NIN* activates the expression of pectate lyase gene that degraded plant cell wall and helped in the formation of infection thread in *L. japonicus*. Near the tip of growing infection threads, rhizobia proliferate and, after reaching the nodule primordium, are discharged into plant cells (Skorupska et al. 2010). At the same time, they get covered by a membrane (derived from the host cell) – peribacteroid membrane (PBM) – and subsequently form symbiosome where they get differentiated into bacteroids (Brewin 2004; Skorupska et al. 2010) which are capable of nitrogen fixation. It appears that H_2O_2 controls the process of differentiation of bacteria into a nitrogen-fixing, symbiotic form (Puppo et al. 2013; Kaur and Kaur 2018). Further, cell divisions increase in root cortex cells; nodule primordium undergoes incessant mitotic activity to form the root nodule, where symbiotic nitrogen fixation occurs (Oldroyd and Downie 2008). A number of authors have documented the role of *ENOD40* in this plant meristematic program (Crespi et al. 1994; Garg and Geetanjali 2007) that leads to the establishment of the nodule as entity. The chief function of the nodule is to protect nitrogenase enzyme from aerobic conditions, thus maintaining the compartment with low oxygen concentrations (Garg and Geetanjali 2007; Kaur and Kaur 2018) and ensuring proper function of the fixation process.

Besides PGPRs, mycorrhizal fungi constitute another significant portion of soil microflora which are found to positively influence plant growth and development in nutrient-stressed soils. Around 50,000 fungal species have been documented to establish effective symbiosis with about 250,000 host plant species (Barea et al. 2017). Among the several types of mycorrhiza recognized, AM fungi, belonging to the phylum *Glomeromycota*, are the most widespread fungi that establish effective symbiosis with over 90% of the plant species belonging to pteridophytes, gymnosperms, and angiosperms (Prasad et al. 2017; Bhandari and Garg 2017). However, plants belonging to families Brassicaceae, Chenopodiaceae, Cyperaceae, Juncaceae, Caryophyllaceae, and Proteaceae are unable to form effective symbiosis with AM fungi (Vierheilig et al. 2003; Kaur and Kaur 2018). In evolutionary terms, it is an ancient type of interaction that has been documented to expedite the colonization of land even over 460 million years ago (Smith and Read 2008; Garg and Chandel 2010). Mycorrhizal symbiosis is a mutualistic association where a "fair trade" between the host plant and mycorrhiza occurs that involves the exchange of N and P fixed by mycorrhiza and carbohydrates by the plant (Fellbaum et al. 2012; Nadarajah 2016; Kaur and Kaur 2018). This mutualistic interaction leads to the formation of tree-shaped subcellular structures – *arbuscules* (meaning bush or little trees) – within root cortical cells, thus ascertaining chief interface for symbiotic nutrient transfer (Parniske 2008; Gutjahr and Parniske 2013; Bhandari and Garg 2017). Several cross talks ranging from AM propagule initiation to intracellular fungal accommodation that regulate the processes of AM establishment have been reviewed by a plethora of researchers worldwide (Gutjahr and Parniske 2013; Bonfante and Desirò 2015; Bhandari and Garg 2017).

During AM establishment, the first step of symbiosis includes the *asymbiotic hyphal growth stage*, where spores germinate and develop hyphae unconventionally but for a partial period due to their obligate biotrophic nature. Generally, it has been observed that AM fungi use spores, fragments of mycorrhizal roots, and AM intra-radical mycelium (IRM) as well as extra-radical mycelium (ERM, developing in the rhizosphere) as propagules for an effective root colonization process (Smith and Read 2008). In order to support the growth of their mycelium, mycorrhizal fungi use triacylglyceride (TAG) and glycogen reserves present in the spore. This phase is followed by the *pre-symbiotic growth stage*, where a cross talk between the host plant and fungi occurs via diffusible signaling molecules, where plant produces strigolactones (SL) (Akiyama et al. 2005), while fungi exude LCOs – *Myc factors*. These Myc factors are perceived at the plant plasma membrane due to lysine motif (LysM) receptor kinases (Broghammer et al. 2012; Oldroyd 2013) that dynamically prepare the root intracellular milieu and induce symbiosis-specific responses, even in the absence of any physical contact (Bhandari and Garg 2017). In addition, auxin has been documented to play a critical role in the regulation of root development (Foo et al. 2013; Diédhiou and Diouf 2018).

The pre-symbiotic phase is straggled by the *symbiotic phase*, where the fungus penetrates the plant root, leading to the materialization of an extremely branched, swollen, and flattened characteristic fungal structure – *appressorium* – on the root epidermal cells (Smith and Read 2008; Genre 2012; Bhandari and Garg 2017). Numerous genes including *VAPYRIN* and *CYCLOPS* have been reported to be involved in this process (Singh et al. 2014; Carbonnel and Gutjahr 2014). Several studies have documented the implication of *VAPYRIN* gene in early and later steps of AM symbiosis (Reddy et al. 2007; Murray et al. 2011). However, the exact function of this gene in root endosymbiosis is still unclear (Feddermann and Reinhardt 2011; Diédhiou and Diouf 2018). With the formation of appressorium, root cortical cells reposition their nucleus as well as remodel their cytoplasm, thus preparing themselves for fungal penetration (Genre et al. 2005). Subsequently, fungal hyphae penetrate the root cortical cells by producing wall-degrading hydrolytic enzymes, thus triggering colonization surface and leading to the formation of *arbuscules* (Parniske 2008; Gutjahr and Parniske 2013; Barea et al. 2014). Arbuscule formation has been reported to induce many changes in gene expression patterns in mycorrhizal roots such as in *Pt4*, *SbtM1*, *Cel1*, *SCP*, *STR1*, and *STR* genes of colonized cortical cells of *Medicago truncatula* (Pumplin and Harrison 2009; Gutjahr et al. 2012; Rech et al. 2013). Correspondingly, studies have revealed that suppression of the *DELLA* and *RAM1* genes led to arbuscule branching and induced the anomalous formation of arbuscules in *M. truncatula* (Floss et al. 2013; Rich et al. 2014; as reviewed recently by Diédhiou and Diouf 2018). Thus, formation of this *tree-like structure* inside the root inner cortical cells symbolizes the establishment of symbiosis amid the two symbiotic partners (Genre 2012; Bhandari and Garg 2017). This stage is followed by the formation of intra-radical and extra-radical mycelium (IRM and ERM) that aids in the acquisition of immobile mineral nutrients from the rhizospheric strata (Barea et al. 2005, 2014). In addition, ERM have been documented to form novel chlamydospores that help in the proliferation of fungus, thus assisting in the completion of its life cycle (Bhandari and Garg 2017).

3 Microbes for Sustainable Agriculture Under Abiotic Stress

Microbial communications are of utmost importance in the rhizosphere especially under stressful conditions. During their life cycle, plants face numerous adverse abiotic constraints including salt stress, drought stress, heat stress, and HM toxicity that severely impede plant growth characteristics and lessen their overall yield by about 70% (Saxena et al. 2013; Talaat and Shawky 2017) by altering their cellular metabolism, promoting membrane disorganization, and generating reactive oxygen species (ROS), thereby reducing the potential for nutrient uptake.

Among various abiotic factors, soil salinization has become a threat for the sustainability of agriculture in arid and semiarid regions of the world (Parvaiz and Satyawati 2008) where extensive usage of artificial irrigation in amalgamation with the extended dry seasons rapidly turns formerly constructive areas virtually into deserts. More than 800 million hectares of land have been assessed to be afflicted either by sodicity (434 million ha) or by salinity (376 million ha), accounting for more than 6% of the world's total area (FAO 2008; Munns and Tester 2008; Zhang et al. 2010a). In general, soils are considered to be saline if their electrical conductivity (ECe) at saturation is greater than 40 mM NaCl (4 dS/m; Munns and Tester 2008). Salt stress has been documented to occur due to the presence of high levels of soluble salts such as sulfates, chlorides, carbonates, and bicarbonates of calcium, magnesium, and sodium ions around the root zone. Accumulations of such ions beyond their normal limit obstruct plant growth and development by causing primarily (1) osmotic stress that alters net assimilation rate and photosynthesis, (2) specific ion toxicity that disturbs ionic homeostasis, and (3) oxidative stress that causes damage to lipids, proteins, and nucleic acids (Munns and Tester 2008; Garg and Bhandari 2016a, b; Bhandari and Garg 2017). Similar to salt stress, water deficit is considered another important abiotic factor that alters water relations of a plant at both cellular and whole plant levels (Talaat and Shawky 2017) due to limitations in available water supply, thus limiting crop productivity. Consequently, water-stressed plants wilt and are incapable of sequestering assimilates in their plant organs (Etesami and Beattie 2017).

Besides salinity stress, presence of HMs including arsenic (As), cadmium (Cd), lead (Pb), and zinc (Zn) are considered harmful beyond their normal concentration in soils, as they hasten mortality rate and lessen survival, thus inducing toxicity symptoms (Talaat and Shawky 2017). However, concentration of toxic pollutants in soil depends on several factors including the sources of origin, their passage to the accumulation site, and their maintenance and fixation with soil components (Saif et al. 2017). Once accrued in the soil, they get adsorbed by soil particles and are reallocated into several chemical forms that vary in their mobility, bioavailability, and toxicity (Zhang et al. 2015; Alamgir 2016; Saif et al. 2017). Studies demonstrated that HMs when present in soil alter the structure and function of enzymes; affect permeability and functionality of the plasma membrane; interfere in the acquisition and dissemination of macro- and micronutrients, photosynthetic process, and nodulation and nitrogen fixation process; and cause overproduction of ROS (Garg and Chandel 2011; Garg and Bhandari 2012; Garg and Singla 2012; Islam et al. 2016). However, plant sensitivity toward metal toxicity varies from plant to plant and also depends on an interrelated network of molecular as well as physiological mechanisms (Chakraborty et al. 2018).

Among the several methods available for remediation of contaminated soils, one of the most promising ones includes usage of symbiotic organisms that are widely used to modulate stress factors by enhancing nutrient availability, thus sustaining productivity under stressful conditions (Dodd and Perez-Alfocea 2012; Pellegrino et al. 2015; Dey et al. 2018). A growing body of evidence exists that underlines the role of beneficial microorganism(s) including PGPRs (Grover et al. 2011; Han et al. 2014; Habib et al. 2016; Talaat and Shawky 2017; Etesami and Beattie 2017) and AM fungi (Garg and Manchanda 2009; Ruiz-Lozano et al. 2012a, b; Porcel et al. 2012; Garg and Bhandari 2016a, b; Garg and Pandey 2015, 2016; Garg and Singla 2015, 2016, 2017; Garg and Kashyap 2017) as elicitors of tolerance to stress which varies in their mode of action as discussed in the following subsections.

3.1 Alleviation by PGPRs

A plethora of evidence exists which highlights the role of PGPRs in improving plant health under stressful conditions (Choudhary 2012; Etesami and Alikhani 2016a, b; Vurukonda et al. 2016; Etesami and Beattie 2017). Among the bacterial genera, several species of *Achromobacter*, *Azospirillum*, *Bacillus*, *Enterobacter*, *Microbacterium*, *Paenibacillus*, *Pseudomonas*, *Rhizobium*, and *Variovorax* have been found to be beneficial under stress as well as unstressed conditions (Upadhyay et al. 2009; Grover et al. 2011; Dodd and Perez-Alfocea 2012; as reviewed by Etesami and Beattie 2017). Several mechanisms have been explored and identified that depict the ameliorating role of PGPRs. One of the prime mechanisms is production of phytohormones (plant growth regulators) including auxin (indole-3-acetic acid, IAA) which have been implicated to play an important role in several plant physiological processes such as cell division and differentiation, germination, and root growth. Production of auxins and biological nitrogen fixation have been implicated to be the major required traits for PGPRs that further trigger plant developmental patterns especially the root system (Bitla et al. 2017).

Plants allocate more of their resources toward the underground part than the foliar region, thus regulating the size of their root system which is essential for optimum nutrient and water uptake under stressful environmental conditions (Etesami and Beattie 2017). Several studies have revealed that inoculation of plants with different PGPRs enhanced the formation of lateral roots and root hairs, thus improving the absorptive surface area that further improved water as well as nutrient uptake (Paul and Sarma 2006; Paul and Lade 2014). This enhancement in lateral root formation has been attributed to PGPR-induced IAA production by several researchers (Ludwig-Müller 2004; Remans et al. 2008; Hayat et al. 2010; Etesami and Beattie 2017). In one of the studies, Junghans et al. (2006) documented that reduced inhibition of root elongation was found to be associated with overexpression of an auxin amidohydrolase that imparted salt resistance in *Arabidopsis*. Similarly, Sadeghi et al. (2012) revealed that a *Streptomyces* isolate promoted germination rate, dry weight, and shoot length and enhanced macro- as well as micronutrient levels in NaCl-stressed wheat via enhancing the production of IAA. Inoculation with *Azospirillum* strain induced morphological variations in coleoptile xylem vessels, upregulated the activity of indole-3-pyruvate decarboxylase gene, and stimulated bacterial IAA synthesis that ultimately improved water status of drought-stressed wheat seedlings (Pereyra et al. 2012).

Besides IAA, gibberellins (GAs), abscisic acid (ABA), cytokinins, and ethylene (ET) are the other well-known and widely explored plant hormones that have been reported to be exuded by several bacteria (Khalid et al. 2006; Maheswari et al. 2013; Chimwamurombe et al. 2016; Naveed et al. 2017; Selim and Zayed 2017). Among them, ET is a gaseous hydrocarbon that regulates many plant physiological processes such as seed germination, vegetative and reproductive development, ripening of fruits, and senescence of plant organs (Kasotia et al. 2016). In plants, ET is synthesized by converting S-adenosyl-L-methionine (SAM) by 1-aminocyclopropane-1-carboxylate synthase (ACS/ACC oxidase) to 1-aminocyclopropane-1-carboxylate (ACC), which is the immediate precursor of ET production (Grover et al. 2011; Miliute et al. 2015; Selim and Zayed 2017). When present below 6.25 ppm concentration in plants, this gaseous hormone has been documented to stimulate the expression of stress-related genes, while at higher concentration of 25 ppm (that usually occurs under stressful conditions), ET causes chlorosis, leaf senescence, flower wilting, etc. that alter plant growth and development (Czarny et al. 2006; Zahir et al. 2009; Etesami et al. 2015; Jha and Saraf 2015). Thus, by reducing ET level in plants, some of the effects of stress could be alleviated (Glick 2005).

It has been seen that levels of ET could be modulated by PGPRs secreting ACC deaminase enzyme that catalyzes the cleavage of ACC into ammonia and α-ketobutyrate, which can be further assimilated by soil microbial population (Glick et al. 2007; Etesami and Beattie 2017). An ample range of PGPRs such as *Azospirillum* (Li et al. 2005), *Burkholderia* (Jiang et al. 2008), *Pantoea* (Zhang et al. 2011), *Bacillus* (Barnawal et al. 2013), *Pseudomonas* (Akhgar et al. 2014), *Arthrobacter* (Barnawal et al. 2014), *Serratia* (Carlos et al. 2016), and *Enterobacter* (Li et al. 2016) exist in nature (reviewed by Naveed et al. 2017) that have been reported to activate ACC deaminase activity, which could limit ET production, thus providing tolerance against stress-induced growth inhibition. In one of the studies, presence of ACC deaminase-containing bacteria reduced the toxic effects of Ni and Cd in *Alyssum murale* (Abou-Shanab et al. 2006) and *Brassica napus* (Sheng and Xia 2006) plants, respectively. Inoculation with bacteria having ACC deaminase restored nodulation potential in pea plants subjected to drought stress (Arshad et al. 2008). In another study, Singh et al. (2015) reported the alleviative role of ACC deaminase-producing *Klebsiella* sp. SBP-8 in improving fresh and dry weight of wheat plants grown under NaCl-stressed conditions. Similarly, inoculation with ACC deaminase-producing *Variovorax paradoxus* strain 5C-2 has been documented to substantially improve plant biomass and photosynthetic efficiency and decrease toxic ion accumulation in NaCl-stressed pea plants (Wang et al. 2016).

Sustenance of intracellular ionic homeostasis is considered a crucial aspect for effective functioning of a cell especially under stressed conditions. Studies have depicted the promising role of PGPRs in restricting the uptake of toxic ions and improving contents of macro- as well as microelement, thus conserving ionic homeostasis under environmentally stressed conditions (Kohler et al. 2009; Etesami and Beattie 2017). For instance, *Herbaspirillum* sp. strain GW103 augmented salt tolerance in Chinese cabbage by reducing Na^+ uptake and improving root K^+/Na^+, thereby maintaining ionic homeostasis which contributed toward improved plant biomass (Lee et al. 2016a). Although the exact mechanism is still unknown, Zhang et al. (2008) in one of their studies revealed the augmenting role of *Bacillus subtilis* in regulating the K^+ transporter HKT1, thus imparting salt tolerance in *Arabidopsis thaliana*. Similarly, PGPR-mediated enhancement in K^+ levels has been proposed to prevent salt-induced stomatal closure (Talaat and Shawky 2017).

In case of metal toxicity, the recombinant strain KT2440-spPCS obtained via cloning of phytochelatin synthase (PCS) gene from *Schizosaccharomyces pombe* expressed in *Pseudomonas putida* KT2440 augmented the levels of Cd, Hg, and Ag up to three- to fivefold, thus imparting metal tolerance (Yong et al. 2014). Besides maintaining cellular homeostasis, certain PGPRs have been documented to alleviate stress via producing volatile organic compounds (VOC such as alcohols, alkenes, benzenoids, esters, ketones, and terpenes) and secreting osmoprotectants. VOC are mobile in nature, thus acting as signaling molecules in (1) short- and long-distance intra- and intercellular pathways and (2) between the host and colonizing microbes (Effmert et al. 2012; Bitla et al. 2017). In a study, VOCs produced by *Trichoderma* substantially improved plant biomass and chlorophyll content of *A. thaliana* (Lee et al. 2016b). In addition, bacteria have been documented to synthesize low molecular weight, hydrophilic molecules that assist them to counteract osmolarity particularly under drought and salt stress (Ullah et al. 2017). For instance, inoculation with *B. subtilis* GB03 conferred dehydration tolerance in *Arabidopsis* plants by accumulating higher levels of osmoprotectants such as choline and glycine betaine (GB) (Zhang et al. 2010b). Similarly, under hypo-osmotic stress, bacteria – *Agrobacterium tumefaciens* and *R. meliloti* – have been reported to release cyclic β-(1,2)-glucans which imparted stress resilience (Ullah et al. 2017).

Among osmoprotectants, trehalose is one of the prevalent sugars that has been described to offer resistance to plants against several abiotic factors including drought, salt, and extreme temperature (Selim and Zayed 2017). Several microorganisms including *Rhizobium* sp. and free-living PGPRs display the ability to accumulate trehalose in stressed plants (Grover et al. 2011). Proline (Pro) is another important osmolyte that displays the ability to stabilize the subcellular structures by protecting protein configuration against denaturation (Ashraf and Foolad 2007), thus helping in combating abiotic stress. Certain species of PGPRs have been identified that modulate the levels of this imino acid under stressful conditions. For instance, co-inoculation with *Rhizobium* sp. and *Pseudomonas* sp. enhanced Pro levels in *Zea mays* when plants were exposed to high salt levels (Bano and Fatima 2009). Similarly, inoculation with different species of PGPRs has been correlated with the increase in the synthesis of Pro in several stressed plants (Zarea et al. 2013; Paul and Lade 2014; Jha and Saraf 2015; Manaf and Zayed 2015; reviewed by Selim and Zayed 2017).

Several soil microbes including species of *Bacillus*, *Pseudomonas*, *Sinorhizobium*, and *Escherichia* have been reported to synthesize and secrete extracellular polymeric substances (EPSs) in their surrounding environment that not only influence soil physical, chemical, and biological composition but also confer an extensive range of benefits to plants when subjected to stressful environments (Paul and Lade 2014; Etesami and Beattie 2017). These high-molecular-weight compounds form a sort of sheath or biofilm between roots and soil and support colonization of several soil microbes, thus providing functional and physical protection to the inhabiting bacteria (Naveed et al. 2017; Ullah et al. 2017). A breadth of reports exists in nature where EPS-producing PGPRs have been documented to bind cations including Na^+, thus causing a decrement in the availability of toxic ion, thereby potentially relieving the negative effects of salinity on plant vegetative and reproductive attributes (Ashraf et al. 2004; Grover et al. 2011; Upadhyay et al. 2009). Investigations have further revealed that by maintaining water status in the rhizosphere (Sandhya et al. 2009), EPS further assist in the mobilization of essential nutrients in soil to be available to plants particularly under abiotic stressful environments (Naveed et al. 2017). PGPRs such as *Rhizobium* and *Azotobacter* not only provide plants with nitrogenous compounds (via nitrogen fixation) but also assist in solubilizing inorganic phosphate, thereby reducing the input of chemical fertilizers, for instance, endomycorrhizae and *B. megaterium* (Ogut et al. 2010; Selim and Zayed 2017). Besides N and P, several soil-borne microbes such as *Pseudomonas* sp. have been reported to produce iron-loving compounds – siderophores (Prasad and Dagar 2014) – that help in acquisition of Fe^{3+} (due to their high affinity) and enhance its availability to plants (Abbamondi et al. 2016; Naveed et al. 2017; Etesami and Beattie 2017) particularly under metal stress conditions. By chelating with toxic metal in the rhizosphere, siderophores lower the formation of ROS, thus exerting a bioprotective effect in plants subjected to stressful environments (Dimkpa et al. 2009; Talaat and Shawky 2017). Another important aspect of siderophore production in the rhizospheric region is augmentation in the colonizing ability of microbes, which otherwise become problematic under iron-starved conditions due to the insufficient availability of Fe that further reduces the surface hydrophobicity of the microbial cells (Simoes et al. 2007; Bitla et al. 2017).

PGPRs have been reported to alleviate stress-induced oxidative damage in plants by upregulating their antioxidant defense response including antioxidant enzymes – superoxide dismutase (SOD), catalase (CAT), ascorbate peroxidase (APOX), glutathione reductase (GR), etc. At low concentrations, ROS including peroxides act as signaling molecule, while at high levels, they become detrimental to the cell. Oxidative damage to cellular membranes, proteins, etc. occurs when the equilibrium between production and scavenging of ROS such as singlet oxygen, superoxide, peroxide, and hydroxyl radical gets disturbed. A breadth of PGPRs have been reported to induce plant synthesis of antioxidant enzymes in response to abiotic factors such as salt stress (Kohler et al. 2009; Chakraborty et al. 2013; Singh et al. 2013; Paul and Lade 2014) and metal stress (Gururani et al. 2013; Oves et al. 2013). In one of the studies, inoculation with *Enterobacter* sp. UPMR18 upregulated and boosted the production of SOD, APOX, and CAT in +NaCl-treated okra plants (Habib et al. 2016). On the contrary, Tiwari et al. (2016) described comparatively less activity of antioxidant enzymes in +PGPR-inoculated stressed chick pea plants, thus lowering the level of oxidative stress when compared with –PGPR-stressed ones which they ascribed to the alleviative role of bacteria in relieving stress.

3.2 Alleviation by AM Fungi

Mycorrhizal fungi play an important role in the existence and progression of plants by improving growth and productivity, nutrient cycling, etc. that determine ecosystem stability and multifunctionality, particularly under stressful environments (Singh et al. 2018). In nature, AM fungi have been reported to exist in the form of spores, propagules, and fragments. However, several stages of the AM development cycle including spore germination, colonization, and hyphal growth and development have been seen to be affected by the presence of different abiotic stressors (Juniper and Abbott 2006; Bhandari and Garg 2017). Studies have further documented that fungal population tends to increase in disturbed regions such as saline soils (Allen and Cunningam 1983; Jahromi et al. 2008; Estrada et al. 2013; Bhandari and Garg 2017; Chakraborty et al. 2018), thereby imparting stress resilience to stressed host plants. Moreover, fungal species differ in their colonizing ability, thus varying in their inherent ability to lessen stress effects even among the plant genotypes. For example, one of our studies (Garg and Pandey 2015) established greater efficiency of *Rhizophagus irregularis* in imparting stress resilience than *Funneliformis mosseae* in differentially tolerant pigeon pea genotypes which was ascribed to the colonizing ability and thus responsiveness of the tolerant genotype with *R. irregularis*.

Improvement in growth attributes and productivity as seen under stressed conditions in +AM plants could not be a consequence of a distinct mechanism but a number of diverse mechanisms operating concurrently under stressful environments. One of the prominent mechanisms is maintenance of plant water status when exposed to water deficit conditions. AM fungi have been reported to impart stress resilience by enhancing plant root hydraulic conductivity (Kapoor et al. 2013) and modifying root morphology under water deficit conditions (Kapoor et al. 2008) that not only generates extensive root system (Hajiboland 2013) but also permits exploration of a substantial soil volume (Bhandari and Garg 2017) (Smith et al. 2010). The fungal hyphal network of inoculated plants is likely to be between 1 and 100 m/g of soil, which not only intensifies the ability for soil exploration (López-Ráez 2016) but also results in uptake of water that further improves plant water status. Numerous researchers have attributed the role of aquaporins (AQP) in modulating root hydraulic conductance (Ruiz-Lozano and Aroca 2010; Li et al. 2013) and water movements. AM fungi have been proposed to induce the expression of AQP genes in host plants that facilitate passive movement of water and some neutral molecules including glycerol, urea, and CO_2 across the biological membranes (López-Ráez 2016). In one of the studies, expression of *GintAQPF1* and *GintAQPF2* improved substantially in root cortical cells of drought-exposed *Z. mays* plants when inoculated with *R. intraradices* (Li et al. 2013). Similarly, in another study, the expression of *LsPIP1* and *LsPIP2* genes was found to be upregulated by mycorrhization in +NaCl *Lactuca sativa* plants which helped in regulating root water permeability, thus alleviating NaCl-induced osmotic stress (Jahromi et al. 2008). This AM-mediated alteration in host AQP not only modulates water relations but also other water-related physiological mechanisms, thus imparting stress resistance. Besides, AM fungi have been documented to influence genes encoding late embryogenesis-abundant protein (*LsLea*) and ABA (*Lsnced*) (Kapoor et al. 2013) under stressed conditions.

Another AM-mediated stress-alleviative mechanism is prevention of toxic ion uptake and/or its sequestration or compartmentalization in different plant organs. Fungal hyphae have been validated to retain toxic ions in their IRM, thus preventing them from translocating to aerial parts (Hajiboland 2013). Mycorrhizal fungi exert a buffering effect on toxic ion uptake via enhancing nutrient but avoiding Na^+ uptake (Hammer et al. 2011). Moreover, fungus has been described to regulate the expression and activity of transporters which are involved in the acquisition of macro- as well as micronutrients and of H^+ pumps that generate the driving force for their transport (Kapoor et al. 2013). AM fungi have been reported to regulate differential expression levels for root Na^+ and K^+ transporters of *Z. mays* implicated in sustaining ionic homeostasis (Estrada et al. 2013). In one of the study, AM symbiosis upregulated the expression of *OsSOS1*, *OsNHX3*, *OsHKT2;1*, and *OsHKT1;5* which favored the extrusion of Na^+ from the cytoplasm and confiscation into the vacuole, leading to unloading of Na^+ from the xylem and its recirculation from photosynthetic organs to roots,thereby enhancing salinity tolerance (Porcel et al. 2016). Under metal toxicity, mycorrhizal fungi have been recorded to produce glycoprotein *Glomalin* that functions as a metal chelating agent, thereby restricting the uptake of metal by plants (Kapoor et al. 2013). At molecular level, several genes such as *GrosMT1*, *GinZnT1*, *GmarMT1*, and *GintABC1* have been explored and identified that have been reported to play a crucial role in maintenance of the cellular homeostasis under metal toxicity (Talaat and Shawky 2017). Among them, *GinZnT1* is a Zn transporter that assists in compartmentalization of Zn in the vacuole; *GmarMT1* is a metal chelator that has been identified to code for metallothioneins (MTs) that protect fungi as well as the host against oxidative stress; and *GintABC1* is actively involved in Cu and Zn detoxification (González-Guerrero et al. 2010).

Accumulation of compatible solute (osmolytes) such as Pro, soluble sugars (SS), and GB is considered as another AM-mediated underlying protective mechanism that has been validated to alleviate osmotic stress (Garg and Chandel 2011; Evelin and Kapoor 2014; Bhandari and Garg 2017). Besides lowering down the osmotic potential, these compatible solutes tend to regulate turgor-related processes (Ruiz-Lozano et al. 2012a). Under As stress, inoculation with *Glomus mosseae* has been described to improve cellular sucrose and GB accumulation in stressed pea plants, thus signifying the protective role of fungi in maintaining higher turgor (Garg and Singla 2012). In one of our lab studies, *F. mosseae*-inoculated chickpea plants displayed higher NaCl-induced osmotic stress resistance which authors ascribed to higher augmentation witnessed in the activities of P5CS and GDH (Pro anabolic enzymes), with a simultaneous regression observed in the activity of ProDH (Pro catabolic enzyme), thus leading to higher Pro accumulation (Garg and Baher 2013). Parallel results were witnessed by Garg and Singh (2017) in pigeon pea plants subjected to Cd and Zn stress. On the contrary, at the molecular level, expression of δ1-pyrroline-5-carboxylate synthetase (*LsP5CS*) in + NaCl +AM-inoculated *L. sativa* plants was found to be downregulated when compared with +NaCl –AM counterparts indicating lower intensity of stress in the former than the latter (Jahromi et al. 2008). Several studies have depicted high rate of accumulation of SS in +AM plants which might be due to the enhanced rates of photosynthesis and/or hydrolysis of sugars and higher concentrations of organic acids that might help fungi in their sustenance (Kapoor et al. 2013). Besides, several researchers have authenticated that mycorrhiza induces higher accrual of trehalose, thus complementing legume-*Rhizobium* symbiosis under several abiotic conditions (Ocón et al. 2007; Garg and Chandel 2011; Garg and Pandey 2016; Garg and Singla 2016; Bhandari and Garg 2017).

Among the multiple benefits of mycorrhizae, enhancement in the uptake of nutrients from the rhizosphere and their translocation to the foliar tissue is considered another important mechanism via which AM fungi progress plant growth and development under stressed environment. As discussed earlier, the fungus after exchange of signals colonizes the host roots and forms a tree-shaped structure – *arbuscule* – where exchange of nutrients especially P, N, and C has been reported to occur. In one of the studies, Talaat and Shawky (2011) documented that mycorrhizal colonization effectively alleviated NaCl-induced injuries on growth attributes by augmenting uptake of N, P, K^+, Ca^{2+}, and Mg^{2+} in wheat plants. Similarly, inoculation with *R. irregularis* and *F. mosseae* not only improved plant nutrient endogenous profile but also imparted plant resistance in pigeon pea and chickpea plants, respectively, when subjected to Cd and Zn stress (Garg and Singh 2017) and salinity (Garg and Bhandari 2016a), respectively. Further, it has been observed that by sustaining higher levels of Ca^{2+}, AM fungi maintain higher membrane stability and cellular ion homeostasis under saline environments (Evelin et al. 2009; Hajiboland et al. 2010; Abdel Latef and Miransari 2014). Recently, fungal inoculations have been reported to enhance the content of Si – a beneficial element in otherwise low accumulating legumes particularly under stressful conditions (Garg and Bhandari 2016a, Garg and Kashyap 2017; Garg and Singh 2017) ultimately improving their productivity.

Besides, several researchers have authenticated a promising role of mycorrhiza in facilitating nitrogen fixation in legumes, by providing P and other immobile nutrients including Cu and Zn which are prerequisites for the nodulation and fixation process (Clark and Zeto 2000; Garg and Manchanda 2008). At molecular level, competence to transfer N has been investigated and identified in *F. mosseae*, where AMT (GmAMT4.1) has been recorded to transfer the same inside the roots of *Glycine max* during arbuscule development (López-Pedrosa et al. 2006; Behie and Bidochka 2014). By facilitating the use of different N forms that are otherwise difficult for non-inoculated plants to exploit, AM fungi assist in enhancing N in crop plants (Pellegrino and Bedini 2014). Studies from our lab have also authenticated the fact that by enhancing trehalose levels and modulating leghemoglobin as well as nitrogenase activity, colonization with AM fungi enhanced the N_2-fixing potential of *C. cajan* (Garg and Pandey 2016) and *C. arietinum* (Garg and Singla 2016) genotypes when exposed to varying degrees of salt levels. In order to alleviate oxidative stress, mycorrhizal fungi display their ability to enhance antioxidant defense responses in stressed host plants. For instance, *Trichoderma harzianum* T22 enhanced seedling vigor by improving the levels of GSH-dependent enzymes such as GR and glutathione S-transferase in different crop species, thus inducing physiological protection against oxidative damage (Shoresh and Harman 2008; Mastouri et al. 2010). Our lab studies (Garg and Singla 2015; Garg and Bhandari 2016b) have also validated the crucial role of different mycorrhizae in improving the efficacy of antioxidant machinery that regulated redox balance in +AM-stressed chickpea plants. Besides, in their study, Garg and Kaur (2013) established that *F. mosseae* inoculations alleviated the toxic effects of Zn and Cd by improving defense responses in pigeon pea nodules that consequently augmented the N_2-fixing ability under stressful environments. Similarly, inoculation with *Piriformospora indica* upregulated several drought-responsive genes, thereby enhancing drought tolerance in *Arabidopsis thaliana* (Sheremati et al. 2009; Venkadasamy et al. 2018).

4 Conclusion and Future Perspectives

Increase in human population coupled with urbanization and industrialization has led to the reduction in the cultivable land. In addition, the problem has been further aggravated by increasing adverse environmental vagaries including salinity, contamination of soil with HMs, drought stress, etc. In order to promote sustainable agriculture, practices that minimize the impact of environmental constrain and impart food security should be taken into consideration. Rhizospheric microbes including PGPRs and AM fungi with multifaceted traits have been validated to play a vital role in enhancing efficiency of the cropping system. Currently, usage of multiple microbial consortia such as co-inoculation of rhizobia along with fungi and/or different fungal isolates has now been considered as a sustainable approach so as to reclamate contaminated soils, thus opening a new and emerging usage of microbes under stressful conditions. However, selection of appropriate microbial inoculants is considered as important criteria before adopting a particular approach.

Therefore, future research should be focused on studying the useful interactions (i.e., whether cumulative, synergistic, or antagonistic) that occur among the different microbes as well as host plant species before implementing them under field conditions. Further, identification of their genetic traits (mechanisms) related to stress alleviation is the prerequisite which must be explored and identified using different molecular approaches, thus contributing to food security worldwide.

Acknowledgements The authors are grateful to the Department of Biotechnology (DBT), Government of India, for providing financial assistance for undertaking related research.

References

Abbamondi GR, Tommonaro G, Weyens N, Thijs S, Sillen W, Gkorezis P, Iodice C, Rangel WDM, Nicolaus B, Vangronsveld J (2016) Plant growth-promoting effects of rhizospheric and endophytic bacteria associated with different tomato cultivars and new tomato hybrids. Chem Biol Technol Agric 3:1–10

Abdel Latef AAH, Miransari M (2014) The role of arbuscular mycorrhizal Fungi in alleviation of salt stress. In: Miransari M (ed) Use of microbes for the alleviation of soil stresses. Springer, New York

Abou-Shanab RA, Angle JS, Chaney RL (2006) Bacterial inoculants affecting nickel uptake by *Alyssum murale* from low, moderate and high Ni soils. Soil Biol Biochem 38:2882–2889

Akhgar M, Arzanlou R, Bakker PAHM, Hamidpour M (2014) Characterization of 1-Aminocyclopropane-1-carboxylate (ACC) deaminase-containing *Pseudomonas* spp. in the rhizosphere of salt-stressed canola. Pedosphere 24:461–468

Akiyama K, Matsuzaki K, Hayashi H (2005) Plant sesquiterpenes induce hyphal branching in arbuscular mycorrhizal fungi. Nature 435:824–827

Alamgir M (2016) The effects of soil properties to the extent of soil contamination with metals. In: Environmental remediation technologies for metal-contaminated soils. Springer, Tokyo, pp 1–19

Allen EB, Cunningham GL (1983) Effects of vesicular-arbuscular mycorrhizae on *Distichlis spicata* under three salinity levels. New Phytol 93:227–236

Arshad M, Shaharoona B, Mahmood T (2008) Inoculation with *Pseudomonas* spp. Containing ACC-deaminase partially eliminates the effects of drought stress on growth, yield, and ripening of pea (*Pisum sativum* L.). Pedosphere 18:611–620

Ashraf M, Foolad M (2007) Roles of glycine betaine and proline in improving plant abiotic stress resistance. Environ Exp Bot 59(2):206–216

Ashraf M, Hasnain S, Berge O, Mahmood T (2004) Inoculating wheat seedlings with exopolysaccharide-producing bacteria restricts sodium uptake and stimulates plant growth under salt stress. Biol Fertil Soils 40:157–162

Badri DV, Loyola-Vargas VM, Broeckling CD, De-La-Peña C, Jasinaski M, Santelia D, Martinoia E, Sumner LW, Banta LM, Stermitz F, Vivanco JM (2008) Altered profile of secondary metabolites in the root exudates of Arabidopsis ATP-binding cassette transporter mutants. Plant Physiol 146(2):762–771

Bagyaraj DJ, Rangaswami G (2005) Microorganisms in soil. In: Agricultural microbiology, 2nd edn. Prentice Hall of India Private Limited, New Delhi, p 254

Bano A, Fatima M (2009) Salt tolerance in *Zea mays* (L). Following inoculation with *Rhizobium* and *Pseudomonas*. Biol Fertil Soils 45(4):405–413

Barea JM, Azcón R, Azcón-Aguilar C (2005) Interactions between mycorrhizal fungi and bacteria to improve plant nutrient cycling and soil structure. In: Buscot F, Varma A (eds) Microorganisms in soils: roles in genesis and functions. Springer, Berlin/Heidelberg, pp 195–212

Barea JM, Pozo MJ, López-Ráez JA, Aroca R, Ruíz-Lozano JM, Ferrol N, Azcón R, Azcón-Aguilar C (2014) Arbuscular mycorrhizas and their significance in promoting soil-plant system sustainability against environmental stresses. In: Rodelas MB, González-López J (eds) Beneficial plant-microbial interactions ecology and applications. CRC Press/Taylor & Francis, Boca Raton, pp 353–387

Barea JM, Azcón R, Azcón-Aguilar C (2017) Mycorrhizosphere interactions to improve a sustainable production of legumes. In: Zaidi A, Khan M, Musarrat J (eds) Microbes for legume improvement. Springer, Cham, pp 199–225

Barnawal D, Maji D, Bharti N, Chanotiya CS, Kalra A (2013) ACC deaminase-containing *Bacillus subtilis* reduces stress ethylene-induced damage and improves mycorrhizal colonization and rhizobial nodulation in *Trigonella foenum-graecum* under drought stress. J Plant Growth Regul 32:809–822

Barnawal D, Bharti N, Maji D, Chanotiya CS, Kalra A (2014) ACC deaminase-containing Arthrobacter protophormiae induces NaCl stress tolerance through reduced ACC oxidase activity and ethylene production resulting in improved nodulation and mycorrhization in *Pisum sativum*. J Plant Physiol 171(11):884–894

Behie SW, Bidochka MJ (2014) Nutrient transfer in plant – fungal symbioses. Trends Plant Sci 19:734–740

Besset-Manzoni Y, Rieusset L, Joly P, Comte G, Prigent-Combaret C (2018) Exploiting rhizosphere microbial cooperation for developing sustainable agriculture strategies. Environ Sci Pollut Res Int 25(30):29953–29970

Bhandari P, Garg N (2017) Dynamics of arbuscular mycorrhizal symbiosis and its role in nutrient acquisition: an overview. In: Varma A, Prasad R, Tuteja N (eds) Mycorrhiza – nutrient uptake, biocontrol, ecorestoration. Springer, Cham, pp 21–43

Bharti N, Barnawal D, Wasnik K, Tewari SK, Kalra A (2016) Co-inoculation of *Dietzia natronolimnaea* and *Glomus intraradices* with vermicompost positively influences *Ocimum basilicum* growth and resident microbial community structure in salt affected low fertility soils. Appl Soil Ecol 100:211–225

Bitla UM, Sorty AM, Meena KK, Singh NP (2017) Rhizosphere signaling cascades: fundamentals and determinants. In: Singh D, Singh H, Prabha R (eds) Plant-microbe interactions in agro-ecological perspectives. Springer, Singapore, pp 211–226

Bonfante P, Desiró A (2015) Arbuscular mycorrhizas: the lives of beneficial fungi and their plant host. In: Lugtenberg B (ed) Principles of plant-microbe interactions. Springer, Cham, pp 235–245

Brewin NJ (2004) Plant cell wall remodeling in the rhizobium-legume symbiosis. Crit Rev Plant Sci 23:293–316

Broghammer A, Krusell L, Blaise M, Sauer J, Sullivan JT, Maolanon N, Vinther M, Lorentzen A, Madsen EB, Jensen KJ, Roepstorff P, Thirup S, Ronson CW, Thygesen MB, Stougaard J (2012) Legume receptors perceive the rhizobial lipochitin oligosaccharide signal molecules by direct binding. Proc Natl Acad Sci U S A 109:13859–13864

Carbonnel S, Gutjahr C (2014) Control of arbuscular mycorrhiza development by nutrient signals. Front Plant Sci 5:462

Carlos MH, Stefani PV, Janette AM, Melani MS, Gabriela PO (2016) Assessing the effects of heavy metals in ACC deaminase and IAA production on plant growth-promoting bacteria. Microbiol Res 189:53–61

Chakraborty U, Chakraborty BN, Chakraborty AP, Dey PL (2013) Water stress amelioration and plant growth promotion in wheat plants by osmotic stress tolerant bacteria. World J Microbiol Biotechnol 29:789–803

Chakraborty U, Chakraborty B, Sarkar J (2018) Amelioration of abiotic stresses in plants through multi-faceted beneficial microorganisms. In: Kashyap PK, Srivastava AK, Tiwari SP, Kumar S (eds) Microbes for climate resilient agriculture. Wiley, Hoboken, pp 105–148

Chandra P, Singh E (2016) Applications and mechanisms of plant growth-stimulating rhizobacteria. In: Choudhary D, Varma A, Tuteja N (eds) Plant-microbe interaction: an approach to sustainable agriculture. Springer, Singapore, pp 37–62

Chimwamurombe PM, Grönemeyer JL, Reinhold-Hurek B (2016) Isolation and characterization of culturable seed-associated bacterial endophytes from gnotobiotically grown Marama bean seedlings. FEMS Microbiol Ecol 92(6):fiw083

Choudhary D (2012) Microbial rescue to plant under habitat-imposed abiotic and biotic stresses. Appl Microbiol Biotechnol 96:1137–1155

Clark RB, Zeto SK (2000) Mineral acquisition by arbuscular mycorrhizal plants. J Plant Nutr 23:867–902

Coats VC, Rumpho ME (2014) The rhizosphere microbiota of plant invaders: an overview of recent advances in the microbiomics of invasive plants. Front Microbiol 5:368. https://doi.org/10.3389/fmicb.2014.00368

Crespi MD, Jurkevitch E, Poiret M, D'Aubenton-Carafa Y, Petrovics G, Kondorosi E, Kondorosi A (1994) Enod40, a gene expressed during nodule organogenesis, codes for a nontranslatable RNA involved in plant growth. EMBO J 13:5099–5112

Czarny JC, Grichko VP, Glick BR (2006) Genetic modulation of ethylene biosynthesis and signalling in plants. Biotechnol Adv 24(4):410–419

D'Haeze W, Holsters M (2002) Nod factor structures, responses and perception during initiation of nodule development. Glycobiology 12:79–105

Dey RK, Pal KM, Thomas DN, Sherathia VB, Mandaliya RA, Bhadania MB, Patel P, Maida DH, Mehta BD, Nawade S, Patel V (2018) Endophytic microorganisms: future tools for climate resilient agriculture. In: Kashyap PL, Srivastava AK, Tiwari SP, Kumar S (eds) Microbes for climate resilient agriculture. Wiley, Hoboken, pp 235–253

Diédhiou I, Diouf D (2018) Transcription factors network in root endosymbiosis establishment and development. World J Microbiol Biotechnol 34(3):37

Dimkpa C, Weinand T, Asch F (2009) Plant-rhizobacteria interactions alleviate abiotic stresses conditions. Plant Cell Environ 32:1682–1694

Dobert RC, Breil BT, Triplett EW (1994) DNA sequence of the common nodulation genes of *Bradyrhizobium elkanii* and their phylogenetic relationships to those of other nodulating bacteria. Mol Plant-Microbe Interact 7:564–572

Dodd IC, Pérez-Alfocea F (2012) Microbial amelioration of crop salinity stress. J Exp Bot 63:3415–3428

Downie JA, Rossen L, Knight CD, Robertson JG, Wells B, Johnston AW (1985) *Rhizobium leguminosarum* genes involved in early stages of nodulation. J Cell Sci Suppl 2:347–354

Effmert U, Kalderás J, Warnke R (2012) Volatile mediated interactions between bacteria and fungi in the soil. J Chem Ecol 38:665–703

Ehrhardt DW, Wais R, Long SR (1996) Calcium spiking in plant root hairs responding to rhizobium nodulation signals. Cell 85:673–681

Estrada B, Barea JM, Aroca R, Ruiz-Lozano JM (2013) A native *Glomus intraradices* strain from a Mediterranean saline area exhibits salt tolerance and enhanced symbiotic efficiency with maize plants under salt stress conditions. Plant Soil 366:333–349

Etesami H, Alikhani HA (2016a) Co-inoculation with endophytic and rhizosphere bacteria allows reduced application rates of N-fertilizer for rice plant. Rhizosphere 2:15. https://doi.org/10.1016/j.rhisph.2016.09.003

Etesami H, Alikhani HA (2016b) Rhizosphere and endorhiza of oilseed rape (*Brassica napus* L.) plant harbor bacteria with multifaceted beneficial effects. Biol Control 94:11–24

Etesami H, Beattie GA (2017) Plant-microbe interactions in adaptation of agricultural crops to abiotic stress conditions. In: Kumar V, Kumar M, Sharma S, Prasad R (eds) Probiotics and plant health. Springer, Singapore, pp 163–200

Etesami H, Alikhani HA, Hosseini HM (2015) Indole-3-acetic acid and 1-aminocyclopropane-1-carboxylate deaminase: bacterial traits required in rhizosphere, rhizoplane and/or endophytic competence by beneficial bacteria. In: Maheshwari DK (ed) Bacterial metabolites in sustainable agroecosystem. Springer, Cham, pp 183–258

Evelin H, Kapoor R (2014) Arbuscular mycorrhizal symbiosis modulates antioxidant response in salt-stressed *Trigonella foenum-graecum* plants. Mycorrhiza 24(3):197–208

Evelin H, Kapoor R, Giri B (2009) Arbuscular mycorrhizal fungi in alleviation of salt stress: a review. Ann Bot 104(7):1263–1280

FAO (2008) FAO land and plant nutrition management service. http://www.fao.org/ag/agl/ahll/spush

Feddermann N, Reinhardt D (2011) Conserved residues in the ankyrin domain of VAPYRIN indicate potential protein-protein interaction surfaces. Plant Signal Behav 6:680–684

Fellbaum CR, Gachomo EW, Beesetty Y, Choudhari S, Strahan GD, Pfeffer PE, Kiers ET, Bücking H (2012) Carbon availability triggers fungal nitrogen uptake and transport in arbuscular mycorrhizal symbiosis. Proc Natl Acad Sci 109(7):2666–2671

Felle HH, Kondorosi E, Kondorosi A, Schultze M (1996) Rapid alkalization of root hairs in response to rhizobial lipochitooligosaccharide signals. Plant J 10:295–301

Filho JAC, Sobrinho RR, Pascholati SF (2017) Arbuscular mycorrhizal symbiosis and its role in plant nutrition in sustainable agriculture. In: Meena V, Mishra P, Bisht J, Pattanayak A (eds) Agriculturally important microbes for sustainable agriculture. Springer, Singapore, pp 129–164

Floss DS, Levy JG, Lévesque-Tremblay V et al (2013) DELLA proteins regulate arbuscule formation in arbuscular mycorrhizal symbiosis. Proc Natl Acad Sci 110:E5025–E5034

Foo E, Ross JJ, Jones WT, Reid JB (2013) Plant hormones in arbuscular mycorrhizal symbioses: an emerging role for gibberellins. Ann Bot 111:769–779

Gans J, Wolinsky M, Dunbar J (2005) Computational improvements reveal great bacterial diversity and high metal toxicity in soil. Science 309:1387–1390

Garg N, Baher N (2013) Role of arbuscular mycorrhizal symbiosis in proline biosynthesis and metabolism of *Cicer arietinum* L. (chickpea) genotypes under salt stress. J Plant Growth Regul 32:767–778

Garg N, Bhandari P (2012) Influence of cadmium stress and arbuscular mycorrhizal fungi on nodule senescence in *Cajanus cajan* (L.) Millsp. Int J Phytoremediation 14(1):62–74

Garg N, Bhandari P (2016a) Silicon nutrition and mycorrhizal inoculations improve growth, nutrient status, K^+/Na^+ ratio and yield of *Cicer arietinum* L. genotypes under salinity stress. Plant Growth Regul 78:371–387

Garg N, Bhandari P (2016b) Interactive effects of silicon and arbuscular mycorrhiza in modulating ascorbate-glutathione cycle and antioxidant scavenging capacity in differentially salt-tolerant *Cicer arietinum* L. genotypes subjected to long-term salinity. Protoplasma 253(5):1325–1345

Garg N, Chandel S (2010) Arbuscular mycorrhizal networks: process and functions. A review. Agron Sustain Dev 30:581–591

Garg N, Chandel S (2011) The effects of salinity on nitrogen fixation and trehalose metabolism in mycorrhizal *Cajanus cajan* (L.) millsp. plants. J Plant Growth Regul 30(4):490–503

Garg N, Geetanjali (2007) Symbiotic nitrogen fixation in legume nodules: process and signaling. A review. Agron Sustain Dev 27:59–68

Garg N, Kashyap L (2017) Silicon and *Rhizophagus irregularis*: potential candidates for ameliorating negative impacts of arsenate and arsenite stress on growth, nutrient acquisition and productivity in *Cajanus cajan* (L.) Millsp. genotypes. Environ Sci Pollut Res 24:18520–18535

Garg N, Kaur H (2013) Response of antioxidant enzymes, phytochelatins and glutathione production towards Cd and Zn stresses in *Cajanus cajan* (L.) Millsp. genotypes colonized by arbuscular mycorrhizal fungi. J Agron Crop Sci 199(2):118–133

Garg N, Manchanda G (2008) Effect of arbuscular mycorrhizal inoculation on salt-induced nodule senescence in *Cajanus cajan* (pigeonpea). J Plant Growth Regul 27(2):115

Garg N, Manchanda G (2009) Role of arbuscular mycorrhizae in the alleviation of ionic, osmotic and oxidative stresses induced by salinity in *Cajanus cajan* (L.) Millsp. (pigeonpea). J Agron Crop Sci 195:110–123

Garg N, Pandey R (2015) Effectiveness of native and exotic arbuscular mycorrhizal fungi on nutrient uptake and ion homeostasis in salt-stressed *Cajanus cajan* L. (Millsp.) genotypes. Mycorrhiza 25(3):165–180

Garg N, Pandey R (2016) High effectiveness of exotic arbuscular mycorrhizal fungi is reflected in improved rhizobial symbiosis and trehalose turnover in *Cajanus cajan* genotypes grown under salinity stress. Fungal Ecol 21:57–67

Garg N, Singh S (2017) Arbuscular mycorrhiza *Rhizophagus irregularis* and silicon modulate growth, proline biosynthesis and yield in *Cajanus cajan* L. Millsp. (pigeonpea) genotypes under cadmium and zinc stress. J Plant Growth Regul 37:46–63

Garg N, Singla P (2012) The role of Glomus mosseae on key physiological and biochemical parameters of pea plants grown in arsenic contaminated soil. Sci Hortic 143:92–101

Garg N, Singla P (2015) Naringenin-and *Funneliformis mosseae*-mediated alterations in redox state synchronize antioxidant network to alleviate oxidative stress in *Cicer arietinum* L. genotypes under salt stress. J Plant Growth Regul 34(3):595–610

Garg N, Singla P (2016) Stimulation of nitrogen fixation and trehalose biosynthesis by naringenin (Nar) and arbuscular mycorrhiza (AM) in chickpea under salinity stress. Plant Growth Regul 80(1):5–22

Genre A (2012) Signalling and the re-structuring of plant cell architecture in am symbiosis. In: Perotto S, Baluška F (eds) Signaling and communication in plant symbiosis, signaling and communication in plants, vol 11. Springer, Berlin, pp 51–71

Genre A, Chabaud M, Timmers T et al (2005) Arbuscular mycorrhizal fungi elicit a novel intracellular apparatus in *Medicago truncatula* root epidermal cells before infection. Plant Cell 17:3489–3499

Gibson KE, Kobayashi H, Walker GC (2008) Molecular determinants of a symbiotic chronic infection. Annu Rev Genet 42:413–441

Giraud E, Moulin L, Vallenet D, Barbe V, Cytryn E, Avarre JC, Jaubert M, Simon D, Cartieaux F, Prin Y, Bena G, Hannibal L, Fardoux J, Kojadinovic M, Vuillet L, Lajus A, Cruveiller S, Rouy Z, Mangenot S, Segurens B, Dossat C, Franck WL, Chang WS, Saunders E, Bruce D, Richardson P, Normand P, Dreyfus B, Pignol D, Stacey G, Emerich D, Vermeglio A, Medigue C, Sadovsky M (2007) Legume symbioses: absence of nod genes in photosynthetic bradyrhizobia. Science 316:1307–1312

Glick BR (2005) Modulation of plant ethylene levels by the bacterial enzyme ACC deaminase. FEMS Microbiol Lett 251:1–7

Glick BR, Todorovic B, Czarny J, Cheng Z, Duan J, McConkey B (2007) Promotion of plant growth by bacterial ACC Deaminase. Crit Rev Plant Sci 26:227–242

Gonzalez-Guerrero M, Benabdellah K, Valderas A, Azcon-Aguilar C, Ferrol N (2010) GintABC1 encodes a putative ABC transprter of the MRP subfamily induced by Cu, Cd, and oxidative stress in *Glomus intraradices*. Mycorrhiza 20:137–146

Grover M, Ali SZ, Sandhya V, Rasul A, Venkateswarlu B (2011) Role of microorganisms in adaptation of agriculture crops to abiotic stresses. World J Microbiol Biotechnol 27:1231–1240

Gururani MA, Upadhyaya CP, Baskar V, Venkatesh J, Nookaraju A, Park SW (2013) Plant growth-promoting rhizobacteria enhance abiotic stress tolerance in *Solanum tuberosum* through inducing changes in the expression of ROS-scavenging enzymes and improved photosynthetic performance. J Plant Growth Regul 32:245–258

Gutjahr C, Parniske M (2013) Cell and developmental biology of the arbuscular mycorrhiza symbiosis. Annu Rev Cell Dev Biol 29:593–617

Gutjahr C, Radovanovic D, Geoffroy J et al (2012) The half-size ABC transporters STR1 and STR2 are indispensable for mycorrhizal arbuscule formation in rice. Plant J 69:906–920

Habib SH, Kausar H, Saud HM (2016) Plant growth-promoting rhizobacteria enhance salinity stress tolerance in Okra through ROS-scavenging enzymes. Biomed Res Int 2016:6284547

Hajiboland R (2013) Role of arbuscular mycorrhiza in amelioration of salinity. In: Ahmad P, Azooz MA, MNV P (eds) Salt stress in plants: signalling, omics and adaptations. Springer Science+Business Media, New York, pp 301–354

Hajiboland R, Aliasgharzadeh N, Laiegh SF, Poschenrieder C (2010) Colonization with arbuscular mycorrhizal fungi improves salinity tolerance of tomato (*Solanum lycopersicum* L.) plants. Plant Soil 331:313–327

Haldar S, Sengupta S (2016) Microbial ecology at rhizosphere: bioengineering and future prospective. In: Choudhary D, Varma A, Tuteja N (eds) Plant-microbe interaction: an approach to sustainable agriculture. Springer, Singapore, pp 63–96

Hammer EC, Nasr H, Pallon J, Olsson PA, Wallander H (2011) Elemental composition of arbuscular mycorrhizal fungi at high salinity. Mycorrhiza 21:117–129
Han Q, Lü X, Bai J, Qiao Y, Paré PW, Wang S, Zhang J, Wu Y, Pang X, Xu W, Wang Z (2014) Beneficial soil bacterium Bacillus subtilis (GB03) augments salt tolerance of white clover. Front Plant Sci 5:1–8
Hartmann A, Rothballer M, Schmid M (2008) Lorenz Hiltner, a pioneer in rhizosphere microbial ecology and soil bacteriology research. Plant Soil 312(1–2):7–14
Hayat R, Ali S, Amara U, Khalid R, Ahmed I (2010) Soil beneficial bacteria and their role in plant growth promotion: a review. Ann Microbiol 60:579–598
Islam F, Yasmeen T, Ali Q, Mubin M, Ali S, Arif MS, Hussain S, Riaz M, Abbas F (2016) Copper-resistant bacteria reduces oxidative stress and uptake of copper in lentil plants: potential for bacterial bioremediation. Environ Sci Pollut Res 23:220–233
Jahromi F, Aroca R, Porcel R, Ruiz-Lozano JM (2008) Influence of salinity on the in vitro development of *Glomus intraradices* and on the in vivo physiological and molecular responses of mycorrhizal lettuce plants. Microb Ecol 55:45–53
Jha CK, Saraf M (2015) Plant growth promoting rhizobacteria (PGPR): a review. E3 J Agric Res Dev 5:108–119
Jiang C, Sheng X, Qian M, Wang Q (2008) Isolation and characterization of a heavy metal-resistant *Burkholderia* sp. from heavy metal-contaminated paddy field soil and its potential in promoting plant growth and heavy metal accumulation in metal-polluted soil. Chemosphere 72:157–164
Jones KM (2012) Increased production of the exopolysaccharide succinoglycan enhances *Sinorhizobium meliloti* 1021 symbiosis with the host plant *Medicago truncatula*. J Bacteriol 194:4322–4331
Jones KM, Kobayashi H, Davies BW, Taga ME, Walker GC (2007) How rhizobial symbionts invade plants: the *Sinorhizobium–Medicago* model. Nat Rev Microbiol 5:619–633
Junghans U, Polee A, Duchting P, Weiler E, Kuhlman B, Grubber F, Teichmann T (2006) Adaptation to high salinity in poplar involves changes in xylem anatomy and auxin physiology. Plant Cell Environ 29:1519–1531
Juniper S, Abbott LK (2006) Soil salinity delays germination and limits growth of hyphae from propagules of arbuscular mycorrhizal fungi. Mycorrhiza 16:371–379
Kapoor R, Sharma D, Bhatnagar AK (2008) Arbuscular mycorrhizae in micropropagation systems and their potential applications. Sci Hortic 116:227–239
Kapoor R, Evelin H, Mathur P, Giri B (2013) Arbuscular mycorrhiza: approaches for abiotic stress tolerance in crop plants for sustainable agriculture. In: Tuteja N, Gill SS (eds) Plant acclimation to environmental stress. Springer Science+Business Media, New York, pp 359–401
Kasotia A, Varma A, Tuteja N, Choudhary DK (2016) Microbial-mediated amelioration of plants under abiotic stress: an emphasis on arid and semiarid climate. In: Choudhary D, Varma A, Tuteja N (eds) Plant-microbe interaction: an approach to sustainable agriculture. Springer, Singapore, pp 155–163
Kaur S, Kaur G (2018) Morphological and physiological aspects of symbiotic plant–microbe interactions and their significance. In: Giri B, Prasad R, Varma A (eds) Root biology. Soil biology, vol 52. Springer, Cham, pp 367–407
Kawaharada Y, Kelly S, Nielsen MW, Hjuler CT, Gysel K, Muszynski A (2015) Receptor-mediated exopolysaccharide perception controls bacterial infection. Nature 523:308–312
Kawaharada Y, Nielsen MW, Kelly S, James EK, Andersen KR, Rasmussen SR (2017) Differential regulation of the Epr3 receptor coordinates membrane-restricted rhizobial colonization of root nodule primordia. Nat Commun 8:14534
Kelly SJ, Muszŷnski A, Kawaharada Y, Hubber AM, Sullivan JT, Sandal N (2013) Conditional requirement for exopolysaccharide in the *Mesorhizobium–Lotus* symbiosis. Mol Plant-Microbe Interact 26:319–329
Khalid A, Arshad M, Zahir ZA (2006) Phytohormones: microbial production and applications. In: Uphoff N, Ball AS, Fernandes E, Herren H, Husson O, Laing M, Palm C, Pretty J, Sanchez P,

Sanginga N, Thies J (eds) Biological approaches to sustainable soil systems. Taylor & Francis/ CRC Press, Boca Raton, pp 207–220

Kohler J, Hernández JA, Caravaca F, Roldán A (2009) Induction of antioxidant enzymes is involved in the greater effectiveness of a PGPR versus AM fungi with respect to increasing the tolerance of lettuce to severe salt stress. Environ Exp Bot 65(2):245–252

Kondorosi E, Banfalvi Z, Kondorosi A (1984) Physical and genetic analysis of a symbiotic region of *Rhizobium meliloti*: identification of nodulation genes. Mol Gen Genet 193:445–452

Lee GW, Lee K, Chae J (2016a) *Herbaspirillum* sp. strain GW103 alleviates salt stress in *Brassica rapa* L. ssp. pekinensis. Protoplasma 253:655–661

Lee S, Yap M, Behringer G, Hung R (2016b) Volatile organic compounds emitted by *Trichoderma* species mediate plant growth. Fungal Biol Biotechnol 3:7

Lerouge P, Roche P, Faucher C (1990) Symbiotic host-specificity of *Rhizobium meliloti* is determined by a sulphated and acylated glucosamine oligosaccharide signal. Nature 344:781–784

Li Q, Saleh-Lakha S, Glick BR (2005) The effect of native and ACC deaminase containing *Azospirillum brasilense* Cd1843 on the rooting of carnation cuttings. Can J Microbiol 51:511–514

Li T, Hu Y, Hao Z, Li H, Wang Y, Chen B (2013) First cloning and characterization of two functional aquaporin genes from an Arbuscular mycorrhizal fungus *Glomus intraradices*. New Phytol 197:617–630

Li L, Nagai K, Yin F (2016) Progress in cold roll bonding of metals. Sci Technol Adv Mater 9:023001 (11pp). https://doi.org/10.1088/1468-6996/9/2/023001

Lopez-Pedrosa A, Gonzalez-Guerrero M, Valderas A, Azcon-Aguilar C, Ferrol N (2006) GintAMT1 encodes a functional high-affinity ammonium transporter that is expressed in the extraradical mycelium of *Glomus intraradices*. Fungal Genet Biol 43:102–110

López-Ráez JA (2016) How drought and salinity affect arbuscular mycorrhizal symbiosis and strigolactone biosynthesis? Planta 243(6):1375–1385

Ludwig-Müller J (2004) From auxin homeostasis to understanding plant pathogen and plant symbiont interaction: editor's research interests. J Plant Growth Regul 23:1–8

Maheswari TU, Anbukkarasi K, Hemalatha T, Chendrayan K (2013) Studies on phytohormone producing ability of indigenous endophytic bacteria isolated from tropical legume crops. Int J Curr Microbiol Appl Sci 2:127–136

Manaf HH, Zayed MS (2015) Productivity of cowpea as affected by salt stress in presence of endomycorrhizae and *Pseudomonas fluorescens*. Ann Agric Sci 60(2):219–226

Mastouri F, Björkman T, Harman GE (2010) Seed treatment with *Trichoderma harzianum* alleviates biotic, abiotic, and physiological stresses in germinating seeds and seedlings. Phytopathology 100:1213–1221

Meena RS, Lal R (2018) Legumes and sustainable use of soils. In: Meena R, Das A, Yadav G, Lal R (eds) Legumes for soil health and sustainable management. Springer, Singapore, pp 1–32

Meena H, Ahmed MA, Prakash P (2015) Amelioration of heat stress in wheat, *Triticum aestivum* by PGPR (*Pseudomonas aeruginosa* strain 2CpS1). Biosci Biotech Res Commun 8:171–174

Mendes R, Kruijt M, de Bruijn I, Dekkers E, van der Voort M, Schneider JH, Piceno YM, De Santis TZ, Andersen GL, Bakker PA, Raaijmakers JM (2011) Deciphering the rhizosphere microbiome for disease suppressive bacteria. Science 332(6033):1097–1100. https://doi.org/10.1126/science.1203980. PMID: 21551032

Miliute I, Buzaite O, Stanys V (2015) Bacterial endophytes in agricultural crops and their role in stress tolerance: a review. Zemdirbyste-Agriculture 102(4):465–478

Munns R, Tester M (2008) Mechanisms of salinity tolerance. Annu Rev Plant Biol 59:651–681

Murray JD, Muni RRD, Torres-Jerez I, Tang Y, Allen S, Andriankaja M, Li G, Laxmi A, Cheng X, Wen J, Vaughan D, Schultze M, Sun J, Charpentier M, Oldroyd G, Tadege M, Ratet P, Mysore KS, Chen R, Udvardi MK (2011) *Vapyrin*, a gene essential for intracellular progression of arbuscular mycorrhizal symbiosis, is also essential for infection by rhizobia in the nodule symbiosis of *Medicago truncatula*. Plant J 65:244–252

Nadarajah KK (2016) Rhizosphere interactions: life below ground. In: Choudhary D, Varma A, Tuteja N (eds) Plant-microbe interaction: an approach to sustainable agriculture. Springer, Singapore, pp 3–23

Nadeem SM, Naveed M, Ahmad M, Zahir ZA (2015) Rhizosphere bacteria for crop production and improvement of stress tolerance: mechanisms of action, applications, and future prospects. In: Arora N (ed) Plant microbes symbiosis: applied facets. Springer, New Delhi, pp 1–36

Naveed M, Aziz MZ, Yaseen M (2017) Perspectives of using endophytic microbes for legume improvement. In: Zaidi A, Khan M, Musarrat J (eds) Microbes for legume improvement. Springer, Cham, pp 277–299

Ocón A, Hampp R, Requena N (2007) Trehalose turnover during abiotic stress in arbuscular mycorrhizal fungi. New Phytol 174:879–891

Ogut M, Er F, Kandemir N (2010) Phosphate solubilization potentials of soil *Acinetobacter* strains. Biol Fertil Soils 46(7):707–715

Oldroyd GED (2013) Speak, friend, and enter: signalling systems that promote beneficial symbiotic associations in plants. Nat Rev Microbiol 11:252–263

Oldroyd GE, Downie JA (2008) Coordinating nodule morphogenesis with rhizobial infection in legumes. Annu Rev Plant Biol 59:519–546

Oldroyd GE, Murray JD, Poole PS, Downie JA (2011) The rules of engagement in the legume-rhizobial symbiosis. Annu Rev Genet 45:119–144

Ortíz-Castro R, Contreras-Cornejo HA, Macías-Rodríguez L, López-Bucio J (2009) The role of microbial signals in plant growth and development. Plant Signal Behav 4:701–712

Ovchinnikova E, Journet E-P, Chabaud M (2011) IPD3 controls the formation of nitrogen-fixing symbiosomes in pea and *Medicago* Spp. Mol Plant-Microbe Interact 24:1333–1344

Oves M, Khan M, Zaidi A (2013) Chromium reducing and plant growth promoting novel strain Pseudomonas aeruginosa OSG41 enhance chickpea growth in chromium amended soils. Eur J Soil Biol 56:72–83

Parniske M (2008) Arbuscular mycorrhiza: the mother of plant root endosymbioses. Nat Rev Microbiol 6:763–775

Parniske M, Schmidt P, Kosch K, Muller P (1994) Plant defense responses of host plants with determinate nodules induced by EPS-defective exoB mutants of *Bradyrhizobium japonicum*. Mol Plant-Microbe Interact 7:631–638

Parvaiz A, Satyawati S (2008) Salt stress and phyto-biochemical responses of plants – a review. Plant Soil Environ 54(3):89–99

Paul D, Lade H (2014) Plant-growth-promoting rhizobacteria to improve crop growth in saline soils: a review. Agron Sustain Dev 34:737–752

Paul D, Sarma YR (2006) Plant growth promoting rhizhobacteria (PGPR)-mediated root proliferation in black pepper (*Piper nigrum* L.) as evidenced through GS root software. Arch Phytopathol Plant Protect 39:311–314

Pellegrino E, Bedini S (2014) Enhancing ecosystem services in sustainable agriculture: biofertilization and biofortification of chickpea (*Cicer arietinum* L.) by arbuscular mycorrhizal fungi. Soil Biol Biochem 68:429–439

Pellegrino E, Opik M, Bonari E, Ercoli L (2015) Responses of wheat to arbuscular mycorrhizal fungi: a meta-analysis of field studies from 1975 to 2013. Soil Biol Biochem 84:210–217

Pellock BJ, Cheng H-P, Walker GC (2000) Alfalfa root nodule invasion efficiency is dependent on *Sinorhizobium meliloti* polysaccharides. J Bacteriol 182:4310–4318

Pereyra MA, Garcia P, Colabelli MN, Barassi CA, Creus CM (2012) A better water status in wheat seedlings induced by *Azospirillum* under osmotic stress is related to morphological changes in xylem vessels of the coleoptile. Appl Soil Ecol 53:94–97

Porcel R, Aroca R, Ruiz-Lozano JM (2012) Salinity stress alleviation using arbuscular mycorrhizal fungi. A review. Agron Sustain Dev 32:181–200

Porcel R, Aroca R, Azcon R, Ruiz-Lozano JM (2016) Regulation of cation transporter genes by the arbuscular mycorrhizal symbiosis in rice plants subjected to salinity suggests improved salt tolerance due to reduced Na^+ root-to-shoot distribution. Mycorrhiza 26:673–684

Prasad MP, Dagar S (2014) Identification and characterization of endophytic bacteria from fruits like avocado and black grapes. Int J Curr Microbiol App Sci 3(8):937–947

Prasad R, Bhola D, Akdi K, Cruz C, Sairam KVSS, Tuteja N, Varma A (2017) Introduction to mycorrhiza: historical development. In: Varma A, Prasad R, Tuteja N (eds) Mycorrhiza. Springer, Cham, pp 1–7

Pumplin N, Harrison MJ (2009) Live-cell imaging reveals periarbuscular membrane domains and organelle location in *Medicago truncatula* roots during arbuscular mycorrhizal symbiosis. Plant Physiol 151:809–819

Puppo A, Pauly N, Boscari A, Mandon K, Brouquisse R (2013) Hydrogen peroxide and nitric oxide: key regulators of the legume *Rhizobium* and mycorrhizal symbioses. Antioxid Redox Signal 18:2202–2219

Radutoiu S, Madsen LH, Madsen EB, Felle HH, Umehara Y, Grønlund M (2003) Plant recognition of symbiotic bacteria requires two LysM receptor-like kinases. Nature 425:585–592

Radutoiu S, Madsen LH, Madsen EB, Jurkiewicz A, Fukai E, Quistgaard EM, Albrektsen AS, James EK, Thirup S, Stougaard J (2007) LysM domains mediate lipochitin–oligosaccharide recognition and Nfr genes extend the symbiotic host range. EMBO J 26:3923–3935

Rech SS, Heidt S, Requena N (2013) A tandem Kunitz protease inhibitor (KPI106)-serine carboxy-peptidase (SCP1) controls mycorrhiza establishment and arbuscule development in *Medicago truncatula*. Plant J 75:711–725

Reddy S, Schorderet M, Feller U, Reinhardt D (2007) A petunia mutant affected in intracellular accommodation and morphogenesis of arbuscular mycorrhizal fungi. Plant J 51:739–750

Remans R, Beebe S, Blair M, Manrique G, Tovar E, Rao I, Croonenborghs A, Torres-Gutierrez R, El-Howeity M, Michiels J, Vanderleyden (2008) Physiological and genetic analysis of root responsiveness to auxin-producing plant growth-promoting bacteria in common bean (*Phaseolus vulgaris* L.). Plant Soil 302:149–161

Rich MK, Schorderet M, Reinhardt D (2014) The role of the cell wall compartment in mutualistic symbioses of plants. Plant-Microbe Interact 5:238

Ruiz-Lozano JM, Aroca R (2010) Host response to osmotic stresses: stomatal behaviour and water use efficiency of arbuscular mycorrhizal plants. In: Koltai H, Kapulnik Y (eds) Arbuscular mycorrhizas: physiology and function, 2nd edn. Springer Science +Business Media BV, Dordrecht, pp 239–256

Ruiz-Lozano JM, Porcel R, Azcón C, Aroca R (2012a) Regulation by arbuscular mycorrhizae of the integrated physiological response to salinity in plants: new challenges in physiological and molecular studies. J Exp Bot 63(11):4033–4044

Ruiz-Lozano JM, Porcel R, Bárzana G, Azcón C, Aroca R (2012b) Contribution of arbuscular mycorrhizal symbiosis to plant drought tolerance: state of the art. In: Aroca R (ed) Plant responses to drought stress, from morphological to molecular features. Springer, Heidelberg, pp 335–362

Sadeghi A, Karimi E, Dahaji PA, Javid MG, Dalvand Y, Askari H (2012) Plant growth promoting activity of an auxin and siderophore producing isolate of *Streptomyces* under saline soil conditions. World J Microbiol Biotechnol 28:1503–1509

Saif S, Zaidi A, Khan MS, Rizvi A (2017) Metal-legume-microbe interactions: toxicity and remediation. In: Zaidi A, Khan M, Musarrat J (eds) Microbes for legume improvement. Springer, Cham, pp 367–385

Sandhya V, Ali SZ, Grover M, Reddy G, Venkateswarlu B (2009) Alleviation of drought stress effects in sunflower seedlings by exopolysaccharides producing *Pseudomonas putida* strain P45. Biol Fertil Soils 46:17–26

Saxena S, Kaur H, Verma P, Petla BP, Andugula VR, Majee M (2013) Osmoprotectants: potential for crop improvement under adverse conditions. In: Tuteja N, Gill SS (eds) Plant acclimation to environmental stress. Springer, New York, pp 197–232

Schnitzer SA, Klironomos JN, HilleRisLambers J, Kinkel LL, Reich PB, Xiao K, Rillig MC, Sikes BA, Callaway RM, Mangan SA, van Nes EH, Scheffer M (2011) Soil microbes drive the classic plant diversity-productivity pattern. Ecology 92(2):296–303

Selim SM, Zayed MS (2017) Role of biofertilizers in sustainable agriculture under abiotic stresses. In: Panpatte D, Jhala Y, Vyas R, Shelat H (eds) Microorganisms for green revolution. Microorganisms for sustainability, vol 6. Springer, Singapore, pp 281–301

Sengupta A, Gunri SK, Biswas T (2017) Microbial interactions and plant health. In: Singh D, Singh H, Prabha R (eds) Plant-microbe interactions in agro-ecological perspectives. Springer, Singapore, pp 61–84

Sharma P, Khanna V, Kumari S (2016) Abiotic stress mitigation through plant-growth-promoting rhizobacteria. In: Choudhary D, Varma A, Tuteja N (eds) Plant-microbe interaction: an approach to sustainable agriculture. Springer, Singapore, pp 327–342

Shelden MC, Roessner U (2013) Advances in functional genomics for investigating salinity stress tolerance mechanisms in cereals. Front Plant Sci 4:123

Sheng XF, Xia JJ (2006) Improvement of rape (*Brassica napus*) plant growth and cadmium uptake by cadmium-resistant bacteria. Chemosphere 64:1036–1042

Sherameti I, Tripathi S, Varma A, Oelmüller R (2009) The root-colonizing endophyte *Pirifomospora indica* confers drought tolerance in Arabidopsis by stimulating the expression of drought stress–related genes in leaves. Mol Plant-Microbe Interact 21:799–807

Shoresh M, Harman GE (2008) The molecular basis of shoot responses of maize seedlings to *Trichoderma harzianum* T22 inoculation of the root: a proteomic approach. Plant Physiol 147:2147–2163

Simões M, Simões LC, Cleto S, Machado I, Pereira MO, Vieira MJ (2007) Antimicrobial mechanisms ofortho – phthalaldehyde action. J Basic Microbiol 47(3):230–242

Singh A, Sarma BK, Upadhyay RS, Singh HB (2013) Compatible rhizosphere microbes mediated alleviation of biotic stress in chickpea through enhanced antioxidant and phenylpropanoid activities. Microbiol Res 168:33–40

Singh S, Katzer K, Lambert J, Cerri M, Parniske M (2014) CYCLOPS, a DNA-binding transcriptional activator, orchestrates symbiotic root nodule development. Cell Host Microbe 15:139–152

Singh RP, Shelke GM, Kumar A, Jha PN (2015) Biochemistry and genetics of ACC deaminase: a weapon to "stress ethylene" produced in plants. Front Microbial. https://doi.org/10.3389/fmicb.2015.00937

Singh RK, Singh P, Li HB, Yang LT, Li YR (2017) Soil–plant–microbe interactions: use of nitrogen-fixing bacteria for plant growth and development in sugarcane. In: Singh D, Singh H, Prabha R (eds) Plant-microbe interactions in agro-ecological perspectives. Springer, Singapore, pp 35–59

Singh RP, Manchanda G, Anwar MN, Zhang JJ, Li YZ (2018) Mycorrhiza – helping plants to navigate environmental stresses. In: Kashyap PK, Srivastava AK, Tiwari SP, Kumar S (eds) Microbes for climate resilient agriculture. Wiley, Hoboken, pp 205–233

Skorupska A, Wielbo J, Kidaj D, Marek-Kozaczuk M (2010) Enhancing rhizobium–legume symbiosis using signaling factors. In: Khan MS, Musarrat J, Zaidi A (eds) Microbes for legume improvement. Springer, Vienna, pp 27–54

Smith SE, Read DJ (2008) Mycorrhizal symbiosis, 3rd edn. Academic, San Diego

Smith SE, Facelli E, Pope S, Smith FA (2010) Plant performance in stressful environments: interpreting new and established knowledge of the roles of arbuscular mycorrhizas. Plant Soil 326(1–2):3–20

Spaink HP (2000) Root nodulation and infection factors produced by rhizobial bacteria. Annu Rev Microbiol 54:257–288

Spaink HP, Sheeley DM, van Brussel AAN et al (1991) A novel highly unsaturated fatty acid moiety of lipooligosaccharide signals determines host specificity of *Rhizobium*. Nature 354:125–130

Sugiyama A, Shitan N, Yazaki K (2008) Signaling from soybean roots to rhizobium: an ATP-binding cassette-type transporter mediates genistein secretion. Plant Signal Behav 3(1):38–40

Sundaramoorthy S, Balabaskar P (2012) Consortial effect of endophytic and plant growth promoting rhizobacteria for the management of early blight of tomato incited by *Alternaria solani*. J Plant Pathol Microbiol 3:145. https://doi.org/10.4172/2157-7471.1000145

Talaat NB, Shawky BT (2011) Influence of arbuscular mycorrhizae on yield, nutrients, organic solutes, and antioxidant enzymes of two wheat cultivars under salt stress. J Plant Nutr Soil Sci 174:283–291

Talaat NB, Shawky BT (2017) Microbe-mediated induced abiotic stress tolerance responses in plants. In: Singh D, Singh H, Prabha R (eds) Plant-microbe interactions in agro-ecological perspectives. Springer, Singapore, pp 101–133

Tiwari S, Lata C, Chauhan PS, Nautiyal CS (2016) *Pseudomonas putida* attunes morphophysiological, biochemical and molecular responses in *Cicer arietinum* L. during drought stress and recovery. Plant Physiol Biochem 99:108–117

UN (2016) United Nations proclaims 2016 as "International year of pulses". Global pulse confederation. Retrieved 24 Jan 2016, (A/RES/68/231)

Ullah S, Hussain MB, Khan MY, Asghar HN (2017) Ameliorating salt stress in crops through plant growth-promoting Bacteria. In: Singh D, Singh H, Prabha R (eds) Plant-microbe interactions in agro-ecological perspectives. Springer, Singapore, pp 549–575

Upadhyay S, Singh D, Saikia R (2009) Genetic diversity of plant growth promoting rhizobacteria isolated from rhizospheric soil of wheat under saline condition. Curr Microbiol 59:489–496

Vacheron J, Desbrosses G, Bouffaud ML, Touraine B, Moënne-Loccoz Y, Muller D, Legendre L, Wisniewski-Dyé F, Prigent-Combaret C (2013) Plant growth promoting rhizobacteria and root system functioning. Front Plant Sci. https://doi.org/10.3389/fpls.2013.00356

Venkadasamy GP, George S, Raina SK, Kumar M, Rane J, Kannepalli A (2018) Plant-associated microbial interactions in the soil environment: role of endophytes in imparting abiotic stress tolerance to crops. In: Bal S, Mukherjee J, Choudhury B, Dhawan A (eds) Advances in crop environment interaction. Springer, Singapore, pp 245–284

Venkateswarlu B, Shanker AK (2009) Climate change and agriculture: adaptation and mitigation strategies. Indian J Agron 54:226–230

Via VD, Zanetti ME, Blanco F (2015) How legumes recognize rhizobia. Plant Signal Behav. https://doi.org/10.1080/15592324.2015.1120396

Vierheilig H, Lerat S, Piché Y (2003) Systemic inhibition of arbuscular mycorrhiza development by root exudates of cucumber plants colonized by *Glomus mosseae*. Mycorrhiza 13:167–170

Vimal SR, Singh JS, Arora NK, Singh SK (2017) Soil-plant-microbe interactions in stressed agriculture management: a review. Pedosphere 27(2):177–192

Vurukonda SSKP, Vardharajula S, Shrivastava M, SkZ A (2016) Enhancement of drought stress tolerance in crops by plant growth promoting rhizobacteria. Microbiol Res 184:13–24

Wang Q, Dodd IC, Belimov AA et al (2016) Rhizosphere bacteria containing 1-aminocyclopropa ne-1-carboxylate deaminase increase growth and photosynthesis of pea plants under salt stress by limiting Na^+ accumulation. Funct Plant Biol 43:161–172

Wang Q, Liu J, Zhu H (2018) Genetic and molecular mechanisms underlying symbiotic specificity in legume-rhizobium interactions. Front Plant Sci 9:313

Werner D (2008) Signalling in the rhizobia–legumes symbiosis. In: Varma A, Abbott L, Werner D, Hampp R (eds) Plant surface microbiology. Springer, Berlin/Heidelberg, pp 99–119

Weston LA, Ryan PR, Watt M (2012) Mechanisms for cellular transport and release of allelochemicals from plant roots into the rhizosphere. J Exp Bot 63(9):3445

Xie F, Murray JD, Kim J, Heckmann AB, Edwards A, Oldroyd GED, Downie JA (2012) Legume pectate lyase required for root infection by rhizobia. Proceed Nat Acad Sci 109(2):633–638

Yano K, Yoshida S, Müller J, Singh S, Banba M, Vickers K, Markmann K, White C, Schuller B, Sato S, Asamizu E, Tabata S, Murooka Y, Perry J, Wang TL, Kawaguchi M, Imaizumi-Anraku H, Hayashi M, Parniske M (2008) CYCLOPS, a mediator of symbiotic intracellular accommodation. Proc Natl Acad Sci 105:20540–20545

Yong X, Chen Y, Liu W, Xu L, Zhou J, Wang S, Chen P, Ouyang P, Zheng T (2014) Enhanced cadmium resistance and accumulation in *Pseudomonas putida* KT2440 expressing the phytochelatin synthase gene of *Schizosaccharomyces pombe*. Lett Appl Microbiol 58:255–261

Yuan J, Raza W, Shen Q (2018) Root exudates dominate the colonization of pathogen and plant growth-promoting rhizobacteria. In: Giri B, Prasad R, Varma A (eds) Root biology. Soil biology, vol 52. Springer, Cham, pp 167–180

Zahir ZA, Ghani U, Naveed M, Nadeem SM, Arshad M (2009) Comparative effectiveness of *Pseudomonas* and *Serratia* sp. containing ACC-deaminase for improving growth and yield of wheat under salt-stressed conditions. Arch Microbiol 191:415–424

Zarea MJ, Goltapeh EM, Karimi N, Varma A (2013) Sustainable agriculture in saline-arid and semiarid by use potential of AM fungi on mitigates NaCl effects. In: Goltapeh EM, Danesh YR, Varma A (eds) Fungi as bioremediators. Soil biology 32. Springer, Berlin, pp 347–369

Zhang H, Irving LJ, McGill C, Matthew C, Zhou D, Kemp P (2010a) The effects of salinity and osmotic stress on barley germination rate: sodium as an osmotic regulator. Ann Bot 106:1027–1035

Zhang H, Murzello C, Kim MS, Xie X, Jeter RM, Zak JC et al (2010b) Choline and osmotic-stress tolerance induced in *Arabidopsis* by the soil microbe *Bacillus subtilis* (GB03). Mol Plant-Microbe Interact 23:1097–1104

Zhang YF, He LY, Chen ZJ, Wang QY, Qian M, Sheng XF (2011) Characterization of ACC deaminase-producing endophytic bacteria isolated from copper-tolerant plants and their potential in promoting the growth and copper accumulation of *Brassica napus*. Chemosphere 83:57–62

Zhang Y, Miró M, Kolev SD (2015) Hybrid flow system for automatic dynamic fractionation and speciation of inorganic arsenic in environmental solids. Environ Sci Technol 49:2733–2740

Zhang H, Kim MS, Sun Y, Dowd SE, Shi H, Paré PW (2008) Soil Bacteria confer plant salt tolerance by tissue-specific regulation of the sodium transporter. Mol Plant-Microbe Interact 21(6):737–744

MIX
Papier aus verantwortungsvollen Quellen
Paper from responsible sources
FSC® C105338

If you have any concerns about our products,
you can contact us on
ProductSafety@springernature.com

In case Publisher is established outside the EU,
the EU authorized representative is:
Springer Nature Customer Service Center GmbH
Europaplatz 3, 69115 Heidelberg, Germany

Printed by Libri Plureos GmbH
in Hamburg, Germany